Using AutoCAD® 2008:
Basics

Using AutoCAD® 2008:
Basics

RALPH GRABOWSKI

Autodesk

THOMSON
DELMAR LEARNING

Australia • Canada • Mexico • Singapore • Spain • United Kingdom • United States

 Autodesk

Using AutoCAD® 2008: Basics
Ralph Grabowski

Vice President, Technology and Trades SBU:
Dave Garza

Senior Acquisitions Editor:
Jim Gish

Marketing Director:
Debbie Yarnell

Channel Manager:
Kevin Rivenburg

Marketing Coordinator:
Mark Pierro

Production Director:
Patty Stephan

Content Project Manager:
Betsy Hough

Editorial Assistant:
Sarah Timm

Senior Project Manager:
John Fisher

Production Manager:
Stacy Masucci

Book Design and Typesetting:
Ralph Grabowski

Cover Images:
Getty Images

Publication Data:

ISBN: 1-4283-1159-9

NOTICE TO THE READER

Library of Congress Cataloging-in-

BRIEF CONTENTS

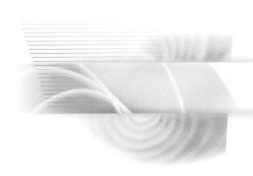

CONTENTS

The icon indicates the command is new to AutoCAD 2008.

2 Understanding CAD Concepts 33

5   Drawing with Precision 181

6   Drawing with Efficiency 239

8   Correcting Mistakes 381

9   Direct Editing of Objects 401

11 Additional Editing Options 477

13   Placing Dimensions 603

15   Geometric Dimensions and Tolerances 719

16   Tables and Fields 743

22   Plotting Drawings 983

INTRODUCTION

With more than 4 million users around the world, AutoCAD offers engineers, architects, drafters, interior designers, and many others, a fast, accurate, and extremely versatile drawing tool.

Now in its 16th edition, *Using AutoCAD 2008* makes using AutoCAD a snap, by presenting easy-to-master, step-by-step tutorials covering all of AutoCAD's commands.

BENEFITS OF COMPUTER-AIDED DRAFTING

AutoCAD is more efficient and versatile than traditional drafting. Some advantages include the following:

Accuracy — AutoCAD drawings are created and plotted to an accuracy of up to fourteen decimal places. The numerical entry of critical dimensions and tolerances is thus more reliable than traditional manual scaling.

Speed — the ability of the AutoCAD operator to copy and array objects, and edit the work on the screen speeds up the drawing process. When the operator customizes the system for specific tasks, the speed of work increases even more markedly.

Neatness and Legibility — plotting produces more exact drawings than traditional methods of hand drafting. AutoCAD drawings are uniform: they contain lines of constant thickness, evenly-spaced hatch patterns, and print-quality lettering. CAD drawings are clean: they are free of smudges and other editing marks.

Consistency — because the system is consistent in its methodology, the problem of individual style is eliminated. A company can have a number of drafters working on the same project and produce a consistent set of graphics.

CAD APPLICATIONS

AutoCAD is being applied to many industries. The flexibility of this software has a major impact on how tasks are performed in architecture, engineering, interior design, manufacturing, mapping, piping design, and entertainment. A brief discussion of each of these applications follows.

Architecture

CAD allows architects to formulate designs in a shorter period of time than by traditional techniques. The work is neater and more uniform. The designer can use 3D modeling to help the client better visualize the finished design. Changes can be made and resubmitted in a very short time. Architects can assemble construction drawings using stored details.

AutoCAD's database allows architects to extract information from the drawings, perform cost estimates, and prepare bills of materials.

Many third-party applications help architects customize AutoCAD for their discipline. Software is available for doing quick 3D conceptual designs, providing building details, generating automatic stairs, and much more. www.autodesk.com/autocad

Revit

Autodesk promotes its Revit product as the future of architectural design. Architects create the building in 3D, and then let Revit generate 2D section views. Worksets allow team members to work together on the same model, fully coordinating their work. Detailed graphic control and view-specific graphics permit drawings to show more or fewer details.

The software can be configured to office standards for graphic style, data and layer export, as well as to additional drafting and CAD standards. The software package includes thousands of building components, and can output its data to ODBC-compliant databases. www.autodesk.com/revit

AutoCAD Architecture

Autodesk has an AutoCAD add-on called Architecture (formerly Architectural Desktop), which customizes AutoCAD for architectural design. Its Content Browser accesses catalogs, tool palettes, and design content in the form of blocks and multiview blocks. With direct manipulation, you can modify designs directly, using either grip manipulation or in-place object editing.

Materials provide graphic and nongraphic

All figures in this chapter are courtesy of Autodesk, Inc.

attributes for design, documentation, and visualization. You can attach materials to building model objects and their respective components for greater visual detail.

Scheduling allows tracking any object in drawings. Because schedules are linked to data, they automatically update as the design changes. You can create schedule tables, objects tagged through xrefs, and schedules of external drawings.

Third-party developers have add-on software for roadway design, digital terrain modeling, and more. www.autodesk.com/autocadarchitecture

Engineering

Engineers use AutoCAD for many different kinds of design; in addition, they use programs that interact with AutoCAD to perform calculations that would take more time using traditional techniques. Among the many engineering disciplines benefitting from AutoCAD are civil, structural, and mechanical.

Building Systems

Autodesk's Building Systems software is meant for designing mechanical, electrical, and plumbing systems for buildings, such as offices. In this case, "mechanical" means heating, cooling, and fire protection. The Building Systems add-on helps you design splined flex ducts with editable grips; single-line, 2D, or 3D pipes for creating chilled water, hot water, steam, and other piping systems; rise/drop symbology for connecting ductwork; and fire protection content, such as sprinklers, deluge valves, and retard chambers.

The electrical design capabilities of Building Systems software include automatic calculation of service and feeder sizes, as well as panel schedules and single-, two-, and three-pole intelligent circuit objects that work across all referenced files.

For plumbing, the software includes schematic tools for piping layout; built-in plumbing code systems for automatically calculating pipe sizing and flow requirements; as well as 3D plumbing content and 3D piping tools. www.autodesk.com/buildingsystems

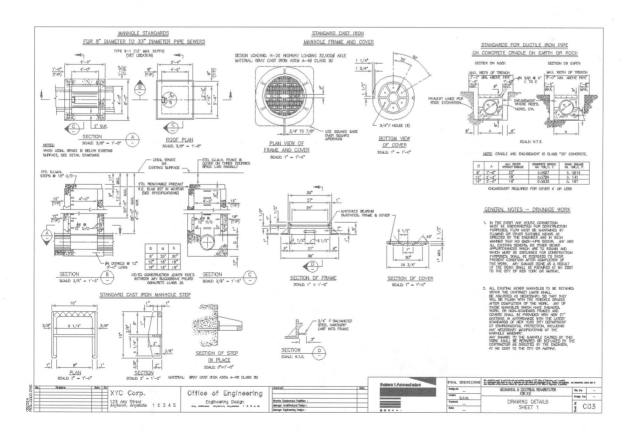

Civil 3D

Autodesk's Civil 3D (formerly Land Desktop) performs COGO (coordinate geometry) tasks, and creates maps, models terrain, designs alignments, and parcels for land planning. Other features include topographic analysis, real-world coordinate systems, volume totals, and roadway geometry.

Civil 3D integrates parcels in a single topology. Changes to one parcel update neighboring parcels. It extracts profiles of multiple surfaces based on alignment geometry, and then automatically creates the profile based on styles. The software generates dynamic models of any road, rail, or corridor project based on design elements, including alignments, profiles, superelevation, and criteria included in design subassemblies. Changes to any element update the corridor volumes, surfaces, and sections.

www.autodesk.com/civil3d

The civil engineering drawing on the previous page illustrates manholes and covers designed with AutoCAD.

Interior Design

AutoCAD is a valuable tool for interior designers. Its 3D capabilities can model interiors for clients, allowing 3D layouts, such as kitchens and baths, to be drawn and modified quickly with third-party add-ons.

VIZ

Autodesk's VIZ software creates high quality renderings of buildings and their interiors. Third-party developers have additional images and symbols for populating scenes.

VIZ has a "modeless" modeling environment with a unified workspace; surface finishes and lighting systems can be created, mapped, and manipulated on-the-fly. Global illumination rendering technology creates more accurate simulations of lighting — either natural or artificial. VIZ comes with a ready-to-use library of common lighting fixtures.

"DWG linking" lets VIZ interact with drawings created by AutoCAD, Architectural Desktop, Mechanical Desktop, and so on. For example, after designing a 3D floor plan in Architectural Desktop, you can link the file to VIZ to study materials and lighting. After modifying the floor plan in Architectural Desktop, the changes can be updated in a VIZ session, preserving materials and lighting defined in the linked model.

Substitution makes it easier to deal with complex details. For example, a chair represented as a simple 2D symbol in the production drawing is substituted in VIZ by a 3D equivalent that contains all the material information. www.autodesk.com/viz

The lifelike rendering shown below was generated by VIZ from an AutoCAD drawing.

Manufacturing

Manufacturing uses for CAD are legion. One main advantage is integrating the program with a database for record keeping and tracking. Maintaining information in a central database simplifies much of the work required in manufacturing. Technical drawings used in manufacturing are constructed quickly and legibly.

The AutoCAD drawing below shows how to fit the components of a truck cab around the driver.

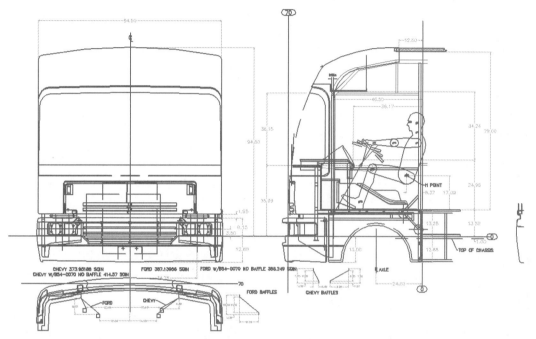

Autodesk has sevearl applications for mechanical engineers. For 2D drafting, there is AutoCAD Mechanical, and for 3D design, there is Mechanical Desktop and Inventor Series.

AutoCAD Mechanical and Mechanical Desktop

AutoCAD Mechanical and Mechanical Desktop are add-on programs that run on AutoCAD. Mechanical is meant for 2D drafting of mechanical parts, while Mechanical Desktop assists in creating 3D mechanical drawings inside of AutoCAD. Their parts tracking feature and large pre-drawn content make it easy for mechanical engineers to create, manage, and reuse their drawing data. www.autodesk.com/autocadmechanical

Inventor Series

The Inventor software uses a paradigm different from "regular" AutoCAD, because mechanical parts can be very complex. You begin by drawing *parts* that make up *assemblies*. The assembly is the finished device, whether an MP3 music player or a plastic injection molding machine. The parts, as the name suggests, make up the assembly. Programs like Inventor are designed to handle tens of thousands of parts.

Once the assembly is created, Inventor can generate 2D plans, parts lists, and assembly instructions. Commands help place standard bolts, route wires and tubing, and determine whether the parts can withstand stresses from pressure and temperature. www.autodesk.com/inventor

The Inventor Series includes AutoCAD, Mechanical, Mechanical Desktop, and Inventor. (Inventor is independent of AutoCAD, but can read and write its drawings.) Inventor Professional adds discipline-specific functions, such as cable routing and PCB (printed circuit board) design. Inventor LT is a low-cost version that leaves out assemblies and add-on applications. www.autodesk.com/inventorlt

Facilities Management

After tenants move into the newly-constructed building, facilities managers can track how the building is used. When staff move to different offices, for example, they need to know whether the correct desks, chairs, cupboards, and telecommunications connections are available in their new office, or whether their existing furniture has to move with them. Tracking hundreds of offices is made easier using FMdesktop from Autodesk. www.fmdesktop.com

Shopping malls and airports use AutoCAD linked to a database to maximize the profit from leased stores. Owners of these complexes need to track the mix of stores, and the revenue they produce. Some stores don't want to be located near competitors; others want to be in prime locations, such as near the entrance or at a corner. Other data stored in the database include agreements with cleaning and maintenance contractors, even details such as floor coverings and air conditioner capacities.

Mapping

Map makers can use CAD to construct maps that store huge amounts of data. For example, a municipal map can contain information about all buildings, transportation systems, and underground services (such as storm sewer and water pipes). By using search criteria, the CAD system returns a map showing, for example, all water pipes older than 20 years. This map then shows where municipal engineers should replace old piping.

GPS (global positioning system) generates the data used to create maps in CAD. This system consists of 24 satellites that always circle the earth, broadcasting data about their location. A handheld GPS receiver fixes your location by analyzing data broadcast from at least three satellites.

Map 3D

Autodesk has a program specifically for AutoCAD users called Map 3D. It assigns properties to objects according to classification, and then identifies, manipulates, and selects objects by their description (e.g., road, river) rather than by their simple primitives (lines and arcs). Topology functions define how nodes, links, and polygons connect to each other. This permits network tracing, shortest-path routing, polygon overlaying, and polygon buffer generation.

Map has tools that help correct errors due to incorrect surveying, digitizing, or scanning. It connects to an external database, and can import data from other GIS products. www.autodesk.com/map3d

The GIS drawing above uses layers and colors to differentiate amongst geographic elements, such as homes, properties, streets, and underground services.

Entertainment

The entertainment industry is largely based on digital media, so CAD graphics are a perfect match. What you view on television and in the newspaper is produced by a form of CAD graphics. Movie and theater set designs are often done in CAD. Movie makers are turning to forms of electronic graphics for manipulations and additions to their filmed and animated work. Autodesk also sells media and entertainment software, such as Maya and MotionBuilder.

OTHER BENEFITS

Using a computer to draft and design also opens a new world of information technology. Many CAD systems now access email and the Web. There are Web sites specific to AutoCAD, as well as discussion groups and email newsletters.

Autodesk maintains many kinds of information at its Web site, www.autodesk.com. You can learn about updates to AutoCAD, download useful utilities, read up tips on for using AutoCAD, and access large libraries of symbols.

For help with bugs in AutoCAD, visit Autodesk's support site at support.autodesk.com.

AUTOCAD OVERVIEW

AutoCAD displays your drawings on the monitor or screen. The screen takes the place of paper in traditional drafting. All the additions and changes to your work are shown on the screen as you perform them.

Drawings are made up of *objects*, such as lines, arcs, circles, text, and others.

You place objects in drawings by means of *commands* and their options. You issue commands by several methods: typing the names of commands on the keyboard, selecting commands from menus, or choosing icons from toolbars or the Dashboard. Some commands start when you select or double-click objects in drawings.

Once a command starts, you are often asked to specify its options. After you identify all the information that AutoCAD requests, the new objects, or changes to objects, are shown on the screen.

TERMINOLOGY

This book contains terms and concepts that you need to understand to use AutoCAD properly. Some of the terms are explained briefly in this chapter. If you need additional help, refer to Appendix B, "AutoCAD Command Summary." In addition, you may consult the index, which contains many terms.

Coordinates

AutoCAD uses the Cartesian coordinate system. The horizontal is represented by the x axis; the vertical, by the y axis. Any point on the graph can be represented by an x and y value shown in the form of x,y. For example 2,10 represents a point 2 units in the x direction and 10 units in the y direction.

The intersection of the x and y axis is the 0,0 point. This point, called the "origin," is normally the lower left corner of the screen. (You can, however, specify a different point for the lower left corner.)

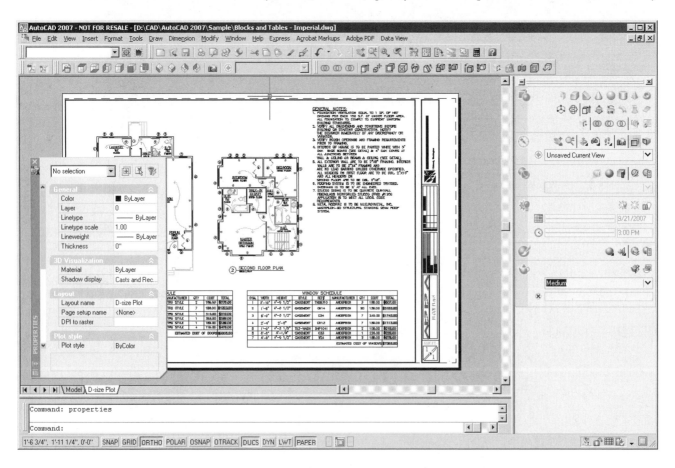

AutoCAD also works with 3D (three-dimensional) drawings. The third axis is called the z axis, which normally points out of the screen at you.

Drawing Files

Drawing files contain the information used to store the objects you draft. Drawing files automatically have the file extension of *.dwg* added to them.

USING THIS BOOK

Each chapter in *Using AutoCAD 2008: Basics* builds on the previous chapters. The basic use for each command is given, along with one or more tutorials that help you understand how the command works. This is followed by a comprehensive look at the command and its many variations.

The problems at the end of a chapter are specifically designed to use the commands covered in the book to that point. Review questions reinforce the concepts explored in the chapter. This method allows you to pace learning. Remember, not everyone grasps each command in the same amount of time.

CONVENTIONS

This book uses the following conventions.

Keys

Several references are made to keyboard keystrokes in this book, such as **ENTER**, **CTRL**, **TAB**, and function keys (**F1**, **F2**, and so on). These keys might be found in different locations on different keyboards.

Control and Alternate Keys

Some commands are executed by holding down one key, while pressing a second key. The control key is labeled **CTRL**, and is always used in conjunction with another key or a mouse button.

To access menu commands from the keyboard, hold down the **ALT** key while pressing the underlined letter. The **ALT** key is also used in conjunction with other keys.

Flip Screen

The text and graphics windows can be alternately displayed by using the Flip Screen key. The **F2** key is used for this function.

Command Nomenclature

When a command sequence is shown, the following notations are used:

> Command: **line** *(Press ENTER.)*
>
> Specify first point: *(Pick a point.)*
>
> Specify next point or [Undo]: *(Pick point 1.)*

Boldface text designates user input. This is what you type at the keyboard. If the command and its options can be entered by other methods, these are described in the book. For example, commands can be selected from menus and toolbars, or entered as keyboard shortcuts.

(Press ENTER) means you press the **ENTER** key; you do not type "enter."

(Pick a point) means that you enter a point in the drawing to show AutoCAD where to place the object. To pick the point, you can either click on the point, or type the x, y, z coordinates.

(Pick point 1) The book often shows you a point on a drawing. The points are designated, such as "Point 1."

<Default> is how AutoCAD indicates *default* values. The numbers or text are enclosed in angle

brackets. The default value is executed when you press ENTER.

[Undo] is how AutoCAD lists command *options*. The words are surrounded by square brackets. At least one letter is always capitalized. This means you enter the capital letter as the response to select the option. For example, if the option is **Undo**, simply entering U chooses the option.

Additional options are separated by the slash mark. When two options start with the same letter, then two letters are capitalized, and you must enter both.

WHAT'S NEW IN AUTOCAD 2008

Fully updated to AutoCAD 2008, this book includes information on these new features. In drawing and editing:

- The new annotation scaling property displays only correctly-scaled objects in model space.

- The Dashboard is expanded to provide access to 2D drawing, editing, and view commands.

- The **HATCH** and **QSELECT** commands now support the annotative scaling property. Commands like **CHANGE** and **PROPERTIES** now support annotative scale factors. The **DBLIST** and **LIST** commands now report on annotative scale factors, when applicable.

- Blocks now support orientation and annotative scaling.

- The new **LAYERSTATE** command provides direct access to layer states.

- Layer properties can now be overridden (and reset) on a per-viewport basis.

- The new **SETBYLAYER** command resets layer properties overridden in viewports.

- The **LAYISO** command now fades isolated layers, as well as freezing them.

- The **RENAME** and **PURGE** commands now support renaming and purging mleader styles.

- The new **RECOVERALL** command recovers and converts drawings and attached reference files.

- The new **COPYMODE** system variable determines whether the **COPY** command repeats automatically.

- The newly-documented **NEWVIEW** command directly accesses the **VIEW** command's New View dialog box.

In text, dimensioning, and tables:

- All text can now have the annotative scaling property, including mtext, single-line text, and leaders.

- The **MTEXT** command now supports multiple columns, forced justification, and new paste special options, as well as sports a redesigned paragraph dialog box.

- The **SPELL** command now begins spell checking immediately, and zooms into unrecognized words.

- The **MLEADER** command now allows multiple leader lines and multiple blocks per leader.

- The new **MLEADERALIGN, MLEADERCOLLECT, MLEADEREDIT, AIMLEADEREDITREMOVE,** and **AIMLEADEREDITADD** commands edit multiline leaders.

- The new **MLEADERSTYLE** command defines mleader styles.

- The **DIMANGULAR** command now locks angular dimensions to specific quadrants.

- The **DIMDIAMETER, DIMRADIUS,** and **DIMJOGGED** commands now have a new **Extension** option.

- The **DIMBREAK**, **DIMJOGLINE**, **DIMINSPECT**, and **DIMSPACE** commands are new.

- The **DIMSTYLE** dialog box has new options.

- The **DIMANNO** system variable now reports whether the dimension style is annotative.

- Dimensions and tolerances now have annotative scaling.

- Tables can now be auto-filled, created from external data sources (an advanced topic), and split into columns.

- Tables now have additional formatting options, including formatting for individual cells.

- The Table and TableStyle dialog boxes are redesigned.

- The new **TABLETOOLBAR** system variable toggles the display of the table toolbar.

- Table cells can now contain more than one piece of content; the order of the content can be changed.

In viewports and plotting:

- Viewport scaling can now be set through the new drawing status bar, also controlling the visibility of objects with the new Annotative property.

- The new **VPLAYEROVERRIDES** system variable reports on overridden layer properties in viewports.

- The **PLOT** command now emulates rendering and 3D effects not supported by the graphics board.

- The new **AUTOPUBLISH** command determines whether drawings are simultaneously saved in DWF format during the **SAVE** and **CLOSE** commands.

- The **SCALELISTEDIT** command now also determines the scale factors used with annotative scaling.

ONLINE COMPANION

The Online Companion™ is your link to AutoCAD on the Internet. We've compiled supporting resources with links to a variety of sites.

In addition, there is information of special interest to readers of *Using AutoCAD 2008*. This includes updates, information about the author, and a page where you can send your comments. You can find the Online Companion at www.autodeskpress.com/resources/olcs/index.aspx. When you reach the Online Companion page, click on All AutoCAD 2008 Titles.

E.RESOURCE

e.Resource™ is an educational resource that creates a truly electronic classroom. It is a CD containing tools and instructional resources to enrich the classroom and reduce preparation time. The elements of e.Resource link directly to the text, and combine to provide a unified instructional system. With e.Resource, you can spend your time teaching, not preparing to teach.

Features contained in e.Resource include: :

- **Instructor Syllabus:** This lists goals, topics covered, reading materials, required texts, lab materials, grading plan and terms. Additional resources and web sites covering descriptive geometry, manual drafting, AutoCAD, and Inventor are also listed.

- **Student Syllabus:** This lists goals, topics covered, required texts, grading plans and terms.

- **Lesson Plans:** These contain goals, discussion topics, suggested reading, and suggested homework assignments. You have the option of using these lesson plans with your own course information.

- **Answers to Review Questions:** These solutions help you grade and evaluate end-of-chapter tests.

- **PowerPoint® Presentation:** These slides help you present concepts and material. Key points and concepts can be graphically highlighted for student retention.

- **Exam View Computerized Test Bank:** This includes over 800 questions of varying levels of difficulty in true/false and multiple-choice formats that assess student comprehension.

- **AVI Files:** These movies, listed by topic, illustrate and explain key concepts.

- **DWG Files:** This list of *.dwg* files matches many of the figures in the textbook. These files can be used to stylize the PowerPoint presentations.

WE WANT TO HEAR FROM YOU!

Many of the changes to the look and feel of this new edition came by way of requests from and reviews by users of our previous editions. We'd like to hear from you as well! If you have any questions or comments, please contact:

> The CADD Team
>
> c/o Autodesk Press
>
> 5 Maxwell Drive
>
> Clifton Park NY 12065-8007

or visit our Web site at www.autodeskpress.com

ACKNOWLEDGMENTS

We would like to thank and acknowledge the professionals who reviewed the manuscript to help us publish this AutoCAD text:

Technical Editor: Bill Fane, British Columbia Institute of Technology, Burnaby BC, Canada

Copy Editor: Stephen Dunning, Douglas College, Coquitlam BC, Canada

ABOUT THE AUTHOR

Ralph Grabowski has been writing about AutoCAD since 1985, and is the author of over 100 books on computer-aided design. He received his B.A.Sc.'80 degree in Civil Engineering from the University of British Columbia.

Mr. Grabowski publishes *upFront.eZine*, the weekly email newsletter about the business of CAD. He is the author of a series of CAD e-books under the eBooks.onLine imprint. Mr. Grabowski is the former Senior Editor of *CADalyst* magazine, the original magazine for AutoCAD users. You can visit his Web site at www.upfrontezine.com and his Weblog at worldcadaccess.typepad.com.

lx

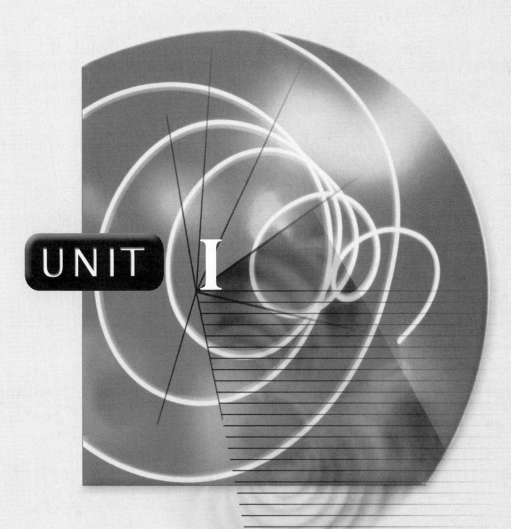

UNIT **I**

Setting Up
Drawings

CHAPTER 1

Quick Tour of AutoCAD 2008

Welcome to *Using AutoCAD 2008: Basics*!

As a new user, you need to experience the feel of AutoCAD firsthand. The concepts behind computer-aided design software can be disorienting to first-time user, and so this quick-start chapter gets your feet wet. It introduces you to features discussed in detail in later chapters.

This chapter is specifically designed to acquaint you with AutoCAD. Subsequent chapters cover the subject more thoroughly. In this chapter, you learn these commands:

>**NEW** starts fresh drawings.
>
>**LINE** draws straight line segments.
>
>**U** undoes mistakes.
>
>**QUIT** exits AutoCAD.
>
>**CLOSE** and **CLOSEALL** close drawings.
>
>**SAVEAS** renames drawings, and then saves them.
>
>**PLOT** prints the drawing on paper.

Let's get started!

STARTING AUTOCAD

Before you can install and run AutoCAD for the first time, your computer must have Microsoft® Windows™ 2000, XP, TabletPC, Server 2003, or Vista installed and running.

TUTORIAL: STARTING AUTOCAD

1. To start AutoCAD 2008, use one of these methods:
   - On the Windows desktop, double-click the icon labeled **AutoCAD 2008**.
     (Double-click means to press the left mouse button twice, quickly.)

AutoCAD 2008

   - Alternatively, you can also start AutoCAD from the Windows taskbar. Click the **Start** button, and then choose **Programs**. Next, choose **Autodesk | AutoCAD 2008 | AutoCAD 2008**. (Your copy of Windows may show a display that differs from the one illustrated below.)

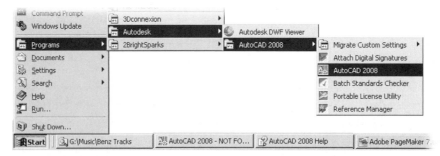

   - In the Windows Explorer (a.k.a File Manager), double-click a *.dwg* drawing file. (If AutoCAD add-ons, such as Mechanical Desktop or Architectural Desktop, are installed, they may start instead.)

   - In the **Run** dialog box, enter *acad.exe*.

     Run: **acad.exe** *(Press* ENTER.*)*

   (The above procedure bypasses command-line startup switches associated with the icon.)
2. In all cases, Windows opens the AutoCAD software.
   An opening screen is displayed, called the "splash screen."

After the splash screen disappears, you see AutoCAD.

3. AutoCAD sometimes shows several dialog boxes before it finishes starting up. Here is what to do with them:

   • If the **New Features Workshop** appears, choose **Maybe Later**, and then click **OK**.

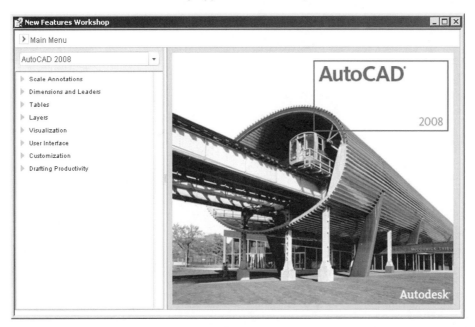

   • If the **Startup** dialog box appears, choose the **Start from Scratch** button, and then select **Imperial (feet and inches)**. Click **OK**.

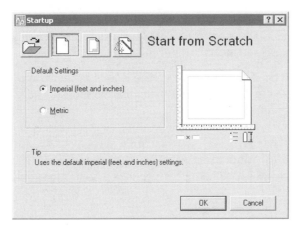

4. The AutoCAD window looks similar to the figure below, except that the background color of the drawing area may be black, instead of white. (The Sheet Set Manager palette, Tool palette, and Express Tools toolbars may appear; you can close them or ignore them.)

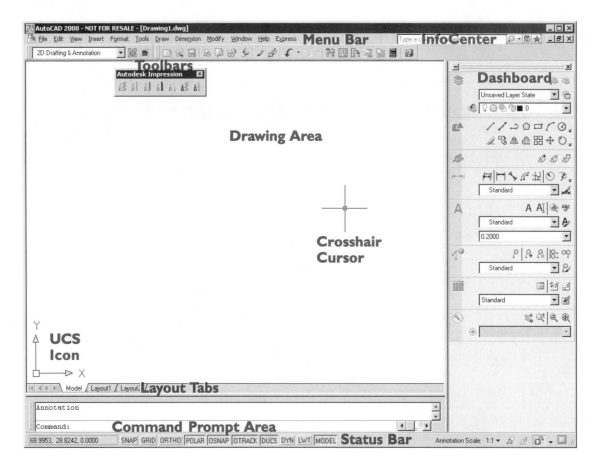

DRAWING IN AUTOCAD

You make drawings in AutoCAD by creating and editing objects. You instruct AutoCAD through *commands*, which are often simple words such as LINE, CIRCLE, MOVE, and ERASE. Commands describe the objects to be drawn — LINE or CIRCLE — and the operation to perform — MOVE or ERASE.

Commands can be executed by several different methods: typing them on the keyboard, selecting them from menus and toolbars, double-clicking or right-clicking the mouse, and by using shortcut keystrokes. Use whichever method you prefer.

Command Prompt Area

At the bottom of the AutoCAD window is the *command prompt area*. This is where you enter typed commands, and one of the places where AutoCAD responds to you. You should see the word "Command:" on that line now.

        Command:

(If you do not see the command prompt area, press CTRL+9 to make it appear. AutoCAD has a larger version of the command prompt area called the "Text window," which displays only the text from commands. You can see it by pressing function key F2; press F2 again to bring back the graphics window.)

Dynamic Input

The drawing area is a second place where you can type commands and receive feedback from AutoCAD. When *dynamic input* is turned on, the command prompts and options appear near the cursor.

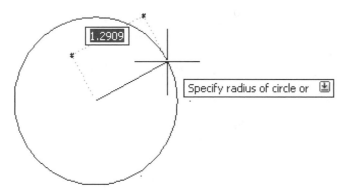

(To turn on dynamic input, click the **DYN** button on the status bar, and ensure the cursor is in the drawing area.) You learn more about dynamic input in Chapter 2, "Understanding CAD Concepts."

Mouse

As an alternative to typing commands, you can use a mouse, digitizing tablet, or other input device to control AutoCAD. (From this point forward, we refer to the mouse only, being the most common input device.) To see how the mouse interacts with AutoCAD, follow these steps:

1. Move the mouse, and notice how the crosshair cursor moves around the screen.
   When the cursor looks like a crosshair, it specifies points in drawings.

2. Move the crosshair cursor out of the drawing area. Notice that the cursor changes to an arrow. This lets you select commands from menus and toolbars.

USING MENUS

A common method for selecting commands is through the menus — just as in almost all other Windows software. The menus contain many of AutoCAD's commands, which can be selected with the mouse. To see how they work, select a command from the menu — choose the **Line** item from the **Draw** menu, as follows:

TUTORIAL: SELECTING MENU ITEMS

1. To select a command from the menu, move the crosshair cursor to the menu bar. As the cursor moves out of the drawing area, it changes from the crosshair to the slanted arrow.
2. As you move the arrow across the words on the menu bar, each is highlighted: **File, Edit, View,** and so on.

3. Select the **Draw** menu: when **Draw** is highlighted by the cursor, click the left mouse button. Notice that a menu drops down into the drawing area.

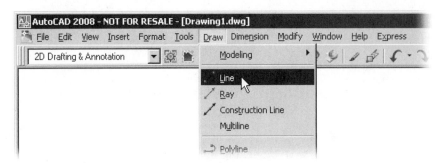

4. Move the pointer down the menu past **Modeling**, **Line**, **Ray**, and so on. As you do, notice that each word is highlighted.

5. Move back up, and then pause the cursor on **Line**. Notice the line of words that appears on the status line — "Creates straight line segments: LINE." This tells you the purpose of the menu item.

6. With the cursor on **Line**, click the left mouse button.

   Notice that text appears in the command bar at the bottom of the window:

   Command: _line Specify first point: *(Pick a point.)*

   This is called a *prompt.* In the command bar, AutoCAD tells you what it expects from you. Here, AutoCAD is asking you to specify the point from which the line starts, called the "first point."

7. Move the crosshair cursor into the drawing area, and then enter a *point* (click the left mouse button).

8. Continue to move the crosshair cursor around the screen. Notice how a line "sticks" to the intersection of the cursor crosshairs. This is called "rubber banding": the line stretches and follows your movement.

   If dynamic input is turned on, then the prompt appears next to the cursor, as illustrated below.

9. The prompt has changed slightly with the addition of "or [Undo]":

   Specify next point or [Undo]: *(Pick another point.)*

   Move the cursor, and pick a point approximately at the point shown in figure below.

The word **Undo** is an *option*. AutoCAD shows options in square brackets, such as [Undo]. Options are alternative actions that can be performed by the command. (**Undo** "undraws" the previous line segment, which is useful when you make mistakes.)

10. The **LINE** command remains active, allowing you to draw as many line segments as you need without reselecting the command. Continue to enter lines as shown in the figure below. As you do, the prompt changes to:

    Specify next point or [Close/Undo]:

    Close is another option of the **LINE** command. When you enter **Close**, it draws a line segment from the end of the last segment to the start of the first segment.

11. To end the command, click the right mouse button. From the shortcut menu, choose **Enter**, as illustrated by the figure below. (Alternately, press **ESC** on the keyboard; for some commands, you need to press **ESC** twice before they end.)

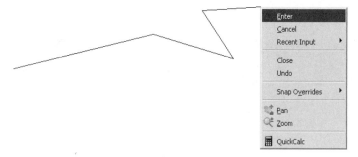

12. Enter the **U** command (short for "undo") to erase the lines you drew.
    Command: **u** *(Press* ENTER.*)*

    AutoCAD confirms:
    Everything has been undone.

WORKSPACES

The way that AutoCAD looks — its menus, toolbars, Dashboard panels, and palettes — is controlled by *workspaces*. Selecting a different workspace changes the number and position of these user interface items. You can select from three workspaces predesigned by Autodesk from the Workspaces toolbar, as illustrated below.

Workspaces are described in greater detail in Chapter 3, "Setting Up Drawings." For now, it is sufficient to know that if you do not see a toolbar or menu item, your workspace may have changed.

This book works with AutoCAD's default workspace, "2D Drafting & Annotation." To return AutoCAD to its default look, select this name from the Workspace toolbar.

TOOLBARS

Toolbars consist of buttons; each button is labeled by a small picture called an "icon." Icons represent commands carried out by buttons. Similar commands are grouped together; for example, most drawing commands appear on the Draw toolbar. AutoCAD has nearly three dozen toolbars, most of which are not initially displayed

AutoCAD 2008 reduces the number of toolbars displayed, depending on the workspace selected. The default workspace, "2D Drafting & Annotation," shows just two toolbars, as illustrated below.

Workspaces toolbar   Standard-Annotation toolbar

The two toolbars are:

- **Workspaces** selects the workspace to be displayed.

- **Standard-Annotation** contains some of the same commands found in other Windows applications, plus a number specific to AutoCAD.

The "3D Modeling" workspace also displays the Layers toolbar, while the "AutoCAD Classic" workspace displays all the toolbars you might expect, if you are used to AutoCAD 2006 and earlier.

The added toolbars include:

- **Draw** toolbar draws many different objects, such as lines, arcs, and hatch patterns.

- **Layers** toolbar controls layers in the drawing.

- **Properties** toolbar changes the properties of objects.

- **Styles** toolbar accesses text, dimension, and table styles.

- **Modify** toolbar edits objects to change their size, location, and other properties.

Using Toolbar Buttons

Every button on every toolbar activates a command or series of commands. For example, the first button of the **Draw** toolbar is **Line**. It starts the LINE command. The icon for LINE is the diagonal line with a dot at either end.

When you cannot remember the purpose of a button, move the cursor over the button and wait a second or two. AutoCAD then displays a *tooltip*, a one- or two-word description of the button's purpose, as well as shortcut keystrokes, if available. Tooltips also appear in the Dashboard, but not in the menus.

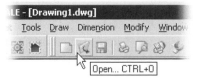

The status bar (at the bottom of the AutoCAD window) displays a one-sentence description of the command's purpose. In the case of the **Open** button, the status bar reads, "Opens an existing drawing file: OPEN."

The same help text appears on the status bar when you access commands from the menus, but not from the Dashboard panels.

DASHBOARD PANELS

Dashboard panels are like toolbars, but tend to be fixed to the right end of the AutoCAD window. Like toolbars, panels group similar commands together.

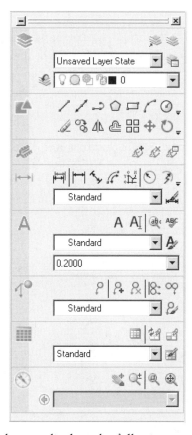

The 2D Drafting & Annotation workspace displays the following panels, from top to bottom:

- **Layers** controls layers and layer states; see Chapter 7.
- **2D Draw** draws many objects, such as lines and circles; see Chapter 3.
- **Annotation Scaling** ensures scale-dependent objects plot at the correct size.
- **Dimensions** places dimensions in drawings; see Chapter 13.
- **Text** places text, checks spelling, selects styles, and so on; see Chapter 12.
- **Multileaders** places and edits multiline leaders; see Chapter 15.
- **Tables** inserts and edits tables, and creates links to external data files; see Chapter 16.
- **2D Navigate** zooms and pans the drawing; see Chapter 5.

TUTORIAL: DRAWING WITH BUTTONS

To draw a line with a button on the Dashboard or a toolbar:

1. Position the cursor over the **Line** button in the 2D Drafting Dashboard panel.

2. Click the left mouse button.

At the command line, you see the familiar prompt:

Command: _line Specify first point:

3. Draw some lines, and then press **ESC** to exit the command.

Flyouts

Sometimes buttons represent secondary toolbars. You recognize these buttons on toolbars and Dashboard panels by the tiny black triangle in the corner of their icons. The triangle indicates the presence of a *flyout*, a group of buttons that "flies out" from the button.

TUTORIAL: ACCESSING FLYOUTS

1. To access flyouts:

Position the cursor over any tiny triangle next to a button, such as the **Circle** button illustrated below.

2. Hold down the left mouse button. Notice that AutoCAD displays a flyout toolbar.

3. Without letting go of the mouse button, drag the cursor over the buttons on the flyout. As the cursor passes over a button, it changes slightly to give the illusion of being depressed.

4. When you reach the button you want, release the mouse button. AutoCAD starts the command associated with the button.

In some cases, the button you selected moves to the top of the toolbar where it is the default next time.

Manipulating Toolbars

Toolbars have many controls hidden in them. You can move, resize, float, dock, and dismiss toolbars. Here's how:

To move a docked toolbar, position the cursor over the two "bars" at the left end of horizontal toolbars (or the top end of vertical toolbars).

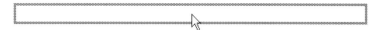

Hold down the mouse button, and drag the toolbar away from the edge of the window.

Release the mouse button. The toolbar now *floats* in the drawing window, and sports a title bar (as shown below).

To move a floating toolbar, position the cursor over the title bar. Hold down the mouse button and drag the toolbar to another location.

To resize (or stretch) a toolbar, position the cursor over one of its four edges. Hold down the left mouse button, and then drag the toolbox to achieve a new shape.

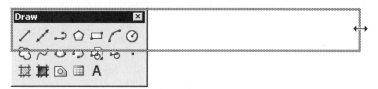

To *dock* a toolbar, drag it against one of the four sides of the drawing area. *Docking* means to move it to the side or top of the drawing area, like the docked toolbars below the menu bar, illustrated in the figure below.

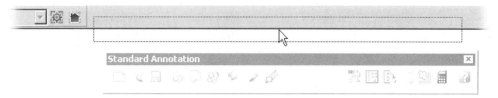

To dismiss (get rid of) a floating toolbar, click the tiny **x** in the upper right corner. The toolbar disappears.

To get back the toolbar, right-click any toolbar button. From the shortcut menu, select its name from the list.

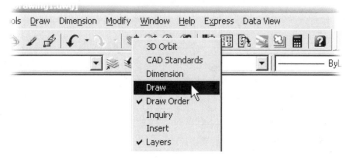

Leave at least one toolbar visible so that you can easily access the other toolbars. If you do close them all, you can use the View | Toolbars command to reopen them.

EXITING DRAWINGS

AutoCAD provides several different ways to close drawings and exit AutoCAD. The following sections outline the possibilities. Choose the option you want, and follow the instructions.

DISCARD THE DRAWING, AND EXIT AUTOCAD

When you don't want to keep your drawing, and wish to stop working:

1. Enter the **QUIT** or **EXIT** commands (from the keyboard), or choose **Exit** from the **File** menu. Notice that AutoCAD displays a dialog box to confirm whether you want to save the drawing — it's your last chance!

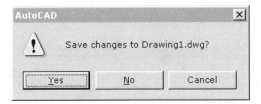

2. Move the cursor to choose the **No** button to exit without saving your work. AutoCAD does not record your work to disk, and exits to the Windows desktop.

Save the Drawing, and Exit AutoCAD

To save your work and exit AutoCAD:

1. Enter the same **QUIT** command in the command window (or select **Exit** from the **File** menu).
2. This time choose **Yes** to save your work. Notice that AutoCAD displays the Save Drawing As dialog box.

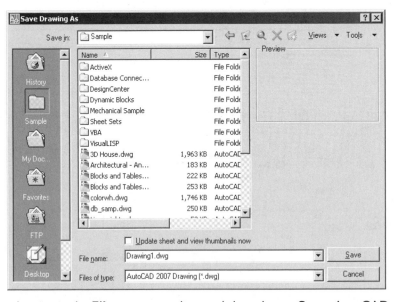

3. Name the drawing in the **File name** text box, and then choose **Save**. AutoCAD saves your work under the name, and exits to the Windows desktop.

Save the Drawing, and Remain in AutoCAD

If you want to keep your work *and* remain in AutoCAD:

1. Enter the **QSAVE** command (or select **File | Save** from the menu).

2. If the drawing has not yet been named, the same Save Drawing As dialog box appears; name the file, and then choose **Save**.

Remain in AutoCAD, and Start a New Drawing

If you would like to start a new AutoCAD drawing:

1. Use the **NEW** command.
2. AutoCAD displays the Select Template dialog box, which looks like the dialog box shown above. Select a template file, and then click **Open**.

Close the Drawing and Remain in AutoCAD

You can have more than one drawing open in AutoCAD. To close one or more drawings, follow one of these options:

- To close the current drawing, enter the **CLOSE** command (or from the **File** menu, select **Close**). If necessary, AutoCAD asks if you wish to save the drawing. AutoCAD then closes the drawing.

- To close all drawings at once, enter the **CLOSEALL** command; from the **Window** menu, select **Close All**.

With the basics of navigating AutoCAD behind you, practice drawing in AutoCAD. In the following tutorial, you draw a two-dimensional object.

2D TUTORIAL

The figure below illustrates the drawing you construct in this tutorial. It is the plan view of a base with four holes, and a shaft with rounded edges.

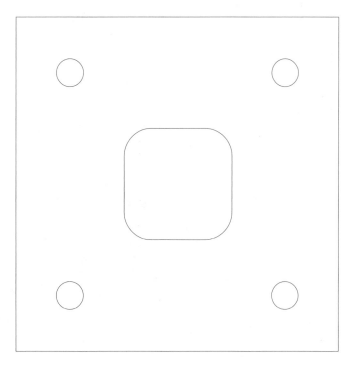

The following tutorial describes the commands to complete the drawing. The commands you type appear on the command line. Your responses are shown by **boldface** in this tutorial. (You can type the items in either uppercase or lowercase.)

Items enclosed in parentheses (*such as these*) are instructions and are not typed. If ENTER is shown, press the ENTER key on the keyboard. For example:

> Command: **line** *(Press ENTER.)*

Type the LINE command, and then press ENTER.

If you make a mistake, just type **U**, and then press ENTER. (Or, you can click the **Undo** button on the toolbar, or press CTRL+Z.) This undoes the previous step. You may use it several times to undo each step in reverse order.

Before entering a command, the prompt on the command line must be "empty," with nothing following the colon:

> Command:

If not, press ESC. This cancels the current command, and makes AutoCAD ready for your command.

SETTING UP NEW DRAWINGS

1. To make sure AutoCAD displays the Create New Drawing dialog box, enter the following command:

   > Command: **startup**

   > Enter new value for STARTUP <0>: **1** *(Press ENTER.)*

2. Start a new drawing. From the **File** menu, choose **New**.
3. In the Create New Drawing dialog box, choose the **Use a Wizard** button.

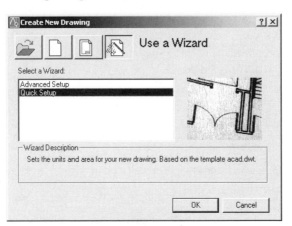

4. Select **Quick Setup**, and then choose the **OK** button.

   Notice that AutoCAD displays the Quick Setup dialog box.

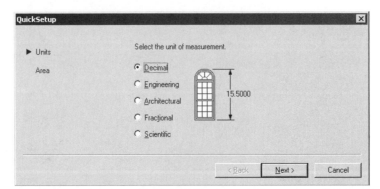

Quick Setup presents a short series of setup dialog boxes that lead you through the steps of quickly setting up a new drawing with units and limits.

5. In the first dialog box, you specify the units. Accept the default of **Decimal** by choosing the **Next** button. Notice that AutoCAD displays the Area dialog box.

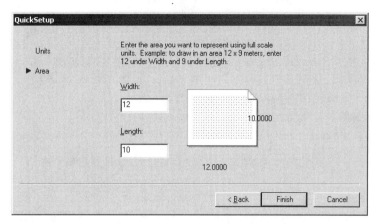

6. Here you specify the limits of the drawing: 12 units wide by 10 units tall (long). In the Area dialog box, define the area by clicking the **Length** text box.

    Clear the current value using the **DEL** or **BACKSPACE** key.

7. Type **10**, and then press the **TAB** key to see the effect in the preview image.

    Keep the **Width** at 12.

8. Choose **Finish**. AutoCAD dismisses the dialog box.

SETTING UP DRAWING AIDS

To draw accurately, you need to turn on drawing aids like grid and snap.

The *grid* is a visual guide that shows distances. Think of it as graph paper with dots, instead of lines. When the grid is set to 1 unit, for example, the drawing is covered with dots spaced one unit apart. The grid is only displayed on the screen; it is not printed.

Snap is like drawing resolution. When on, snap causes the cursor to move in precise increments. When the snap is set to 1 unit, for example, the cursor moves in one-unit increments. The grid and snap can have either the same or different spacing).

1. From the **Tools** menu, choose **Drafting Settings**. Notice that AutoCAD displays the Drafting Settings dialog box.

2. Select the **Snap and Grid** tab.

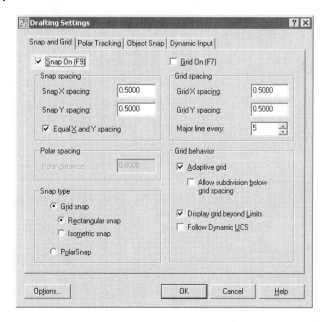

3. Turn on snap by clicking the check box next to **Snap On**.

    ☑ Check mark indicates the option is turned on.

    ☐ No check mark means option is turned off.

The **F9** next to **Snap On** reminds you to press function key **F9** to turn snap on and off. This is handy when you need to *toggle* (turn on and off) snap during commands. (Alternatively, click the word **SNAP** on the status bar at the bottom of the AutoCAD window.)

4. Change both the **Snap X spacing** and **Snap Y spacing** to 1.

5. Do the same for the grid: turn it on, and set the grid spacing to 1.

The settings in the dialog box should look like the highlighted area in the figure below.

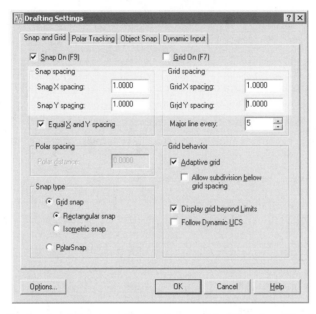

6. Choose **OK** to dismiss the dialog box. Notice that a grid of fine dots fills the drawing area.

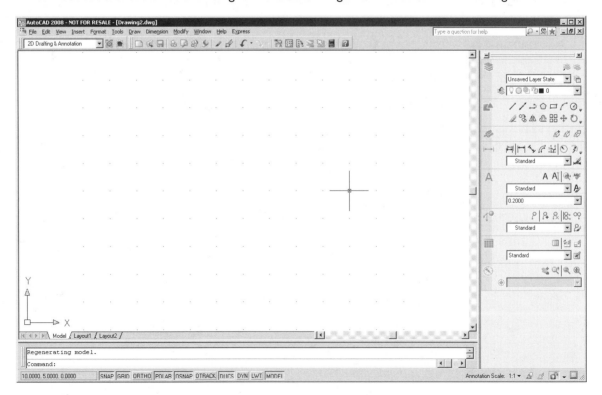

If the dots don't fill the drawing area, use the **ZOOM** command with the **All** option to see all of the drawing:

Command: **zoom**

Specify corner of window, enter a scale factor (nX or nXP), or

[All/Center/Dynamic/Extents/Previous/Scale/Window/Object] <real time>: **all**

7. Move the mouse to see the crosshairs cursor move in increments of one unit.

DRAWING THE BASEPLATE

The baseplate is a square. You can draw it with the RECTANGLE command, which is AutoCAD's easiest command for drawing rectangles and squares.

AutoCAD needs to know two things to draw a square: the position of two opposite corners.

1. From the **Draw** menu bar, select **Rectangle**.

    As an alternative, enter the **RECTANGLE** command at the keyboard:

    Command: **rectangle** *(Press ENTER.)*

2. AutoCAD next needs to know where to place the square. One corner is at the x,y coordinates of 3,2:

    Specify first corner point or [Chamfer/Elevation/Fillet/Thickness/Width]: **3,2** *(Press ENTER.)*

3. The opposite corner is located at x,y coordinates of 9,8:

    Specify other corner point or [Dimensions]: **9,8** *(Press ENTER.)*

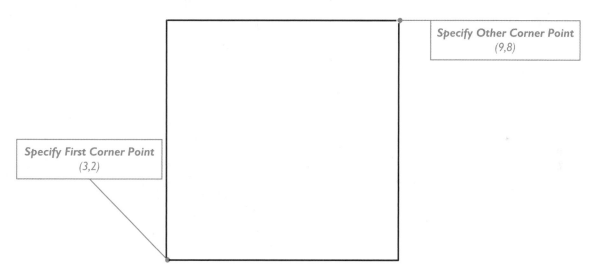

Specify Other Corner Point
(9,8)

Specify First Corner Point
(3,2)

Unlike the **LINE** command, the **RECTANGLE** command stops by itself; there is no need to press **ESC**.

4. Your drawing should look similar to the figure above. If necessary, enter the **ZOOM** command, and select the **All** option:

    Command: **zoom** *(Press ENTER.)*

    Specify corner of window ... <real time>: **all** *(Press ENTER.)*

DRAWING THE SHAFT

The shaft is also a square, but with rounded corners. The RECTANGLE command can draw squares with corners that are square (as above), rounded (also known as "filleted"), or cutoff ("chamfered").

Draw the square shaft by repeating the RECTANGLE command. In addition to specifying the two corners, you tell AutoCAD the radius of the rounded corners.

1. In AutoCAD, pressing the SPACEBAR repeats the previous command — this is a handy shortcut that saves you typing the full name a second time.

   Command: *(Press SPACEBAR.)*

   RECTANGLE

2. Rounded corners are called "fillets," and their size is specified by radius. To specify the radius of the fillets, enter the **Fillet** option:

   Specify first corner point or [Chamfer/Elevation/Fillet/Thickness/Width]: **fillet** *(Press ENTER.)*

3. The fillets are 0.5 inches in radius:

   Specify fillet radius for rectangles <0.0000>: **.5** *(Press ENTER.)*

4. With the fillet radius set, specify the corners of the square using x,y coordinates:

   Specify first corner point or [Chamfer/Elevation/Fillet/Thickness/Width]: **5,4** *(Press ENTER.)*

   Specify other corner point or [Dimensions]: **7,6** *(Press ENTER.)*

   Your drawing should look similar to the figure.

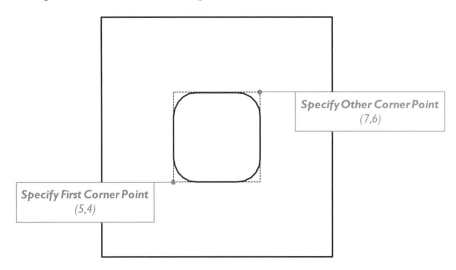

Specify Other Corner Point (7,6)

Specify First Corner Point (5,4)

DRAWING HOLES

Now draw the "holes" in the baseplate. Holes can be drawn with circles.

To draw a circle, AutoCAD needs to know two things: (1) the position of the circle in the drawing, based on its center point; and (2) the radius (or diameter) of the circle.

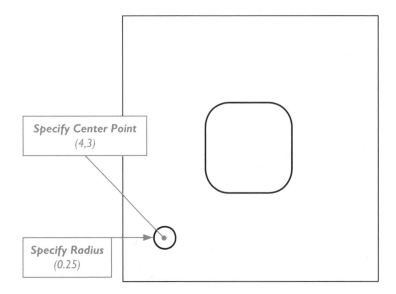

Specify Center Point (4,3)

Specify Radius (0.25)

1. From the **Draw** menu, select **Circle**. Notice that AutoCAD enters the command name for you, and immediately moves on to the first prompt:

    Command: _circle

2. One circle is located at 4,3 in the drawing:

    Specify center point for circle or [3P/2P/Ttr (tan tan radius)]: **4,3** *(Press* ENTER.*)*

3. The circle has a radius of 0.25 units:

    Specify radius of circle or [Diameter]: **0.25** *(Press* ENTER.*)*

COPYING CIRCLES

To draw the other three circles, you don't need to repeat the CIRCLE command; instead, you can copy the first one to the other locations on the baseplate using the COPY command.

AutoCAD needs to know two things for making copies: (1) which object(s) is to be copied; (2) the location to place the copies. Often, the location is specified as the distance from the original to the copies.

1. Select the object to be copied using AutoCAD's object selection method called "Last" (or "L" for short). It select the last-drawn object visible on the screen, as follows:

    Command: **copy** *(Press* ENTER.*)*

    Select objects: **L** *(Press* ENTER.*)*

    Notice that the circle changes its look: it is made of dashed lines.

    This effect is called "highlighting," and is how AutoCAD indicates selected objects.

2. Press ENTER to tell AutoCAD you have finished selecting objects.

    1 found Select objects: *(Press* ENTER *to end object selection.)*

3. To specify the distance that the circle should be copied, you tell AutoCAD the base point and the displacement. The *base point* is the point from which the copying takes place:

    Specify base point or [Displacement] <Displacement>: **4,3** *(Press* ENTER.*)*

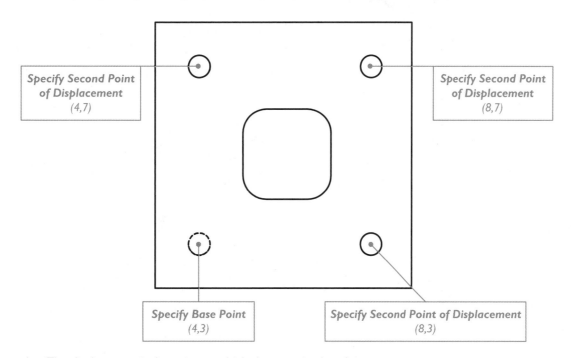

4. The *displacement* is the point at which the copy is placed:

    Specify second point of displacement or <use first point as displacement>: **8,3** *(Press* ENTER.*)*

5. The copy command repeats its displacement prompt so that you can make multiple copies:

Specify second point of displacement...: **8,7** *(Press* ENTER.*)*

Specify second point of displacement...: **4,7** *(Press* ENTER.*)*

6. Press **ESC** to tell AutoCAD you have finished making copies:

Specify second point of displacement...: *(Press* ESC *to end the command.)*

Your drawing should look similar to the figure illustrated above.

SAVING THE DRAWING

With the drawing complete, it is time to save your valuable work.

1. On the Standard toolbar, click the **Save** button (looks like a diskette, if you can recognize it!).

2. In the Save Drawing As dialog box, enter "tutorial" for the **File name**.
3. Click **Save** to save the drawing.

On the title bar, notice that the name changes from the generic [Drawing1.dwg] to [tutorial.dwg], perhaps prefixed by the path name.

PLOTTING THE DRAWING

With the drawing saved, print out a copy to show your fine work to your instructor, family, and friends! This is done with the **PLOT** command.

Before printing, AutoCAD needs to know: (1) the printer on which the drawing will be plotted, (2) the view to plot, and (3) the size. There are many other options, but these three are the most important.

1. From the **File** menu, select **Plot**. Notice that AutoCAD displays the Plot dialog box.
2. In the Printer/plotter area, select a printer from the **Name** droplist. "Default Windows System printer.pc3" is usually a safe choice.

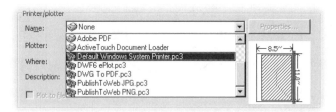

3. In the Plot Area section, select **Extents** from the What to Plot droplist. This option ensures the entire drawing is plotted.

4. In the Plot Offset section, select **Center the plot** to center the drawing on the paper.

5. In the Plot Scale area, select **Fit to Paper**. This option ensure the drawing fits the paper, no matter the size of paper.

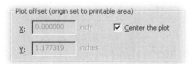

6. To ensure the plot works out correctly, and to save paper, click the **Preview** button.

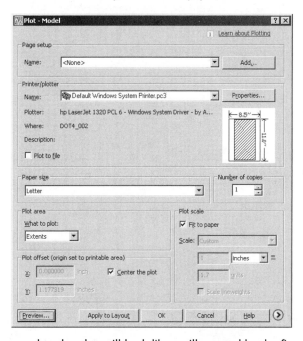

AutoCAD shows you what the plot will look like, as illustrated by the figure below.

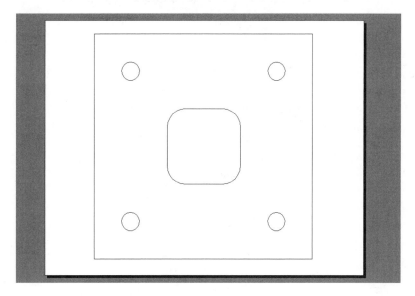

7. Press **ESC** to exit the print preview mode.

   *Press ESC or ENTER to exit, or right-click to display shortcut menu. (Press **ESC**.)*

8. Back in the Plot dialog box, choose **OK** to start the plot.

   AutoCAD displays a dialog box showing the progress of the plot. When done, a yellow alert balloon appears at the right end of the status bar, reporting on the success of the plot (or lack thereof).

   After a moment, the drawing should emerge from your printer.

9. Save the drawing with the **QSAVE** command, and then exit AutoCAD, as you learned earlier in this chapter.

3D TUTORIAL

Let's repeat the tutorial and draw the same part in three dimensions. The figure below illustrates the baseplate with its four holes and shaft with rounded edges.

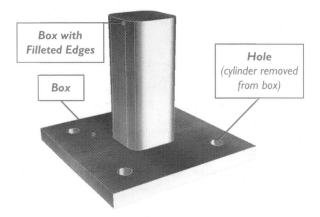

Modeling objects in 3D is quite different from 2D drafting, because 3D objects are made from *solid primitives*, parts like boxes and cylinders. For example, the base is made of a box, the shaft is another box with edges filleted, and the holes are cylinders removed from the base.

As well, you work in a different modeling workspace.

1. Start AutoCAD with the **3D Modeling** workspace. (If AutoCAD is already running, choose the **Tool** menu, **Workspaces**, and then **3D Modeling**. Start a new drawing with the *acad3d.dwt* template file.)

   Notice that AutoCAD's environment changes significantly.

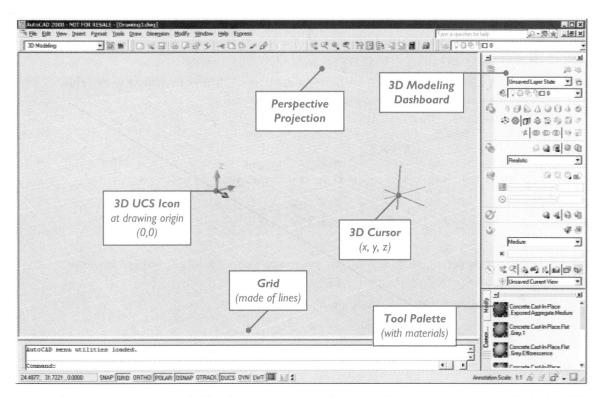

Ensure dynamic input is turned off; otherwise your results may differ from this tutorial. Click the **DYN** button on the status bar.

2. From the **Draw** menu bar, select **Modeling**, and then choose **Box**.

   As an alternative, enter the **BOX** command at the keyboard:

   Command: **box** *(Press* ENTER.*)*

3. AutoCAD next needs to know where to place the box. One corner is at the x,y coordinates of 3,2:

   Specify first corner or [Center]: **3,2** *(Press* ENTER.*)*

4. The opposite corner is located at x,y coordinates of 9,8:

   Specify other corner or [Cube/Length]: **9,8** *(Press* ENTER.*)*

5. The height is 0.5:

   Specify height or [2Point]: **0.5** *(Press* ENTER.*)*

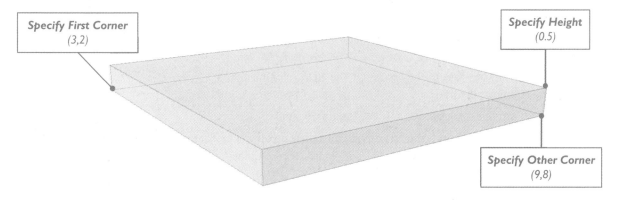

6. Your drawing should look similar to the figure above. If necessary, enter the **ZOOM** command, and select the **All** option:

   Command: **zoom** *(Press* ENTER.*)*

   Enter option [All/Extents/Window/Previous] <real time>: **all** *(Press* ENTER.*)*

MODELING THE SHAFT

The shaft is also a box, but with rounded edges. Unlike RECTANGLE, the BOX command doesn't have a built-in filleting option, so you'll need to use the FILLET command later.

1. Recall that the **SPACEBAR** repeats the previous command — without requiring you to type its name a second time. Press it to repeat the **BOX** command.

    Command: *(Press* SPACEBAR.*)*

    BOX

2. To make the boxes easier to see and work with, turn on transparency. In the Dashboard, click the **X-ray Mode** button (found in the Visual Styles control panel).

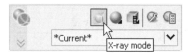

3. Specify the corners and height of the box by typing x,y coordinates, as follows:

    Specify first corner or [Center]: **5,4** *(Press* ENTER.*)*

    Specify other corner or [Cube/Length]: **7,6** *(Press* ENTER.*)*

    Specify height or [2Point]: **6** *(Press* ENTER.*)*

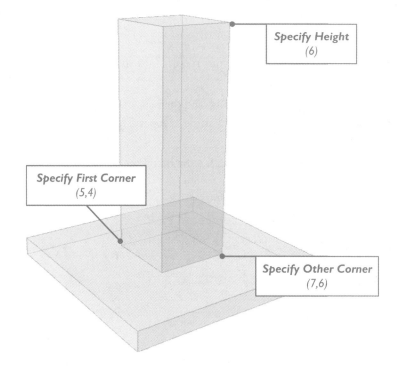

4. To round the edges of the box, enter the **FILLET** command:

    Command: **fillet**

    Current settings: Mode = TRIM, Radius = 0.0000

5. Enter the "m" option (short for multiple), because you will be filleting multiple (four) edges:

    Select first object or [Undo/Polyline/Radius/Trim/Multiple]: **m**

6. The fillets are 0.5 inches in radius:

    Enter fillet radius <0.0000>: **0.5** *(Press* ENTER.*)*

    Notice that the cursor changes to a small square; this is called the "pick" cursor.

7. Now tell AutoCAD which edges to fillet by picking each one: you must position the square pick cursor over each edge, and then click the left mouse button:

Select an edge or [Chain/Radius]: *(Pick an edge.)*

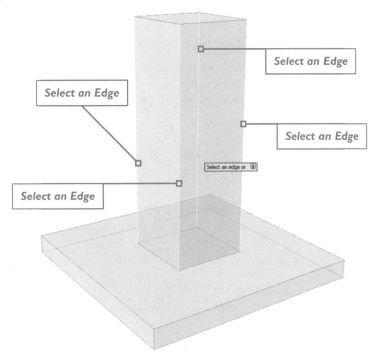

As you do, the edge turns white. Repeat for the other three edges:

Select an edge or [Chain/Radius]: *(Pick another edge.)*

Select an edge or [Chain/Radius]: *(Pick a third edge.)*

Select an edge or [Chain/Radius]: *(Pick the last edge.)*

8. With the four edges selected, press **ENTER** to exit selection mode:

Select an edge or [Chain/Radius]: *(Press **ENTER** to exit selection mode.)*

4 edge(s) selected for fillet.

9. Press **ENTER** a second time to exit the command.

Select first object or [Undo/Polyline/Radius/Trim/Multiple]: *(Press **ENTER** to exit the command.)*

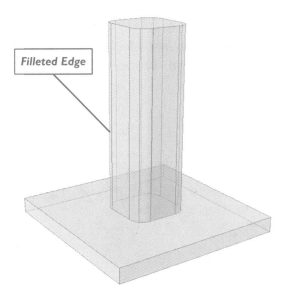

Notice that the column has rounded edges.

MODELING HOLES

The next step is to draw the holes in the baseplate. In AutoCAD, holes are constructed by drawing, then removing, cylinder shapes.

1. From the **Draw** menu, select **Modeling**, and then choose **Cylinder**.

   Command: _cylinder

2. The cylinder is located at 4,3 in the drawing:

   Specify center point of base or [3P/2P/Ttr/Elliptical]: **4,3** *(Press ENTER.)*

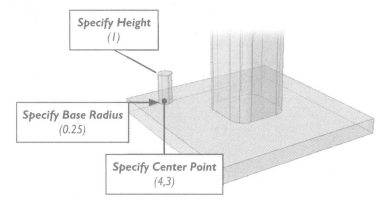

Specify Height
(1)

Specify Base Radius
(0.25)

Specify Center Point
(4,3)

3. The circle has a radius of 0.25 units:

   Specify base radius or [Diameter]: **.25** *(Press ENTER.)*

4. The height does not matter, just as long as it is taller than the base:

   Specify height or [2Point/Axis endpoint] <6.0>: **1** *(Press ENTER.)*

COPYING CYLINDER

To draw the other three cylinders, use the COPY command to copy the first cylinder to the other locations on the baseplate.

1. Start the **COPY** command, and then select the last-drawn cylinder with the **L** object selection option, as follows:

   Command: **copy** *(Press ENTER.)*

   Select objects: **L** *(Press ENTER.)*

   1 found Select objects: *(Press ENTER to end object selection.)*

2. Recall that the *base point* is the point from which the copying takes place:

   Specify base point or [Displacement] <Displacement>: **4,3** *(Press ENTER.)*

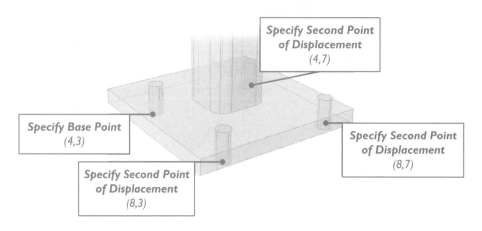

Specify Second Point
of Displacement
(4,7)

Specify Base Point
(4,3)

Specify Second Point
of Displacement
(8,7)

Specify Second Point
of Displacement
(8,3)

3.   The *displacement* is the point at which each copied cylinder is placed:

   Specify second point of displacement or <use first point as displacement>: **8,3** *(Press* ENTER.*)*

   Specify second point of displacement...: **8,7** *(Press* ENTER.*)*

   Specify second point of displacement...: **4,7** *(Press* ENTER.*)*

   Specify second point of displacement...: *(Press* ESC *to end the command.)*

SUBTRACTING CYLINDERS TO CREATE HOLES

With the four cylinders in place, you now use the SUBTRACT command to turn them into holes.

1.   From the **Modify** menu, select **Solid Editing**, and then choose **Subtract**.

   AutoCAD first asks you for the object(s) to subtract *from*; that would be the baseplate:

   Select solids and regions to subtract from ..

   Select objects: *(Select the baseplate.)*

   Select objects: *(Press* ENTER *to end object selection.)*

2.   Now AutoCAD needs to know which objects to subtract; that would be the four cylinders:

   Select solids and regions to subtract ..

   Select objects: *(Select one cylinder.)*

   Select objects: *(Select another cylinder.)*

   Select objects: *(Select a third cylinder.)*

   Select objects: *(Select the fourth cylinder.)*

   Select objects: *(Press* ENTER *to end object selection.)*

   Notice that AutoCAD removes the cylinders from the baseplate, creating four holes.

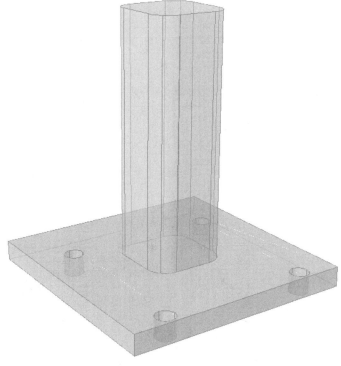

3.   The final step is to join the shaft with the baseplate. This is done with the UNION command. From the **Modify** menu, select **Solid Editing**, and then choose **Union**. AutoCAD asks you for the objects to bring together:

Select objects: *(Select the base plate.)*

Select objects: *(Select the shaft.)*

Select objects: *(Press **ENTER** to end object selection.)*

The model does not look any different, but it is now a single object.

4. Save the drawing as "3d Tutorial," and then print, if you wish.

So, that was a quick start to using AutoCAD.

Welcome to the rest of *Using AutoCAD 2008: Basics*. The remaining chapters take you step by step through the program. Before long, you'll be using AutoCAD to create drawings like a pro.

Enjoy!

EXERCISES

1. In this exercise, you start drawing in AutoCAD.
   Start AutoCAD. (If the Start New Drawing dialog box appears, click **Cancel**.)
   With the **LINE** command, draw several shapes:
   a. Rectangle.
   b. Triangle.
   c. Irregular polygon (any shape you like).

2. Use the **U** command. What happens to the last object you drew?

3. Press function key **F9**, and then look at the status line.
   Does the button look depressed (pressed in)?

4. Repeat the **LINE** command, and again try drawing these shapes:
   a. Rectangle.
   b. Triangle.
   c. Irregular polygon.
   Do you find it easier?

5. Draw a rectangle with the **RECTANGLE** command using these parameters:
   Corner   **2,3**
   Other corner   **6,7**

6. Draw a circle with the **CIRCLE** command using these parameters:
   Center point   **4,5**
   Radius   **0.75**
   Is the circle drawn "inside" the box?

7. Print your drawing with the **PLOT** command.
   Does the plot look like the drawing on your computer screen?

CHAPTER REVIEW

1. What are AutoCAD drawings constructed with?
2. Name three areas in which commands are entered:
   a.
   b.
   c.
3. Describe the purpose of the 'Command:' prompt.
4. How do you exit from print preview mode?
5. What is the purpose of the **LINE** command?
6. How is the **U** command helpful?
7. Describe how to cancel commands.
8. What does the mouse control?
9. What is an *icon*?
10. Name two methods by which you can determine the function of toolbar buttons:
    a.
    b.

11. What are *tooltips*?

12. What are *flyouts*?

13. Can toolbars be moved around the AutoCAD window?

14. Which commands close drawings without exiting AutoCAD?

15. Which commands exit AutoCAD?

16. Which command saves drawings?

17. Which command starts new drawings?

18. Describe how snap and grid are useful:

    Snap

    Grid

19. Is the grid plotted?

20. List three things AutoCAD needs to know, as a minimum, before plotting drawings:

    a.

    b.

    c.

21. Explain the advantage of the **PLOT** command's **Fit to Scale** option.

22. Why is the **Preview** option environmentally friendly?

23. How do you exit from print preview mode?

24. Identify the user interface elements illustrated below:

    a.

    b.

    c.

    d.

    e.

    f.

    g.

    h.

    i.

    j.

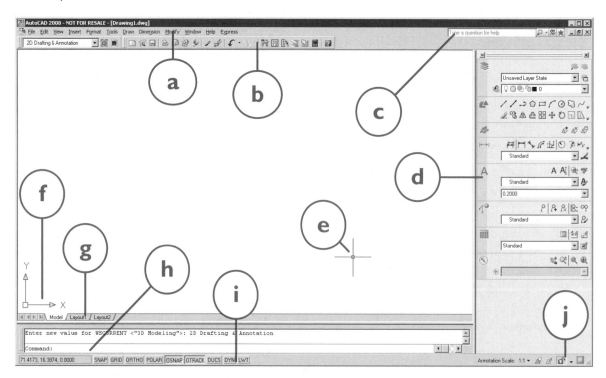

CHAPTER 2

Understanding CAD Concepts

CAD software contains concepts that distinguish them from other software programs. In part, the uniqueness stems from CAD being based on *vectors*, geometric objects that have length and direction. Using AutoCAD means knowing how to manipulate vectors; before you can manipulate vectors successfully, you need to understand how AutoCAD operates.

This chapter covers the following activities:

Touring the AutoCAD user interface.

Saving and restoring user interface configurations.

Understanding the information AutoCAD presents to you.

Entering commands.

Learning shortcut keystrokes and command aliases.

Picking points on the screen.

Using mouse buttons and menus.

Identifying the elements of the Dashboard, toolbars, and dialog boxes.

Understanding the draft-edit-plot cycle.

Recognizing AutoCAD's coordinate systems.

Entering coordinates from the keyboard and on the screen.

NEW TO AUTOCAD 2008 IN THIS CHAPTER

- AutoCAD has a new default workspace named "2D Drafting & Annotation."
- The Infobar (a.k.a. Drawing Status bar) is a new user interface element.
- The InfoCenter replaces the Communications Center.

BEGINNING A DRAWING SESSION

Your drawing sessions begin by starting the AutoCAD program. As described in Chapter 1, "Quick Start in AutoCAD," double click the AutoCAD icon to start the program.

AutoCAD 2008

(If there is no icon on the Windows desktop, click the Windows **Start** button, and then choose **Programs**. Choose the **Autodesk** folder, the **AutoCAD 2008** folder, and then the **AutoCAD 2008** program.)

As an alternative, double-click a *.dwg* file in Windows Explorer. (Avoid this method when another Autodesk application, such as Mechanical Desktop, is installed.)

When AutoCAD loads, you see either a blank drawing or one or more dialog boxes — depending on how it is configured. Follow the instructions in the boxed text to close them. You now see the *drawing area*, the window in which you draw.

AutoCAD's "2D Drafting & Annotation" workspace looks like this:

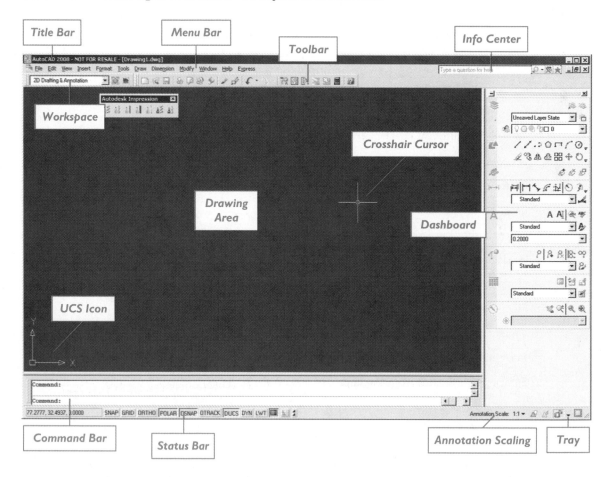

ENCOUNTERS WHILE STARTING AUTOCAD

When you first start AutoCAD, you will probably encounter numerous dialog boxes. Here is how to navigate through them:

New Features Workshop
The New Features Workshop tells you about features new to AutoCAD. For this book, decline the offer, and then click **OK**. (If you wish to pursue the workshop later, you can access it from the **Help** menu: select **New Features Workshop**.)

Startup
In some cases, the Startup dialog box may offer to start new drawings from scratch, from templates, or through wizards. You learn more about Startup later in this and following chapters. For now, click **Cancel** to close it.

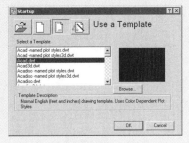

Impression Toolbar
Once AutoCAD finishes loading, it may display the Autodesk Impression toolbar. It is not required for this book. Close it by clicking the **X** at the right end of its title bar.

Workspaces
AutoCAD provides workspaces that rearrange toolbars, Dashbaord panels, and so on. For example, the **AutoCAD Classic** workspace makes AutoCAD look like it has in previous releases, as shown below.

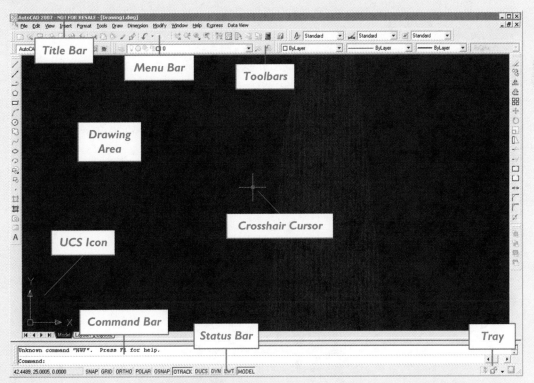

For most of this book, you use the **2D Drafting & Annotation** workspace. Select it, and then click **OK**.

AUTOCAD USER INTERFACE

Notice that AutoCAD displays information in several areas. Let's look at these areas, starting at the top and moving down the window.

Title Bar

At the very top of the AutoCAD window is the *title bar*. The title bar tells you the name of the software, AutoCAD 2008, followed in square brackets by the name of the current drawing, such as [Drawing1.dwg].

Menu Bar

Below the title bar is the menu bar, which categorizes commands, such as File, Edit, and View. For example, file-related commands appear under **File**, and many drawing command are under **Draw**.

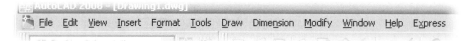

But not every part of the menu is logical; most editing commands are under **Modify** rather than **Edit**. And not all commands are found in the menus; thus, you may resort to using toolbars, the Dashboard, or typing command.

 Info Center

To the right of the menu bar is the Info Center. (It replaces the Info palette from earlier releases of AutoCAD.) When you enter terms, it accesses the online help.

The "satellite" button replaces the Communication Center from earlier releases; Autodesk uses it to provide information about bug patches and other news.

Toolbars

Just below the menu bar, AutoCAD displays toolbars. The toolbars provide access to basic services, such as saving drawings, accessing workspaces, and opening palettes.

Workspace Toolbar

The Workspace droplist allows you to switch quickly between workspaces. For more information, see Section V of this book, "Introduction to 3D Modeling."

Drawing Area

The drawing area is where you draw, but it also displays information. The most important is the *crosshair cursor*, because it shows you where you are in the drawing.

The small square at the center of the cursor indicates that you can *select* (or "pick") objects in the drawing for editing. (Pressing the left mouse button selects objects.) The square is called the "pickbox." Its size can be made larger or smaller through a system variable named **PICKBOX**.

When you move the crosshair cursor out of the drawing area, it changes to an *arrow cursor*, with which you are probably familiar from other Windows software.

The arrow cursor lets you select items outside of the drawing area.

UCS Icon

In the lower-left corner is an L-shaped arrow. It is called the *UCS icon*; UCS is short for "user coordinate system."

The UCS icon is useful primarily when you draw in three dimensions; it helps you orient yourself in three-dimensional space. Sometimes, the UCS icon changes to other shapes. You learn how to use UCSs and the icon in volume 2 of this book, *Using AutoCAD: Advanced*.

The UCS icon is not needed for 2D drafting. You can turn it off with the **UCSICON** command. Type the command name, and then type the **Off** option, as follows:

> Command: **ucsicon** *(Press ENTER.)*
>
> Enter an option [ON/OFF/All/Noorigin/ORigin/Properties] <ON>: **off** *(Press ENTER.)*

Command Prompt

Below the layout tabs is the *command prompt*. This is the traditional area for entering commands, and one place where AutoCAD responds with prompts and messages. (Other places include message boxes, tooltips, the status bar, and dynamic input prompts.) If you become confused as to what AutoCAD expects of you, the prompt area is one place to look.

The command prompt area can be dragged away from the main AutoCAD window. With the mouse, drag the two vertical bars (at the left end of the 'Command:' prompt) away from the AutoCAD window. Release the mouse button.

The floating command bar looks like this:

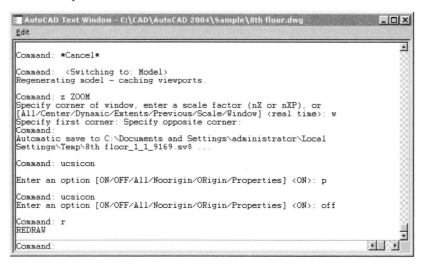

To dock it, drag the Command Line title bar against an edge of the AutoCAD window. If you close it, you can open it again by pressing CTRL+9.

Text Window

The command bar displays just 2-3 lines of command history. You can see a much longer history of commands by pressing function key F2. AutoCAD displays the *text window*.

Here you see a listing of about 20-40 of your most recent command inputs and AutoCAD's responses. Use the scroll bar to see up to 400 lines of command text.

Press F2 a second time to return to the drawing window.

Status Bar

At the bottom of the AutoCAD window is the *status bar*. This line reports the coordinates of the cursor location, and the status of numerous drawing settings.

X, Y Coordinates and Elevation

At the left of the status bar are three sets of numbers that represent the *coordinates* of the crosshair cursor. The figure above shows x, y coordinates, plus z as the current elevation. (The format of the numbers is controlled by the UNITS command, as you learn in Chapter 3, "Setting Up Drawings.")

The coordinates constantly update as you move the cursor. You can change the display to *relative* (distance<angle) readout by clicking the coordinate display.

39.2206<340, 0.0000

If the coordinates look gray and do not update as you move the cursor, then they are turned off; click the coordinate display to turn them on.

OTHER USER INTERFACE ELEMENTS

AutoCAD has a very flexible user interface, and so you don't always see all aspects at the same time. Here are some other user interface elements you may encounter:

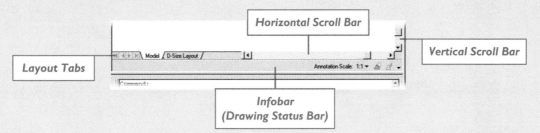

Layout Tabs

Sometimes tabs appear at the bottom of the drawing area, labeled Model, Layout1, and so on. These tabs allow you to switch between model space and layout mode. You learn more about layout mode later in this book. To turn on layout tabs, right-click the layout icons on the toolbar, and then select **Display Layout and Model Tabs**.

 Infobar (a.k.a. Drawing Status Bar)

The icons related to annotation scaling normally appear on the status bar, but you can move it to a new bar named "Infobar." Confusingly, Autodesk also refers to this as the "Drawing Status Bar," yet it is different from the status bar at the bottom of the AutoCAD drawing window. When turned on, this new bar appears just above the Command bar.

To turn on the Infobar, enter the **OPTIONS** command: choose the Display tab, and then select the **Display Drawing Status Bar** option.

Scroll Bars

Scroll bars let you move around the drawing area. If these appear, you can click or drag them to move the drawing horizontally. Moving the drawing is called "panning." To the right of the drawing area is the vertical scroll bar, which pans the drawing up and down.

To turn on scroll bars, enter the **OPTIONS** command: choose the Display tab, and then select the **Display Scroll Bars in Drawing Window** option.

Additional Toolbars

AutoCAD has many toolbars, most of which are not displayed. You can see more of them when you change the workspace to "AutoCAD Classic."

To turn on a toolbar, right-click any toolbar, and then select the toolbar name from the shortcut menu.

Mode Indicators

Next to the coordinates on the status bar are *mode indicators* that look like buttons. These are "toggles," because they turn the modes on and off — like light switches. Clicking a button turns the mode on or off; when on, the mode's button looks as if it is pressed in. In the figure below, OSNAP and MODEL are on; the other modes are turned off.

The modes have the following meaning:

Mode	Shortcut Keystroke	Meaning
SNAP	F9	Cursor snap is on or off.
GRID	F7	Grid display is on or off.
ORTHO	F8	Orthogonal mode is on or off.
POLAR	F10	Polar snap mode is on or off.
OSNAP	F3	Object snap modes are on or off.
OTRACK	F11	Object tracking mode is on or off.
DUCS	F6	Dynamic user-coordinate system.
DYN	F12	Dynamic input mode is on or off.
LWT	...	Lineweights are displayed or not.
MODEL	...	Model or paper space (layout mode) is current.

As an alternative to clicking buttons on the status bar, you can press the shortcut keystrokes listed by the table above.

NEW IN 2008 Annotation Scaling

Most objects in AutoCAD can be drawn and plotted at any size, but some cannot. Text, hatching, and linetypes must look the correct size when plotted. Annotation scaling ensures that scale-dependent objects, like text and hatches, are displayed and plotted at the correct scale factor.

Next to the status toggles, AutoCAD displays several buttons for controlling annotation scaling, as illustrated below. Annotative scaling is discussed in detail in Chapter 12, "Placing and Editing Text."

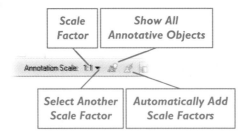

Tray

At the right end of the status bar are several icons, some of which appear only when AutoCAD needs to tell you information. This area is called the "tray." Some of the icons are displayed below:

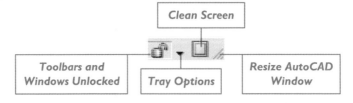

The following icons are usually present, unless turned off via Tray Options:

Toolbars and Windows Unlocked reports whether any toolbars and/or windows are locked into position, or if they are all unlocked.

Tray Options displays a shortcut menu with options for the status bar and tray.

Clean Screen maximizes the AutoCAD window, and turns off the title bar and toolbars; click the button a second time to return the window to its former size.

Resize AutoCAD changes the size of the AutoCAD window; drag the corner with the mouse; not present when AutoCAD is in full-screen mode.

The following icons appear when AutoCAD needs to communicate with you:

Associated Standards File(s) warns you when CAD standards have been breached; available only when the standards checker is active.

Attribute Extraction indicates the drawing has tables filled with linked attribute data.

Manage Xrefs displays the Xref Manager dialog box; available only when an externally-referenced drawing is attached.

Signed by Digital Signature displays the Digital Signature dialog box; available only when the drawing has been signed digitally to warn against changes.

DWG Check reports whether the drawing was created by Autodesk software.

Plot/Publish Details reports when AutoCAD is plotting, and when plots are complete. During plotting, the icon appears to move a sheet of paper back and forth — even though few people use such a plotter any more.

New to AutoCAD 2008 are the following:

Unreconciled Layers warns that layers have been added to the drawing, such as from externally-referenced drawings.

Data Link indicates that tables in the drawing are linked to external data files.

Sometimes, the icons display yellow alert balloons to warn you of changes and problems.

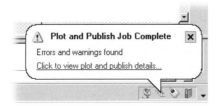

Click the blue underlined text for more information. Or click the **x** in the upper-right corner to dismiss the balloon.

Locking Toolbars and Windows

To keep the toolbars and windows from shifting on the screen, you can lock them in place. ("Windows" is the old term for palettes, such as the Properties palette — and not the AutoCAD window itself.) From the **Window** menu, select **Lock Location**, and then choose which user interface elements to lock: floating and docked toolbars, floating and docked windows, or all.

The padlock icon in the tray indicates whether any element is locked, or whether all are unlocked. Hold down the **CTRL** key to move locked items temporarily.

 **DASHBOARD**

The Dashboard is like a group of toolbars; but instead of toolbars, commands are grouped into "panels." Just as different workspaces display different toolbars, they also display different groups of panels in the Dashboard. Like toolbars, panels have buttons, droplists, and flyouts; unlike toolbars, panels can also contain sliders and can trigger specific Tool palettes.

As illustrated above, the "2D Drafting & Annotation" workspace displays panels (at left) related to 2D drafting and dimensioning, while the "3D Modeling" workspace displays panels (at right) for 3D modeling and rendering.

Some panels can expand to show more elements. When you pass the cursor over a panel, a chevron icon appears. Click the chevron, and the panel expands in size — and, curiously enough, turns orange.

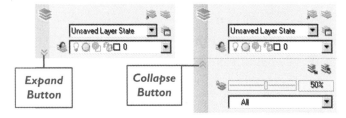

To condense a panel, click its chevron button again. The panel condenses automatically when another panel is expanded.

Some of the elements of the Dashboard are worth examining in greater detail immediately.

Layer

In the Layers panel, the layer droplist reports the status, color, and name of the current layer.

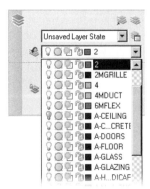

Every object resides on one layer. Every new drawing contains at least one layer called "0" (zero). Layers usually have one or more objects; each object is assigned to just one layer. The status of the layer determines the color, linetype, lineweight, plot style, visibility, and plotability of the objects — unless overridden. See Chapter 7, "Changing Object Properties," for details on controlling layers.

The small square prefixing the layer name indicates the color of the layer. The icons that prefix layer names indicate their status:

Lightbulb means the layer is on or off.

Shining sun means the layer is thawed (can be seen and edited).

Snowflake means the layer is frozen (cannot be seen or edited).

Sun with rectangle means the layer is thawed in the current paper space viewport.

Padlock means the layer is locked (can be viewed but not edited) or unlocked.

ENTERING COMMANDS

Commands can be entered in AutoCAD by several means, such at the keyboard, through dynamic input, and by keyboard shortcuts. Let's look at each.

USING THE KEYBOARD

The original method for entering commands is through the keyboard. The earliest versions of AutoCAD did not rely on a mouse, which was expensive and rare at the time. (The technical editor notes that he used a joystick until it fell out of favor due to its association with games.) Every command had to be entered with the keyboard.

Over the years, menus, toolbars, and right-click shortcut menus were slowly introduced to the user interface. Despite these changes to AutoCAD, you still need to deal with the keyboard, such as when entering coordinates or using commands that work only with the keyboard.

When typing commands at the keyboard, your input is displayed either in the command prompt area (illustrated below) or on the screen — depending on where the cursor is located.

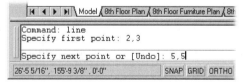

After typing the command, you must press **ENTER** or the spacebar to activate the command.

Command Input Considerations

Before typing a command, the command line must be "clear." That is, another command must not be in progress. If the 'Command' prompt is not clear, you can clear it by pressing **ESC**.

> Specify next point or [Undo]: *(Press* **ESC**.*)*
>
> Command:

Transparent commands are the exception. These commands are a selection of regular commands that, when prefixed with the ' (apostrophe) character, operate during other commands.

When you enter a command, AutoCAD doesn't care if you use UPPERCASE or lowercase characters, or a combination of either. If you make a mistake typing on the command line, press the **BACKSPACE** key to correct, or edit the text on the command line using these keys:

Keystroke	Meaning
ESC	Cancels the current command.
←	Moves the cursor one character to the left.
→	Moves the cursor one character to the right.
HOME	Moves the cursor to the beginning of the command text.
END	Moves the cursor to the end of the line.
DEL	Deletes the character to the right of the cursor.
BACKSPACE	Deletes the character to the left of the cursor.
INS	Switches between insert and typeover modes.
↑	Displays the previous line in the command history.
↓	Displays the next line in the command history.
PGUP	Displays the previous screen of command text.
PGDN	Displays the next screen of command text.
CTRL+V	Pastes text from the Clipboard into the command line.
TAB	Cycles through command names.

Press **ESC** at any time. If you press **ESC** while an operation is in progress, it terminates the command.

Repeating Commands

To repeat the command you previously entered in AutoCAD, press **ENTER** (or the spacebar) at the 'Command:' prompt. As an alternative, right-click to display the cursor menu, and then select the **Repeat** option.

Some commands repeat automatically until you press **ESC**, but most commands do not. To force a command to repeat automatically, type **MULTIPLE** before entering the command name, as in:

> Command: **multiple**

> Enter command name to repeat: **circle**

This forces the **CIRCLE** command to repeat itself until you press **ESC**.

Recent Input

AutoCAD remembers the last twenty commands and options you enter at the 'Command:' prompt, or select from a menu or toolbar. Autodesk calls this "Recent Input."

To access command history, press the ↑ key at any time, except in a dialog box. AutoCAD shows the previous commands you entered. When you see the one you need, press the **ENTER** key, and the command is executed.

During commands, you can also press the ↑ key. Instead of listing previous commands, AutoCAD shows you the coordinates you previously entered, whether at the keyboard or picked with the cursor.

If you prefer to see a list of the command history, right-click, and then select **Recent Input** from the shortcut menu. AutoCAD lists the previous commands and/or coordinate inputs in a submenu.

Dynamic Input

A second way to view commands typed at the keyboard is in the drawing area. Autodesk calls this "dynamic input," because it displays tooltips, distances, angles, and tracking lines dynamically on the screen during drawing and editing commands. (*History*: The "dynamic" name goes back to Release 9, when Autodesk introduced dialog boxes to AutoCAD as "dynamic dialogs.")

There are three aspects to dynamic input, any of which you can turn on or off:

> **Dynamic Prompts** — display commands and prompts in tooltips near the cursor.

> **Pointer Input** — displays coordinates and angles in tooltips near the cursor. They can be edited.

> **Dimension Input** — displays distances and angles in tooltips near the cursor. These appear only at the 'Specify next point:' prompt.

You turn on overall dynamic input like this:

1. On the status bar, turn on dynamic mode by clicking the **DYN** button.

2. Move the cursor into the drawing area, and then type commands, as described below.

To hide the command bar and force all command input to be dynamic, use the COMMANDLINEHIDE command. (Alternatively, press CTRL+9 to toggle the display of the command prompt bar.)

Entering Commands

When entering commands during dynamic input, the command name appears in the drawing area near the cursor, such as "circle" in the figure below.

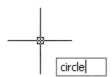

You can edit the command name by using the same keys as at the 'Command:' prompt. For example, you can press the [↑] key to scroll through the names of previous commands, or use the TAB key to step through the names of commands that start with the same letter(s). Press BACKSPACE to erase the name, or ESC to exit the command.

If you enter an incorrect command name or invalid value, the input box is outlined in red.

Entering Coordinates

Upon entering the command name and pressing ENTER, the command's prompts appear at the cursor. For the CIRCLE command, the first prompt is 'Specify center point for circle:'. Pick a point in the drawing, or interact with the dynamic prompt, as described next.

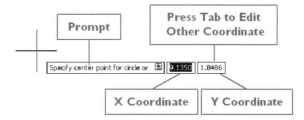

A pair of numbers appears next to the prompt. In many cases, the number represents the x and y coordinates of the cursor's current position. As the cursor moves, the values update.

The x coordinate of 9.1350 is highlighted, because it can be edited. To edit the y coordinate, press the **TAB** key; press the **TAB** key a second time to return to the x coordinate.

You can press the ↑ key repeatedly to cycle through earlier inputs. As you do, older x, y coordinates appear next to the prompt. In addition, an orange marker appears in the drawing showing the location of the coordinate.

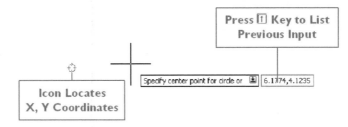

Press **ENTER** to fix the x, y-coordinate.

Selecting Options

Next to the prompt text is a small icon that looks like an arrow on a keycap.

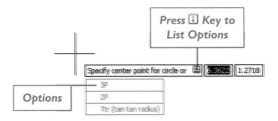

This is a reminder to press the ↓ key to see the available options. (When the arrow-key icon does not appear, the command has no options at this time.) To select an option, move the cursor to it, and then click, such as **2P** illustrated below. (The **2P** option draws two-point circles.)

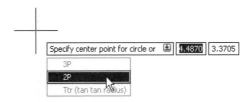

Dynamic Dimensions

After placing the first point, AutoCAD displays the **CIRCLE** command's next prompt, 'Specify radius of circle:'. Also shown is the *dynamic dimension*, which displays the size of the circle's radius.

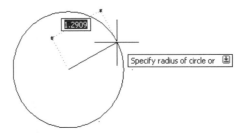

Technically, dynamic dimensions show the relative distance from the previous point to the current cursor location. The same distance is shown on the status bar, if relative coordinates are selected.

Selecting the Default Value

As before, you can press the ⬇ key to select an option. This time, the previous radius (or diameter) is shown with a black dot next to it. The black dot, as in • 0.9509, indicates this is the default value. (This is the equivalent to seeing '<0.9509>:' in the command prompt area.)

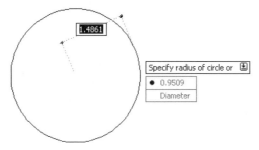

To select the default, press **ENTER** or pick the value with the cursor.

Dynamic Angles

Dynamic dimensions also display angles. (The angle was not shown for constructing the circle, because the angle doesn't matter for circles specified by a radius.) To see the angle, start the **LINE** command, and then pick two points. After picking the first point, notice that the dynamic angle appears:

The angle of 37° is measured from the positive x axis.

Error Messages and Warnings

Dynamic input also displays error messages and warnings at the cursor.

When you enter an incorrect value for coordinates, AutoCAD displays a red box around the problem, such as the "w" below:

Modifying Dynamic Input

AutoCAD allows you to modify many aspects of dynamic input. To make changes, right-click the **DYN** button, and then select **Settings** from the shortcut menu.

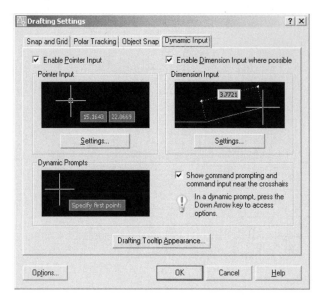

Notice that you can toggle pointer input, dimension input, and command prompting. Click the **Settings** buttons to determine how and when dynamic input appears.

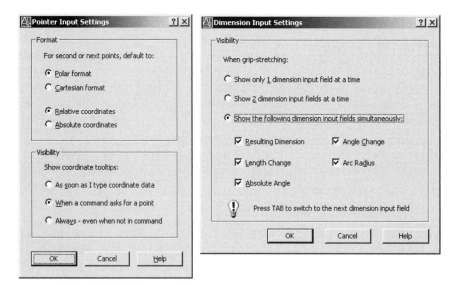

To change the color and size of the tooltips, click the **Drafting Tooltip Appearance** button.

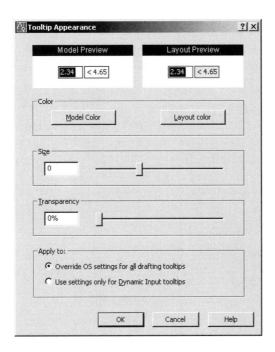

Keyboard Shortcuts

In addition to accepting the full name of commands, AutoCAD allows you to type shortcuts. AutoCAD recognizes several types of shortcuts, including *aliases,* *function keys*, and *control keys*. The advantage to shortcuts is that you execute commands more quickly.

AutoComplete

AutoCAD has around one thousand commands and system variables, and it can be difficult to remember the correct spelling of each. For example, do you enter **POLYLINE** or **PLINE**? **POLYGON** or **PGON**? One way to deal with the confusion is to select command names from the menu; however, you have to know the menu in which the command resides.

Another issue: the commands in menus and toolbars change from release to release.

An alternative is to type the start of a command, such as "p," and then press the **TAB** key. Each time you press **TAB**, AutoCAD displays the next command, system variable, or alias in alphabetical order. Autodesk calls this "AutoComplete."

For example, enter p, and then press **TAB**:

> Command: **p** *(Press* **TAB**.*)*
>
> Command: P *(Alias for* **PLOT** *command; press* **TAB**.*)*
>
> Command: PA *(Alias for* **PASTESPEC** *command; press* **TAB**.*)*
>
> Command: PAGESETUP *(Press* **ENTER** *to accept this command.)*

Press **SHIFT+TAB** to go backwards through the list. For example, enter "p" and then press **SHIFT+TAB**. The next to appear is the **PURGE** command, and then the **PUCSBASE** system variable.

When you reach the command you need, press **ENTER** to execute it.

Function Keys

Function keys appear on the top row of your keyboard; they are all prefixed with the letter F (short for "function"). AutoCAD assigns meanings to many of the function keys:

Function Key	Meaning
F1	Calls up the help window.
F2	Toggles between the graphics and text windows.
F3	Toggles object snap on and off.
F4	Toggles tablet mode on and off; you must first calibrate the tablet before toggling tablet mode.
F5	Switches to the next isometric plane when in iso mode; the planes are displayed in order of left, top, right, and then repeated.
F6	Toggles dynamic UCS on and off.
F7	Toggles grid display on and off.
F8	Toggles ortho mode on and off.
F9	Toggles snap mode on and off.
F10	Toggles polar tracking on and off.
F11	Toggles object snap tracking on and off.
F12	Toggles dynamic input mode.
CTRL+F4	Closes the current drawing.
ALT+F4	Exits AutoCAD.
ALT+F8	Displays the Macros dialog box.
ALT+F11	Starts the Visual Basic for Applications editor.
ALT+TAB	Switches to the next application.
CTRL+TAB	Switches to the next drawing (window).

Aliases

Aliases are abbreviations of full command names. For example, the alias of the LINE command is L. To execute the LINE command more quickly, press L, followed by ENTER. AutoCAD executes the LINE command just as if you had entered the full command name. Examples of common aliases include:

Alias	Command
a	Draws arcs.
aa	Finds the area.
adc	Opens the Design Center.
b	Creates blocks (symbols).
c	Draws circles
co	Copies objects.
h	Opens the Hatch dialog box.
i	Inserts blocks.
l	Draws lines.
la	Opens the Layers dialog box.
le	Draws leaders.
m	Moves objects.
t	Places text in drawings.
z	Executes the ZOOM command.

There are many other command aliases, and you can create your own. The full list provided with AutoCAD is described in Appendix B.

Shift Keys

AutoCAD uses the **SHIFT** key to override drafting settings temporarily. For some, there are two shortcuts, one for left-hand and another for right-handed mouse users.

Shift-key		Meaning
SHIFT		Toggles orthogonal mode.
Right-handed	**Left-handed**	
SHIFT+A	SHIFT+'	Toggles object snap mode.
SHIFT+C	SHIFT+,	Overrides CENter osnap.
SHIFT+D	SHIFT+L	Disables all snap modes and tracking.
SHIFT+E	SHIFT+P	Overrides ENDpoint osnap.
SHIFT+Q	SHIFT+]	Toggles osnap tracking mode.
SHIFT+S	SHIFT+;	Enables osnap enforcement.
SHIFT+V	SHIFT+M	Overrides MIDpoint osnap.
SHIFT+X	SHIFT+.	Toggles polar mode.

Control Keys

You use control-key shortcuts by holding down the **CTRL** key, and then pressing another key. You may already be familiar with some of them, such as **CTRL+C** to copy to Clipboard, and **CTRL+Z** for undo. AutoCAD displays its control-key shortcuts in the drop-down menus, next to the related command. These keystrokes can also be customized (changed) with the **CUI** command.

Ctrl-key	Meaning
CTRL	Unlocks the position of toolbars and windows temporarily, and prevents toolbars and windows from docking.
	Cycles through overlapping objects during object selection.
	Selects faces, during 3D editing.
CTRL+0	Toggles clean screen.
CTRL+1	Toggles the Properties palette.
CTRL+2	Toggles the AutoCAD DesignCenter palette.
CTRL+4	Toggles the Sheet Set Manager palette.
CTRL+6	Launches dbConnect.
CTRL+7	Toggles the Markup Set Manager palette.
CTRL+8	Toggles the Quick Calc palette.
CTRL+9	Toggles the command bar.
CTRL+A	Selects all unfrozen objects in the drawing.
CTRL+B	Turns snap mode on or off.
CTRL+C	Copies selected objects to the Clipboard.
CTRL+D	Changes the coordinate display mode.
CTRL+E	Switches to the next isoplane.
CTRL+F	Toggles object snap on and off.
CTRL+G	Turns the grid on and off.
CTRL+H	Toggles pickstyle mode.
CTRL+K	Creates hyperlinks.
CTRL+L	Turns ortho mode on and off.
CTRL+N	Starts a new drawing.
CTRL+O	Opens drawings.
CTRL+P	Prints drawings.
CTRL+Q	Quits AutoCAD.

Ctrl-key	Meaning
CTRL+R	Switches to the next viewport.
CTRL+S	Saves drawings.
CTRL+T	Toggles tablet mode.
CTRL+V	Pastes from the Clipboard into the drawing or command prompt.
CTRL+X	Cuts selected objects to the Clipboard.
CTRL+Y	Performs the REDO command.
CTRL+Z	Invokes the U command.
CTRL+SHIFT+A	Toggles group selection mode.
CTRL+SHIFT+C	Copies selected objects with a base point to the Clipboard.
CTRL+SHIFT+S	Displays Save Drawing As dialog box.
CTRL+SHIFT+V	Pastes with an insertion point.

USING MOUSE BUTTONS

AutoCAD recognizes as many as sixteen buttons on mice and digitizer pucks. Most mouse devices, of course, are limited to two, three, or four buttons, but digitizer pucks commonly have four, twelve, or sixteen buttons.

By default, AutoCAD defines the first ten buttons found on input devices, as follows:

Button #	Mouse Button	Meaning
1	Left	Selects commands or objects.
2	Right	Displays shortcut menus.
3	Center	Displays object snap menu (when MBUTTONPAN = 0).
SHIFT+1	SHIFT+Left	Toggles cycle mode.
SHIFT+2	SHIFT+Left	Displays object snap shortcut menu.
CTRL+2	CTRL+Right	Displays object snap shortcut menu.
...	Wheel	Zooms or pans.
4		Cancels command.
5		Toggles snap mode.
6		Toggles orthographic mode.
7		Toggles grid display.
8		Toggles coordinate display.
9		Switches isometric plane.
10		Toggles tablet mode.

Buttons 11 through 16 are not predefined by AutoCAD. It is possible to change the meaning of all the buttons except the first (left) button, which is always the Select button. (The Windows mouse driver can reverse the meaning of the left and right buttons for left-handed users.) You learn more about button customization in *Using AutoCAD: Advanced*.

Left Mouse Button

Earlier, I noted that you can pick points in the drawing with the cursor. By moving the mouse, you control the movement of the cursor. To pick a point, press the mouse's left button.

Right Mouse Button

In many cases, pressing the right mouse button (when inside the AutoCAD window) displays a *context-sensitive shortcut menu*. "Context-sensitive" means that the content of the menu changes, depending on *where* you right-click and *which* command is in effect.

The shortcut menu (cursor menu) appears at the cursor when you press the right-mouse button.

Mouse Wheel

The wheel on a mouse can zoom and pan the drawing without invoking the ZOOM and PAN commands.

To zoom in or out, roll the wheel forward (zoom in) and backward (zoom out). To zoom the drawing to its extents, double-click the wheel button.

To pan about the drawing, hold down the wheel and drag the mouse.

In some cases, you might prefer that the wheel act as a button. You can do this by setting the value of the **MBUTTONPAN** system variable to 0. Now when you click the wheel, it acts like the middle button of a three-button mouse. Some mouse brands come with more than three buttons. This allows you to access additional AutoCAD functions.

Selecting Commands from Menus

Click an item on the menu bar, and a menu appears on the screen. You then choose a command from the menu. (If you pull down a menu and do not want to select an item, you can exit the menu by clicking the menu name a second time.) To access items on the dropdown menu from the keyboard, press ALT, and then press the underlined letter. For example, to start the **Line** command, which is found in the **Draw** menu:

1. Press and hold **ALT**.
2. Notice that the <u>D</u>raw menu has the D underlined. Press **d**.
3. Notice that the <u>L</u>ine command has the L underlined. Press **l**.

(A bug in AutoCAD causes numerous commands to lack underlined letters, making keyboard access more difficult.)

Menus use punctuation marks as shorthand to indicate special meaning. Look for these marks:

▶ Arrowheads indicate menu items that contain submenus. A menu item can have one, two, or more submenus. In the figure above, the **Display** item displays two submenus.

✓ Check marks indicate the menu item is turned on. The check mark next to **On** means the UCS icon is displayed.

... Ellipses indicate menu items that display dialog boxes; think of the ellipses as "more to come." In the figure, **Named Views...** displays the Viewport dialog box.

Gray Text indicates menu items are not available at the current time. In the figure, 3D Views is not available in paper space.

Highlighting indicates the menu item will be selected when you click the mouse's left button.

USING DIALOG BOXES

Many commands require you to specify options in dialog boxes. The benefit of dialog boxes is that they display all your options at the same time, unlike the command-line, which typically displays one option at a time. The dialog box and the command-line for creating arrays are compared below.

Command: **array**

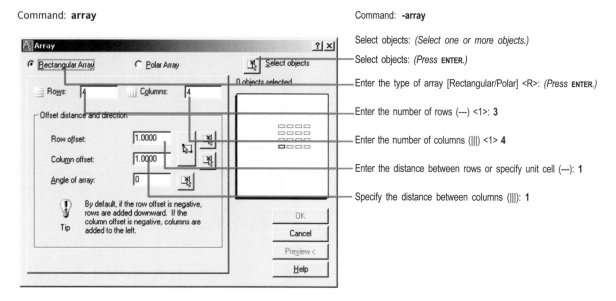

Command: **-array**

Select objects: *(Select one or more objects.)*

Select objects: *(Press* ENTER.*)*

Enter the type of array [Rectangular/Polar] <R>: *(Press* ENTER.*)*

Enter the number of rows (---) <1>: **3**

Enter the number of columns (||||) <1> **4**

Enter the distance between rows or specify unit cell (---): **1**

Specify the distance between columns (||||): **1**

The drawback to dialog boxes is that they obscure drawings. On the other hand, dialog boxes present options grouped logically; command-line prompts guide you through the steps needed to complete the action, but lack the previews and hints found in dialog boxes. The command-line steps you through options serially; dialog boxes allow you to jump to options directly.

A number of commands can display a dialog box or display prompts on the command line. There are several reasons for this. One is historical: until Release 9, AutoCAD didn't use dialog boxes at all, so all commands had to be operated at the command prompt. Also, many commands don't need a dialog box, such as drawing and editing commands like **LINE** and **COPY**. A third reason is that you may prefer using the command line, because it can be faster than using dialog boxes. For example, I find it faster to type the **-VPORTS** command and its options than to use the Viewports dialog box. A final reason is technical: script files and AutoLISP routines cannot control dialog boxes, and so must use the command-line version. There are two ways to suppress dialog boxes.

Hyphen Prefix

Prefixing some commands with a hyphen (–) suppresses the related dialog box, displaying prompts at the command line instead. For example, the **ARRAY** command normally displays the Array dialog box:

> Command: **array**
>
> *(AutoCAD displays dialog box, shown above.)*

Prefixing the command with a hyphen displays the command line options:

> Command: **-array**
>
> Select objects: *(Select one or more objects.)*
>
> Select objects: *(Press* ENTER.*)*
>
> *(Command continues with the prompts listed above.)*

Not all commands have the hyphen option; the following table lists some of those that do:

Commands	Hyphenated	Commands	Hyphenated
ARCHIVE	-ARCHIVE	PAN	-PAN
ARRAY	-ARRAY	PARTIALOAD	-PARTIALOAD
ATTDEF	-ATTDEF	PLOT	-PLOT
ATTEDIT	-ATTEDIT	PLOTSTAMP	-PLOTSTAMP
ATTEXT	-ATTEXT	PLOTSTYLE	-PLOTSTYLE
BEDIT	-BEDIT	PSETUPIN	-PSETUPIN
BLOCK	-BLOCK	PUBLISH	-PUBLISH
BOUNDARY	-BOUNDARY	PURGE	-PURGE
COLOR	-COLOR	REFEDIT	-REFEDIT
DIMSTYLE	-DIMSTYLE	RENAME	-RENAME
EATTEXT	-EATTEXT	SCALELISTEDIT	-SCALELISTEDIT
GROUP	-GROUP	STYLE	-STYLE
HATCH	-HATCH	TABLE	-TABLE
HATCHEDIT	-HATCHEDIT	TEXT	-TEXT
IMAGE	-IMAGE	TOOLBAR	-TOOLBAR
IMAGEADJUST	-IMAGEADJUST	UNITS	-UNITS
INSERT	-INSERT	VBARUN	-VBARUN
LAYER	-LAYER	VIEW	-VIEW
LINETYPE	-LINETYPE	VPORTS	-VPORTS
MTEXT	-MTEXT	WBLOCK	-WBLOCK
OPENSHEETSET	-OPENSHEETSET	XBIND	-XBIND
OSNAP	-OSNAP	XREF	-XREF

There are some curious exceptions: **-PARTIALOPEN** is the true name, while **PARTIALOPEN** is its alias. Similarly, **-SHADEMODE** is the true name, while **SHADEMODE** is an alias for the **VSCURRENT** command.

System Variables

Whether or not certain commands display dialog boxes is controlled by system variables. *System variables* store the current state of AutoCAD and its many settings.

FileDia

When the **FILEDIA** system variable is set to 0, AutoCAD suppresses the display of file-related dialog boxes, such as those associated with the **NEW**, **OPEN**, **SAVEAS**, and **VSLIDE** commands. Prompts are displayed at the command line:

> Command: **filedia**
>
> Enter new value for FILEDIA <1>: **0**
>
> Command: **open**
>
> Enter name of drawing to open: *(Enter path and name of .dwg drawing file.)*

To force the display of the dialog box, type the tilde (~) character when AutoCAD prompts for the file name:

> Command: **open**
>
> Enter name of drawing to open: **~**
>
> *(AutoCAD displays the Select File dialog box.)*

Expert

When the **EXPERT** system variable is set to a value other than zero, it suppresses warning dialog boxes displayed by AutoCAD. Examples include "About to regen, proceed?" and "Really want to turn the current layer off?"

AttDia

When the **ATTDIA** system variable is set to 0, the **INSERT** command displays prompts for attribute data at the command line; when set to 1, a dialog box is displayed instead. (Curiously, though, in this age of dialog boxes and palettes, 0 continues to be the default.)

Alternate Commands

In a very few cases, AutoCAD uses one command name for the command line, and a different one for dialog boxes. Examples include:

Command Line	Dialog Box
FILEOPEN	OPEN
CHANGE	PROPERTIES
UCS	UCSMAN

Controlling Dialog Boxes

When a dialog box opens, the arrow cursor replaces the crosshair cursor. Use the cursor to choose options in the dialog box. Then choose the **OK** button to accept the changes and dismiss the dialog box.

Some dialog boxes have an **Apply** button to apply changes you made, yet keep the dialog box open. Others have a **Preview** button that temporarily dismisses the dialog box, so that you can see the changes. To dismiss the dialog box without affecting the drawing, choose **Cancel**.

Almost all dialog boxes contain a **Help** button. Selecting the button displays a help box that explains the purpose and use of the dialog box. Many also have a **?** button in the upper-right corner: click it, and then click an item in the dialog box for a context-sensitive explanation.

You can move dialog boxes around the screen by dragging them by their *title bar*. This means that you position the cursor over the title bar, and holding down the left mouse button, move the dialog box. Moving the dialog box is useful if you need to see objects under the box.

Most dialog boxes are *modal*, which means you must dismiss them before you can continue working in AutoCAD. A few are *non-modal*, which means they hover on your Windows desktop as you continue working in AutoCAD; these are called "palettes." An example is the Properties palette. To dismiss a non-modal dialog box, click the small **x** in the upper corner.

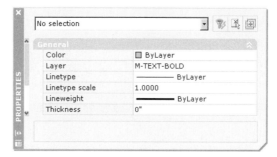

Other dialog boxes contain *tabs*, which allow a single dialog box to display a lot of information, separated into similar-looking boxes. For example, the **OPTIONS** command displays a dialog box with nine tabs!

Some dialog boxes contain additional or sub-dialog boxes. If a sub-dialog box is displayed, you must choose **OK** or **Cancel** to return to the original dialog box. In the figure below, New Text Style is the sub-dialog box displayed by the **New** button.

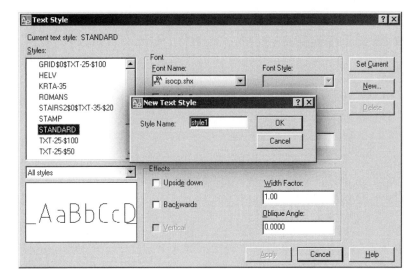

Buttons in Dialog Boxes

Buttons are used to select items. Select buttons with the cursor or by pressing keys on the keyboard. To move from button to button, press the TAB key.

- The button with the second border is the default. Pressing **ENTER** is like choosing the button with the cursor. In the figure above, the **OK** button has the second border.

- Three periods (...) after a button's name opens a sub-dialog box. An example is the **New...** button in the dialog box illustrated above.

- Buttons with a single arrow pointing left (**<**) indicate action is required in the drawing, such as selecting an object. An example is the **Preview <** button illustrated below.

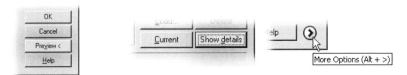

- Buttons with a round arrow button expand the dialog box. Alternatively, the dialog box may have a **Show Details** button to expand it.

- If the text on the button is gray, the button's function is not available at this time, such as the **Delete** button in the Text Style dialog box, above.

The following is a summary of the buttons used in dialog boxes.

Check Boxes

Check boxes *toggle* options, which means the option is either on or off.

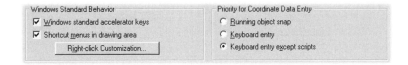

- ☑ A check mark indicates the option is selected (or turned on).

- ☐ No check mark means the option is turned off.

Radio Buttons

Two (or more) round radio buttons force you to select one option out of several. Only one option is active at a time.

⊙ The black dot in the button indicates it is selected or turned on.

○ No dot indicates it is unselected or turned off.

Select Objects Button

Some dialog boxes allow you to pick objects or points in the drawing. When you click the ⌧ **Select Objects** button, AutoCAD temporarily dismisses the dialog box, and prompts you to pick the points, such as:

Pick a point: *(Pick a point.)*

Pick a point (or press ENTER to accept): *(Press ENTER to return to the dialog box.)*

Once done, the dialog box returns.

List Boxes

List boxes contain a list from which you choose one or more items. When the list is too long for the box, a scroll bar appears automatically, allowing you to scroll to choices not visible. To choose more than one item, hold down the CTRL key as you select items; not all list boxes allow this.

Scroll Bars

Some list boxes contain more entries than can be displayed at a time. Scroll bars exist on some list boxes to bring additional items or options into view by scrolling up or down. To display another entry below the current field of view, click on the down arrow of the scroll bar.

Droplists

Droplists normally display a single item. They have an arrow button; when you click the button, a list "drops down" to show the entire list. An example is shown below from the Hatch dialog box.

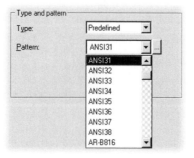

You can also move through the scroll list by clicking on the bar, and while continuing to hold the mouse button down, sliding the bar up or down with the cursor. When you click again, the entries are redisplayed at the new location. Note that the position of the scroll bar is relative to the position of the displayed items. Thus, if there are many items in the list, a relatively small movement of the scroll bar moves by several items.

Text Edit Boxes

Text edit boxes contain a single line of text (or, in some cases, a paragraph of text) to be edited. To edit the text, click in the box. A cursor bar appears at the text. You move the cursor bar with the arrow keys on the keyboard.

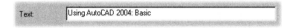

Image Tiles

Image tiles are small windows that display graphical images. Select the image to select the option, such as the 3D surface objects illustrated below.

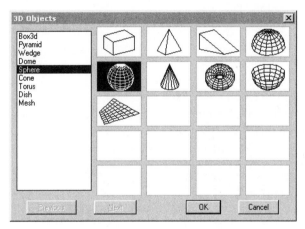

This style of dialog box was common in the early days of graphical user interfaces, but now is rarely used by AutoCAD.

THE DRAFT-EDIT-PLOT CYCLE

Once you set up a drawing, working with AutoCAD involves a cycle, consisting of drafting, editing, and plotting. You draft drawings, edit, and then plot them. This book is arranged in a similar manner:

Cycle	Chapters
Setup	1: Quick Tour of AutoCAD 2008
	2: Understanding CAD Concepts
	3: Setting Up Drawings
Drafting	4: Drawing with Basic Objects
	5: Drawing with Precision
	6: Drawing with Efficiency
Editing	7: Changing Object Properties
	8: Correcting Mistakes
	9: Direct Editing of Objects
	10: Constructing Objects
	11: Additional Editing Options
Text	12: Placing and Editing Text
	13: Placing Dimensions
	14: Editing Dimensions
	15: Geometric Dimensioning and Tolerancing
	16: Tables and Fields
	17: Reporting on Drawings
Intro to 3D	18: 3D Modeling and Viewing
	19: 3D Editing
	20: 3D Rendering
Plotting	21: Layouts
	22: Plotting Drawings
Appendices	A: Computers and Windows
	B: AutoCAD Commands, Aliases, and Keyboard Shortcuts
	C: AutoCAD System Variables
	D: AutoCAD Toolbars and Menus
	E: AutoCAD Fonts
	F: AutoCAD Linetypes
	G: AutoCAD Hatch Patterns
	H: AutoCAD Lineweights
	I: DesignCenter Symbols & Dynamic Block Libraries

This book concentrates on two-dimensional drafting, editing, and plotting — with an introduction to 3D modeling and rendering. Volume 2 of this book, *Using AutoCAD: Advanced*, covers three-dimensional drafting and rendering, customizing AutoCAD, and working with multiple drawings (sheet sets, raster images, and externally-referenced drawings).

DRAFTING

Drafting consists of placing and constructing objects in the drawing. AutoCAD comes with a set of objects, from which everything else in the drawing must be made:

Objects	Drawn with Command(s)
Points	POINT
Lines	LINE, TRACE, PLINE, *and* 3DPLINE
Parallel lines	MLINE
Construction lines	RAY *and* XLINE
Circles	CIRCLE *and* DONUT
Ellipses	ELLIPSE
Arcs	ARC *and* PLINE
Elliptical arcs	ELLIPSE
Regular Polygons	POLYGON
Rectangles	RECTANG
Splines	SPLINE *and* HELIX
Irregular boundaries	BOUNDARY
Revision clouds	REVCLOUD
3D meshes	AI_BOX, AI_DOME, AI_DISH, AI_PYRAMID, AI_CONE, CYLINDER, AI_SPHERE, AI_TORUS, AI_WEDGE, AI_ MESH, SURFTAB, RULETAB, *and* 3DFACE
3D surfaces	PLANESURF, LOFT, SWEEP, THICKEN
3D solids	BOX, CONE, CYLINDER, SPHERE, TORUS, PYRAMID, POLYSOLID, *and* WEDGE
Text	TEXT, MTEXT, *and* FIELD
Tables	TABLE
Dimensions	*All commands starting with* DIM, *plus* LEADER, QDIM, QLEADER, TOLERANCE, *and* MLEADER*

* New to AutoCAD 2008

Constructing Objects

When you cannot draw an object with one of AutoCAD's commands, you must *construct* it. For example, AutoCAD has no commands for drawing right-angle triangles and irregular polygons, so you construct them with the **LINE** or **PLINE** commands.

In addition, AutoCAD has a collection of commands to construct copies of objects:

Copy Method	Copied with Command(s)
Copy objects	COPY
Copy blocks	INSERT, XBIND *and* ADCENTER
Offset copies	OFFSET
Mirror copies	MIRROR *and* MIRROR3D
Array copies	ARRAY *and* 3DARRAY
Array as block	MINSERT
Copy properties	MATCHPROP *and* MATCHCELL
Copy faces, edges	SOLIDEDIT, FLATSHOT *and* IMPRINT
Copy along paths	DIVIDE *and* MEASURE

Blocks and Templates

One of Autodesk's cofounders, John Walker, said that you should never have to draw the same object twice in CAD. The best way to reuse drawing details is to turn them into *blocks*. Blocks can be used repeatedly in drawings, shared between drawings and offices, and can also hold database information, called "attributes."

Sometimes you start drawings with *templates*, other drawing files that already contain previously-created settings and objects. The most common templates consist of a drawing border, title block, and perhaps layers. Many firms also use templates for standard drawings of details. Illustrated below is a new drawing begun as a template, in this case the "Tutorial-iArch" drawing border and title block.

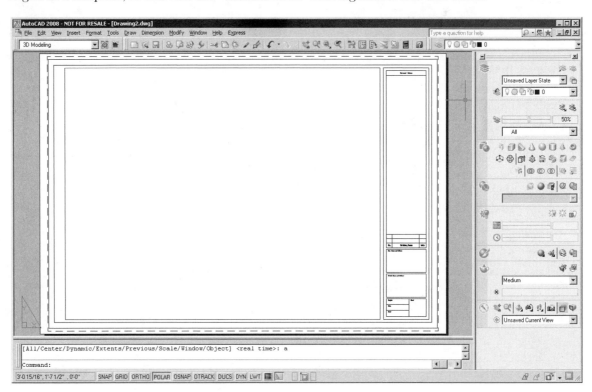

EDITING

Editing consists of modifying objects in the drawing, correcting mistakes, moving objects, changing your mind, your boss changing his mind, and so on.

Editing Method	Edited with Command(s)
Move objects	MOVE, STRETCH, ALIGN, 3DMOVE
Rotate objects	ROTATE, ROTATE3D, 3DROTATE
Resize objects	SCALE
Stretch objects	STRETCH, LENGTHEN, EXTEND
Mirror objects	MIRROR
Erase objects	ERASE
Trim objects	TRIM, BREAK
Fillet objects	FILLET
Chamfer objects	CHAMFER
Undo editing	U, UNDO, REDO, MREDO
Edit splines	SPLINEDIT
Edit multilines	MLEDIT
Edit text	DDEDIT
Change text size	SCALETEXT, STYLE
Align text	JUSTIFYTEXT, STYLE

Editing Method	Edited with Command(s)
Edit attributes	EATTEDIT, ATTIPEDIT*
Edit polylines	PEDIT
Edit 3D solids	UNION, INTERSECT, SUBTRACT, SOLIDEDIT
Edit 3D meshes	PEDIT
Reduce objects	EXPLODE, XPLODE
Change properties	CHANGE, PROPERTIES

* New to AutoCAD 2008

In addition to using commands to edit, you can edit objects directly. For example, double-clicking most objects brings up the Properties palette, which lets you change their properties; double-clicking text brings up the appropriate text editor.

Single-clicking objects produces *grips* (handles), allowing you to stretch, move, copy, rotate, mirror, and resize objects.

In addition to editing objects, AutoCAD has a collection of commands that *construct* objects:

Construction Type	Constructed with Command(s)
Hatch patterns	HATCH, GRADIENT, HATCHEDIT
Filled areas	SOLID, HATCH
Blank areas	WIPEOUT
3D meshes	REVSURF, TABSURF, RULESURF, EDGESURF
3D revolutions	REVOLVE
3D extrusions	EXTRUDE
3D solids	INTERFERE, SLICE, SECTION, IMPRINT, PRESSPULL, SWEEP, LOFT

PLOTTING

When your drawings are complete, you print them on paper (a process called "plotting"), or send them through the Internet, called an "e-plot," short for "electronic plot." When plots are created before the drawing is complete, they are called "check plots."

AutoCAD's **PREVIEW** command displays plots before they are committed to paper, while the **PLOTSTAMP** command labels plots with the drawing file name, date and time of plot, and other information.

AutoCAD supports two kinds of plots: *color based* and *style based*.

- **Color-based** plotting is the older method, whereby the colors of objects determine the pens (or inkjets) with which they are plotted.

- **Style-based** plotting applies *styles* to layers and individual objects; these styles affect every aspect of how objects can be plotted: color of lines, width of lines, screen percentage, style of end caps, and even whether the objects are plotted at all.

ePlots

In addition to plotting drawings on paper, you can also plot drawings to cyberspace. To send copies of drawings — but not the original *.dwg* drawing file — you save the drawing as *.dwf* files (short for "design web format") or *.pdf* files (short for "portable document format").

This is sometimes called an "electronic plot." The *.dwf* and *.pdf* files look just like the original, but cannot be edited. The "DWF ePlot" plotter configuration of the **PLOT** command creates *.dwf* eplots, while "PDF ePlot" creates *.pdf* plots.

To view *.dwf* files on your computer and over the Internet, Autodesk provides the DWF Viewer software with AutoCAD; to mark up drawings, download the free Design Review from www.autodesk.com/dwf.

To view *.pdf* files, most computers include the Acrobat Reader from Adobe; you can download it free from www.adobe.com/acrobat. Viewing software allows you to zoom in and out, select predefined views, and print drawings.

The **PUBLISHTOWEB** command generates Web pages and images of the drawings, while the **3DDWF** command creates 3D versions of DWFs.

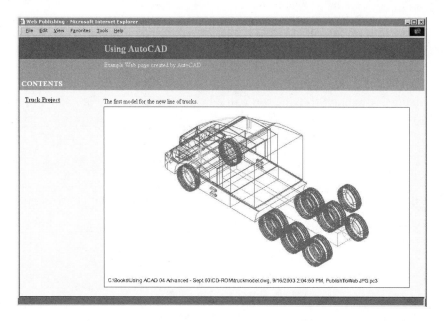

PROTECTING DRAWINGS

To protect drawings, AutoCAD includes *password* protection and digital signatures. When you save drawings with passwords (through the **SECURITYOPTIONS** command), they cannot be opened unless the correct passwords are entered.

Be careful: If you forget the password, the drawing is rendered useless. In this regard, it's best not to use passwords at all.

A different level of security is the *digital signature*. When drawings are "signed" digitally, a warning appears after they have been edited or altered. This alerts you that the drawings may not be originals; the system does not, however, tell you *what* has been changed. You must purchase a digital ID from a certificate authority.

SCREEN POINTING

You enter points, distances, and angles in two ways: by "showing" AutoCAD the information on the screen, or by entering coordinates and angles at the keyboard.

Picking two points in the drawing could indicate, for example, a distance, an angle, or both — depending on the command and what it expects as input. With AutoCAD, you press the left mouse button to pick points in the drawing.

Showing Points by Window Corners

Some commands require input of both a horizontal and a vertical displacement. This is shown by a rectangle on the screen, called the "window." You pick a point in the drawing, and then move the cursor to form the rectangle. AutoCAD determines the x and y distances by measuring from the lower left corner to the upper right corner.

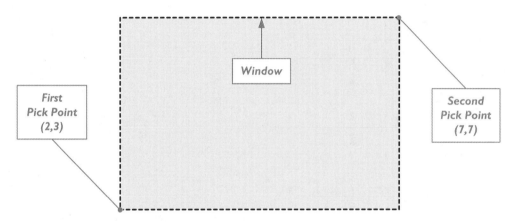

For example, the window in the figure above starts at (2,3) and rises to (7,7). To AutoCAD, this indicates a displacement of (5,4) by using subtraction:

Horizontal displacement (x)   = 7 - 2 = 5

Vertical displacement (y)   = 7 - 3 = 4

COORDINATE SYSTEMS

AutoCAD works with *coordinate* systems. Coordinates locate objects in the drawing. (Technically, AutoCAD does not record lines and other objects, but rather their endpoints and properties.) All 2D drafting is performed on the x,y-plane; for 3D drafting, you can redefine this plane to any orientation in space through UCSs.

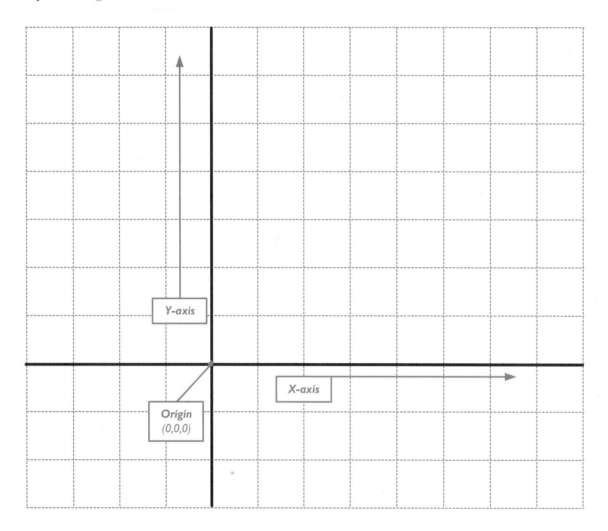

AUTOCAD COORDINATES

AutoCAD works with several coordinate systems, most of which measure distances relative to the *origin*, which is located at 0,0,0:

- **Cartesian** define 2D and 3D points by two or three distances along the x, y, and (sometimes) z axes.

- **Polar** define 2D points by a distance and an angle.

- **Cylindrical** define 3D points by two distances and an angle.

- **Spherical** define 3D points by a distance and two angles.

- **Relative** define 2D and 3D distances (and optionally angles) relative to the last point.

- **User-defined coordinate systems** (UCS) define an x,y-plane anywhere in 3D space.

You enter coordinates in the format of your choice: decimal (metric), engineering (decimal inches), architectural (Imperial units), fractional (inches only), and scientific notation (exponents). Accuracy ranges from zero to eight decimal places (whole inches to 1/256th of an inch).

Angles can be entered in degrees-minutes-seconds (360 degrees in a circle), radians (2*pi), grads (400), and in surveyor's units (N E).

In new drawings, angles are measured counterclockwise, with 0 degrees pointing from the origin to the East, or 3 o'clock. Using this convention, 90 degrees points North (12 o'clock), 180 degrees points West (9 o'clock), and 270 degrees points South (6 o'clock).

To measure an angle in the clockwise direction, use negative angles. Prefix the angle with a dash, as in -45 degrees. Because there are 360 degrees in a circle, -45 degrees is the same as 315 degrees.

In the same way, you specify negative distances by prefixing numbers with the dash. For example, -2,-3 means two units in the negative x-direction, and three units in the negative y-direction.

Recall that *absolute* coordinates are measured relative to the origin (0,0,0), while *relative* coordinates are measured relative to the last point. Similarly, absolute angles are measured from the x-axis counterclockwise, while relative angles are measured from the value stored in the **LASTANGLE** system variable.

If nothing else, AutoCAD is flexible in its measurement systems. You can make changes with the **UNITS** command so that angles are measured clockwise, and that 0 degrees points in any direction.

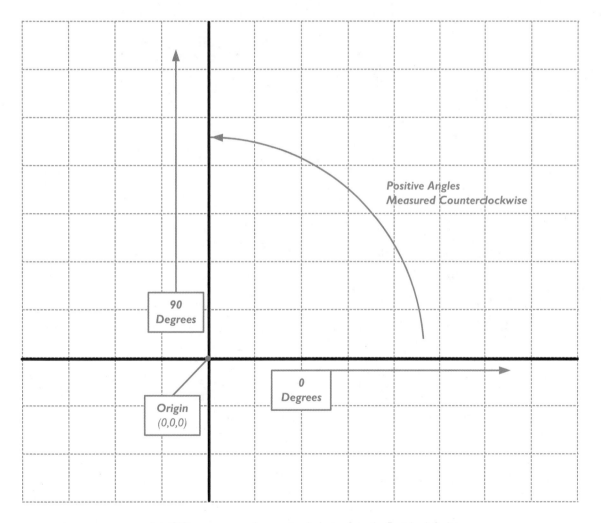

AutoCAD measures angles counterclockwise from the East, by default.

Cartesian Coordinates

Cartesian coordinates define 2D and 3D points by measuring two or three distances: each distance is along one of the three axes: x, y, and z, either positive or negative. In AutoCAD, the x-direction is horizontal, at 0 degrees; positive x is measured to the right. The y-direction is vertical along 90 degrees; positive y is measured upward. The z-direction comes out of the screen; positive z points at your face.

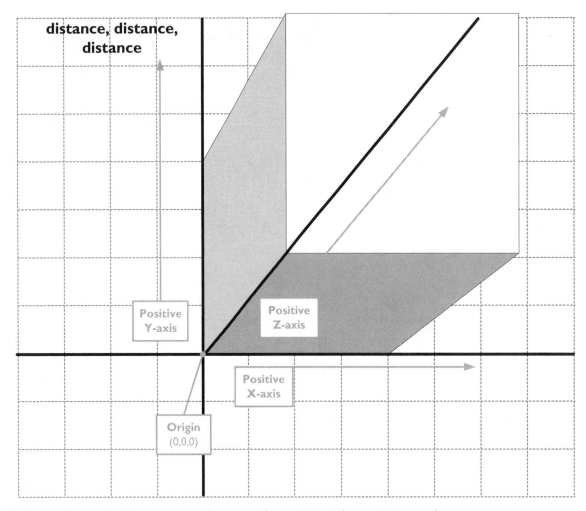

To draw a line using Cartesian coordinates with x,y (2D) and x,y,z (3D) coordinates:

> Command: **line**
>
> Specify first point: **1,2**
>
> Specify next point or [Undo]: **3,4,5**

Polar Coordinates

Polar coordinates define 2D points by measuring one distance and one angle in the x,y-plane. The distance is the straight-line distance measured from the origin to the point, while the angle is measured from 0 degrees. To indicate the angle to AutoCAD, you prefix the angle with an angle bracket (<), such as <45 for 45 degrees.

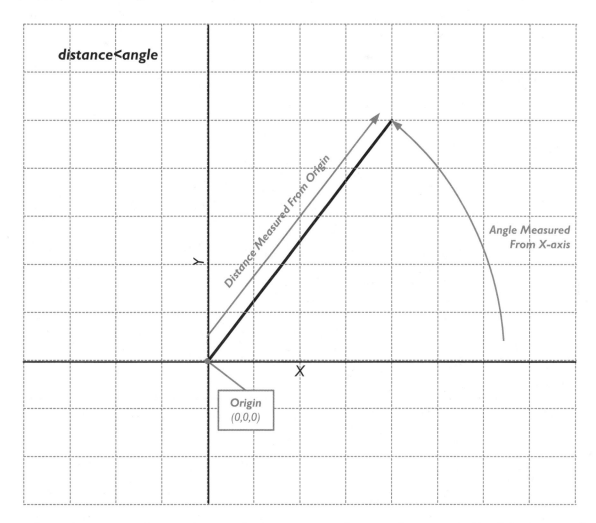

To draw a line using polar coordinates, the first point is relative to the origin (0,0), while the second is relative to the first point:

> Command: **line**
>
> Specify first point: **1<23**
>
> Specify next point or [Undo]: **4<55**

Cylindrical Coordinates

Cylindrical coordinates define 3D points by measuring two distances and one angle in 3D space. They are like polar coordinates, but with the z-distance added. The distances are measured from the origin to the point, while the angle is measured from 0 degrees.

The first distance lies in the x,y-plane; the second distance measures the distance along the z-axis. To indicate the measurement to AutoCAD, you use this notation: 1<23,4 — where 1 is the polar distance, <23 is angle from 0 degrees, and 4 is the height in the z direction.

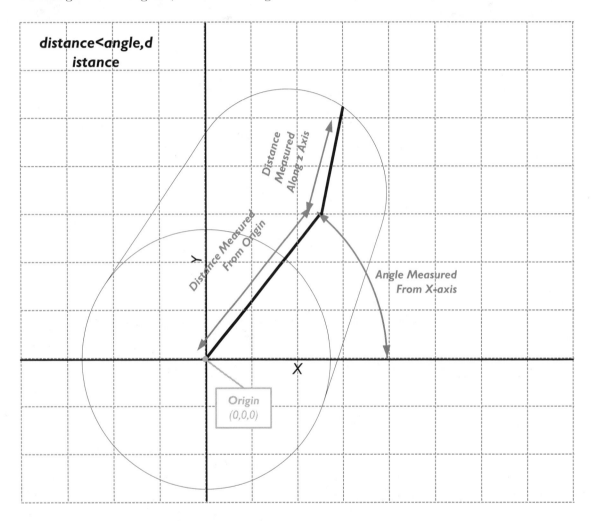

To draw a line using cylindrical coordinates:

     Command: **line**

     Specify first point: **1<23,4**

     Specify next point or [Undo]: **5<67,8**

Spherical Coordinates

Spherical coordinates define 3D points by measuring a distance and two angles from the origin. The first angle lies in the x,y-plane, while the second angle is up (or down) from the x,y-plane. The distance lies, of necessity, in the x,y,z-plane.

To indicate the angles to AutoCAD, you prefix each with angle brackets, such as <15<45 for 15 and 45 degrees.

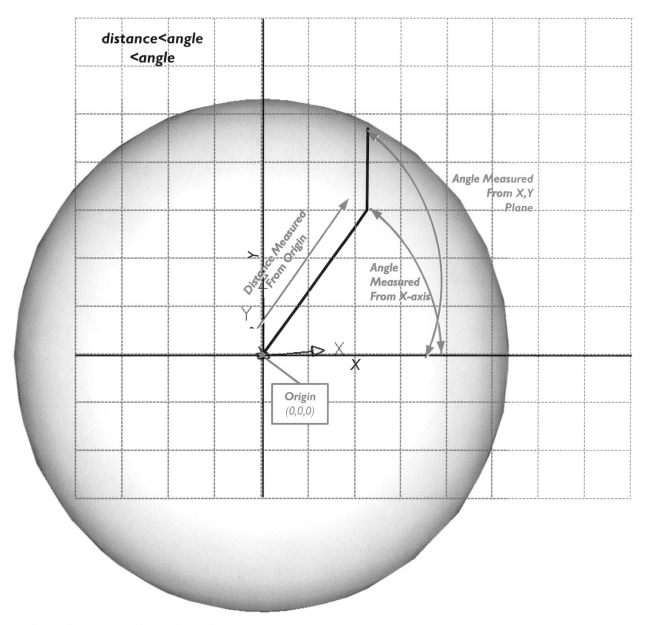

To draw a line using spherical coordinates:

>   Command: **line**
>   Specify first point: **1<23<45**
>   Specify next point or [Undo]: **6<78<90**

Relative Coordinates

All coordinate systems can use *relative* coordinates, which measure the distance (and angle) from the last point. (The opposite of relative is *absolute*, which measures distances from the origin.) To indicate relative coordinates to AutoCAD, you prefix them with the at symbol (@), such as @2,3.

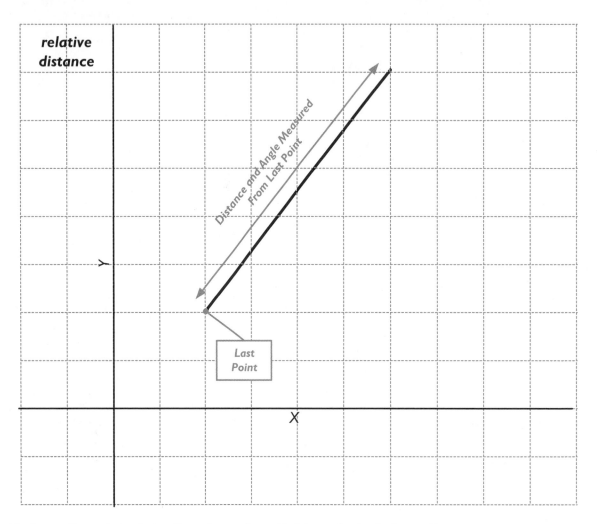

To draw a line using relative Cartesian, polar, cylindric, and spherical coordinates

> Command: **line**
>
> Specify first point: **@1,2**
>
> Specify next point or [Undo]: **@3,4,5**
>
> Specify next point or [Undo]: **@6<78**
>
> Specify next point or [Undo]: **@9<10<11**
>
> Specify next point or [Undo]: **@9<10<11**

Note that the first point can also be relative. It is measured from whatever last point was specified in the drawing with a previous drawing command. AutoCAD stores the coordinates of the last point in a system variable called **LASTPOINT**.

Similarly, relative angles are measured from the value stored in the **LASTANGLE** system variable.

User-defined Coordinate Systems

As I noted earlier, all drawing in AutoCAD takes place in the x,y-plane, also called the "working plane." How, then, do you draw in 3D space, which involves the z axis, along with the x,z and y,z planes? You could specify the z-coordinate, but this is not possible for all commands. In any case, specifying z coordinates becomes cumbersome when drafting details, say, on the side of a slanted roof, where the value of z is constantly changing.

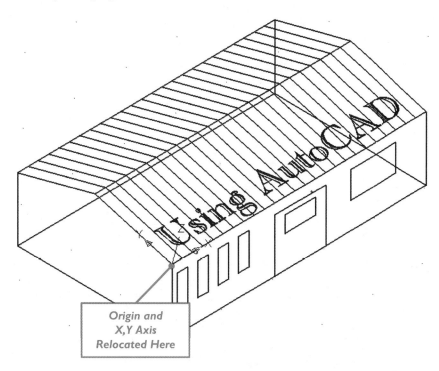

Origin and
X,Y Axis
Relocated Here

AutoCAD takes care of that through user-defined coordinate systems, or "UCS" for short. Through UCSs, you can define the working plane anywhere in 3D space, such as on the sloped roof of a barn. The UCS command lets you set the working plane relative to the current view, to three points picked in the drawing, and by several other methods. UCSs can be saved by name, so that you can return to previously-created working planes.

ENTERING COORDINATES

Much of working with AutoCAD involves specifying points on the screen. Typical are the prompts of the LINE command:

> Command: **line**
>
> Specify first point: *(Pick a point, or enter x,y,z coordinates.)*
>
> Specify next point or [Undo]: *(Pick another point, or enter more coordinates.)*

You can do this in three ways:

- Picking points on the screen.

- Entering coordinates at the keyboard.

- Using object snaps, tracking, direct distance entry, and point filters.

Screen Picks

When AutoCAD prompts you to specify a point, you can use the crosshair cursor to pick a point in the drawing. Move the mouse to move the cursor; press the left mouse button to pick the point. AutoCAD determines the x,y coordinates of the pick point.

Screen picks are strictly 2D (two dimensional), even when working in 3D (three dimensional) drawings. "2D" means AutoCAD records the x and y coordinates only, and assumes the z coordinate is 0. If you wish to specify a z coordinate, you may do this with the ELEVATION command (which sets the distance along the z axis), or with the .XY *point filter* (which forces AutoCAD to prompt you for the z distance), or by entering an x,y,z triplet.

Keyboard Coordinates

You can enter coordinates at the keyboard in a variety of methods, as I described earlier. The formats permitted by AutoCAD are:

Absolute	Relative	Meaning
x,y	@x,y	2D Cartesian coordinates.
x,y,z	@x,y,z	3D Cartesian coordinates.
d<a	@d<a	2D polar coordinates.
d<a,z	@d<a,z	3D cylindrical coordinates.
d<a<r	@d<a<r	3D polar coordinates.

x, y, and z = distance along the x, y, and z axis, respectively.
d = distance in x,y plane.
a = angle from x axis (in x,y plane).
r = angle from x,y plane (in z direction).

Recall that *absolute* coordinates are measured relative to the origin (0,0,0), while *relative* coordinates are measured relative to the last point. Similarly, absolute angles are measured from the x-axis counterclockwise, while relative angles are measured from the value stored in the LASTANGLE system variable.

Snaps, Etc.

Entering coordinates at the keyboard can get tedious. Thus, there are alternatives and aids:

- **Point filters** combine screen picks with keyboard entry.

- **Snap spacing** restricts cursor movement to increments.

- **Polar snap** and **ortho mode** restrict cursor movement to specific angles.

- **Direct distance entry** allows you to show angles and enter distances.

- **Tracking** shows distances.

- **Object snaps** snap the cursor to the nearest geometric feature.

- **Object tracking** shows geometric relationships to nearby objects.

You learn more about snap mode, polar snaps, objects snaps, and object tracking in Chapter 5, "Drawing with Precision."

Point Filters

Point filters combine screen picks with keyboard entry to specify, for example, the x coordinate with the keyboard and the y coordinate on the screen. AutoCAD recognizes five point filters:

Point Filter	AutoCAD Asks For...
.x	y and z coordinates.
.y	z and z coordinates.
.z	x and y coordinates.
.xy	z coordinate.
.xz	y coordinate.
.yz	x coordinate.

Here is an example of using point filters with the LINE command. Notice that you can enter coordinates at the keyboard or on the screen at any time:

> Command: **line**
>
> Specify first point: **.x**
>
> of *(Pick a point; AutoCAD reads just the x-coordinate.)*
>
> (need YZ): **2,3** *(AutoCAD reads these as the y and z coordinates.)*
>
> Specify next point or [Undo]: **.y**
>
> of **2** *(AutoCAD reads this as the y coordinate.)*
>
> (need XZ): *(Pick a point; AutoCAD reads the x and z coordinates.)*

Snap Spacing

Snap mode restricts cursor movement. To restrict the cursor to moving in increments of 0.5, for example, set the snap spacing to 0.5 with the SNAP command. This allows you to pick points easily on the screen, accurate to the nearest 0.5 units. If you measured the rooms in your house to the nearest one inch, when drawing the floor plan in AutoCAD, you would set the snap spacing to 1".

Snap is often used in conjunction with the grid; when the grid is set to the same spacing as the snap, then you can "see" the snap points.

Polar Snap and Ortho Mode

Polar snap snaps the cursor to specific angles, while ortho mode restricts the cursor to 90-degree increments. (Ortho is short for "orthographic.") You can specify the polar angle, although increments of 15 degrees are common. In addition, you can specify polar distances, such as 1.

While drawing, AutoCAD displays a tooltip showing relative distance and angle from the last point.

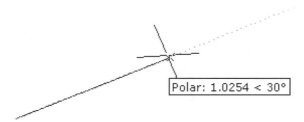

You turn on polar mode by clicking **POLAR** on the status bar; turn on ortho mode by clicking **ORTHO**.

Direct Distance Entry

Direct distance entry shows angles, and types distances. It allows you to draw and edit by showing a relative distance: you move the cursor to indicate the angle, and then type a number to specify the distance. If you need a precise angle, then ensure that polar tracking or ortho mode are first turned on.

> Specify first point: *(Move the cursor, and then enter a distance, such as* **2.5***.)*

Tracking

Tracking shows distances with the "pen up" during drawing and editing commands. You can think of it as the opposite of direct distance entry, which draws; tracking moves the cursor without drawing. During a prompt, such as "Specify first point:", enter **tk** to enter tracking mode. Move the cursor, in the direction to track, and then pick a point or enter a distance, such as **2.5**. You can continue tracking as far as you need.

> Specify first point: **tk**

First tracking point: *(Move cursor, and then enter a distance or pick a point.)*

Next point (Press ENTER to end tracking): *(Press ENTER.)*

Object Snaps

Object snaps snap the cursor to the nearest geometric feature, such as the endpoints of lines, the center points of circles and arcs, and the insertion points of text. To help you find these geometric features, AutoCAD displays icons and tooltips.

The figure below illustrates the line being drawn to the center of the circle. Because object snapping is turned on, AutoCAD displays the center icon (a circle), as well as the tooltip, "Center."

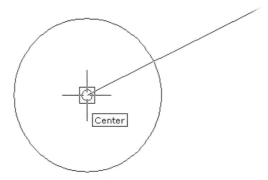

To use objects snaps, you can turn them on with the **OSNAP** command, or enter them at prompts, such as:

Specify first point: **center**

of *(Pick a circle or arc.)*

Object Tracking

Object tracking shows geometric relationships to nearby objects. When turned on with at least one object snap mode, object snap tracking shows *alignment paths* (dashed lines) based on object snap locations. The figure below illustrates the alignment path showing the quadrant object snap.

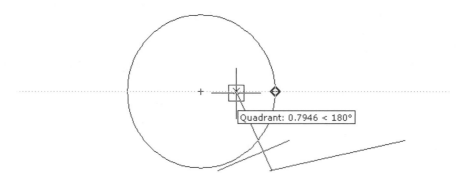

DISPLAYING COORDINATES

AutoCAD reports the cursor position by displaying the x,y,z coordinates on the status line (at the extreme left end, as illustrated below).

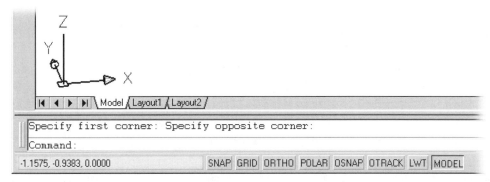

In most cases, the display shows absolute x,y,z coordinates; the z coordinate, however, always reads 0.0000 — unless changed by the ELEV command.

If you wish, you can turn off the display, although there is no point to doing that, at least not that I can think of. In addition, you can change the display to relative d<r coordinates. To change the coordinate display, press **CTRL+D** or function key **F4** — or, simply click the coordinate display on the status line. In the figure below, notice that AutoCAD displays relative coordinates during line drawing:

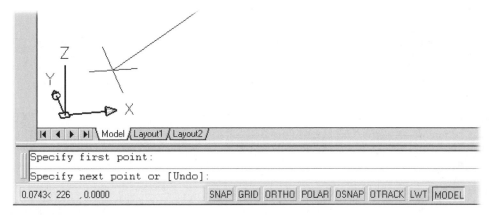

0.0743 is the distance.

<226 is the angle

0.0000 is the z coordinate (elevation).

In addition to coordinates on the status line, AutoCAD also displays coordinate information through tooltips.

EXERCISES

1. In the following exercises, watch how the coordinates and cursor react to different modes.

   Move the cursor around with your mouse or digitizer. Do the coordinates at the lower left corner of the screen move?

   Press **F6**. Move the cursor again, and watch the coordinates. What is the difference?

2. Press **F8**. Check the status bar: is ortho mode turned on?

   Move the cursor. Do you notice a difference?

3. Press **F9** to turn on snap mode.

   Move the cursor. Now is there a difference in cursor movement?

4. In this exercise, you practice using menus and keystrokes that affect commands.

   Move the crosshairs to the top of the drawing area and into the menu bar. Does the cursor change its shape?

   Move the arrow from left to right over the menu bar. What happens as the cursor moves over each word?

   Position the cursor over **Draw**, and then click. Does a menu drop down into the drawing area?

   Click on **Draw** a second time. What happens?

5. Choose the **Draw** menu again, and then **Line**.

   Draw some lines.

   To exit the command, right-click and select **Cancel** from the shortcut menu.

6. Press the **ENTER** key. Did the Line command repeat?

   Clear the command line with **ESC**.

   Press the spacebar. What happens?

7. Press **F2**. Do you see the Text window?

   Press **F2** again to return to the drawing window.

8. Press **F1**. Do you see the Help window? Familiarize yourself with the Help window.

9. Use the **LINE** command to draw a line graph of insurance premiums for nonsmoking men. Enter the coordinates listed by the table below:

X (Age)	Y (Rate $)
25	14
30	15
35	19
40	23
45	37
50	59
55	88
60	130
65	194

   Are these coordinates relative or absolute? (If you cannot see the drawing, use the **ZOOM Extents** command.)

10. Using the figure below, label each axis and the origin.

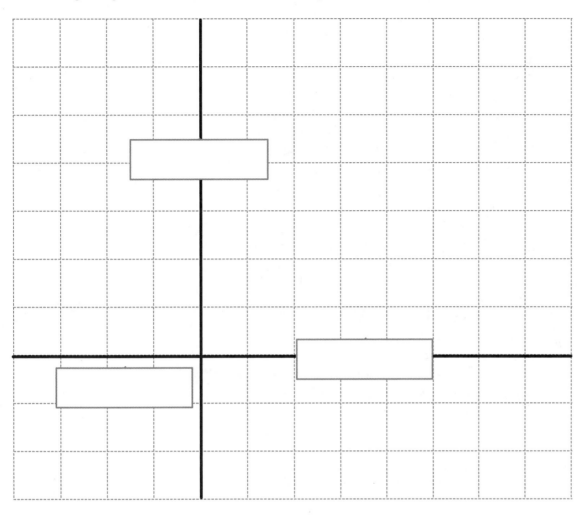

CHAPTER REVIEW

1. List the information found on AutoCAD's title bar.

    a.

    b.

2. What are the benefits of using *template* drawings?

3. Where are the mode indicators located?

4. How does a menu indicate the presence of submenus? Of dialog boxes?

5. Can you copy and delete files from within AutoCAD?

6. What happens when you press the **F2** key?

7. List the control and/or function key associated with the following actions:

    Turns ortho mode on and off.

    Selects all objects in the drawing.

    Toggles between graphic and text screens.

    Toggles tablet mode on and off.

    Cancels all the characters on the command line.

    Toggles to the next isometric plane in iso mode.

    Toggles the screen coordinate display modes.

    Toggles the grid on and off.

    Toggles snap on and off.

8. After a command ends, what happens when you press **ENTER**?

9. Name three ways to invoke the on-line help facility:

    a.

    b.

    c.

10. What is the purpose of the *prompt* area?

11. Explain the meaning of the icons in the layer droplist:

    Lightbulb

    Shining sun

    Closed padlock

    Color square

    Gray layer name

12. What does the color BYLAYER mean?

13. What are *linetypes*?

14. Explain the purpose of the small square at the center of the crosshair cursor.

15. "UCS" is short for:

16. Can the UCS icon be turned off? If so, why?

17. Explain the meaning of these numbers on the status line:

32'-11 3/8", 1'-2 1/2", 0'-0"

18. What does it mean when the following buttons on the status line look depressed?

    SNAP

    OSNAP

    MODEL

19. Where is the *tray* located?

20. Can you "show" AutoCAD an angle by picking points in the drawing?

21. How do you show AutoCAD x and y displacement simultaneously?

22. Describe how *transparent* commands work.

23. Can any command be transparent?

24. What is an *alias*?

25. Write out the alias for each of the following commands:

    ARC

    LINE

    BLOCK

    CIRCLE

    COPY

    ZOOM

26. Where are *function keys* located?

    How can you identify function keys?

27. Write out the function key you would use for the following actions:

    Toggle snap mode

    Toggle ortho mode

    Toggle grid display

    Exit AutoCAD

28. What does *toggle* mean?

29. Write out the **CTRL** key you would press for the following actions:

    Display Properties window

    Toggle clean screen mode

    Copy objects to the Clipboard

    Save the drawing

    Switch to the next drawing window

30. Describe two methods to repeat commands.

31. Explain in two or three words the function of each of these mouse buttons:

    Left button

    Right button

    Scroll wheel

    SHIFT+Left button

32. What happens when you click an item on the menu bar a second time?

33. Describe two methods to access commands on menus.

34. List some pros and cons to using dialog boxes.

    Pro:

    Con:

35. Do all commands use dialog boxes?

36. Explain the difference between the **ARRAY** and the **-ARRAY** commands.

37. How do you force file-related commands to display their dialog boxes, when the function has been turned off?

38. Describe the purpose of the **Preview** button on dialog boxes.

39. What is a *non-modal* dialog box?

40. What is the purpose of *tabs* in dialog boxes?

41. Describe the difference between:

    Check boxes

    Radio buttons

42. What happens in dialog boxes when lists are too long?

43. How can you select more than one item from a list?
44. Describe the function of these commands:

    LINE

    CIRCLE

    ARC

    POLYGON

    TEXT

    Are these commands for drawing or editing?
45. Describe the function of these commands:

    COPY

    MIRROR

    INSERT

    MATCHPROP
46. Give two reasons why you might need to edit drawings?
47. Describe the function of these commands:

    MOVE

    ERASE

    ROTATE

    DDEDIT

    PEDIT

    PROPERTIES
48. What happens when you double-click a line of text?

    Objects other than text?
49. Explain the difference between the two types of plotting:

    Color-based.

    Styles-based.
50. Describe the function of *plot styles*.
51. List one way that the **PREVIEW** command is useful.
52. What is an *e-plot*?
53. How would you email a drawing without sending the *.dwg* file itself?
54. What is *.dwf* short for?

    Name a benefit to *.dwf* files.
55. What happens if you forget the password to drawings?
56. Briefly explain these coordinate systems:

    Cartesian

    Polar

    Spherical

    Cylindrical
57. Describe the meaning of the following symbols as used by AutoCAD:

    <

    @

    , (comma)
58. Name the coordinate system used by the following:

    25<45

    25,45,60

    25<45<60

    25<45,60

59. Write out relative coordinates for a distance of 5 units and an angle of 75 degrees.

60. Can coordinates be negative?

    If yes, when?

61. What are the coordinates of the *origin*?

62. Are relative coordinates measured relative to the origin?

    Are absolute coordinates measured relative to the origin?

63. In which direction are positive angles measured in AutoCAD, usually?

    Can the direction be changed?

64. In which direction is 0 degrees in AutoCAD, usually?

    Can the direction be changed?

65. Fill in the missing Cartesian coordinates to draw a line from the origin to 10,15.

    Command: **line**

    Specify first point:

    Specify next point:

66. Fill in the missing relative coordinates to draw a vertical line 10 units long from 20,15.

    Command: **line**

    Specify first point:

    Specify next point:

    What are the coordinates of the line's endpoint?

67. What happens when you press **ENTER** at the **LINE** command's "Specify first point" prompt?

68. What is the likely z coordinate when you pick a point in the drawing?

    When might the z coordinate change?

69. When might you use *cylindrical* and *spherical* coordinates?

70. When might you use UCSs?

71. Describe the distances and angles employed by spherical coordinates:

    Distance

    Angle 1

    Angle 2

    Write down an example of a spherical coordinate.

72. How are *point filters* helpful?

73. When you enter **.xy**, what does AutoCAD ask for?

74. Explain the purpose of:

    Snap spacing

    Object snaps

    Polar snaps.

75. What is the difference between *direct distance entry* and *tracking*?

76. How do you enter tracking mode?

77. Can angles be negative?

    If so, when?

78. Describe what happens when you press the ⬆ key at the 'Command:' prompt.

79. List five ways to enter commands:

    a.

    b.

    c.

    d.

    e.

80. What happens when you enter the first few letters of a command name, and then press the **TAB** key?

81. What is *dynamic input*?

82. How is dynamic input activated?

83. Can the x, y coordinates be changed during dynamic input? If so, how?

84. How do you access command options during dynamic input?

85. What is the meaning of the dot (•) next to an option in dynamic input?

86. From where are dynamic angles measured?

87. From where are dynamic dimensions measured?

CHAPTER 3

Setting Up Drawings

After starting a new drawing, you must prepare it for use. This chapter covers the following items to help you set up drawings:

NEW starts new drawings from scratch, or through the Startup dialog box.

QNEW starts new drawings from templates.

Advanced Wizard sets up units and angles.

OPEN opens existing drawings, and helps manage files.

MVSETUP sets up units, and creates scaled borders.

OPTIONS changes user interface colors, specifies backup options, and much more.

UNITS specifies unit and angle formats.

Scale factors affect the size of text, linetypes, hatch patterns, and dimensions.

SAVE and **SAVEAS** save drawings by other names, in other formats, and to other computers.

QSAVE saves drawings quickly.

QUIT exits AutoCAD.

NEW TO AUTOCAD 2008 IN THIS CHAPTER

- Annotative scaling affects scale-dependent objects.
- Modest changes to TrustedDWG.
- New "2D Drafting & Annotation" workspace.

FINDING THE COMMANDS

On the **Standard Annotation** toolbar:

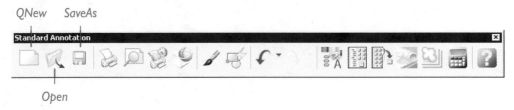

On the **Standard** toolbar:

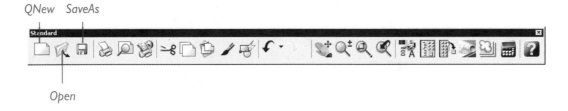

NEW AND 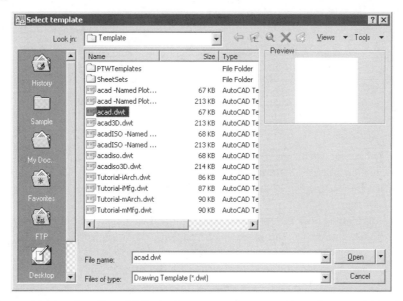 QNEW

When you start AutoCAD, you either begin with new drawings or call up previously-saved drawings. Let's begin with starting new drawings.

After AutoCAD starts up, it displays a new blank drawing. (At the same time, it may also display a variety of startup dialog boxes. If the New Features Workshop dialog box appears, cancel it.)

You use the NEW or QNEW commands at any time to start additional new drawings, as described below. You can have many drawings open in AutoCAD at the same time.

TUTORIAL: STARTING NEW DRAWINGS

1. To open a new drawing when AutoCAD is already running, start the **NEW** command by one of the following methods:
   * From the **File** menu, choose **New**.

   * Alternatively, press the **CTRL+N** keyboard shortcut.

   * Or, at the 'Command:' prompt, enter the **new** command.

      Command: **new** *(Press* ENTER.*)*

   In some cases, AutoCAD may display the Startup dialog box. But usually, AutoCAD prompts you to select a template file.

2. Select a template file, which determines what new drawings look like. Some are blank, while others contain drawing borders and other elements. If you are unsure, select *acad.dwt*.

3. Click **Open**. Notice that AutoCAD displays a new drawing; if based on *acad.dwt*, the new drawing is blank.

About Templates

Each time you begin a new drawing, AutoCAD copies a list of settings from a *template* drawing. (This is a drawing file with the extension of *.dwt*.) AutoCAD copies the settings from template files, which can have names such as *acad.dwt*, *acadiso.dwt*, and *acad3d.dwt*.

Template drawings contain preset settings that new drawings use; sometimes, templates also include drawing elements. New drawings are identical to the template, except for the file name. You can change any setting in the drawings, regardless of how it is set in the template.

AutoCAD stores its list of template drawings in a folder with the long path of */documents and settings/<user login name>/local settings/application data/autodesk/autocad 2008/r17.1/enu/template*. AutoCAD includes a number of template drawings for a variety of sizes and standards. As you select a file name, AutoCAD displays its preview image. Clicking **Open** opens the template, and then this file becomes the starting point for your new drawing.

QNew

The QNEW command is a more flexible version of NEW: depending on how AutoCAD is set up, the command either starts with a preselected *.dwt* file, or else prompts you to select the template. (QNEW is short for "quick new.")

 Notes: Whether AutoCAD displays the Select Template dialog box is determined by a setting in the Options dialog box. From the **Tools** menu, select **Options**, choose the **Files** tab, and then look for the **Template Settings** section.

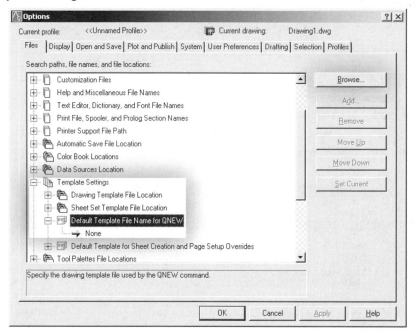

The **Default Template File Name for QNEW** item has two options:
* If **None** is shown, then the QNEW command displays the Select Template dialog box, just like the NEW command.
* If a file name with a *.dwt* extension is shown, then the QNEW command starts new drawings with the specified *.dwt* template file. This is the "quick" version of the command.

To select a specific template or another drawing, click the **Browse** button, and then choose the file.

TUTORIAL: STARTING NEW DRAWINGS WITHOUT TEMPLATES

1. To open a new drawing based on a template, start the QNEW command:
   * From the **Standard** toolbar, choose **QNew**.

   * Or, at the 'Command:' prompt, enter the **qnew** command.

     Command: **qnew** *(Press ENTER.)*

   If a template file is defined as described above, then AutoCAD displays a new, blank drawing based on a preselected *.dwt* file.

STARTUP DIALOG BOX

If AutoCAD displays the Startup dialog box, then you are faced with the following options:

> **Start from Scratch** starts new drawings in metric or English units.
>
> **Use a Template** starts new drawings based on *.dwt* template files.
>
> **Use a Wizard** starts new drawings based on two wizards, Quick and Advanced.
>
> **Open a Drawing** opens existing drawings.

(The dialog box discussed here is displayed when AutoCAD first starts; the nearly-identical Create New Drawing dialog box is displayed when you enter the NEW command. The only difference between the two, other than the title, is that the Create New Drawing dialog box grays out the **Open a Drawing** button, because this function is handled by the OPEN command.)

Note: Whether AutoCAD displays the Create New Drawing dialog box is determined by the value of the STARTUP system variable:
1 — displays the Startup dialog box.
0 — does not display the dialog box (default setting).

Starting From Scratch

The **Start from Scratch** option is one way to start new drawings. It presents just two options: measuring the drawing in either Imperial (English) or metric units.

Imperial

Select **Imperial** to use feet and inches. The drawing limits are set to 12" x 9", and dimension styles to inches. (You can change these settings later.) AutoCAD copies the settings stored in the *acad.dwt* template file.

Metric

Select **Metric** for metric units. Limits are set to 429 mm x 297 mm, and the dimension styles to metric. AutoCAD copies the settings stored in the *acadiso.dwt* template file.

MeasureInit

Behind the scenes, AutoCAD stores your selection in the MEASUREINIT system variable, which has a value of 0 or 1. The value determines AutoCAD's actions, as follows:

> **0** — (English) AutoCAD reads the hatch patterns and linetypes defined by the **ANSIHatch** and **ANSILinetype** setting in the Windows registry.

> **1** — (Metric) AutoCAD reads the hatch patterns and linetypes defined by the **ISOHatch** and **ISOLinetype** registry settings.

Using Templates

In the Setup dialog box, choose the **Use a Template** button. AutoCAD displays the same list of template drawings as it did with the NEW command. Select a file name, and AutoCAD displays its preview image. Click **Open** to open the template, which is the starting point for your new drawing.

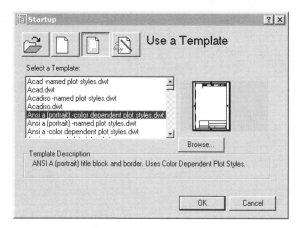

Although template drawings have the *.dwt* extension, you are free to use *any* AutoCAD drawing as a template. Choose **Browse** to find other files. Or, in the Select a Template dialog box, click the **Files of type** droplist, and select *.dwg*.

Using Wizards

AutoCAD includes two "wizards" that step you through the stages of setting up some of the many parameters that define a drawing: Quick and Advanced.

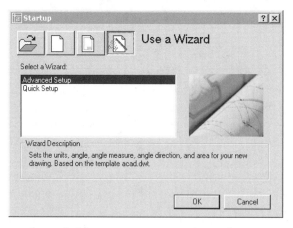

The Quick Wizard takes you through just two steps — units and area — in creating new drawings. You worked through the Quick Wizard in Chapter 1, "Quick Start in AutoCAD." Below, you work through the Advanced Wizard, which takes you through five steps to set up units, angles, and area for new drawings.

TUTORIAL: USING THE ADVANCED WIZARD

1. To start a new drawing with the advanced wizard, choose the **Use a Wizard** button.

2. Select **Advanced Wizard**.

   Advanced Setup presents a series of dialog boxes that lead you through the steps of setting up new drawings. Curiously, no online help is available for this early encounter with AutoCAD.

3. The Units dialog box asks you to specify the units.

   Select a measurement unit. If you are not sure, select **Decimal**. (If you change your mind later, you can change the units with the **UNITS** command.)

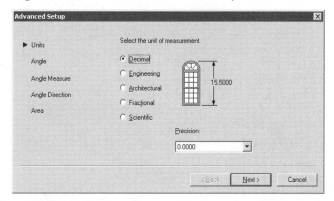

Technically, AutoCAD does not work with Imperial or metric units; instead, the software uses scientific units internally. To help human operators, however, AutoCAD accepts and displays units in the formats listed below, converting to and from scientific format on the fly.

AutoCAD Units	Example	Comments
Decimal	14.5	Decimal and metric units.
Engineering	1'-2.5"	Feet and decimal inches.
Architectural	1'-2 1/2"	Feet and fractional inches.
Fractional	14 1/2	Fractional inches, no feet.
Scientific	1.45E+01	Exponent notation.

4. *Precision* refers to the number of decimal places or fractions of an inch. From the **Precision** droplist, select the number of decimal places (or fractional accuracy). If you are not sure, select 2 decimal places (0.00) or 1/8 fractional.

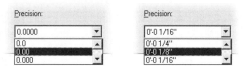

Left: *Selecting decimal precision.*
Right: *Selecting fractional precision.*

This affects the display of units only; no matter which precision you select, AutoCAD calculates to 16 decimal places of accuracy internally, and then displays the result rounded to the maximum of eight places or 1/256th.

- Decimal, engineering, and scientific units: 0 to 8 decimal places.

- Architectural and fractional units: 1/1, 1/2, 1/4, 1/8, 1/16, 1/32, 1/64, 1/128, or 1/256.

5. Click **Next.**

6. The Angle dialog box specifies the style of angle measurement.
Select one of them. If you are not sure, select decimal degrees.

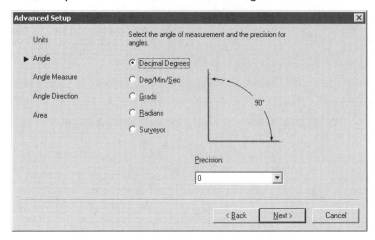

AutoCAD works internally with radians, but accepts and displays angles in the following formats:

Angle Formats	Example	Comments
Decimal Degrees	45.00	360 degrees in a circle.
Degrees-minutes-seconds	45d00'00.00"	Fractions of seconds in decimals.
Grads	50.00g	400 grads in a circle.
Radians	0.79r	2pi radian in a circle.
Surveyor	N45d0'0.00"E	Fractions of seconds in decimals.

7. As with units, Precision affects only the *display* of angles:

   • Decimal degrees, grads, and radians: 0 to 8 decimal places.

   • Deg/min/sec and surveyor angles: 45d, 45d00' through to 45d00'00.00000".

   From the **Precision** droplist, select the number of decimal places. (There is no fractional accuracy for angles.) If you are not sure, select 1 decimal place.

8. Click **Next.**

9. The Angle Measure dialog box asks you to "Select the direction for angle measurement."
The dialog box would be more accurate if it read, "Select the direction for 0 degrees."

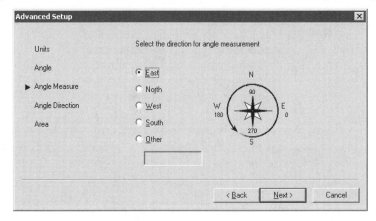

The default is **East**, which is the correct setting for most drawings. Measuring angles from East is the same as measuring from the positive x axis.

If your drawings use a direction other than 0 degrees, select from **North**, **West**, **South**, or **Other**. In the **Other** text box, you enter an angle in degrees between 0 and 360 — even when you selected grads or radians in the previous dialog box.

10. Click **Next**.

11. The Angle Direction dialog box asks you the direction in which to measure positive angles — counterclockwise or clockwise. For most drawings, you keep the direction counterclockwise.

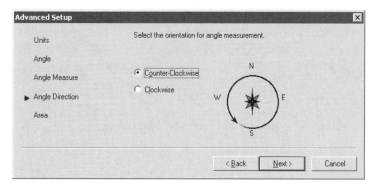

12. Click **Next**.

13. The Area dialog box asks you to specify the limits of the drawing. The default is 12 units wide (x direction) by 9 units tall (long, y direction).

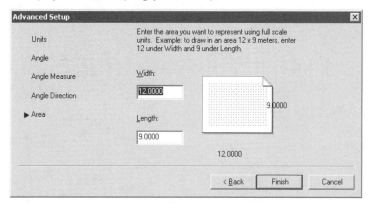

If you know how large the drawing will be, enter values in the **Width** and **Length** text boxes. Clear the current values using the DEL or BACKSPACE key.

AutoCAD uses these values to set the limits of the drawing, which affect the extent of the grid display (as shown in the preview image above) and the ZOOM All command. You can always change the limits later with the LIMITS command.

(*History*: AutoCAD prevented you from drawing outside the area defined by the limits; this didn't make much sense, and Autodesk soon dropped the restriction with 1986's version 2.17g, adding the LIMCHECK system variable.)

14. Click **Finish**.

AutoCAD opens the new drawing with the settings you specified.

OPEN

The OPEN command opens existing drawings.

This command opens drawings for editing. It can open files stored in a couple of formats; there are other commands in AutoCAD for opening files in non-Autodesk formats.

TUTORIAL: OPENING DRAWINGS

1. To open drawings, start the OPEN command:
   * From the **File** menu, choose **Open**.
   * From the Standard toolbar, choose **Open**.
   * Alternatively, press the **CTRL+O** keyboard shortcut.
   * Or, at the 'Command:' prompt, enter the **open** command.

     Command: **open** *(Press* ENTER.*)*

   In most cases, AutoCAD displays the Select File dialog box.

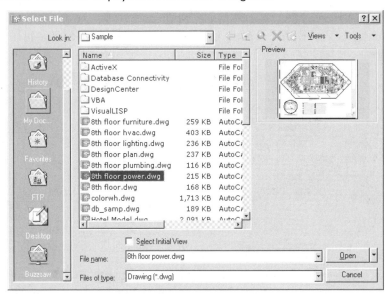

2. Select a *.dwg* file from the list, and then choose **Open**.
   Notice that AutoCAD opens the drawing.

The Select File dialog box accesses the following functions:

   * Opens *.dwg* (drawing), *.dxf* (interchange), *.dws* (standards) and *.dwt* (template) files.
   * Selects one or more files to open at the same time.
   * Previews drawings before opening, via thumbnail representation.
   * Sorts files by name, date, type, and size.
   * Views names of drawings recently opened (History), and most often used (Favorites).
   * Opens drawings protected against change (their status set to "read-only").
   * Loads all or portions of drawings.
   * Accesses drawings stored on your computer, on any other computer located on your network, or from any location accessible through the Internet (FTP).

- Searches for drawings anywhere on your computer (Find).
- Searches the Internet for information.
- Creates a list of frequently-accessed folders.
- Creates new folders and manipulates files and folders, including renaming and deleting files.
- Sends drawings as email messages, or to removable drives.
- Resizes the Select Drawing dialog box.
- Accesses Autodesk's Buzzsaw Web site for managing construction projects.

Moving counterclockwise, the Select File dialog box consists of five primary sections:

Files List lists the names of files and folders.

Preview displays a thumbnail-size image of the drawing.

Files of Type specifies the name and type of file to be opened.

Standard Folders Sidebar provides shortcuts to folders and Web locations.

Toolbar holds tools useful for manipulating files and folders.

Note: While the **OPEN** command is the most common method to open files, it is not the only way:

INSERT inserts other drawings as blocks into the current drawing.
PARTIALOAD loads additional portions of partially-loaded drawings.
FILEOPEN opens drawings without a dialog box; useful for scripts and macros, but is limited to AutoCAD running in SDI (single drawing interface) mode.

OPENSHEETSET opens *.dst* sheet set data files and related drawings.
DWFATTACH attaches *.dwf* files for viewing, while **OPENDWFMARKUP** attaches marked-up ones.
DXBIN opens *.dxb* (drawing exchange binary) files created by CAD\camera (obsolete).
DXFIN opens *.dxf* (drawing interchange format) files created by AutoCAD and other CAD programs.
ACISIN imports *.sat* format solid models created by ACIS-compatible CAD programs.

IMAGEATTACH attaches raster (bitmap) images to drawings.
WMFIN inserts *.wmf* (Windows meta format) files.
XOPEN opens externally-referenced drawings in their own windows.
3DSIN opens *.3ds* models created by 3D Studio.
DGNIN opens MicroStation V8 *.dgn* design files (new to AutoCAD 2008).

And the Clipboard pastes vector and raster graphics and text into drawings from almost any program.

Files List

To open drawings:

1. Select the name from the Files list. Notice that the name appears in the **File name** field.
2. Click **Open**.

As a shortcut, you can double-click the file name. This is equivalent to selecting the file name, and then clicking the **Open** button.

Preview

When you select a drawing in the Files list, AutoCAD displays its preview image. The image is small, but provides sufficient detail to distinguish it from most drawings. The preview image, also called a

TRUSTED DWG & DWGCHECK

When AutoCAD opens a drawing, it reports on its source. When the drawing was last saved by Autodesk software, the following prompt appears in the command bar:

Autodesk DWG. This file is a TrustedDWG last saved by an Autodesk application or Autodesk licensed application.

When the drawing comes from a non-Autodesk source, a warning appears in a dialog box:

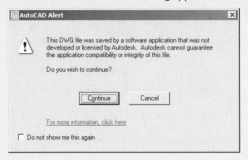

- **Continue** opens the drawing, and then repeats the warning at the command line:

    Non Autodesk DWG. This DWG file was saved by a software application that was not developed or licensed by Autodesk. Autodesk cannot guarantee the application compatibility or integrity of this file.

- **Cancel** does not open the drawing.
- **For more information click here** opens Internet Explorer with an Autodesk Web page that explains Trusted DWG (www.autodesk.com/trusteddwg).
- ☐ **Do not show me this again** stops the dialog box from appearing in the future.

AutoCAD 2007 introduced *Trusted DWG*, because Autodesk felt that drawings edited and saved by non-Autodesk software may corrupt data and destabilize applications used with AutoCAD. Trusted DWG does not repair defects in files, but allows you to choose whether to load these "foreign" *.dwg* files.

The stern wording of the dialog box was changed in AutoCAD 2008 after Autodesk launched a law suit against the Open Design Alliance. The alliance had used the registered term "Trusted DWG." As part of the settlement, the alliance agreed to stop using the term, and Autodesk agreed to tone down the wording of the warning.

DWGCHECK

The **DWGCHECK** system variable determines how and when these warnings are displayed.

- **0** — When the drawing has a possible problem, the warning appears in a dialog box.
- **1** — When the drawing was saved by software other than AutoCAD or AutoCAD LT or when it has a possible problem, the warning appears in a dialog box.
- **2** — When the drawing has a possible problem, the warning appears on the command line.
- **3** — When the drawing was saved by software other than AutoCAD or AutoCAD LT or when it has a possible problem, the warning appears on the command line.

If you do not want the warnings to appear, turn off **DWGCHECK** by setting its value to 0, as follows:

Command: **dwgcheck**
Enter new value for DWGCHECK <1>: **0**

"thumbnail," is generated automatically by AutoCAD when you save the drawing. Because the thumbnail shows the view when the drawing was last saved, the view may not include the entire drawing.

(*History:* The first several releases of AutoCAD presented a numbered *menu* of choices. To open a drawing, users pressed **2**, and then **ENTER**. Option 2 was labeled "Edit an EXISTING drawing." AutoCAD then prompted for the name of the drawing; users had to memorize the drawing's file name, because they could not select file names from a list. Release 9 introduced the basic dialog box that allowed users to select drawing names, as well as to choose the drive and folder names. AutoCAD 2000i greatly expanded the options through the dialog box you see today.)

Notes: If you see no preview image, the **RASTERPREVIEW** system variable was turned off (set to 0) when the drawing was saved. Turn it on by changing the value to 1:

Command: **rasterpreview**

Enter new value for RASTERPREVIEW <0>: **1** *(Press* ENTER.*)*

The other possibility is that the drawing files were created by old versions of AutoCAD or other types of CAD software that don't include preview images.

Drawings in 3D are shown rendered when the drawing was saved with a visual style turned on.

Files of Type

By default, the dialog box displays the names of *.dwg* files, which are AutoCAD drawings. In addition, you can select other kinds of files to open:

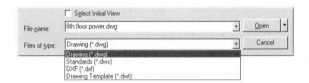

To select a file type different from *.dwg*, click the **Files of type** list.

DWG — (drawing) for creating AutoCAD drawing files.

DWS — (drawing standards) for ensuring drawings use standardized elements, such as linetypes and layers, against which drawings are checked.

DXF — (drawing interchange format) for creating drawings that can be exchanged between incompatible programs.

DWT — (drawing template) for creating preset frameworks of new drawings, such as dimension styles and drawing borders. When you select "Drawing Template (*.DWT)", the dialog box automatically switches to the *template* folder.

If you are not sure, select *.dwg*.

There is more than one way to open a drawing file. Click the down-arrow next to the **Open** button.

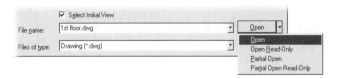

Read-Only

Notice the menu that appears:

Open — loads all of the drawing; you can edit the drawing, and save the changes.

Open Read-Only — loads all of the drawing, and does not allow you to save changes unless saved by to a different file name.

Partial Open — loads parts of the drawing, based on layer and view names.

Partial Open Read-Only — loads parts of the drawing, and does not save editing changes.

When you select one of the read-only options, AutoCAD displays the "Read Only" message on the title bar, as illustrated below.

If you attempt to save read-only drawings, a warning dialog box is displayed, as illustrated below. You are prevented from saving the drawing over the original. (You can save the drawing when you give it a different file name, as described later in this chapter.)

Partial Open

When you select one of the partial-open options, AutoCAD displays the Partial Open dialog box. (*Partial-open* means that only part of the drawing is loaded, useful for very large drawings.)

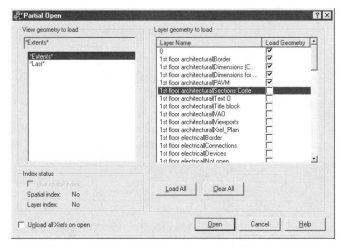

Select a view from the left hand column and the layers to be opened in the right hand column. When done, click **Open**. The partial opening of drawings is described in detail in *Using AutoCAD: Advanced*.

Select Initial View

Getting back to the Select File dialog box, the **Select Initial View** check box determines whether the drawing is opened with a saved view:

    ☐ (Off) When the drawing is opened, AutoCAD displays the same view as when the drawing was last saved.

    ☑ (On) Before the drawing is opened, AutoCAD displays the Select Initial View dialog box, which lets you select a named view.

If no views have been saved, then the only option is *Last View*, the view of the drawing when it was last saved. Views are described in Chapter 5, "Drawing with Precision."

Making the Dialog Box Bigger

When you first see the dialog box, it is at its smallest. You can make it larger to see more file names at a time. With the cursor, grab an edge of the dialog box, and stretch the dialog box larger, revealing more file names.

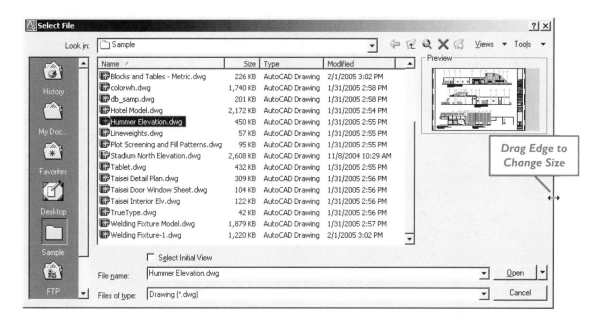

 Note: Press function key **F5** to update the files list. You may need to do this when files appear to be missing or incorrectly named.

Standard Folders Sidebar

A vertical list of folder names, called the "sidebar," is located on the left of the dialog box. It directly accesses folders on your computer, networks, and the Internet. From top to bottom, these are:

> **History** displays a list of the most-recently opened drawings.
>
> **My Documents** displays files stored in your computer's *my documents* folder.
>
> **Favorites** displays files stored in the *windows**favorites* folder; add files to this folder with **Tools | Add to Favorites**.
>
> **FTP** browses FTP (file transfer protocol) sites on the Internet; add more sites with **Tools | Add/Modify FTP Locations**.
>
> **Desktop** displays the files and folders found on the Windows "desktop."
>
> **Buzzsaw** connects you to Autodesk's www.buzzsaw.com Web site, after you install their ProjectPoint client software.

To add folders to the list, simply drag them from the Files list. For example, I like to have AutoCAD's *sample* folder on the list for fast access to the sample drawings, so I dragged it onto the folder.

Commands for editing the sidebar are available by right-clicking the folder sidebar, as illustrated by the figure below.

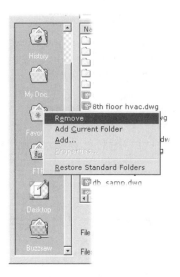

Toolbar

The toolbar is located at the top of the Select File dialog box. It contains a number of utility commands that are useful in AutoCAD. (These functions may vary, depending on which version of Windows is running on your computer.)

From left to right, the toolbar droplist and icons perform the following functions:

Look in selects the folder, drive, and network location from the droplist.

Back moves back to the previous folder (keyboard shortcut is **ALT+1**).

Up One Level moves up one level in the folder structure (**ALT+2**).

Search the Web displays a simple Web browser (**ALT+3**).

Delete erases selected files and folders (**DEL**); Windows asks if you are sure.

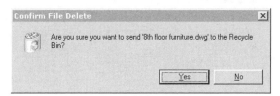

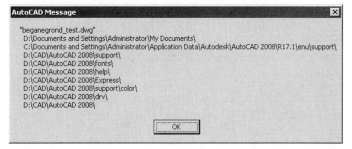 **Create New Folder** creates new folders (**ALT+5**).

The **Views** menu changes how files are listed:

List lists files by name only.

Details lists files by name, size, type, and date last modified.

Thumbnails displays files by images of drawings.

Preview toggles the display of the preview image; the check mark indicates that preview images are displayed.

The **Tools** menu provides additional utility commands:

Find displays the Windows Find dialog box that lets you search for files.

Locate displays a dialog box listing the names of folders used by AutoCAD's search paths:

```
AutoCAD Message                                                    [x]

"beganegrond_test.dwg"
  D:\Documents and Settings\Administrator\My Documents\
  C:\Documents and Settings\Administrator\Application Data\Autodesk\AutoCAD 2008\R17.1\enu\support\
  D:\CAD\AutoCAD 2008\support\
  D:\CAD\AutoCAD 2008\fonts\
  D:\CAD\AutoCAD 2008\help\
  D:\CAD\AutoCAD 2008\Express\
  D:\CAD\AutoCAD 2008\support\color\
  D:\CAD\AutoCAD 2008\drv\
  D:\CAD\AutoCAD 2008\

                          [      OK      ]
```

Add/Modify FTP Location displays the Add/Modify FTP Locations, which allows you to specify the URL, user name, and password for accessing FTP sites (more later).

Add to Favorites adds selected files and folders to the Favorites list, which quickly accesses frequently-used drawings.

FINDING FILES

Sometimes you cannot remember the names of drawing files or where they are located. To help you locate them, AutoCAD includes the **Find** option in the **Tools** item on the dialog box's toolbar. When you choose **Find**, AutoCAD displays a Windows dialog box for finding files.

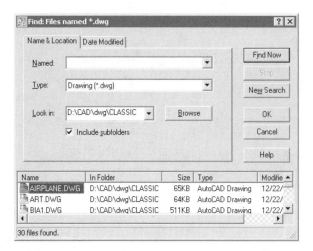

It is able to look at every folder on every drive in your computer. Drives include the hard drives, external drives, CD/DVD drives, and other storage devices. In addition, if your computer is connected to a network of other computers, the Find feature can search every folder of every networked drive your computer has permission to read.

Named

The **Named** text box allows you to enter as much of the name as you can remember. Use wildcard characters to specify part of a name.

Recall that ? is a placeholder for any character, while * searches for all files that match the rest of the criteria you supply. For example, *door** searches for all files that start with **Door**:

> door   door36   doorstop

Types

The **Type** droplist provides a filter for selecting *.dwg* drawing, interchange *.dxf*, template *.dwt*, and standards *.dws* files.

Look In

The **Look In** droplist restricts the search to specific drives, folders, and paths.

Date Modified

The **Date Modified** item narrows the search to specific dates. This is useful when you know roughly the date the drawing was created. For example, if the drawing was created last month, specify "Before this date," and then enter this month.

After you enter the search parameters, choose **Find Now**. Windows spends a bit of time rummaging through your computer's drives, and produces a list of matching files. At any time, choose **Stop** to bring the process to a premature halt.

Searching with windows Explorer

Windows Explorer can also search for AutoCAD drawing files, as well as for text in *.dwg* files, such as text and attributes.

MVSETUP

The Basic and Advanced Wizards are both quite basic. They do not initiate many of the tasks we require for setting up new drawings. It is unfortunate that AutoCAD "hides" a useful command called MVSETUP (short for "model view setup"). It sets up drawings with predrawn borders, sets the scale, creates multiple viewports, and orients the view in each viewport.

TUTORIAL: SETTING UP NEW DRAWINGS

1. To set up a new drawing, start the **MVSETUP** command:
   - At the 'Command:' prompt, enter the **mvsetup** command.

     Command: **mvsetup** *(Press* ENTER.*)*

2. AutoCAD asks if you wish to enable paper space. (Paper space is dealt with in volume 2 of this book.) Answer "N":

   Enable paper space? [No/Yes] <Y>: **n**

3. Specify the type of units:

   Enter units type [Scientific/Decimal/Engineering/Architectural/Metric]: *(Enter an option.)*

4. Depending on the units you specified, AutoCAD displays different scale factors.

   Enter the scale factor: *(Enter a number.)*

   AutoCAD expects you to enter the number in parentheses next to each scale factor. For example, enter 4 for "4 Times" decimal scale; enter 120 for 1"=10' engineering scale.

- Scientific and Decimal scale factors:

   (4.0)   4 TIMES
   (2.0)   2 TIMES
   (1.0)   FULL
   (0.5)   HALF
   (0.25) QUARTER

- Metric scale factors:

   (5000) 1:5000
   (2000) 1:2000
   (1000) 1:1000
   (500)   1:500
   (200)   1:200
   (100)   1:100
   (75)   1:75
   (50)   1:50
   (20)   1:20
   (10)   1:10
   (5)   1:5
   (1)   FULL

- Architectural scale factors:

   (480) 1/40"=1'
   (240) 1/20"=1'
   (192) 1/16"=1'
   (96)   1/8"=1'
   (48)   1/4"=1'
   (24)   1/2"=1'
   (16)   3/4"=1'
   (12)   1"=1'
   (4)   3"=1'
   (2)   6"=1'
   (1)   FULL

- Engineering scale factors:

   (120) 1"=10'
   (240) 1"=20'
   (360) 1"=30'
   (480) 1"=40'
   (600) 1"=50'
   (720) 1"=60'
   (960) 1"=80'
   (1200) 1"=100'

5. As a last step, AutoCAD prompts

you to enter the width and height of the paper.

Enter the paper width: *(Enter a value.)*

Enter the paper height: *(Enter a value.)*

Paper means the media the drawing is plotted on. Below are sizes for standard sheets.

Sheet Name	Size (width x height)
ANSI Engineering:	
A	11" x 8.5"
B	17" x 11"
C	22" x 17"
D	34" x 22"
E	44" x 34"
US Architectural:	
A	12" x 9"
B	18" x 12"
C	24" x 18"
D	36" x 24"
E	48" x 36"
ISO/DIN Metric:	
A4	278mm x 198mm
A3	408mm x 285mm
A2	582mm x 285mm
A1	819mm x 582mm
A0	1167mm x 819mm

6. AutoCAD draws the outline of the sheet as a rectangle. The rectangle is multiplied by the scale factor. For example, an A-size sheet at scale factor 4 is drawn at 34" x 44". And that's all it does: it does not scale linetypes, dimensions, or text.

CHANGING SCREEN COLORS

When first installed on your computer, AutoCAD is designed to start up in a workspace called "2D Drafting & Annotation." Other workspace names include AutoCAD Classic and 3D Modeling:

- **2D Drafting & Annotation** is the default workspace for AutoCAD 2008.

- **AutoCAD Classic** makes AutoCAD look like it did in 2006 and older releases.

- **3D Modeling** sets up interface elements for 3D modeling.

You can change workspaces through the Workspaces toolbar.

2D Drafting & Annotation Workspace

In the "2D Drafting & Annotation" workspace, AutoCAD displays the drawing area of model space in black. Notice that the cursor and UCS icon show as white on the black background.

History: In this way, today's AutoCAD mimics history. In the past, computer screens had black backgrounds, displaying the text and graphics in monochrome colors, such as green, amber, or white — hence the name, "monochrome displays."

With the advent of Macintosh and its paper metaphor, the background color of choice became white, because it is the color of paper and other media upon which drawings are printed. The white background was then mimicked by Windows. Still, some drafters prefer the black background, because it makes lighter colors, such as yellow and green, stand out better, while others like it because it creates less strain on their eyes.

AutoCAD Classic Workspace

In the "AutoCAD Classic" workspace, AutoCAD also displays the drawing area of model space in black.

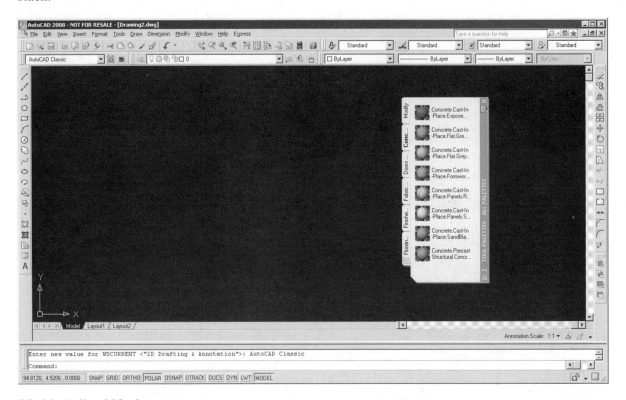

3D Modeling Workspace

In the "3D Modeling" workspace, AutoCAD displays the drawing area of model space in shades of gray to provide a sense of depth. Notice that the cursor and UCS icon are 3D and colored.

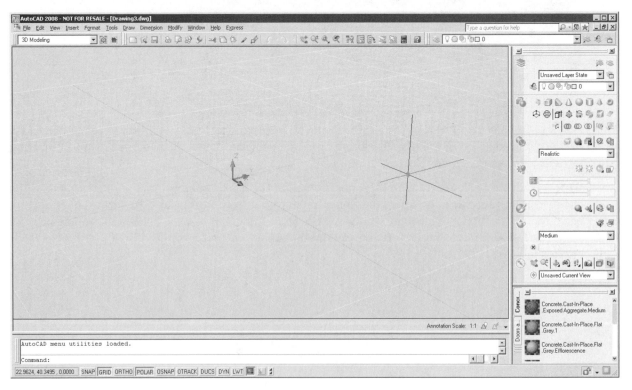

(If your copy of AutoCAD does not look like the figure above, start a new drawing with the *acad3d.dwt* template file.)

Other differences in the 3D worksapce include grid lines instead of dots, perspective mode instead of parallel projection, appearance of the Dashboard, fewer toolbars, and no scroll bars. Unit V of this book, "Introduction to 3D," has more information.

Changing Interface Colors

AutoCAD allows you to change the background color of the drawing area and many other parts of its user interface.

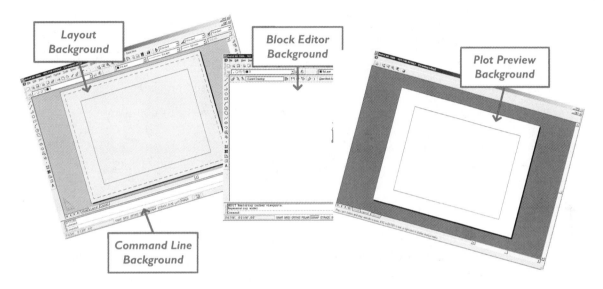

TUTORIAL: CHANGING COLORS

1. To change the colors of the AutoCAD user interface, start the **OPTIONS** command:
   - From the **Tools** menu bar, choose **Options**.
   - At the 'Command:' prompt, enter the **options** command.

      Command: **options** *(Press ENTER.)*

2. In the Options dialog box, choose the **Display** tab, and then click **Colors**. AutoCAD displays the Drawing Window Colors dialog box.

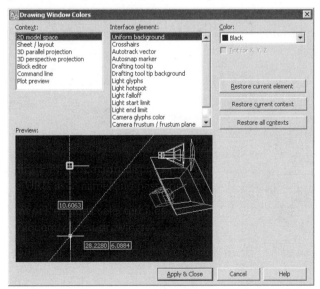

3. From the **Context** area, select a user interface "context," such as **2D model space**.

4. From the **Interface Element** area, select an element, such as **Uniform background**.

5. From the **Color** droplist, assign a color to the element. If the basic eight colors are insufficient, click **Select Color**. AutoCAD displays the Select Color dialog box, which accesses the full palette of 16.7 million colors.

6. If you mess up, click **Restore current element** to change the element back to its default color. Or, choose **Restore current context** to change all elements back to their default colors.

7. Click **Apply & Close** to see the change in colors.

8. Click **OK** to exit the Options dialog box.

UNITS

The **UNITS** command specifies the units of measurement, angles, and precision.

If you choose not to use the Advanced Wizard, set drawing units with this command.

TUTORIAL: SETTING UNITS

1. To set the units for drawings, start the **UNITS** command:

   • From the **Format** menu, choose **Units**.

   • At the 'Command:' prompt, enter the **units** command.

   Command: **units** *(Press ENTER.)*

   • Alternatively, enter the aliases **un** or **ddunits** (the old name for this command) at the 'Command:' prompt.

   In all cases, AutoCAD displays the Units dialog box.

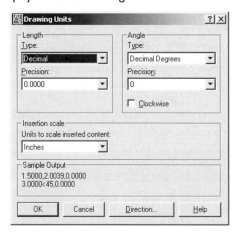

2. In the **Length** area, select the type of unit and precision. To learn more about the units used by AutoCAD, see the discussion earlier in this chapter for the Advanced Wizard. Watch the Sample Output area for examples of the units you select.

3. In the **Angle** area, select the type and precision of angles.

4. In the **Insertion scale** area (named "Drag-and-drop scale" in earlier releases of AutoCAD), from the following possibilities, select the units for blocks dragged into drawings from the DesignCenter, Tool Palettes, or from a Web site using i-drop:

Unit	Comment
Microinches	0.000001 inches
Mils	0.001 inches
Yards	3 feet
Inches	25.4mm
Feet	12 inches
Miles	5,280 feet
Angstroms	0.1 nanometers
Nanometers	10E-9 meters
Microns	10E-6 meters
Millimeters	0.0393 inches
Centimeters	10 mm
Decimeters	0.1 meter
Meters	100 cm
Kilometers	1000 m
Decameters	10 meters
Hectometers	100 meters
Gigameters	10E+9 meters
Astronomical Units	149.597E+8 kilometers
Light Years	9.4605E+9 kilometers
Parsecs	3.26 light years
Unitless	Inserted blocks are not scaled; units match the drawing's units.

5. To change the direction of angle measurement, click **Direction**. AutoCAD displays the Direction Control dialog box.

6. Select a direction for the angle. If you wish to specify an angle that isn't listed, select **Other**, and then enter the angle in the **Angle** text box.

   Alternatively, choose the **Pick an Angle** button to pick the angle in the drawing:

   Pick angle: *(Pick a point in the drawing.)*

   Specify second point: *(Pick a second point.)*

7. Click **OK** to dismiss the Direction Control dialog box.

   Click **OK** to dismiss the Drawing Units dialog box.

8. Move the cursor, and watch the coordinate display on the status bar. It should match the units you selected. The figure below illustrates scientific units and grad angles:

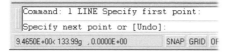

SETTING SCALE FACTORS

One of the hardest concepts for the new CAD user to grasp is that of *scale*. Think of when you sketch a picture of your house on a piece of paper. You draw the house small enough to fit the paper. (There isn't a piece of paper big enough for you to draw the house full size!).

A typical house is 50 feet long. To fit a drawing of it on a 10"-wide sheet of paper, you sketch the house about 60 times smaller. Where does "60" come from? The math works out like this:

Convert house width from feet to inches: 50 feet x 12 inches/foot = 600 inches.

Divide the house size (600 inches) by the paper size (10 inches) = 60.

The scale factor is 1:60. One inch on the drawing is 60 inches on the house.

Here are additional examples of scale factors:

A truck is 20 feet long. The paper is 10 inches wide. The scale factor works out to 1:24.

The sailboat is 30 m long. The paper is 1m wide. The scale factor is 1:30. (It's much easier in metric!)

The gear is 1 inch in diameter. The paper is 8 inches wide. The scale factor is 8:1. (Scale factor numbers are "reversed" when objects are smaller than the paper.)

In CAD, scaling is done late in the process. Instead of drawing the house to scale, you create the CAD drawing full-size. That means the 50-foot long house is drawn 50 feet long; the one-inch gear is drawn one inch in diameter. That's much easier, isn't it? (AutoCAD can handle drawings of extreme size. The technical editor notes that the largest object would be a circle with a radius of 1 000 000 000 000 000 000 E+99 units.)

You could draw the entire known universe — full size — if you want. Full size is shown as 1:1. There is no need to work out scale factors — or is there?

Scale comes into the scene when you plot your drawing. That big drawing, whether of your house or the solar system, must be made small enough to fit the paper it is plotted on. Thus, scale factor comes into play during the **PLOT** command, where you specify that AutoCAD should print the house drawing 60 times smaller to fit the paper.

Scale-dependent Objects

There is, however, a catch. Some items in the drawing cannot be 60 times smaller. These items include text, hatch patterns, linetypes, and dimensions. These are called *scale-dependent*. If AutoCAD plots the text 60 times smaller, you would not be able to read it. The solution is to do the inverse: draw the text 60 times larger. When it is plotted, the text becomes the correct size.

You may be puzzled. How big should text be when plotted on a drawing? The standard in drafting is to draw text 1/8" tall for "normal" sizes. When you place text in your house drawing, specify a height of 7.5" tall (=1/8" x 60). This may seem much too high to you, but trust me: when plotted, it looks correct.

You must apply the scale factor to other scale-dependent objects as well, including dimensions, linetypes, and hatch patterns. Scale factors for hatch patterns, linetypes, and dimensions are stored in these system variables:

System Variable	Scale Factor
HPSCALE	Hatch patterns.
LTSCALE	Linetype patterns.
DIMSCALE	Dimension text height, arrowhead size, and leaders; does not affect dimension values.

To change the scale factor, enter the name of the related system variable. For example, to set the hatch pattern scale to 60:

> Command: **hpscale**
>
> Enter new value for HPSCALE <1.0000>: **60**

Fortunately, you can use the same scale factor for all. For text, you specify a scaled height during the **TEXT** command:

> Command: **text**
>
> Current text style: "TECHNICLIGHT" Text height: 0'-2"
>
> Specify start point of text or [Justify/Style]: *(Pick a point.)*
>
> Specify height <0'-2">: **7.5**

Alternatively, use the **STYLE** command to create a text style with a fixed height of 7.5". If the text is the wrong size, you can use the **SCALETEXT** command to change it.

> **Note:** Be aware of this inconsistency: **HPSCALE** and **DIMSCALE** do not apply to objects already in the drawing, but **LTSCALE** does.

 ## Annotation Scaling

Until recently, there was no master scale factor that applied to all scale-dependent objects, and there was no global scale factor for text at all. AutoCAD 2008 introduces *annotation scaling*, a way to ensure that text, dimensions, linetypes, and hatch patterns are displayed and plotted at the correct size.

Annotation scaling works like this: when the annotative scale of objects matches the view and plot scales, then the objects appear; when not, they do not appear. If the drawing is plotted at several sizes, then you need to assign several annotation scale factors to each object. (Fortunately, AutoCAD makes this automatic.) When multiple annotation scales are assigned to objects, AutoCAD takes care of automatically selecting the representation that appears at the correct scale.

Annotation is a property of the following objects: text, mtext, fields, dimensions, tolerances, attributes, leaders, qleaders, multileaders, and hatch patterns. In addition, blocks and dynamic blocks are aware of annotation scaling.

Linetypes are handled differently. They don't have individual annotative scales; instead, the new **MSLTYPE** system variable forces all linetypes in the drawing to take on the current annotative scale factor.

You learn more about annotation scaling in Chapter 12, "Placing and Editing Text," as well as in other chapters that deal with scale-depending objects (hatches, dimensions, and so on).

SAVE AND SAVEAS

The **SAVE** and **SAVEAS** commands operate identically — both display the Save Drawing As dialog box.

The purpose of the commands is to save drawings by other names, and in a few other formats. You may want to give drawings different names when you open them for editing, but don't want to save the changes to the original file. These commands also allow you to save the drawings in other folders, on other drives, and over networks on other computers.

Drawing files created by AutoCAD 2008 cannot be read by earlier versions of AutoCAD (except for AutoCAD 2007), unless the drawings are *translated*. In contrast, AutoCAD 2008 reads just about any drawing created by earlier releases of AutoCAD. For this reason, the **SAVEAS** command includes options to translate drawings to certain earlier formats.

TUTORIAL: SAVING DRAWINGS BY OTHER NAMES

1. To save drawings under different names, start the **SAVEAS** command:
   * From the **File** menu, choose **Save As**.

   * Alternatively, use the **CTRTL+SHIFT+S** keyboard shortcut.

   * Or, at the 'Command:' prompt, enter the **saveas** or **save** commands.

     Command: **saveas** *(Press* ENTER.*)*

   In all cases, AutoCAD displays the Save Drawing As dialog box.

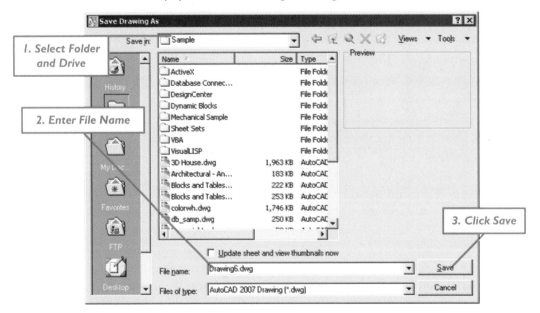

2. In the **Save in** droplist, select the drive and folder in which to save the drawing.
3. In the **File name** text box, enter a different name for the drawing. There is no need to include the .*dwg* extension; AutoCAD adds it automatically.
4. Click **Save**.

When you enter the name of a drawing that already exists in the folder, AutoCAD displays the message box shown below.

5. To replace the existing drawing with the new one, choose **Yes**. If not, choose **No**, and then rename the drawing.

SAVING DRAWINGS: ADDITIONAL METHODS

The SAVEAS command allows you to save drawings on computers connected to networks, including the Internet. It also saves drawings in other formats. The dialog box has provides these options:

• **My Network Places** saves drawings over a network.

• **FTP** saves drawings on the Internet.

• **Files of type** saves drawings in other file formats.

• **Options** specifies options for saving *.dwg* and *.dxf* files.

Let's look at each option.

My Network Places

The **My Network Places** option saves drawings to other computers, provided two conditions are met: (1) the computers are connected via a network and (2) you have access rights.

TUTORIAL: SAVING DRAWINGS ON OTHER COMPUTERS

1. In the Save Drawing As dialog box, click the **Save in** drop list.
2. From the list, select **My Network Places**. Windows displays the drives of computers connected to yours, as well as generic network groupings.
   • *Networked drives* are identified by names similar to "C on Downstairs," where "C" is the name of the hard drive, and "Downstairs" is the name of the computer.

   • *Network groupings* are identified by generic names, such as "Entire Network" and "Computers Near Me." If you regularly access another computer, click **Add Network Place** item to add it to the list.

3. Select a network drive, and then choose the folder in which to save the drawing.

4. Click **Save**.

   Depending on the speed of the network, you may notice it takes longer to save drawings through the network.

FTP

FTP is a method of sending and receiving files over the Internet (short for "file transfer protocol"). You may be familiar with sending files through email as *attachments*. The drawback to using email is that it was not designed to handle very large files. Some email providers limit the attachment size to 5MB or smaller.

FTP allows you to send files of any size, even entire CDs of files — 650MB or more. (FTP is the method used to upload Web pages to Web sites.) The drawback is that FTP first needs to be set up with a user name, password, and other data.

Clicking on **FTP** in the **Places** list results in an empty list. You first need to add a site.

TUTORIAL: SENDING DRAWINGS BY FTP

1. In the Save Drawing As dialog box, choose **Tools**, and then **Add/Modify FTP Locations**.

   AutoCAD displays the Add/Modify FTP Locations dialog box.

   There are two types of FTP sites: anonymous and password.

   - *Anonymous* sites allow anyone to access them (also known as "public FTP sites"). You need only your email address as the password; a username is unnecessary.

   - *Password* sites require a username and a password, usually provided by the person running the FTP site.

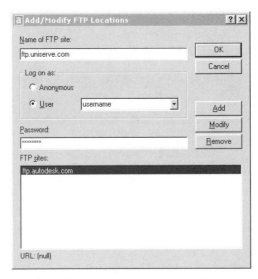

2. Enter the details for the ftp site. For example, enter the following parameters to access Autodesk's public FTP site:

Name of FTP site:	**ftp.autodesk.com**
Logon as:	**Anonymous**
Password:	**email@email.com**

3. Click **Add**.

4. Click **OK**. The FTP item has to be created. Notice that AutoCAD adds the FTP site to the list in the dialog box.

5. To save the drawing, double-click the FTP site.

    Notice that the dialog box displays a list of folders.

6. Select a folder, and then click **Save**.

    A dialog box shows the progress as AutoCAD transfers the file to the site. Expect it to be slower than saving to disk.

7. Some sites require a username and password before you can save files. If so, a dialog box appears requesting the information. Fill in the missing information, and click **OK**.

Files of Type

To save the drawing in an earlier release of AutoCAD, or in other AutoCAD formats, choose the **File of type** list box.

TUTORIAL: SAVING DRAWINGS IN OTHER FORMATS

1. To save drawings in other formats, start the **SAVEAS** command:

    Command: **saveas** *(Press* ENTER.*)*

    AutoCAD displays the Save Drawing As dialog box.

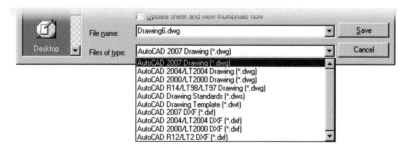

2. In the **Files of type** droplist, select a file format:

    *   **AutoCAD 2007 (*.dwg)** saves drawings in 2007 format, which can also be read by AutoCAD LT 2007.

    *   **AutoCAD 2004/LT 2004 Drawing (*.dwg)** saves drawings in 2004 format, which can also be read by AutoCAD 2005 and 2006.

    *   **AutoCAD 2000/LT 2000 Drawing (*.dwg)** saves drawings in 2000 format, which can also be read by AutoCAD 2000i and 2002, as well as by AutoCAD LT 2000i and 2002.

- **AutoCAD Drawing Standards (*.dws)** saves drawings as standards files.

- **AutoCAD Drawing Template (*.dwt)** saves drawings as templates.

- **AutoCAD 2007 DXF (*.dxf)** saves drawings in DXF format compatible with AutoCAD 2007 and 2008. This saves the drawing in a format called DXF (short for "drawing interchange format"). To open the drawing in Release 12, Release 11, and AutoCAD LT Release 1 and 2, use the **DXFIN** command.

- **AutoCAD 2004 DXF (*.dxf)** saves drawings in DXF format compatible with AutoCAD 2004 - 2006.

- **AutoCAD 2000/LT 2000 DXF (*.dxf)** saves drawings in DXF format compatible with AutoCAD 2000.

- **AutoCAD R12/LT R2 DXF (*.dxf)** saves the drawing for use with AutoCAD Release 12.

3. Click **Save**.

   Caution! Some objects may be lost or changed in translation to older formats.

Note: To save drawings in other formats — *.wmf, .sat, .stl, .bmp, .dgn,* and *.3ds* — use the **EXPORT** command.

To save drawings in *.tif, .jpg,* and *.png* formats, use the **TIFOUT**, **JPGOUT**, and **PNGOUT** commands, respectively.

To save drawings in PostScript (*.eps*) format, use the **PSOUT** command.

To save drawings in *.sld* slide format, use the **MSLIDE** command.

To save drawings in *.dwf* format, use the **ePlot** option of the **PLOT** command, the **PUBLISH** command, or the **3DDWF** command.

To save drawings in MicroStation V8 format, use the **DGNOUT** command.

Options

The **Options** option allows you to specify how *.dwg* and *.dxf* files are saved.

1. In the Save Drawing As dialog box, choose **Tools**, and then **Options**.

   AutoCAD displays the Saveas Options dialog box.

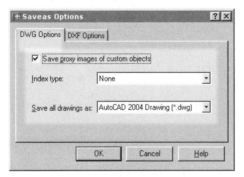

2. The **DWG Options** tab has the following settings:

   ☑ **Save proxy images of custom objects** stores images of *custom objects* in the drawing file. Custom objects are created by ObjectARX programs, and are not understood by AutoCAD when the programs are not present. *Proxy images* are graphical representations of custom objects, which allow you to see, but not edit them.

   ☐ When off, AutoCAD displays rectangles in place of custom objects. It usually makes sense to leave this option turned on.

   Index type specifies whether AutoCAD creates indices when saving drawings. *Indices* improve performance during demand loading, but may increase save time. The options are:

None creates no indices.

Layer loads layers that are on and thawed.

Spatial loads only drawing parts within clipped boundaries.

Layer & Spatial combines both options for optimal partial loading performance.

Save all drawings as determines the default format when drawings are saved with the SAVEAS and QSAVE commands. Keep this set to "AutoCAD 2007 Drawing (*.dwg)" to preserve all features unique to AutoCAD 2007 and 2008. If, however, you regularly exchange drawings with clients who use earlier releases, you may want to change the setting here.

3. The **DXF Options** tab has these settings:

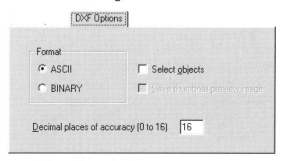

Format selects either ASCII or binary format:

⊙ **ASCII** *.dxf* files are readable by a larger number of non-Autodesk software applications than binary, and also by humans when opened in a text editor.

○ **Binary** files are created and read faster than ASCII, and have smaller file sizes.

☑ **Select Objects** prompts you to select objects from the drawing.

☐ When option is off, AutoCAD outputs the entire drawing to the *.dxf* file.

☑ **Save Thumbnail Preview Image** includes a preview image in the *.dxf* file.

Decimal Places of Accuracy (0 to 16) specifies the accuracy of *.dxf* files. Higher values, such as 16, create larger files. CNC machines typically work to four decimal places, and get confused when files contain five or more decimal places. This option is available only for ASCII *.dxf* files; binary files are always saved with 16 decimal places of accuracy.

 QSAVE

The QSAVE command (short for "quick save") saves drawings as *.dwg* files on disc.

When you start a new drawing, AutoCAD gives it the generic name of *drawing1.dwg*. The first time you use QSAVE, AutoCAD displays the Save Drawing As dialog box, so that you can name the drawing, and file it in the correct folder. Subsequent uses of QSAVE unobtrusively save the drawing without the dialog box.

TUTORIAL: SAVING DRAWINGS

1. To save drawings to disc, start the **QSAVE** command:

   • From the **File** menu, choose **Save**.

   • On the Standard toolbar, select the **Save** button.

   • As an alternative, press the **CTRL+S** keyboard shortcut.

   • Or, at the 'Command:' prompt, enter the **qsave** command.

   Command: **qsave** *(Press ENTER.)*

2. AutoCAD saves the drawing.

   If the drawing's name is *drawing1.dwg*, then AutoCAD displays the Save Drawing As dialog box.

AUTOMATIC BACKUPS

Computers are notorious for *crashing*, where the software stops working for no apparent reason. Even with improvements to Windows XP and Vista, AutoCAD can crash unexpectedly due to *bugs* (accidental errors in the programming code) and conflicts with the operating system. (Technically, computers do not crash; the application software conflicting with the operating system causes the crash.) When these happen, you can lose work.

It is therefore wise to save your work periodically, even if AutoCAD automatically saves to a temporary file. (If you choose to discard your work with the **QUIT** command, only the part of the drawing changed since the last file save is discarded.)

To guard against loss of work, AutoCAD includes two functions to back up drawings: one makes backup copies of drawing files; the other creates a second set of backup copies at set time intervals. Autodesk warns that it is possible to lose drawing data when the power fails during a save. I recommend plugging your computers into uninterruptable power supplies that provide ten minutes of emergency power, giving you time to save drawings during power outages.

TUTORIAL: MAKING BACKUP COPIES

1. To ensure your drawings are backed up, open the Options dialog box:
   * From the **Tools** menu, choose **Options**, and then the **Open and Save** tab.
   * At the 'Command:' prompt, enter the **options** command.

     Command: **options** *(Press* ENTER.*)*

   In either case, AutoCAD displays the Options dialog box.

2. Notice the items in the **File Safety Precautions** section:

   ☑ **Automatic save** turns on the automatic backup facility. *This setting should always be turned on!* The only time to turn it off is when your computer's hard drive is low on free space — which is rare in this day of huge, cheap hard drives.

   10 Minutes between saves means that AutoCAD automatically backs up the drawing every ten minutes. You can change this value to any number you like. If your drawings are small, and your computer fast, then you may want to set this value to five minutes or less. That way, at most, only five minutes' work is lost.

   A setting of 0 (zero) means that no backups are made.

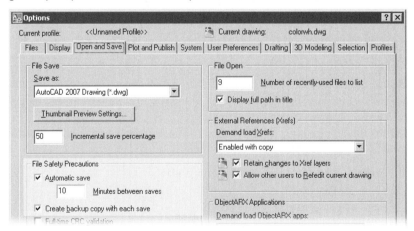

   ☑ **Create backup copy with each save** creates a *.bak* backup copy each time you use **QSAVE** and **CTRL+S**. *This option also should always be turned on,* unless the hard drive is low on free space.

3. Click **OK** to save the settings, and exit the dialog box.

AutoCAD starts the **SAVETIME** timer when you first change the drawing, not when you first open the drawing. The timer resets each time you use the **QSAVE**, **SAVE**, and **SAVEAS** commands. This prevents unnecessary hard drive activity.

MAKING BACKUPS: ADDITIONAL METHODS

In addition to instructing AutoCAD to back up copies of your drawings, you can specify the folder in which backup files are stored.

- **SAVEFILEPATH** specifies the folder for storing backup copies.

Let's look at how it works.

SaveFilePath

AutoCAD normally saves backup files in the following folder:

C:\Documents and Settings*username*\Local Settings\Temp\

This is the same folder used by all other software on your computer. On my computer, for example, this folder holds 1,873 backup files from a variety of sources — word processors, Web browsers, paint programs, and AutoCAD. With all that clutter, you may prefer to store AutoCAD's backup files in a more suitable location, such as *autocad 2008**dwgbackups*, so that the backup files are easier to access.

1. Use Windows Explorer (or File Manager) to create the *dwgbackups* folder in *autocad 2008*. (AutoCAD 2008 already has a folder called *backup* for its own purposes.)
2. In AutoCAD, change the setting of the **SAVEFILEPATH** system variable, as follows:

   Command: **savefilepath**

   Enter new value for SAVEFILEPATH, or . for none <>: *(Enter a new path, such as* **"c:\autocad 2008\dwgbackups".***)*

   Backup files are now stored in the new folder.

When the path name includes spaces, as does the one shown above, you must surround it with quotation marks.

If the path name is not valid, AutoCAD complains, "Cannot set SAVEFILEPATH to that value. *Invalid*." Ensure the folders exist, and that the full path is correct — *full path* means *all* folder names, from the drive name down to the file folder.

Accessing Backup Files

The first time you save a drawing, AutoCAD adds the *.dwg* extension to the file name. With the second save, AutoCAD renames the file with the *.bak* file extension. Each time you use the **QSAVE** command, AutoCAD updates the backup file.

When AutoCAD crashes, it attempts to rename the *.bak* file as *.bk1*. (If *.bk1* exists, then AutoCAD renames it *.bk2*, and, if necessary, continues with *.bka* and then all the way through to *.bkz*.) This prevents AutoCAD from replacing previous backup files.

Backup files are full drawing files, just with a different file extension. If necessary, you can open the backup files in AutoCAD, following these steps:

1. Use Windows Explorer to copy the *.bak* files to a different folder.
2. Rename the *.bak* files to *.dwg* files.
3. Open the renamed *.dwg* files in AutoCAD.

(These steps are also carried out by the **DRAWINGRECOVERY** command.)

When AutoCAD crashes, it sometimes leaves behind temporary files; if your computer's hard drive is getting low on free space, you can erase these leftover files — provided AutoCAD is not running at the time.

Drawings saved by the automatic backup process are given the extension *.ac$*.

When drawings are in use, AutoCAD *locks* the drawings to prevent other AutoCAD users from editing them. The lock status is indicated by the presence of *.dwl* and *.dwl2* (drawing lock) files. These files are used by the **WHOHAS** command to determine which user is editing specific drawings.

The *acminidump.dmp* file is created by AutoCAD when it crashes, to help programmers determine the reason.

These files can be erased with Windows Explorer when AutoCAD is not running.

QUIT

Occasionally you create drawings that you do not wish to keep. To exit the current drawing and discard the changes, use the **QUIT** command.

TUTORIAL: QUITTING AUTOCAD

1. To exit AutoCAD, without saving changes to drawings, start the **QUIT** command:
   * From the **File** menu, choose **Exit**.

   * At the 'Command:' prompt, enter the **quit** or **exit** commands.

   * Alternatively, use the **CTRTL+Q** or **ALT+F4** keyboard shortcuts.

     Command: **quit** *(Press* ENTER.*)*

2. If nothing has changed in the drawing since it was opened, AutoCAD exits.
   In all other cases, AutoCAD displays the strangely-named AutoCAD dialog box.

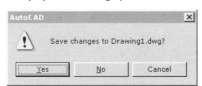

3. To save the drawing, choose **Yes**. AutoCAD displays the Save Drawing As dialog box.
   Enter a name, and then click **Save**.
   To not save the drawing, choose **No**. AutoCAD exits.
   To not save the drawing and return to AutoCAD, choose **Cancel**.

 Note: The **CLOSEALL** command closes all drawings. **QUIT** closes all drawings *and* exits AutoCAD.

EXERCISES

1. In the following exercises, you start new drawings by a variety of methods.
   Start AutoCAD.

   Which do you see: a blank drawing, or the Startup dialog box?

   If you do not see the Startup dialog box, follow the instruction given in this chapter to turn it on.

2. Once the Startup dialog box is on the screen, click the **Use a Wizard** button.

   Select **Advanced Setup**, and then enter the settings for a engineering drawing with four decimal places precision.

   Write down a sample length expressed in the units you set: _____ .

   Select Degrees-minutes-seconds with 0 decimal places of precision.

   Write down a sample angle expressed in the units you set: _____ .

   Select the 0 angle direction as North and clockwise.

   Draw a sketch below showing the direction of North and the direction of positive angle measurement:

   Specify a drawing size of 11" wide and 17" long.

3. On the **Standard** toolbar, click the **New** icon. What happens?

   If AutoCAD displays the Create New Drawing dialog box, select the **Use a Template** button.

   In either the Create New Drawing dialog box, or the Select Template dialog box, watch the preview window while clicking on the names of template drawings.

   When you see a template that looks interesting, open it.

4. In the following exercises, you change the units of measurement.

   a. From the **Format** menu, select **Units**.

   b. In the **Length** droplist of the dialog box, select **Fractional Units**.

   c. Set the precision to **1/16**.

   d. Click **OK** to exit the dialog box.

   e. Move the cursor, and watch the coordinates on the status line.

   f. What do they look like?

5. Press the **SPACEBAR** to return to the Units dialog box.

   In the **Angle** droplist, select **Surveyor's Units**.

   Change the precision to two decimal places.

   After exiting the dialog box, start the **LINE** command, and then click a point.

   Move the cursor, and watch the status line. What do the coordinates look like?

   Move the cursor to the coordinates, and click. Does the display change?

   Press **CTRL+D**, and then move the cursor. Is the coordinate display different?

6. In this exercise, you set the scale factor for hatch patterns to match the drawing.

   Start the **MVSETUP** command in a new drawing, with the following settings:

Units:	**Architectural**
Scale factor:	**240**
Paper width:	**34"**
Paper height:	**22"**

   Does the command draw a rectangle?

   Watching the coordinate display on the status line, move the cursor to the corners of the rectangle, and write down the x, y coordinates to the nearest feet and inches.

   Upper-left
   corner:

   _____ , _____

   Upper-right
   corner:

   _____ , _____

   Lower-left
   corner:

   _____ , _____

   Lower-right
   corner:

   _____ , _____

   Are the upper-right coordinates similar to the scale factor times the width and height dimensions?

   Change the **HPSCALE** system variable to **240**.

   Start the **HATCH** command.

   In the **Hatch** tab of the dialog box, look at the Scale setting. What do you see?

   Click the **Pick Points** button, and pick a point inside the border rectangle.

   Press **ENTER**, and then click **OK**.

   Does the hatch pattern appear to be sized correctly (neither too large nor too small)?

7. Continuing from the previous exercise, practice using the save commands.

    Use the **QSAVE** command to save the drawing.

    Does AutoCAD ask for a name? If so, enter *mvsetup.dwg*.

    Repeat the **QSAVE** command. Does AutoCAD ask for a name?

    This time, use the **SAVE** command. Does AutoCAD ask for a name? If so, press **Cancel**.

    Finally, use the **SAVEAS** command. Does this work identically to the **SAVE** command?

8. If your computer is connected to others through a network, ask your instructor for the following information:

    Name of another computer.

    Name of the drive and folder in which to store drawings.

    Use the **SAVEAS** command to save the drawing from exercise #6 using the network.

9. If your computer has a connection to the Internet, ask your instructor for the following information:

    Address of the school's FTP site.

    Username and password (or email address, if an anonymous FTP site).

    Name of a folder in which to store drawings.

    Use the **SAVEAS** command to save the drawing from exercise #6 to the Web site using FTP transfer.

CHAPTER REVIEW

1. Is it necessary to add a *.dwg* extension to the drawing name when you save drawing files?

2. What is the danger of saving an AutoCAD 2008 drawing to a version earlier than 2004?

3. Which file types does the **OPEN** command open?

4. Write out the meaning of the following file extensions:

    DWG

    DXF

    DWF

    DWT

5. Does the Select Drawing File dialog box allow you to access drawing files located on the hard drive of your coworkers' computers and at sites located on the Internet?

6. Can the **OPEN** command access Web pages?

7. Is AutoCAD limited to working with just one drawing at a time?

8. Describe how to open more than one drawing in the Select Files dialog box.

9. How can you open a 3D model provided to you in SAT format?

10. Is AutoCAD able to open files other than DWG and DXF?

    If so, how?

11. Describe how to rename, move, and delete files with the **OPEN** command.

12. When can you erase a *.ac$* file?

13. AutoCAD starts up, but does not display the Startup dialog box. Describe how to enable this dialog box.

14. Explain the difference between the **NEW** and **QNEW** commands.

15. What is the importance of the *acad.dwt* file?

16. Describe how *template* drawings are useful.

17. What is the purpose of *wizards* in AutoCAD?

18. Write out the unit of measurement for each example shown below:

    1'-2 1/2"

    12.3456

    2'-2.2"

    6.54E+03

    43 3/4

19. Which unit of measurement is used for metric drawings?

20. Which unit of measurement does AutoCAD use internally?

21. How many decimal places can AutoCAD display?

    When would you not want to work with that many decimal places?

22. When you specify two decimal places, does this restrict the accuracy of AutoCAD's calculations?

23. Write out the angle of measurement for each example shown below:

    45.00

    45d00'00.00"

    50.00g

    0.79r

    N45d0'0.00"E

24. How many *degrees* are there in a circle?

    How many *grads*?

    How many *radians*?

25. In which direction does AutoCAD think of 0 degrees, by default?

26. In which direction are *positive* angles measured, by default?

27. Can you change the direction of 0 degrees?

28. What does AutoCAD use *limits* for?

29. Does the Select File dialog box allow you to preview drawings?

30. How would you open a drawing, but prevent changes from being saved?

31. What happens when you double-click a drawing file name in the Select Files dialog box?

    What happens when you click a drawing file name twice?

32. Explain the meaning of:

    Write-protected

    Read-only

    Read-write

33. Describe how to save drawings that have been opened *read-only*.

34. What are *.bak* files?

    Why are they useful?

35. Describe how to open a *.bak* file.

36. How can you guard against losing drawing data from a power outage?

37. What happens when drawings are "saved"?

38. Which command saves drawings quickly?

    Can AutoCAD save drawings on its own?

39. Describe the difference between the **QSAVE** and **SAVE** commands.

    And the difference between the **SAVE** and **SAVEAS** commands.

40. In the Select File dialog box, displayed by the **OPEN** command, does the **Folders** sidebar provides quick access to:

    My Documents

    History

    FTP

    Desktop

    Buzzsaw

41. Can you create new folders with the Select Files dialog box?

42. Explain when you would use the **Find** tool?

43. What functions does the **MVSETUP** command perform?

44. What are the dimensions of a B-size sheet of ANSI Engineering paper?

45. Which metric sheet is larger — A4 or A0?

    Which imperial sheet is larger — A or E?

46. Can you change the background color of the drawing area?

    If so, why might you want to?

47. Describe the purpose of the **UNITS** command.

48. How does the **UNIT** command's **Unitless** setting help the operator insert blocks into drawings?

49. Work out the scale factors for drawing these objects on a sheet of paper that is 10 inches wide. Show all your calculations.

    a. An automobile and trailer 12 feet long.

    b. A tool shed 10 feet wide.

    c. A telephone 9 inches long.

50. Work out the scale factors for the following objects drawn on a sheet of paper that is 100 cm wide. Show all your calculations.

    a. A pen holder 7 cm in diameter.

    b. A stereo system 42 cm wide.

    c. A house 16m long.

51. Which system variable sets the scale factor for the following objects:

    Hatch patterns

    Linetypes

    Dimensions

    Text

52. Explain why text cannot be drawn full-size in drawings scaled at 1:50.

53. How tall should text be drawn in drawings with the following scale factors. (Show all your calculations.)

    1:50

    1:1

    10:1

    1:2500

54. If the scale factor for a drawing is 1:50, what should the scale factor be for hatch patterns?

55. If the scale factor for linetypes is 10:10, what should the scale factor be for hatch patterns?

56. Under what conditions can drawings be saved to others' computers?

57. What happens when you use the **SAVE** command to save a drawing, when another drawing of the same name already exists?

58. Describe what happens when you quit AutoCAD without saving the drawing.

59. Explain a benefit and drawback to FTP.

    What is FTP short for?

    When might you use FTP?

60. What is the difference between *anonymous* and *private* FTP sites?

61. Describe the difference between the two types of *.dxf* file:

    ASCII

    Binary

62. Name the command that belongs to each keyboard shortcut:

    CTRL+S

    CTRL+SHIFT+S

    CTRL+O

    CTRL+N

    CTRL+Q

UNIT II

Drafting
Essentials

CHAPTER 4

Drawing with Basic Objects

Creating drawings with AutoCAD involves a primary set of commands. These drawing commands construct the objects basic to most drawings — lines, arcs, circles, and so on. Now that you know how to set up new drawings, let's begin creating them.

In this chapter, you learn about these basic drawing commands:

LINE draws line segments.

RECTANG constructs squares and rectangles.

POLYGON constructs regular polygons, from 3 to 1,024 sides.

CIRCLE constructs circles by several methods.

ARC constructs arcs by many methods.

DONUT draws thick and solid-filled circles.

PLINE draws connected lines, arcs, and curves.

ELLIPSE constructs ellipses and elliptical arcs.

POINT draws point objects.

FINDING THE COMMANDS

On the Dashboard's **2D Draw** panel:

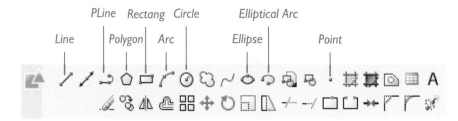

On the **Draw** toolbar:

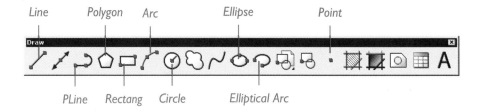

LINE

The **LINE** command draws line segments.

Constructing lines in drawings is the most basic CAD operation. In AutoCAD, you can draw different types of lines (as listed below) and apply a variety of options to them, but many lines are drawn with the **LINE** command.

A single line is sometimes called a "segment." At each end, the line has *endpoints*.

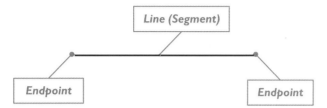

Let's talk about "first points" and "next points." To draw lines, you must show AutoCAD the point from which to start. AutoCAD calls this the "first point." Think of this as the place where you initially put your pencil to the paper.

Secondly, determine the point where the line segment should end. AutoCAD calls this is the "next point."

AutoCAD repeats the 'Specify next point:' prompt so that you can draw a series of lines without the inconvenience of restarting the **LINE** command.

AutoCAD repeats the 'Specify next point:' prompt until you press **ESC**, or press **ENTER** twice to end the command.

Note: AutoCAD has commands that draw different types of lines, some of which you encounter in this chapter:

- **MLINE** (multiline) draws as many as 16 parallel lines as a single object.

- **PLINE** (polyline) draws lines in the same manner as the **LINE** command, but a single polyline can include arcs, splines, and variable widths. The **3DPOLY** command draws three-dimensional polylines.

- **SKETCH** draws freehand lines.

- **SPLINE** draws splined curves.

- **XLINE** (construction line) and **RAY** draw infinite construction lines (xlines) and semi-infinite construction lines (rays).

TUTORIAL: DRAWING LINE SEGMENTS

1. To draw line segments, start the **LINE** command with one of these methods:
    *   From the **Draw** menu, choose **Line**.
    *   From the Draw toolbar, choose the **Line** button.
    *   From the Dashboard's 2D Draw panel, choose the **Line** button.
    *   Or, at the 'Command:' prompt, enter the **line** command.

        Command: **line** *(Press* ENTER.*)*
    *   Alternatively, enter the **l** alias at the 'Command:' prompt.

2. Specify the starting point of the line segment:

    Specify first point: *(Pick point 1, or specify coordinates.)*
    *   With the cursor, pick a point on the screen.
    *   Or, at the keyboard, enter x,y coordinates, such as **2,3**.

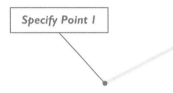

3. Specify the next point(s):

    Specify next point or [Undo]: *(Pick point 2, or specify coordinates.)*
    *   Move the cursor, and then pick another point.
    *   Or, enter another set of x,y coordinates, such as **3,4**.

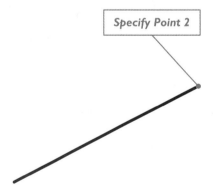

4. Press **ENTER** to end the command.

    Specify next point or [Undo]: *(Press* ENTER *to exit the command.)*

Alternatives to pressing **ENTER** include:
*   Pressing **ESC**.
*   Right-clicking, and selecting **Enter** from the shortcut menu.

5. To repeat the **LINE** command, press **ENTER**.

    AutoCAD restarts the **LINE** command and displays "LINE" at the 'Command:' prompt.

    　Command: LINE Specify first point:

6. This time, draw several lines. Notice that each new line connects precisely at the endpoint of the previous line. This connection is called a "vertex."

7. Press **ENTER** to exit the command.

    In AutoCAD, pressing **ENTER** ends (most) commands, as well as restarting the same command.

DRAWING LINE SEGMENTS: ADDITIONAL METHODS

The **LINE** command provides additional options for drawing lines and ending the command:

- **Close** automatically draws another line to the start of the first one.

- **Undo** undraws the last segment.

- **@** and **<** draw with relative coordinates and angles.

- **ENTER** continues from the last segment.

Some (but not all) of the command's options are shown on the shortcut menu (below, at left) and in the drafting tooltip (at right).

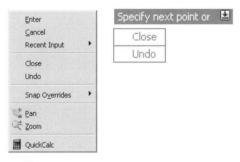

Note: Shortcut menus contain options specific to the command, as well as options that apply to any command. In the shortcut menu illustrated above, four options belong to the **LINE** command: **Enter**, **Cancel**, **Close**, and **Undo**.

The remaining five options apply to all commands:
Recent Input — lists the last 20 commands and options you entered.
Snap Overrides — displays a submenu of object snap modes (see Chapter 5).
Pan — enters realtime pan, allowing you to shift the drawing around the viewport (see Chapter 5).
Zoom — enters realtime zoom, allowing you to change the apparent size of the drawing.
QuickCalc — displays the Quick Calc window, which allows you to carry out calculations, the results of which are used as input for the active command (see *Using AutoCAD: Advanced*).

Let's look at each option specific to the **LINE** command.

Close

When you construct polygons with the **LINE** command, the last segment connects to the first segment. Aligning the ends of segments can be tedious, so the **Close** option automatically does this for you. The **Close** option does not appear until you pick three points, because a minimum of two lines is needed before they can be closed with a third line to make the polygon.

To see how this works, consider drawing a right angle triangle. The illustration shows how the line at the final intersection is connected with the **Close** option.

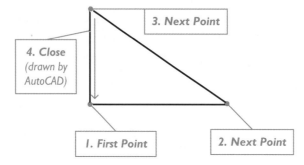

Command: **line**

Specify first point: *(Pick point 1.)*

Specify next point or [Undo]: *(Pick point 2.)*

Specify next point or [Undo]: *(Pick point 3.)*

Specify next point or [Close/Undo]: **c**

Undo

As you draw line segments with the LINE command, you may sometimes make a mistake. Instead of canceling the command and starting over, use the **Undo** option to "undraw" the last segment. You may undo all the way back to the first segment.

Relative Coordinates and Angles

The @ and < symbols allow you to specify relative coordinates and angles. When you prefix x,y-coordinates with @, AutoCAD reads the numbers as distances, not coordinates. When you include the < symbol, AutoCAD reads the number following it as an angle. Here are two examples of using relative coordinates:

Command: **line**

Specify first point: @2,3   *Draws the line 2 units right and 3 units up from the previous point.*

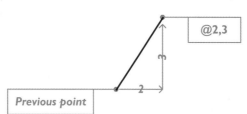

Specify next point or [Undo]: 10<45   *Draws the line 10 units long at a 45-degree angle from the last point.*

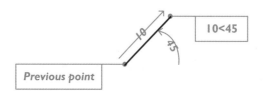

Continue

When you terminate the LINE command, you can begin a new line anywhere else in the drawing. If, however, you wish to go back and continue from the last endpoint, reconnect with the LINE command's **Continue** option.

This option is hidden by AutoCAD: the next time you start the LINE command, press ENTER at the "Specify first point:" prompt.

 ## RECTANG

The RECTANG command draws rectangles and squares.

The command has options for drawing rectangles by a variety of methods and in a variety of styles. It can draw them with thin or fat lines, tilted at an angle, and with rounded or cutoff corners. The primary parts of a rectangle are its *length* and *width*:

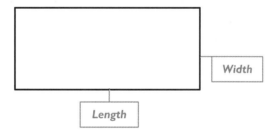

This command does not draw geometric variations, such as parallelograms or rhomboids.

TUTORIAL: DRAWING RECTANGLES

To draw a basic rectangle, pick two points that define the opposite corners of the rectangle.

1. Start the RECTANG command by one of these methods:
   * From the **Draw** menu, choose **Rectangle**.

   * From the Draw toolbar, choose the **Rectangle** button.

   * From the Dashboard's 2D Draw panel, choose the **Rectangle** button.

   * Or, at the 'Command:' prompt, enter the **rectang** command.

     Command: **rectang** *(Press ENTER.)*

   * Alternatively, enter the aliases **rec** or **rectangle** at the 'Command:' prompt.

2. Pick a point for one corner of the rectangle, such as the lower-left corner:
   Specify first corner point or [Chamfer/Elevation/Fillet/Thickness/Width]: *(Pick point 1.)*

3. And pick a point for the opposite corner:
   Specify other corner point or [Area/Dimensions/Rotation]: *(Pick point 2.)*

Notes: The sides of the rectangle are parallel to the x and y axis of the current UCS (user-defined coordinate system). To draw rectangles at an angle, use the command's **Rotation** option.

To draw squares, pick two points that create the square. As an alternative, you can use the **POLYGON** command to draw squares.

ADDITIONAL METHODS FOR DRAWING RECTANGLES

The **RECTANG** command provides additional options for drawing rectangles:

- **Dimensions** specifies the width and height of the rectangle.
- **Area** specifies the area and one side of the rectangle.
- **Rotation** specifies the angle of the rectangle.
- **Width** specifies the width of the four lines making up the rectangle.
- **Elevation** draws the rectangle a specific height above the x,y-plane.
- **Thickness** draws the rectangle with a thickness in the z-direction.
- **Fillet** rounds off the corners of the rectangle.
- **Chamfer** cuts off the corners of the rectangle.

Let's look at each option.

Dimensions

The basic way to draw a rectangle is to specify the points of opposite corners. If you prefer, you can specify the size of the rectangle by the length and width to draw, for example, a 3" x 2" box.

1. Start the **RECTANG** command, pick a starting point, and then enter the **Dimension** option:

    Command: **rectang**

    Specify first corner point or [Chamfer/Elevation/Fillet/Thickness/Width]: *(Pick point 1.)*

    Specify other corner point or [Dimensions]: **d**

2. Enter the *length*, which is in the x-direction (to the left):

    Specify length for rectangles <0.0000>: **3**

3. And enter the *width*, which is in the y-direction (up):

    Specify width for rectangles <0.0000>: **2**

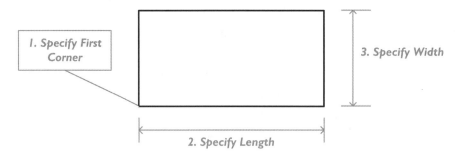

4. At this point, AutoCAD asks you to specify the orientation of the rectangle. Depending on where you pick the "other corner point," you place the rectangle in one of four positions around the first corner point:

    Specify other corner point or [Dimensions]: *(Move the cursor to position the rectangle, and then pick point 4.)*

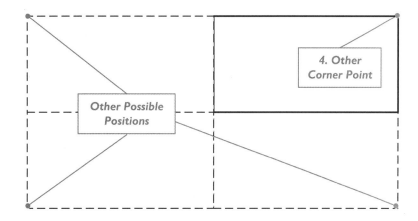

Area

The **Area** option specifies the area and one side of the rectangle. After entering the **Area** option, the command prompts you:

> Enter area of rectangle in current units <100>: *(Enter a value.)*

> Calculate rectangle dimensions based on [Length/Width] <Length>: *(Type **L** or **W**.)*

When you enter **L**, AutoCAD prompts:

> Enter rectangle length <10>: *(Enter a value smaller than the area.)*

If you enter **W**, AutoCAD prompts:

> Enter rectangle width <10>: *(Enter a value smaller than the area.)*

AutoCAD then repeats the earlier prompt:

> Specify other corner point or [Area/Dimensions/Rotation]: *(Pick a point or enter an option.)*

You can define the other corner of the rectangle, or specify an option, such as the rotation angle.

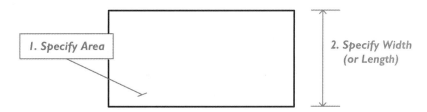

When the **Chamfer** or **Fillet** option is active, AutoCAD accounts for the subtracting effect of the chamfers or fillets on the area; that is, the area is made "larger" to compensate. Both rectangles illustrated below have a length of 10 units and an area of 50 square units. The gray rectangle is wider, because the fillets at each corner would otherwise reduce the area.

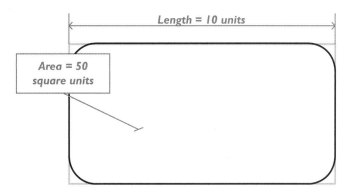

Rotation

The **Rotation** option specifies the angle of the rectangle. AutoCAD prompts you:

Specify rotation angle or [Points] <0>: *(Enter the angle or type **P**.)*

When you enter the angle, AutoCAD ghosts the rotated rectangle, and then prompts you for the other corner point:

Specify other corner point or [Area/Dimensions/Rotation]: *(Pick a point, or enter an option.)*

When you enter **P**, AutoCAD prompts you to pick two points:

Specify first point: *(Pick a point.)*

Specify second point: *(Pick another point.)*

And then asks for the location of the other corner.

Width

You can specify the width of the four lines making up the rectangle. (Technically, the "lines" are *polylines*, which are discussed later in this chapter.) Once you set the width, the same value is used by subsequent RECTANG commands, until you change it.

To change the width, enter the **Width** option, and then specify a width in units:

Command: **rectang**

Specify first corner point or [Chamfer/Elevation/Fillet/Thickness/Width]: **w**

Specify line width for rectangles <0.0000>: **.1**

Width changed to 0.1 units.

Elevation

You can specify the height of the rectangle above (or below) the x,y-plane — in the z-direction. Once you set the height, the same value is used by subsequent RECTANG commands until you change it. To change the elevation, enter the **Elevation** option, and then specify an elevation in units:

Command: **rectang**

Specify first corner point or [Chamfer/Elevation/Fillet/Thickness/Width]: **e**

Specify the elevation for rectangles <0.0000>: **2**

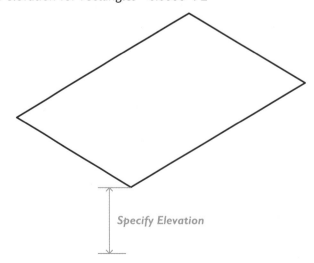

Specify Elevation

A positive value for elevation draws the rectangle above the x,y plane. A negative value pushes the rectangle below the x,y-plane.

Thickness

You can specify a thickness for the rectangle, which turns it into a 3D box-looking object that lacks a top and bottom. Once you set the thickness, the same value is used by subsequent RECTANG commands, until you change it. To change the thickness, enter the **Thickness** option, and then specify a thickness in units:

> Command: **rectang**
>
> Specify first corner point or [Chamfer/Elevation/Fillet/Thickness/Width]: **t**
>
> Specify thickness for rectangles <0.0000>: **2**

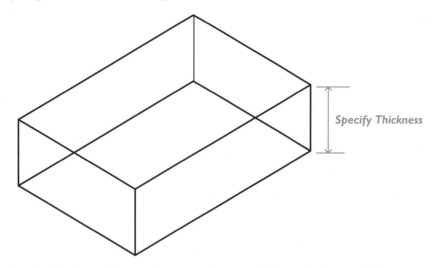

Specify Thickness

A positive value for thickness draws the rectangle upward from the base, while a negative value draws the rectangle downward.

Fillet

To round the corners of the rectangle, enter the **Fillet** option:

> Command: **rectang**
>
> Current rectangle modes: Elevation=2.0000 Thickness=2.0000 Width=0.1000
>
> Specify first corner point or [Chamfer/Elevation/Fillet/Thickness/Width]: **f**
>
> Specify fillet radius for rectangles <0.0000>: **.25**

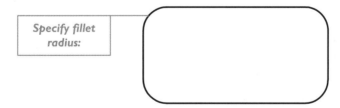

Specify fillet radius:

The single fillet radius applies to all four corners; you cannot specify a different radius for each corner. If the fillet radius is too large for the rectangle, AutoCAD does not draw the fillets; instead, the rectangle has square corners. Once you set the fillet, the same value is used until you change it.

You can enter a negative radius, which produces this interesting effect:

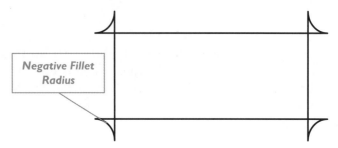

Negative Fillet
Radius

After the rectangle is drawn, you can use grips to move the fillets around, such as illustrated below.

Chamfer

To cut off the corners of the rectangle, enter the **Chamfer** option:

> Command: **rectang**
>
> Specify first corner point or [Chamfer/Elevation/Fillet/Thickness/Width]: **c**
>
> Specify first chamfer distance for rectangles <0.2500>: *(Press* ENTER.*)*
>
> Specify second chamfer distance for rectangles <0.2500>: *(Press* ENTER.*)*

The chamfer distances are measured from the corners of the rectangle. You can enter a different value for the first and second chamfer distances, as illustrated below.

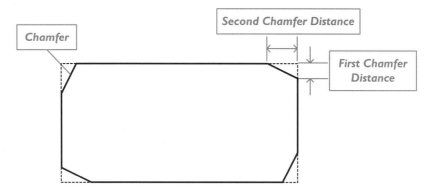

Chamfer

Second Chamfer Distance

First Chamfer
Distance

The pair of chamfer distances applies to all four corners; you cannot specify a different chamfer for each corner. If the chamfer distances are too large for the rectangle, AutoCAD does not draw the chamfers; instead, the rectangle has square corners. Once you set the chamfers, the same value is used by subsequent RECTANG commands, until you change it.

You can enter negative and positive distances, which produces this pinwheel effect:

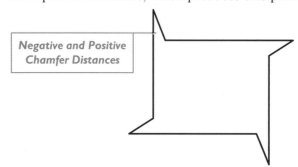

Negative and Positive
Chamfer Distances

POLYGON

The **POLYGON** command draws regular polygons.

AutoCAD draws regular polygons, from 3 to 1,024 sides, through the **POLYGON** command. A *regular polygon* has all the sides the same length; in contrast, an *irregular polygon* has sides of different lengths. With this command, you draw equilateral triangles, squares, pentagons, hexagons, octagons, and so on.

The polygons are constructed by any of three methods: on the basis of the length of one edge, or by fitting inside ("inscribed" within) or outside ("circumscribed" by) a circle. AutoCAD draws polygons from polylines, so you can change their widths with the **PEDIT** command.

The primary parts of polygons are the vertices, edges, radius of inscribed circle, and center point:

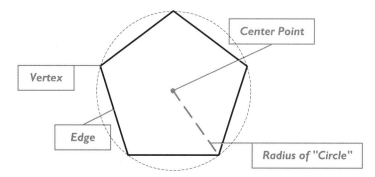

You can use either this command to draw four-sided polygons, more commonly called "squares," or you can use the **RECTANG** command. Neither command draws irregular polygons, such as acute triangles and parallelograms; you need to construct these using AutoCAD's other commands, as described at the end of this chapter.

In drawing regular polygons, the word "circle" is used, because each vertex is the same distance from the center point — just like every point on a circle's circumference.

TUTORIAL: DRAWING POLYGONS

1. To draw regular polygons, start the **POLYGON** command by one of these methods:
    *   From the **Draw** menu, choose **Polygon**.
    *   From the Draw toolbar, choose the **Polygon** button.
    *   From the Dashboard's 2D Draw panel, choose the **Polygon** button.
    *   Or, at the 'Command:' prompt, enter the **polygon** command.

        Command: **polygon** *(Press* ENTER.*)*
    *   Alternatively, enter the **pol** alias at the 'Command:' prompt.

2. Specify the number of sides. For example, enter 3 for a triangle, 4 for a square, 5 for a pentagon, and so on.

    Enter number of sides <4>: *(Enter a value between 3 and 1024.)*

3. Pick a point for the center of the polygon:

    Specify center of polygon or [Edge]: *(Pick a point.)*

4. Decide if the polygon fits inside (is inscribed within) or outside (circumscribes) an imaginary circle:

    Enter an option [Inscribed in circle/Circumscribed about circle] <I>: *(Type* I *or* **C**.*)*

5. Specify the radius of the circle, which determines the size of the polygon:

    Specify radius of circle: *(Enter a radius, or pick a point.)*

ADDITIONAL METHODS FOR DRAWING POLYGONS

The **POLYGON** command provides options for drawing polygons:

- **Edge** defines the polygon by the length of one of its edges.

- **Inscribed in circle** fits the polygon inside an imaginary circle.

- **Circumscribed about circle** fits the imaginary circle inside the polygon.

Let's look at each option.

Edge

Perhaps the easiest method is to construct polygons by specifying the length of one edge. Pick two points that define the length of the edge. Here is how to draw a pentagon with the **Edge** option.

1. Start the **POLYGON** command:
   Command: **polygon**

2. To draw a pentagon, enter **5** for the number of sides:
   Enter number of sides <4>: **5**

3. Select the **Edge** option by typing **e**:
   Specify center of polygon or [Edge]: **e**

4. Pick a point (or enter x,y coordinates) for the start of the edge:
   Specify first endpoint of edge: *(Pick a point.)*

5. And pick another point for the end of the edge:
   Specify second endpoint of edge: *(Pick another point.)*

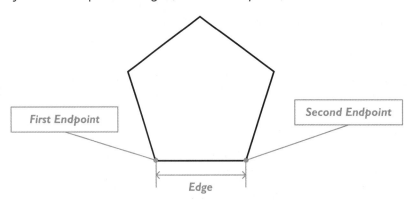

The angle of the two edge points determines the angle of the polygon. For example, to draw a diamond shape, pick the two points at a 45-degree angle:

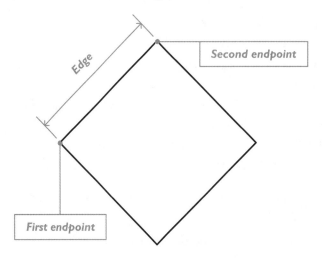

Inscribed in Circle

Inscribed polygons are constructed inside a circle of a specified radius. (The circle itself is not drawn, but is imaginary.) The vertices of the polygon fall on the circle. Here is an example of how to draw a triangle inside a circle:

> Command: **polygon**
>
> Enter number of sides <4>: **3**
>
> Specify center of polygon or [Edge]: *(Pick a point.)*
>
> Enter an option [Inscribed in circle/Circumscribed about circle] <I>: **i**
>
> Specify radius of circle: *(Pick a point, or enter a radius.)*

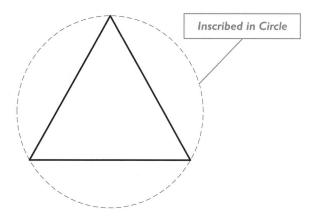

Inscribed in Circle

Just as the two points for the **Edge** option determine the rotation of the polygon, so too does the radius specification. As you move the cursor during the "Specify radius of circle" prompt, notice that one vertex moves with the cursor:

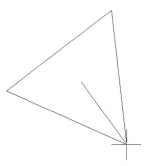

Circumscribed About Circle

Circumscribed polygons are constructed outside a circle of a specified radius. The midpoints of the polygon's edges are placed on the circle's circumference. As with the inscribed option, the rotation of the polygon is determined at the "Specify radius of circle" prompt.

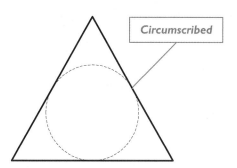

Circumscribed

 **CIRCLE**

The **CIRCLE** command draws circles.

AutoCAD can draws circles by several methods. As you progress in your drafting, you will find that some methods work more easily than others. The important parts of the circle are the center point, the radius or diameter, and the circumference.

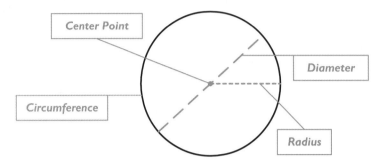

(To draw isometric circles, use the **Isocircle** option of the **ELLIPSE** command, after turning on **Iso** mode with the **SNAP** command.)

TUTORIAL: DRAWING CIRCLES

1. To draw circles, start the **CIRCLE** command by one of these methods:
   * From the **Draw** menu, choose **Circle**, and then choose **Center, Radius**.
   * From the Dashboard's 2D Draw panel, choose the **Circle** button.
   * From the Draw toolbar, choose the **Circle** button.
   * Or, at the 'Command:' prompt, enter the **circle** command:

     Command: **circle** *(Press ENTER.)*
   * Alternatively, enter the **c** alias at the 'Command:' prompt.

2. Pick a point for the center of the circle:

   Specify center point for circle or [3P/2P/Ttr (tan tan radius)]: *(Pick point 1, or enter x,y coordinates.)*

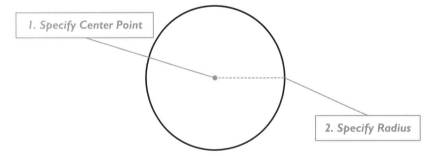

3. Enter the radius of the circle:

   Specify radius of circle or [Diameter]: *(Specify the radius, or pick point 2.)*

ADDITIONAL METHODS FOR DRAWING CIRCLES

The CIRCLE command provides additional options for drawing circles:

- **Diameter** draws circles based on a center point and diameter.

- **2P** draws circles based on two diameter points.

- **3P** draws circles based on three points along the circumference.

- **Ttr (tan tan radius)** draws circles touching two tangent points and a radius.

- **Tan, Tan, Tan** draws circles touching three tangent points.

Let's look at each option.

Diameter

Instead of specifying the circle's center point and radius, you can specify the diameter. (The diameter is twice the radius.) From the **Draw** menu bar, select **Circle | Center, Diameter.**

> Command: **circle**
>
> Specify center point for circle or [3P/2P/Ttr (tan tan radius)]: *(Pick point 1.)*
>
> Specify radius of circle or [Diameter] <1.0>: **d**
>
> Specify diameter of circle <2.0>: *(Pick point 2, or enter a value for the diameter.)*

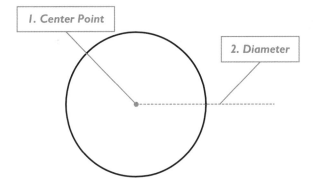

2P

Instead of specifying a center point and diameter, you can specify only the diameter by picking two points on the circumference. These two points define the diameter. From the **Draw** menu bar, select **Circle | 2 Points.**

> Command: **circle**
>
> Specify center point for circle or [3P/2P/Ttr (tan tan radius)]: **2p**
>
> Specify first end point of circle's diameter: *(Pick point 1.)*
>
> Specify second end point of circle's diameter: *(Pick point 2.)*

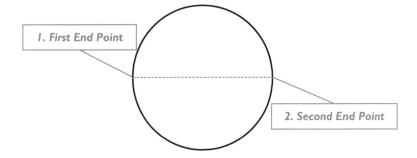

3P

As an alternative, you can pick any three points on the circumference, and AutoCAD constructs the circle. From the **Draw** menu bar, select **Circle | 3 Points.**

> Command: **circle**
>
> Specify center point for circle or [3P/2P/Ttr (tan tan radius)]: **3p**
>
> Specify first point on circle: *(Pick point 1.)*
>
> Specify second point on circle: *(Pick point 2.)*
>
> Specify third point on circle: *(Pick point 3.)*

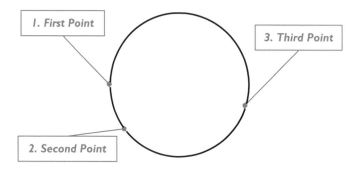

Ttr (tan tan radius)

Often, you need to draw a circle that's precisely tangent to other objects. The **Ttr** option automatically invokes **Tangent** object snap. (A *tangent* is the point at which the circle touches another object.) You can draw circles tangent to other circles, arcs, lines, polylines, and so on. From the **Draw** menu bar, select **Circle | Tan Tan Radius.**

> Command: **circle**
>
> Specify center point for circle or [3P/2P/Ttr (tan tan radius)]: **t**
>
> Specify point on object for first tangent of circle: *(Pick point 1.)*
>
> Specify point on object for second tangent of circle: *(Pick point 2.)*
>
> Specify radius of circle <0.9980>: *(Pick point 3, or enter a value for the radius.)*

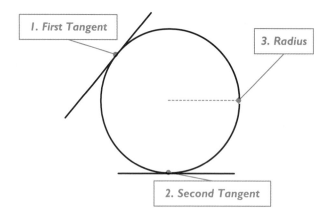

You may find it tricky placing the circle with tangents, because the radius determines where the tangents are placed.

Tan, Tan, Tan

AutoCAD has a sixth way of drawing circles that's not available from the CIRCLE command. Instead, you have to access it from the **Draw** menu: select **Circle | Tan Tan Tan.** This "secret" option constructs circles tangent to three objects. After you select the command from the menu, notice that AutoCAD fills in several options for you, and automatically invokes **Tangent** object snap (displayed as "_tan").

All you do is pick three objects, as follows:

> Command: _circle Specify center point for circle or [3P/2P/Ttr (tan tan radius)]: _3p
> Specify first point on circle: _tan to *(Pick point 1.)*

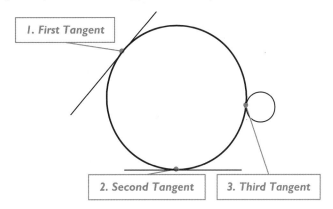

1. First Tangent

2. Second Tangent   3. Third Tangent

Specify second point on circle: _tan to *(Pick point 2.)*

Specify third point on circle: _tan to *(Pick point 3.)*

 ## ARC

The ARC command draws arcs by a variety of methods.

An *arc* is portion of a circle. Like circles, arcs have a center point and a radius (or diameter). Like lines, arcs have starting and ending points.

It can be difficult for beginners to remember that AutoCAD draws arcs in the *counterclockwise* direction, with a couple of exceptions: the three-point arc, and the start-end-direction arc. If AutoCAD does not draw the arc as you expect, it may be because you are trying to draw it clockwise. Sometimes, you may find it easier to draw a circle, and then use the BREAK command to create an arc.

Some disciplines, such as railroad track design, define arcs by *chords*, the straight distance between the start and endpoints. AutoCAD calls this distance the "length." The parts of an arc are shown below:

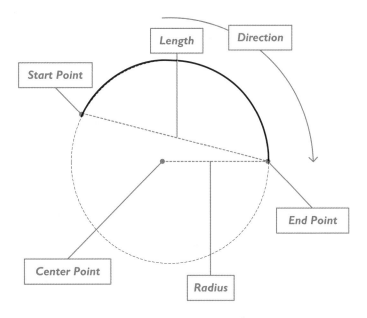

Length   Direction

Start Point

End Point

Center Point

Radius

AutoCAD constructs arcs by many methods — too many, you might think by the end of this section. This flexibility allows you to place arcs in many different situations.

TUTORIAL: DRAWING ARCS

1. To draw arcs, start the **ARC** command by one of these methods:
    - From the **Draw** menu, choose **Arc**, and then choose **Start, Center, End**.

    - From the Draw toolbar, choose the **Arc** button.

    - From the Dashboard's 2D Draw panel, choose the **Arc** button.

    - Or, at the 'Command:' prompt, enter the **arc** command:

        Command: **arc** *(Press* ENTER.*)*

    - Alternatively, enter the **a** alias at the 'Command:' prompt.

2. Pick the starting point of the arc:

    Specify start point of arc or [Center]: *(Pick point 1.)*

3. Pick a point that lies on the arc:

    Specify second point of arc or [Center/End]: *(Pick point 2.)*

4. Pick a point at the end of the arc:

    Specify end point of arc: *(Pick point 3.)*

    You can pick points in the drawing, or enter x,y coordinates. The arc drawn in this tutorial is called the "three-point arc." Unlike other arcs, this one is drawn in the direction determined by the first and second points.

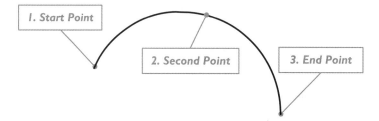

ADDITIONAL METHODS FOR DRAWING ARCS

The ARC command provides additional options for drawing arcs. Several of the variations are similar; thus I have grouped them together:

- **Start, Center, End** constructs arcs from the start, center, and endpoints.
- **Center, Start, End** constructs arcs from the center, start, and endpoints.

- **Start, Center, Angle** constructs arcs from the start and center points, and the included angle.
- **Center, Start, Angle** constructs arcs from the center and start points, and the included angle.
- **Start, End, Angle** constructs arcs from the start and endpoints, and the included angle.

- **Start, Center, Length** constructs arcs from the start and center points, and the chord length.
- **Center, Start, Length** constructs arcs from the start, center, and endpoints.

- **Start, End, Radius** constructs arcs from the start and end points, and the radius.
- **Start, End, Direction** constructs arcs from the start and endpoints, and the direction.
- **Continue** continues the arc tangent to a line or another arc.

Let's look at each option.

START, CENTER, END & CENTER, START, END

The most common method of drawing arcs is to specify their start, center, and endpoints. AutoCAD constructs the arc in the counterclockwise direction, starting from the start and drawing to the endpoint. AutoCAD calculates the arc's radius as the distance from the start point to the center point.

If the arc is too big or looks wrong, it could be you started the arc at the point where it should end. From the **Draw** menu, select **Arcs | Start, Center, End** or **Center, Start, End**.

> Command: **arc**
>
> Specify start point of arc or [Center]: *(Pick point 1, or enter x,y coordinates.)*
>
> Specify second point of arc or [Center/End]: **c**
>
> Specify center point of arc: *(Pick point 2, or enter x,y coordinates.)*
>
> Specify end point of arc or [Angle/chord Length]: *(Pick point 3, or enter x,y coordinates.)*

After you pick the center point, notice that AutoCAD ghosts the endpoint as you move the cursor. The endpoint that you pick need not lie on the arc.

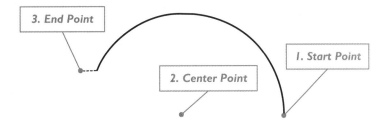

START, CENTER, ANGLE & START, END, ANGLE & CENTER, START, ANGLE

Arcs are sometimes specified by their *included angle*. This is the angle formed between the start and endpoints, with the angle's vertex at the arc's center point.

Keep in mind that AutoCAD measures the arc in the counterclockwise direction, starting from the start to the endpoint. When you enter a negative angle, AutoCAD draws the arc clockwise from the start point.

From the **Draw** menu, select **Arc | Start, Center, Angle** or **Start, End, Angle** or **Center, Start, Angle**.

> Command: **arc**
>
> Specify start point of arc or [Center]: *(Pick point 1.)*
>
> Specify second point of arc or [Center/End]: **c**
>
> Specify center point of arc: *(Pick point 2.)*
>
> Specify end point of arc or [Angle/chord Length]: **a**
>
> Specify included angle: *(Enter an angle, 3.)*

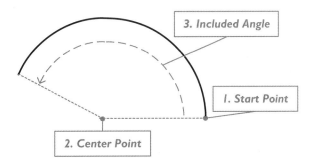

START, CENTER, LENGTH & CENTER, START, LENGTH

When you need to draw an arc with a specific chord, use the **Length** option. The length is the distance from the start point to the endpoint. From the **Draw** menu, select **Arc | Start, Center, Length** or **Center, Start, Length**.

> Command: **arc**
>
> Specify start point of arc or [Center]: *(Pick point 1.)*
>
> Specify second point of arc or [Center/End]: **c**
>
> Specify center point of arc: *(Pick point 2.)*
>
> Specify end point of arc or [Angle/chord Length]: **l**
>
> Specify length of chord: *(Enter a length, 3.)*

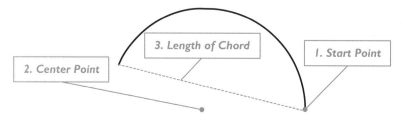

AutoCAD does not draw the arc if the chord length does not work with the start and center points you pick. When you see "*Invalid*" on the command prompt area, the chord is too long or too short. Note that the center point is *not* usually on the chord.

AutoCAD constructs a *major arc* or a *minor arc*, depending on the chord length. (Given that the chord divides a circle into two parts, the major arc is the larger arc, while the minor arc is the smaller one.) When the chord is positive, AutoCAD draws the minor arc; when negative, the major arc.

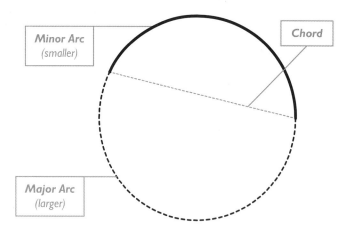

START, END, RADIUS

This option is most like drawing a circle: you specify two points and the radius. From the **Draw** menu, select **Arc | Start, End, Radius.**

> Command: **arc**
>
> Specify start point of arc or [Center]: *(Pick point 1.)*
>
> Specify second point of arc or [Center/End]: **e**
>
> Specify end point of arc: *(Pick point 2.)*

Specify center point of arc or [Angle/Direction/Radius]: **r**

Specify radius of arc: *(Enter a positive or negative radius, or pick point 3.)*

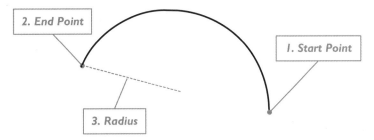

By default, AutoCAD draws the minor arc. If, however, you enter a negative value for the radius, AutoCAD draws the major arc.

START, END, DIRECTION

This option is perhaps the trickiest to understand. AutoCAD starts the arc tangent in the direction you specify. From the **Draw** menu, select **Arc | Start, End, Direction**.

Command: **arc**

Specify start point of arc or [Center]: *(Pick point 1.)*

Specify second point of arc or [Center/End]: **e**

Specify end point of arc: *(Pick point 2.)*

Specify center point of arc or [Angle/Direction/Radius]: **d**

Specify tangent direction for the start point of arc: *(Pick point 3.)*

At the "Specify tangent direction" prompt, move the cursor. Notice how AutoCAD ghosts the arc, depending in the cursor's location.

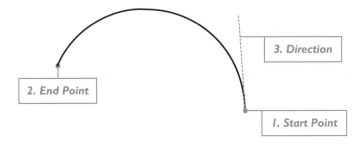

CONTINUE

Among the ARC command's many options, one manages to stay hidden. You find it on the menu under **Draw | Arc | Continue**, but it is not visible at any of the ARC command's prompts. The **Continue** option is designed to draw an arc tangent to the last-drawn object. Here's how it works:

Command: **arc**

Specify start point of arc or [Center]: *(Press* ENTER.*)*

Notice that AutoCAD automatically selects the end of the last line, polyline, or arc as the starting point for the new arc, which is drawn tangent to the last-drawn object. As you move the cursor, AutoCAD ghosts it in the drawing. There is just one option:

Specify end point of arc: *(Pick point 1.)*

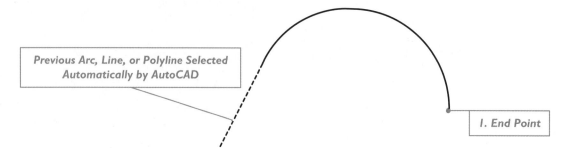

Previous Arc, Line, or Polyline Selected Automatically by AutoCAD

1. End Point

 Note: AutoCAD can convert arcs into circles, and back again. Here's how:
- To convert arcs to circles, use the **JOIN** command.
- To convert circles to arcs, use the **BREAK** command to remove a portion of the circle.

DONUT

The **DONUT** command draws circles with thick walls and solid-filled circles. These kinds of circles are useful for PCB (printed circuit board) designs. Donuts are drawn as polylines.

This is one of AutoCAD's few drawing commands that keeps on repeating itself until you press ESC or ENTER.

TUTORIAL: DRAWING DONUTS

1. To draw donuts, start the **DONUT** command by one of these methods:
   - From the Draw menu, choose **Donut**.
   - Or, at the 'Command:' prompt, enter the **donut** command:

     Command: **donut** *(Press ENTER.)*
   - Alternatively, enter the **doughnut** alias at the 'Command:' prompt.

2. Enter a value for the donut's "hole," its inside diameter:
   Specify inside diameter of donut <0.5000>: *(Specify value, or pick two points, 1.)*

3. Enter a value for the outside of the donut:
   Specify outside diameter of donut <1.0000>: *(Specify value, or pick two points, 2.)*

4. Pick a point to place the donut.
   Specify center of donut or <exit>: *(Pick point 3.)*

5. This command repeats until you exit the command. Press **ENTER** to end it:
   Specify center of donut or <exit>: *(Press ENTER to exit command.)*

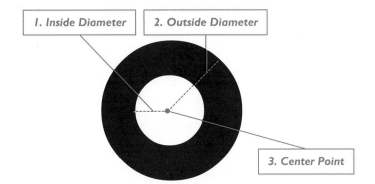

 Note: When the inside diameter equals the outside diameter, AutoCAD draws donuts as circles, albeit ones made from polylines. This is a "circle" that can be edited with the **PEdit** command. When the inside diameter is zero, AutoCAD draws solid-filled donuts, as illustrated below.

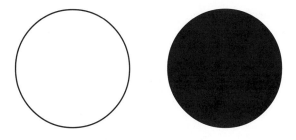

 PLINE

The **PLINE** command draws polylines.

We have referred to the "polyline" in some commands earlier, specifically **POLYGON** and **DONUT**. The **PLINE** command draws polylines, perhaps the most unique and flexible object created by any CAD program.

A polyline is a single object that consists of connected lines and arcs. It can be curved, splined, and open or closed (like an irregular polygon). You can specify the width of each segment, or give each segment a tapered width.

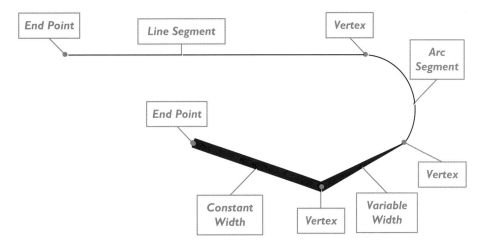

(*History*: Until lineweights were introduced to AutoCAD, polylines were the primary way to draw objects with width.)

TUTORIAL: DRAWING POLYLINES

1. To draw polylines, start the **PLINE** command.

   - From the **Draw** menu, choose **Polyline**.

   - From the Draw toolbar, choose the **Polyline** button.

   - From the Dashboard's 2D Draw panel, choose the **Polyline** button.

   - Or, at the 'Command:' prompt, enter the **pline** command.

        Command: **pline** *(Press* ENTER.*)*

   - Alternatively, enter the **pl** alias at the 'Command:' prompt.

2. Pick a point from which to start drawing the polyline:

        Specify start point: *(Pick a point.)*

   AutoCAD reminds you of the current line width, which is 0, unless you change it with the **Width** option:

        Current line-width is 0.0000

3. Pick the next point, or select an option:

        Specify next point or [Arc/Halfwidth/Length/Undo/Width]: *(Pick a point, or enter an option.)*

4. Continue picking points, and then press ENTER to exit the command:

        Specify next point or [Arc/Close/Halfwidth/Length/Undo/Width]: *(Press* ENTER *to exit the command.)*

ADDITIONAL METHODS FOR DRAWING POLYLINES

The **PLINE** command provides additional options for drawing polylines:

- **Undo** removes the previous segment.

- **Close** draws a segment to the start point.

- **Arc** switches to arc-drawing mode.

- **Width** specifies the width of the polyline.

- **Halfwidth** specifies the halfwidth.

- **Length** draws a tangent segment of specific length.

Let's look at each option.

Undo

As you draw polyline segments, you may sometimes make a mistake. Instead of canceling the command and starting over, use the **Undo** option to "undraw" the last segment. You can undo all the way back to the first segment.

Close

When you construct irregular polygons using the **PLINE** command, the last segment connects to the first segment. Aligning the ends of segments can be tedious, so the **Close** option automatically performs this for you. The **Close** option appears after you pick three points, because a minimum of two segments is needed before the polygon can be closed with a third segment.

Arc

The **Arc** option switches the **PLINE** command to arc-drawing mode. This mode has options similar to those of the **ARC** command, with some additions:

        Specify next point or [Arc/Halfwidth/Length/Undo/Width]: **a**

        [Angle/CEnter/Direction/Halfwidth/Line/Radius/Second pt/Undo/Width]:

The **Width** and **Halfwidth** options specify the width of the arc segments, as described below.

The **Line** option switches out of arc-drawing mode and back to line drawing mode.

Width

The **Width** option allows you to specify the width of each segment. In addition, you can specify a different starting and ending width, which creates tapers. By default, the width is 0 in new drawings. When changed, AutoCAD remembers the previous width setting.

1. Start the **PLINE** command, and pick a starting point:

   Command: **pline**

   Specify start point: *(Pick a point.)*

2. Notice the current line width. Enter **w** to access the **Width** option:

   Current line-width is 0.1000

   Specify next point or [Arc/Halfwidth/Length/Undo/Width]: **w**

3. Enter a value for the "starting width." This is the width at the starting end of the polyline.

   Specify starting width <0.1000>: *(Press **ENTER** to accept the default, or enter another value.)*

4. Press **ENTER** if you want the ending width to be the same as the starting width. For a taper, enter a different value for the ending width:

   Specify ending width <0.1000>: **.25**

5. Pick the next point, or select another option:

   Specify next point or [Arc/Halfwidth/Length/Undo/Width]:

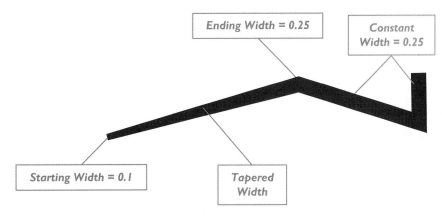

Notes: When you provide different values for the starting and ending widths, AutoCAD draws a taper but for that segment only! The ending width becomes the next starting width.

When two polyline segments have a width other than 0, AutoCAD automatically bevels the vertex between them. To see the beveled vertices, set the **FILLMODE** system variable to 0, followed by the **REGEN** command.

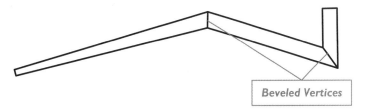

To select a polyline of width other than 0, you must pick its edge. Picking the centerline of the polyline fails. For some reason, AutoCAD is unable to detect the inside of polylines.

Halfwidth

The **Halfwidth** option is identical to **Width**, except that you specify the width from the edge of the polyline to its centerline.

> Specify next point or [Arc/Halfwidth/Length/Undo/Width]: **h**
>
> Specify starting half-width <0.1250>: *(Enter a value, or press* ENTER *to keep the default.)*
>
> Specify ending half-width <0.1250>: *(Press* ENTER *for constant width; enter a different value for tapered width.)*

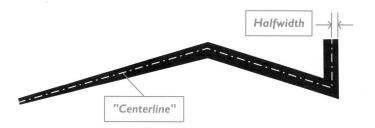

Length

The **Length** option draws the next segment at the same angle as the previous one. You specify the length. This option is most useful when the previous segment is an arc, because the segment is drawn tangentially to the arc's endpoint.

> Specify next point or [Arc/Halfwidth/Length/Undo/Width]: **l**
>
> Specify length of line: **3**

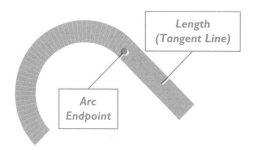

ELLIPSE

The **ELLIPSE** command draws ellipses and elliptical arcs. When isometric mode is turned on, this command adds the **Isocircle** option for drawing isometric circles.

Ellipses are elongated circles drawn with two diameters called "axes." The *major axis* is the longer axis; the *minor axis*, the shorter.

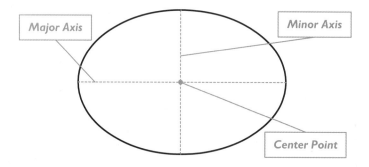

An ellipse is sometimes referred to by its *rotation angle*. A circle is an ellipse that has not been rotated: its rotation is zero degrees; the major and minor axes have the same length.

An ellipse is a circle that is rotated, or viewed at an angle. For instance, a 40-degree ellipse is a circle that has been rotated by 40 degrees about the major axis. AutoCAD allows you to specify ellipse rotation between 0.0 (a circle) and 89.4 degrees (a very thin ellipse).

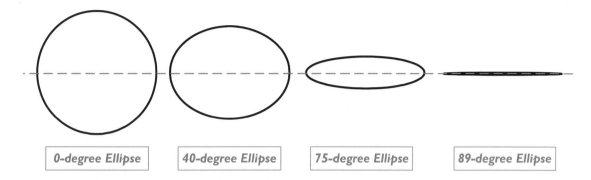

| 0-degree Ellipse | 40-degree Ellipse | 75-degree Ellipse | 89-degree Ellipse |

TUTORIAL: DRAWING ELLIPSES

1. To draw ellipses, start the **ELLIPSE** command.
   * From the **Draw** menu, choose **Ellipse**, and then **Center**.
   * From the Draw toolbar, choose the **Ellipse** button.
   * From the Dashboard's 2D Draw panel, choose the **Ellipse** button.
   * Or, at the 'Command:' prompt, enter the **ellipse** command.

     Command: **ellipse** *(Press* ENTER.*)*
   * Alternatively, enter the **el** alias at the 'Command:' prompt.

2. Pick a point to indicate an endpoint of one axis.
   Specify axis endpoint of ellipse or [Arc/Center]: *(Pick point 1.)*

3. Pick another point for the other end of the axis.
   Specify other endpoint of axis: *(Pick point 2.)*

4. Show the half-distance to the other axis.
   Specify distance to other axis or [Rotation]: *(Pick point 3.)*

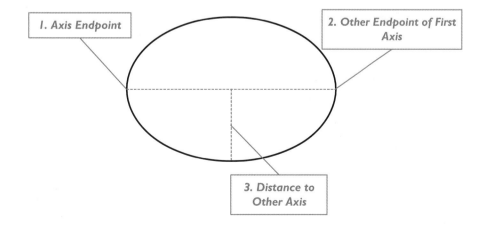

The "distance to other axis" does not need to be perpendicular to the first axis.

ADDITIONAL METHODS FOR DRAWING ELLIPSES

The **ELLIPSE** command contains options that provide additional methods for drawing ellipses:

- **Center** starts with the center point of the ellipse, followed by the axes.

- **Rotation** starts with the rotation angle about the major axis.

- **Arc** constructs an elliptical arc.

Center

The **Center** option allows you to pick the center of the ellipse, and then lets you specify distance from the center point to the endpoints of the axes.

> Command: **ellipse**
>
> Specify axis endpoint of ellipse or [Arc/Center]: **c**
>
> Specify center of ellipse: *(Pick point 1.)*
>
> Specify endpoint of axis: *(Pick point 2.)*
>
> Specify distance to other axis or [Rotation]: *(Pick point 3.)*

As you move the cursor during the "Specify distance to other axis" prompt, AutoCAD ghosts in the ellipse. Note that AutoCAD doesn't care which axis you create first.

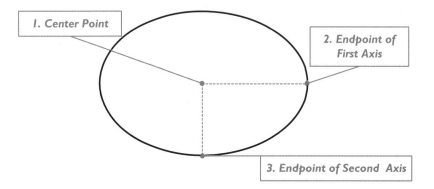

Rotation

The **Rotation** option rotates the ellipse about the major axis; it replaces the prompts for drawing the minor axis. (This option is also available when constructing ellipses with the **Center** option.) As you move the cursor during the "Specify rotation" prompt, AutoCAD ghosts the image of the ellipse.

> Command: **ellipse**
>
> Specify axis endpoint of ellipse or [Arc/Center]: *(Pick point 1.)*
>
> Specify other endpoint of axis: *(Pick point 2.)*
>
> Specify distance to other axis or [Rotation]: **r**
>
> Specify rotation around major axis: *(Pick point 3, or enter an angle.)*

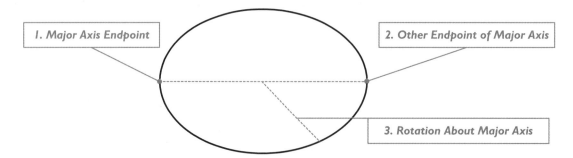

Arc

The **Arc** option draws elliptical arcs. The first few times you try drawing an elliptical arc, it might not turn out as desired. It takes some practice! Or, draw an ellipse and then trim or break it to an arc.

> Command: **ellipse**
>
> Specify axis endpoint of ellipse or [Arc/Center]: **a**
>
> Specify axis endpoint of elliptical arc or [Center]: *(Pick point 1.)*
>
> Specify other endpoint of axis: *(Pick point 2.)*
>
> Specify distance to other axis or [Rotation]: *(Pick point 3.)*
>
> Specify start angle or [Parameter]: *(Pick point 4.)*
>
> Specify end angle or [Parameter/Included angle]: *(Pick point 5.)*

As you move the cursor during the "Start angle" and "End angle" prompts, AutoCAD ghosts a preview of the elliptical arc. The figure below shows an arc that consists of the upper half of an ellipse.

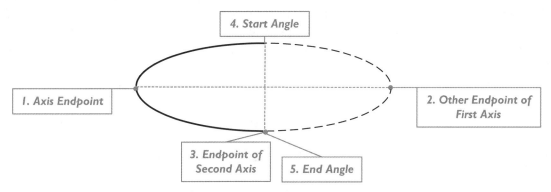

The **Arc** option has three methods of determining the arc: **Angle**, **Included Angle**, and **Parameter**.

The **Angle** option specifies the start angle and the end angle of the arc. The two angles are measured relative to the first axis endpoint. The *start angle* determines where the arc starts; the *end angle* where it ends. When you specify 0 degrees for the start angle, the arc starts at the first axis endpoint. AutoCAD draws the arc in the counterclockwise direction. If you run into trouble, it's probably because you are trying to draw the arc clockwise.

The **Included Angle** option defines the arc by an angle that starts from the start angle (described above). The angle is relative to the start angle. AutoCAD prompts:

> Specify end angle or [Parameter/Included angle]: **i**
>
> Specify included angle for arc <180>: *(Enter an angle, or pick a point.)*

The **Parameter** option uses this formula to construct the elliptical arc :

> $p(u) = \mathbf{c} + \mathbf{a} \times \cos(u) + \mathbf{b} \times \sin(u)$

where:

> a is the major axis.
>
> b is the minor axis.
>
> c is the center point of the ellipse.

Does anyone use this formula in the real world? Probably not. In any case, AutoCAD prompts you:

> Specify start angle or [Parameter]: **p**
>
> Specify start parameter or [Angle]: *(Pick a point.)*
>
> Specify end parameter or [Angle/Included angle]: *(Pick a point.)*

Notice how you can switch back and forth between the three arc definitions – Angle, Included Angle, and Parameter.

Isocircle

The **Isocircle** option appears in the **ELLIPSE** command's prompts only when isometric drafting mode is turned on with the **DSETTINGS** command. Isometric drafting is discussed in *Using AutoCAD: Advanced*.

▪ POINT

The **POINT** command constructs *point* objects.

Points have no height or width: they are dots. Points are the smallest object that output devices can produce: single pixels on the screen, single dots on printed paper.

Because points are so small, they can be hard to see. Use the **PDMODE** and **PDSIZE** system variables to change the look and size of points. As illustrated below, points can be invisible, or combinations of lines, circles, and squares. The effect of the two system variables is retroactive, meaning changing their values changes the look and size of all previously-drawn points.

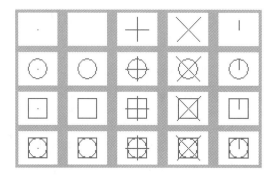

TUTORIAL: DRAWING POINTS

1. To draw points, start the **POINT** command.
   • From the menu bar, choose **Draw**, then **Points**, and then **Single Point**.

   • From the Draw toolbar, choose the **Point** button.

   • From the Dashboard's 2D Draw panel, choose the **Point** button.

   • Or, at the 'Command:' prompt, enter the **point** command:

     Command: **point** *(Press ENTER.)*

   • Alternatively, enter the **po** alias at the 'Command:' prompt.

2. Notice that AutoCAD displays the values of the **PDMODE** and **PDSIZE** system variables.
   Current point modes: PDMODE=0 PDSIZE=0.0000

3. Pick a point to place the point.
   Specify a point: *(Pick a point.)*

ADDITIONAL METHODS FOR DRAWING POINTS

The **POINT** command does not contain any options, but there are related commands that provide additional methods for drawing points:

   • **MULTIPLE** command repeats the **POINT** command.

   • **DDPTYPE** command changes the look of points.

Multiple

The **MULTIPLE** command repeats the **POINT** command, which is handy when you want to place more than one point at a time quickly. Indeed, the **MULTIPLE** command works with any command. From the **Draw** menu, select **Point | Multiple Point**. At the command prompt, enter:

> Command: **multiple**
>
> Enter command name to repeat: **point**
>
> Current point modes: PDMODE=0 PDSIZE=0.0000
>
> Specify a point: *(Pick a point.)*
>
> POINT Current point modes: PDMODE=0 PDSIZE=0.0000
>
> Specify a point: *(Press* ESC *or* ENTER *to exit repeating command.)*

The command repeats until you press ESC or ENTER.

DdPType

The **DDPTYPE** command displays a dialog box that allows you to select the style and size of point. There are limitations: you can only choose from the styles listed in the dialog box, which excludes custom point styles; all points in the drawing take on the same style and size, which means different points cannot have different styles. (**DDPTYPE** is short for "dynamic dialog point type.")

From the **Format** menu, select **Point Style** to display the dialog box:

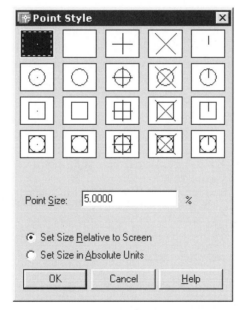

The upper half of the dialog box displays the point styles you can choose from. Note that the first style (the single dot) cannot be resized; the second style (the blank square) creates invisible points, which also cannot be sized.

The lower half changes the size of the point. The **Point Size** option sets the size as a percentage relative to the screen size or in units. The defaults are 5% and 5 units. I recommend you stick with the **Set Size Relative to Screen** (percentage) setting, because the **Absolute Units** setting can create points that are too large to see (when zoomed in) and too small to see (when zoomed out).

Click **OK** to accept the changes; AutoCAD automatically regenerates the drawing so that you can see the changes to the point display.

CONSTRUCTING OTHER OBJECTS

This chapter has introduced you to AutoCAD's basic drawing commands. These allow you easily to draw lines, circles, rectangles, regular polygons, arcs, ellipses, elliptical arcs, polylines, and points. They don't, however, make it easy to draw other common 2D geometric shapes, such as isosceles and right triangles, spirals, parallelograms; as well as 3D objects, such as cubes and cylinders. You learn about 3D objects in chapter 18 of this book.

To draw other kinds of 2D shapes, for example the four kinds of triangle, you need to *construct* them, using one or more commands available in AutoCAD:

> **Equilateral** — all three sides the same length.
>
> **Isosceles** — two sides the same length.
>
> **Scalene** — all three sides at different lengths.
>
> **Right Angle** — one angle at 90 degrees.

As you may recall from your high school geometry class, triangles are drawn with a combination of lengths and angles, and sometimes require the use of sine, cosine, and tangent calculations. Although it's not obvious, AutoCAD provides all the tools you need to draw any kind of triangle — without getting out the scientific calculator.

EQUILATERAL TRIANGLES

All three sides of equilateral triangles have the same length. (Again, if you recall your high school geometry class, the three interior angles are all 60 degrees.) This kind of triangle is most easily drawn with the **POLYGON** command set to draw 3 sides. When you know the length of the sides, use the **Edge** option. For example, if the sides are 2.5 units long, use *direct distance entry* to specify the length, as follows:

1. Start the **POLYGON** command:
   Command: **polygon**

2. To draw triangles, specify 3 sides:
   Enter number of sides <4>: **3**

3. By using the **Edge** option, you specify the length of one side; AutoCAD draws the other two sides to match.
   Specify center of polygon or [Edge]: **e**
   Specify first endpoint of edge: *(Pick point 1.)*
   Specify second endpoint of edge: *(Move cursor in a direction, and then enter length.)* **2.5**

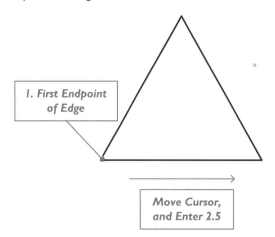

1. First Endpoint of Edge

Move Cursor, and Enter 2.5

AutoCAD draws the equilateral triangle for you.

ISOSCELES TRIANGLES

Two sides of isosceles triangles have the same length (the two opposite angles also being the same). One method to draw this triangle is to use the **LINE**, **MIRROR**, and **TRIM** commands. For example, consider a triangle with a base of 3 units and isosceles angles of 63 degrees.

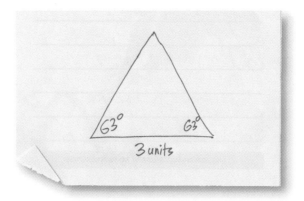

We don't know the length of the other two sides, so we draw them an arbitrary length, and then trim or extend to fit. For the best result, turn on polar mode.

1. Start the **LINE** command to draw the base and one angled leg:

    Command: **line**

    Specify first point: *(Pick point 1.)*

2. Use direct distance entry to draw the base 3 units long:

    Specify next point or [Undo]: *(Move cursor to the right, and then enter length.)* **3**

3. Use relative coordinates to draw one leg 10 units long at 63 degrees. The 10 units is arbitrary, because we do not know its length; we edit the length later.

    Specify next point or [Undo]: **10<63**

    Specify next point or [Close/Undo]: *(Press ENTER to exit command.)*

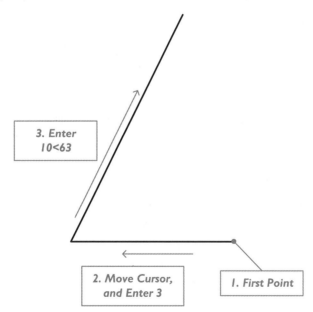

4. Start the **MIRROR** command to make a mirror copy of the angled leg:

    Command: **mirror**

5. Enter **L** to select the last-drawn object:

   Select objects: **L**

   1 found Select objects: *(Press* ENTER *to end object selection.)*

6. Use MIDpoint object snap to create an imaginary vertical mirroring line:

   Specify first point of mirror line: **mid**

   of *(Pick point 4.)*

   Specify second point of mirror line: *(Pick point 5.)*

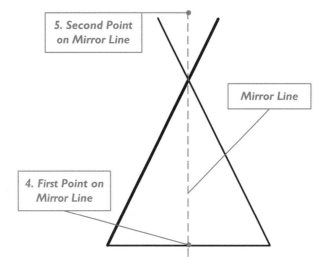

7. Keep the source object, and exit the command:

   Delete source objects? [Yes/No] <N>: *(Press* ENTER *to accept default,* **No***, and to exit the command.)*

8. Use the TRIM command to trim back the two angled lines to their intersection point:

   Command: **trim**

   Current settings: Projection=UCS, Edge=None

   Select cutting edges ...

   (Filleting with radius = 0 also works.)

9. Select all the lines. They become each other's cutting edge:

   Select objects or <select all>: *(Press* ENTER *to select all objects.)*

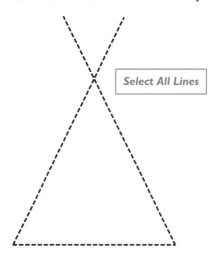

10. Select the two line extensions to trim them. Be sure to pick the portion of the lines to be trimmed away:

> Select object to trim or shift-select to extend or
>
> [Fence/Crossing/Project/Edge/eRase/Undo]: *(Select end of one angled line, 6.)*
>
> Select object to trim or shift-select to extend or
>
> [Fence/Crossing/Project/Edge/eRase/Undo]: *(Select end of the other line, 7.)*

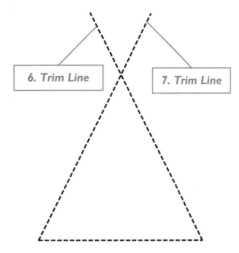

> Select object to trim or shift-select to extend or
>
> [Fence/Crossing/Project/Edge/eRase/Undo]: *(Press **ENTER** to exit command.)*

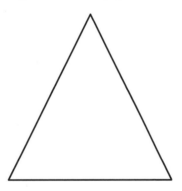

The isosceles triangle is complete.

SCALENE TRIANGLES

All three sides and corners of scalene triangles have different lengths and angles. This kind of triangle is best drawn with the LINE command, using a combination of direct distance entry and polar coordinates, as required. Take, for example, a triangle with the following specifications:

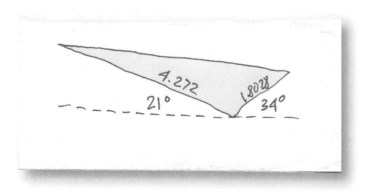

1. Start the **LINE** command:

   Command: **line**

   Specify first point: *(Pick point 1).*

2. Use polar coordinates (which are also relative) to draw the short leg at right:

   Specify next point or [Undo]: **1.8028<34**

3. End the **LINE** command, and start over to draw the next leg (at left):

   Specify next point or [Undo]: *(Press* **ENTER.***)*

   Command: *(Press* **ENTER.***)*

   LINE Specify first point: **endp**

   of *(Again, pick point 1).*

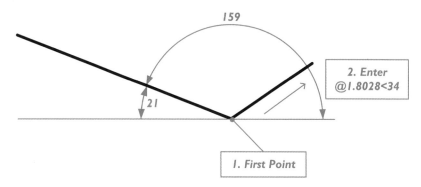

4. Use polar coordinates to draw the other leg. The angle of 159 degrees = 180 - 21=159.

   Specify next point or [Undo]: **4.272<159**

5. Use ENDpoint object snap to draw the third leg:

   Specify next point or [Undo]: **end**

   of *(Pick point 4.)*

   Specify next point or [Close/Undo]: *(Press* **ENTER.***)*

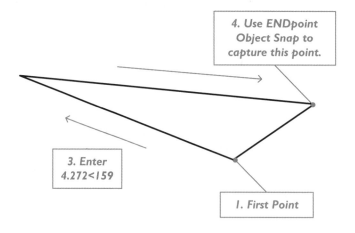

RIGHT ANGLE TRIANGLES

One corner of right angle triangles is at 90 degrees; the two remaining corners have other angle that add up to 90 degrees, such as the same angle (45 degrees each, making them isosceles triangles), or two different angles, making them scalene.

This kind of triangle is easily drawn with the LINE command. When you know the length of two sides, such as 3 and 2 units, follow these steps:

1. Turning on ortho mode helps you draw the 90-degree angle easily:

    (Press **F8** *to turn on ortho mode.)*

2. Start the LINE command:

    Command: **line**

    Specify first point: *(Pick point 1.)*

3. Notice that it is easier to draw this kind of triangle when you start at a corner away from the right angle, and then draw toward the right angle.

    Specify next point or [Undo]: *(Move cursor to the left, and then enter length.)* **3**

    Specify next point or [Undo]: *(Move cursor up, and then enter length.)* **2**

4. By using the **Close** option, you don't have to work out the length of angle of the third side.

    Specify next point or [Close/Undo]: **c**

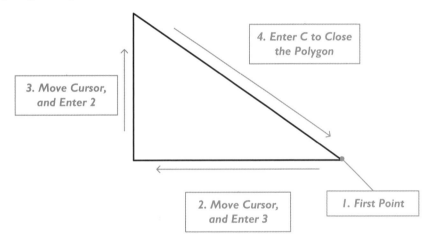

The right angle triangle is complete.

EXERCISES

1. With the **LINE** command, draw the house shape shown below:

   Command: **line**

   Specify first point: *(Enter point 1.)*

   Specify next point: *(Enter point 2.)*

   Specify next point: *(Enter point 3.)*

   Specify next point: *(Enter point 4.)*

   Specify next point: *(Enter point 5.)*

   Specify next point: *(Enter point 1.)*

   Notice that the line stretched behind the crosshairs. This is called "rubber banding."

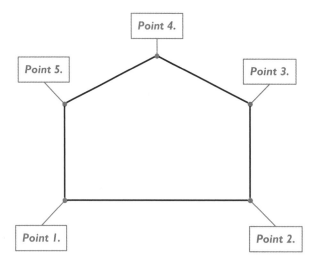

2. Using the figure below, fill in the missing absolute coordinates. Each side of the floor plan is dimensioned. Place the answers in the boxes provided. Assume the origin is at the lower left corner.

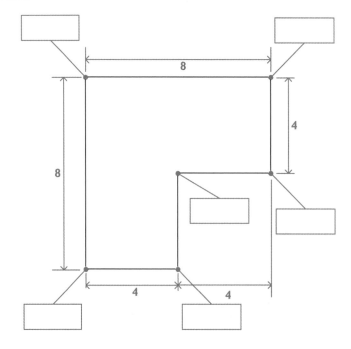

3. Using the **LINE** command, connect the following points designated by absolute coordinates.

    Point 1: **1,1**

    Point 2: **5,1**

    Point 3: **5,5**

    Point 4: **1,5**

    Point 5: **1,1**

    What shape did you draw?

4. Determine the length of each side of the floor plan in the figure below. (Calculate the lengths from the absolute coordinates given.)

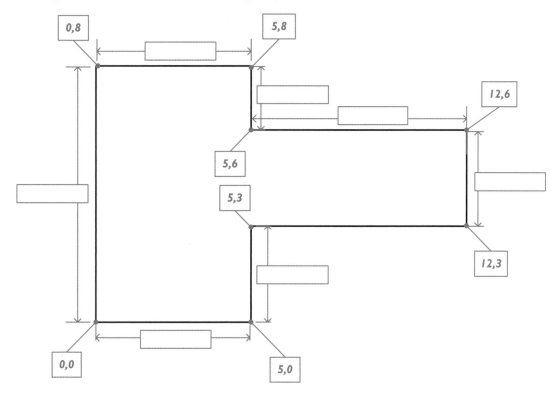

5. Use the following absolute and relative coordinates to draw an object.

    Point 1: **0,0**

    Point 2: **@3,0**

    Point 3: **@0,1**

    Point 4: **@–2,0**

    Point 5: **@0,2**

    Point 6: **@–1,0**

    Point 7: **0,0**

6. List the relative coordinates used to draw the following baseplate.

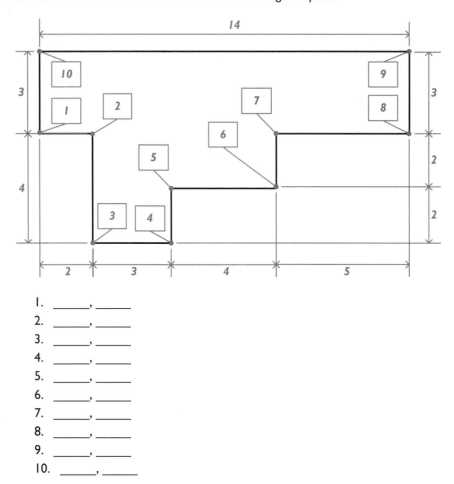

1. _____, _____
2. _____, _____
3. _____, _____
4. _____, _____
5. _____, _____
6. _____, _____
7. _____, _____
8. _____, _____
9. _____, _____
10. _____, _____

7. List the polar coordinates used to draw the baseplate above.

8. Write a list of the absolute coordinates used to construct the shim shown in the figure below. The origin is at the lower left corner.

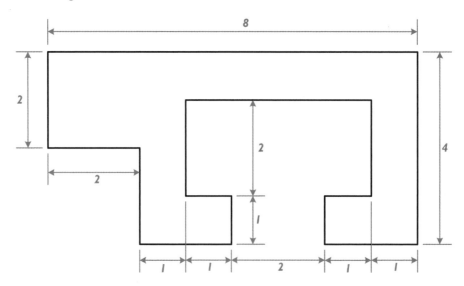

9. Use the following relative distances and angles to draw an object.

    Point 1: **0,0**
    Point 2: **@4<0**
    Point 3: **@4<120**
    Point 4: **@4<240**

10. Write a list of the relative coordinates used to construct the support plate shown in the figure below. The origin is at the lower left corner.

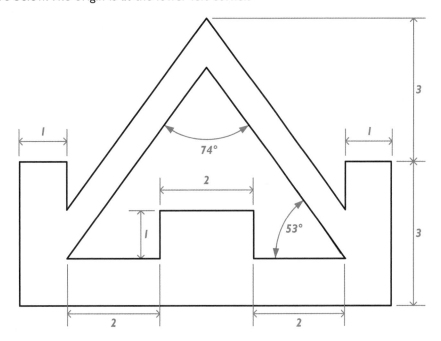

11. Write a list of the polar coordinates used to construct the corner shelf shown in the figure below.

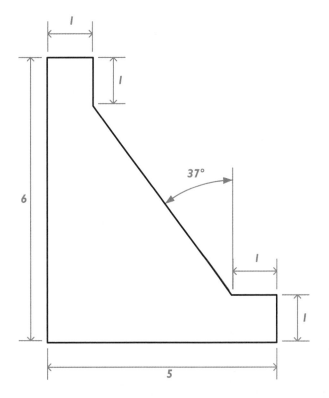

12. Set up a point type of 66, and then place it in the drawing:

    Command: **pdmode**

    New value for PDMODE <default>: **66**

    Recall that **PDMODE** affects all points in the drawing. Points placed previously are updated with the next command that causes a regeneration, such as **REGEN**. To place the point in the drawing, enter:

    Command: **point**

    Current point modes: PDMODE=66 PDSIZE=0.0000

    Specify a point: *(Pick a point.)*

13. Construct a circle with a *radius* of 5. From the **Draw** menu, choose **Circle | Center, Radius**.

    Command: _circle

    Specify center point for circle or [3P/2P/Ttr (tan tan radius)]: *(Pick a point.)*

    Specify radius of circle or [Diameter]: **5**

14. Construct a circle using a center point and a *diameter* of 3.

    Command: **circle**

    Specify center point for circle or [3P/2P/Ttr (tan tan radius)]: *(Enter point 1.)*

    Specify radius of circle or [Diameter]: **d**

    Specify diameter of circle <5.0000>: **3**

15. With the **RECTANG** command, draw two rectangles. One rectangle represents a B-size drawing sheet (17" x 11"); the second represents the title block in the lower right corner (4" x 2").

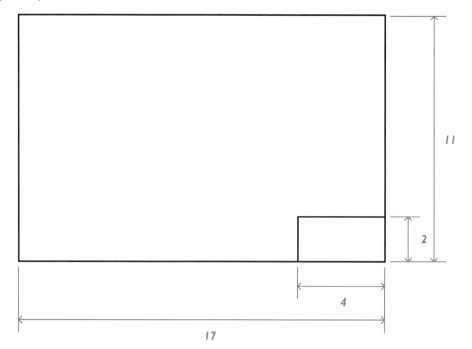

16. With the **RECTANGLE** command, draw the outline of a standard 24" x 36" speed limit sign; do not draw the text.

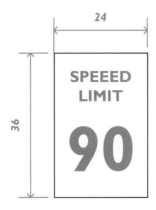

17. Use the **POLYGON** command to draw the outline of a standard 30" warning sign; do not draw the text:

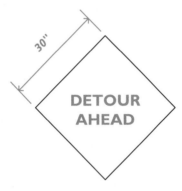

18. Use the **POLYGON** command to draw the outline of a standard 30" Stop sign; do not draw the text:

19. With the **POLYGON** command, draw the outline of a standard 36" Yield sign:

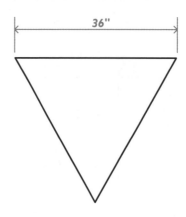

20. Draw donuts with the following diameters:
    a. ID= 1.0; OD = 1.5
    b. ID = OD = 2.5
    c. ID = 0.0; OD = 1.0

21. Use drawing commands discussed in this chapter to draw the following baseplates:
    a. Rectangular baseplate with two holes:

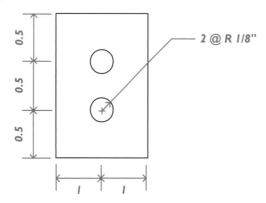

    b. Rectangular baseplate with four holes:

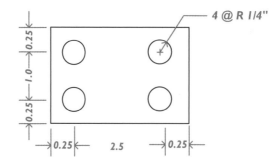

22. Use drawing commands discussed in this chapter to draw the following baseplates:

    a. Circular baseplate with two square holes:

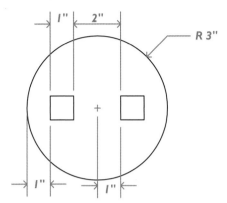

    b. Elliptical baseplate with two square holes. The minor radius is 3", and the major radius is 4.5":

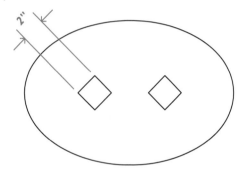

23. Use the **ELLIPSE** command to draw the face of an alien. Use the figure below to help you. Don't worry about the size.

24. Draw a compact disc, which has an outer diameter of 4.75" and an inner diameter of 0.6". What are the equivalent measurements as radii?

25. Draw arcs with the following specifications:

    a. Start point   **7,5**
       Second point **9.5,3.5**
       Endpoint   **6,2**

    b. Center point **4,4**
       Start point   **4,2**
       Angle   **270**

    c. Start point   **10.5,5.5**
       Center point **9,6**
       Length   **3**

26. Draw circles with the following specifications:

    a. Center point   **5,5**
       Radius   **3.5**

    b. Center point   **5,5**
       Diameter   **3.5**

    c. 2P: first point   **10,5**
       Second point   **3,2**

    d. 3P: first point   **4,7**
       Second point   **6,5**
       Third point   **5,2**

27. Draw the 4.5" x 2.75" electrical cover plate shown below. The screw holes are 1/8" in diameter, and 2-5/16" apart. The light switch cover plate has an opening of 0.95" x 0.4".

28. With the **Line** and **Arc** options of the **PLINE** command, draw the outline of a standard credit card — 3.35" wide by 2.15" high, with a 0.1"-radius arc at each corner.

29. Draw the inserts for CD cases:
    a. The cover insert is 12cm x 12cm.
    b. The back cover insert is 15cm wide x 11.7 cm tall; the two spine strips are 6.4 mm wide.

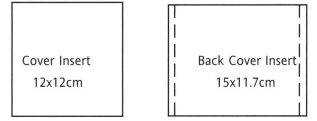

Cover Insert
12x12cm

Back Cover Insert
15x11.7cm

30. Draw the bicycle shown in the figure below.
    Use polylines for the frame, lines for the spokes, and donuts for the tires.

CHAPTER REVIEW

1. What is another name for a single line?
2. Is using the **Close** option of the **LINE** command more accurate at closing the polygon than attempting to line up endpoints manually?
3. What does the @ symbol mean?
   The < symbol?
4. How do you continue drawing a line *tangent* to a previously-drawn arc?
5. Name the best command for drawing the following triangles:
   Equilateral triangle
   Scalene triangle
6. What is AutoCAD's default method of constructing arcs?
7. What command can be used to change the look and size of points?
8. When rectangles are drawn with the **RECTANG** command, can each corner have a different fillet radius?
9. The **POLYGON** command uses three methods to determine the size of polygons. Explain the meaning of the methods:
   Edge
   Inscribed
   Circumscribed
10. The **PDMODE** and **PDSIZE** system variables are *retroactive*. What does this mean?
11. There are six methods of constructing circles. What four circle properties are used in various combinations to comprise these methods?
    a.
    b.
    c.
    d.
12. Describe the meaning of these options for drawing circles:
    2P
    3P
    Ttr
13. Describe a method of drawing arcs that does not involve the **ARC** command.
14. List three methods for ending the **LINE** command?
    a.
    b.
    c.
15. In which direction does AutoCAD draw arcs (by default).
16. What is the difference between a *minor* arc and a *major* arc?
17. What is the *chord* of an arc?
18. How do you continue drawing an arc tangent to a previously-drawn arc or line?
19. What does the **ARC** command's **Direction** option specify?
20. What must you do to end the **DONUT** command?
21. What is the alias for the following commands:
    ARC
    LINE
    CIRCLE
    PLINE

22. Where on the menu bar do you find commands for drawing objects?
23. Name the three kinds of circles drawn by the **DONUT** command:
    a.
    b.
    c.
24. What do the **POLYLINE**, **DONUT**, and **POLYGON** commands have in common?
25. Can polylines have varying width?
26. What is the purpose of the **PLINE** command's **Arc** option?
27. When selecting a wide polyline, where must you pick it?
28. Describe the purpose of the **PLINE** command's **Length** option.
29. What is the difference between the *minor* axis and the *major* axis of an ellipse?
30. What is another name for an ellipse with a rotation of 0 degrees?
31. Which must you specify first, the minor axis or the major axis of an ellipse?
32. Describe the purpose of the **MULTIPLE** command.
33. List two methods of drawing circles without using the **CIRCLE** command:
    a.
    b.
34. What is the effect of chamfers and fillets on the **RECTANGLE** command's **Area** option?
35. Does the **POLYGON** command draw scalene triangles?
36. Does the **RECTANGLE** command draw rhomboids?

CHAPTER **5**

Drawing with Precision

One major advantage of CAD over hand drafting is precision. Instead of drawing with an accuracy to the nearest pencil (or pen) width, AutoCAD draws with an accuracy to 14 decimal places. In this chapter, you learn to use some of AutoCAD's commands for drawing with precision, and viewing the drawing in different ways. They are:

DSETTINGS controls drafting settings.

GRID displays a grid of dots or lines.

SNAP specifies the cursor increment and additional drafting options.

ORTHO constrains cursor movement to the horizontal and vertical.

POLAR helps draw at specific angles.

OSNAP helps accurately select geometric features.

OTRACK implements object snap tracking.

RAY and **XLINE** place construction lines in the drawing.

REDRAW and **REGEN** clean up and update the drawing.

ZOOM enlarges and reduces the view of the drawing.

PAN moves the view around.

CLEANSCREENON minimizes the user interface to maximize the drawing.

DSVIEWER provides a "bird's-eye view" of the drawing in an independent window.

VIEW stores and recalls views of drawings by name and controls viewport backgrounds.

VIEWRES controls the roundness of circular objects.

NEW TO AUTOCAD 2008 **IN THIS CHAPTER**

- The **NEWVIEW** command directly accesses the **VIEW** command's New View dialog box.
- The Dashboard provides access to zoom, pan, and view options.

FINDING THE COMMANDS

On the Dashboard's **2D Navigate** panel:

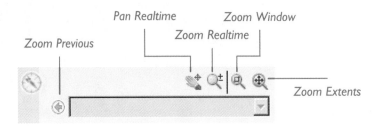

On the Dashboard's **2D Draw** panel:

On the status bar:

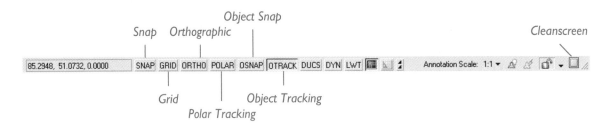

GRID

The **GRID** command toggles (turns on and off) the *grid*, a display of dots or lines. The grid is like a sheet of graph paper that helps you see the horizontal and vertical frame of reference. In 2D wireframe mode, AutoCAD displays dots (the intersections of the lines); in other modes, it displays a grid of lines.

The grid gives you a feeling for the size of the drawing, and of distances within the drawing. Because it is a visual aid, the grid is not plotted.

You can change the spacing of grid dots and lines, the subspacing between grid lines, the angle of the grid, the style (standard or isometric), the extent of the grid, and the color of grid lines.

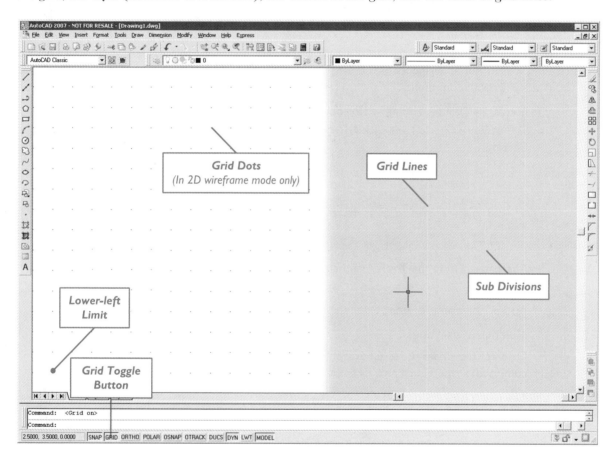

TUTORIAL: TOGGLING THE GRID

1. To turn on the grid, use the **GRID** command by one of these methods:
   - From the status bar, choose **GRID** (turned on when buttons looks depressed).
   - On the keyboard, press function key **F7** or **CTRL+G**.
   - Or, at the 'Command:' prompt, enter the **grid** command:

     Command: **grid** *(Press* ENTER.*)*

2. Use the **ON** option to turn on the grid:

   Specify grid spacing(X) or [ON/OFF/Snap/Major/aDaptive/Limits/Follow/Aspect] <0.5>: **on**

   Notice that the grid is displayed.

3. To turn *off* the grid, repeat the command with the **OFF** option:

   Command: **grid** *(Press* ENTER.*)*

   Specify grid spacing(X) or [ON/OFF/Snap/Major/aDaptive/Limits/Follow/Aspect] <0.5>: **off**

CONTROLLING THE GRID DISPLAY: ADDITIONAL METHODS

When you start new drawings in AutoCAD, the grid spacing is 0.5 units. You can increase and decrease the spacing, make the spacing the same as the snap distance (as detailed later in this chapter), and change the spacing in the x and y directions separately.

To effect these changes, you can use a dialog box or the command line:

- **DSETTINGS** displays a dialog box for changing grid settings.
- **GRID** specifies the grid spacing and aspect ratio via the command line.
- **LIMITS** controls the extent of the grid.

Let's look at how each command affects the grid display.

DSettings Command

The **DSETTINGS** command (short for "drafting settings") displays a dialog box with settings that control many (but not all) aspects of the grid display. The dialog box lists some of the options found in the **GRID** and **SNAP** commands.

1. To access the Drafting Settings dialog box:
    - The easiest way is to right-click **GRID** on the status line. From the shortcut menu, select **Settings**.

- Alternatively, from the **Tools** menu, choose **Drafting Settings**, and then the **Snap and Grid** tab.
- At the 'Command:' prompt, enter the **dsettings** command, and then the **Snap and Grid** tab.

    Command: **dsettings**

- Or, enter the aliases **ds**, **se** (short for "settings"), or **ddrmodes** (the old name for this command) at the 'Command:' prompt.

Notice that AutoCAD displays the Drafting Settings dialog box. (If necessary, choose the **Snap and Grid** tab.)

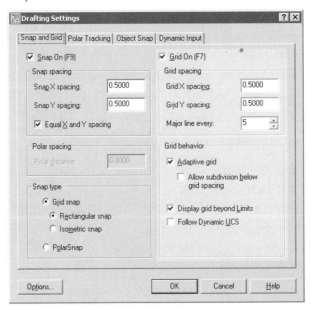

2. In this dialog box, you can make the following changes to the grid settings:

- **Grid On (F7)** toggles the grid display. The "(F7)" notation reminds you that you can also press function key **F7** to turn the grid on and off.

- **Grid X spacing** specifies the horizontal spacing between grid dots or lines. When you enter a value for the x spacing, AutoCAD automatically changes the y-spacing to match, if the **Equal X and Y spacing** option is turned on (a check mark shows).

- **Grid Y spacing** specifies the vertical spacing between dots or lines. Some disciplines, such as highway design, use a different scale in the x-direction than the y-direction. To make the vertical spacing different from the horizontal, you must turn off the **Equal X and Y spacing** option (under Snap spacing).

- **Major line every** specifies the number of subdivision lines per major line; subdivisions are not seen in 2D wireframe mode.

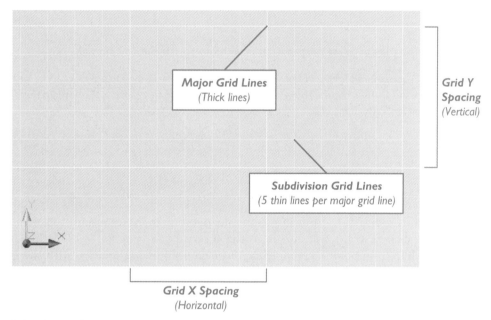

Major Grid Lines (Thick lines)

Grid Y Spacing (Vertical)

Subdivision Grid Lines (5 thin lines per major grid line)

Grid X Spacing (Horizontal)

- **Adaptive grid** makes the grid spacing wider when zoomed out. (AutoCAD no longer complains that the grid is too dense to display.)

- **Allow subdivision below grid spacing** displays more and more minor grid lines as you zoom into the drawing.

- **Display grid beyond limits** allows the grid display to go beyond the limits set by the **LIMITS** command.

- **Follow Dynamic UCS** forces the grid to match the x,y-orientation of the dynamic UCS, when activated.

3. To accept the changes, click **OK**.

Grid Spacing

The **GRID** command lets you change the spacing between grid lines:

To make the grid spacing the same as the snap distance, use the **Snap** option:
Command: **grid** *(Press* ENTER.*)*

Specify grid spacing(X) or [ON/OFF/Snap/Major/aDaptive/Limits/Follow/Aspect] <0.5>: **s**

To set the grid spacing as a multiple of the snap distance, add an "x" to the **GRID** command's default option.

For example, to make the grid spacing twice that of the snap spacing, enter **2x**, as follows:

Command: **grid** *(Press* **ENTER.***)*

Specify grid spacing(X) or [ON/OFF/Snap/Major/aDaptive/Limits/Follow/Aspect] <0.5>: **2x**

LIMITS

The grid could extend infinitely in all directions, and so the extent of the grid is limited by the **LIMITS** command. Typically, you want the grid covering the area in which you are drafting. Often, the lower-left corner is kept at 0,0 and only the location of upper-right corner is changed, as follows:

Command: **limits**

Reset Model space limits:

Specify lower left corner or [ON/OFF] <0.0000,0.0000>: *(Press* **ENTER,** *or enter x,y coordinates.)*

Specify upper right corner <12.0000,9.0000>: *(Enter x,y coordinates.)*

Notes: The grid can be set independently in each viewport, layout, and drawing.

Grid lines are colored light gray, but you can change the color in the **Display** tab of the **OPTIONS** dialog box: click the **Colors** button. The grid lines that lay on the x and y axes are colored differently: red for x and green for y, which matches the color coding of the UCS icon.

SNAP

The **SNAP** command specifies the cursor increment.

For example, if you are drafting a drawing to the nearest one inch, it's useful to set the snap to 1". When turned on, snap restricts the cursor to the interval you specify — 1" in this example. The cursor seems to jump from point to point. Snap is often used in conjunction with the grid.

As with the grid, you can change the snap's spacing, aspect ratio, and so on. Similarly, the snap can be set individually for each viewport, layout, and drawing. (The **LIMITS** command does not apply to snap.)

TUTORIAL: TOGGLING SNAP MODE

1. To turn on snap, invoke the **SNAP** command by one of these methods:
   - On the status bar, click **SNAP** (turned on when buttons looks depressed).
   - On the keyboard, press function key **F9** or **CTRL+B**.
   - Or, at the 'Command:' prompt, enter the **snap** command.

     Command: **snap** *(Press* **ENTER.***)*
   - Alternatively, enter the **sn** alias at the 'Command:' prompt.

2. Use the **ON** option to turn on snap:

   Specify snap spacing or [ON/OFF/Aspect/Rotate/Style/Type] <0.5000>: **on**

   When you move the mouse, notice that the cursor jumps.

3. To turn off snap, repeat the command with the **OFF** option:

   Command: **grid** *(Press* **ENTER.***)*

   Specify snap spacing or [ON/OFF/Aspect/Rotate/Style/Type] <0.5000>: **off**

   Notice that the cursor no longer jumps.

CONTROLLING THE SNAP: ADDITIONAL METHODS

When you start new drawings in AutoCAD, the default snap spacing is set to 0.5 units — the same as the grid. You can increase and decrease the spacing, make the spacing the same as the grid distance, or specify different spacing in the x and y directions. To make these changes, you can use a dialog box, or enter the changes at the command line, using these commands:

- **DSETTINGS** displays a dialog box for changing snap settings.

- **SNAP** specifies the snap spacing, aspect ratio, and modes via the command line.

Let's look at how each command affects the snap.

DSettings

The **DSETTINGS** command displays a dialog box that controls all settings affecting the snap. (The **Snap and Grid** tab combines the **SNAP** and **GRID** commands.)

1. To access the Drafting Settings dialog box, right-click **SNAP** on the status line.
    From the shortcut menu, select **Settings**.

 Note: The shortcut menu (shown above) has settings that you can access without going through the dialog box:
> **Polar Snap On** turns on polar snap mode, where the cursor aligns with distances and angles.
> **Grid Snap On** turns on both the snap and the grid.
> **Off** turns off snap mode.

AutoCAD displays the Drafting Settings dialog box, showing the **Snap and Grid** tab.

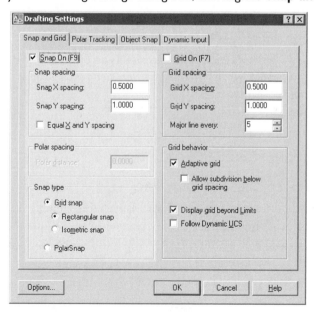

2. In this dialog box, you can make the following changes to the snap settings:
    - **Snap On (F9)** toggles the snap on and off. The "**(F9)**" reminds you that you can press **F9** to turn snap mode on and off.

- **Snap X spacing** specifies the horizontal snap increment. When you enter a value for the x-spacing, AutoCAD automatically changes the y-spacing to match, if **Equal X and Y spacing** is turned on. (The old press-the-**TAB**-key trick no longer works.)

- **Snap Y spacing** specifies the vertical snap increment.

- **Grid Snap** sets the snap spacing equal to the grid spacing, and ignores the values set earlier in the **Snap X and Y spacing** fields.

- **Rectangular Snap** keeps standard snap, as opposed to isometric snap.

- **Isometric Snap** turns on isometric snap, which is useful for drawing isometric objects. (See volume 2 of this book, *Using AutoCAD: Advanced.*)

- **Polar Snap** sets snap to polar angles.

3. To accept the changes, click **OK**.

Note: The **Angle**, **X Base**, and **Y Base** options were removed from this dialog box in AutoCAD 2007, as was the **Rotate** option of the **SNAP** command. Autodesk notes that the snap angle and origin match the current UCS; relocate the UCS to relocate the snap origin. As an alternative, use the **SNAPANGLE** and **SNAPBASE** system variables to change the values from 0 and 0,0, respectively.

Snap

The SNAP command has these options for changing the snap spacing:

1. To set the snap increment, use the **SNAP** command's default option. For example, to change the increment to 12 units (or inches), enter 12, as follows:

   Command: **snap** *(Press ENTER.)*

   Specify snap spacing or [ON/OFF/Aspect/Style/Type] <A>: **12**

2. To make the vertical (y) snap increment 10 times that of the horizontal (x) direction, use the **Aspect** option:

   Command: **snap** *(Press ENTER.)*

   Specify snap spacing or [ON/OFF/Aspect/Style/Type] <A>:: **a**

   Specify the horizontal spacing <0.5000>: **1**

   Specify the vertical spacing <0.5000>: **10**

3. To select the type of snap mode, use the **Style** and **Type** options. **Style** selects between regular and isometric modes, while **Type** selects between rectangular and polar modes. Here is how to enter polar mode:

   Command: **snap** *(Press ENTER.)*

   Specify snap spacing or [ON/OFF/Aspect/Style/Type] <A>: **t**

   Enter snap type [Polar/Grid] <Grid>: **p**

TUTORIAL: DRAWING WITH GRID AND SNAP MODES

Although the grid display is often used together with the snap spacing, the two can be independent of each together. While it is common to set the snap spacing to the drawing resolution (such as 1/8" or 1 cm), it's better set the grid spacing to a larger value (such as 1' or 1 m), so that the grid doesn't clutter the screen.

In the first part of the tutorial, you turn on and off the grid display, and change its spacing.

1. On AutoCAD's status line, right-click **GRID**, and then choose **Settings** from the shortcut menu. Notice that the Snap and Grid tab of the Drafting Settings dialog box appears.

2. Select **Grid On** to turn on the grid display, so that a check mark appears in the box. (The grid does not appear until after the dialog box closes.)

3. Set the **Grid X** and **Y Spacing** to 1.0.
4. Click **OK**. Notice the grid on the screen. If AutoCAD is in 2D wireframe mode, the grid appears as an array of dots; otherwise, it appears as vertical and horizontal lines.
5. Press **F7**. This key toggles the grid off and on.
    Press **F7** again.
6. Return to the Snap and Grid dialog box, and then turn off **Equal X and Y spacing**.
7. Set the **X Grid Spacing** to 10, and then the **Y Grid Spacing** to 1. Notice that it is not necessary for the vertical and horizontal values to be the same.
8. Click **OK**, and notice the grid spacing has changed, as illustrated below.
    (Depending on the mode AutoCAD is in, you may see grid dots, instead of lines.)

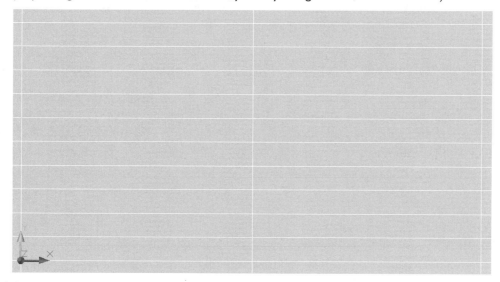

In the following steps, you draw a few lines with snap mode turned on.

9. Right-click the word **SNAP** on the status line, and then choose **Settings**.
10. In the Snap and Grid tab, choose **Snap On** to turn on snap mode.
    Ensure the **Equal X and Y spacing** option is turned on, and then change the **Snap X** and **Y Spacing** to 0.25.
    Click **OK**. You won't notice any difference in the display until you start drawing.
11. Start the **LINE** command, and then pick endpoints on the screen. Notice the crosshair cursor "snapping" to specific points on the screen.

Here, you make the grid spacing equal to the snap spacing.

12. Right-click **GRID** on the status line, and then choose **Settings**.
    Set the grid spacing to 0.25. The grid display has the same spacing as the snap.
    Click **OK** to exit the dialog box.
13. Use the **LINE** command to draw some lines. Notice that the crosshair cursor now lines up with the grid points.

Now you practice drawing rectangle, one with snap turned off, and another on.

14. Use **F9** to turn the snap mode off and on.
15. Draw two rectangles with the **LINE** command — one with snap mode on and one with snap mode off. Notice how the endpoints are easier to line up with snap mode on.

Finally, rotate the snap angle with the SNAPANG system variable to see how much easier it is for drawing at angles. (Even though you enter system variables at the 'Command:' prompt just like commands, system variables are different from commands: they contain settings and do not execute commands.)

16. At the 'Command:' prompt, change the snap angle to 45 degrees, as follows:

    Command: **snapang**

    Enter new value for SNAPANG <0>: **45**

    Notice that the **SNAPANG** system variable affects the snap and crosshair cursor. If you are drawing in 2D wireframe mode, the grid dots are also rotated. Grid lines do not rotate.

17. Draw a rectangle with the rotated snap grid and crosshairs. This is an excellent method of drawing objects that have many lines at the same angle.

ORTHO

When turned on, ortho mode (short for "orthographic") constrains cursor movement to the horizontal and vertical directions. This mode is very useful, because much drafting takes place at right angles — 0, 90, 180, and 270 degrees — as evidenced by the T-squares and right-triangles used in manual drafting.

Ortho mode can be changed from the horizontal and vertical. Use the SNAPANG system variable to change the ortho angle, as detailed earlier. Alternatively, use the polar option, described later.

TUTORIAL: TOGGLING ORTHO MODE

1. To turn on ortho mode, use the **ORTHO** command by one of these methods:
   - From the status bar, click **ORTHO** (turned on when button looks depressed).

   - On the keyboard, press function key **F8** or **CTRL+L**.

   - At the 'Command:' prompt, enter the **ortho** command.

     Command: **ortho** (Press ENTER.)

2. Use the **ON** option to turn on ortho:

   Enter mode [ON/OFF] <ON>: **on**

   You don't notice any difference in the cursor movement until you use drawing and editing commands. Then the cursor moves only vertically or horizontally.

3. To turn off ortho mode, repeat the command with the **OFF** option:

   Command: **ortho** (Press ENTER.)

   Enter mode [ON/OFF] <ON>: **off**

Unlike the GRID and SNAP commands, the ORTHO command has no additional options. Ortho mode is, however, affected by these snap options: angle, base point, and isometric style.

 Notes: While you can rotate the ortho angle, drawing still takes place at 90 degree increments in most cases. The exception is when isometric mode is turned on: then the ortho angle is 120 degrees.

To draw at other angles, use polar tracking together with PolarSnap. Ortho mode and polar tracking, however, cannot both be on at the same time: turning on one turns off the other. Ortho mode is ignored in 3D perspective views.

TUTORIAL: DRAWING IN ORTHO MODE

1. If ortho mode is on, click **ORTHO** on the status line to turn it off.
2. Draw a rectangle with the **LINE** command. Notice that is hard to draw lines that are perfectly horizontal and vertical.
3. Press **F8** to turn on ortho mode again.
4. Draw another rectangle with the **LINE** command. After picking the first endpoint, notice how the cursor moves precisely horizontally or vertically until you pick the next point.

POLAR

If ortho mode is like manual drafting with a T-square and right-triangle, then polar mode is like drafting with 30-60-90 and 45-45-90 triangles. Polar mode is more flexible than ortho mode, allowing you to draw at 90-degree increments as well as any other angle, such as 45 degrees or 22.5 degrees to mimic a Vemco drafting machine.

When you turn *on* polar mode, AutoCAD automatically turns *off* ortho mode. That's because ortho limits cursor movement to 90 degrees, while polar is any angle,

TUTORIAL: TOGGLING POLAR MODE

Curiously, there is no "POLAR" command even though AutoCAD has shortcut keystrokes to turn polar mode on and off. (As you later find out, polar mode can be controlled by system variables and dialog boxes.)

1. To turn on polar mode, use one of these methods:
   - From the status bar, click **POLAR** (turned on when button looks depressed).
   - On the keyboard, press function key **F10** or **CTRL+U**.
2. To initiate polar mode, start the line command, and then pick a point.
3. At the "Specify next point" prompt, move the cursor around. Notice that every so often a *tooltip* appears, together with an *x-marker* and an *alignment path*, as illustrated below:

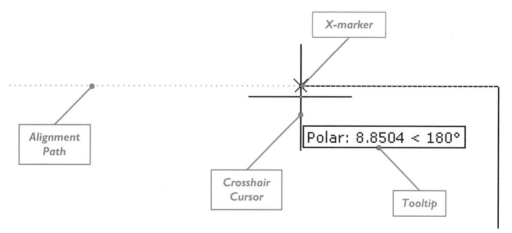

Toolips are small rectangles with explanatory text. In this case, the tooltip displays the polar distance, such as

Polar: 8.8504 < 180°

This means that the line is 8.8504 units long, at an angle of 180 degrees.

X-marker emphasizes the end of the line and the start of the alignment path. It marks the point where the line would end if you were to click the mouse button.

Alignment path is the thin dotted line that shows you where the line would be drawn, if you were to continue in that direction — almost like a preview.

CONTROLLING POLAR MODE: ADDITIONAL METHODS

In new drawings, the default polar angle is 90 degrees, the same as ortho mode. You can add angles and specify whether angle measurements are absolute, or relative to the last angle. The changes are made through a dialog box or at the command line using system variables.

- **DSETTINGS** command displays a dialog box for changing snap settings.

- **POLARANG** system variable specifies the increment for polar angles.

- **POLARDIST** system variable specifies the polar snap increment, but only when the **SNAPSTYL** system variable is set to 1.

- **POLARMODE** system variable specifies a number of variables for polar snap.

DSettings

The DSETTINGS command displays a dialog box that controls polar mode, because there is no "POLAR" command.

1. To access the Drafting Settings dialog box:
   - From the **Tools** menu, choose **Drafting Settings**. Choose the **Polar Tracking** tab.

   - At the 'Command:' prompt, enter the **dsettings** command, and then choose the **Polar Tracking** tab.

   - Alternatively, right-click **POLAR** on the status line. From the shortcut menu, select **Settings**.

In all cases, AutoCAD displays the Drafting Settings dialog box with the **Polar Settings** tab.

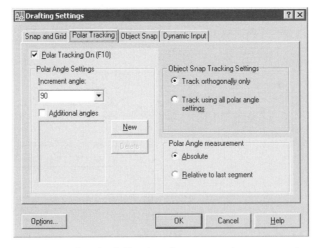

2. In this dialog box, you can make the following changes to the snap settings:
   - **Polar Tracking On (F10)** toggles the snap on and off. The "(F10)" reminds you that you can press function key **F10** to turn on and off polar mode.

   - **Increment Angle** specifies the polar tracking angle. AutoCAD presets a number of common angles, ranging from 5 to 90 degrees.

   - **Additional Angles** allows you to add any angle not provided by the Increment Angle list, such as 7.5. Adding angles is a bit awkward: click the **New** button, enter a value, and then press **ENTER**. Repeat for additional angles.

- **Object Snap Tracking Settings** is an option described later in this chapter.

Polar Angle Measurement switches between absolute and relative angle measurements:

- **Absolute** means that the angle is measured from AutoCAD's 0-degree, usually the positive x axis.

- **Relative to last segment** means the angle is measured relative to the last-drawn segment. The tooltip changes to read "Relative Polar."

3. Not all polar settings are listed in this tab of the dialog box. Return to the **Snap and Grid** tab to uncover additional settings.

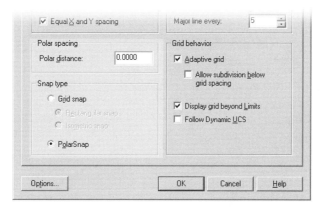

4. Under Snap Type, select **PolarSnap**. This action causes all the Snap settings to gray out, meaning they are unavailable.

5. Instead, you now specify a distance next to **Polar distance**, such as 2. This sets a snap spacing in the polar direction, i.e., in the direction of the angle specified in the previous tab. (Snap must be turned on.)

6. Click **OK** to exit the dialog box.

7. With the LINE command, draw some lines, noticing the action of the cursor: you draw lines at specific angles, and the lines have lengths that are multiples of 2.

In general, you use the Drafting Settings dialog box to control polar mode, but you may prefer the command line at times.

PolarAng

The **POLARANG** system variable specifies the increments for polar angles.

    Command: **polarang**

    Enter new value for POLARANG <90>: *(Enter a new value, such as 15.)*

PolarDist

The **POLARDIST** system variable specifies the polar snap increment (distance), but only when the **SNAPTYPE** system variable is set to 1. Recall that polar measurements consist of a distance and an angle; this system variable specifies the distance.

    Command: **polardist**

    Enter new value for POLARDIST <0.0>: *(Enter a new value, such as 1.0.)*

PolarMode

The **POLARMODE** system variable specifies a number of variables for polar snap. It handles eight situations using a single number by means of *bitcodes*. Bitcodes are added up to create a single number.

When all options are turned on, for example, the value of **POLARMODE** is 15 (which comes from 1 + 2 + 4 + 8). Zero means *no*, *off*, or *not*.

PolarMode	Meaning
Measurement Mode	
0	Absolute mode: Polar angle measurements are based on the current UCS.
1	Relative mode: Polar angles are measured from the last-drawn object.
Object Snap Tracking	
0	Tracking uses orthogonal angles only.
2	Tracking uses polar angle settings.
Additional Polar Tracking Angles	
0	Additional angles are not used.
4	Additional angles are used.
Acquire Object Snap Tracking Points	
0	Acquire automatically.
8	Press SHIFT to acquire.

Some of the values used by this system variable are meant for otrack'ing, covered later in this chapter.

TUTORIAL: DRAWING WITH POLAR MODE

In the previous chapter, you saw that the **POLYGON** command draws regular polygons but not irregular ones. An example of an irregular polygon is the 45-degree isosceles triangle, which has two angles at 45 degrees and the third at 90 degrees. In this tutorial, you draw such a triangle with two sides 2.5 units long.

1. Before starting to draw, set up polar mode using the **DSETTINGS** command.

   In the Drafting Settings dialog box, make these changes:

Polar Tracking tab:	
Polar Tracking On	**On**
Increment Angle	**45 degrees**
Polar Angle Measurement	**Absolute**
Snap and Grid tab:	
Snap On	**On**
Snap Type	**PolarSnap**
Polar Distance	**2.5**

   Click **OK** to exit the dialog box.

2. Start the **LINE** command, and then pick a point anywhere in the drawing:

   Command: **line**

   Specify first point: *(Pick point 1.)*

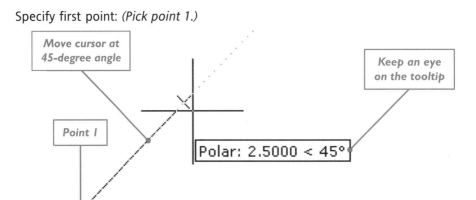

Move cursor at 45-degree angle

Keep an eye on the tooltip

Point 1

Polar: 2.5000 < 45°

3. Move the cursor at a 45-degree angle to the upper-right. You may need to move the cursor around until the tooltip appears, reporting **Polar: 2.5000 < 45°**.

    Specify next point or [Undo]: *(Pick point 2.)*

4. Now move the cursor at a 45-degree angle to the lower-right. Click when the tooltip reports **Polar: 2.5000 < 315°**. (The angle of 315 comes from 45 + 270.)

    Specify next point or [Undo]: *(Pick point 3.)*

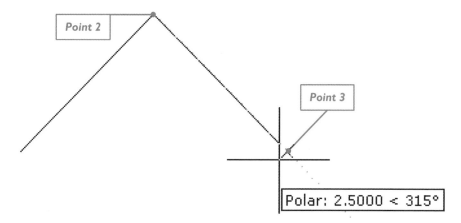

5. Close the triangle using the **Close** option:

    Specify next point or [Close/Undo]: **c**

    The isosceles triangle is complete. AutoCAD determines the length of the third side.

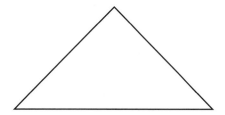

 ## OSNAP

Object snap is *the most important tool* for drawing accurately in AutoCAD. You might be able to do without the precise drawing aids described up to this point, but not object snap. As the name implies, object snap causes AutoCAD to snap to objects. More precisely, AutoCAD's cursor selects geometric features of objects, such as the end of an arc, the center of a circle, or the intersection of two lines.

In all, AutoCAD has 13 object snaps, which are displayed in drawings as temporary icons. (See figure at right.) Object snaps are often called "osnaps" for short.

In addition, AutoCAD has an entire toolbar devoted to object snaps, and provides the names of object snaps in shortcut menus:

APERTURE

A powerful feature of object snap is that you don't have to be right on the geometric feature — you just have to be close enough (as in horseshoes, hand grenades, and dancing) — and AutoCAD finds a point to which to snap.

□ Endpoint

△ Midpoint

○ Center

⊠ Node

◇ Quadrant

✕ Intersection

--- Extension

⤵ Insertion

⊥ Perpendicular

○ Tangent

⊠ Nearest

⊠ Apparent intersection

∥ Parallel

The close-enough distance is called the *aperture*, which is a square ten pixels in size. You can change the size of the aperture, from as large as 50 pixels to as small as 1 pixel with the **APERTURE** command.

The aperture appears as a square around the crosshair cursor. AutoCAD examines the objects that lie within or cross the aperture for likely object snap connections.

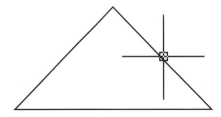

RUNNING AND TEMPORARY OSNAPS

Object snaps are used in two ways during drawing and editing commands: running and temporary. *Running* object snaps are set with the **OSNAP** and **-OSNAP** commands; in effect, the object snaps run on and on, until you turn them off.

Temporary object snaps are in effect for the next object selection only. At prompts such as "Select first point:" or "Specify next point", enter the names of one or more object snap modes. (When entering more than one mode, separate them with commas.) Here a line is drawn from the endpoint of one object to the midpoint or center point of another. "End" and "mid,cen" are abbreviations for the endpoint, midpoint, and center object snaps.

    Command: **line**

    Specify first point: **end**

    of *(Pick a point.)*

    Specify next point or [Undo]: **mid,cen**

    of *(Pick a point.)*

    Specify next point or [Undo]: *(Press* **ENTER** *to end command.)*

During commands, you can select object snap modes from a shortcut menu: hold down the **CTRL** key, and then press the right mouse button (or **SHIFT**+right-click). The shortcut menu is illustrated below:

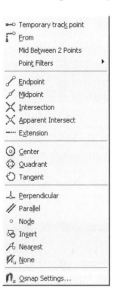

SUMMARY OF ALL OBJECT SNAP MODES

In the review of object snap modes below, notice that the first three letters of each mode are capitalized. These three letters are the abbreviation for each mode. When you enter object snaps at the command line, you need enter only the three letters, not the entire name.

In alphabetical order, AutoCAD's object snap modes are:

APParent intersection

The apparent intersection object snap (called "APPint" for short) works differently, depending on whether you are working with z coordinates.

In 2D drawings, APPint turns on *extended intersection* mode, so that AutoCAD snaps to the intersection of two objects that don't physically intersect. AutoCAD creates an imaginary extension to the two objects, and then determines if an intersection could occur.

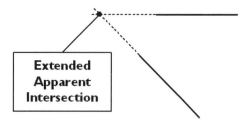

In 3D drawings, APPint snaps to the point where two objects *appear* to intersect from your viewpoint; the objects do not actually need to intersect. The objects can be arcs, circles, ellipses, elliptical arcs, lines, multilines, polylines, rays, splines, and xlines.

CENter

The center object snap snaps to the center point of arcs, circles, ellipses, and elliptical arcs.

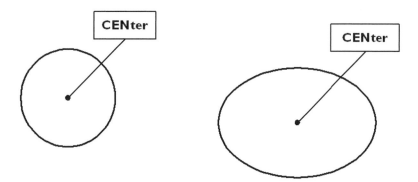

ENDpoint

The endpoint object snap snaps to the closest endpoint of arcs, elliptical arcs, lines, multilines, polyline segments, splines, regions, rays, and the closest corner of traces and 2D solids. Use ENDpoint for the corners of 3D faces and solids, because INTersection and EXTended intersection don't work with them.

— ··· EXTension

The extension object snap displays an extension line when the cursor passes over the endpoints of objects. This allows you to draw objects that start and end on the extension line.

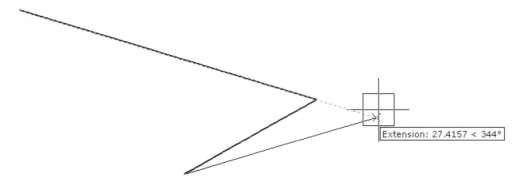

In the figure above, AutoCAD displays the extension line (shown dotted) from the endpoint of the line. The tooltip reports the distance and angle. (*Extensions* are similar to, but a more limited form of, object snap tracking, discussed later in this chapter.)

INSERTION

The insertion object snap snaps to the insertion point of attributes, blocks, shapes, and text.

INTersection

The intersection object snap snaps to the intersection of arcs, circles, ellipses, elliptical arcs, lines, multilines, polylines, rays, regions, splines, and xlines. Sometimes, more than one intersection is possible, such as when two circles intersect.

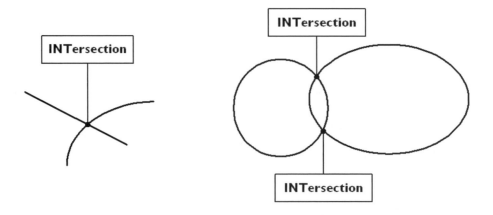

If extended intersection mode is turned on, AutoCAD can snap to the intersection of two objects that don't physically intersect. An imaginary extension to the two objects is created to determine if an intersection could occur.

 ## MIDpoint

The midpoint object snap snaps to the midpoint of arcs, ellipses, elliptical arcs, lines, multilines, polyline segments, regions, solids, splines, and xlines.

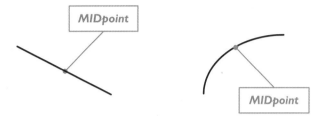

 ## NEArest

The nearest object snap snaps to the nearest point on the nearest arc, circle, ellipse, elliptical arc, line, multiline, point, polyline, ray, spline, or xline.

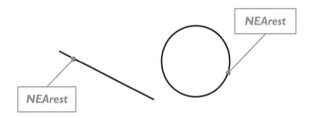

 ## NODe

The node object snap snaps to points, dimension definition points, and dimension text origins.

 ## PARallel

The parallel object snap displays an alignment path parallel to a straight line segment, such as a line or polyline.

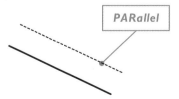

 **PERpendicular**

The perpendicular object snap snaps to a point perpendicular to an arc, circle, ellipse, elliptical arc, line, multiline, polyline, ray, region, solid, spline, or xline.

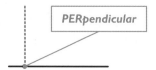

If necessary, AutoCAD turns on *deferred perpendicular* mode automatically when you need to pick more than one point to determine perpendicularity.

 **QUAdrant**

The quadrant object snap snaps to the *quadrant points* of arcs, circles, ellipses, and elliptical arcs.

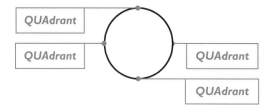

The quadrant points are located at the 0-, 90-, 180-, and 270-degree points of these curves. Thus, circles and ellipses always have four possible quadrant points, while arcs and elliptical arcs can have anywhere from one to four quadrant points.

 **TANgent**

The tangent object snap snaps to the tangent points of arcs, circles, ellipses, elliptical arcs, and splines.

If necessary, AutoCAD turns on *deferred tangent* mode automatically when you draw more than one tangent, such as a line tangent to two arcs.

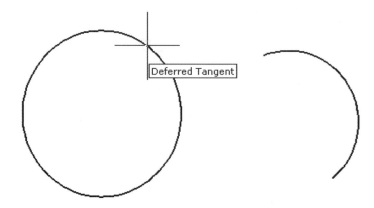

ADDITIONAL OBJECT SNAP MODES

There are several additional object snap modes that don't work with geometry. These find distances, or else turn off snap modes.

 From

The from mode allows you to specify a temporary reference or base point, from which to locate the next point.

Command: **line**

Specify first point: **from**

Base point: *(Pick point 1.)*

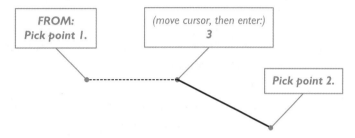

<Offset>: *(Move the cursor to show the distance, or enter a distance:)* **3**

Specify next point or [Undo]: *(Pick point 2.)*

Specify next point or [Undo]: *(Press ENTER to end command.)*

 NONe

The none mode turns off all object snap modes temporarily for the next pick point.

 TT

The temporary tracking mode allows you to specify a *temporary tracking* point. AutoCAD places a + marker at the point, and then, as you move the crosshair cursor, it displays alignment paths relative to the + marker.

Command: **line**

Specify first point: **tt**

Specify temporary OTRACK point: *(Pick point 1.)*

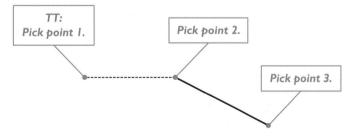

Specify first point: *(Pick point 2.)*

Specify next point or [Undo]: *(Pick point 3.)*

Specify next point or [Undo]: *(Press ENTER to end command.)*

QUIck

The quick object snap snaps to the first snap point AutoCAD finds. Quick mode works only when you have two or more other snap modes turned on. For example, when you have intersection and endpoint turned on, AutoCAD returns the first object snap point that is either an intersection or endpoint. This is significant in large, complex drawings.

OFF

The Off option turns off all object snap modes.

TUTORIAL: TOGGLING AND SELECTING OSNAP MODE

Object snap probably has the most complete set of command variations:

1. To toggle ortho mode, use one of these methods:
   - On the status bar, click **OSNAP** (turned on when button looks depressed).
   - At the keyboard, press function key **F3** or **CTRL+F**.

2. To toggle ortho mode *and* specify one or more object snaps, use one of these methods:
   - From the **Tools** menu, choose **Drafting Settings**. In the dialog box, choose the Object Snap tab, and then select object snap modes.
   - During another command, hold down the **CTRL** (or **SHIFT**) key, and then press the right mouse button to display a shortcut menu of object snap modes.
   - Similarly, you can use the Object Snap toolbar during a command: from the toolbar, select a button corresponding to the object snap mode you wish to employ.
   - On the keyboard, enter the **-osnap** command, followed by one or more object snap mode names. (The hyphen prefix forces AutoCAD to display the command at the 'Command:' prompt; without the hyphen, the **OSNAP** command displays the dialog box.)
   - Alternatively, type the aliases **os** or **ddosnap** (the old command name) for the dialog box, and the **-os** alias for the command line.
   - And finally, the **OSMODE** system variable can be used to set object snap modes.

CONTROLLING OSNAP MODES: ADDITIONAL METHODS

In new drawings, AutoCAD remembers the object snap modes set previously. You change the modes through a dialog box, or at the command-line:

- **OSNAP** command displays a dialog box for changing osnap settings.
- **-OSNAP** command changes osnap settings at the command line.
- Several system variables affect object snaps.

OSnap (DSettings)

The OSNAP command displays a dialog box that allows you to select running object snap modes. (Strictly speaking, the OSNAP command is a shortcut for using the DSETTINGS command, followed by selecting the Object Snap tab in the dialog box.)

1. To access the Drafting Settings dialog box:
   - From the **Tools** menu, choose **Drafting Settings**. Click on the Object Snap tab.
   - At the 'Command:' prompt, enter the **osnap** command.
   - Alternatively, right-click **OSNAP** on the status line. From the shortcut menu, select **Settings**.

   In all cases, AutoCAD displays the Drafting Settings dialog box with the Object Snap tab.

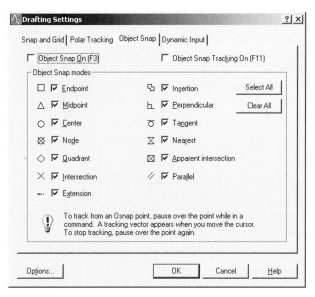

2. In this dialog box, you can make the following changes to the snap settings:

    *   **Object Snap On (F3)** toggles the snap on and off. The "(F3)" reminds you that you can press **F3** to turn on and off osnaps.

    *   **Object Snap Modes** lists most of the modes available. Notice the icon next to each mode; this is the same icon AutoCAD displays in the drawing when it finds a geometric feature matching the selected mode(s).

3. Select one or more object snap modes. You can turn on all osnap modes with the **Select All** button, and turn them all off with the **Clear All** button.

4. Not all object snap options are listed in this dialog box. Click the **Options** button to view more. AutoCAD displays the Drafting tab of the Options dialog box:

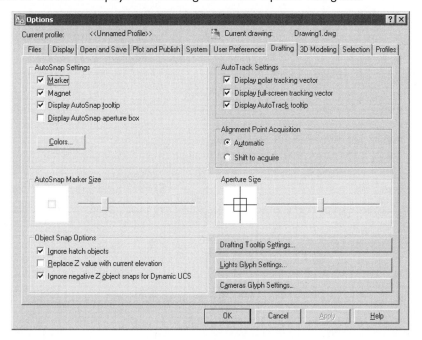

Here you select from the many options that affect how object snaps show up in the drawing. For example, if you changed AutoCAD's background color to white, you may want to change the marker color from the hard-to-see yellow to blue or to another higher-visibility color. The options have the following meanings:

TEMPORARY OVERRIDES

Temporary overrides temporarily turn on several object snap and drawing modes. For example, if you are drawing with all osnap modes off, you can hold down **SHIFT+E** to turn on the ENDpoint osnap mode temporarily.

The override does not work in reverse, unfortunately: when ENDpoint is on, holding down **SHIFT+E** does not turn it off temporarily. The workaround is to press **SHIFT+D** to disable all osnaps temporarily.

Temporary Override	Keyboard Shortcut(s)	
Disable osnap enforcement	SHIFT+A	SHIFT+'
Enable osnap enforcement	SHIFT+S	SHIFT+;
Enable CENter osnap	SHIFT+C	SHIFT+,
Enable ENDpoint osnap	SHIFT+E	SHIFT+P
Enable MIDpoint osnap	SHIFT+V	SHIFT+M
Toggle ortho mode	SHIFT	...
Toggle osnap tracking mode	SHIFT+Q	SHIFT+]
Toggle polar mode	SHIFT+X	SHIFT+.
Disable tracking and all osnap modes	SHIFT+D	SHIFT+L

There are two shortcuts for each, one for left-hand people and the other for right-handed. If none of these temporary overrides works, then the **TEMPOVERRIDE** system variable has been turned off — or, the shortcuts were changed with the **CUI** command.

In addition, the following shortcuts toggle osnap and other modes:

Toggles	Keyboard Shortcut(s)	
Toggle osnap mode	F3	CTRL+F
Toggles ortho mode	F8	CTRL+L
Toggle snap mode	F9	CTRL+B
Toggle polar mode	F10	CTRL+U
Toggle osnap tracking mode	F11	...

Marker toggles the display of the osnap icons.

Magnet toggles whether the cursor automatically moves to the osnap point.

Display AutoSnap Tooltip toggles whether the tooltip appears, which labels the object snap by name.

Display AutoSnap Aperture Box indicates AutoSnap mode is on, through a small square appearing at the center of the cursor.

Colors selects any color for the marker.

AutoSnap Marker Size makes the marker (icon) larger and smaller.

Ignore Hatch Objects determines whether osnaps snap to the lines that make up hatch patterns.

Replace Z values with current elevation determines whether osnaps use the z coordinate or the value of the **ELEVATION** system variable.

Ignore negative Z object snaps for Dynamic UCS snaps only to positive z-axis locations when dynamic UCS is turned on.

Aperture Size makes the object snap cursor larger and smaller.

5. Click **OK** to exit the dialog box.

6. AutoCAD returns you to the Drafting Settings dialog box. Click **OK** to exit this dialog box.

-OSnap

The **-OSNAP** command displays a prompt at the command line. It lists the current osnap modes, and then prompts you to enter additional modes. You can enter a single osnap mode, such as **Int**, or several modes separated by commas, as shown below:

> Command: **-osnap**
>
> Current osnap modes: End,Mid,Cen
>
> Enter list of object snap modes: **int,qua**

You need only type the first three letters of each mode name. For example, "int" means **INTersection** and "qua" means **QUAdrant**.

You cannot selectively turn off osnap modes. Instead, you turn off other modes by entering new modes. In the example above, entering **int** and **qua** turns off **end**, **mid**, and **cen**.

OSnap System Variables

Several system variables affect how AutoCAD performs its object snaps.

Back in AutoCAD 2004, Autodesk gave NODe osnap the ability to snap to grips on mtext; customers complained, because this made it difficult to avoid snapping to text. In AutoCAD 2005, Autodesk added the **OSNAPNODELEGACY** system variable to toggle this behavior on and off:

OSnapNodeLegacy	Comments
0	NODe snaps to mtext grips (default).
1	NODe ignores mtext grips.

It can be difficult to avoid selecting hatch objects, so Autodesk added the **OSNAPHATCH** system variable. When off, object snap ignores hatch objects:

OSnapHatch	Comments
0	Ignores hatch objects (default).
1	Snaps to hatch objects.

The **DWFOSNAP** system variable determines if you can snap to geometry in DWF files attached to drawings:

DwfOsnap	Comments
0	Ignores DWF entities.
1	Snaps to entities in attached DWF files (default).

NEW IN 2008 Similarly, the **DGNOSNAP** system variable determines if you can snap to geometry in imported MicroStation DGN V8 design files:

DgnOsnap	Comments
0	Ignores MicroStation elements.
1	Snaps to elements in imported MicroStation files (default).

The **OSNAPZ** system variable determines how osnaps work in 3D:

OSnapZ	Comments
0	Uses the z coordinate of the current point (default).
1	Uses the value stored in the ELEVATION system variable.

The **TEMPOVERRIDE** system variable determines whether temporary overrides are available.

TempOverride	Comments
0	Does not allow temporary overrides (off).
1	Allows temporary overrides (on).

The **OSOPTIONS** system variable determines how osnaps operate during dynamic UCS mode:

OsOptions	Comments
0	Operates normally during dynamic UCS.
1	Ignores hatch objects.
2	Ignores geometry with negative Z values.

The curious **APSTATE** system variable, undocumented by Autodesk, reports whether the aperture box cursor is on or off. I suspect it is meant for programmers instead of users.

TUTORIAL: USING OBJECT SNAP MODES

1. Draw a circle and arc of any size.
2. From the **Draw** menu, choose the **Line** command. Notice that AutoCAD prompts you "Specify first point:"
3. Do not pick a point. Instead, hold down the **CTRL** key, and then press the right mouse button to access the shortcut menu.
4. Choose **Endpoint** from the menu. Notice that the aperture square appears at the intersection of the crosshair cursor.
5. Place the cursor over one end of the arc, and click. The line should snap precisely to the endpoint of the existing line.
6. Before placing the next point of the line, type **cen**, and press **ENTER**, as follows:

   Specify next point or [Undo]: **cen** *(Press* ENTER.*)*

   of *(Place the cursor over any part of the circle's circumference, and then click.)*

   Notice how the line snaps to the precise center of the circle.

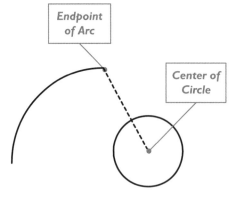

7. Continue the line, using the **Center** object snap on the arc. The line snaps to the center point of the arc.

8. Try other object snap modes on the line, arc, and circle. For example, use the **Tangent** object snap to construct lines tangent to circles and arcs.

OTRACK

Otrack is short for "object snap tracking"; Autodesk also calls it AutoTrack™, complete with the trademark symbol. Otracking adds *alignment paths* to object snapping.

Alignment paths are thin dotted lines that show the osnap relationships to other geometry. Think of otrack as "super extension osnap." At least one object snap mode must be turned on for otracking to work.

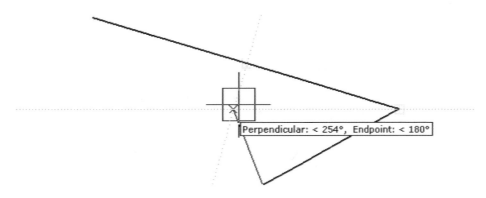

TUTORIAL: TOGGLING OTRACK MODE

Like polar mode, there is no "OTRACK" command, so there are shortcuts for turning it on and off. In addition, otrack mode is controlled by system variables and dialog boxes.

To turn on otrack mode, use one of these methods:

*   From the status bar, click **OTRACK** (turned on when button looks depressed).
*   On the keyboard, press function key **F11**.

At any prompt that asks you to 'Select objects:', move the cursor around the drawing to see geometric relationships. In the figure above, AutoTrack has placed a marker (the **x**) at a point perpendicular to one segment and in-line with the endpoint of another segment.

You can use otracking together with temporary tracking. Enter **tt** for *tracking* to show alignment paths relative to the temporary tracking point. The technical editor notes that OTRACK and TT together eliminate the need for the rays, xlines, and other construction lines discussed next.

 **RAY** AND  **XLINE**

The **RAY** and **XLINE** commands place *construction lines* in the drawing. Construction lines are useful for creating drawings, helping to reference things like center lines and offset lines. The construction lines are, unfortunately, plotted by AutoCAD, so you should place them on frozen or no-plot layers.

RAY draws "semi-infinite" lines, which have start points, but no endpoints. **XLINE** draws infinitely-long lines, which have no start or endpoints.

TUTORIAL: DRAWING CONSTRUCTION LINES

1. To draw construction lines, start the **RAY** or **XLINE** command with one of these methods:
   * From the **Draw** menu, choose **Ray** or **Xline**.
   * From the **Draw** toolbar, choose the **Construction Line** (xline) button.
   * From the Dashboard's **2D Draw** panel, choose the **Construction Line** button.
   * Or, at the 'Command:' prompt, enter the **ray** or **xline** commands.

     Command: **xline** *(Press ENTER.)*
   * Alternatively, enter the **xl** alias at the 'Command:' prompt.

2. Specify two points that lie along the construction line:
     Specify a point or [Hor/Ver/Ang/Bisect/Offset]: *(Pick point 1.)*
     Specify through point: *(Pick point 2.)*

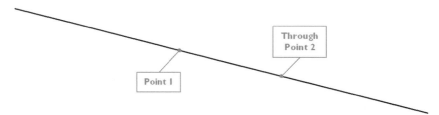

3. Because the **RAY** and **XLINE** commands automatically repeat, press **ENTER** or **ESC** to exit them:
     Specify through point: *(Press ENTER to exit command.)*

The **RAY** command differs from **XLINE** by first asking for the starting point:

     Command: **ray**
     Specify start point: *(Pick point 1.)*
     Specify through point: *(Pick point 2.)*
     Specify through point: *(Press ENTER to exit command.)*

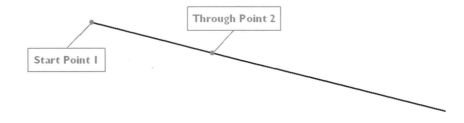

DRAWING CONSTRUCTION LINES: ADDITIONAL METHODS

The XLINE command contains these options that provide additional methods for drawing construction lines.

- **Hor** draws horizontal construction lines:

    Command: **xline**

    Specify a point or [Hor/Ver/Ang/Bisect/Offset]: **h**

    Specify through point: *(Pick a point.)*

    Specify through point: *(Press* ENTER *to exit command.)*

- **Ver** draws vertical construction lines:

    Command: **xline**

    Specify a point or [Hor/Ver/Ang/Bisect/Offset]: **v**

    Specify through point: *(Pick a point.)*

    Specify through point: *(Press* ENTER *to exit command.)*

- **Ang** draws construction lines at a specified angle:

    Command: **xline**

    Specify a point or [Hor/Ver/Ang/Bisect/Offset]: **a**

    Enter angle of xline (0) or [Reference]: *(Enter an angle, or type* **r** *for the reference option.)*

    Specify through point: *(Pick a point.)*

    Specify through point: *(Pick another point.)*

    Specify through point: *(Press* ENTER *to exit command.)*

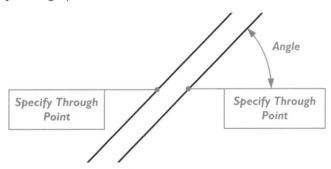

- **Bisect** draws construction lines that bisects an angle defined by a vertex and two endpoints:

    Command: **xline**

    Specify a point or [Hor/Ver/Ang/Bisect/Offset]: **b**

    Specify angle vertex point: *(Pick a point.)*

    Specify angle start point: *(Pick a point.)*

    Specify angle end point: *(Pick a point.)*

    Specify angle end point: *(Press* ENTER *to exit command.)*

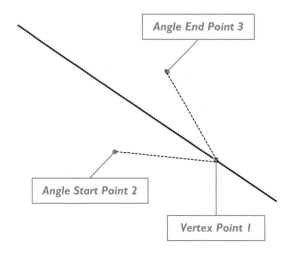

- **Offset** draws construction lines parallel to existing lines and construction lines:

    Command: **xline**

    Specify a point or [Hor/Ver/Ang/Bisect/Offset]: **o**

    Specify offset distance or [Through] <Through>: *(Enter a distance, or type **t** for the through option.)*

    Select a line object: *(Pick a line.)*

    Specify side to offset: *(Pick a point.)*

    Select a line object: *(Press **ENTER** to exit command.)*

TUTORIAL: DRAWING WITH CONSTRUCTION LINES

In this tutorial, you draw the true surface of a rectangular inclined face. You use much of what you have learned so far in this chapter: grid, snap, ortho, object snaps, and construction lines.

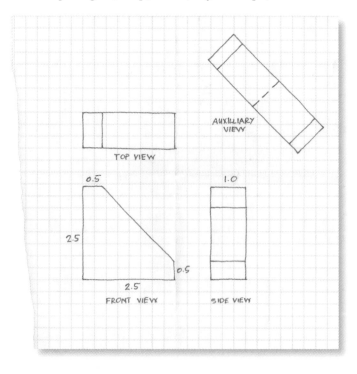

1. Start AutoCAD with a new drawing.

2. Change the following settings:
    Turn on **SNAP**; ensure the snap spacing is 0.5 units.
    Turn on **GRID**.
    Turn on **ORTHO**.
    Turn on **OSNAP**; ensure that object snap modes **Intersection** and **Perpendicular** are turned on.

3. With the LINE command (menu: **Draw | Line**), create the multiview drawing shown below.

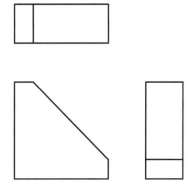

Because the top view is a mirror image of the side view, consider using the MIRROR command (menu: **Modify | Mirror**) to create it. *Hint:* Turn off **ORTHO** mode, and use a 45-degree mirror line.

4. Use the SNAP command's **Rotate** option to rotate the snap and grid to align with the sloping edge of the front view:
    Command: **snap**
    Specify snap spacing or [ON/OFF/Aspect/Rotate/Style/Type] <0.5000>: **r**
    Specify base point <0.0000,0.0000>: *(Pick point 1.)*

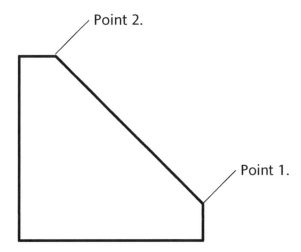

Because you set the object snap modes in Step 2, you should have no difficulty snapping to the ends of the sloped edge.
    Specify rotation angle <0>: *(Pick point 2.)*
    Angle adjusted to 315
Notice that the grid changes to align itself with the sloped edge.

5. Turn off **SNAP**.

   With the RAY command (menu: **Draw | Ray**), draw a series of parallel construction lines.

   > Command: **ray**
   >
   > Specify start point: *(Pick point 1.)*
   >
   > Specify through point: *(Pick point 2.)*
   >
   > Specify through point: *(Press ESC.)*

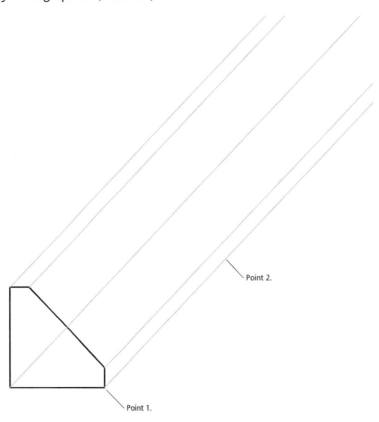

Point 2.

Point 1.

Repeat the command to draw four more construction lines.

Hint: Press ESC to end a command that repeats itself; press the spacebar to repeat the command.

6. Now draw the auxiliary view with the LINE command. Remember that the object is 1" wide.

   Hint: Use the **NEArest** object snap to start drawing from the construction line.

   > Command: **line**
   >
   > Specify first point: **nea**
   >
   > to *(Pick point 1.)*
   >
   > Specify next point or [Undo]: *(Pick point 2.)*
   >
   > Specify next point or [Undo]: *(Move cursor toward point 3.)* **1**
   >
   > Specify next point or [Close/Undo]: *(Pick point 4.)*
   >
   > Specify next point or [Close/Undo]: **c**

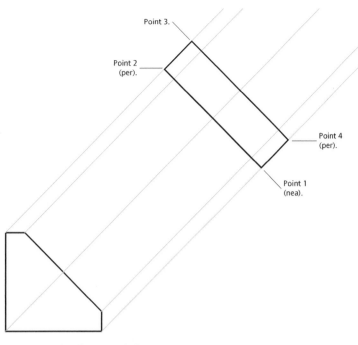

7.   Draw the two edge lines, as follows:

Command: **line**

Specify first point: *(Pick point 1.)*

Specify next point or [Undo]: *(Pick point 2.)*

Specify next point or [Undo]: *(Press ESC.)*

Command: *(Press ENTER to repeat the command.)*

LINE Specify first point: *(Pick point 3.)*

Specify next point or [Undo]: *(Pick point 4.)*

Specify next point or [Undo]: *(Press ESC.)*

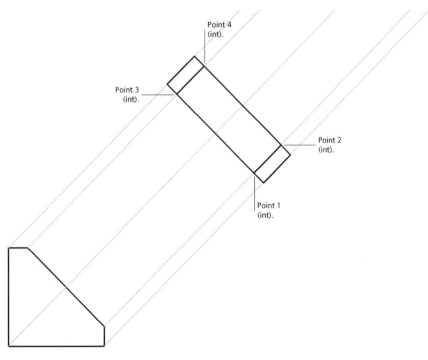

8.   Save the drawing as *auxview5.dwg*.

REDRAW AND REGEN

The REDRAW and REGEN commands clean up and update the drawing. REDRAW cleans up the display by removing blip marks (if any) and blanked out areas. REGEN is short for "regeneration," and updates the display by recalculating all the vectors making up the drawing. During a regeneration, AutoCAD recalculates line endpoints, hatched areas, and so on. A redraw is always faster than a regen.

If you have more than one viewport displayed, REDRAW and REGEN operate on the current viewport only. Use the REDRAWALL and REGENALL commands to clean up and update all viewports.

TUTORIAL: USING THE REDRAW COMMAND

1. To clean up the display, start the REDRAW command with one of these methods:
   * At the 'Command:' prompt, enter the **redraw** command.

     Command: **redraw** *(Press ENTER.)*
   * Alternatively, enter the **r** alias at the 'Command:' prompt.

   The screen may flicker as AutoCAD cleans up the drawing.

The commands associated with REDRAW work similarly:

Command	View Menu	Alias
RedrawAll	Redraw	ra
Regen	Regen	re
RegenAll	Regen All	rea

The "REDRAW" command found on the menu actually executes the REDRAWALL command.

TRANSPARENT REDRAWS

A redraw can be executed while another command is active. This is called a *transparent* redraw. To perform this, enter an apostrophe (') before the command. For example, to perform a transparent redraw during the LINE command:

> Command: **line**
>
> Specify first point: *(Pick a point.)*
>
> Specify next point or [Undo]: **'redraw**
>
> Resuming LINE command.
>
> Specify next point or [Undo]: *(Pick another point.)*
>
> Specify next point or [Undo]: *(Press ENTER.)*

When the transparent redraw is completed, the previously-active command resumes. REGEN is not a transparent command.

REGENAUTO

Some commands cause an automatic regeneration under certain circumstances. Among these are ZOOM, PAN, and VIEW Restore. The regeneration occurs if the new display contains areas not within the currently-generated area. The REGENAUTO command controls automatic regeneration in AutoCAD.

Because each regeneration can take a long time on a slow computer or with a very complex drawing, AutoCAD provides a command to warn you before regenerations are performed.

Command: **regenauto**

Enter mode [ON/OFF] <ON>: **OFF**

When REGENAUTO is turned off and a regeneration is required, AutoCAD displays a warning dialog box: "About to regen—proceed?" Choose **OK** to permit the drawing regeneration; choose **Cancel** to stop the regen.

As an alternative, press ESC to cancel the process. If the regen is interrupted in this way, some of the drawing might not be redisplayed. To redisplay the drawing, you must reissue the REGEN command.

 ZOOM

The ZOOM command enlarges and reduces the view of the drawing.

Most drawings are too large and too detailed to work with on the computer's 17" — or even 21" — screen. CAD operators routinely zoom in to small areas to work on details. When zoomed out, you see more of the drawing; when zoomed in, you see less of the drawing but more detail. Think of zoom as a magnifier and shrinker.

Let's consider this analogy: imagine that your drawing is the size of the wall in your room. The closer you walk to the wall-size drawing, the more detail you see, but the less you see of the entire drawing. When you move a great distance away from the wall, you see the entire drawing, but not much detail.

The wall does not change size, only your viewing distance. The ZOOM command works in the same way: you enlarge and reduce the drawing size on the screen, but the drawing itself does not change size (or scale).

Because zooming is performed often, AutoCAD has many ways to zoom.

TUTORIAL: USING THE ZOOM COMMAND

1. To change the viewing size of the drawing, start the ZOOM command with one of these methods:
   * From the **View** menu, choose **Zoom**, and then one of its options.
   * From the Dashboard's 2D Navigate panel, choose one of the **Zoom** buttons.
   * On the Zoom toolbar, click any one of the buttons to execute an option.
   * Right-click anywhere in the drawing, and from the shortcut menu select **Zoom**.
   * Or, at the 'Command:' prompt, enter the **zoom** command.

     Command: **zoom** *(Press* ENTER.*)*
   * Alternatively, enter the **z** alias at the 'Command:' prompt.

2. The command displays its many options:

   Specify corner of window, enter a scale factor (nX or nXP), or
       [All/Center/Dynamic/Extents/ Previous/Scale/Window/Object] <real time>:

The prompt indicates there are 12 options, and there are more that are hidden: two are **Left** and **Vmax**, and the third is the mouse wheel.

Many of AutoCAD's commands have a single default option, but Autodesk managed to endow this one with three defaults! These are the three defaults:

   (1) Specify corner of window, **(2)** enter a scale factor (nX or nXP), or **(3)** <real time>:

Here's what they mean:

Specify corner of window — click a point on the screen, and AutoCAD executes the **Window** option.

Enter a scale factor — type a number, and AutoCAD executes the **Scale** option.

<real time> — press **ENTER**, and AutoCAD executes real-time zooms.

ZOOM COMMAND: ADDITIONAL METHODS

The **ZOOM** command contains these options for zooming in and out. The options are shown at the command prompt, in the **View | Zoom** menu, and on the **Standard** and **Zoom** toolbars.

Zoom Option	Meaning
Shows Entire Drawing (zoom out):	
All	Shows the entire drawing, or the limits of the drawing, whichever is larger.
Extents	Shows everything in the drawing.
Vmax	Zooms out to the maximum without invoking a regen *(hidden option)*.
Shows Portion of Drawing (zoom in):	
Center	Zooms about a center point.
Object	Zooms to the extents of the selected objects.
Window	Zooms in to a rectangular area specified by two points.
Left	Zooms relative to a lower left point *(hidden option)*.
Shows All or Portion of Drawing (zoom out or in):	
real time	Zooms in real time with mouse movement.
Previous	Shows the previous view, whether a zoom or pan.
Scale	Zooms by absolute and relative factors in model space and paper space.
Dynamic	Displays a zoom/pan box for interactive zooming *(obsolete)*.

Despite the many options, I find I use only a few frequently — specifically **Extents**, **Window**, and **Previous**. Some CAD users find they don't need the **ZOOM** command at all: they use the mouse wheel to zoom in and out of the drawing transparently. The Dashboard provides access to these Zoom options:

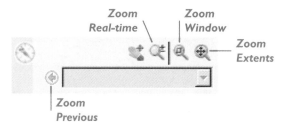

Let's look at all of the **ZOOM** command's options in alphabetical order. To practice zooming, open the *Carson-A.dwg* drawing file from the Companion CD.

Zoom All

The **All** option displays the entire drawing on the screen. This typically displays the entire area of the limits and the extents of the drawing, including the area of the drawing outside the limits.

Note this difference between the **Extents** and **All** options: **Extents** displays the extents of the drawing, while **All** displays either the drawing extents or the drawing limits, depending on which is larger.

> Command: **zoom**
>
> Specify corner of window, enter a scale factor (nX or nXP), or
>   [**All**/Center/Dynamic/Extents/Previous/Scale/Window/Object] <real time>: **a**

Occasionally, ZOOM All has to regenerate the drawing twice. If this is necessary, AutoCAD displays the following message on the prompt line:

> * * Second regeneration caused by change in drawing extents.

When the limits are changed, the entire drawing area is not shown until the next ZOOM All is performed.

Zoom Center

The ZOOM command's **Center** option allows you to choose a center point for the zoom. You then specify the magnification or height for the new views.

> Command: **zoom**
>
> Specify corner of window, enter a scale factor (nX or nXP), or
>   [All/**Center**/Dynamic/Extents/Previous/Scale/Window/Object] <real time>: **c**
>
> Specify center point: *(Select a point.)*
>
> Enter magnification or height <5.0000>: *(Enter a number.)*

If an "x" follows the magnification value, such as 5x, the zoom factor is relative to the current display.

Zoom Dynamic

The **Dynamic** option allows *dynamic* zooming and panning. This option is a predecessor to the Aerial View palette and real-time zoom and pan, and so is now rarely used.

> Command: **zoom**
>
> Specify corner of window, enter a scale factor (nX or nXP), or
>   [All/Center/**Dynamic**/Extents/Previous/Scale/Window/Object] <real time>: **d**

When you specify **Dynamic**, you are presented with a new window displaying information about current and possible view selections. The window shown in the figure represents a typical display.

Each of the view boxes has a specific color and pattern:

> **Drawing Extents** is shown by the blue dotted rectangle. The drawing extents can be thought of as the "sheet of paper" on which the drawing resides.
>
> **Initial View Box** is shown by the green dotted rectangle. This is the view of the drawing when you began the **ZOOM Dynamic** command.
>
> **Desired View Box** is shown by the black solid-line rectangle. This defines the size and location of the desired view. The desired view box is initially the same as the current view box. Move this box to obtain the view you want. It operates in two modes: pan and zoom.
>
> **Pan Mode** moves the view box to another location. A large **X** is initially placed in the center of the box. This denotes panning mode. When the **X** is present, moving the cursor causes the box to move around the screen.
>
> **Zoom Mode** is denoted by the arrow. Moving the cursor right or left increases or decreases the size of the view box. The view box increases and decreases in proportion to your screen dimensions, resulting in a "what you see is what you get" definition of the zoomed area. This differs from the standard window zoom, which works from a stationary window corner and may show more of the screen, depending on the proportions of the defined zoom window. Pressing the pick button causes the **X** to change to an arrow at the right side of the box.

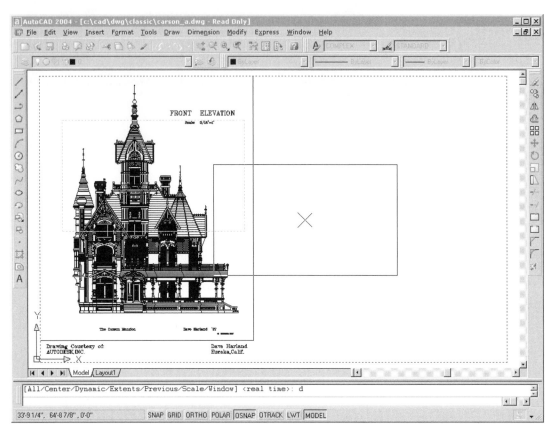

You can toggle between zoom and pan modes as many times as you wish to set the size and location of the view box. When you have windowed the desired area, press ENTER and the zoom is performed. The area defined by the zoom box is now the current screen view.

It is possible to use dynamic zoom without a mouse — something that harkens back to the days when the mouse was not so common. Use the arrow keys to move the view box. Press ENTER to toggle between pan mode and zoom mode.

Zoom Extents

The **Extents** option displays the drawing at its maximum size on the display screen. This results in the largest possible display, while showing the entire drawing.

> Command: **zoom**
>
> Specify corner of window, enter a scale factor (nX or nXP), or
>   [All/Center/Dynamic/**Extents**/Previous/Scale/Window/Object] <real time>: **e**

 Note: The fastest access to ZOOM Extents is through double-clicking the middle button or mouse wheel.

Zoom Object

The **Object** option displays selected objects at their maximum size on the screen. This option is useful for inspecting one or more objects visually.

> Command: **zoom**
>
> Specify corner of window, enter a scale factor (nX or nXP), or
>   [All/Center/Dynamic/Extents/Previous/Scale/Window/**Object**] <real time>: **o**

Zoom Previous

The **Previous** option returns to the last zoom or pan you viewed. This option is useful if you need to move frequently between two areas. AutoCAD remembers the last ten view changes; simply use the **Z P** aliases several times in a row. Since the view coordinates are stored automatically, you do not need any special procedure to access them.

> Command: **zoom**
>
> Specify corner of window, enter a scale factor (nX or nXP), or
>   [All/Center/Dynamic/Extents/**Previous**/Scale/Window/Object] <real time>: **p**

Zoom Scale

The **Scale** option enlarges or reduces the entire drawing (original size) by a numerical factor. For example, entering **5** results in a zoom that increases the drawing to five times its normal size — AutoCAD zooms in. The zoom is centered on the screen's center point.

When an **x** follows the zoom factor, the zoom is computed relative to the current display. For example, entering **5x** makes the drawing five times larger than its current zoom factor.

Only positive values can be used for zoom scales. To zoom out (i.e., display the drawing smaller), use a decimal value smaller than 1. For example, **0.5** results in a view of the drawing that is one-half its normal size. (As of Release 14, ZOOM **All** and **Extents** leave a bit of room around the drawing, effectively doing a zoom 0.95x.)

> Command: **zoom**
>
> Specify corner of window, enter a scale factor (nX or nXP), or
>   [All/Center/Dynamic/Extents/Previous/**Scale**/Window/Object] <real time>: **s**
>
> Enter a scale factor (nX or nXP): **5**

It is not necessary to enter the "s" for the **Scale** option. Enter a number after starting the command. Then AutoCAD assumes you want the scaled zoom.

The **XP** option scales the model viewport relative to the paper space viewport. This is discussed in *Using AutoCAD: Advanced.*

Zoom Window

The **Window** option uses a rectangular window to show the area, which you specify with two screen picks.

> Command: **zoom**
>
> Specify corner of window, enter a scale factor (nX or nXP), or
>   [All/Center/Dynamic/Extents/Previous/Scale/**Window**/Object] <real time>: **w**
>
> Specify first corner: *(Pick point 1.)*
>
> Specify other corner: *(Pick point 2.)*

A box is shown around the area to be zoomed.

It is not necessary to enter the "w" for the **Window** option. Simply pick two points after starting the command; AutoCAD assumes you want a windowed zoom.

Zoom Real Time

With today's computers and fast display drivers providing the horsepower, AutoCAD can zoom in real-time. *Real time* means that the zoom changes continuously as you move the mouse. To enter real-time zoom mode:

> Command: **zoom**
>
> Specify corner of window, enter a scale factor (nX or nXP), or
>   [All/Center/Dynamic/Extents/Previous/Scale/Window/Object] <**real time**>: *(Press* ENTER.*)*
>
> Press ESC or ENTER to exit, or right-click to display shortcut menu.

Once in real-time zoom mode, the cursor changes to a magnifying glass. To zoom, hold down the left mouse button, and then move the mouse up (toward the monitor) and down (away from the monitor). As you move the mouse up, the drawing becomes larger; as you move the mouse down, it becomes smaller.

While in real-time zoom mode, right-click the mouse to see the following shortcut menu (redesigned in AutoCAD 2008):

VIEW TRANSITION

The **ZOOM** command can employ "view transition." When zooming in and out, AutoCAD executes an animation that makes objects appear to move closer and further away.

View transitions are useful for seeing where the view "ends up" in the drawing, particularly in 3D; the drawback is that zooms and pans take longer. Real-time zooms still occur in real-time; they don't have the animation.

View transitions are controlled by these system variables:

VTENABLE specifies when view transitions take place (using bit codes):

VtEnable	Meaning
0	View transitions turned off.
1	Turned on for the ZOOM and -PAN commands.
2	Turned on for view rotation.
4	Turned on for scripts.

VTDURATION specifies the duration of view transitions. The default value is 750 milliseconds; range of values is 0 (no transition) to 5000 (very slow transition).

VTFPS specifies speed at which the view transition takes place. The range is 1 to 30fps (frames per second).

GROUPED VIEW-CHANGE UNDO

Undoing a series of view changes, such as repeated pans, can be annoying. The value of 16 in the **UNDOCTL** system variable groups together all zooms and pans as a single action.

Alternatively, you can turn on the setting in the Options dialog box. From the **Tools** menu, select **Options**, and then choose the **User Preferences** tab.

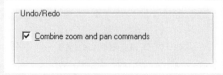

Option	Meaning
Exit	Exits the ZOOM command.
Pan	Switches to real-time pan mode (discussed later in this chapter).
Zoom	Switches to real-time zoom mode (default).
3D Orbit	Switches to interactive 3D rotation.
Zoom Window	Displays a cursor with small rectangle. Select two points, which AutoCAD displays as the new zoomed view. Remains in real-time zoom.
Zoom Original	Displays the drawing as it was when you began real-time zooming.
Zoom Previous	Displays the extents of the drawing's objects.

To dismiss the shortcut menu without choosing one of its options, click anywhere outside the menu. When the drawing is the size you want, press ENTER or ESC to exit real-time zoom mode and the ZOOM command.

Transparent Zoom

A transparent zoom can be executed while another command is active. Enter 'ZOOM at almost any prompt (notice the apostrophe prefix.) There are, however, restrictions:

Fast zoom mode must be turned on, which it is, by default. (This option is set with the VIEWRES command.) You cannot perform a transparent zoom if a regeneration is required, which means you cannot zoom outside the generated area — see **ZOOM Dynamic**.

Transparent zooms cannot be performed when certain commands are in progress. These include the **VPOINT**, **PAN**, **VIEW**, and the **ZOOM** command itself. The drop-down menus and toolbar buttons automatically use transparent zooms.

 PAN

The PAN command moves around the view of the drawing.

Many times, you zoom into an area of the drawing to see more detail. You may want to "move" the screen a short distance, to continue working while remaining at the same zoom magnification. This sideways movement is called "panning."

A pan is similar to placing your eyes at a certain distance from a paper drawing, and then moving your head about the drawing. This would allow you to see all parts of the drawing at the same distance from your eyes.

TUTORIAL: PANNING THE DRAWING

1. To move the view of the drawing, use the **PAN** command by one of these methods:
    *   From the **View** menu, choose **Pan**, and then choose one of its options.
    *   On the Standard toolbar, select the **Pan Realtime** button.
    *   On the Dashboard's 2D Navigate panel, select the **Pan Realtime** button.
    *   Or, at the 'Command:' prompt, enter the **pan** command.

        Command: **pan** *(Press ENTER.)*
    *   Alternatively, enter the **p** alias at the 'Command:' prompt.

2. Hold down the left mouse button; notice the hand cursor.
    Move the mouse; notice that the drawing moves with the mouse.

Right-click to see the same shortcut menu as displayed during realtime zoom.

3. To exit pan mode, press **ENTER** or **ESC**.

Press ESC or ENTER to exit, or right-click to display shortcut menu. *(Press ESC.)*

TRANSPARENT PANNING

A pan can be performed transparently while another command is in progress. To do this, enter 'PAN at any prompt, except those expecting you to enter text. The same restrictions apply as for transparent zooms.

Alternatively, you can press-drag the middle mouse button (or mouse wheel).

SCROLL BARS

Like most other Windows applications, AutoCAD has a pair of scroll bars that allow you to pan the drawing horizontally and vertically (though not diagonally). To pan with a scroll bar, position the cursor over a scroll bar and click. AutoCAD pans the drawing transparently.

(If you don't see the scroll bars, you need to turn them on in the Options dialog box. Enter the **OPTIONS** command, choose the Display tab, and then turn on the **Display Scroll Bars in Drawing Window** option.)

There are three ways to use the scroll bars:

*   Click the arrow at either end of the scroll bar. This pans the drawing in one-tenth increments of the viewport size.

*   Click and drag on the scroll bar button. This pans the drawing interactively as you move the button. This is also the way to pan by a very small amount.

*   Click anywhere on the scroll bar, except on the arrows and button. AutoCAD pans the drawing by 80 percent of the view.

-PAN

The -PAN command is the command-line alternative to panning. It prompts you to pick a pair of points, much like the line command:

Command: **-pan**

Specify base point or displacement: *(Pick a point.)*

As you move the cursor, AutoCAD draws a dragline.

Specify second point: *(Pick another point.)*

After the second pick point, AutoCAD pans by drawing by the distance indicated with the two pick points.

On the **View** menu, **Pan** has several options. The **Realtime** menu pick executes the PAN command, while the **Point** menu pick executes the -PAN command. The **Left**, **Right**, **Up**, and **Down** menu picks pan the view by 25% in the indicated direction.

▢ CLEANSCREENON

The CLEANSCREENON command minimizes the user interface to maximize the drawing. It turns off the title bar, toolbars, and window edges.

 Note: You can make the drawing area even larger by turning off the scroll bars and layout tabs. This is done through the **Display** tab of the Options dialog box. In addition, you can drag the Command Prompt away from the bottom of the screen area to make it float, and then make it transparent.

TUTORIAL: TOGGLING THE DRAWING AREA

1. To make the drawing area as large as possible, start the **CLEANSCREENON** command with one of these methods:
   - From the **View** menu, choose **Clean Screen**.
   - On the keyboard, press **CTRL+0** (zero).
   - At the 'Command:' prompt, enter the **cleanscreenon** command.

   Command: **cleanscreenon** (Press ENTER.)

2. To return to the "normal" screen, use the **CLEANSCREENOFF** command.

   Command: **cleanscreenoff** (Press ENTER.)

You can press CTRL+0 (zero) to toggle between maximized and regular views.

DSVIEWER

The **DSVIEWER** command provides a "bird's-eye view" of the drawing in an independent window called "Aerial View." This window is an alternative to the realtime **ZOOM** and **PAN** commands and scroll bars. The Aerial View window lets you see the entire drawing at all times in an independent window. This is sometimes called the bird's-eye view.

After the drawing appears in the Aerial View window, zooming in is as simple as with the **ZOOM Window** command: pick two points. To zoom to another area, move the cursor. To pan, move the rectangle to the new location.

TUTORIAL: OPENING THE AERIAL VIEW WINDOW

1. To view the entire drawing in an independent window, start the **DSVIEWER** command with one of these methods:
   * From the **View** menu, choose **Aerial View**.
   * At the 'Command:' prompt, enter the **dsviewer** command.

     Command: **dsviewer** *(Press ENTER.)*
   * Alternatively, enter the **av** alias at the 'Command:' prompt.

2. Notice that AutoCAD opens the Aerial View window.
   The Aerial View always shows the entire drawing. The heavy rectangle shows the current view in AutoCAD, while the light rectangle is the proposed view.

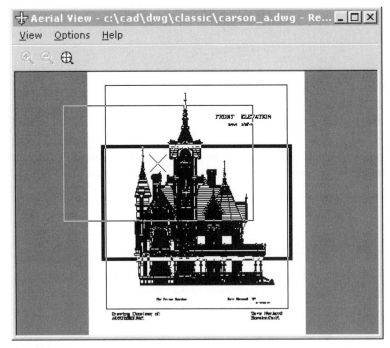

3. As with **ZOOM Dynamic,** you can switch between zoom and pan modes: left-click to switch. The **X** indicates pan mode, while the arrow indicates zoom mode.
   Moving the mouse in pan mode pans the view.
   Moving the mouse in zoom mode zooms in or out.

4. When you have the view you want, right-click in the Aerial View window.

AERIAL VIEW WINDOW: ADDITIONAL METHODS

The Aerial View window contains a menu bar and toolbar that provide options for viewing the drawing.

View Menu

The options in the **View** menu change the magnification of the Aerial View — not the drawing itself. The **Zoom In** option zooms in by a factor of 2, while the **Zoom Out** option zooms out by a factor of 2. The **Global** option displays the entire drawing of the current view in the Aerial View window.

Options Menu

The **Options** menu changes how the Aerial View window operates. The **Auto Viewport** option displays the model space view of the *current* viewport automatically. The **Dynamic Update** option updates the Aerial View window as you edit the drawing. (When off, the image in the window is not updated until you click it). The **Realtime Zoom** option updates the drawing area in real time when you zoom using the Aerial View window.

VIEW

The **VIEW** command stores and recalls views of the drawing by name. This lets you quickly move about a drawing. Think of it as combining the **ZOOM** and **PAN** commands, without needing to specify zoom ratios and pan directions.

You can save the current display as a view, or else window an area to define the view. Naturally, if you are saving the current view, you need to zoom and pan into that view first. Views can be stored relative to the current UCS (user-defined coordinate system), such as when using an architectural plan with an angled wing of the building or when working with 3D drawings.

You take advantage of named views outside of the **VIEW** command in two ways. When AutoCAD starts, you can specify the name of a view to be displayed when the drawing opens. And, when AutoCAD plots, you can specify a named view with the **PLOT** command.

Named views can be connected with layer names. That means you can access a view that automatically turns off certain layers for a more specific view. Views can also defined visual styles and backgrounds.

The name of views can be changed with the **RENAME** command, and views can be deleted from drawings with the **VIEW** command's "hidden" **Delete** option.

PREDEFINED VIEWS

AutoCAD provides several predefined views — six standard orthographic views and four standard isometric views. These are meant for viewing 3D drawings. The orthographic views show the top, bottom, left, right, front, and back of drawings, while the isometric views show the southwest, northwest, northeast, and southeast corners of drawings.

You can access these named views from the **View** menu and the **View** toolbar. From the **View** menu, select **3D Views**.

TUTORIAL: STORING AND RECALLING NAMED VIEWS

1. To create named views, start the **VIEW** command with one of these methods:
   * From the **View** menu, choose **Named Views**.
   * From the View toolbar, choose the **Named Views** button.
   * Or, at the 'Command:' prompt, enter the **view** command.

Command: **view** *(Press ENTER.)*

- Alternatively, enter the aliases **v** or **ddview** (the old name for this command) at the 'Command:' prompt.

Notice the View dialog box.

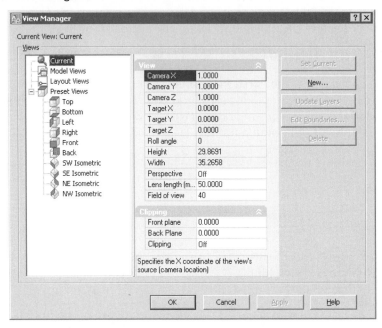

2. Click **New** to create a new named view. Notice the **New View** dialog box. (This dialog box changed slightly in AutoCAD 2008.) You can access this dialog box directly with the **NEWVIEW** command.

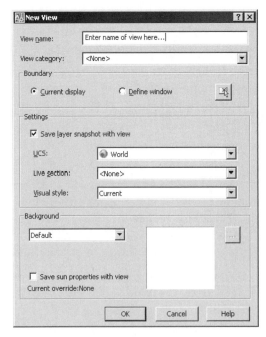

3. Enter a name for the view.

   Later, when it comes time to recall the view, you specify it by its name. That's why it makes sense to give views descriptive names, such as "Titleblock."

4. Determine the view's boundary:

   Current display — the view is the current display.

   Define window — the view is smaller than the current display. Click the **Define**

Window button. Notice that AutoCAD dismisses the dialog boxes temporarily, and prompts you to pick two points that define the rectangular window.

Specify first corner: *(Pick a point.)*

Specify opposite corner: *(Pick another point.)*

5. Decide whether you want the visibility of layers changed with the view.

When the **Save layer snapshot** option is turned on, AutoCAD remembers which layers were frozen, locked, turned off, and so on. When you later restore the view, the same layers are frozen, etc.

You can ignore the other options — **UCS**, **View Category**, **Background**, and so on — because they are meaningful mainly for working with sheet sets and 3D drawings.

6. Click **OK**.

AutoCAD adds the view name to the list.

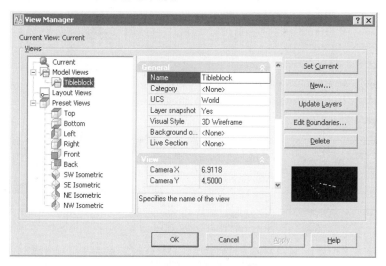

7. To recall the view, select its name (such as "Titleblock"), and then click **Set Current**. A small triangle points to the current view name.

8. Click **OK**, and notice how AutoCAD fills the screen with a closeup of the drawing's titleblock (or whatever you windowed).

You can also access view names from the Dashboard's 2D Navigate panel.

TUTORIAL: RENAMING AND DELETING NAMED VIEWS

You can rename and delete named views.

To delete views: select a name, and then click **Delete**.

To rename views: under the General section, click the **Name** field, and then change the name.

The meaning of the items in this shortcut menu is:

Shortcut Menu	Comments
Set Current	Makes the named view current.
New	Creates new named views.
Update Layers	Changes layer visibility to the current setting.
Edit Boundaries	Shows the windowed view, with non-view areas in gray.
Delete	Removes the named view from the drawing.
	Warning! AutoCAD removes the view without warning.

TUTORIAL: OPENING DRAWINGS WITH NAMED VIEWS

When you use the OPEN command, AutoCAD displays the Select File dialog box. One option you may have overlooked is **Select Initial View**, which is normally turned off. This causes AutoCAD to present a list of view names, from which you select one.

To open a drawing with an initial view:

1. Enter the **OPEN** command.
2. In the Select File dialog box:
   a. Select the *bias1.dwg* drawing file from the Companion CD.
   b. Click **Select Initial View** to turn it on (a check mark shows).
   c. Click **Open**.

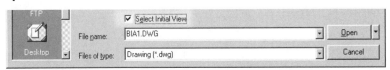

3. AutoCAD starts to open the drawing, and then displays the Select Initial View dialog box. Select the "KYOTO" from the list, and then click **OK**.

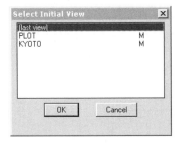

   The "M" means the named view was saved in model space; a "P" means the view was saved in a layout.
4. AutoCAD finishes opening the drawing, and displays the view.

TUTORIAL: STARTING AUTOCAD WITH NAMED VIEWS

AutoCAD originally ran on another operating system called DOS (short for "disk-based operating system, upon which Windows was also designed.) In those days, it was common to start a program with *command-line switches*, which instructed the program how to start. These switches are still available for AutoCAD, and the /v switch specifies the named view to display when AutoCAD opens a drawing. The drawing, naturally, needs to have the named view; otherwise AutoCAD ignores /v, and instead shows the last saved view.

To use the /v switch, edit AutoCAD's command line, as follows:

1. On the Windows desktop, right-click the AutoCAD icon.
2. From the shortcut menu, select **Properties**.
3. In the Properties dialog box, select the **Shortcut** tab.

4. In the Target text box, add the path and file name of the drawing. An example is shown in boldface text, but may differ for your computer:

    "C:\AutoCAD 2008\acad.exe" **"c:\autocad 2008\sample\file name.dwg" /v titleblock**

5. Click **OK** to close the dialog box, and double-click the AutoCAD icon to see if it loads the drawing with the named view.

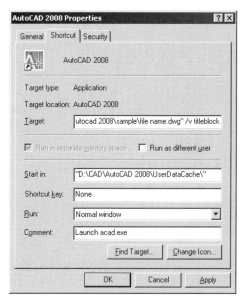

 Notes: A space is required after the switch. Double quotes are needed when there are spaces in the file name.

You can have more than one shortcut icon on the Windows desktop. To make copies of icons, hold down the **CTRL** key, and then drag the icon. When you let go, Windows creates the copy, which you can then rename and edit.

TUTORIAL: CHANGING WINDOWED VIEWS

AutoCAD makes it easier to change window views, by showing the current windowing.

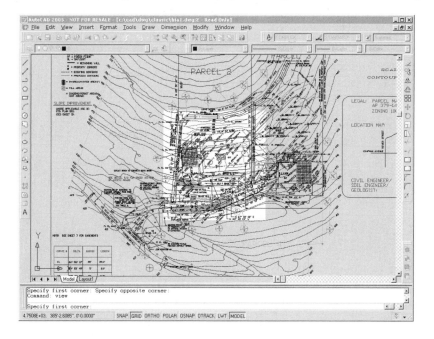

1. From the Companion CD, open the *bias1.dwg* file.
2. Enter the **VIEW** command, and then select the "KYOTO" view.
3. Click the **Edit Boundaries** button.

   Notice that the dialog box disappears. The area outside the window is shown in gray. (See figure on next page.)
4. At the command prompt, AutoCAD asks you:

   Specify first corner: *(Pick a point.)*

   Specify opposite corner: *(Pick another point.)*
5. After you pick two points for the new window, AutoCAD shows the area in white (or black, depending on the background color of the drawing area), and asks again:

   Specify first corner (or press ENTER to accept): *(Press* ENTER.*)*

   Pick new points, or else press **ENTER** to accept the change.

   The dialog box returns.
6. Click **OK** to dismiss the dialog box.

VIEWRES

The **VIEWRES** command controls the roundness of circular objects. This is a technical command that is almost never needed anymore, but it can be useful to know the reason to use it when circles look like polygons.

AutoCAD stores the objects mathematically. When you draw a circle, for example, AutoCAD stores the circle's primary parameters: its center point and radius. If you drew it by specifying its diameter, AutoCAD internally converts the diameter to the equivalent radius. When you use the **LIST** command on a circle, AutoCAD lists the circle's center point and radius, no matter how it was drawn.

AutoCAD displays objects as *vectors*, lines that have direction and length. (Windows and the graphics board convert the vector data to the raster image displayed by the screen.) Everything displayed by AutoCAD is vectors, including circles, which are composed of many very short vectors — so short that circular objects look round to you.

In most cases, you won't notice that circular objects are made of very short lines. Sometimes, however, circles and arcs look like polygons. This occurs when you zoom into the drawing by a large amount, because AutoCAD does not display any more circle segments than it determines are necessary for the current zoom level. If the circle uses less than two screen pixels at the maximum zoom magnification, the circle is displayed as a single pixel.

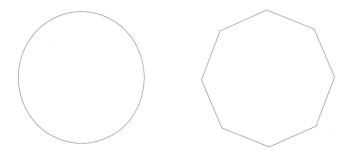

AutoCAD regenerates some zooms, some pans, and view restores, and when entering layout mode (paper space). The "Fast Zoom" mode is retained only for compatibility with scripts and macros; at one time, it allowed AutoCAD to refresh the screen at the faster redraw speed, instead of the slower regen speed. Some extreme zooms still require a regeneration.

TUTORIAL: CHANGING THE VALUE OF VIEWRES

1. Start the **VIEWRES** command, as follows:
   * At the 'Command:' prompt, enter the **viewres** command.

     Command: **viewres** *(Press* ENTER.*)*

2. Entering **n** at the first prompt causes all zooms, pans, and view restores to regenerate. For fastest speed, answer **y**:

   Do you want fast zooms? [Yes/No] <Y>: *(Enter* **y** *or* **n**.*)*

3. The default value for the circle zoom percent is 100. A value less than 100 diminishes the resolution of circles and arcs, but results in faster regeneration times.

   Enter circle zoom percent (1-20000) <1000>: **10000**

Values greater than 100 result in a larger number of vectors than usual being displayed for circles and arcs. For today's fast computers, Autodesk has increased VIEWRES to 1000, resulting in smooth circles and arcs at a zoom factor of 10. Increase the value to 10,000 for smooth circles at zoom factor 100.

EXERCISES

1. Draw the following baseplate from lines and circles.

    Set snap to 0.25 and grid to 1.0. Use the appropriate object snaps to assist your drafting. The dimensions need not be exact, but it may be helpful to know that the circles have a radius of 1.0.

    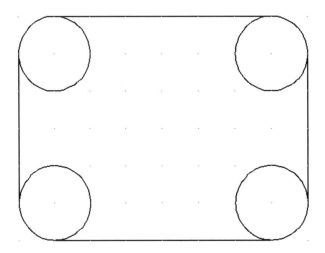

2. Draw the following floor plan using lines.

    Set snap and grid to 12.0. Set limits to 0,0 and 600,400. Use the **ZOOM All** command to see the entire drawing area. Turn on ortho mode. Use the appropriate object snaps to assist your drafting. The dimensions need not be exact, but it may be helpful to know that the longest wall is 500 units long.

    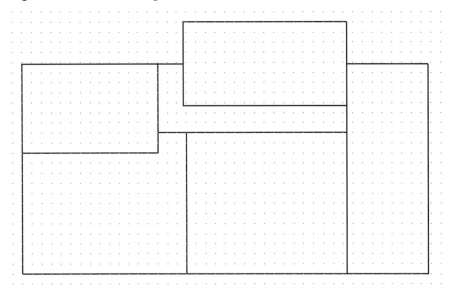

3. Draw the following L-shaped bracket using lines and arcs.

    Set the snap to 1.0 and the grid to 12.0.

    Set the limits to 0,0 and 216,144.

    Use the **ZOOM All** command to see the entire drawing area.

    Turn on ortho mode.

    Use the appropriate object snaps to assist your drafting. Remember that arcs are drawn counterclockwise. The dimensions need not be exact, but it may be helpful to know that the longest edge is 10 units long, and that the metal is 1 unit thick.

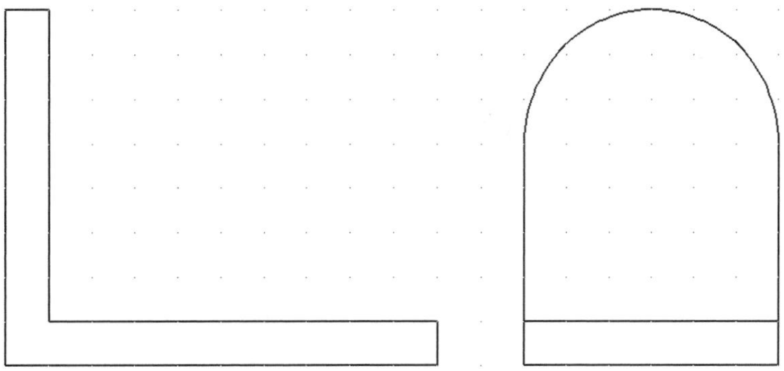

4. Draw an isosceles triangle with lines.

    Its angles are 45-45-90, and the two sides are 5 units. Use polar tracking set to 45 degrees and polar snap set to 5 units.

5. Draw the profile of the electrical switch spring shown below.

    Use polylines, switching between line- and arc-drawing mode, as necessary.

    Set the pline width to 0.1 units. Snap and grid are 1.0 units.

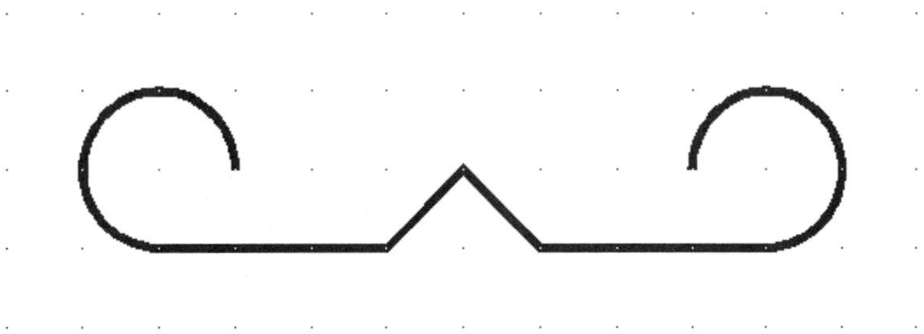

6. Use object snap tracking to assist you in creating the third view (shown in gray) of the industrial strength door wedge. The wedge is 9 units long, 2 units tall, and 3 units wide.

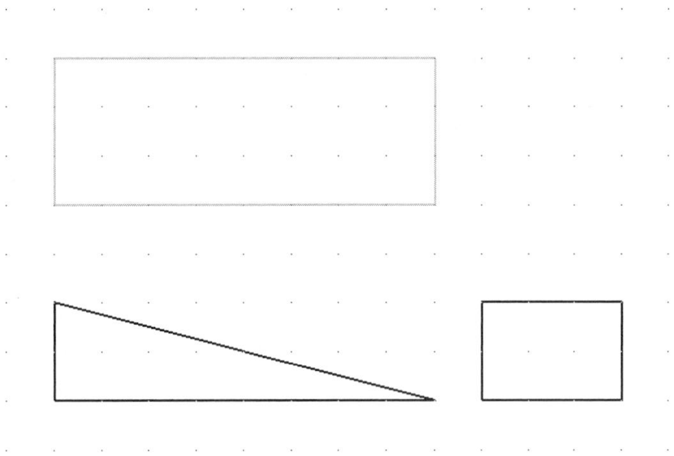

7. Use construction lines to assist you in creating the true view of the object shown below. The units are shown on the sketch.

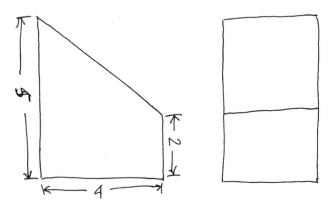

8. Draw the front, side, and top views of the bracket made of 1/4" sheet metal. The units are shown on the sketch.

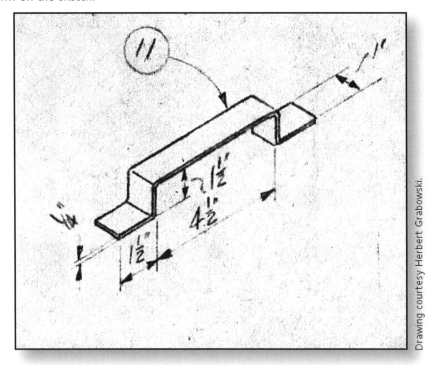

9. Using object snaps, place an arc tangent to the two circles.
   Then, draw the remaining lines, using the object snaps shown.
   The larger circle has a radius of 4 units, the smaller a radius of 2 units.

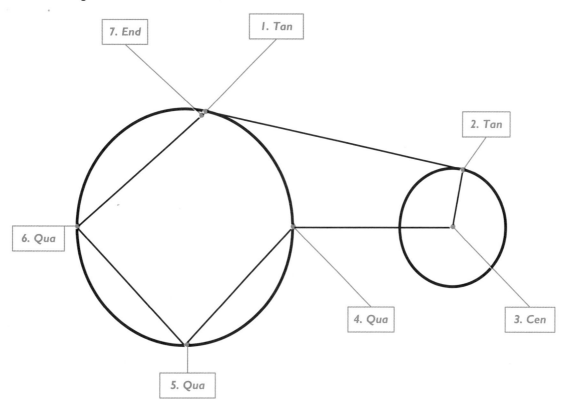

10. Draw the 3.5" wide by 3.7" tall diskette.
    The figure below illustrates the diskette full size, 1:1; take your measurements directly from the photograph.

11. Draw the 3.9" wide by 1.75" tall cross brace for ceiling lamps.

Take your measurements directly from the photograph, which is shown below at full-size, 1:1 scale.

12. Draw the 2.25" wide by 1.0" tall latch face for a security deadbolt lock.

Take your measurements directly from the full-size photograph, below.

CHAPTER REVIEW

1. What command allows you to enlarge and reduce the visual size of the drawing?
2. What is the purpose of the **PAN** command?
3. What is the **Realtime** option under the **ZOOM** command used for?
4. When a real-time zoom is in progress, can you override it?
   If so, how?
5. How do the **PAN** and **ZOOM** commands differ?
6. What does *toggle* mean?
7. What is a *transparent* command?
   How is this option invoked?
8. How do the **REDRAW** and **REGEN** command differ?
9. List four ways by which the grid can be toggled:
   a.
   b.
   c.
   d.
10. What is *temporary tracking* used for?
11. Name two ways to create *construction lines*:
    a.
    b.
12. What is the *aperture*?
13. What object snap mode would you choose to snap to a point object?
14. Can you snap to the midpoint of arcs?
15. When does AutoSnap come into effect?
16. Can different x and y spacings be given to the snap?
    Can it coincide with the grid spacing?
17. What advantages do rays and xlines have over grids?
18. What modes provide great accuracy for creating true horizontal and vertical lines?
19. What purpose does **APPint** mode serve?
20. Which command controls the extent of the grid?
21. Although the grid is not a part of the drawing, can it be plotted?
22. Where does a ray start?
    End?
23. Name the function of the Aerial View.
24. In addition to the snap angle, the **Angle** option of the **SNAP** command also affects:
    a.
    b.
    c.
    d.
25. Which mode does **F8** toggle?
26. What is the difference between *ortho* and *polar* modes?
27. What are *tooltips*?
28. When would you use the **<** symbol in AutoCAD?
29. What is an *alignment path*?
30. Describe the function of *PolarSnap*.

31. Provide the meaning of the following abbreviations:

    osnap

    otrack

    appint

    tt

32. Define the following object snap abbreviations:

    int

    cen

    qua

    end

33. Which part of the object geometry do the following object snaps snap to?

    ins

    nod

    per

    tan

34. When would you use the **From** object snap?

35. Describe how the EXTension object snap operates.

36. Can AutoCAD snap to the intersection of two circles?

    If so, in how many places could the object snap take place?

37. Why does AutoCAD sometimes defer snapping to a tangent?

38. Which command is toggled by the following function keys?

    F3

    F9

    F10

    F11

39. What is the purpose of AutoSnap's magnet?

40. How do the **REDRAW** and **REDRAWALL** commands differ?

41. When might you turn off **REGENAUTO**?

42. Name the command that matches the alias:

    z

    sn

    xl

    ds

43. Which command maximizes the drawing area?

44. What is the function of the **DSVIEWER** command?

45. Name four commands that take advantage of *named views*.

    a.

    b.

    c.

    d.

46. How do you delete named views?

47. When do you need to increase the setting of the **VIEWRES** command?

48. Describe *view transitions*.

49. How would you turn off view transitions?

50. What keystrokes would you press to turn on MIDpoint object snap temporarily?

CHAPTER 6

Drawing with Efficiency

In addition to precision, another advantage of CAD over hand drafting is *efficiency*. "You should never have to draw anything twice," said John Walker, one of the founders of Autodesk, Inc. By drawing efficiently, you complete projects in less time — or more projects in the same time. In this chapter, you learn to use some of AutoCAD's commands for creating and reusing content, such as symbols (blocks), poches (hatch patterns), and groups of objects. The commands are:

BLOCK and **-BLOCK** create reusable components called "blocks."

INSERT and **-INSERT** place blocks in drawings.

BASE changes the base point of drawings.

WBLOCK exports blocks and drawings as *.dwg* drawing files.

MINSERT inserts blocks as rectangular arrays.

BLOCKICON creates icons (preview images) of blocks.

BEDIT edits blocks.

ADCENTER (DesignCenter) provides access to content in other drawings.

TOOLPALETTES provide access to frequently-used content.

HATCH and **-HATCH** place hatch patterns and colored fills.

GRADIENT places gradient fills.

HATCHEDIT edits associative hatch patterns and fills.

BOUNDARY and **-BOUNDARY** create single regions from disparate areas.

continued...

NEW TO AUTOCAD 2008 IN THIS CHAPTER

- **HATCH** and **QSELECT** now support the annotative scaling property.
- Blocks now support orientation and annotative scaling.

SELECT selects objects by their location.

QSELECT selects objects by their properties.

PICKBOX changes the size of the pick cursor.

DDSELECT controls selection options.

FILTER selects objects based on their properties and location.

GROUP creates selectable groups of objects.

DRAWORDER and **TEXTTOFRONT** control the order in which objects are displayed.

FINDING THE COMMANDS

On the Dashboard's **2D Draw** panel:

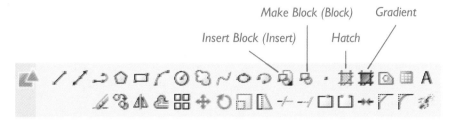

On the **Standard Annotation** toolbar:

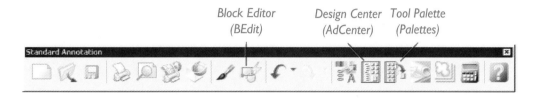

On the **Draw** toolbar:

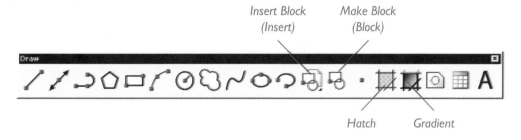

On the **Modify II** and **Draw Order** toolbars:

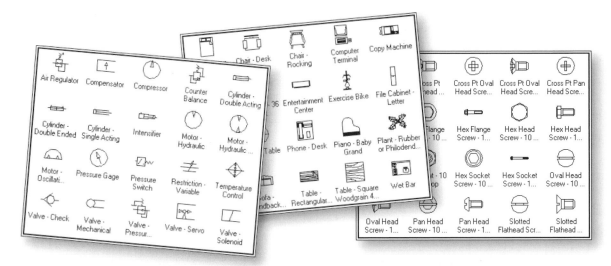 BLOCK AND [INSERT] INSERT

The **BLOCK** command creates symbols that AutoCAD calls "blocks." Its complement is the **INSERT** command, which places blocks in drawings. Using blocks in drawings gives you several distinct advantages.

Entire libraries of blocks can be used over and over again for repetitive details. By using pre-drawn blocks, you reduce the amount of drafting to complete drawings. AutoCAD includes more than a dozen libraries of hundreds of blocks in the *DesignCenter* and *Dynamic Blocks* folders, found in *AutoCAD 2008\Sample*. (See Appendix F for the complete listing.)

Libraries of Standard Blocks (*Sample\DesignCenter*)

Analog Integrated Circuits	Basic Electronics
CMOS Integrated Circuits	Electrical Power
Fasteners - Metric	Fasteners - US
Home - Space Planner	House Designer
HVAC - Heating Ventilation Air Conditioning	Hydraulic - Pneumatic
Kitchens	Landscaping
Pipe Fittings	Plant Process
	Welding

Libraries of Dynamic Blocks (*Sample\Dynamic Blocks*)

Annotation	Architectural
Civil	Electrical
Mechanical	Structural

Blocks and nested blocks are excellent for building drawings from "pieces." (A *nested block* is a block placed within another block.) Several blocks require less space than several copies of the same objects, because AutoCAD stores only the information for the original block *definition*.

 Note: It is more efficient to use blocks than to copy the same objects over and over. The **COPY** command makes a complete copy each time you use it; in contrast, AutoCAD creates a single definition of a block, and then points to it when you make "copies" with the **INSERT** command. This reduces the drawing's file size and improves the display time.

While standard blocks are static, *dynamic blocks* can be modified interactively according to the designer's intentions. Dynamic blocks are easily stretched, rotated, mirrored, and aligned to other objects — if defined that way. In addition, *lookup tables* allow one block to display itself in multiple forms. (Manipulating dynamic blocks is discussed in Chapter 9, "Direct Editing," while the construction of dynamic blocks is covered in *Using AutoCAD: Advanced*.)

Blocks can be used with *attributes*, which are text records that can be visible or invisible. (Attributes can only be attached to blocks.) The data from attributes can be exported to tables in drawings, and to spreadsheet programs for further analysis. This is useful in facilities management, for example, where multiple occurrences of desks, chairs, and computers are found. Each is stored as a block, and has attributes, such as the person's name and telephone number, associated with it. (Attributes are detailed in *Using AutoCAD: Advanced*.)

AutoCAD stores blocks in the drawing in which they were made. You can share them with other drawings through DesignCenter, or by exporting blocks as *.dwg* drawing files to disk with the WBLOCK command; AutoCAD can import any *.dwg* drawing file as a block.

Editing the Content of Blocks

Because a block is a group of objects combined into a single object, you move, scale, copy, and erase a block as though it were a single object. Indeed, blocks are so "tight" that it is difficult to edit their individual members. The three ways to edit the objects in blocks are, in historical order:

- **EXPLODE** command breaks blocks into their individual parts; when you are done editing the parts, you can use the **BLOCK** command to recombine them.

- **REFEDIT** command "checks out" blocks, and then the **REFSET** command edits them by adding and removing objects; when done, the **REFCLOSE** command adds the changes to the block definitions.

- **BEDIT** command opens blocks in the Block Editor, where the objects can be edited; when done, the **BSAVE** command saves the changes to the original blocks. The **BSAVEAS** command saves the changes to new blocks, preserving the old ones.

If you need to work with groups of objects where members are more easily edited, consider using the GROUP command discussed later in this chapter.

TUTORIAL: CREATING BLOCKS

There are three basic pieces of information AutoCAD needs before it creates blocks: (1) a name for the block, so that you can later identify it; (2) x, y, z coordinates for the insertion point, so that AutoCAD knows where to place the block; and (3) the objects that make up the block.

With that in mind, let's create blocks with the **BLOCK** command. (Dynamic blocks are created with the BEDIT command as described in *Using AutoCAD: Advanced*.)

1. To create blocks, start the **BLOCK** command with one of these methods:
   - From the **Draw** menu, choose **Block**, and then **Make**.

   - On the Draw toolbar, pick the **Block** button.

   - In the Dashboard's 2D Draw panel, choose the **Make Block** button.

   - Or, at the 'Command:' prompt, enter the **block** command:

     Command: **block** *(Press* ENTER.*)*

   - Alternatively, enter the aliases **b**, **bmod**, or **bmake** (one of the command's older names, short for "block make") at the keyboard.

Notice the **Block Definition** dialog box (redesigned in AutoCAD 2008):

2. Give the block a name in the **Name** field. You can enter nearly anything, from a single letter to a word 255 characters long. Not permitted are the following characters: < > / \ " ' : ; ? * | = and space.

When you try to give the block a name that already exists in the current drawing, AutoCAD warns you:

> Blockname is already defined. Do you want to redefine it?

Most times you enter **No**, and then change the name.

Notes: Sometimes you deliberately want to redefine an existing block. Perhaps you made an error in creating the block or left out attribute definitions. In this case, answer **Yes** to the "Do you want to redefine it?" question.

If you need help seeing the names of blocks already defined in the drawing, click the down arrow in the **Name** field. AutoCAD displays the names of blocks in the current drawing.

Block names that begin with * (asterisk), such as *X20, are created by AutoCAD. These are called *anonymous blocks*.

3. For the Base Point, enter coordinates in the **X, Y**, and **Z** fields, or else choose the **Pick Point** button to pick the insertion point in the drawing.

When you choose the **Pick Point** button, the dialog box disappears temporarily, and AutoCAD prompts you,

> Specify insertion base point:

I recommend using an object snap mode to make the pick accurate, such as ENDpoint for the end of a line or CENter for the center of a circle or arc.

After you pick the insertion point, the dialog box reappears and AutoCAD fills in the **X, Y**, and **Z** fields. Normally, the **Z** value is 0, unless you have set the elevation or select a point on a 3D object.

Note: The base point defines the *insertion point*, which is where the block is later placed by the **INSERT** command.

The base point is usually at the lower left corner of the block for convenience, but sometimes is located elsewhere, such as the center of the block. In the figure below, the logical location for the sprinkler head's base point is the center of the head. That makes it easy to insert and attach to existing objects, such as lines representing the irrigation piping.

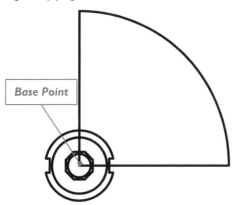

Base Point

4. To select the objects that make up the block, choose the **Select Objects** button. The dialog box disappears and AutoCAD prompts you:

    Select objects: *(Select one or more objects.)*

    Select objects: *(Press ENTER to return to the dialog box.)*

    You can use any method of object selection, such as Window or Fence. After you finish selecting objects, press **ENTER** and the dialog box reappears. AutoCAD reports the number of objects you selected.

5. Click **OK**, and AutoCAD creates the block — even though it's hard to tell it happened. You access it later with the **INSERT** command.

Suggestions for Designing Blocks

When creating blocks, it is best to make them the size you intend for use. For example, office desks are commonly 24" x 36", so it makes sense to draw them that size.

Sometimes, however, you don't what size will be used. For example, the block of a tree symbol could be inserted at 2' or 3' or 5' or whatever size you need. When you don't know how large the inserted block will be, draw the block at *unit size*. This means that the block fits inside a 1-unit square:

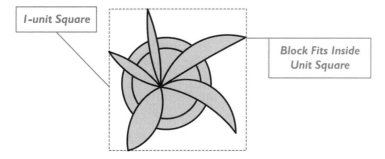

I-unit Square

Block Fits Inside Unit Square

When it comes time to insert the block, the scale factor option sizes the block correctly and automatically.

CREATING BLOCKS: ADDITIONAL METHODS

The Block Definition dialog box contains a number of options that help you create blocks.

Objects

The Objects section of the dialog box has these settings:

- ☐ **Specify On-screen** prompts you to select objects after you click the **OK** button to close the dialog box (new to AutoCAD 2008).

- **Select** objects clears the dialog box, and then prompts you to select objects in the drawing. After pressing **ENTER**, the dialog box returns.

- **Quick Select** displays the Quick Select dialog box, where you can select objects based on their properties; see the **QSELECT** command later in this chapter.

- **Retain** keeps the objects making up the block in the drawing as individual objects, as well as stores them in the new block definition.

- **Convert to Block** erases the objects, and replaces them with the block in the same position in the drawing (default).

- **Delete** erases the objects, while storing the block definition in the drawing. The block seems to disappear, which can freak out new AutoCAD users: "Wha' happened to my drawing!!??" You can insert the block with the **INSERT** command.

The objects selected note reports the number of objects selected for the block. You cannot create a block with no objects.

Behavior

The Behavior section determines how the block reacts during and after insertion. The first two options are new to AutoCAD 2008.

- ☐ **Annotative** gives the block annotative scaling, which means it is displayed only when the viewport scaling matches the block's annotative scale factor. (Clicking the i icon displays online help for this feature.)

- ☐ **Match Block Orientation to Layout** forces the block to remain "upright," even when rotated with the **ROTATE** command or through layout rotation for plotting.

In the figure below, the boxed "Arbour" text is a block, with orientation turned on. When all objects in the drawing are rotated, the block remains in place.

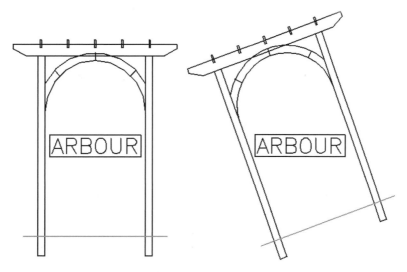

☐ **Scale Uniformly** fixes the scale factor ratio at 1:1:1. This means the block cannot be inserted with different scale factors in the x, y, and/or z directions. It's a good idea to turn on this option when you need the block kept proportional.

☑ **Allow Exploding** allows the block to be reduced to its constituent parts with the **EXPLODE** command. Exploding causes block insertions to sever their link with block definitions, and to lose attribute data, if any. To prevent the block from being exploded, turn off this option.

Settings

The Settings section sets settings.

- **Block Unit** specifies units for the block. Usually, AutoCAD makes one drawing unit equal to the current unit.

   When you drag blocks from DesignCenter or *i-Drop*-enabled Web sites into drawings, AutoCAD automatically scales the block. (i-Drop is Autodesk technology that allows you to drag blocks from Web sites directly into drawings.)

   Select a drawing unit from the droplist:

Inches	Nanometers	Feet	Microns
Miles	Decimeters	Millimeters	Decameters
Centimeters	Hectometers	Meters	Gigameters
Kilometers	Astronomical Units	Microinches	Light Years
Mils	Parsecs	Yards	Angstroms
Unitless			

   (1 parsec = 3.26 light years; 1 astronomical unit = average distance between Earth and Sun; 1 angstrom = typical size of an atom, or 1 ten-billionth of a meter; 1 microinch = 1 millionth of an inch.)

- **Hyperlink** displays the Insert Hyperlink dialog box so that you can attach a hyperlink to the block. (Hyperlinks are discussed in *Using AutoCAD: Advanced.*)

☐ **Open in Block Editor** opens blocks in the Block Editor after you click **OK**. This allows you to edit blocks further, or to turn them into a dynamic blocks.

Description

The description text is displayed by the **INSERT** and **DESIGNCENTER** commands.

- **Description** provides room for you to include a long (or short) description of the block. This item is completely optional.

-Block Command

AutoCAD also provides the **-BLOCK** command, which displays its prompts at the command line. This version of the command is meant for scripts and for power users who prefer to use the keyboard (like myself). The command displays the following prompts:

Command: **-block**

Enter block name or [?]: *(Enter a name, or type* **?** *to list the names of blocks already defined in the drawing.)*

Specify insertion base point or [Annotative]: *(Pick a point, or enter x, y coordinates; type* **A** *to specify the annotative scaling and orientation properties.)*

Select objects: *(Select the objects that make up the block.)*

Select objects: *(Press* ENTER *to end the command.)*

The objects disappear from the drawing, and AutoCAD creates the block definition. To bring them back, use the **OOPS** command:

Command: **oops**

INSERTING BLOCKS

After you define blocks with the **BLOCK** command or copy them from other drawings, you can place them with the **INSERT**, **DESIGNCENTER**, or **TOOLPALETTE** commands.

INSERT needs at least two pieces of information to place blocks: (1) the name, and (2) the insertion point. In addition, you can optionally specify the scale factors, normally 1.0, the rotation angle, usually 0 degrees, and other options. (**DESIGNCENTER** and **TOOLPALETTE** need only the insertion point.)

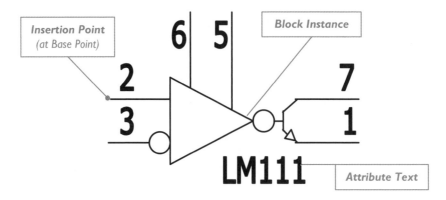

The *insertion point* is the x, y-coordinate in the drawing where the block is inserted, as illustrated below. More specifically, the block is inserted at its base point, which was defined early during the block's creation. The base point becomes the insertion point.

After blocks are inserted, you can use the INSertion object snap to snap to their insertion points.

Note: The inserted block resides on the layer that is current when the block is placed. Objects within the block retain the color, linetype, and so on of their original layer (i.e., the layer they were on before they were made part of the block). The properties are called "ByBlock."

There is one exception: if objects were on layer 0, then they take on the properties of layer in which they are inserted.

TUTORIAL: PLACING BLOCKS

1. To place blocks in drawings with the **INSERT** command, start the command with one of these methods:
   - From the **Insert** menu, choose **Block**.
   - On the Draw or Insert toolbars, pick the **Insert** button.
   - In the Dashboard's 2D Draw panel, choose the **Insert** button.
   - Or, at the 'Command:' prompt, enter the **insert** command:

     Command: **insert** *(Press ENTER.)*
   - Alternatively, enter the aliases **i**, **ddinsert** (the command's old name), or **inserturl** at the keyboard.

   Notice the **Insert** dialog box.

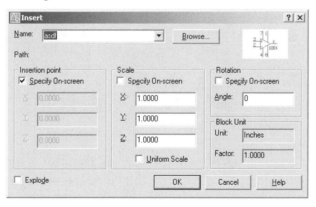

2. In order to insert a block, you select it by its name:
   - If the block resides in the drawing, enter its name in the **Name** field.
   - If you don't know the name of the block, click the arrow at the end of the **Name** field, and AutoCAD displays a list of all blocks defined in the drawing. Select one of them.

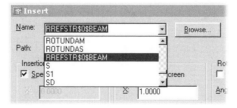

   - If the block is not in the drawing, click **Browse** to select any *.dwg* drawing file to insert as a block. (Alternatively, you can use DesignCenter to search for and insert blocks.)

3. To specify the insertion point, you can enter the x, y, and z coordinates in the dialog box or in the drawing. The **Specify on Screen** option determines where this takes place:

☐ Specify the insertion point in the **X**, **Y**, and **Z** text entry boxes.

☑ Specify the insertion point in the drawing. After clicking the **OK** button, AutoCAD prompts you:

Specify insertion point or [Scale/X/Y/Z/Rotate]: *(Pick a point, enter an option, or type the x, y, z coordinates.)*

4. After you specify the insertion point, AutoCAD places the block in the drawing. Technically, this is called "inserting an instance of the block definition."

INSERTING BLOCKS: ADDITIONAL METHODS

The Insert dialog box contains several options:

• **Scale** changes the size of the block instance.

• **Rotation** rotates the block instance.

• **Explode** inserts the block with its original components, not as an instance.

In addition, the command-line version of the command lists a number of options, including some that are hidden from you:

Insert Option	Meaning
Basepoint	Relocates the block's base point.
Scale	Specifies a single scale factor for the x, y, and z-directions.
X	Scales the block in the x-direction.
Y	Scales the block in the y-direction.
Z	Scales the block in the z-direction.
Rotate	Rotates the block in the counterclockwise direction.
The following options are hidden:	
PScale	Sets the preview scale factor for the x, y, and z-directions.
PX	Sets the preview scale factor and insertion scale for the x-direction.
PY	Sets the preview scale factor and insertion scale for the y-direction.
PZ	Sets the preview scale factor and insertion scale for the z-direction.
PRotate	Sets the preview rotation angle and insertion scale factor.

Basepoint

The **Basepoint** option allows you to specify a new base point, the point at which the block is inserted. This can make it easier inserting blocks relative to other objects, such as placing a bathtub symbol into the corner of the bathroom. AutoCAD temporarily places the block in the drawing, and then prompts you:

Specify base point: *(Pick a point to relocate the base point.)*

Pick a point, and AutoCAD proceeds with the 'Insertion point:' prompt, allowing you to place the block in its intended location.

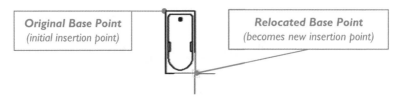

Original Base Point
(initial insertion point)

Relocated Base Point
(becomes new insertion point)

Changing the base point affects only this block insertion; it does not affect the block reference's base point. In other words, you need to reuse this option every time you insert the block and need a different base point.

Scale

AutoCAD normally uses a *unit* scale factor for the inserted block. ("Unit" means that the scale factor is 1.0 so that the block is inserted at its original size.) There are times, however, when you might want the block to be a different size.

When the scale factor is more than 1.0, the block is made larger; when the scale factor is less than 1.0, the block is smaller.

The three scale factors — X, Y, and Z — can be the same or different, negative or positive. (When the **Scale Uniformly** option is turned on during block creation, only the X Scale factor is available.) When the scale factors are different from each other, the block is stretched; when negative, the block is mirrored, including its text; when positive, the block is inserted the right way around.

The illustration below shows the effect of some different scale factors on inserted blocks:

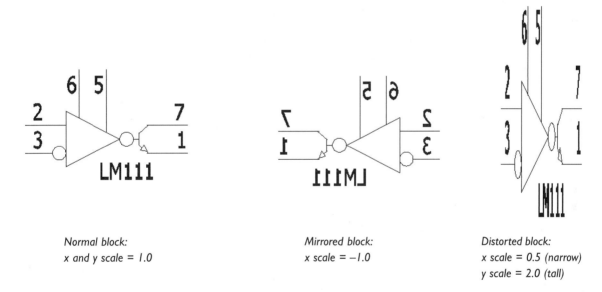

Normal block:
x and y scale = 1.0

Mirrored block:
x scale = −1.0

Distorted block:
x scale = 0.5 (narrow)
y scale = 2.0 (tall)

The **Specify On-screen** option determines whether you specify the scale factor in the dialog box or in the drawing:

☐ Specify the scale factor in the **X**, **Y**, and **Z** text entry boxes.

☑ Specify the factor in the drawing. After you click the Insert dialog box's **OK** button, AutoCAD will prompt you:

> Enter X scale factor, specify opposite corner, or [Corner/XYZ] <1>: *(Press ENTER to accept the default factor, enter a different x-scale factor, pick a point, or specify an option.)*

> Enter Y scale factor <use X scale factor>: *(Press ENTER to make the y scale factor the same as x, or enter the y-scale factor.)*

AutoCAD presents a large number of options at the command line. Here's what they mean:

Scale Option	Meaning
X scale factor	Scales the block in the x-direction.
Opposite corner	Specifies a rectangle that defines the x and y scale factors; insertion point is the first corner.
Corner	Same effect as the above option.
XYZ	Scales the block independently in the x, y, and z-directions.
Y scale factor	Scales the block in the y-direction.
Use X scale factor	Makes the y scale factor equal to the x scale factor.

The **Uniform Scale** option makes the y and z scale factors the same as the x factor.

Rotation Angle

You can have AutoCAD rotate the block about its insertion point relative to the original orientation of the block. The **Specify On-screen** option determines when the specification takes place:

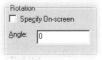

☐ Specify the rotation angle in the **Angle** text entry box.

☑ Specify the angle in the drawing. After clicking **OK**, AutoCAD prompts you:

Specify rotation angle <0>: *(Enter the rotation angle, or pick two points to indicate the angle.)*

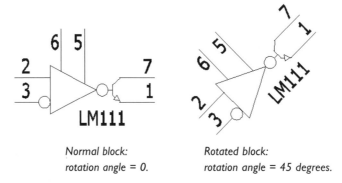

Normal block:
rotation angle = 0.

Rotated block:
rotation angle = 45 degrees.

If you choose to indicate the angle, move the cursor. AutoCAD ghosts a rubber band cursor between the previously-set insertion point and the cursor. Move the cursor until the rubber-banded line shows the desired angle, and then click. The distance between the two points is irrelevant. The angle, shown by rubber-banded line between the insertion point and the angle point, determines the angle of insertion.

Explode

When you turn on the **Explode** option, AutoCAD inserts the block's constituent parts in the drawing — the lines, arcs, and text that make up the block definition.

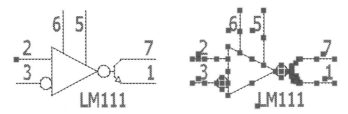

Normal block: a single grip.

Exploded block: grips for every constituent part.

This option is not available when the **Allow Exploding** option was turned off during block creation. Note that AutoCAD limits you to specifying a single scale factor for exploded blocks, which applies equally to the x, y, and z directions.

Preview

The Preview window shows you what the block looks like.

Icons indicate special blocks:

 dynamic blocks.

annotatively-scaled blocks (new to AutoCAD 2008).

DESIGNCENTER AND TOOL PALETTES

The drawback to the **INSERT** command is that it provides previews of just one block at a time. Sometimes, it is easier to pick out blocks from a large group by their look, rather than by name. Also, you may not be 100% sure of the block's location: which drawing or folder is it stored in? For these reasons, you may find it easier to use Design Center or Tool palettes to insert blocks, which provide the added convenience of drag'n drop insertion.

(The following sections describe working just with blocks and Design Center and Tool palettes, with additional details later in the chapter.)

Inserting Blocks with DesignCenter

The advantage of the DesignCenter is that it locates and displays blocks stored in the current drawing, as well as in drawings on your computer, on the local network, and the Internet.

After finding the block, simply drag its icon into your drawing; there are no prompts to answer. The place where you let go of the mouse button becomes the insertion point. Using snaps and object snaps helps you place blocks accurately.

When you want more control over placing block in drawings, right-click the icon in DesignCenter, and then select **Insert Block** from the shortcut menu. AutoCAD displays the familiar Insert dialog box, which lets you change any option you wish.

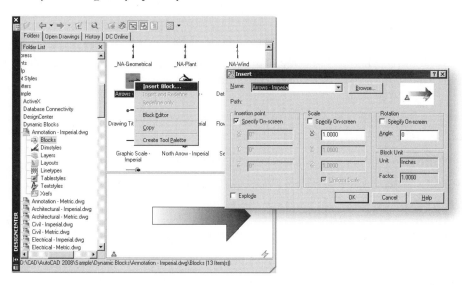

Inserting Blocks with Tool Palettes

The advantage of the Tool palette is that it can contain the blocks you use the most. However, it does not search for blocks as does DesignCenter; you add blocks to the palette by dragging them from the drawing or DesignCenter. Once on the palette, the blocks are available to all drawings.

As with DesignCenter, you add blocks to drawings by dragging their icons from the palette and into the drawing; there are no prompts to answer.

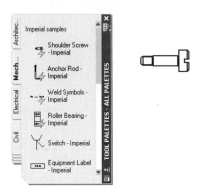

When you want to control block placement, right-click the icon, and then select **Properties** from the shortcut menu. This displays the Tool Properties dialog box, where you can change the scale factor, rotation angle, and so on. The changes made are semipermanent: they affect future insertions of the block until you again change the properties.

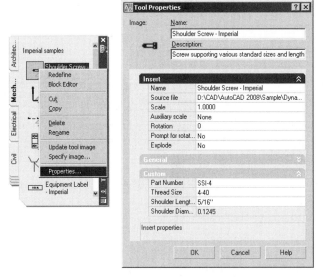

(A drawback to Tool palettes is that they display images of blocks, even when a block no longer exists. Dragging the icon of the non-existant block into the drawing results in nothing.)

CREATING AND INSERTING BLOCKS: ADDITIONAL METHODS

AutoCAD has several commands that are useful for working with blocks:

- **-INSERT** command places blocks using the command line.

- **BASE** command specifies the base point (origin) of the drawing.

- **WBLOCK** command exports blocks as *.dwg* files on disk.

- **MINSERT** command inserts blocks as rectangular arrays.

- **BLOCKICON** command creates icons of blocks.

Let's look at each command.

-Insert Command

AutoCAD also provides the **-INSERT** command, which displays its prompts at the command line. This version of the command is meant for scripts and for power users who prefer to use the keyboard. The prompts displayed by this command vary depending on the properties of the block being inserted. A typical variation looks like this:

> Command: **-insert**
>
> Enter block name or [?]: *(Enter a name, or type* **?** *to list the names of blocks already defined in the drawing.)*
>
> Units: Inches
>
> Specify insertion point or
>
> [Basepoint/Scale/X/Y/Z/Rotate]: *(Pick a point, or enter an option.)*
>
> Enter X scale factor, specify opposite corner, or [Corner/XYZ] <1>: *(Press* **ENTER**, *or enter an option).*
>
> Enter Y scale factor <use X scale factor>: *(Press* **ENTER**, *or enter a scale factor).*
>
> Specify rotation angle <0>: *(Press* **ENTER**, *or enter an angle).*

Unlike the Insert dialog box, which displays all available options, the **-INSERT** command "hides" many options. Here's how to access them:

Inserting Exploded Blocks

To insert an exploded block, prefix the block's name with an asterisk (*****):

> Enter block name or [?]: ***blockname**

Redefining Blocks

To redefine a block definition, suffix the block's name with an equals sign (**=**):

> Enter block name or [?]: **blockname=**

Be careful, because this forces AutoCAD to replace the existing block definition with the new one. The new definition will be applied to all block insertions of the same name in this drawing. AutoCAD then asks:

> Block "blockname" already exists. Redefine it? [Yes/No] <No>: **y**

Replacing Blocks with Drawings

To replace a block definition with an external *.dwg* drawing file, use the equals sign, as follows:

> Enter block name or [?]: **blockname=filename.dwg**

Base

The **INSERT** command can insert other drawings as blocks into the current drawing. The inserted drawing is normally inserted with coordinates 0,0 as its base point. If you wish to change the base point, use the **BASE** command on the drawing *before* inserting it.

> Command: **base**
>
> Specify base point <default>: *(Pick the point for the new base point.)*

Save the drawing to save the new base point, and then insert it in the other drawing.

Note: You can drag *.dwg* files from Windows Explorer into drawings. AutoCAD treats them differently, depending on where you drag them.

Drag files to the title bar — AutoCAD opens the *.dwg* files as drawings.
Drag files into the current viewport — AutoCAD inserts the *.dwg* files as blocks.

WBlock

The **WBLOCK** command is the complement to the **INSERT** command. Whereas **INSERT** inserts *.dwg* drawing files as blocks, **WBLOCK** saves all or part of the current drawing as another *.dwg* file on disk. This is an alternative to using the DesignCenter for sharing blocks.

- From the **File** menu, choose **Export**. In the **Export Data** dialog box, choose "Block (*.dwg)" from the **File of type** list box.

- At the 'Command:' prompt, enter the **wblock** command.

- Alternatively, enter the **w** alias at the keyboard.

    Command: **wblock** *(Press ENTER.)*

AutoCAD displays the Write Block dialog box, which looks similar to the **BLOCK** command's dialog box. The primary difference is in the **Source** area, where you choose what you want written to disk.

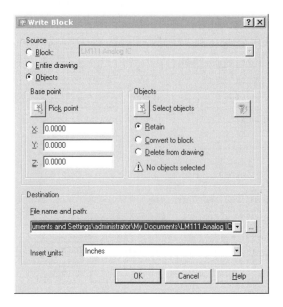

Source

Block saves blocks as *.dwg* files; the blocks must exist in the current drawing.

Entire Drawing saves the whole drawing with an exception: unused layers, blocks, styles, and dimension styles are not saved. This is a quick way to clean up a drawing.

Objects saves parts of the drawing; you specify the base point, and then select the objects.

Destination

In the **Destination** area, specify the file name and location (drive and folder.)

When you change the file name extension from *.dwg* to *.dxf*, AutoCAD saves the drawing or block in DXF format. DXF is short for "drawing interchange format."

(With AutoCAD 2004, Autodesk removed **WBLOCK**'s ability to save the drawing as an *.xml* file — extended markup language.

MInsert

The **MINSERT** (short for "multiple insert") command combines the **INSERT** command with the **Rectangular** option of the **ARRAY** command.

AutoCAD doesn't use a dialog box for this command, instead displaying all prompts at the command line. The sequence starts by issuing prompts in the same manner as the **-INSERT** command, followed by the prompts for constructing the rectangular array.

- At the 'Command:' prompt, enter the **minsert** command.

    Command: **minsert** *(Press ENTER.)*

AutoCAD displays prompts on the command lines:

> Enter block name or [?]: *(Enter a name, or type **?** to list the names of blocks in the drawing.)*
>
> Specify insertion point or [Scale/X/Y/Z/Rotate]: *(Pick a point.)*
>
> Enter X scale factor, specify opposite corner, or [Corner/XYZ] <1>: *(Press ENTER.)*
>
> Enter Y scale factor <use X scale factor>: *(Press ENTER.)*
>
> Specify rotation angle <0>: *(Press ENTER.)*
>
> Enter number of rows (---) <1>: *(Enter a number for the rows).*
>
> Enter number of columns (| | |) <1>: *(Enter a number for the columns).*
>
> Enter distance between rows or specify unit cell (---): *(Enter a distance between rows, or pick two points that show the row and column distance.)*
>
> Specify distance between columns (| | |): *(Enter a distance between columns.)*

The blocked arrays created by this command have many of the same properties as blocks, with some exceptions.

Create arrays using **MINSERT** when you need to edit the array as a whole. For example, a seating arrangement consisting of several rows of chairs may need to be moved around for design purposes. Creating the arrangement with the **MINSERT** command allows you to move and rotate the seating as a whole.

The following qualities apply to multiple inserts:

- The array reacts to editing commands as if it were one block. You cannot edit the individual blocks making up the array. When you select one object to move or copy, the entire array is affected.

- Annotative blocks cannot be used with this command.

- Dynamic blocks can be array-inserted, but act like non-dynamic blocks; you cannot access their specialized grips.

- You cannot "minsert" the block as exploded. If you need to do this, use the **INSERT** command followed by the **ARRAY** command. Nor can you explode the block into individual objects following insertion.

- When the initial block is inserted with a rotation angle, the entire array is rotated around the insertion point of the initial block (see figure below). This creates an array in which the original object appears to have been inserted at a standard zero angle, with the entire array rotated.

The illustration below shows a block inserted at a 30-degree angle through the **MINSERT** command, with three rows and four columns.

History: The technical editor recalls that Autodesk once touted **MINSERT** as a way of making drawings uneditable: convert the entire drawing to a block, they suggested, and then minsert it. Users quickly found a workaround to make the drawing editable again: erase the minsert, reinsert the block, and then explode it.

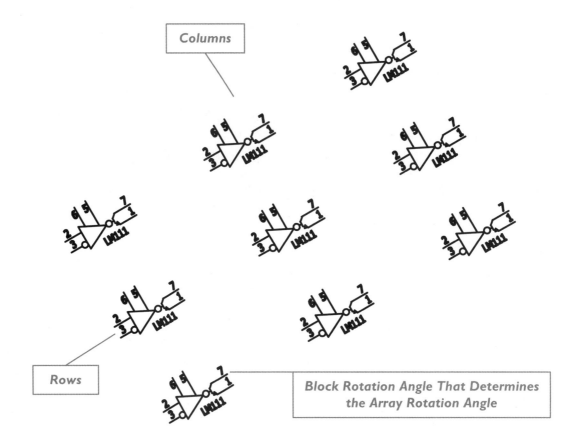

Columns

Rows

Block Rotation Angle That Determines the Array Rotation Angle

BlockIcon

The Block Definition dialog box automatically creates preview images of blocks, as illustrated below in DesignCenter.

When you have access to old blocks created without icons, you can use the **BLOCKICON** command to add icons to all blocks in the drawing.

> Command: **blockicon**
>
> Enter block names <*>: *(Press* **ENTER** *to iconize all blocks.)*
>
> *nnn* blocks updated.

After the **BLOCKICON** command finishes, the drawing looks no different, but the icons now show up in DesignCenter and the Insert dialog box.

REDEFINING BLOCKS

After you insert the same block many times in drawings, you may need to change them. This is done by redefining one block, which changes all inserted blocks of the same definition. This is an especially powerful feature: imagine being able to change 100 identical drawing parts in a single operation!

Here's how:

1. Explode one block that has been inserted. This action breaks the link to its block definition and reduces the block to its constituent parts.

2. Edit the exploded block.

3. Use the **BLOCK** command to convert the edited parts back into a block. Give this "new" block the same name. You may select the name from the **Name** drop list.

4. Click **OK**. AutoCAD displays the following dialog box:

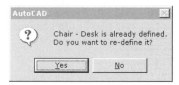

5. Click **Yes**. Notice that all other blocks of the same name change, including those inserted with **MINSERT**.

If the block is another drawing that was inserted whole, edit the original drawing, and then reinsert it as described above.

 BEDIT

The **BEDIT** command invokes the Block Editor for editing blocks and making them dynamic.

The Block Editor is a separate editing environment. Many (not all) AutoCAD commands are available for drawing and editing the blocks. In addition, the editor includes unique commands for constructing dynamic blocks. (The **-BEDIT** command prompts for the block's name at the command line before entering the Block Editor.)

The Block Editor performs these functions:

- Creates new blocks.

- Edits existing blocks.

- Adds dynamic actions to blocks.

Before entering the Block Editor, select an existing block, the entire drawing, or provide the name for a new block.

Once in the editor, there are several special commands for all block definitions (those operating only inside the Block Editor):

BATTORDER — specifies the order of attributes.

BSAVE — saves the edited block definition.

BSAVEAS — saves the edited block definition by another name.

BCLOSE — closes the Block Editor

A second group of commands that works only in the Block Editor is meant for dynamic blocks, and is described in *Using AutoCAD: Advanced*. In addition, these system variables affect the Block Editor:

BLOCKEDITLOCK — toggles the editing of blocks by preventing the Block Editor from opening.

BLOCKEDITOR — reports whether the BlockEditor is open.

(Additional system variables affect the display of dynamic blocks only while in the Block Editor.)

TUTORIAL: EDITING BLOCKS WITH THE BLOCK EDITOR

1. From the CD, open *bedit.dwg*, a drawing that contains a block.
2. To edit the block with the Block Editor, invoke the **BEDIT** command by one of these methods:
   - From the **Tools** menu, choose **Block Editor**.
   - On the Standard toolbar, pick the **Block Editor** button.
   - Or, at the 'Command:' prompt, enter the **bedit** command:

     Command: **bedit** *(Press ENTER.)*
   - Alternatively, enter the **be** alias at the keyboard.

 Note: There are two alternate, faster methods to start the Block Editor: selecting a block, and then entering the **BEDIT** command opens the block in the Block Editor. Even faster is double-clicking the block, which opens it in the Block Editor.

3. (Click **Cancel** if AutoCAD asks, "Do you want to see how dynamic blocks are created?") Before entering the Block Editor, AutoCAD displays the Edit Block Definition dialog box.

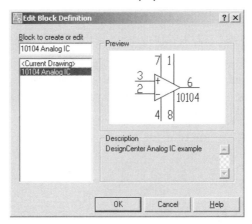

The dialog box provides you with three options:
- To edit an existing block, select its name from the list.
- To edit the entire drawing, select **<Current Drawing>**.
- To create a new block from scratch, enter a new name in the **Block to create or edit** field.

(When a yellow lightning icon appears in the corner of the preview window, this indicates a dynamic block is selected.)

4. Select a block to be edited. For this tutorial, select "10104 Analog IC," and then click **OK**. Notice the Block Editor. It looks much like regular AutoCAD, but the background color is pale yellow, and there is a new toolbar and palette specific to block editing. In addition, the UCS icon is shifted to the block's insertion point.

 Note: To change the background color of the Block Editor, use the **Tools | Options | Display | Color** menu selection. In the Window Element list, select "Block editor background", and then select a color.

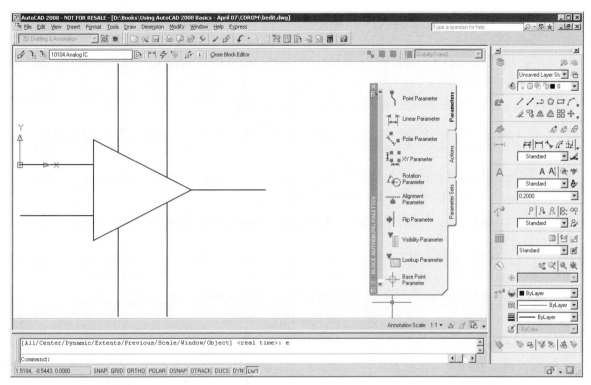

5. The palette is useful only for dynamic blocks, so close it with the **BAUTHORPALETTECLOSE** command, or click the **x** in its upper left corner.

   The toolbar contains some commands useful for block editing, as illustrated below; the remaining commands are specific to editing dynamic blocks, and are not discussed here.

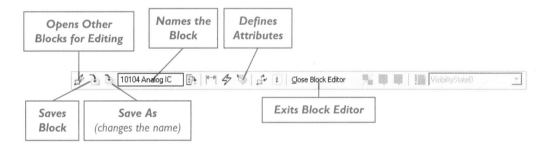

6. Make some changes to the block. For instance:

   Select the lines making up the triangle, and thicken them with the **Lineweights** droplist.

   Use the **STYLE** command to change the font to Arial.

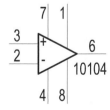

7. Use the **BSAVE** command to save the changes.

8. On the toolbar, click the **Close Block Editor** button.

   Back in AutoCAD's drawing editor, notice that previous insertions of the block take on the changes you made.

ADCENTER

The **ADCENTER** command (short for "AutoCAD Design Center") displays a palette for accessing content in the current and other drawings. DesignCenter provides these services:

- Previews the content of drawing file and raster images.

- Displays layers, linetypes, text styles, table styles, blocks, dimension styles, external references, and layouts (if any) of drawings.

- Accesses blocks from Autodesk's Web site.

- Copies content from one drawing to another.

- Views the content of folders and files on your computer and network.

TUTORIAL: DISPLAYING DESIGNCENTER

1. To display the DesignCenter palette, start the **ADCENTER** command by one of these methods:
   - From the **Tools** menu, choose **Palettes**, and then **DesignCenter**.

   - On the Standard toolbar, pick the **DesignCenter** button.

   - On the keyboard, press **CTRL+2**.

   - Or, at the 'Command:' prompt, enter the **adcenter** command:

      Command: **adcenter** *(Press ENTER.)*

   - Alternatively, enter the aliases **adc**, **dc**, **dcenter**, or **content** at the keyboard.

2. Notice the DesignCenter palette.

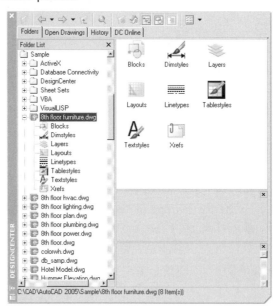

To turn off the DesignCenter palette, use the **ADCCLOSE** command, or press **CTRL+2**. As an alternative, click the **x** in the palette's upper corner.

The **ADCNAVIGATE** command specifies the initial path for DesignCenter. This determines which folder is displayed when DesignCenter is first opened:

Command: **adcnavigate**

Enter pathname <>: *(Enter a path, such as c:\drawings.)*

NAVIGATING DESIGNCENTER: ADDITIONAL METHODS

DesignCenter is a collapsable palette that can float independently of AutoCAD, be docked to the edge of the drawing area, or minimized to just its title bar.

Docking DesignCenter

To dock it, drag DesignCenter by its title bar to an edge of the AutoCAD window. Notice that DesignCenter attaches itself to the edge. The drawback to docking is that the drawing area becomes significantly smaller.

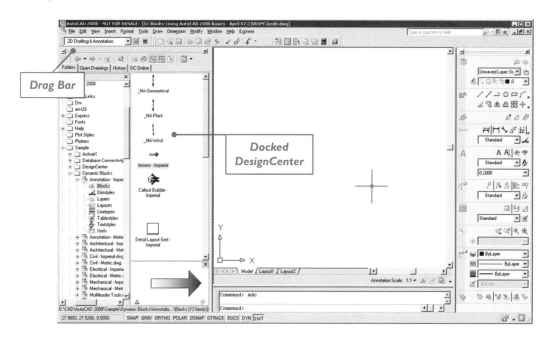

Floating DesignCenter

To float, drag the double-line (at the top of DesignCenter) away from the edge of the AutoCAD window. Notice that DesignCenter changes its shape.

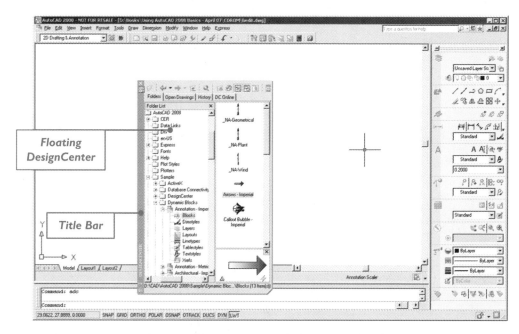

You can move DesignCenter anywhere on the screen, including outside of the AutoCAD window. If you have a two monitor system, consider dragging it to the second monitor.

Minimizing DesignCenter

To minimize, click the double-arrow **AutoHide** button on DesignCenter's title bar. Notice that DesignCenter collapses to just its title bar. To restore, pass the cursor over DesignCenter, and it springs back to its original size.

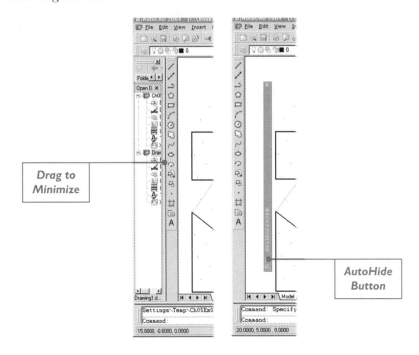

UNDERSTANDING DESIGNCENTER'S USER INTERFACE

The DesignCenter consists of many user interface elements, including toolbar, tabs, title bar, and several panes.

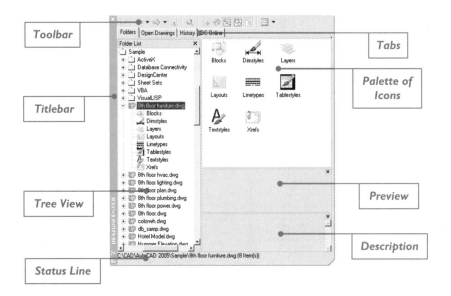

Tabs

DesignCenter employs tabs to display content in several contexts.

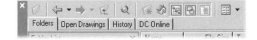

Folders

Click the **Folders** tab to see the names of folders (sub directories) and files on the drives of your computer and network — similar to Windows Explorer.

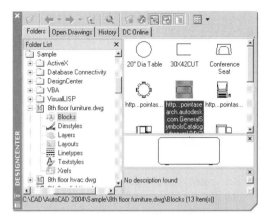

Open Drawings

The **Open Drawings** tab displays a list of the drawings currently open in AutoCAD. The named content of each drawing is also displayed: blocks, xrefs, layouts, layers, linetypes, table styles, dimension styles, and text styles.

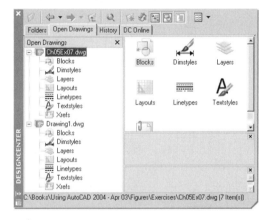

Custom Content

The **Custom Content** tab is present only if a drawing contains proxy data generated by registered (third-party) applications. Either the third-party application must be running or the application must be registered with DesignCenter.

History

The **History** tab displays the last 120 documents that the DesignCenter viewed. Double-click a file name to display its named content (if an AutoCAD drawing) or a thumbnail (if a raster image).

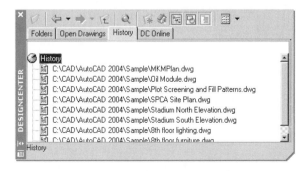

DC Online

The **DC Online** tab displays content available for downloading from Autodesk's Web site.

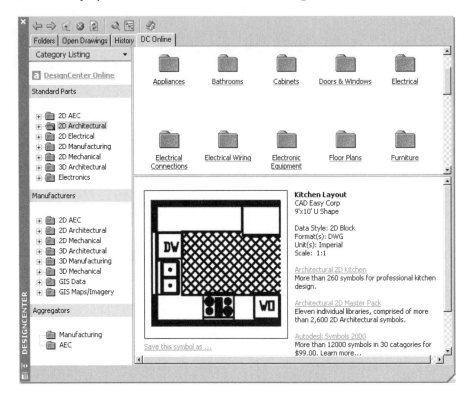

Toolbar

The toolbar changes, depending on the tab you select. For most tabs, the toolbar contains the following buttons:

Load — displays the Load dialog box for selecting content to load into DesignCenter. You can select content from files located on your computer, from other computers connected with a network, and from Web sites on the Internet.

Back — returns to the previous location in the history list. Click the arrow next to this button for the history list.

Forward — lists the next location in the history list.

Up — moves up the tree view by one level.

Search — displays the Search dialog box, so that you can locate content by specific criteria.

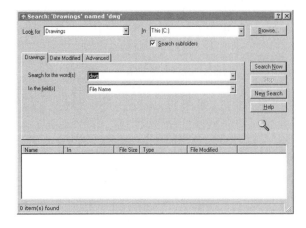

Favorites — displays the Windows' Favorites folder. Various software applications, including AutoCAD, automatically add files to the Favorites folder. AutoCAD, for example, provides several drawings with blocks that you may find useful for your own drawings. See Appendix I in this book.

Home — sends DesignCenter to the designated home folder, which is *C:\Program Files\AutoCAD 2008\ Sample\DesignCenter* by default.

Tree View Toggle — shows and hides the tree view (the left-hand pane).

Preview — shows and hides the preview pane.

Description button — shows and hides the description pane.

Views button — changes the display format of the content: large icons, small icons, list view (names only), and detail view (names and file information).

> **Note**: You can force DesignCenter to update its content by *refreshing*. To refresh the display, right-click in the content area, and then select **Refresh** from the shortcut menu.

DC Online

The **DC Online** tab (short for "DesignCenter Online") adds Web browser-like buttons to the toolbar:

Stop — stops the current transfer of data.

Reload — reloads the current page.

Palette

Right-click the palette (the area displaying the icons) to display a shortcut menu.

Copying Content

All palette shortcut menus have one action in common: select **Copy** to copy the content from one drawing to another. For example, to copy a layer name from one drawing to another:

1. Navigate to a *.dwg* drawing file with DesignCenter, and then open the *Layers* folder.
2. Right-click the layer you want to copy.
3. From the shortcut menu, select **Copy**.

4. In the drawing to which you want the layer name copied, choose **Paste** from the **Edit** menu. The layer and its properties are added to the drawing.

As an alternative, drag content, such as a group of layers, from DesignCenter into the drawing:

5. Select the first layer name.

6. Hold down the **SHIFT** key, and then select another layer name. Notice that DesignCenter highlights all layer names between your two selections. (To select layers in a nonconsecutive order, hold down the **CTRL** key, and then pick them.)

7. Drag the group of highlighted layer names into the drawing. AutoCAD copies only the layer names and properties, not objects located on the layers. Sometimes AutoCAD warns:

> Duplicate names ignored.

This means that if a layer of that name already exists, AutoCAD does not duplicate it.

Adding Content

The second item on the shortcut menu varies, depending on the content; many have **Add**. This option adds the content to the current drawing.

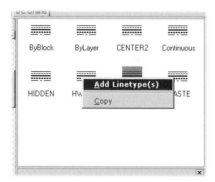

It may seem to you that the **Copy** and **Add** options are similar in purpose. The difference is that **Copy** copies the content to the Clipboard. From there you have to use the **PASTE** command to get the data into the drawing. (This allows you to move content to another CAD system, such as AutoCAD LT.) In contrast, **Add** immediately adds the content to the current drawing.

For example, to add a linetype to the current drawing, right-click the linetype, and then select **Add** from the shortcut menu. As before, AutoCAD warns that duplicate names are ignored.

Working with Blocks and Xrefs

The shortcut menus for blocks and external references are different:

Blocks. When right-clicking blocks, the shortcut menu includes the **Insert Block** option, which displays the **Insert** dialog box. It is identical to the **Insert** dialog box described later in this chapter.

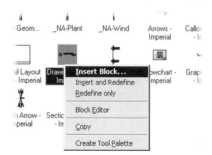

Insert and Redefine. When you try to add a layer or linetype that already exists in the drawing, AutoCAD ignores the command, and doesn't add it; blocks are handled differently, however. When you add a block from DesignCenter to a drawing that already contains a block of the same name, the **Insert and Redefine** and **Redefine Only** options update blocks in the drawing with the definition you are adding.

Block Editor. Selecting the **Block Editor** option opens the block in the Block Editor. This allows you to edit the block.

External References. Selecting the **Attach Xref** option displays the External Reference dialog box, which also is identical to AutoCAD's External Reference dialog box. See *Using AutoCAD: Advanced*.

Create Tool Palette

The **Create Tool Palette** option adds the selected object to the Tool palette.

Tree View

Right-click the tree view to display a shortcut menu that mimics the functions of some buttons on the toolbar: **Find** and **Favorites**.

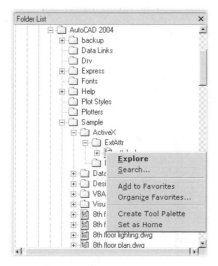

Explore has the same effect as clicking the small + symbol next to a folder or file name.

Search brings up the Search dialog box to hunt down content.

Add to Favorites adds the drawing or folder to the list of "favorites," which acts like a bookmark.

Organize Favorites displays the Windows folder showing the *Documents and Settings*\\<username>\\ *Autodesk**Favorites* folder.

Create Tool Palette adds *all* of the drawing's blocks to the Tool palette.

Set as Home is the same as using the ADCNAVIGATE command.

Description

Right-click the description area to display a cursor menu with commands suitable for copying the text to the Clipboard.

Title Bar

Double-click the title bar to dock DesignCenter on the right side of the AutoCAD window.

 TOOLPALETTES

The **TOOLPALETTES** command displays palettes that provide access to frequently-used content.

The Tool palettes is a centralized collection of blocks, hatches, objects, materials, and commands that you use often. It replaces the need to hunt through DesignCenter, the Hatch and Gradient dialog box, the Materials palette, and the Insert Block dialog box for specific blocks, hatch patterns, and materials. The Tool palette can also include programming routines and scripts.

TUTORIAL: DISPLAYING AND USING THE TOOLPALETTE

1. To display the Tool palette, use the **TOOLPALETTES** command with one of these methods:
   - From the **Tools** menu, choose **Palettes**, and the **Tool Palettes**.
   - On the Standard toolbar, pick the **Tool Palettes** button.
   - On the keyboard, press **CTRL+3**.
   - Or, at the 'Command:' prompt, enter the **toolpalettes** command:

     Command: **toolpalettes** *(Press* ENTER.*)*
   - Alternatively, enter the **tp** alias at the keyboard.

   Notice the Tool palette has numerous tabs. (The number and names of tabs depend on how the palette was last used.)

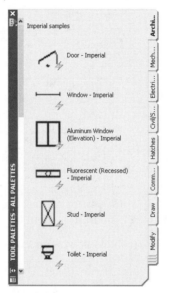

2. The *tabs* along the side access the palettes. Click a tab to view the associated palette. Each palette contains one or more *tools*. A tool can be a hatch pattern, a block, a material, an object, or a command.

3. To place a block (or any other object) in the drawing, drag it from the palette into the drawing. AutoCAD places the block at the point where you let go of the mouse button; use object snaps to place the block accurately. There are no prompts to answer; the block is scaled automatically to its preset units.

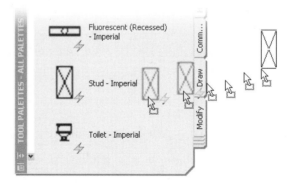

To turn off the Tool palette, press **CTRL+3** or enter the **TOOLPALETTESCLOSE** command. As an alternative, click the **x** in the palette's upper corner.

NAVIGATING THE TOOL PALETTES: ADDITIONAL METHODS

AutoCAD hides many of the Tool palette's options in shortcut menus, which differ depending on which part of the palette you right-click.

Right-click Commands: Tabs

Right-click the tab of the current palette (the one "on top") to see the following commands:

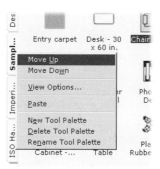

Command	Comment
Move Up	Moves up the palette by one position.
Move Down	Moves down the palette by one position.
View Options	Displays the View Options dialog box for changing the size and style of icons and text.
Paste	Pastes data from the Clipboard into the palette. This option is available only when the Clipboard contains appropriate data, such as an AutoCAD block. I can copy a block in DesignCenter for pasting in the tool palette, but not from AutoCAD itself, because in AutoCAD you never see the block *definition*; you only see insertions of it.
New Tool Palette	Creates a new blank palette, and prompts you to name the tab.
Delete Tool Palette	Warns against deleting the palette, and then deletes it when you answer **OK**.
Rename Tool Palette	Renames the tab.

View Options Dialog Box

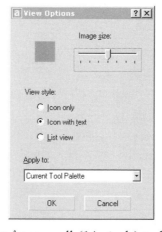

Image Size changes the size of icons from small (14 pixels) to large (54 pixels).

View Style changes the display of icons and text:

- **Icon only** displays icons only.
- **Icon with text** displays icons and text.

- **List view** displays small icons with text.

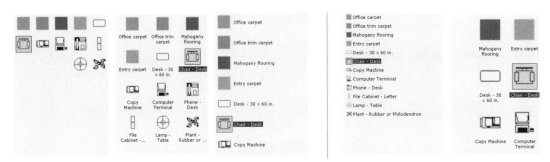

At left: Icons only, icons with text, and list view.
At right: Smallest images, and largest images.

Apply to determines whether the changes apply to the current palette (tab), or to all palettes (tabs).

Right-click Commands: Icons

Right-click an icon (called "tool" by Autodesk) to get the following shortcut menu:

Command	Comment
Redefine	(*Blocks only*) Redefines the block, if the definition has changed.
Block Editor	(*Blocks only*) Opens the block in the Block Editor.
Cut	Removes the tool and places it in the Clipboard.
Copy	Copies the tool to the Clipboard.
Delete	Removes the tool.
Rename	Renames the tool.
Update Tool Image	(*Blocks only*) Forces AutoCAD to regenerate images, should the block have changed.
Specify Image	Selects a raster image for the tool's icon.
Properties	Displays the Properties dialog box for changing the properties of the hatch and block tools. This palette has content similar to the one displayed by the PROPERTIES command.

Tool Properties Dialog Box

The Tool Properties dialog box displays everything AutoCAD knows about the object. Shown below is the list of properties for the selected block. For example, you can change the name of the block, its description, scale, and whether it is inserted exploded (**Explode**).

In particular, you can define a target layer through the **Layer** property: change "-- use current --" to another name, and the block is inserted on that layer, instead of the current one.

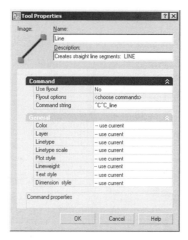

Properties shown in white can be changed; gray ones cannot. Those properties listed under **Custom** are specific to dynamic blocks.

Right-click Commands: Palettes

Right-click anywhere on an unused area of the palette — not on an icon or a tab — to get the following shortcut menu:

Command	Comment
Allow Docking	Determines whether the tool palette docks against the side of the AutoCAD window when it gets close enough.
Auto-Hide	Toggles whether the tool palette reduces itself to just the title bar when the cursor leaves the palette.
Transparency	Displays the Transparency dialog box to make the tool palette "see thru."
View Options	Displays the same dialog box as on the tab's shortcut menu.
Sort By	Sorts icons by name or type of object.
Paste	Pastes the tool; command is available only when a tool has been copied or cut to the Clipboard.
Add Text	Adds a line of text to describe a group of icons.
Add Separator	Adds a horizontal line to separate groups of icons visually.
New Palette	Adds a new blank palette.
Delete Palette	Removes the current palette.
Rename Palette	Changes the name of the current palette.
Customize Palettes	Displays the Customize dialog box with the Tool Palettes tab.
Customize Commands	Displays the Customize User Interface dialog box (CUI).

Title Bar

When you right-click the title bar, you can access all the palette groups and "covered" tabs:

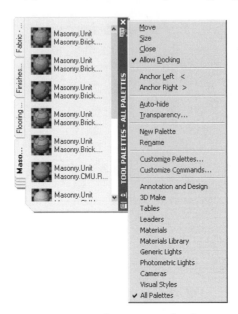

In the lower portion of the shortcut menu are the names of palette groups, such as Dynamic Blocks and Samples. If you cannot find a group, this is the place to look.

Transparency Dialog Box

You can change the transparency of tool palettes:

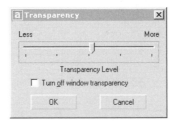

Moving the pointer toward **Less** makes the tool palettes more opaque, while moving it toward **More** makes them more transparent. You can turn off transparency by clicking the **Turn off window transparency** check box. (This option is unavailable when AutoCAD graphics are hardware-accelerated.)

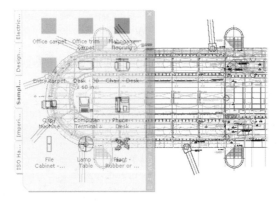

Transparency lets you see both the tool palettes and the drawing, as illustrated below. Whether or not you use transparency is a matter of personal preference. I find transparency makes it harder to see both the palette and the drawing, so I keep it turned off; others like being able to see both at once.

Customize

The **CUSTOMIZE** command organizes palettes by order and by group.

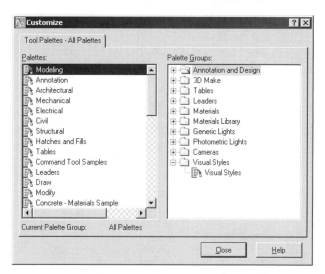

The left pane shows all palettes; the right pane allows you to segregate palettes into groups. When you make a group current, only the palettes in the group are displayed by the Tool palette. Right-click a group, and then select **Make Current** from the shortcut menu.

Drag palettes up and down to change their order. To create a new (blank) palette, right-click, and then select **New Palette** from the shortcut menu. To delete or rename a palette, right-click it and then select the appropriate command from the shortcut menu. Deleting palettes is permanent; there is no undo.

You can export selected palettes to share with other AutoCAD users. Right-click a palette name, and then select **Export**. Notice that AutoCAD prompts you to save the palette as a *.xtp* file. Use the **Import** button to read tool palettes created by others. The source drawing files must accompany blocks sent with this export and import process.

COMMAND TOOLS BY EXAMPLE

Command tools are made from geometric objects, such as lines, polylines, and dimensions — any object other than blocks, xrefs, materials, images, hatches, or gradient fills. You create tools by dragging objects onto the Tool palette. Autodesk calls this feature "tools by example," because the tools have the same properties as the original objects, including layer, color, and lineweight.

(When you drag the command tool from the Tool palette into the drawing, AutoCAD executes the command associated with the tool.)

To create command tools, drag objects onto the palette. Follow these steps:

1. Select the object; notice the handles.

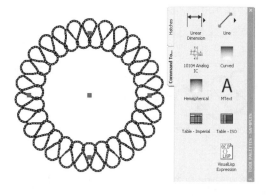

2. Hold down the **right** mouse button; notice that the handles disappear.

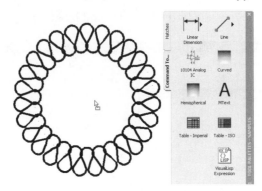

3. Drag the object onto the palette. The thick line helps you position the tool.

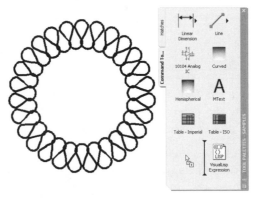

Once the object is on the palette, AutoCAD automatically adds a flyout that shows all other drawing commands. Flyouts are created for commands that create objects and dimensions only. The flyout lets you place other objects on the same layer.

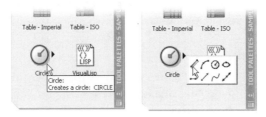

Left: *The arrowhead (▶) indicating the Circle tool has a flyout.*
Right: *Clicking the arrowhead to display the flyout of additional tools.*

4. Select an item from the flyout, such as the line. Notice that the icon changes to Line.
5. Click the **Line** icon. Notice that the LINE command appears at the 'Command:' prompt:

   Command: _line Specify first point: *(Pick a point.)*

   Specify next point or [Undo]: *(Pick a point, and so on.)*

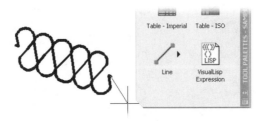

Notice that the line has the same properties as the circle you dragged onto the palette.

In addition to storing objects as tools, palettes remember the object's properties, such as layer, color, and linetype. The tools have the same properties as the objects selected from drawings, including the creating object's layers if they do not exist in the drawing.

To change their properties, select one or more tool icons. (To select more than one, hold down the CTRL key while selecting them.) Right-click them, and then select **Properties** from the shortcut menu.

The Tool Properties dialog box has a number of commands that are accessed through shortcut menus:

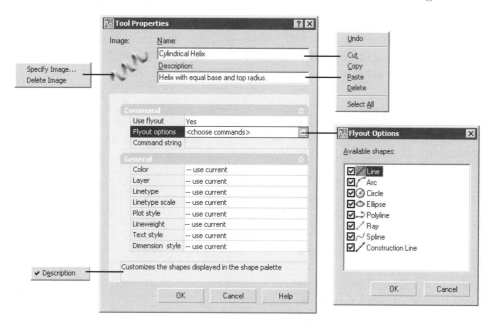

Changing the Image

AutoCAD automatically includes an image appropriate to the object. To change the image, right click the icon, and then select **Specify Image**. In the Specify Image File dialog box, select a bitmap (*.bmp*), JPEG (*.jpg*), PNG (*.png*), GIF (*.gif*), or TIFF (*.tif*) file, and then click **Open**. I have found that the size does not matter: AutoCAD resizes the image to fit.

You cannot change the images for hatches, tables, or blocks.

Changing the Flyout (Objects and Commands)

In most cases, AutoCAD automatically adds the flyout. If it doesn't, change the **Use Flyout** option from **No** to **Yes**.

The flyout consists of the objects illustrated by the figure above. Click **Flyout Options** to reduce the objects displayed on the flyout. You cannot add flyouts to hatches, tables, materials, or blocks.

Auxiliary Scale (Blocks and Hatches)

The **Auxiliary Scale** option applies to blocks and hatches. This is a scale factor that multiplies the dimension scale (as set by the DIMSCALE system variable) and the plot scale (as set by the PAGESETUP command) as follows:

Auxiliary Scale	Scale Factor is...
None	Unchanged.
Dimscale	Multiplied by the value stored in DIMSCALE sysvar.
Plot Scale	Multiplied by the value set by the PAGESETUP command's Plot Scale factor.

Prompt for Rotation (Blocks)

The **Prompt for Rotation** option prompts you for the rotation angle when you drag blocks into the drawing, when on. When off, it does not; blocks are placed at 0 degrees rotation.

ADDITIONAL CONTENT

Tool palettes also handle additional content, as described here.

Drawings and Images

In addition to dragging objects from drawings onto palettes, you can also drag drawing files (*.dwg*) and raster image files (such as *.jpg* and *.tif*) onto palettes from Windows Explorer. Drawing files are placed as blocks, while raster files are placed as images.

Command Tools

Commands, macros, AutoLISP and Visual LISP expressions, VBA macros, and scripts can be placed on palettes. The process, however, is not as simple as dragging and dropping objects. For programs and macros, you need to follow this indirect method:

1. Place a geometric object (tool) on a palette.
2. Right-click the tool, and then select **Properties** from the shortcut menu.
3. If necessary, turn off **Use Flyout** (set to **No**).
4. In the **Command String** area, replace the command with the programming code or macro. You can either type the new code, or copy and paste it from another source.

To change the icon for programs like this: (1) right-click the image; (2) from the shortcut menu, select **Specify Image**; (3) select another image.

Sharing Tool Palettes

In addition to dragging objects from DesignCenter into drawings, you can also drag them onto the Tool Palettes. Here is a summary of the dragging operations you can perform:

- From DesignCenter onto the Tool Palettes (but not from the Tool Palettes onto DesignCenter).
- From the Tool Palettes onto the drawing, and in AutoCAD from the drawing onto the Tool Palettes.
- From DesignCenter onto the drawing (but not from the drawing onto the DesignCenter).

 HATCH

The HATCH command places hatch and fill patterns in drawings, while the HATCHEDIT command edits the patterns.

ABOUT HATCHING

Many times, parts and assemblies cannot be fully described by orthographic projection. It can be helpful to view the object as if it were cut apart, a view called a "section." Mechanical designers use sections to show interior details. Civil engineers detail profiles by showing sections along roadways and railways. Architects use sections through entire structures to show how buildings are designed.

Crosshatching is used to show solid parts of the section. The spacing of the crosshatching should be relative to the scale of the section. The angle of the crosshatching should be oriented 45 degrees from the main lines of the cut area whenever possible. AutoCAD crosshatches sectional views easily. After drawing the section, apply the hatch pattern(s) to the cut areas with the HATCH command.

Full Sections

Full sections cut across the entire object, and are usually cut wherever the view needs to be clarified.

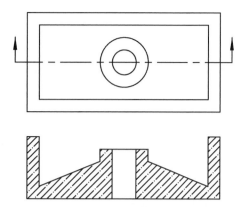

Parts that are "cut" are shown with crosshatching.

Revolved, Offset, and Removed Sections

It is often helpful to view a section of a part of an object that is transposed on top of the point where the section was cut. Such a section is referred to as a *revolved section*.

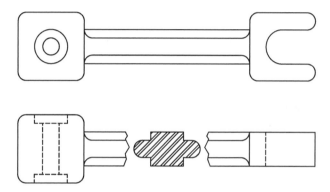

Offset sections are cut along an uneven line. Offset sections should be used carefully; change the cutting plane only to show essential elements.

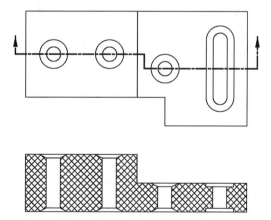

Removed sections are similar to revolved sections, except the section is not placed at the point where the section was cut.

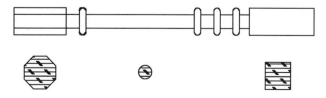

How Hatching Works

The HATCH command generates a boundary around an area, and then places an associative hatch pattern within that area. *Associative hatch patterns* automatically update themselves when their boundaries change. If you prefer that patterns not update themselves, then place non-associative hatches.

Hatches differ from other objects in that the lines making up the pattern are treated as one object. When you select one line of a pattern, the entire hatch is selected.

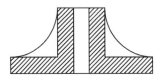

To place a hatch pattern in the drawing, AutoCAD needs to know at least two parameters: (1) the name of the hatch pattern, (2) the area to be hatched, and (3) if it is annotative. Optionally, you can also specify the scale, rotation angle, and style of area detection. You can also determine whether the boundary should be retained, and create simple custom patterns.

BASIC TUTORIAL: PLACING HATCH PATTERNS

1. To place hatch patterns in drawings, use the HATCH command with one of these methods:
   - From the **Draw** menu, choose **Hatch**.
   - On the Draw toolbar, pick the **Hatch** button.
   - In the Dashbaord's 2D Draw panel, choose the **Hatch** button.
   - Or, at the 'Command:' prompt, enter the **hatch** command:

     Command: **hatch** (*Press* ENTER.)
   - Alternatively, enter the aliases **h**, **bh**, or **bhatch** (one of its old names, short for "boundary hatch") at the keyboard.

   Notice the Hatch and Gradient dialog box.

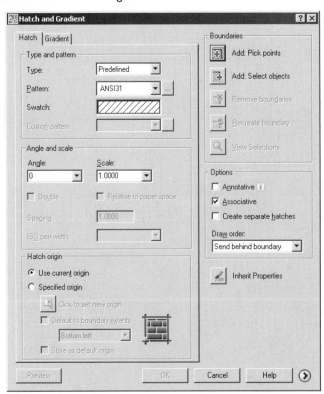

2. From the **Pattern** droplist, select a hatch pattern name. If you prefer to select the pattern visually, click **Swatch** to display the Hatch Pattern Palette dialog box. (See figure next page.)

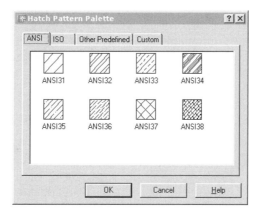

- **ANSI** — patterns defined by the ANSI (American) standard.

- **ISO** — patterns defined by the ISO (international) standard.

- **Other Predefined** — additional hatch patterns, ones that are the most useful for drawings.

- **Custom** — patterns defined by you.

   For a solid color fill, select the "Solid" pattern found in the **Other Predefined** tab. Select a pattern, and then click **OK**.

3. Select the area to be hatched by clicking the **Pick Points** button. Notice that the dialog box disappears, and that AutoCAD prompts you on the command line:

    Pick internal point or [Select objects/remove Boundaries]: *(Pick a point.)*

    Selecting everything... Selecting everything visible... Analyzing the selected data...
        Analyzing internal islands...

    Pick internal point or [Select objects/remove Boundaries]: *(Press ENTER to return to dialog box.)*

4. Click **OK**. AutoCAD immediately hatches the area.

PLACING HATCH PATTERNS: ADDITIONAL METHODS

The Hatch and Gradient dialog box has two tabs, **Hatch** and **Gradient**, plus an area common to both tabs. The **More Options** button displays advanced options. The many options can be confusing, so let us step through them.

Select a Type of Hatch

The first step is to select the type of pattern. The **Type** droplist supports three classes of pattern.

Predefined selects a predefined hatch pattern listed by the **Pattern** droplist. Hatch patterns provided with AutoCAD are stored in the *acad.pat* and *acadiso.pat* files. (ISO patterns permit a pen width to be assigned to the pattern; "ISO" is short for the International Organization of Standards.)

User Defined specifies a simple pattern of lines based on the angle and spacing. To create a unique pattern, apply linetypes to the hatching.

Custom selects patterns in *.pat* files typically provided by third party or in-house developers.

In most cases, you select one of the **Predefined** patterns.

Select a Pattern Name

Second, select the pattern to use from the **Pattern** droplist. The default pattern is made of diagonal lines, and is called "ANSI31." To select a different pattern, click the down arrow, and select from the available hatch patterns.

If you don't recognize the pattern by name, the alternative is to click the pattern next to **Swatch**. AutoCAD displays a tabbed dialog box of patterns grouped together. Select a pattern swatch, and choose **OK**.

Specify the Angle and Scale

Hatch patterns can be drawn at varying angles and scales. The hatch pattern samples shown in the dialog box have an angle of 0 degrees, even if the lines of the hatch are drawn at a different angle.

To change the angle in the **Angle** text box, select one of the preset angles, or else use the keyboard to enter a different angle.

Left: *Selecting an angle for the hatch pattern.*
Right: *Selecting a scale factor.*

The scale factor multiplies the hatch pattern by a factor. Changing the value in **Scale** to 2, for example, doubles the spacing between lines, while changing the scale to 0.5 reduces the spacing to half-size. To change the thickness of the lines, use lineweights.

To change the scale in the **Scale** box, select one of the preset scales, or use the keyboard to enter another scale factor.

Scaling hatch patterns is similar to scaling text and linetypes: enter the inverse scale factor in the **Scale** box. For example, if the drawing is to be plotted at a scale of 1:100, then the hatch pattern must be applied at a scale of 100.

 Note: The **OSNAPHATCH** system variable performs the useful function of determining whether object snap modes will snap to hatch patterns and gradient fills. When set to **0**, the default, then all the osnap modes ignore hatch and fill patterns; when set to **1**, osnap modes select component objects within the hatch.

Relative to Paper Space

The **Relative to Paper Space** option scales hatch patterns appropriately for each layout, using the scale of the layout. (This option is grayed out in model space.)

Specifying User Defined Patterns

When you select **User Defined** as the hatch type, you work with a single pattern consisting of simple, parallel lines. You specify three parameters: angle, spacing between lines, and single or double.

The **Angle** option is the same as described earlier. Turning on the **Double** check box causes AutoCAD to draw the pattern twice, the second time at 90 degrees to the first pattern, creating a crosshatch. The **Spacing** option specifies how far apart the parallel lines are drawn. The illustration below shows examples of using linetypes together with user-defined hatch patterns.

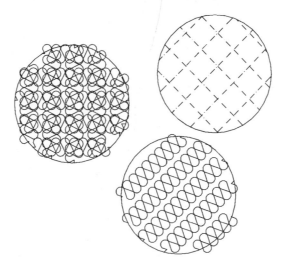

In the circle at left, I used the Batting pattern with double hatching; below it, with single hatching.

The **ISO pen width** option is available only when you select an ISO hatch pattern. It allows you to specify the width of the lines making up the ISO pattern.

Selecting Hatch Boundaries

Hatches must be contained within *boundaries*. Boundaries are made up of one or more objects that form closed polygons. Objects such as lines, polylines, circles, and arcs can form boundaries.

Before placing hatch patterns, you must identify the objects that form the closed boundary. If the boundary is not closed, the hatch pattern might leak out; if the boundary contains gaps, you can use the **Gap Tolerance** option, described later. There are several ways to identify the boundary:

- Select objects that make up the boundary.
- Pick points within areas to be hatched, and then let AutoCAD find the boundaries.
- Remove islands from inside boundaries.

Defining Boundaries by Selecting Objects

You can specify the boundary for the hatch pattern by selecting the objects that surround the area to be hatched. The illustration shows a shape constructed with four lines.

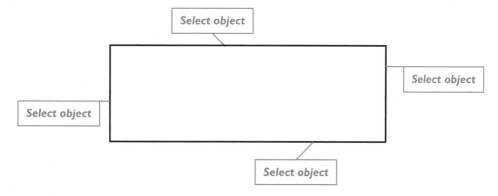

To place a hatch in the rectangle, choose the **Add: Select Objects** button. The dialog box disappears from the screen; notice the command prompt area:

Select objects or [picK internal point/remove Boundaries]: *(Pick one or more objects.)*

Select objects or [picK internal point/remove Boundaries]: *(Press ENTER.)*

AutoCAD prompts you to select the objects that will form the boundary. Place the pickbox over each object, and then select.

 Note: Use the Window or Crossing option to select all the sides of the boundary in one operation.

When you are finished selecting objects, press ENTER. The dialog box returns.

To test the boundary, choose the **Preview** button. If the boundary is inadequate to contain the hatch pattern, AutoCAD attempts to close the boundary.

Press ENTER or right-click to return to the dialog box. To review the objects selected for the boundary, choose **View Selections**. The dialog box again disappears, and AutoCAD highlights the boundary objects. If you need to, change the boundary objects by choosing the **Select Objects** button to select and deselect objects.

Defining Boundaries by Picking Areas

The simplest way to define the boundary is to show AutoCAD by selecting a point within the area to be hatched.

Let's look at an example. The illustration shows an chair with different areas that could be hatched: seat, back, arm rests.

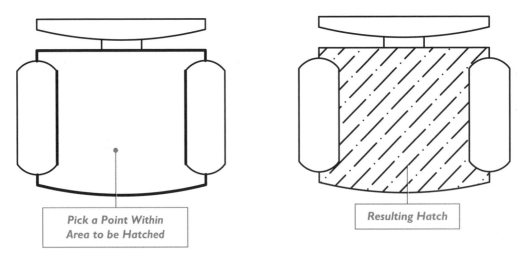

Pick a Point Within Area to be Hatched

Resulting Hatch

In the dialog box, choose the **Add: Pick Points** button. When the dialog box disappears from the screen, place the crosshairs in the area.

Pick internal point or [Select objects/remove Boundaries]: *(Pick point.)*

On the command line, AutoCAD reports:

> Selecting everything...
>
> Selecting everything visible...
>
> Analyzing the selected data...
>
> Analyzing internal islands..

You can think of those statements as AutoCAD muttering to itself as it searches for the boundary. (There is the NOMUTT system variable to turn off muttering.) AutoCAD analyzes the area, looking for leakage. Unknown to you, AutoCAD then places a polyline as the boundary around the inside of the area. (You can place boundaries by themselves with the BOUNDARY command, discussed later in this chapter.)

If you select a point in an area that is not contained, AutoCAD warns:

> Valid hatch boundary not found.

You must reconsider the area you want hatched; perhaps you need to draw another line to close off the area, or else define the maximum size of allowable gap.

You can select more than one area. When you finish picking areas, press ENTER to return to the dialog box. Choose **Preview** to see how the hatch will turn out. When satisfied, choose **OK** and the hatch is completed.

Removing Boundaries

Sometimes the area to be hatched is not as straightforward as the rectangle hatched above. It is common for hatch areas to contain islands. An *island* is another closed area within the boundary. In the illustration below, the two circles are islands.

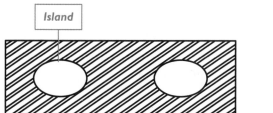
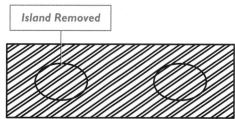

AutoCAD assumes you do *not* want the islands hatched. If, however, you *do* want AutoCAD to draw the pattern right through islands, choose the **Remove Boundary** button. The dialog box disappears, and AutoCAD prompts you to pick the islands to be removed:

> Select objects or [Add boundaries]: *(Pick object.)*

After you select an object, AutoCAD prompts you to pick another:

> Select objects or [Add boundaries/Undo]: *(Pick another object.)*

Or, you can type **U** to deselect the object. When done removing islands, press ENTER. Back in the dialog box, choose the **Preview** button to see the effect of removing the islands.

Hatch Origin

The default origin of hatch patterns is 0,0 — coinciding with the origin of the drawing. In many cases, that's fine, but in some cases you want adjacent hatches to line up, or have the pattern begin at the edge of its boundary.

In the figure, the first hatch pattern has its origin at 0,0; notice that the brick pattern does not line up along the left or bottom edges. For the second hatch pattern, the origin was moved to the lower left corner of the wall; the bricks line up correctly.

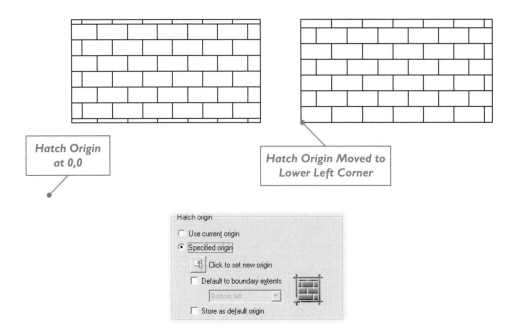

The Hatch Origin options allows you to relocate the origin, which is applied to all subsequent hatches — until you change the origin again. (Prior to AutoCAD 2006, the hatch origin was relocated using the **SNAPBASE** system variable.)

The **Use Current Origin** option reads the value stored in the **HPORIGIN** system variable, which is 0,0 until it is changed.

The **Specified Origin** option allows you to specify a different starting point for the hatch pattern:

> **Click to set new origin** — click the button to pick a point in the drawing to relocate the origin of the hatch pattern.

> **Default to boundary extents** — select one of the points on the extents of the pattern boundary: lower left, lower right, upper left, upper right, or center.

The *extents of the pattern boundary* is an imaginary rectangle that encompasses the hatch object. When the boundary is rectangular, then the extents match up; otherwise, they do not, as illustrated by the figure.

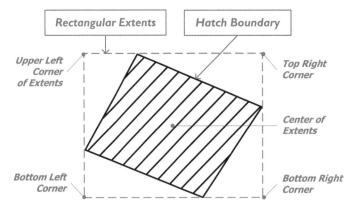

Additional Options

When AutoCAD draws hatch patterns, they are *associative*, which means they associate with their boundaries: change a boundary, and the hatch pattern updates by enlarging and shrinking itself to fit. AutoCAD draws the associative hatch pattern as a single object.

In some rare cases, you may want to place *non-associative* hatch patterns. This pattern is "dumb," not knowing its boundary, or its parameters. AutoCAD draws the non-associative hatch pattern as a collection of independent lines collected into an anonymous block.

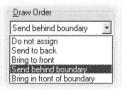

The **Annotation** option turns on the annotative scaling property. When on, the hatch pattern does not appear in model space unless its scale factor matches that of the viewport. (Click the **i** button for online help for annotation.)

The **Associative** option lets you decide if the hatch pattern should be associative or not. If you are not sure, keep the option set to **Associative**, because you can later use the **EXPLODE** command to change associative hatch patterns to non-associative; going the opposite direction is not possible. When you select two or more boundaries, AutoCAD can fill them with one hatch, or each with its own. The **Create Separate Hatches** option determines the number of patterns created when more than one boundary is selected: one for all, or one for each.

Draw Order

The **Draw Order** option determines whether patterns are displayed in front of or behind their boundary and other objects. Hatch patterns and filled areas can overwhelm the drawing when they are displayed over top of other objects. This occurs when the hatches are placed later in the drawing process.

One solution is to freeze the layer containing the pattern; a better solution is to select **Send to Back**, which ensures the hatch patterns are visually underneath all other objects in the drawing.

This option is particularly useful for areas filled with solid color or gradients. You can use this option, together with the **DRAWORDER** and **TEXTTOFRONT** commands, to control the display order of all objects in drawings.

Inherit Properties

Associative hatch patterns are also self-aware. They know their parameters, such as pattern name, scale, and origin. Click the **Inherit Properties** button to copy the properties from an existing hatch pattern:

> Select hatch object: *(Pick another hatch pattern.)*

The dialog box then copies the properties for use with the current pattern.

Advanced Hatch Options

As you have read, the **Add: Pick Points** button effectively automates the process of finding the hatch area. In some cases, the hatch area might be too complicated for AutoCAD to determine automatically.

AutoCAD keeps hidden a collection of advanced (read: rarely used) options that are revealed when you click the **More Options** button. These options were formerly on the dialog box's Advanced tab.

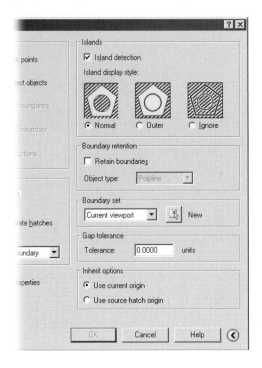

Islands

When the area to be hatched contains other objects, the **Island Detection** option lets you choose which get hatched.

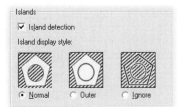

Let's look at how island detection works. The figure below illustrates a part with text.

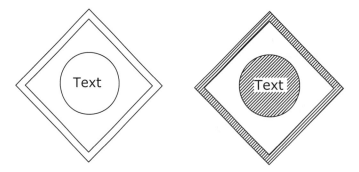

Left: Objects before being hatched.
Right: Islands and text hatched using Normal style.

The **Normal** is the default detection style, which hatches inward from the outermost boundary, skips the next boundary, and hatches the next. In the figure above, notice how the text is bordered by the hatch. With this style, an invisible window protects text from being obscured by hatching.

The **Outer** style hatches only the outermost enclosed boundary. The hatch continues only until it reaches the first inner boundary, and continues no further. Text is not hatched.

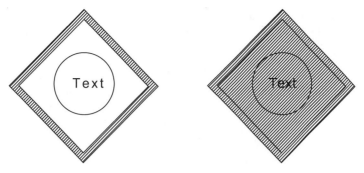

Left: *Outer island hatched with Outer style.*
Right: *Islands and text hatched using Ignore style.*

The **Ignore** style hatches all areas defined by the outer boundary — with no exceptions. This style even hatches through text.

Boundary Retention

Boundaries are drawn as polylines or region objects. AutoCAD normally erases the boundary after the HATCH command is finished. To retain the polyline boundary in the drawing, check the **Retain Boundaries** box.

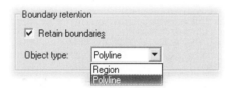

When the **Retain Boundaries** option is on, you can request that AutoCAD use a polyline for compatibility with older versions of AutoCAD; otherwise, use the **Region** option.

Boundary Set

When you use the **Add: Pick Points** option to define boundaries, AutoCAD analyzes all objects visible in the current viewport. You can, however, change the set of objects AutoCAD examines. In large drawings, reducing the set lets AutoCAD operate faster.

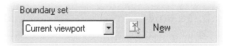

The **Current Viewport** option is the default, and examines all objects visible in the current viewport to create the boundary.

The **New** button prompts you to select objects from which to create the boundary set. (AutoCAD includes only objects that can be hatched.)

The **Existing Set** option creates the boundary from the objects selected with the **New** option. (You must use the **New** option *before* the **Existing Set** option.)

Gap Tolerance

In the early days of AutoCAD, if a boundary had a gap, the hatch pattern could "leak" out, covering much of the rest of the drawing. Drawing hatches is a processing-intensive operation, and leaky hatches would bog down the slow computers of the day. Hence, CAD operators were careful to ensure hatches would not leak.

In more recent releases of AutoCAD, Autodesk programmers wrote code that checked for leaks before applying the hatching. If it found a gap, AutoCAD would warn, "Valid hatch boundary not found." But CAD operators still had to fix gaps manually.

As of AutoCAD 2005, the **Gap Tolerance** option allows gaps in the boundary, ranging from 0 (the default value) to 5000 units. (*Warning!* Hatching to a gapped boundary leaves associative hatching turned off until you turn it back on again.)

 Note: Hatches can be handled more easily if they are put on their own layer. They can also be turned off and frozen to speed redraw time. Be sure that the layer linetype is continuous. Although the hatch pattern may contain dashed lines and dots, the linetype should be continuous to ensure a proper hatch.

Inherit Options

The **Inherit Options** option determines whether the origin is considered when properties are inherited from another hatch pattern.

Use Current Origin ignores the origin of the source pattern.

Use Source Hatch Origin copies the origin from the source pattern.

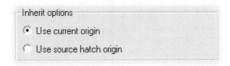

CREATING HATCHES: COMMAND LINE

The **-HATCH** command creates hatches after you enter options at the command line:

> Command: **-hatch**
>
> Current hatch pattern: ANSI31
>
> Specify internal point or [Properties/Select objects/draW boundary/remove
>
> Boundaries/Advanced/DRaw order/Origin]: *(Pick a point, or enter an option.)*

If you pick a point within an area, AutoCAD continues:

> Analyzing the selected data... Analyzing internal islands...
>
> Current hatch pattern: ANSI31
>
> Specify internal point or [Properties/Select objects/draW boundary/remove Boundaries/
> Advanced/DRaw order/Origin/ANnotative]: *(Press* ENTER *to complete the hatch and exit the command.)*

Properties

To control the hatch pattern (and to see prompts familiar from earlier releases of AutoCAD), enter the **Properties** option:

> Enter a pattern name or [?/Solid/User defined] <ANSI31>: *(Enter a hatch pattern name, or enter an option.)*
>
> Specify a scale for the pattern <1.0000>: *(Enter a scale factor, or press* ENTER *to accept the default value.)*
>
> Specify an angle for the pattern <0>: *(Enter an angle, or press* ENTER *to accept the default value.)*

Current hatch pattern: ANSI31

Specify internal point or [Properties/.../Origin]: *(Pick a point in the area to be hatched.)*

Specify internal point or [Properties/.../Origin]: *(Press* ENTER *to complete the hatch and exit the command.)*

The **?** option lists the names of hatch patterns loaded into AutoCAD. If you know letter(s) that the hatch pattern begins with, you can enter those few letters to narrow down the search, such as:

Enter a pattern name or [?/Solid/User defined] <ANSI31>: **?**

Enter pattern(s) to list <*>: **b***

This lists all hatch patterns that begin with 'b'. Pressing ENTER lists all hatch pattern names, like this:

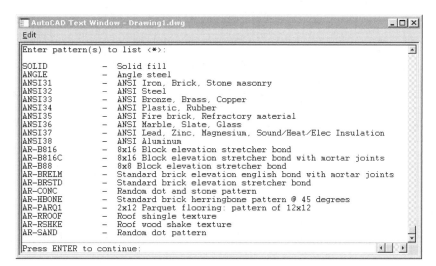

Press ENTER to continue the listing (which can go on for a long time), or else press ESC to return to the 'Command:' prompt.

Hatch styles are not displayed by the command. Append the pattern name with a comma and a code for the style. For example:

Enter a pattern name or [?/Solid/User defined] <ANSI31>: **ansi32,o**

The style codes are:

Style Code	Meaning
,i	Ignore
,n	Normal
,o	Outer
,?	List styles

The **Solid** option is a shortcut to the solid fill.

The **User defined** option takes you through the steps of defining the distance between lines and their angle:

Specify angle for crosshatch lines <0>: *(Enter an angle, or press* ENTER.*)*

Specify spacing between the lines <1.0000>: *(Enter a distance, or press* ENTER.*)*

Double hatch area? [Yes/No] <N>: *(Type* **y** *or* **n**.*)*

AutoCAD fills the area with the hatch pattern.

Select objects

The **Select Objects** option prompts you to select objects, and then generates a boundary for the hatch pattern.

draW boundary

The **draW boundary** option prompts you to pick points in a manner similar to the **PLINE** command. The points determine the boundary for the hatch pattern.

remove Boundaries

The **remove Boundaries** option allows you to add and remove objects from the boundary:

> Select Objects or [Add boundaries]: *(Select objects to remove, or type **A** to add objects.)*

Advanced

The **Advanced** option displays a prompt that allows you to control the hatch in a manner similar to the **More Options** section of the Hatch and Gradient dialog box:

> Enter an option [Boundary set/Retain boundary/Island detection/Style/Associativity/Gap tolerance/separate Hatches]: *(Enter an option.)*

DRaw order

The **DRaw order** option allows you to change the display order of hatches relative to other objects:

> Enter draw order [do Not assign/send to Back/bring to Front/send beHind boundary/ bring in front of bounDary] <send beHind boundary>: *(Enter an option.)*

Origin

The **Origin** option relocates the origin of the hatch pattern:

> [Use current origin/Set new origin/Default to boundary extents] <current>: *(Enter an option.)*

 Notes: You can use the **TRIM** command to trim hatch and fill patterns. The process is similar to trimming other objects. (See Chapter 11, "Additional Editing Options.") Start the **TRIM** command, and then select an object as the cutting edge. At the "Select object to trim" prompt, pick the portion of the hatch you want trimmed (removed).

Conversely, the hatch pattern can be used to trim other objects. Trimmed hatches are associated with the new boundary.

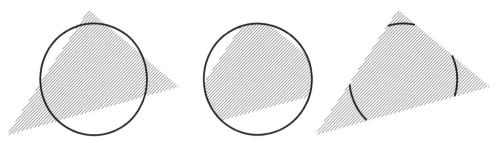

Left: *Circle and hatch patterns — before trimming.*
Center: *The circle trimming the hatch.*
Right: *The hatch boundary trimming the circle.*

GRADIENT

The **GRADIENT** command places *gradient* fills.

Gradients are colors that change in intensity (more white or more black), or change from one color to another. The **GRADIENT** command displays the **Gradient** tab of the Hatch and Gradient dialog box:

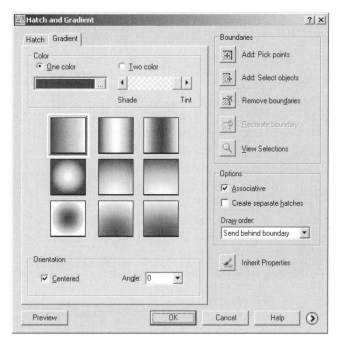

One Color means just one color is used with white or black to provide the second "color." To select the color, click the **...** button, which displays the Select Color dialog box. Pick a color, and then click **OK**.

Move the **Shade-Tint** slider to make the "second" color more black (Shade) or more white (Tint). Moving the slider to the center makes the gradient subtle.

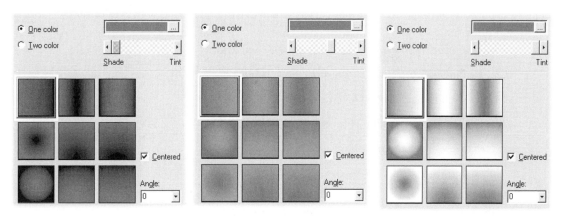

Color gradients near black.   *Color with little gradient.*   *Color gradients near white.*

Two Color creates a gradient from two colors. Pick the **...** button next to each color sample to select the colors. You don't have the slider to help you create subtle gradients; instead, select two similar colors.

Centered toggles the gradient between being centered in the fill area, or starting at the left edge.

Angle rotates the gradient inside the fill area.

Using gradients allows you to create 3D-like effects in 2D drawings, as the example below illustrates.

 HATCHEDIT

The **HATCHEDIT** command changes the properties of patterns and fills.

The command displays a dialog box that looks exactly like that of the **HATCH** command, with some options grayed out (unavailable).

TUTORIAL: CHANGING HATCH PATTERNS

1. To change hatch patterns in drawings, use the **HATCHEDIT** command with one of these methods:

   - From the **Modify** menu, choose **Object**, and then **Hatch**.

   - On the Modify II toolbar, pick the **Edit Hatch** button.

   - Double-click the associative hatch object.

   - At the 'Command:' prompt, enter the **hatchedit** command.

   - Alternatively, enter the **he** alias at the keyboard.

     Command: **hatchedit** (*Press* ENTER.)

2. AutoCAD prompts you to select the hatch pattern to modify:

   Select associative hatch object: (*Pick a hatch pattern.*)

   Notice the Hatch Edit dialog box.

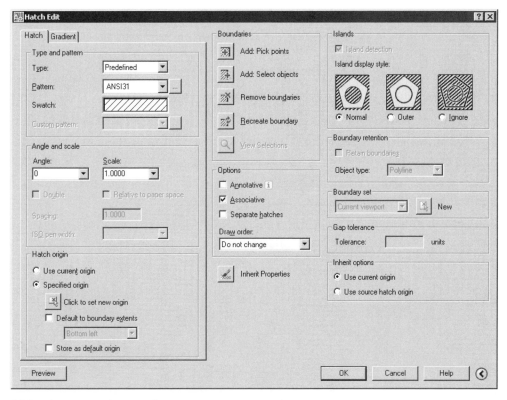

3. Make changes, such as to the pattern name or scale.

4. Click **Preview** to see the effect of the changes. Notice that the dialog box disappears, and that AutoCAD prompts you:

   Pick or press Esc to return to dialog or <Right-click to accept hatch>: *(Enter an option.)*

5. To accept the changes, press **ENTER** or right-click the mouse.

   To return to the dialog box, press **ESC**, or left-click the mouse.

As an alternative, click the **Inherit Properties** button. This allows you to match the pattern to another in the drawing. AutoCAD temporarily dismisses the dialog box, and then prompts you:

   Select associative hatch object: *(Pick a hatch pattern.)*

AutoCAD copies the properties of the other hatch pattern, such as its name, rotation angle, and scale factor. When the Hatch Edit dialog box returns, click **OK**. The pattern changes to match.

BOUNDARY

The **BOUNDARY** command creates single regions from disparate areas.

Earlier, you read how AutoCAD creates a temporary boundary out of a polyline or region to hold the hatching. To work with boundaries independent of hatch patterns, use the **BOUNDARY** command. The boundary is selected in the same manner, and is constructed as a polyline or region. The difference is that no hatch is placed within the boundary.

Note: The **BOUNDARY** command converts the outlines of objects to polylines; this allows you to select them more easily for editing.

TUTORIAL: CREATING BOUNDARIES

1. To place boundaries in the drawing, use the **BOUNDARY** command with one of these methods:

    *   From the **Draw** menu, choose **Boundary**.

    *   At the 'Command:' prompt, enter the **boundary** command.

        Command: **boundary** *(Press ENTER.)*

    *   Alternatively, enter the aliases **bo** or **bpoly** (the command's original name, short for "boundary polyline") at the keyboard.

    Notice the Boundary Creation dialog box.

2. Select the object you want used for the boundary: a polyline or a region.

    Polyline boundaries are more easily edited, while region boundaries can be analyzed for properties, such as the centroid (geometrically-weighted center), using the **MASSPROP** command.

3. Click **Pick Points** to pick a point within the area to be bounded. If the area is enclosed, AutoCAD creates the boundary, makes the following report, and exits the command:

    BOUNDARY created 1 polyline

    Command:

    If AutoCAD finds the area is not closed, a dialog box complains in somewhat misleading terms, "Valid hatch boundary not found."

You probably won't notice the new boundary in the drawing, because AutoCAD traces over the objects defining the boundary. To edit it, for example to move or color it, use the **Last** object selection mode, which selects the last-drawn object in the drawing.

CREATING BOUNDARIES: COMMAND LINE

The **-BOUNDARY** command creates boundaries after you specify them at the command line. If you don't need to change options, the command-line is faster than the dialog box. The command runs like this:

Command: **-boundary**

Specify internal point or [Advanced options]: *(Pick a point.)*

Selecting everything... Selecting everything visible... Analyzing the selected data...

Analyzing internal islands...

Specify internal point or [Advanced options]: *(Press ENTER.)*

BOUNDARY created 1 polyline

The **Advanced options** include **Boundary set**, **Island detection**, and **Object type**:

> Enter an option [Boundary set/Island detection/Object type]: *(Enter an option.)*

Boundary Set

The **Boundary set** option selects the objects to be considered for determining the boundary:

> Specify candidate set for boundary [New/Everything] <Everything>

The **New** option prompts you to select objects, while the **Everything** option includes all objects in the drawing.

Island Detection

The **Island detection** option toggles island detection:

> Do you want island detection? [Yes/No] <Y>: *(Enter an option.)*

When off, islands are ignored; when on, islands are taken into account.

Object Type

The **Object type** option selects the object from which the boundary is made, a region or a polyline:

> Enter type of boundary object [Region/Polyline] <Polyline>: *(Enter an option.)*

SELECT

The **SELECT** command selects objects by their location. For example, you can select all objects that are inside a selection window, or all objects in the entire drawing.

More commonly, objects are not selected with **SELECT**, but during other commands that present the 'Select objects:' prompt. For example, you can start the **MOVE** command, and then select the objects you wish to move. Both the **SELECT** command and the 'Select objects:' prompt have the same options.

In addition, AutoCAD allows you to select objects *without* the **SELECT** command or its options. With no command active, pick one or more objects: AutoCAD highlights them, and displays *grips* (called "handles" in other software applications). Edit the objects by manipulating their grips, until you press **ESC** to exit this direct editing mode. Editing with grips is described in Chapter 9, "Direct Editing."

AutoCAD identifies selected objects by displaying them with *highlighting*, as if they were drawn with a dashed line. As the cursor passes over them, objects take on highlighting and thicken their lines. When you click the mouse button, the object is selected.

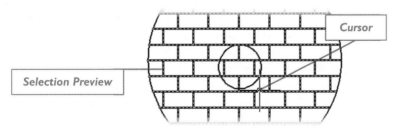

Selection preview is controlled by options found in the **Selection** tab of the Options dialog box, described later in this chapter.

TUTORIAL: SELECTING OBJECTS BY LOCATION

1. To select objects in the drawing, use the **SELECT** command with one of these methods:
   * At the 'Command:' prompt, enter the **select** command.

- Alternatively, enter a selection option at any "Select objects" prompt.

    Command: **select** *(Press ENTER.)*

2. AutoCAD prompts you to select objects:

    Select objects: *(Select an object.)*

    1 found Select objects:

    Notice that AutoCAD highlights the objects you select, and keeps a running tally of the number of objects selected, as in "1 found."

3. You can keep selecting objects until you press **ENTER** to exit the command:

    Select objects: *(Select more objects.)*

    3 found Select objects: *(Press ENTER.)*

4. After pressing **ENTER** to exit the **SELECT** command, the highlighting disappears. The objects you selected are added to AutoCAD's *selection set*.

So, you may wonder, what good is the **SELECT** command? Not a lot, which is why many CAD operators never use it. You can, however, access selection sets with the **Previous** option of the **SELECT** command, or at any 'Select objects:' prompt. In this way, the **SELECT** command is good for creating a selection set used later by other commands.

Note: The **SELECT** command fails to list its options. To force it to display options, enter **?**, as follows:

Command: **select**
Select objects: **?**
Invalid selection
Expects a point or Window/Last/Crossing/BOX/ALL/Fence/WPolygon/CPolygon/Group/
    Add/Remove/Multiple/Previous/Undo/AUto/SIngle/SUbobject/Object
Select objects: *(Enter an option.)*

Select Options

The **SELECT** command's options have the following meaning:

Select Option	Meaning
Expects a point	Selects one object under the cursor.
Window	Selects all objects fully within a rectangle defined by two points. Window mode is indicated by a solid rectangle filled with blue color.
Last	Selects the most recently created visible object.
Crossing	Selects all objects within and crossing a rectangle defined by two points. Crossing mode is indicated by a dashed rectangle filled with green color.
BOX	Selects all objects within and/or crossing a rectangle specified by two points. If the selection rectangle is picked from: • Right to left, performs Crossing selection. • Left to right, Window selection.
All	Selects all objects in the drawing, except those residing on frozen and locked layers.
Fence	Selects all objects crossing a selection line. The fence line can cross itself.
WPolygon	Selects all objects completely within a selection polygon, which can be any shape but cannot cross itself; AutoCAD closes polygon.
CPolygon	Selects all objects within and crossing a selection polygon, which can be any shape but cannot cross itself; AutoCAD closes polygon.
Group	Selects all objects comprising a named group.
CLass	Selects object classes defined by add-on software, such as Autodesk Map.

Add	Switches selection mode to Add, after being in Remove mode. Objects selected by any means listed above are added to the selection set.
Remove	Switches selection mode to Remove. Objects selected by any means listed above are removed from the selection set. As an alternative, you can hold down the SHIFT key to remove objects from the selection set.
Multiple	Selects objects without highlighting them. Also selects two intersecting objects when the intersecting point is selected twice.
Previous	Adds objects that were previously selected to the selection set. The Previous selection set is ignored when you switch between model and paper space.
Undo	Removes the object most recently added to the selection set.
Auto	Selects objects by three methods. Picking a point in a blank area of the drawing starts Window or Crossing mode, depending on how you move the cursor: • Right to left performs Crossing selection. • Left to right performs Window selection. • Picking an object selects it.
Single	Selects the first object(s) picked, and then does not repeat the "Select objects:" prompt.
SUbobject	Select subobjects on 3D models: vertices, edges, and faces; equivalent to holding down the CTRL key during object selection.
Object	Exits subobject selection mode.

That probably seems like too many options for you! It is useful to know about all of them, but in practice you probably use just a half dozen — All, Previous, Last, Window, Crossing, and point are the ones I use most frequently, while the technical editor uses Auto the most. "Because," he says, "Auto is the most efficient. Most of the other options are left over from older versions of AutoCAD."

Pick (Select Single Object)

To select a single object, click on it with the cursor. There is no need to enter any options at the "Select objects:" prompt:

> Select objects: *(Pick an object.)*

In this mode, AutoCAD selects a single object. To select another object, pick again; to "unselect" the object, hold down the SHIFT key. To select from two or more overlapping objects, follow this procedure:

1. Hold down the **CTRL** key while selecting an object. If the object overlaps another, AutoCAD displays <Cycle on> at the 'Command:' prompt.
2. Release the **CTRL** key.
3. Click repeatedly (cycle) until AutoCAD selects the object you want.
4. Press the **ENTER** key to exit cycle mode and return to the 'Select objects:' prompt.

To select more than one object at a time, use one of the modes described next.

ALL (Select All Objects)

The **All** option selects all the objects in the drawing, except those on frozen and locked layers — most of the time. For some commands, such as COPYCLIP, the **All** option selects only those objects visible in the viewport; in model space, it selects all objects, visible or not.

 Notes: As a shortcut, you can press the **CTRL+A** keys, which selects all objects in the drawing (subject to the restrictions noted above). To select nearly all objects in the drawing, first select all objects with the **All** option, and then use the **Remove** (explanation follows) option to "deselect" the exceptions.

W (Select Within a Window)

One of the most common forms of object selection is the **W** option (short for "window"). It places a rectangle around the objects to be selected. You define the rectangle by picking its opposite corners, as follows:

> Select objects: **w**
>
> Specify first corner: (Pick point 1.)
>
> Specify opposite corner: (Pick point 2.)

Objects *entirely* within the window rectangle are selected. If an object crosses the rectangle, it is not selected (see Crossing mode). Only objects currently visible on the screen are selected.

You can place the selection rectangle so that all parts of those objects you want to choose are contained in the windows, while those that you don't want are not entirely contained. With this method, you select objects in an area of your drawing where objects overlap.

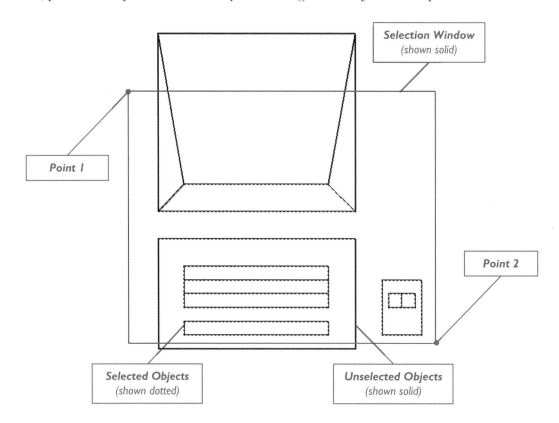

Selection Window (shown solid)

Point 1

Point 2

Selected Objects (shown dotted)

Unselected Objects (shown solid)

C (Select with a Crossing Window)

The **C** option (short for "crossing") is similar to the **Window** option, with an important difference: objects crossing the rectangle are included in the selection set — as well as those objects entirely within the rectangle. Define the rectangle by picking its opposite corners.

> Select objects: **c**
>
> Specify first corner: *(Pick point 1.)*
>
> Specify opposite corner: *(Pick point 2.)*

Notice that the crossing window is dashed and filled with a transparent green color. This distinguishes it from the Window rectangle, which is made of solid lines and blue color.

The color and opacity of area selections can be changed with the Options dialog box: choose the Selection tab, and then click the **Visual Effect Settings** button.

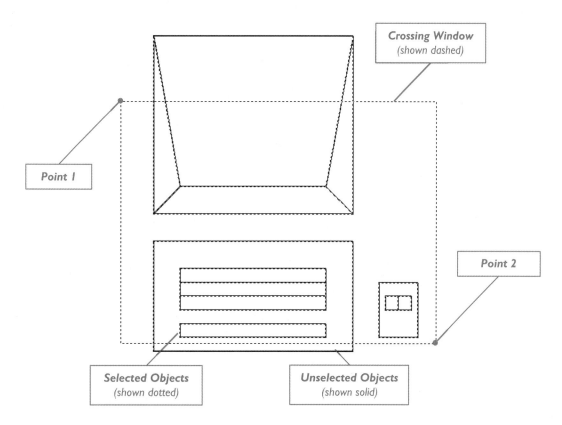

BOX (Select with a Box)

The **BOX** option allows you to use either the crossing or window rectangle to select objects. Define the rectangle by picking its opposite corners.

After you pick the first corner of the box, move the cursor either to the right or to the left. If you move to the right, the result is the **Window** selection: AutoCAD selects objects completely within the blue selection rectangle.

When you move to the left, the result is the crossing selection: AutoCAD selects objects within the rectangle, as well as those crossing it. In addition, the rectangle is drawn with dashed lines and filled with green color.

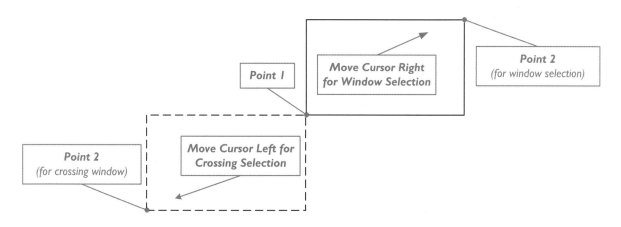

F (Select with a Fence)

The **F** option (short for "fence") uses a polyline to select objects. The fence is displayed as a dashed line; all objects crossing the fence are selected. You can construct as many fence segments as you wish, and can use the **Undo** option to undo a fence line segment.

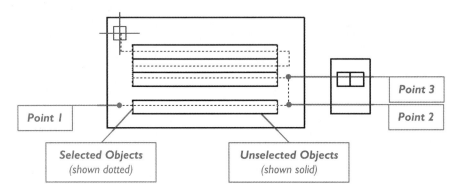

Select objects: **f**

First fence point: *(Pick point 1.)*

Specify endpoint of line or [Undo]: *(Pick point 2.)*

Specify endpoint of line or [Undo]: *(Pick point 3.)*

et cetera

AU (Select with the Automatic Option)

The **AU** option (short for "automatic") combines the pick and **Box** options. It is the default for most selection operations. After you enter **AU** in response to the "Select objects" prompt, you select a point with the pickbox. If an object is found, the selection is made.

If an object is not found, the selection point becomes the first corner of the **Box** option. Move the box to the right for Window, or to the left for Crossing. The **AU** option is excellent for advanced users who wish to reduce the number of modifier selections.

WP (Select with a Windowed Polygon)

The **WP** option (short for "windowed polygon") selects objects by placing a polygon window around them. The polygon window selects in the same manner as the **Window** option: all objects completely within the polygon are selected. Objects that cross or are outside the polygon are not selected.

The difference is that the **WP** option creates a multisided window, instead of a rectangle. Let's look at a sample command sequence:

Select objects: **wp**

First polygon point: *(Pick point 1.)*

Specify endpoint of line or [Undo]: *(Pick point 2.)*

Specify endpoint of line or [Undo]: *(Pick point 3.)*

After you enter the first polygon point, build the window by placing one or more endpoints. The polygon rubber-bands to the cursor intersection, always creating a closed polygon window filled with the color blue.

Undo the last point entered by typing **U** in response to the prompt. Pressing **ENTER** closes the polygon window and completes the process.

Note that the polygon window must not cross itself or rest directly on a polygon object. If it does, AutoCAD warns "Invalid point, polygon segments cannot intersect," and refuses to place the vertex.

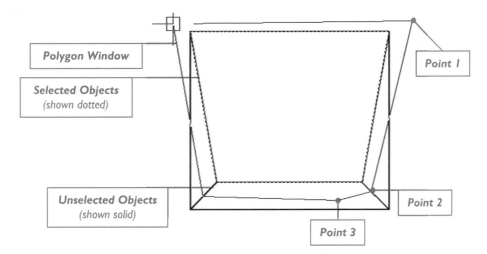

CP (Select with a Crossing Polygon)

The **CP** option (short for "crossing polygon") works in the same manner as the **WP** option, except that the polygon functions in the same manner as a crossing window. All objects within or crossing the polygon are selected. The crossing polygon is displayed with dashed lines filled with green, similar to a crossing window.

SI (Select a Single Object)

The **SI** option (short for "single") forces AutoCAD to issue a single "Select objects:" prompt. (All other selection modes repeat the "Select objects:" prompt until you press ENTER.) You can use other selection options in conjunction with **SI** mode, such as the Crossing selection shown in this example:

> Select objects: **si**
>
> Select objects: **c**
>
> Specify first corner: *(Pick point 1.)*
>
> Specify opposite corner: *(Pick point 2.)*
>
> *nnn* found

The single option is useful for efficient single object selection in macros, because it does not require an ENTER to end the object selection process.

SU and O (Select Subobjects and Objects)

The **SU** option (short for "subobjects") selects faces, edges, and vertices of 3D models. This is equivalent to holding down the CTRL key during object selection. The **O** option (short for "objects") exits subobject selection mode.

M (Select Through the Multiple Option)

Each time you select an object, AutoCAD scans the entire drawing to find the object. If you are selecting an object in a drawing that contains a large number of objects, there can be a noticeable delay.

The **M** option (short for "multiple") forces AutoCAD to scan the drawing just once. This results in shorter selection times. Press ENTER to finish the object selection and begin the scan. Note that AutoCAD does not highlight the objects selected until you press ENTER .

> Select objects: **m**
>
> Select objects: *(Pick three times.)*
>
> *3* selected, *3* found

Special Selection Modes

P (Select the Previous Selection Set)

The **P** option (short for "previous") uses the previous selection set. This very useful option allows you to perform several editing commands on the same set, without reselecting them. In addition, you can add to and remove from the **Previous** selection set.

L (Select the Last Object)

The **L** option (short for "last") selects the last object drawn still visible on the screen. When the command is repeated, and **Last** is used a second time, AutoCAD chooses the same last object and reports "1 found (1 duplicate), 1 total." This option is useful for immediately editing an object just drawn.

G (Select the Group)

The **G** option (short for "group") adds the members of a named group to the selection set. (Groups are covered later in this chapter.) The option prompts:

> Enter group name: *(Enter a group name.)*

Changing Selected Items

You can add and remove objects from the group of selected objects by using modifiers. Modifiers must be entered after you select at least one object, and before you press ENTER to end object selection.

U (Undo the Selected Option)

The **U** option (short for "undo") removes the most recent addition to the set of selections. If the undo is repeated, you will step back through the selection set. This shortcut replaces the two-step process of using the **Remove** option, and then remembering the objects to pick.

R (Remove Objects from the Selection Set)

The **R** option (short for "remove") removes objects from the selection set by any object selection method.

> Select objects: **r**
>
> Remove objects: *(Pick an object.)*
>
> 1 found, 1 removed, 2 total

 Notes: The **Remove** option is useful when you need to select a large number of objects with the exception of one or two objects located in the area. Select all the objects, then use the **R** option to remove the excess objects.

SHIFT+Select removes objects in **Add** mode, but adds objects in **Remove** mode.

A (Add Objects to the Selection Set)

The **A** option (short for "add") adds objects to the set. **Add** is usually used after **Remove**. **Add** changes the prompt back to 'Select objects:' so you may add objects to the selection set. Hold down the SHIFT key to remove objects.

Canceling the Selection Process.

Pressing ESC at any time cancels the selection process and removes the selected objects from the selection set. The prompt line returns to the 'Command:' prompt.

Pressing ENTER ends the selection process, and continues with the editing command's other options.

Selecting Objects During Commands

To use the selection options at any 'Select objects:' prompt, enter one of the abbreviations listed in the table on the earlier page (abbreviations shown in uppercase letters). Alternatively, just pick two points, because the **Auto** mode is the default.

For example, to move all the objects within the selection rectangle, do the following

> Command: **move**
>
> Select objects: **w**
>
> Specify first corner: *(Pick point 1.)*
>
> Specify opposite corner: *(Pick point 2.)*
>
> 5 found Select objects: *(Press* ENTER *to end object selection.)*
>
> Specify base point or displacement: *(Pick a point.)*
>
> Specify second point of displacement or <use first point as displacement>: *(Pick another point.)*

Notice that the "Select objects:" prompt repeats until you press ENTER. This means you can keep selecting (and deselecting) objects until you are satisfied with the selection set.

QSELECT

The QSELECT command allows you to select objects based on their properties, instead of on their location. For example, use the command to select all objects with hidden linetypes on a specific layer, or all circles with a radius larger than one inch.

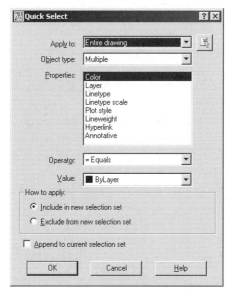

The dialog box is also accessible through the **Quick Select** button in several other commands, such as **BLOCK** and **PROPERTIES**.

TUTORIAL: SELECTING OBJECTS BY PROPERTIES

1. To select objects by their common properties, start the QSELECT command with one of these methods:
   - From the **Tools** menu, choose **Quick Select**.
   - Right-click, and from the shortcut menu choose **Quick Select**.
   - At the 'Command:' prompt, enter the **qselect** command:

Command: **qselect** *(Press* ENTER.*)*

Notice the Quick Select dialog box.

2. The selection can be made from the entire drawing or a subset of the drawing, called the "current selection."

   To select a subset, click the **Select Objects** button next to the **Apply to** droplist. AutoCAD prompts you:

   > Select objects: *(Select one or more objects.)*

   > Select objects: *(Press* ENTER *to return to the dialog box.)*

   Notice that the **Apply to** drop list now has two options. Select one:
   - Entire Drawing.
   - Current Selection.

3. The **Object Type** droplist allows you to narrow down the selection to specific objects. Only the objects found in the drawing or the current selection are listed here. To include all objects, select "Multiple."

4. Narrow down your selection by picking one of the items in the **Properties** list. The object you selected earlier affects the content of this list, which contains only properties specific to the object. You can select only one property from this list.

5. The **Operator** list further narrows the range AutoCAD searches for. Depending on the property you select, the choice of operator includes:

Operator	Meaning
=	Equals
< >	Not Equal To
>	Greater Than
<	Less Than
*	Wildcard Match (selects all).

   Use the ***Wildcard Match** operator with text fields that can be edited, such as the names of blocks.

6. The **Value** field goes with the **Operator** list: if AutoCAD can determine values, it lists them here; if not, you type the value.

7. Under **How to Apply**, include or exclude objects matching these parameters from the existing selection set.

8. The **Append to Current Selection Set** check box determines whether these are added to (on) or replace (off) the current selection set.

9. Choose **OK** to exit the dialog box. AutoCAD highlights the objects selected, and reports the number selected:

   > nnn item(s) selected.

CONTROLLING THE SELECTION: ADDITIONAL METHODS

In addition to selection modes and direct selection, AutoCAD has these commands:

- **PICKBOX** changes the size of the square pick cursor.
- **DDSELECT** changes selection options; displays the **Selection** tab of the Options dialog box.
- **FILTER** selects objects based on their properties.
- **SELECTURL** highlights all objects that contain hyperlinks. See *Using AutoCAD: Advanced* for details on using hyperlinks in drawings.

Let's look at how these commands work.

Pickbox

When an editing command displays the "Select objects:" prompt, the cursor is supplemented by a small square called the "pickbox." When you place the pickbox over the object, and then click (press the left mouse button), AutoCAD scans the drawing, and selects the object under by the pickbox.

The pickbox can be made larger and smaller. The **Selection** tab of the Options dialog box allows you to change the pickbox size; or, you just might prefer to use the PICKBOX system variable:

> Command: **pickbox**
>
> Enter new value for PICKBOX <3>: *(Enter a value between 0 and 50.)*

The size of the pickbox is measured in pixels. A large pickbox forces AutoCAD to scan through more objects, which can take longer in complex drawings on slow computers.

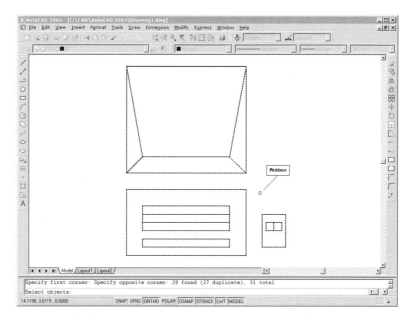

DdSelect

The DDSELECT command allows you to change selection modes. (The command is actually an alias for displaying the **Selection** tab of the Options dialog box.)

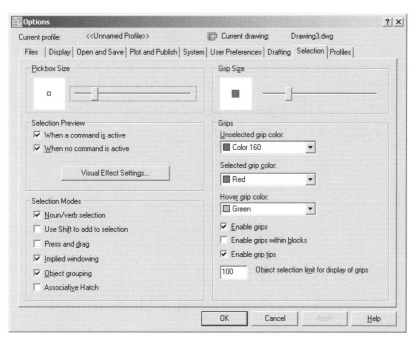

- From the **Tools** menu, choose **Options** , and then the **Selection** tab.

- Right-click, and from the shortcut menu choose **Options**.

- At the 'Command:' prompt, enter the **ddselect** command.

    Command: **ddselect** *(Press* ENTER.*)*

Let's look at each option.

Pickbox Size

The pickbox is the square cursor that appears at the 'Select objects:' prompt. Here, you can change its size from small to large.

Selection Preview

Selection preview highlights objects as the pickbox cursor moves over them. You can turn the mode on and off:

When a Command Is Active means that selection preview works only when a command displays the 'Select objects:' prompt.

When **No Command Is Active** means that selection preview works any other time.

The **Visual Effect Settings** button displays a dialog box that controls the visual effects of selection previews and windowed selections. For selection previews, you can choose whether dashed lines, thickened lines, or both are displayed. For windowed and crossing selections, you can choose the color and opacity. (Autodesk documentation is incorrect in suggesting these affect the appearance of windowed areas only during selection preview.)

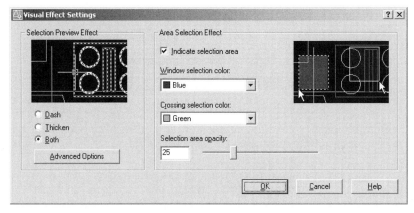

The **Advanced Options** button accesses further options, specifically which objects should be excluded from selection previewing.

Noun/Verb Selection

The "normal" way to edit drawings is to (1) first select the editing **command**, and then (2) select the **objects** to modify. Technically, this is choosing the *verb* (the action represented by the edit command), and then the *noun* (the object of the action represented by the selection set).

As an alternative, AutoCAD allows you to reverse the procedure: (1) first select the **object(s)** to edit, and then (2) choose the editing **command**. This is called "noun/verb selection."

When the **Noun/Verb Selection** option is turned on (as it is by default), AutoCAD places the pickbox at the intersection of the crosshairs. The presence of the pickbox indicates that you can select objects before entering one of these editing commands:

ALIGN   LIST   ARRAY   MIRROR   BLOCK   MOVE   CHANGE   PROPERTIES   CHPROP

ROTATE   COPY   SCALE   DVIEW   STRETCH   ERASE   WBLOCK   EXPLODE   BEDIT

Here is the command sequence to erase a single object, for example:

> Command: *(Select the object.)*
> Command: **erase**
> 1 found

Use Shift to Add to Selection

When selecting, you normally choose the object(s), and then press ENTER when you are finished. As you select each object, it is automatically added to the selection set. To remove objects from the selection set, hold down the SHIFT key while picking them.

As an alternative, turn on the **Use Shift to Add to Selection** option. Now you must hold down the SHIFT key to add objects to the selection set; this is similar to the method used by many other Windows programs. When you make more than one selection without holding the SHIFT key, the previous selections are removed from the selection set.

Press and Drag

The traditional AutoCAD method for selecting objects through windowing is to (1) click one corner, (2) move the cursor to the other corner, and then (3) click the other corner.

As an alternative, turn on the **Press and Drag** option. You build a window by (1) clicking one corner, and then (2) dragging (by holding down the mouse button) the cursor to the other corner. Like the **Shift to Add** option, this is used by most other Windows programs.

Implied Windowing

You previously learned how to use a selection window or a crossing window to select objects. All you have to do is enter either a **W** or a **C** to invoke the window mode.

The **Implied Windowing** option does this automatically (the option is on by default). You "imply" the window by clicking in an empty area of the drawing. When AutoCAD does not find an object within the area of the pickbox, it assumes that you want to use windowing. When you move the cursor to the right, **Window** selection mode is entered; move to the left, **Crossing** selection mode is entered.

Because the first point entered describes the first corner of the window, be sure to select a desirable position.

Object Grouping

When the **Object Grouping** option is turned on, AutoCAD selects the entire group when you pick one object in the group. When off, just the picked object is selected. (More on groups follows in this chapter.)

As a shortcut, you can toggle object grouping mode with CTRL+SHIFT+A. Each time you press the key combination, AutoCAD comments:

> Command: *(Press* CTRL+SHIFT+A.*)* <Group off>

"Group off" means objects are selected, while "Group on" means the entire group is selected:

Command: *(Press* CTRL+SHIFT+A.*)* <Group on>

Curiously, CTRL+H does the same thing. Technically, it toggles the **PICKSTYLE** system variable between **0** (individual objects selected from the group) and **1** (entire group is selected). **PICKSTYLE** can also have the values of **2** and **3** for toggling the selection of associative hatch patterns.

Associative Hatch

When the **Associative Hatch** option is turned on, AutoCAD selects the boundary object(s) when you pick an associative hatch pattern. When off, only the hatch is selected.

Filter

The **FILTER** command displays a dialog box that selects objects based on their properties. You can save the resulting selection set by name, and then access it at the 'Select objects:' prompt.

- At the 'Command:' prompt, enter the **filter** command.

- Alternatively, enter the **f** alias at the keyboard.

  Command: **filter** *(Press* ENTER.*)*

- Because **FILTER** is a *transparent* command, it can be invoked during another command to filter the selection set:

  Command: **erase**

  Select objects: **'filter**

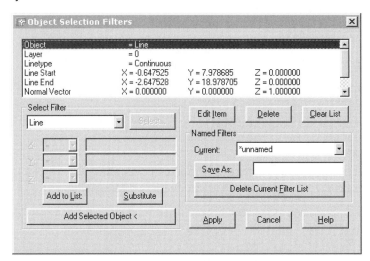

The selection set created by **FILTER** can be accessed via the **P** (previous) selection option. In the following tutorial, you erase all construction lines from a drawing:

1. Start the **ERASE** command:

   Command: **erase**

2. Then invoke the **FILTER** command transparently (prefix the command with the quotation mark):

   Select objects: **'filter**

   Notice that AutoCAD displays the Object Selection Filters dialog box.

3. In the **Select Filter** droplist, select "Xline."

   Xline is near the end of the list. Here's a quick way to get to any item in an alphabetical list: after clicking on the droplist, press **x** on the keyboard. You are taken to the first word starting with "x".

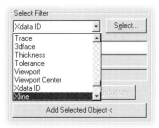

4. Click the **Add to List** button. Notice that AutoCAD adds this text to the "list" (the large white area at the top of the dialog box):

    Object   =Xline

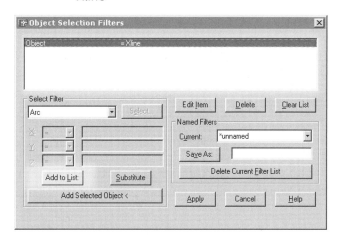

5. Click **Apply**. Notice that the dialog box disappears, and that the **FILTER** command displays prompts on the command line:

    Applying filter to selection.

    This means that AutoCAD will apply the filter (search for all xlines) to the objects you now select.

6. Specify that the filter should apply to the entire drawing with the All option:

    >>Select objects: **all**

    9405 found 9401 were filtered out.

    The double angle bracket (**>>**) indicates AutoCAD is currently in a transparent command. (The **ERASE** command continues later.)

7. Return to the **ERASE** command by exiting the **FILTER** command. Press **ENTER**:

    >>Select objects: *(Press* **ENTER** *to exit the* **FILTER** *command.)*

    Exiting filtered selection.

    Resuming ERASE command.

    AutoCAD returns to the **ERASE** command, and picks up the four objects selected by the **FILTER** command:

    Select objects: 4 found

8. Press **ENTER** to erase the four objects and exit the **ERASE** command:

    Select objects: *(Press* **ENTER** *to exit the* **ERASE** *command.)*

    AutoCAD uses the selection set created by the **FILTER** command to erase the xlines.

In addition to filtering objects, this command also filters x, y, z coordinates, such as the center point of circles. Other options become available in the dialog box as you select them. For example, when you select **Elevation**, the X text entry box allows you to specify the elevation using the following operators:

Operator	Meaning
<	Less than.
<=	Less than or equal to.
=	Equal to.
!=	Not equal to.
>	Greater than.
>=	Greater than or equal to.
*	All values.

Specifying elevation **> 100**, for example, means that all objects with an elevation greater than 100 will be added to the filtered selection set.

In addition, you can group filter sequences using these operators:

Operator	Meaning
Begin OR with **End OR	Include *any* of these items.
Begin AND with **End AND	Include *all* of these items.
Begin NOT with **End NOT	Include *none* of these items.
Begin XOR with **End XOR	Include none of these items if one item is found.

You can save filter definitions by name to *.nfl* (short for "named filter") files on disk for use in other drawings or editing sessions.

GROUP

The **GROUP** command creates selectable groups of objects.

A selection set lasts only until a new selection set is created. When you select a circle, for example, and then later a line, AutoCAD "forgets" about the circle you selected first. (There is a workaround, but it involves writing code with the AutoLISP programming language.)

To overcome the limitation, AutoCAD lets you create "groups." Each group has a name and consists of any selection set of objects. Members of one group can be members of other groups. Unlike blocks, however, groups cannot be shared with other drawings.

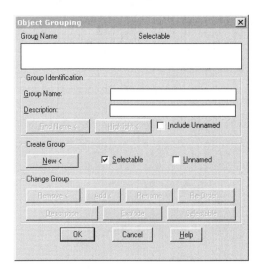

TUTORIAL: CREATING GROUPS

1. To create groups, start the **GROUP** command with one of these methods:
   - At the 'Command:' prompt, enter the **group** command.

      Command: **group** *(Press* **ENTER.***)*
   - Alternatively, enter the **g** alias at the keyboard.

2. Notice the Object Grouping dialog box.
3. In creating a group, you first give it a descriptive name, such as "Linkage." The name can consist of up to 255 letters and numbers.

      Group Name: **Linkage**
4. Optionally, you can describe the group with a label of up to 448 characters:

      Description: **The left end of the linkage**
5. Next, select the objects that become part of the group. Choose the **New** button. AutoCAD dismisses the dialog box temporarily so that you can select objects in the drawing. (Use any object selection mode.)

      Select objects for grouping:

      Select objects: *(Pick one or more objects.)*

      5 found Select objects: *(Press* **ENTER** *to end object selection, and return to the dialog box.)*
6. Decide whether the group should be *selectable*:

      When selectable, selecting one member of the group selects the entire group.

      When not selectable, selecting a member of the group selects the member only.

      The **Selectable** setting is not crucial, because you can toggle selectable mode on and off at any time by pressing **CTRL+SHIFT+A**.
7. As you create groups, AutoCAD adds their names to the list at the top of the dialog box. When done creating groups, click **OK**.

Notes: The **Unnamed** check box determines whether the group is named. When selected (turned on), AutoCAD automatically assigns a name to the group. The first assigned name is ***A0**; the next is ***A1**, and so on. When **Unnamed** is not selected (turned off), you must name the group.

Groups can contain groups, or members of other groups. Selecting one member of a group selects all.

WORKING WITH GROUPS: ADDITIONAL METHODS

Once a group is created, AutoCAD treats all members of the group as a single object, much like a block. Select one member of the group, and AutoCAD selects the entire group; drag one member, and the entire group moves.

Unlike blocks, however, you can quickly work with the individual members. Press **CTRL+SHIFT+A**, and AutoCAD turns off group mode:

      Command: *(Press* **CTRL+SHIFT+A.***)*

      <Group off>

Now you can select a single member of the group — not the entire group. To return to group mode, press **CTRL+SHIFT+A** again:

      Command: *(Press* **CTRL+SHIFT+A.***)*

      <Group on>

Highlighting Groups

AutoCAD unfortunately has no easy way of identifying groups in drawings. (Curiously enough, AutoCAD LT has a different — and, in my opinion, better — user interface for handling groups.) The workaround is to return to the Object Grouping dialog box:

1. In the list under **Group Name**, select the name of a group. Notice that the **Highlight** button becomes available.
2. Choose the **Highlight** button. The dialog box disappears, and AutoCAD highlights the objects in the group.
3. Click the **Continue** button to return to the dialog box.

The method works in reverse: select an object to find out the name of its group(s).

4. Choose the **Find Name** button.
5. The dialog box disappears, and AutoCAD prompts you to pick an object:
   Pick a member of a group. *(Select one object).*
6. Select the object you are curious about, and AutoCAD displays a dialog box listing the name(s) of the group(s) to which the object belongs.

If the object is not a member of a group, AutoCAD complains, "Not a group member."

Changing Groups

After the groups are created, you can change their descriptions, as well as the selection of objects comprising the group. You can change these by choosing the appropriate button in the Object Grouping dialog box:

- **Remove** removes objects from the group. AutoCAD temporarily dismisses the dialog box, and then highlights members of the group so that you can select the ones to remove.

  Select objects to remove from group...

  Remove objects: *(Select one or more objects.)*

  Remove objects: *(Press* ENTER *to return to dialog box.)*

- **Add** adds objects to the group. AutoCAD again dismisses the dialog box, highlighting members of the group so that you don't accidentally select current members.

  Select objects to add to group...

  Select objects: *(Select one or more objects.)*

  Select objects: *(Press* ENTER *to return to dialog box.)*

- **Rename** renames the group. Enter a new name in **Group Name** text box, and then choose the **Rename** button.

- **Re-order** changes the "order" in which the objects are handled for selection, tool paths, and so on. AutoCAD numbers the objects in the group as you add them, the first object being numbered 0. Other than reversing the order, the renumbering process is, unfortunately, painful to execute.

- **Description** changes the description of the group. Enter a new description in the **Description** text box, and then choose the **Description** button.

- **Explode** deletes the selected group from the drawing.

DRAWORDER

The **DRAWORDER** command determines the order in which overlapping objects are displayed. In the figure below, the text "Using AutoCAD" is obscured by the rectangle. By changing the display order — the order in which AutoCAD redraws objects — you can make the text visible on top of the inappropriately placed rectangle.

Left: *Text above the rectangle.*
Right: *Text below the rectangle.*

TUTORIAL: SPECIFYING DRAWORDER

1. To specify the display order of overlapping objects, start the **DRAWORDER** command with one of these methods:
   * From the **Tools** menu, choose **Draw Order**.
   * On the Modify II toolbar, pick the **Draworder** button.
   * At the 'Command:' prompt, enter the **draworder** command.

      Command: **draworder** *(Press* ENTER.*)*
   * Alternatively, enter the **dr** alias at the keyboard.

2. Select one or more objects:

      Select objects: *(Select one or more objects.)*

      1 found Select objects: *(Press* ENTER *to end object selection.)*

3. Specify the display order for the selected objects(s)

      Enter object ordering option [Above object/Under object/Front/Back] <Back>: *(Press* ENTER.*)*

      Regenerating model.

CHANGING THE DRAW ORDER: ADDITIONAL METHODS

The **DRAWORDER** command has four options for controlling the visual overlapping of objects. The first two listed below are useful when three or more objects overlap.

   * **Above object** places the object visually on top of other selected objects.
   * **Under object** places the object under other selected objects.
   * **Front** places the object above all other objects.
   * **Back** places the object below all other objects.

HpDrawOrder

AutoCAD provides additional control over draw order. The **HPDRAWORDER** system variable determines whether hatch and fill patterns are above or below their boundaries and other objects. See the **HATCH** command earlier in this chapter.

TextToFront

The **TEXTTOFRONT** command determines whether text and/or dimensions are always placed on top (or in front) of overlapping objects in the drawing.

>   Command: **texttofront**
>
>   Bring to front [Text/Dimensions/Both] <Both>: *(Press* ENTER.*)*

This command ensures the text is always legible, not obscured by other objects. It applies to all text and dimensions in the drawing. It does not, however, work when model space objects overlay text in a layout, and vice versa.

This command is not a mode; if text and dimensions are obscured by later objects, you must run this command again.

EXERCISES

1. In this exercise, you use some of the options of the **SELECT** command to erase portions of a valve housing drawing.

    From the Companion CD, open the *edit1.dwg* drawing.

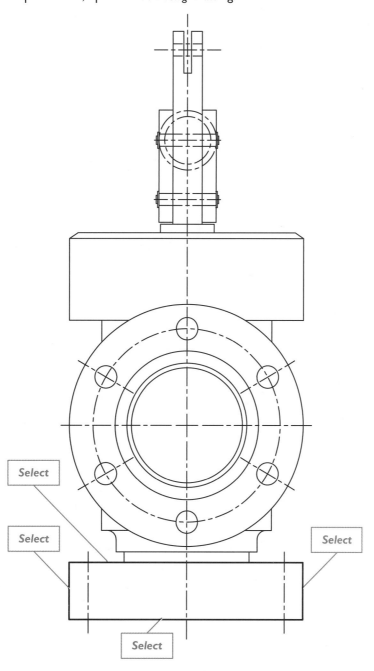

    Use the **ERASE** command to delete some of the objects.

    You should now see a pickbox on the screen. Place the pickbox over the bottom line of the part, as shown in the illustration, and click. The line should be highlighted.

    One by one, select the remaining lines, as shown in the figure.

    Finally, press **ENTER**. The four lines are removed from the drawing.

2. Repeat the above exercise on the same drawing with the **Window** object selection mode. But first, use the **u** command to undo the erasure.

Select **ERASE** again, and this time enter **w** in response to the "Select objects:" prompt.

Consult the figure for the points referenced in the following command sequence.

Command: **erase**

Select objects: **w**

Specify first corner: *(Select point 1.)*

Specify other corner: *(Select point 2.)*

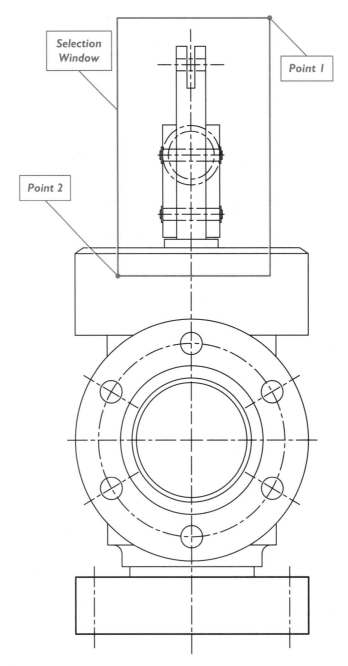

All items within the window are selected and highlighted. Notice that objects that extend outside the window area (but not wholly contained therein) are not selected.

Select objects: *(Press ENTER.)*

Press **ENTER** and the selected objects are deleted.

3. Repeat the above exercise on the same drawing with the **Crossing** window object selection mode. Again, first undo the erasure with the **u** command.

    Select **ERASE** again, entering **C** (for "crossing") as the option. Refer to the following command sequence and the figure.

    Command: **erase**

    Select objects: **c**

    Select first corner: *(Select point 1.)*

    Select other corner: *(Select point 2.)*

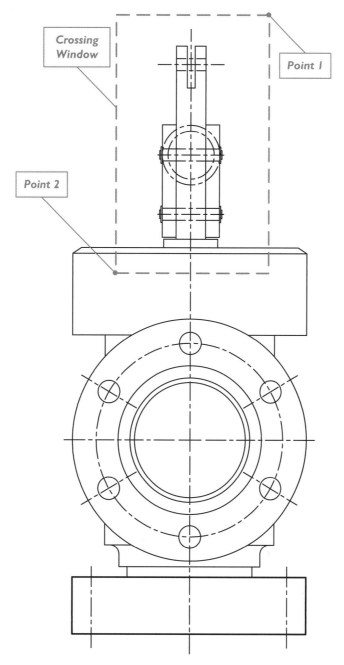

Notice that all the objects within and crossing the window are selected (as noted by the highlighting).

4. Continuing from the above exercise, remove objects from the selection set.

After you placed the crossing window in the previous exercise, AutoCAD asked you to select more objects. This time, enter **R** for remove. The command sequence continues:

Select objects: **r**

Remove objects: *(Select one of the horizontal lines.)*

1 found, 1 removed Remove objects: *(Select the other horizontal line.)*

1 found, 1 removed Remove objects: *(Press ENTER.)*

The objects you removed from the object selection set are not erased.

Exit the drawing, discarding the changes you made so that the edits are not recorded.

5. In this exercise, you create a block from a drawing, and then insert the block in the drawing several times.

From the Companion CD, open the drawing named *office.dwg*.

Window the entire desk, and then start the **BLOCK** command.

Name the block "SDESK", and then pick an insertion point.

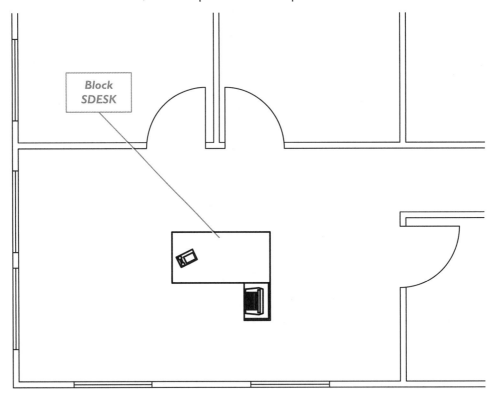

Click **OK**. The block is now stored with the drawing, and can be used as many times as you require.

With the **INSERT** command, place the SDESK block in the drawing three times, using these scales and rotation angles:

Scale = 1.0   Angle = 0 degrees

Scale = 1.5   Angle = 90 degrees

Scale = 0.75   Angle = 45 degrees

6.   In this exercise, use DesignCenter to place blocks in a drawing.
     From the Companion CD, open the site plan drawing named *insert.dwg*. The landscape
     items are drawn for you and stored as blocks in the drawing.

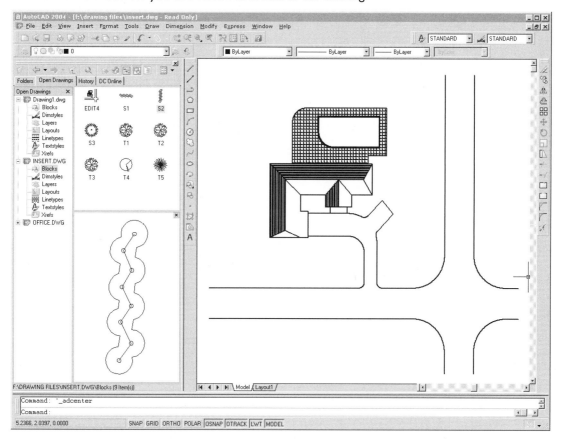

Use **ADCENTER** to view the blocks, and then insert them in the drawing to create a
landscaping design of your own.

7.   In this exercise, you draw an electronic part called a "diode," and then insert it as an array
     of blocks.
     Using the **PLINE** command, construct the semi-conductor symbol shown below.

With the **BLOCK** command, select the diode, and name it "DIODE"; use the left end of the
diode as the insertion point.
You have now created the block DIODE in the drawing.
Use the **MINSERT** command to place an array of diodes, as follows:

    Command: **minsert**

    Enter block name or ?: **diode**

    Specify insertion point or [Scale/X/Y/Z/Rotate]: *(Pick a point.)*

    Enter X scale factor, specify opposite corner, or [Corner/XYZ] <1>: *(Press ENTER.)*

    Enter Y scale factor <use X scale factor>: *(Press ENTER.)*

    Specify rotation angle <0>: *(Press ENTER.)*

Enter number of rows (—-) <1>: **4**

Enter number of columns (| | |) <1>: **6**

Enter distance between rows or specify unit cell (—-): *(Pick point 1.)*

Specify opposite corner: *(Pick point 2.)*

You may need to use the **ZOOM All** command to see the completed array.

Notice the (- - -) and (||||) notations in the rows and columns prompts that make it easier to remember which way rows and columns operate.

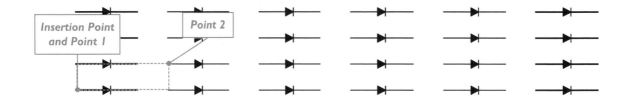

8. In this exercise, you hatch portions of a drawing of several hot air balloons.

    From the Companion CD, open the drawing named *solids.dwg.*

    Use the **HATCH** command to place the solid hatch pattern in a variety of colors in some of the balloon areas.

Now try placing a variety of hatch patterns in the same drawing.

9. In this exercise, you specify the area to be hatched with a pick point. The figure shows an object containing three areas that could be hatched.

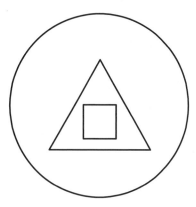

You need to hatch the square and circle, but not the triangle.

Draw the circle, triangle, and square using the **CIRCLE** and **POLYGON** commands.

From the **Draw** menu, select **Hatch**.

In the **Pattern** list box, select **ANSI31**.

Select the **Advanced** tab, and then ensure that **Normal** appears in the **Style** list box and that the **Island Detection** method has **Flood** selected.

Select the **Pick Points<** button, and then click inside the circle, but outside the triangle

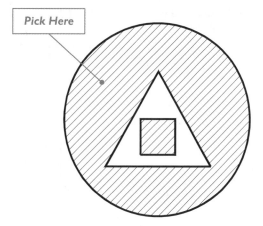

Pick Here

Right-click to return to the dialog box.

Choose the **Preview** button. Your hatch should look similar to the figure above. If not, change the hatch options and/or your pick points.

When the preview hatch looks right, choose **OK** to apply the hatch pattern.

10. In this exercise, you use hatch styles to control the behavior of hatch patterns.

    From the Companion CD, open the *hatch.dwg* drawing file.

    Start the **HATCH** command, and set the following options:

Pattern	**ANSI31**
Scale	**1**
Angle	**0**
Island Detection Style	**Normal**
Select objects	**W** *(or pick a point between the two outer rectangles)*

    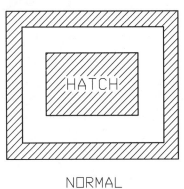

    Place the window around the object, and then press **ENTER**. Your drawing should look similar to the following illustration:

    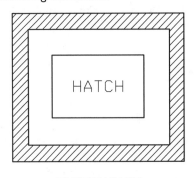

11. Repeat the above exercise on the same drawing, but use the **Outer** style. The drawing should look similar to the following illustration:

    OUTERMOST

12. Repeat the above exercise on the same drawing, but use the **Ignore** style of island detection. Notice that the hatch has ignored the boundaries and the text. Your drawing should look similar to the following illustration:

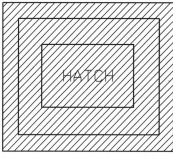

IGNORE

13. Use the Block Editor to design the following symbols. Save them using the name indicated.

 a. Fan   b. TransformerDelta

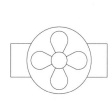   

 c. HexHeadScrew-Top   c. HexHeadScrew-Side

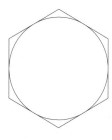   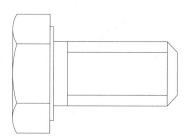

 d. FlangedCross   e. ElbowOutlet-Up

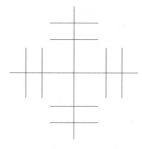

CHAPTER REVIEW

1. Describe the purpose of DesignCenter.
2. What are two ways to insert a block from DesignCenter into a drawing?
   a.
   b.
3. When might you use the History tab?
4. Which key do you hold down to select more than one item?
5. List four types of drawing content displayed by DesignCenter:
   a.
   b.
   c.
   d.
6. Which shortcut keystroke toggles the display of the DesignCenter?
7. What is the purpose of the **ADCNAVIGATE** command?
8. What is the meaning of the message "Duplicate names ignored"?
9. Explain what the **HATCH** command does.
10. If you did not want the solid filled areas in your drawing to plot, what could you do?
11. What are *hatch boundaries*?
12. What are *blocks*?
13. What is the *base point* of a block?
14. When placed in drawings, how do blocks handle their layer definitions?
15. What is a *nested* block?
16. How would you create a separate drawing file from an existing block?
17. How do you place blocks in drawings?
18. How can you place blocks in other drawings?
19. When you place blocks in the drawings, what is the *insertion point*?
20. Can you place one AutoCAD drawing in another AutoCAD drawing?
21. Name two advantages of using blocks.
22. What commands are combined to create the **MINSERT** command?
23. How would you change the origin point of a hatch?
24. When do hatch patterns obscure text?
25. When an editing command is invoked, must a selection option, such as **SIngle** or **Window**, be entered before an object is selected?
26. When choosing a set of objects to edit, other than by using the **U** command, how can you alter an incorrect selection without starting over?
27. If you wish to edit an item that was not drawn last, but was the last item selected, would the **Last** selection option allow you to select the item?
28. How do you increase and decrease the size of the pickbox?
29. When entering **BOX** in response to the "Select objects:" prompt, how are you then allowed to choose objects?
30. What makes it evident that an item has been selected?
31. When a group of items is selected with the **Window** option, does an item become a part of the selection set as long as it is partially inside the window?
32. In the object selection process, how is the **BOX** option different from the **AUtomatic** option?
33. What is the purpose of the tool palette?

34. How do you add blocks from a tool palette to the drawing?
35. Describe the purpose of the **BASE** command.
36. In what situation do you use the **BLOCKICON** command?
37. What kind of hatches can the **HATCHEDIT** command not change?
38. Can the **HATCHEDIT** command copy the properties of one hatch pattern and apply them to another?
39. What is the difference between the **SELECT** and **QSELECT** commands?
40. Can you change the size of the *pick* cursor?
41. What are the two keyboard shortcuts for toggling group mode?
42. What is the purpose of the **DRAWORDER** command?
43. Name two differences between blocks and groups:

    a.

    b.

44. Explain how *auxiliary scale* differs from the normal scale factor.
45. Describe how to add objects to the Tool Palette.
46. Why might you create *groups* of palettes?
47. What is the purpose of the **BATCH** command's **Gap Tolerance** option?
48. When might you use the Hatch and Gradient dialog box's **Draw Order** option?
49. Explain the effect of turning off the **OSNAPHATCH** system variable.
50. Can hatch patterns be trimmed?

    Can hatch patterns be used as trim boundaries?

51. How would you force text and dimensions to display in front of other overlapping objects in drawings?
52. Name three uses for the Block Editor:

    a.

    b.

    c.

53. Describe two ways to change blocks to a new definition.

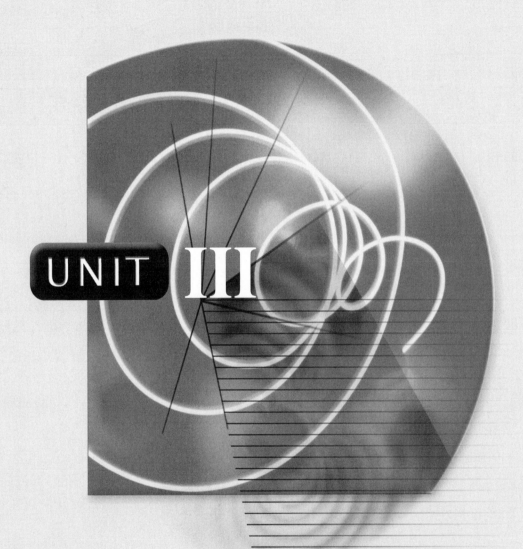

UNIT III

Editing
Drawings

CHAPTER 7

Changing Object Properties

As you create drawings, you sometimes need to change the properties of objects, such as their color, layer assignment, and linetype pattern.

In some earlier chapters, you learned how to modify specific objects, for example hatch patterns with the HATCHEDIT command and point styles with the DDPDMODE command. These commands are specific to certain objects; in this chapter, however, you learn how to change the properties of *any* object using the following commands:

COLOR and **-COLOR** change the colors of objects.

LWEIGHT and **-LWEIGHT** change the display width of lines.

LINETYPE and **-LINETYPE** change the patterns of lines.

LTSCALE and **CELTSCALE** change the scale factor of linetypes.

LAYER and **-LAYER** apply properties to all objects on layers.

LAYERSTATE accesses the Layer State Properties dialog box directly (new to AutoCAD 2008).

SETBYLAYER resets layer properties in overridden viewports (new to AutoCAD 2008).

AI_MOLC and **CLAYER** set the current layer.

LAYERP and **LAYERPMODE** set the layers to their previous state.

LAYISO isolates selected layers by fading out all other layers (new to AutoCAD 2008).

PROPERTIES palette and **CHPROP** change nearly all properties of objects.

MATCHPROP matches the properties of one object to other objects.

NEW TO AUTOCAD 2008 IN THIS CHAPTER

- **LAYERSTATE** and **SETBYLAYER** are new commands.
- Layer properties can be overridden (and reset) on a per-viewport basis.
- **LAYISO** now fades isolated layers.
- Commands like **CHANGE** and **PROPERTIES** now support annotative scale factors.

FINDING THE COMMANDS

On the Dashboard's **Layers** panel:

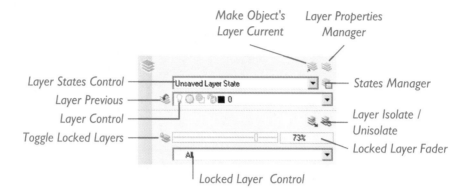

On the Dashboard's **Object Properties** panel:

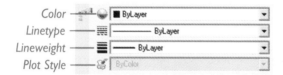

On the **Layers** and **Properties** toolbars:

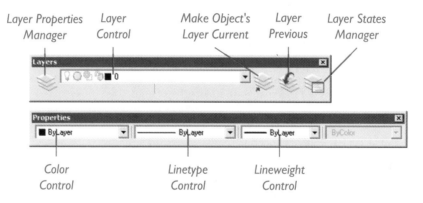

COLOR

Colors are used to identify objects, to segregate them visually, and to set lineweights for plotting.

In AutoCAD, the default color of 2D objects is white or black, depending on the background color of the viewport. (If the background color is black, then objects are displayed in white; if write, then AutoCAD automatically switches to black objects.) In 3D modeling, AutoCAD uses pale blue as the default color (RGB = 231,240,255).

AutoCAD has a number of methods of setting and changing the color of objects:

- **Preset the color.** For objects you plan to draw, the best method is to specify colors through the Color option of the **LAYER** command; less preferred is to set the working color through the **COLOR** command.

- **Change the color.** For objects already in the drawing, the fastest method is to use the Color Control droplist on the Object Properties panel; alternatives are to use the **PROPERTIES** or **CHPROP** commands. (The best method is to move the objects to the appropriate layer, so that they take on the color of the layer.)

Here is the difference: the LAYER command sets the color for objects drawn on each layer; the COLOR command sets the color for *subsequently* drawn objects. It is important to understand that the COLOR command overrides the colors set by the LAYER command. Thus, it is possible for a layer to contain objects of different colors, regardless of the color set for the layer.

TUTORIAL: CHANGING COLORS

1. To change the color setting, start the **COLOR** command with one of these methods:
   - From the **Format** menu, choose **Color**.
   - From the Properties toolbar, choose the Color Control droplist, and then **Select Color**.
   - In the Dashboard's **Object Properties** panel, click the Color Control droplist, and then choose **Select Color**.
   - Or, at the 'Command:' prompt, enter the **color** command:

     Command: **color** *(Press* ENTER.*)*
   - Alternatively, enter the aliases **col**, **colour** (the British spelling), or **ddcolor** (an old name) at the 'Command:' prompt.

2. In all cases, AutoCAD displays the Select Color dialog box.

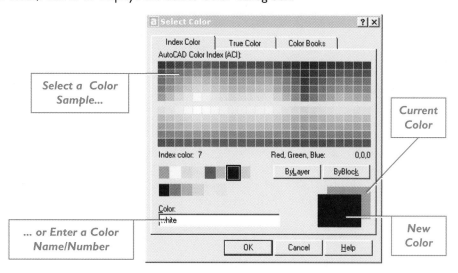

3. Select a color sample (one of the colored squares), or enter a color name/number.

4. Click **OK**. All the objects you now draw take on the new color — until you change color again.

 Note: The Select Color dialog box is used by more than just the **COLOR** command; it is also accessed by many other commands that select colors, such as **LAYER, OPTIONS,** and **PROPERTIES.**

(*History*: Some drafters were upset when Autodesk introduced the **COLOR** command with AutoCAD v2.5. They felt that layer settings should always determine the colors of objects. Like this book's technical editor, these drafters feel it is wrong to override the layer's color setting with the **COLOR** command. Overriding colors is a poor method of working with CAD, which you may come across in your place of work. AutoCAD 2008 partially solves the problem by adding the **SETBYLAYER** command, which forces objects to take the color, linetype, lineweight, and plot style of their assigned layers.)

WORKING WITH AUTOCAD COLORS

For many drafters, color is extremely important; it's more than the hues and shades of objects. In old releases of AutoCAD, color was the only way to control pens in plotters. The color red, for example, was assigned to pen #1, which in turn could be the color red, or a black pen with a width of 0.1".

Today in AutoCAD, you control the plotter by color (the old method) or *styles* (the current method). More details follow in Chapter 22, "Plotting Drawings."

You can specify colors by one of several methods: select a color swatch from the dialog box, specify a number between 1 to 255, enter an RGB value, a color book name and number, or (for the first seven colors) type a color name, such as "blue."

By the way, *hue* refers to differences in a color, while *shade* refers to differences in gray.

Color Numbers and Names

Of the 16.7 million colors available on computers, AutoCAD identifies the first seven by name and number, and the first 255 by number only. Color numbers are called the "AutoCAD Color Index," or ACI for short. Beyond that, colors are identified by RGB value (amounts of red, green, and blue) or by color book number.

The first seven color names and associated ACI numbers are liste below in numerical order:

ACI	Name	Abbreviation*
1	Red	R
2	Yellow	Y
3	Green	G
4	Cyan	C
5	Blue	B
6	Magenta	M
7	White (black)	W

* These abbreviations are used at the command line during commands like -COLOR, CHANGE, and -LAYER.

You may be wondering about my reference to the *first* 255 colors? Before AutoCAD 2004, AutoCAD worked with just 255 colors. These are the ones you see in the Index Color tab of the Select Color dialog box. (Colors 250 through 255 are shades of gray.) Every CAD package adopts a different order for colors, and so Autodesk identified AutoCAD's as the ACI — *AutoCAD* Color Index.

Background Color 0

Color 0 is a special ACI color number that you never use in drawings. It identifies the *background* color of the drawing area (a.k.a viewport color).

To change the background color, use the Colors dialog box accessed through the **Tools | Options | Display | Colors** menu sequence. This dialog box lets you change the colors of the background of the model layout, layout viewports, plot preview, block editor, and both 3D projections, as described in Chapter 3, "Setting Up Drawings."

Color 7

The background can be any color, but typically is white or black. Because white lines are invisible against a white background, AutoCAD automatically changes color 7 (white) as to be the opposite of the background color (color #0) — so here is a case in life when white is black, and black is white.

Officially, however, color 7 is white, because AutoCAD's original background color was black. For this reason, you may encounter color 7 being referred to as "white," even though it appears black on your screen, as well as black when plotted on paper.

Colors ByLayer, ByBlock, and ByEntity

In addition to the special colors 0 and 7, AutoCAD has three more special "colors": the oft-seen ByLayer, less-commonly seen ByBlock, and rarely-seen ByEntity. These solve the problem of what to name colors when they are controlled by layers, blocks, or entities (the old term for objects).

ByLayer means that the color of objects is determined by the layer color setting. Choosing ByLayer causes subsequently drawn objects to inherit the layer's color; you return control of objects' colors to the layer's color setting. This color is also known as color #256.

ByBlock means that objects in blocks take on the color of the block. Choosing ByBlock causes objects to be drawn in white until they are turned into a block. When the block is inserted, the objects inherit the color of the block insertion. This color is also known as entity color #0.

ByEntity means that each object carries its own color, rather than one assigned by the layer or block. ByEntity is only used by system variables, such as **OBSCUREDCOLOR** and **INTERSECTIONCOLOR**, although programmers can make use of it in their software programs. This color has a value of 257.

True Color

AutoCAD now displays millions of colors. Technically, computer software works with three sets of 8-bit colors, which translates into 16.7 million hues. It can get confusing trying to select a single color out of millions, and so AutoCAD provides several methods to help you make your selection. The methods are through RGB, HSL, and color books.

RGB

RGB is short for "red, green, blue," a common system of selecting colors that uses three sets of color numbers ranging from 0 to 255. Each number represents the amount of red, green, or blue, ranging from black (0) to full color (255).

As you may recall from elementary school physics, all colors of light can be represented by varying amounts of red, green, and blue. Yellow is made from mixing red and green, orange is made from full red (255) and half green (127), and no blue (0), and so on.

The hue of **blue** used in this book is represented by R=38, G=133, and B=187. Black consists of no colors (0,0,0), and white is all colors (255,255,255).

To specify a color in AutoCAD with the RGB system, follow these steps:

1. In the Select Color dialog box, select the **True Color** tab.
2. Under Color Model, select **RGB**.
3. Drag the sliders for Red, Green, and Blue.

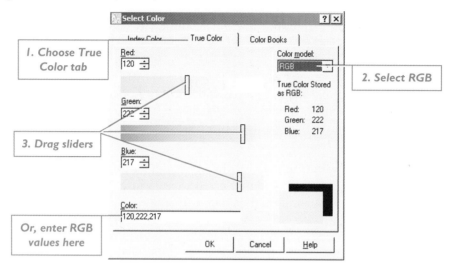

Alternatively, enter an R,G,B color triplet, (such as 120,222,217) in the **Color** text box.

4. Click **OK**.

HSL

HSL is short for "hue, saturation, luminance." It is a second system for specifying colors, which varies the *hue* (color), *saturation* (amount of color, ranging from gray to full color), and *luminance* (brightness of color, ranging from darkened to bright).

Hue ranges from 0 to 360. Starting at 0 = red, the colors progress through the rainbow: yellow is 60, green is 120, cyan is 180, blue is 240, violet is 300, and then back to red at 360.

Saturation and luminance each range from 0 to 100 percent, where 0 is none (black) and 100 is full (white). The HSL value for the blue used in this book is H=202, S=66, and L=44.

To specify a color through HSL:

1. In the Select Color dialog box, select the **True Color** tab.
2. Under Color Model, select **HSL**.

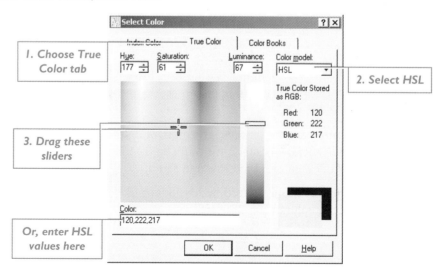

3. Move the cross slider around to select the hue and saturation. Move the horizontal slider up and down to choose luminance.

   Alternatively, In the **Color** text box, enter a color triplet, such as 120,222,217.

4. Click **OK**.

 Note: AutoCAD includes a drawing that shows all its colors in the form of *color wheels*. In the *\autocad 2008\samples* folder, open *colorwh.dwg*.

On the left, looking like a smoothly shaded donut, is the true color wheel made of 35,840 tiny tiles. (The coloring makes it appear like a 3D torus.)

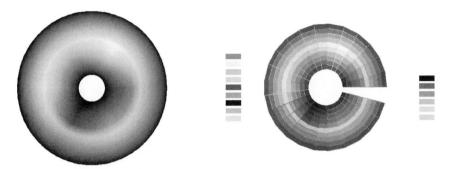

On the right is the original 256-color ACI color wheel. It is flanked by the named colors (to the left) and shades of gray (to the right).

You can set precise colors in your drawings using RGB or HSL. For example, using a raster editor, such as PaintShop Pro, you can determine the color of a client's logo.

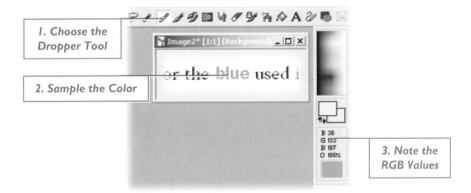

Here's how I do it: get a sample of the color, either through a screen grab or by scanning. Open the sample in the raster editor. Use the color dropper tool to sample the color; note the RGB values, and then enter them in AutoCAD's Select Color dialog box.

Color Books

AutoCAD provides a third naming system used by publishing and design industries, called "color books." Colors are seen differently by individuals, and so a standardized system for specifying a select number of colors was needed. Color books consist of swatches and numbers.

For example, the name of the blue color used in this book is named Pantone 2925. When I specify this name, the identical hue of blue shows up in my desktop publishing software as in the printed book, and in AutoCAD.

AutoCAD supports three of systems: the American Pantone, the European RAL, and the Japanese DIC systems.

The Pantone Matching System®(www.pantone.com) was designed in 1963 to specify colors for the graphic arts, textiles, and plastics industries. Today, designers typically work with Pantone's fan-format book of standardized colors. This system is used primarily in North America.

The RAL color system (www.ral.de) was designed back in 1927 to standardize colors by limiting the number of color gradations, first to just 30 and now to over 1,600. RAL (Reichs Ausschuß für Lieferbedingungen – German for "Imperial Committee for Supply Conditions") is administered by the German Institute for Quality Assurance and Labeling. This system is used primarily in Europe.

The DIC Color Guide (www.dic.co.jp, Japanese language only) is the Japanese standard for specifying colors. DIC is the abbreviation for the ink company that came up with the system, Dainippon Ink and Chemicals.

Here's how to specify colors by Color Books in AutoCAD.

1. In the Select Color dialog box, choose the **Color Books** tab.

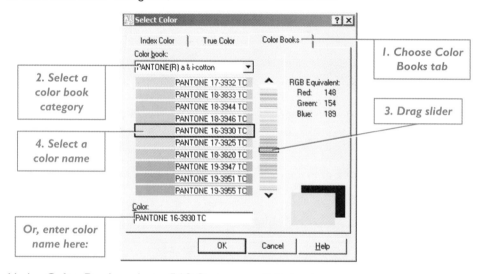

2. Under **Color Books**, select a DIC, Pantone, or RAL book name.
3. Move the slider up and down to choose a color group, and then pick a specific color name.

    As an alternative, in the **Color** text box, enter a color book specification.
4. Click **OK**.

CHANGING COLORS: ADDITIONAL METHODS

AutoCAD provides several other ways of changing the color of objects:

*   **-COLOR** command changes the working color at the command line.
*   **Color Control** droplist on the Properties toolbar changes the color of selected objects.

Later in this chapter, you learn about changing color with the **PROPERTIES** and **CHPROP** commands.

-Color

The **-COLOR** command changes the working color at the command line. Power users prefer this over the dialog box, because it can be faster. *Real* power users, however, always set all colors to ByLayer.

    Command: **-color**

    Enter default object color [Truecolor/COlorbook] <1 (red)>: *(Enter a color number, name, or abbreviation, or else enter an option.)*

The color names, numbers, and abbreviations were discussed earlier in this section.

The **Truecolor** option prompts you for a color specified by red, green, and blue. (The HSL method is not available.)

> Red,Green,Blue: **38,133,187**

The **COlorbook** option prompts you to enter a color book name:

> Enter Color Book name: (*Enter* **pantone** *or* **ral** *or* **dic**.)
>
> Enter color name: (*Enter a color name, such as* **11-0604 TC**.)

Color Control Droplist

As an alternative to the color commands, you can use the **Color Control** droplist in the Properties toolbar or the Object Properties panel of the Dashboard. The controls affect color by two methods, depending on the order in which you use them. Method 1: To *set* the working color (subsequent objects taking on the color), select a color from the droplist. Method 2: To *change* the color of objects(s), first select the objects, and then a color from the droplist.

In the following tutorial, we look at both methods.

TUTORIAL: SETTING AND CHANGING COLORS

1. Start AutoCAD with a new drawing, and then draw a few lines. Notice that they are colored white or black — depending on the background color of your screen.

2. From the Properties toolbar or the Dashboard's Object Properties panel, click on the **Color Control** droplist. (It can be hard sometimes to figure out the location of the droplist on the toolbar, because it is not labeled; I look for the one with the color square.)

    When you click the Color Control droplist, notice that it lists a basic set of colors: ByBlock, ByLayer, and the first seven colors, red through white. (White has a black/white icon indicating the ambiguity of the color's name.)

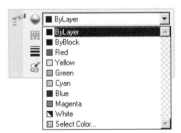

3. Select **Red**.
4. Draw some more lines. Notice that they are colored red.
5. To change the red lines to blue, do this: select the red lines by picking them. They become highlighted and show grips (squares).

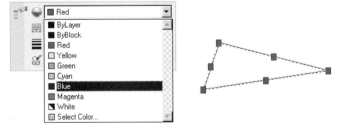

6. From the Color Control droplist, select **Blue**. Notice that the lines turn blue.

    Press **ESC** to "unselect" the lines (remove the highlighting and grips).

LWEIGHT

The **LWEIGHT** command (short for "lineweight") changes the apparent width of lines and other objects.

In old releases of AutoCAD, only polylines and traces could have a width (weight); all other objects were drawn one pixel wide, as thin as the display. Widths were assigned at plot time based on the object's color. As of AutoCAD 2000, almost any object can have width.

Using a variety of weights (heavy and thin lines) helps make drawings clearer. Weights can be assigned to layers or to individual objects. In the figure below, the upper part of the drawing has lineweights turned off; in the lower part, lineweights are turned on.

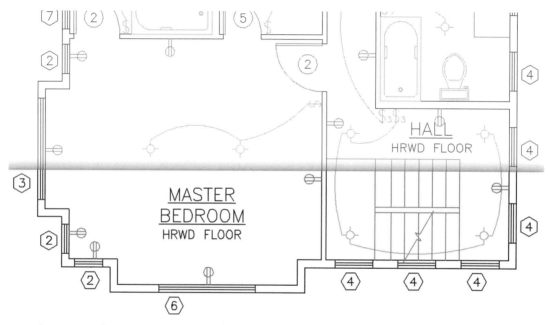

Lineweights range from 0.05 mm (0.002") to 2.11 mm (0.083"). AutoCAD displays the values in millimeters or inches. The lineweight of 0 is compatible with earlier versions of AutoCAD, and displays as one pixel in model space. It is plotted at the thinnest width the plotter is capable of.

When on, the weight appears when both displayed and plotted. The display is constant, in pixels. This means that as you zoom out, the lines appear to get wide — "WYSINQWYG," what you see is not quite what you get, in the words of the technical editor.

There is an exception: objects with lineweights under a certain value are displayed one pixel wide in model space. The default is 0.025mm but can be adjusted by the **Display Scale** slider in the Linetype Settings dialog box, described below.

At plot time, objects with weight are plotted at the exact same widths. Objects copied to the Clipboard (through **CTRL+C**) retain their lineweight data when pasted (**CTRL+V**) back into AutoCAD, but may lose their weight when pasted into other Windows applications.

Some drafters prefer to define lineweight through layers, as with colors; they do not approve of overriding the layer's lineweight ByLayer setting.

AutoCAD includes three named lineweights:

ByLayer — objects take on the lineweight defined by the layer.

ByBlock — objects take on their block's lineweight.

Default — displays the default value of 0.25 mm, or another default value.

Note: Lineweights are meant as a visual aid. Do not use lineweights to represent the actual width of objects. If printed circuit board traces are 0.05 inches wide, use polylines with width of 0.05 inches; don't draw lines with a weight of 0.05 inches.

TUTORIAL: CHANGING LINEWEIGHTS

1. In new drawings, lineweights are not displayed. To turn on their display, choose the **LWT** button on the status bar. Lineweights are on when the button looks depressed.

2. To change the lineweights of objects, start the **LWEIGHT** command with one of these methods:

   • From the **Format** menu, choose **Lineweight**.

   • On the status line, right-click **LWT**, and then choose **Settings** from the shortcut menu.

   • At the 'Command:' prompt, enter the **lweight** command.

   • Alternatively, enter the aliases **lw** or **lineweight** at the 'Command:' prompt.

   Command: **lweight** *(Press* ENTER.*)*

3. In all cases, AutoCAD displays the Lineweight Settings dialog box.

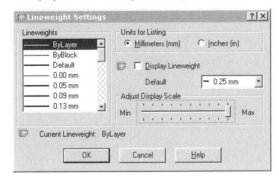

4. Under **Lineweights** is the list of weights available in AutoCAD. (This is the same list displayed by the Lineweights Control droplist on the Properties toolbar.) You cannot customize lineweights.

   Select a lineweight; this becomes the default until you change the lineweight again.

5. Other items of interest in the dialog box include:

   Units for Listing — selects between inches and millimeters (the default).

   Display Lineweight check box — functions identically to the **LWT** button on the status bar.

   Default list box — specifies the lineweight with which all objects are drawn, unless overridden by layer.

   Adjust Display Scale slider — controls the display scale of lineweights in model space. This slider is beneficial if your computer displays AutoCAD on a high-resolution monitor, which tends to make lines look thinner. Experiment with adjusting the lineweight scale to see if you get a better display with different lineweights. As you move the slider, notice that the widths of lines in the Lineweights list change.

6. Choose **OK** to dismiss the dialog box.

CHANGING LINEWEIGHTS: ADDITIONAL METHODS

AutoCAD provides several other ways to change the lineweight of objects:

   • **-LWEIGHT** command changes the working lineweight at the command line.

   • **Lineweight Control** droplist on the Properties toolbar changes the lineweight of selected objects.

AUTOCAD'S DEFAULT LINEWEIGHTS

The table below shows the default lineweight values used by AutoCAD and their equivalent values for industry standards. There are 25.4 mm per inch; 72.72 points per inch.

Millimeters	Inches	Points	Pen Size	ISO	DIN	JIS	ANSI
0.50	0.002						
0.90	0.003	$^1/_4$ pt					
0.13	0.005				✓		
0.15	0.006						
0.18	0.007	$^1/_2$ pt	0000	✓	✓	✓	
0.20	0.008						
0.25	0.010	$^3/_4$ pt	000	✓	✓	✓	
0.30	0.012		00				2H or H
0.35	0.014	1 pt	0	✓	✓	✓	
0.40	0.016						
0.50	0.020		1	✓	✓	✓	
0.53	0.021	$1^1/_2$ pt					
0.60	0.024		2				H, F, or B
0.70	0.028	$2^1/_4$ pt	$2^1/_2$	✓	✓	✓	
0.80	0.031		3				
0.90	0.035						
1.00	0.039		$3^1/_2$	✓	✓	✓	
1.06	0.042	3 pt					
1.20	0.047		4				
1.40	0.056			✓	✓	✓	
1.58	0.062	$4^1/_4$ pt					
2.00	0.078			✓	✓		
2.11	0.083	6 pt					

Let's look at each option. (Later in this chapter, you also learn about changing lineweight with the **PROPERTIES** and **CHPROP** commands.)

-LWeight

The **-LWEIGHT** command changes the lineweights and related options at the command line:

> Command: **-lweight**
>
> Current lineweight: 0.024"
>
> Enter default lineweight for new objects or [?]: *(Enter a value, such as **.042**, or type **?**.)*

The **?** option lists the available lineweights:

Enter default lineweight for new objects or [?]: **?**

ByLayer ByBlock Default

0.000" 0.002" 0.004"   0.005" 0.006" 0.007"

0.008" 0.010" 0.012"   0.014" 0.016" 0.020"

0.021" 0.024" 0.028"   0.031" 0.035" 0.039"

0.042" 0.047" 0.055"   0.062" 0.079" 0.083"

To change the display between imperial and metric units, use the LWUNITS system variable, as follows:

Command: **lwunits**

Enter new value for LWUNITS <0>: **1**

A value of 0 (zero) causes AutoCAD to display lineweights in imperial units (inches), while a value of 1 displays units in metric (millimeters).

Lineweight Control

As an alternative to using commands and dialog boxes, you can change lineweights directly through the **Lineweight Control** droplist. As with color, this control affects lineweights by two methods, depending on the order in which you use it. Method 1: To *set* the working lineweight (subsequent objects taking on the selected lineweight), select a lineweight from the droplist. Method 2: To *change* the lineweight, first select the objects, and then a lineweight from the droplist.

In the following tutorial, we look at both methods.

TUTORIAL: ASSIGNING AND CHANGING LINEWEIGHTS

1. Start AutoCAD with a new drawing, and then draw a few lines. Notice that they are thin looking.
2. From the Dashboard's Object Properties panel (illustrated below) or the Properties toolbar, click on the **Lineweight Control** droplist.

3. Select **0.30mm** (or 0.012").
4. Draw some more lines. Notice that they are thicker.
5. Now, change the weight of the lines, as follows: Select the lines, which become highlighted and show grips.

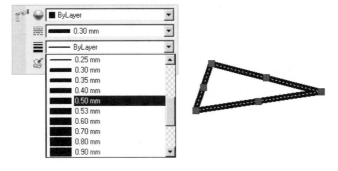

6. From the Lineweight Control droplist, select **0.050mm** or **0.020"**. Notice that the lines become thicker.

7. Press **ESC** to "unselect" the lines.

LINETYPE

The **LINETYPE** command sets the patterns of lines.

Linetypes are used to identify objects in 2D drawings. For example, solid lines represent the edges of objects; AutoCAD refers to solid lines as "continuous lines." Lines marked with "HW" represent hotwater pipes.

Edges hidden from view are shown in a linetype referred to as "hidden," constructed of short segments. The figure shows an object containing edges that are hidden in some views. These edges are defined with the Hidden linetype.

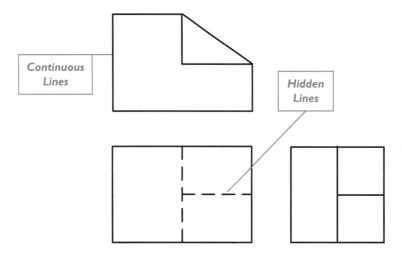

AutoCAD provides many linetypes, and you can add custom ones. AutoCAD provides three classes of linetype:

Simple — consist of line segments, dots, and gaps; these are the most commonly-used linetypes.

ISO — are similar to simple linetypes, except that they conform to standards set by the International Organization of Standards, and can have pen widths assigned to them.

Complex — use dashes and spaces like simple linetypes, but add characters and shapes, such as HW (mentioned above) or squiggles (to indicate insulation).

AutoCAD stores linetype definitions in *acad.lin* and *acadiso.lin*, and the shapes for complex linetypes in *ltypeshp.shp* — all found in the *\support* folder. You can create your own linetypes, as well.

Using linetypes in drawings takes two steps: (1) loading them into the drawing, and (2) selecting for use. Applying them by layer is the preferred method.

TUTORIAL: CHANGING LINETYPES

1. To change the linetype, start the **LINETYPE** command with one of these methods:

   • From the **Format** menu, choose **Linetype**.

   • From the Properties toolbar, choose the Linetype Control droplist, and then **Other**.

Linetype	Description
ACAD_ISO02W100	ISO dash __ __ __ __ __ __ __ __ __ __ __ __
ACAD_ISO03W100	ISO dash space __ __ __ __ __ __
ACAD_ISO04W100	ISO long-dash dot ____ . ____ . ____ . ____
ACAD_ISO05W100	ISO long-dash double-dot ____ .. ____ .. ____ .
ACAD_ISO06W100	ISO long-dash triple-dot ____ ... ____ ... ____
ACAD_ISO07W100	ISO dot . . . .
ACAD_ISO08W100	ISO long-dash short-dash ____ __ ____ __ ____ _
ACAD_ISO09W100	ISO long-dash double-short-dash ____ __ __ ____
ACAD_ISO10W100	ISO dash dot __ . __ . __ . __ . __ . __
ACAD_ISO11W100	ISO double-dash dot __ __ . __ __ . __ __ . __
ACAD_ISO12W100	ISO dash double-dot __ . . __ . . __ . . __ . .
ACAD_ISO13W100	ISO double-dash double-dot __ __ . . __ __ . .
ACAD_ISO14W100	ISO dash triple-dot __ . . . __ . . . __ . . .
ACAD_ISO15W100	ISO double-dash triple-dot __ __ . . . __ __ .
BATTING	Batting SSSSSSSSSSSSSSSSSSSSSSSSSSSSSSSSSSSS
BORDER	Border __ __ . __ __ . __ __ . __ __ .
BORDER2	Border (.5x) __ __ . __ __ . __ __ . __ __ .
BORDERX2	Border (2x) ____ ____ . ____ ____ . ____
CENTER	Center ____ _ ____ _ ____ _ ____ _ ____
CENTER2	Center (.5x) ____ _ ____ _ ____ _ ____ _
CENTERX2	Center (2x) _____ __ _____ __ _____
DASHDOT	Dash dot __ . __ . __ . __ . __ . __ .
DASHDOT2	Dash dot (.5x) _._._._._._._._._._._._.
DASHDOTX2	Dash dot (2x) ____ . ____ . ____ . ____
DASHED	Dashed __ __ __ __ __ __ __ __ __ __ __
DASHED2	Dashed (.5x) _ _ _ _ _ _ _ _ _ _ _ _ _ _
DASHEDX2	Dashed (2x) ____ ____ ____ ____ ____
DIVIDE	Divide ____ . . ____ . . ____ . . ____
DIVIDE2	Divide (.5x) _.._.._.._.._.._.._.._.
DIVIDEX2	Divide (2x) _____ . . _____ . . _
DOT	Dot . . . .
DOT2	Dot (.5x) ....
DOTX2	Dot (2x) . . . .
FENCELINE1	Fenceline circle ----0----0----0----0----0---
FENCELINE2	Fenceline square ----[]----[]----[]----[]----
GAS_LINE	Gas line ----GAS----GAS----GAS----GAS----GAS---
HIDDEN	Hidden __ __ __ __ __ __ __ __ __ __ __
HIDDEN2	Hidden (.5x) _ _ _ _ _ _ _ _ _ _ _ _ _ _
HIDDENX2	Hidden (2x) ____ ____ ____ ____ ____ ____
HOT_WATER_SUPPLY	Hot water supply ---- HW ---- HW ---- HW ----
PHANTOM	Phantom _____ __ __ _____ __ __ _____
PHANTOM2	Phantom (.5x) ____ _ _ ____ _ _ ____ __ _
PHANTOMX2	Phantom (2x) _____ ____ ____ _
TRACKS	Tracks -I-I-I-I-I-I-I-I-I-I-I-I-I-I-I-I-I-I-I-
ZIGZAG	Zig zag /\/\/\/\/\/\/\/\/\/\/\/\/\/\/\

- From the Dashboard's Object Properties panel, select the Linetype Control droplist, and then **Other**.

- At the 'Command:' prompt, enter the **linetype** command:

  Command: **linetype** *(Press* ENTER.*)*

- Alternatively, enter the aliases **lt**, **ltype**, or **ddltype** (the old name) at the 'Command:' prompt.

2. In all cases, AutoCAD displays the Linetype Manager dialog box:

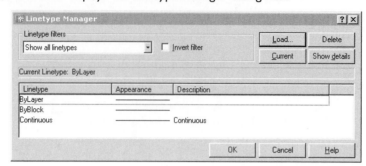

The dialog box lists the linetypes loaded into the current drawing. Just as for colors and lineweights, there are two special linetypes:

ByLayer — assigned by layers.

ByBlock — assigned by blocks.

3. To load additional linetypes, choose the **Load** button.

AutoCAD displays the Load or Reload Linetypes dialog box. (Unless you know you have linetype definitions stored in a file other than *acad.lin*, you can ignore the **File** button.)

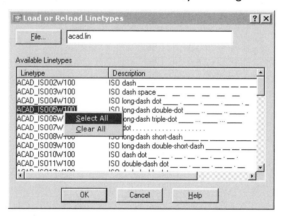

4. I find it easier to load all linetypes at once, and then later remove any that remain unused. To do so, right-click on any linetype. Notice that AutoCAD displays a small pop-up menu with two selections.

Choose **Select All,** and AutoCAD highlights all linetypes.

5. Choose **OK**, and AutoCAD loads all linetypes.

6. Back in the Linetype Manager dialog box, select a linetype, and then click **Current**.

7. Choose **OK**.

From now on, all objects are drawn in this linetype — until you change it again.

Global, Local, and Annotative Linetype Scales

AutoCAD lets you specify linetype scales in three ways:

- **Globally** for all objects in the drawing. The global linetype scale is set by the **LTSCALE** command, described later.

- **Locally** for individual objects. Although it is poor drafting practice to use more than one linetype scale in drawings, local scaling is sometimes needed for complex linetypes, such as Batting and Hot Water Supply.

- **Annotatively** for scale-specific viewports (new to AutoCAD 2008).

Local (Object) Linetype Scaling

To change the linetype scale of individual objects, use either the **PROPERTIES** command (as described later in this chapter) or the **CELTSCALE** system variable (short for "current entity linetype scale"). There is a difference between the two: the **PROPERTIES** command changes the linetype scale of objects that already exist in the drawing, while the **CELTSCALE** system variable changes the scale of objects you plan to draw.

CELTSCALE can be tricky to use, because it sets the linetype scale *relative* to the value set by the **LTSCALE** command setting: when **LTSCALE** is set to 3.0 and **CELTSCALE** is set to 0.75, objects will be drawn with a linetype scale of 2.25 (3.0 x 0.75 = 2.25).

Both global and object linetype scaling are available as features "hidden" in the Linetype Manager dialog box. Click the **Details** button to reveal them.

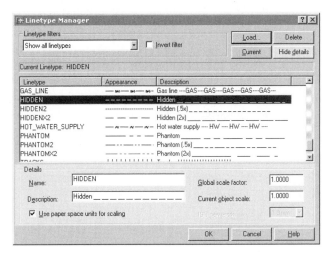

Global scale factor — sets the linetype scale for all objects in the drawing. This scale factor takes effect with the next drawing regeneration.

Current object scale — sets the linetype scale for subsequently-created objects.

ISO pen width — selects from one of the ISO's pen widths, such as 1.00 mm. This option is grayed out when non-ISO linetypes are selected.

Use paper space units for scaling — scales linetypes identically in paper space and model space through the **PSLTSCALE** system variable. More details may be found in *Using AutoCAD: Advanced*.

 ## Annotative Scaling

Annotative scaling makes linetypes appear at the correct size in model space viewports. This includes the Model tab and viewports in Layout tabs. When a scale factor is assigned a viewport (or in Model tab), then the scale factor for all linetypes is adjusted automatically by AutoCAD.

The new **MSLTSCALE** system variable (short for "model space linetype scale") toggles annotative scaling of linetypes. When set to 1, linetypes are scaled to match the current viewport scale; this is the default setting. (The variable is set to 0 in drawings opened from AutoCAD 2007 and earlier.)

To use **MSLTSCALE** correctly, you have to set all linetype scale system variables to 1:

> **CELTSCALE** = 1
> **LTSCALE** = 1 (or another size correct for plotting)
> **MSLTSCALE** = 1
> **PSLTSCALE** = 1

The changes due to annotation scaling appear with the next regeneration (**REGEN** command).

CHANGING LINETYPES: ADDITIONAL METHODS

AutoCAD provides several other ways to handle linetypes:

- **-LINETYPE** command changes the working linetype at the command line.
- **LTSCALE** command changes the scale of linetypes.
- **Linetype Control** droplist changes lineweights of selected objects.
- **DesignCenter** window allows you to import linetypes from other drawings.

Let's look at each option. (Later you also learn about changing linetypes with the **PROPERTIES** and **CHPROP/CHANGE** commands.)

-Linetype

The **-LINETYPE** command loads and sets linetypes at the command line. Use the **Set** option to set the working linetype:

Command: **-linetype**

Enter an option [?/Create/Load/Set]: **s**

Specify linetype name or [?] <ByLayer>: *(Enter a linetype name, such as* **hidden.***)*

Enter an option [?/Create/Load/Set]: *(Press* **ENTER** *to exit the command.)*

The **?** option prompts you to select a *.lin* file, and then lists the available linetypes. The display starts off like this:

Linetypes defined in file C:\Documents and Settings\username\Application

Data\Autodesk\AutoCAD 2008\R16.2\enu\Support\acad.lin:

Name	Description
"BORDER"	Border __ __ . __ __ . __ __ . __ __ . __ __ .
"BORDER2"	Border (.5x) __.__.__.__.__.__.__.__.__.
"BORDERX2"	Border (2x) ____ ____ . ____ ____ . ___
"CENTER"	Center ____ _ ____ _ ____ _ ____ _ ____ _ ____
"CENTER2"	Center (.5x) ____ _ ____ _ ____ _ ____ _ ____ _ ____

Press ENTER to continue: *(Press* **ENTER** *to see more, or else* **ESC** *to exit the listing.)*

The **Load** option is like the **Load** button in the linetype dialog box: it selects a *.lin* file to load into the drawing. The **Create** option creates custom linetypes by defining the pattern of segments, dots, and gaps, and their lengths.

LtScale

Because linetypes are constructed of line segments and dashes, the pattern can look too small or too large for drawings. How do you tell when the scale is wrong? When the pattern lines do not look continuous.

Like hatch patterns, linetypes need to be set to an appropriate scale; fortunately, the same scale factor applies to both hatch patterns and linetypes. When you figure out the correct factor for one, you can use it for the other, as well as for text and dimensions.

The scale of linetypes is adjusted by the **LTSCALE** (short for "linetype scale") command:

Command: **ltscale**

Enter new linetype scale factor <1.0000>: *(Enter a value.)*

Regenerating model.

Numbers larger than 1.0 result in longer line segments, while numbers smaller than 1.0 create shorter line segments.

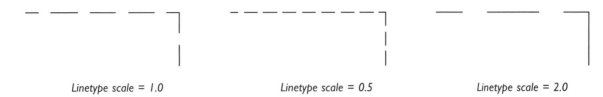

Linetype scale = 1.0   Linetype scale = 0.5   Linetype scale = 2.0

To display the new linetype scale correctly, AutoCAD automatically performs a regeneration. If a regeneration does not occur after you change the linetype scale, force a regeneration with the **REGEN** command.

If, after changing the scale, the linetypes still appear continuous, change **LTSCALE** to yet a different value. It is possible for the linetype scale to be too large or small to display. If lines take longer to draw, it is likely that the scale is too small.

General rule: the linetype scale is the inverse of the plot scale. For example, when the plot scale is 1/8" = 1' (1:96), the linetype scale factor is 96. The exception is when your drawing uses annotative scale factor, in which case AutoCAD uses the viewport scale for linetypes when **MSLTSCALE** = 1.

Linetype Control

As with color and lineweights, you can change linetypes directly through the **Linetype Control** droplist. The control affects linetypes by two methods, depending on the order in which you use it. Method 1: To *set* the working lineweight (objects subsequently taking on the selected linetype), simply select a linetype from the droplist. Method 2: To *change* the linetype of existing objects(s), first select them, and then a linetype from the droplist.

In the following tutorial, we look at both methods.

TUTORIAL: SETTING AND CHANGING LINETYPES

1. Start AutoCAD with a new drawing, and then draw a few lines. Notice that they are continuous lines.
2. From the Dashboard's Object Properties panel (illustrated below), or the Properties toolbar, click on the Linetype Control droplist.

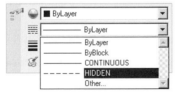

3. Select **Hidden**. (If the linetype name does not appear in the list, return to the earlier tutorial to learn how to load linetypes into drawings.)
4. Draw more lines. Notice that they are dashed lines.
5. Now, change the pattern of the lines: select the lines, which become highlighted and show grips.

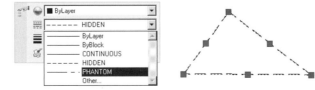

6. From the Linetype Control droplist, select another linetype, such as Phantom. Notice that the lines change, showing dashes and dots.

7. Press **ESC** to "unselect" the lines.

 Note: If you (or someone else) creates custom linetypes, AutoCAD has two ways to share them. One is to exchange a copy of the *.lin* file that stores the custom definitions. The other is to use DesignCenter to drag a copy of the linetype from one drawing to another.

 LAYER

The **LAYER** command applies common sets of properties to objects assigned to layers.

Nearly all CAD drawings are constructed using *layers* that can be turned on and off, and changed. Layers go back to the days of traditional drafting, which sometimes used a method of drawing called "overlay drafting." Drafters overlaid sheets of semitransparent drafting media so that the drawings showed through each other. Multiple sheets could be printed together, resulting in prints that showed the complete design.

The bottom sheet was typically referred to as the "base drawing." Each additional sheet showed different items. For instance, when a drafter prepared a set of floor plans, he typically drafted separate drawings showing the items to be removed, plumbing, electrical, and so on. Often, the floor plan is the base drawing, with each discipline, such as electrical and plumbing, placed on overlay sheets.

AutoCAD's layers are like sheets of glass stacked on top of one another. You draw different aspects of the drawing on the sheets of glass (layers), yet are able to see through all the layers. The work appears as though it were one drawing.

AutoCAD goes two steps further. You can turn layers on or off, so that they are either visible or invisible, as well as change their properties. The properties of a layer affects all objects assigned to the layer, but can be overridden with other commands, such as **COLOR**, **LINETYPE**, and **PROPERTIES** (described elsewhere this chapter). In turn, these overrides can be overridden.

Before adding objects to a layer, you switch to the appropriate layer. The working layer is called the "current layer." AutoCAD can have only one layer current at a time. The current layer cannot be *frozen*. (Frozen layers are invisible and cannot be edited.) A layer named "0" exists in all new drawings; it has some special properties, and cannot be renamed or removed.

TUTORIAL: CREATING LAYERS

1. To create layers, start the **LAYER** command with one of these methods:
    * From the **Format** menu, choose **Layer**.
    * From the Layers toolbar, choose the **Layer Properties Manager** button.
    * In the Dashboard's Layers panel, choose the **Layer Properties Manager** button.
    * Or, at the 'Command:' prompt, enter the **layer** command:

        Command: **layer** *(Press* **ENTER**.*)*
    * Alternatively, enter the aliases **la** or **ddlmodes** (an old name for this command) at the 'Command:' prompt.

2. In all cases, AutoCAD displays the Layer Properties Manager dialog box.

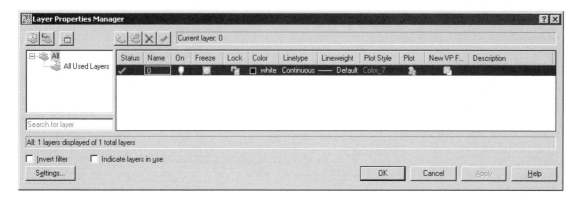

3. Click the **New Layer** icon. Alternatively, press **ALT+N**.

Notice that AutoCAD creates a new layer named "Layer1," which you can rename to anything up to 255 characters long.

To the right of the name are icons indicating that the layer is turned on, thawed (not frozen), unlocked, and colored white/black; has continuous linetype, and default lineweight; and will be plotted/printed. (More about these settings later.)

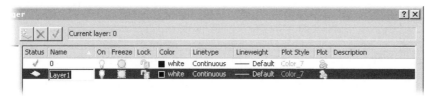

4. To change the layer name, edit the "Layer1" text, type a new name.

5. To make the layer current, click the **Set Current** button. Alternatively, press **ALT+C**.

6. Click **OK** to dismiss the dialog box. Notice that the name also appears on the Layer droplist.

> **Notes**: AutoCAD creates some layers on its own. Two layers you must become acquainted with are **0** and **Defpoints**. Every new drawing contains the layer named 0 (zero). This layer cannot be removed or renamed. Layer 0 has a special property for creating blocks, of which you learn elsewhere in this book.
>
> The first time you draw dimensions, AutoCAD creates a layer named Defpoints (short for "definition points"). This layer contains data that AutoCAD needs to keep its dimensions associative. The layer cannot be removed. You can rename it, but then AutoCAD creates a new Defpoints layer with the next dimension. Anything you draw on this layer, accidentally or otherwise, AutoCAD does not plot. For this reason, some students become frustrated to find that part of their drawing won't plot, because they accidentally drew on layer Defpoints but had overridden colors and linetypes, so everything looked correct. (Use the **No Print** toggle to make *any* layer non-printing.)
>
> A third layer created by AutoCAD is **AShade**, which is used to hold data regarding renderings.

Shortcut Keystrokes

The following shortcut keystrokes operate only while the Layer Properties Manager dialog box is displayed:

Keystroke	Meaning
ALT+N	Creates new layer.
ALT+D	Deletes selected layer(s).
ALT+C	Sets selected layer as current.
ALT+P	Creates new property filter.
ALT+G	Creates new group filter.
ALT+S	Displays layer states manager.

Working with Layers

The Layer Properties Manager dialog box provides you with a fair degree of control over layers.

Status

The **Status** column uses icons to report the status of each layer:

Status	Comment
	Layer is current.
	Layer is in use (has at least one object).
	Layer is empty (can be deleted, unless part of an xref).
	Layer is marked for deletion.
	Layer is part of a group filter.
	Layer is part of a property filter.

Turning Layers On and Off

When layers are *on*, their objects are seen and can be edited; the status is indicated by the yellow light bulb icon in the **On** column. When *off*, the objects on that layer are not seen; the light bulb icon turns blue-gray.

To turn one or more layers on or off, first highlight the layer(s) by selecting their name(s), and then choose the light bulb icon. The "on" column immediately reflects the change.

On and **Off** were used in the earliest versions of AutoCAD to toggle the display of layers, and are rarely used anymore; **Freeze** and **Thaw** are preferred now.

Freezing and Thawing Layers

Frozen layers cannot be seen or edited; their status is shown by the snowflake icon in the **Freeze** column. *Thawed* layers can be seen and edited, just like *on* layers; the symbol for thawed layers is the sun icon.

To freeze or thaw one or more layers, highlight the target layer(s), and then choose the sun or snowflake icon. It is more efficient to freeze layers than to turn them off. Objects on frozen layers are ignored by AutoCAD during regenerations and other compute-intensive operations.

 Notes: To select *all* layers, right-click any layer name. AutoCAD displays a shortcut. Choose **Select All** to highlight all layer names. Choose **Clear All** to deselect all layers.

To change the width of the columns in the layer dialog box, grab the black bar separating column tiles and drag left or right.

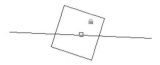 **Locking and Unlocking Layers**

Objects on *locked* layers can be seen, but not edited in most cases. You can still use these objects for osnaps, trimming, and extending.

Locked layers show a padlock icon in the **Lock** column. *Unlocked* layers are like *on* layers: their objects are seen and can be edited; an unlocked layer shows an open padlock.

When you attempt to select objects on locked layers, AutoCAD shows a padlock icon near the cursor.

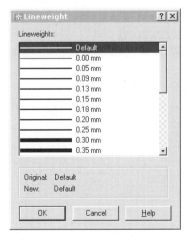

To lock or unlock layers, select the target layer(s), and then choose the padlock icon(s) in the **Lock** column.

■ **Setting Layer Colors**

To set the color for objects on layers, first select the layer names(s), and then click on the color square. AutoCAD displays the Select Color dialog box. Choose a color, and then click **OK**.

From now on, all objects on the layer are displayed in that color, unless overridden by the COLOR command or by a block's properties.

Setting Layer Linetypes

When you select a linetype for a layer, it affects all the objects on that layer. First select the layer names(s) for which you wish to set a linetype, and then choose the linetype. AutoCAD displays the Select Linetype dialog box. If necessary, load linetypes. Select a linetype and choose **OK**.

Setting Layer Lineweights

You can specify the lineweights for objects residing on layers. First, select the layer name(s) for which you wish to set a lineweight, and then choose the lineweight, such as "Default." The Lineweight dialog box is displayed; select a lineweight, and then choose **OK**.

Setting the Layer Plot Style

Plot styles are available only when the feature is turned on, as described in Chapter 19; otherwise, the plot style names are shown in gray and cannot be changed. Plot styles determine how all objects residing on a layer are plotted.

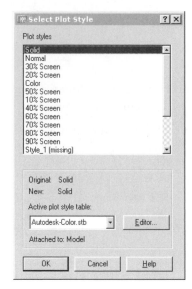

To change the plot style, select the layer names(s) for which to set a plot style. Select a plot style from those available in the Select Plot Style dialog box, and then choose **OK**.

Setting Layer Print Toggles

You can specify that some layers print, while others do not. Under the **Plot** column, choose the printer icon for the layer(s) you don't want to print or plot. To allow the layer to print, simply choose the icon a second time.

Description

The **Description** column permits a descriptive sentence for each layer.

Layer Controls in Layouts

In layout mode (paper space), more columns are added to the layer dialog box. To see them, close the dialog box, and switch to layout mode by selecting any layer tab. Re-open the Layer Properties Manager dialog box, and scroll the layer listing all the way to the right.

As described in *Using AutoCAD: Advanced*, AutoCAD can work with tiled or floating viewports. You can freeze layers in floating viewports — something you cannot do in model view — to display different sets of layers in different viewports.

AutoCAD allows you to do the following to layers in floating viewports: to freeze the active viewport, freeze new viewports, and to override the color, linetype, lineweight, and plot style in specific viewports.

Layout Freeze Modes

The **New VP Freeze** column freezes specified layers when a new viewport is created. Select the layer name(s), and then choose the icon in the New VP Freeze column. When the icon is a shining sun, the layer is thawed; when the icon is a snowflake, the layer is frozen.

The **VP Freeze** column automatically freezes the specified layers in the current viewport. Select the layer name(s), and then choose the icon in the VP Freeze column (formerly named "Current VP Freeze"). As always, when the icon is a shining sun, the layer is thawed; when the icon is a snowflake, the layer is frozen. When the icon is gray, however, AutoCAD cannot display floating viewports, because the drawing is not in layout mode.

Layout Property Modes

The **VP Color** column allows you to override the layer color for the current viewport. You use it like this:

1. In model space, draw an object, like a rectangle.
2. Switch to a layout tab, and then create two viewports. Double-click a viewport to make it current.
3. Start the **LAYER** command.
4. In the layer dialog box, change **VP Color** from "White" to "Red," and then exit the dialog box.

Notice that the rectangle is red in one viewport (shown at left, below), but remains black in the other (at right).

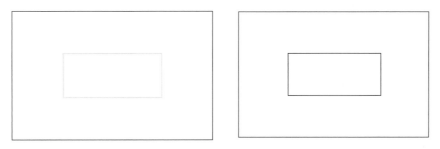

The same trick can be played using linetypes, lineweights, and plot styles. Each viewport can display the drawing in a different manner. The viewport is said to have its layer properties "overridden."

 Resetting Overridden Layout Properties

To revert the viewport properties to the layers' default settings, use the **SETBYLAYER** command. Technically, this command resets the properties of objects to ByLayer. The command applies only to viewports in layouts.

Selecting Layers

Some types of drawings may have many, many layer names. In theory, AutoCAD drawings can have an unlimited number of layers; in practice, drawings may contain thousands. Working your way through long lists of layers can become tedious.

For example, open the *8th floor.dwg* file included on the Companion CD. Start the **LAYER** command, and notice the text on the status line of the dialog box:

> All: 225 layers displayed of 225 total layers.

Scroll through the list of layers to appreciate its length. Two hundred and twenty-five is a lot of layers! The reason is that this drawing includes six other drawings, known as "external references." The original drawing has 27 layers; the xrefs contribute the additional 198 layers.

To help out, AutoCAD allows you to sort and to shorten the list of names using several techniques:

- By sorting names and properties alphabetically in columns.

- By searching for layer names.

- By inverting layer listings.

- By using group and property filters.

- By deleting unused layers.

Sorting by Columns

The names of the columns — **Name**, **On**, **Freeze**, **Lock**, and so on — are actually buttons. For instance, click once on the **Name** column button, and the column sorts layer names in alphabetical order: 0 - 9 followed by A - Z.

Click a second time, and the column sorts in reverse order: Z - A followed by 9 - 0. The figure below illustrates layer names sorted in reverse order starting with Z:

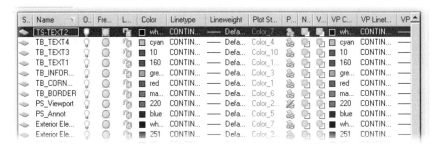

To sort layers by color *number,* click the **Color** column button. The first color is red. Choose it a second time to reverse the order of color numbers, starting with the highest color number used by drawing.

Similarly, clicking the **On**, **Freeze**, and other column titles sorts the layers by state. Click once, and the column sorts by light bulb, sun, or open padlock. Click a second time, and the column reverses the sorting by dim bulb, snowflake, or closed padlock.

Notes You can change the order of the columns by dragging their headers around. For instance, if you want Color next to On, just drag the Color header next to On.

You can also hide the display of columns. Right-click any column header, and then select the column name you want hidden. (The checkmarks indicate the headers that are displayed.)

Searching for Names

The **Search for Layer** field reduces the list of layer names displayed by the dialog box.

Continuing with the *8th floor.dwg* sample drawing:

1. Click in the **Search for Layer** field.

   Notice that an asterisk appears (*). This is called a "wild card" character. It means that all layer names are listed.

   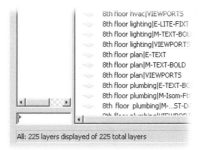

   Let's assume that you want to list all layers starting with the letters VIEW and ending with any characters. The filter would read:

   VIEW*

   This lists layers named VIEWPORTS, VIEW_A, and VIEW2C. It would not list 1PRT, VIW21, or 3RD_VIEW, because the names don't match VIEW*.

   Another wild card is the question mark (?), which matches single characters.

2. Enter "e*". The list is shortened to all layer names starting with the letter E.

   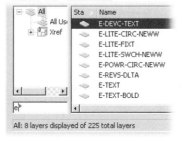

3. Now erase the asterisk, leaving the **e** on its own. Notice that no layer names appear, because none consists of just the letter **e**.

4. Prefix the search term with * to find all layers with a term. For example, ***text*** finds all layer names that contain "text":

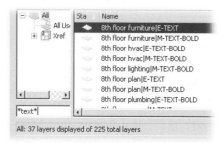

5. Click the **Invert Filter** option to display all layers, *except* those that contain **text**:

The **Apply to layers toolbar** option determines whether the abbreviated list of layer names is also displayed by the droplist on the **Layers** toolbar. When on (check mark shows), the list is abbreviated.

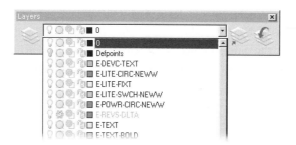

> ***Warning!*** When both **Invert Filter** and **Apply to layers** options are turned on, it is possible for the Layer droplist to be empty, except for layer 0.

Creating Group and Property Filters

The *tree view* at the left of the layer dialog box shows groups of layers in a *hierarchy*. The topmost group is always **All**, and holds the names of all layers. Select it, and the list view (the right half of the dialog box) lists the names of all layers.

Underneath the **All** group is the **All Used Layers** group, which contains all layers that have at least one object. When you click it, you see the list of all used layers in the list view. The list will be shorter when some layers are empty. Click **Invert Filter** to see only the names of empty layers, those that contain no objects.

 Note: A layer may appear to be empty, but sometimes it contains objects from a block definition.

When drawings contains xrefs (short for "externally referenced drawings"), the tree view also includes a group named **Xref**. Click it to see the names of all layers found in xrefs.

Below the **Xref** group are the names of the xref drawings. Each is a group that contains the names of layers found in that particular drawing.

 Note: To view all layers *not* in xrefs, select the **Xref** group, and then turn on the **Invert Filter** option.

TUTORIAL: CREATING GROUP FILTERS

In this tutorial, you create a group filter that lists all text-related layers. Follow these steps:

1. Start AutoCAD with the *8th floor.dwg* drawing, found on the Companion CD.
2. Enter the **LAYER** command to open the Layer Properties Manager dialog box.
3. Click the **New Group Filter** button. (Alternatively, press **ALT+G**.)

    Notice that AutoCAD creates an empty filter with the generic name of "Group Filter 1."

4. Change the name to something more meaningful: **Text**.

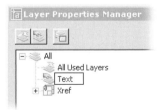

5. Now isolate the layer names that have "text" in them:
    a. Click on **All**.
    b. In the list view, click the **Name** header to sort layer names alphabetically, if necessary.
    c. In the search bar, enter "*text*".

d. Scroll down the list view to find all layers *not* part of an xref. These are layer names that don't start with "8th floor...".

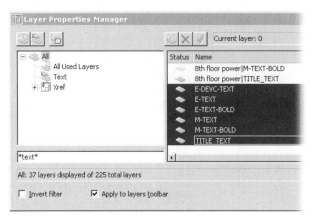

6.   Select the six layer names, and then...

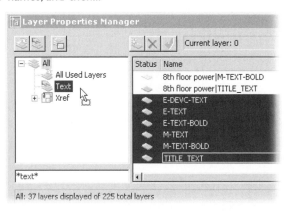

7.   ... drag them over to the tree view, and deposit them on the **Text** group filter.
     Don't worry! When you drag layer names around this dialog box, you do not move or lose them; AutoCAD copies them.

8.   Select the **Text** filter. Notice that the list view shows a shortened list of layer names.

Select the **All** filter to see all names again.

TUTORIAL: CREATING PROPERTY FILTERS

Group filters are groups of layers that you drag from the list view into the group filter name. Group filters are static; they change only when you add and subtract layers to and from the group.

AutoCAD also supports *property filters*, based on the properties of layers. For example, you might want a group that holds all layers with the color blue and linetype Hidden. Property filters are dynamic; as you create (or delete) layers of the same properties, AutoCAD automatically adds and removes them from the filter group.

In this tutorial, you create a group filter that lists all layers that are colored blue and have the Hidden linetype. Follow these steps.

1. Start AutoCAD with the *8th floor.dwg* drawing, found on the Companion CD.

2. Enter the **LAYER** command to open the Layer Properties Manager dialog box.

3.  Click the **New Property Filter** button. (Alternatively, press **ALT+P**.)

   Notice that AutoCAD displays the Layer Properties Filter dialog box. In the upper half, you define the filter(s); the lower half displays the names of layers and their properties.

4. In the **Filter Name** field, replace the generic name "Properties Filter 1" with:

   Filter Name: **Blue Hidden**

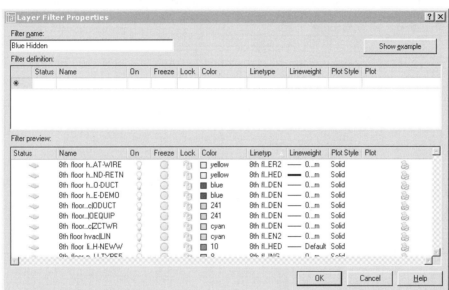

5. Under each heading (Status, Name, On, ...) you can enter a restriction that shortens the list. For example, this filter is supposed to have just layers colored blue.

   a. Under **Color**, click the blank area. Notice the small button that appears.

   b. Click the **...** button. Notice the Select Color dialog box.

   c. Select the blue color (index color 5), and then click **OK**.

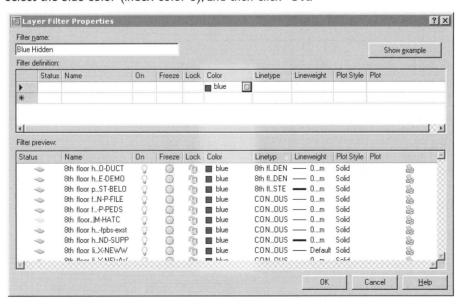

   In the lower half of the dialog box, notice that the list of layers is shortened: just the blue layers are listed.

LAYER FILTERS

Layer filters slowed down AutoCAD, so Autodesk added two system variables that help speed up the display speed of the layers dialog box.

The **SHOWLAYERUSAGE** system variable toggles the display of icons in the Layer dialog box. (The icons indicate whether layers have objects or are empty.) Turn off this system variable (set to **0**) to improve the speed of the dialog box.

When drawings have more than 99 filters and the number of filters exceeds the number of layers, the **LAYERFILTERALERT** system variable determines how to delete layer filters, thereby improving performance:

LayerFilterAlert	Meaning
0	Filters are not deleted.
1	All filters are deleted without warning the next time the layer dialog box is opened.
2	All filters are optionally deleted; prompts 'Do you want to delete all layer filters now?' the next time the Layer dialog box is opened.
3	Filters are selectively deleted; displays dialog box for selecting filters to delete the next time the drawing is opened.

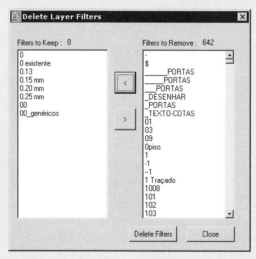

6. Apply a further restriction: just those layers with the Hidden linetype. Repeat the steps of #4, but this time select the Hidden linetype when the Select Linetype dialog box appears.

    In the lower half of the dialog box, the list of layers is again shortened to blue layers with Hidden linetype.

7. Click **OK**.

8. Select the "Blue Hidden" property filter, and only one layer appears.

9. Create a new layer with color blue and linetype Hidden. Notice that it is automatically added to the property filter.

> **Note:** To edit a property filter, double-click its name in the tree view. This causes the Layer Properties Filter dialog box to appear.

 Deleting Layers

AutoCAD allows you to delete *empty* layers, those with no objects. But you cannot delete layers 0 and DefPoints, as well as externally-referenced layers — even when these are empty. Sometimes, empty layers cannot be erased, because they are part of an unused block definition.

Select one or more layer names, and then choose the **Delete Layer** button. Alternatively, press **ALT+D**. If the layer cannot be erased, AutoCAD displays the following dialog box to explain why not:

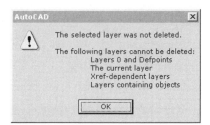

The layer is not erased immediately. Rather, an icon in the **Status** column marks the layer for deletion. This allows you to change your mind.

The layer is deleted when you click **Apply** or **OK**.

 Layer States Manager

The **Layer States Manager** saves the current state of layers, and then restores the state at a later time. The *state* of layers includes names and properties, such as whether they are thawed or frozen, as well as their colors and linetypes. Once you save a layer state by name, you can edit the state, and export states for sharing with others.

TUTORIAL: CREATING AND APPLYING LAYER STATES

1. To create layer states, start the **LAYERSTATES** command with one of these methods:
   - From the **Format** menu, choose **Layer States Manager**.
   - From the Layers toolbar, choose the **Layer States Manager** button.
   - In the Dashboard's Layers panel, choose the **Layer States Manager** button.
   - Or, at the 'Command:' prompt, enter the **layerstates** command:

     Command: **layerstates** *(Press* ENTER.*)*

   - Alternatively, enter the aliases **la** or **ddlmodes** (an old name for this command) at the 'Command:' prompt. You can also access the dialog box through the Layer Properties Manager dialog box.

   In all cases, AutoCAD displays the Layer States Manager dialog box.

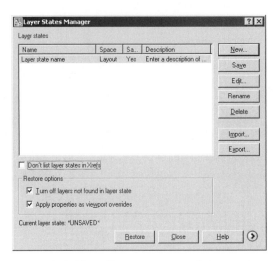

2. To create a new layer state, click **New**. Enter a name and description, and then click **OK**.

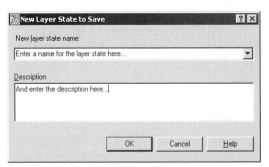

3. Now you can determine which properties will be saved. Click the **More** button to expand the dialog box, and then choose from the list of properties.

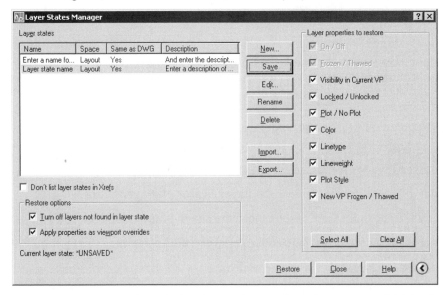

4. To save the layer state to disk, click the **Export** button. Provide a name and folder for the *.lay* file, and then click **Save**.

5. The easy way to apply a layer state is to select its name from the Dashboard's Layer panel.

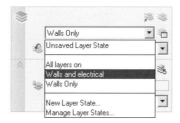

To import a layer state, click the **Import** button. Select the folder and name of a file, and then click **Open**. Duplicate layer states names are not imported. As of AutoCAD 2008, you can also import layer states from *.dwg* (drawing), *.dws* (drawing standards), and *.dwt* (drawing template) files — as well as *.las* (layer state) files.

CONTROLLING LAYERS: ADDITIONAL METHODS

AutoCAD provides several other methods to control layers:

- **-LAYER** command controls layers at the command line.
- **Layers** toolbar and panel control layers from the toolbar and Dashboard.
- **CLAYER** system variable makes a layer current quickly.
- **AI_MOLC** command makes the selected object's layer current.
- **LAYERP** command restores the previous layer state.
- **Layer Tools** manipulate layers through individual commands.
- **LAYISO** command isolates selected layers by freezing or locking all other layers.

Let's look at each of these.

-Layer

The **-LAYER** command creates and changes layers at the command line:

> Command: **-layer**
>
> Current layer: "0"
>
> Enter an option
>
> [?/Make/Set/New/ON/OFF/Color/Ltype/LWeight/Plot/PStyle/Freeze/Thaw/LOck/Unlock/ stAte]: *(Enter an option.)*

Most options prompt you for the name of a layer (or to select an object), and then apply the option. The action is applied to the layer of the selected object. For example, to freeze a layer:

> [?/Make/Set/New/ON/OFF/Color/Ltype/LWeight/Plot/PStyle/Freeze/Thaw/LOck/Unlock/ stAte]: **f**
>
> Enter name list of layer(s) to freeze or <select objects>: *(Enter name of layer, or select an object.)*

To freeze more than one layer at a time, separate their names with commas:

> Enter name list of layer(s) to freeze or <select objects>: **layer1,layer2,layer3**

-Layer Options

The meaning of the options are:

> **?** lists the names of layers in the drawing.
>
> **Make** creates a new layer by prompting you for the name, and then sets it as current.
>
> **Set** sets a layer as current.
>
> **New** creates a new layer by prompting you for the name.

ON turns on layers that have been turned off; you must enter "ON" as the option.

OFF turns off specified layers.

Color changes the color of specified layers.

Ltype changes the linetype of specified layers.

LWeight changes the lineweight of specified layers.

Plot toggles whether the layers will be plotted.

PStyle changes the plot style of specified layers.

Freeze freezes the specified layers.

Thaw thaws frozen layers.

LOck locks specified layers.

Unlock unlocks locked layers.

stAte controls layer states through these options:

> Enter an option [?/Save/Restore/Edit/Name/Delete/Import/EXport]:

> > **?** lists the names of layer states in the drawing, if any.

> > **Save** — saves the current settings of layers as a state; prompts for a name, and then reports the settings.

> > **Restore** restores a named layer state.

> > **Edit** edits the settings of a layer state.

> > **Name** changes the name of a layer state (should be called the Rename option).

> > **Delete** erases a layer state from the drawing.

> > **Import** imports a layer state from *.las*, *.dwg*, *.dws*, and *.dwt* files; enter ~ to force display of the Select Layer State File dialog box.

> > **EXport** exports the layer states to a *.las* file for use in other drawings.

Layer Control

When the Layers toolbar and/or Layers panel are available, they let you select the current layer without needing the **LAYER** command. The Dashboard's Layer panel looks like this:

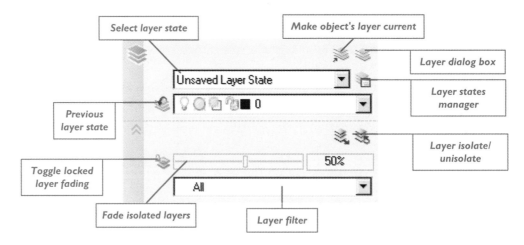

Changing the Current Layer

To change the current layer, click the down arrow at the right end of the layer name on the toolbar. It reveals a list of layer names, along with icons signifying their status.

Select the name of a layer to make it current. It's that simple.

There is, unfortunately, one catch. You cannot make current any layer whose name is shown in gray, because it is frozen (and you can't see or work with frozen layers.)

Changing the Layer's Status

The drop-down box lists a quintet of icons beside each layer name. The colored square icon is the color assigned this layer. The other four icons each have two states:

Lightbulb on or off — layer is on (default) or off.

Sun or **snowflake** — layer is thawed (default) or frozen; frozen layers also show their names in gray.

Sun or **snowflake on square** — layer is thawed (default) or frozen in the current viewport; this icon changes only when the drawing is in paper space (layout mode).

Padlock — open or closed: layer is unlocked (default) or locked.

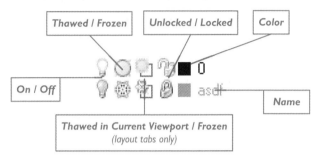

You change the status of each of the four icons simply by clicking them. Note that the thawed/frozen in current viewport icon changes only when the drawing is in layout mode (paper space). The color square performs no action.

 Ai_Molc

An easy way to switch to another layer is with the **Make Object's Layer Current** button. Suppose you are interested in switching to the layer holding a green dotted line whose name you're not sure about. You use the feature as follows:

1. Choose the **Make Object's Layer Current** button. AutoCAD prompts you:

   Select object whose layer will become current:

2. Select the green dotted line. AutoCAD reports:

   layername is now the current layer.

Check the toolbar, and you will see that layer *layername* is now current.

CLayer

If you prefer the keyboard to dialog boxes, then CLAYER system variable (short for "current layer") is the fastest way to switch between layers, as follows:

   Command: **clayer**

   New value for CLAYER <default>: *(Enter the name of the layer, and then press* ENTER.*)*

AutoCAD immediately makes that layer current.

 LayerP

The LAYERP command (short for "layer previous") undoes changes to layer settings, much like the **ZOOM Previous** command restores the previous view.

   Command: **layerp**

   Restored previous layer status.

The command, however, cannot undo the changes to renamed, deleted, purged, and newly-created layers. (Purging is described in the next chapter.) When you rename a layer and change its properties, only the properties are changed back, not the name.

A related command is **LAYERPMODE** (short for "layer previous mode"). It toggles whether AutoCAD tracks changes to layers.

> Command: **layerpmode**
>
> Enter LAYERP mode <ON>: *(Enter* **ON** *or* **OFF**.*)*

When turned off, the **LAYERP** command does not work, and so AutoCAD reminds you:

> Layer-Previous is disabled. Use LAYERPMODE to turn it on.

Layer Tools

The layer tools were formerly part of Express Tools, a collection of commands distributed with AutoCAD but not supported by Autodesk. In AutoCAD 2007, Autodesk changed the layer tools from Express Tools status to fully-supported status. The layer tools are:

LAYCUR changes the layer of selected objects to that of the current layer.

LAYMCH changes the layers of selected objects to that of a selected object.

LAYMCUR makes the selected object's layer current, just like the **AI_MOLC** command.

LAYMRG moves objects to another layer, and then removes the layer of the moved objects.

LAYWALK displays objects on selected layers.

LAYVPI isolates the selected object's layer in the current viewport by freezing its layer in all other viewports. This command works only in paper space with two or more viewports.

LAYISO turns off all layers except those holding selected objects (detailed below). **LAYUNISO** turns on layers that were turned off with the last **LAYISO** command.

LAYDEL erases all objects from the specified layer, and then purges the layer from the drawing. The current layer cannot be deleted. If you make a mistake, you can use the **U** command later to restore the layer name and its objects.

LAYFRZ freezes the layers of the selected objects. **LAYTHW** thaws all layers.

LAYLCK locks the layer of the selected object. **LAYULK** unlocks the layer of a selected object.

LAYOFF turns off the layer of the selected object. **LAYON** turns on all layers, except frozen layers.

LAYISO

The **LAYISO** command (short for "layer isolation") isolates one for more layers, causing the objects on all other layers to fade away. The command can either freeze or fade all other layers. Faded layers are locked, not frozen; this lets you see but not edit them.

The related **LAYUNISO** command returns layers to normal.

The **LAYLOCKFADECTL** system variable determines the amount of fading. You may find it easier to use the Locked Layer Fading slider on the Dashboard's Layer panel. The maximum fade amount is 90%.

1. From the Companion CD, open *property.dwg*.
2. Start the **LAYISO** command:
   * From the **Format** menu, choose **Layer Tools**, and then **Layer Isolate**.
   * From the Layers II toolbar, choose **Layer Isolate**.
   * In the Dashboard's Layers panel, pick the **Layer Isolate** button.

- At the 'Command:' prompt, enter the **properties** command:

   Command: **layiso**

   Current setting: Lock layers, Fade=50

3. Type "s" to specify the Settings options:

   Select objects on the layer(s) to be isolated or [Settings]: **s**

   Enter setting for layers not isolated [Off/Lock and fade] <Lock and fade>:

   The Settings options have the following meaning:

   - **Off** freezes layers.

   - **Lock and fade** locks and fades layers:

      - **Fade value** specifies the percentage to fade locked layers, from 0% to 90%. I find that the maximum of 90 works most effectively.

         Enter fade value (0-90) <50>: **90**

4. Select one or more objects on the layers you want to isolate (display fully, not fade). For this tutorial, select one of the black property lines; this will fade the blue text and dimensions.

   Select objects on the layer(s) to be isolated or [Settings]: *(Pick a black property line.)*

5. Press **ENTER** to exit object selection.

   Select objects on the layer(s) to be isolated or [Settings]: *(Press* **ENTER***.)*

   Layer Property Lines has been isolated.

   AutoCAD reports the name of the layer(s) isolated. Notice how the objects on the other layers are isolated (faded), as illustrated below:

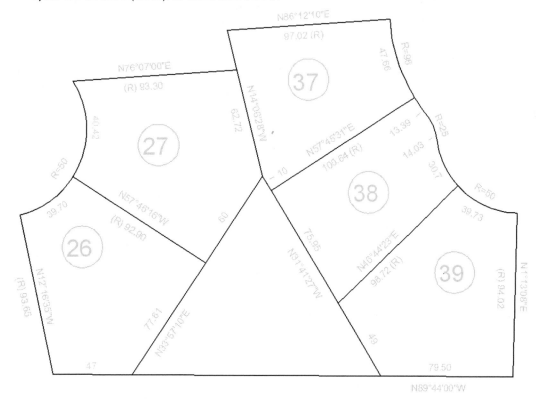

You may find it much easier to use the Dashboard's Layers panel to control layer isolation; for example, the Locked Layer Fading slider lets you interactively control the amount of fading displayed by locked layers.

Follow the steps illustrated below to use the panel's layer isolation controls.

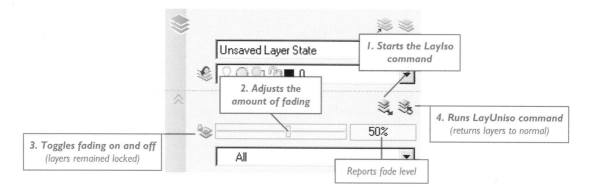

PROPERTIES

The **PROPERTIES** command displays a palette for changing (almost) *all* properties of objects. It's called a *palette*, because you can continue working in AutoCAD without needing to close it, unlike dialog boxes.

The Properties palette displays different sets of properties depending on the object(s) you select: text, mtext, 3D face, 3D solid, multiline, arc, point, attribute definition, polyline, block insertion, ray, body, region, circle, shape, dimension, mleader, 2D solid, ellipse, spline, external reference, hatch, tolerance, image, trace, leader, viewport, line, xline, and more.

That's because each object has a somewhat different set of modifiable properties. For example, you can modify the endpoints of a line through x, y, and z coordinates; you cannot, however, modify the endpoint of an arc; instead, you have to alter its start and end angle.

TUTORIAL: CHANGING ALL PROPERTIES

1. To change all the properties of objects, start the **PROPERTIES** command with one of these methods:

   • From the **Tools** or **Modify** menus, choose **Properties**.

   • From the Standard toolbar, choose **Properties**.

   • At the keyboard, press **CTRL+1**.

   • At the 'Command:' prompt, enter the **properties** command:

     Command: **properties** *(Press* ENTER.*)*

   • Alternatively, enter the aliases **pr**, **props**, **mo** (short for "modify"), **ch**, (short for "change"), **ddchprop**, or **ddmodify** (the old names in AutoCAD) at the 'Command:' prompt.

   In all cases, AutoCAD displays the Properties palette.

2. Notice that there are two colors of field: white and gray. White means you can modify the property, while gray means you cannot.

   The effect of changes you make in this window depends on whether or not objects are selected — much like changing settings in the Properties toolbar. When no objects are selected (as shown above), making changes to properties affects objects drawn from now on; when objects are selected, changing the properties affects the selected objects immediately.

   You can collapse unneeded sections by clicking the chevron (double-arrow) button, which makes the window smaller.

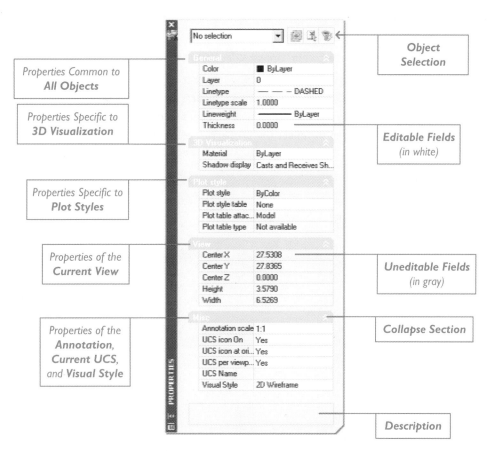

4. Notice the section named "General" located in the upper part of the window. The General section includes these modifiable properties:

> **Color** — selects the color square to change colors.
>
> **Layer** — selects a layer name.
>
> **Linetype** — selects a linetype name.
>
> **Linetype Scale** — specifies a different scale factor.
>
> **Lineweight** — selects a preset lineweight.

5. To select one or more objects, pick an object in the drawing, such as a line. Note that you do not dismiss the Properties palette: simply move the cursor into the drawing, and then pick the objects. (To select all objects in the drawing, press **CTRL+A**.)

As an alternative, choose one of the buttons at the top of the palette:

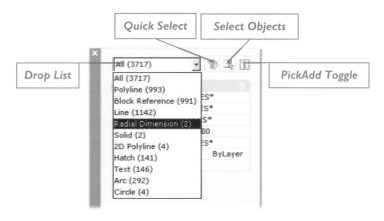

The droplist shows the variety (and, in parentheses, the number) of objects.

Quick Select displays the Quick Select dialog box, described in Chapter 6, "Drawing with Efficiency."

Select Objects causes AutoCAD to prompt you (unnecessarily):

Select objects: *(Select one or more objects.)*

Select objects: *(Press* ENTER *to end object selection.)*

Toggle value of PICKADD **sysvar** changes how you select objects:

- Plus (**+**) sign means you select additional objects by picking them.
- Number **1** means you select additional objects by holding down the SHIFT key.

6. The Properties palette changes to reflect the properties of the selected object. For example, if you select an arc, the palette lists geometry specific to arcs, such as the x, y, and z coordinates of the arc's center point.

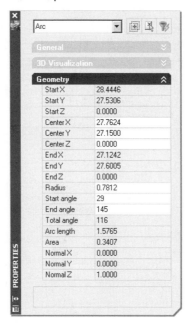

When you pick two objects whose properties differ, AutoCAD displays *VARIES* as the property value.

7. In the **Geometry** section, click in the **Center X** field, and then change the value to a different number. Notice that the line's endpoint moves.

As an alternative, click the ⬚ button. AutoCAD prompts you:

Pick a point in the drawing: *(Move the cursor, and then pick a point.)*

You may use object snaps to assist you in making an accurate pick.

(Clicking the calculator icon brings up **QuickCalc**, described in *Using AutoCAD: Advanced*.)

8. When done with this palette, you can click the **x** or enter the PROPERTIESCLOSE command.

Note: You can easily display the Properties window by double-clicking (almost) any object in the drawing. Exceptions include text, which displays the text editing window.

CHANGING PROPERTIES: ADDITIONAL METHOD

AutoCAD provides another way to change properties:

- **CHPROP** command changes properties at the command line.

ChProp

The **CHPROP** command prompts you to change the properties of selected objects at the command line:

> Command: **chprop**
>
> Select objects: *(Select one or more objects.)*
>
> Select objects: *(Press* ENTER *to end object selection.)*
>
> Enter property to change
>
> [Color/LAyer/LType/ltScale/LWeight/Thickness/Material/Annotative]: *(Enter an option.)*

Depending on the option you select, AutoCAD presents a different set of prompts that match the related command. For example, to change the color, AutoCAD prompts you in a manner similar to the **-COLOR** command:

> Enter property to change
>
> [Color/LAyer/LType/ltScale/LWeight/Thickness/PLotstyle]: **c**
>
> New color [Truecolor/COlorbook] <BYLAYER>: *(Enter a color number, such as* **2**.*)*
>
> Enter property to change
>
> [Color/LAyer/LType/ltScale/LWeight/Thickness/PLotstyle]: *(Press* ENTER *to exit the command.)*

 MATCHPROP

The **MATCHPROP** command "reads" the properties of one object, and then applies them to other objects.

At times, you may want a group of objects to match the properties of another object. For example, you may accidentally place several doors on the Landscape layer, instead of the Door layer. This command lets you make this change. Or, you may realize that some lines drawn with hidden linetype should be in another linetype — whose name you don't recall.

You could use the Properties toolbar to change layers, colors, linetype, and so on, but it doesn't work well if you are not sure of the exact layer and linetype name. Complex drawings have hundreds of layer names, many of which look similar. One sample drawing provided with AutoCAD has the following layer names: ARCC, ARCCLR, ARCDIMR, ARCDSHR, ARCG, ARCM, ARCR, ARCRMNG, ARCRMNR, and ARCTXTG. And that's just the first ten! Similarly, linetypes can look confusingly alike.

The **MATCHPROP** command solves those problems. It lets you copy the properties from one object to a selection set of objects. You must, however, be sure to select objects in the correct order: (1) select the single object whose properties to copy, and then (2) select the object(s) to take on those properties.

TUTORIAL: MATCHING PROPERTIES

1. To match the properties of one object to other objects, start the **MATCHPROP** command with one of these methods:
   - From the Standard, toolbar choose the **Match Properties** button.
   - At the 'Command:' prompt, enter the **matchprop** command:

Command: **matchprop** *(Press* ENTER.*)*

- Alternatively, enter the aliases **ma** or **painter** (the old name in AutoCAD LT) at the 'Command:' prompt.

2. In all cases, AutoCAD prompts you at the command line:

   Select source object: *(Pick a single object.)*

   Notice that the cursor turns into a paintbrush.

3. AutoCAD lists the *active settings*, properties to be "picked up" by the paintbrush.

   Current active settings: Color Layer Ltype Ltscale Lineweight Thickness PlotStyle Dim Text Hatch Polyline Viewport Table Material Shadow display Multileader

   Select the objects to which the properties should be applied.

   Select destination object(s) or [Settings]: *(Pick one or more objects, or enter* **S**.*)*

4. You can continue selecting objects using windows, fences, single picks, and so on.

   Press **ENTER** to exit the command.

   Select destination object(s) or [Settings]: *(Press* ENTER.*)*

Property	Objects Affected
Color	All except OLE objects.
Dimension	Dimension, leader, and tolerance objects; also paints dimension styles.
Hatch	Hatches; also paints the hatch pattern.
Layer	All except OLE objects.
Linetype	All except attributes, hatches, multiline text, OLE objects, points, & viewports.
Linetype Scale	All objects except attributes, hatches, multiline text, OLE objects, points, and viewports.
Lineweight	All.
Material	All.
Multileader	Multileaders (*new to AutoCAD 2008*).
Plot Style	All OLE objects; unavailable when PSTYLEPOLICY is 1.
Polyline	Polylines; also paints width and linetype generation, but not fit/smooth, variable width, or elevation.
Shadow display	All.
Thickness	Arcs, attributes, circles, lines, points, 2D polylines, regions, text, and traces.
Text	Text and multiline text only; also paints the text style.
Viewport	Viewports; also paints on/off, display locking, standard or custom scale, shade plot, snap, grid, as well as UCS icon visibility and location settings, but not clipping, UCS-per-viewport, or layer freeze/thaw state.
Visual style	All.

MATCHING PROPERTIES: ADDITIONAL OPTION

The **MATCHPROP** command includes a single option that lets you decide which properties should be picked up and copied:

- **Settings** determines which properties are painted.

Settings

At the "Select destination object(s) or [Settings]:" prompt, enter **S** to display the Property Settings dialog box.

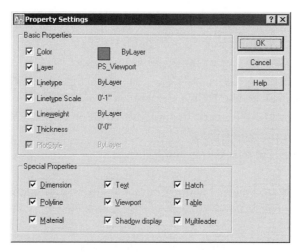

The properties listed by this dialog box are more extensive than those listed by the Properties toolbar. Not all properties, however, work with all objects. For example, it makes no sense to match the hatch properties of text objects, since text cannot be hatched.

By default, all properties are on, meaning they will all be copied.

When you turn off properties, AutoCAD remembers this the next time you use **MATCHPROP**, which is why it has that list of "Current active settings."

EXERCISES

1. In this exercise, you change the colors of objects.
    a. Open AutoCAD to start a new drawing.
    b. Use the **LAYER** command to set the color of layer 0 to cyan (light blue).
    c. Click **OK** to exit the dialog box.
    d. Draw some lines. Do the circles and boxes change color?

    e. Use the **COLOR** command to set the current color to yellow.
    f. Draw three circles. Did the **COLOR** command override the layer's color setting?
    g. Next, set the current color to red.
    h. Draw three boxes. Do they appear in the correct color?

    i. Start the **PROPERTIES** command, and then select all the objects on the screen.
    j. Select blue as the color. Did all the objects change to blue?

2. In this exercise, you create a layer, and then freeze and thaw it.
    a. Create a layer named "Mylayer."
    b. Set **Mylayer** as the current layer. Do you see the layer name in the toolbar at the top of the screen?
    c. Set the layer color to green, and then draw some objects on the layer.
    d. Using the Layer toolbar, freeze layer **Mylayer**, and then set the current layer to **0**. Do the objects you drew disappear?
    e. Use the **-LAYER** command to thaw **Mylayer**. Do the objects reappear?

3. In this exercise, you create several layers, and then change their properties.
    With the **LAYER** command, create seven layers using these names:
    >   Landscape
    >   Roadways
    >   Hydro
    >   ToBeRemoved
    >   StormSewer
    >   CableTV
    >   Building

    Using the layer dialog box:
    a. Change the color of layer **Landscape** from white to green.
    b. Change the lineweight of layer **Roadways** from Default to 0.083" (2.11mm).
    c. Change the linetype of layer **Hydro** from **Continuous to Gas_line**.
    d. Change the **ToBeRemoved** layer from white to red, and from Continuous to Dashed linetype.
    e. Change the linetype of the **CableTV** layer from Continuous to Hidden, and change the lineweight to 0.020" (0.51mm).
    f. Make the **CableTV** layer current by choosing the **Current** button.
    g. Click **OK** to exit the layer dialog box.
    h. Draw some lines. Do they appear in hidden linetype? If the lines do not look thick, ensure the **LWT** button is depressed on the status line.

i. Use the Layer toolbar to change to the **ToBeRemoved** layer, and then draw some lines. Do they appear red and dashed?

j. Again, use the Layer toolbar to change to the **Building** layer, and then draw some lines. Do they appear black and continuous?

4. In this exercise, you copy the properties from one object to all others in the drawing. Continue with the drawing you created in the previous exercise.

   Start the **MATCHPROP** command.

   Select source object: *(Select one of the red, dashed lines drawn on the **CableTV** layer.)*

   Select destination object(s) or [Settings]: **all**

   Select destination object(s) or [Settings]: *(Press* **ENTER.***)*

   Do all the objects become red and dashed?

CHAPTER REVIEW

1. What do hidden lines show in drawings?
2. Name two ways linetypes can be applied to objects in AutoCAD:

   a.

   b.
3. How do you control the length of individual segments in dashed lines?
4. How do you load all linetypes into a drawing?
5. What is the difference between *global* and *local* linetype scaling?
6. Is it possible to save the current state of layers?

   If so, how?
7. Describe five properties that the **PROPERTIES** command changes.
8. What is meant when the color of objects is changed to **ByLayer**?
9. Why would you use the **PROPERTIES** command instead of the Properties toolbar?
10. Can the Properties toolbar change objects?

    If so, how?
11. What is the **MATCHPROP** command used for?
12. Why would a CAD drafter use layers?
13. What layer option do you use to create new layers?
14. How do you turn on frozen layers?
15. What is the difference between *locking* layers and *freezing* layers?
16. How do you obtain a listing of all layers in drawings?
17. Can you have objects of more than one color on the same layer?

    Explain.
18. What do the following layer symbols mean?

    Snowflake

    Open lock

    Lightbulb glowing

    Printer

    Colored square
19. Name a benefit to using lineweights.
20. Is it possible to add custom linetypes to drawings?

    Custom lineweights?
21. What color is designated by R?

    What is color 7?

    What is a *color book*?
22. Describe what is meant by *RGB* in terms of colors.
23. Should you use lineweights to represent objects with width?
24. What is the purpose of the **LWT** button on the status bar?
25. Describe the steps to changing the lineweight of circles using the Properties toolbar:

    a.

    b.
26. Explain how these three classes of linetype differ:

    Simple

    ISO

    Complex
27. Before using linetypes in new drawings, what must you do first?

28. What is the name of the linetype used to show hidden edges?
29. **LTSCALE** is set to 0.5 and **CELTSCALE** is set to 5.0. At what scale is the next linetype drawn?
30. Under what condition can layers be deleted?

    Can layer 0 be deleted?
31. Describe two ways to manage long lists of layer names:

    a.

    b.
32. When drawings are in model space, why do the **Current VP Freeze** and **New VP Freeze** columns not appear in the Layer dialog box?
33. What is the purpose of the **Show all used layers** filter in the layers dialog box?
34. Can you assign lineweights to layers?

    Assign linetypes?

    Assign hatch patterns?
35. When linetypes are assigned to layers, are all objects on those layers displayed with that linetype?

    If not, why not?
36. What do the following wild card characters mean?

    *

    ?
37. What is the purpose of the **CLAYER** system variable?
38. Which keystroke shortcut displays the Properties window?
39. Describe how the **PICKADD** system variable affects object selection.
40. Can the Properties window be used to change the endpoints of line segments?
41. Explain the purpose of the following filters:

    Group

    Properties
42. What are xref layers?

CHAPTER 8

Correcting Mistakes

As you create a drawing, you sometimes make mistakes. One way to correct them is through editing, which is the subject of the chapters following. Sometimes, it is faster to do something "wrong" and then fix it, than it is to do it correctly the first time. Or you can say "Oops!," and reverse the mistake by undoing it, the subject of this chapter.

In all chapters until now, you have been learning how to create things. In this chapter, you learn how to un-create them — how to revert, reverse, and repair — using the following commands:

U and **UNDO** reverse the changes of most commands.

REDO and **MREDO** reverse the effect of the undoing.

RENAME changes the names of linetypes, layers, and so on.

PURGE removes unused layers, blocks, and so on.

RECOVER and **AUDIT** attempt to fix corrupted drawings.

RECOVERALL recovers drawings and related xrefs (new to AutoCAD 2008).

DRAWINGRECOVERY lists the names of damaged drawing files.

File Manager allows you to retrieve drawing backups.

QUIT and **CLOSE** discard editing changes made to drawings.

NEW TO AUTOCAD 2008 IN THIS CHAPTER

- **RENAME** and **PURGE** now support renaming and purging mleader styles.
- **RECOVERALL** recovers and converts drawings and attached reference files.

FINDING THE COMMANDS

On the **Standard** toolbar:

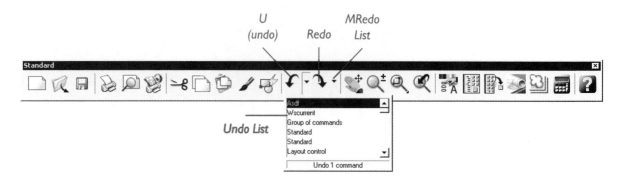

 **U** AND **UNDO**

The U and UNDO commands reverse the effect of many (not all) commands.

- **U** — undoes the last action; this is the quick command to use when just one or two changes need to be undone.

- **UNDO** — provides many options for undoing the effects of commands; this is the advanced command that provides utter control.

You can draw, edit, and then undo your work — then redo it. This is useful if you have just performed an operation that you wish to reverse, either a mistake or a what-if scenario. After undoing one or more operations, you can use the REDO command *once* to reverse the undo, while MREDO redoes multiple undoes.

The U command reverses the effect of the most recent command. You can execute a series of U commands to back up through a string of changes. (The U command is not an alias for the more advanced UNDO command, although it does function identically to the UNDO command's **1** option.)

Undoing a command restores the drawing to the state before the command was executed. For example, erase an object, and then execute the U command: the object is restored. When you scale an object, and then undo it, the object is scaled back to its original size. Undoing a just-completed **BLOCK** command restores the block, and deletes the block definition that was created, leaving the drawing exactly as it was before the block was inserted.

At the command prompt, the U command lists the command that is undone to alert you to the type of command that was affected.

TUTORIAL: UNDOING COMMANDS

1. Open a drawing, and erase a part of it with the **ERASE** command.
2. To reverse the erasure, start the **U** command with one of these methods:
   - From the **Edit** menu, choose **Undo**.
   - At the keyboard, press **CTRL+Z**.
   - At the 'Command:' prompt, enter the **u** command.

     Command: **u** *(Press* ENTER.*)*
3. AutoCAD undoes the effect of the **ERASE** command, and then displays the name of the undone command:

       ERASE

   When you undo all the way back to the first command, AutoCAD reports:

       Everything has been undone

 Note: *Several commands cannot be undone.* **SAVE**, **PLOT**, and **WBLOCK**, for example, are unaffected, because AutoCAD cannot "unsave" or "unplot." If you attempt to use the U or UNDO command after these commands, the name of the command is displayed, but the command's action is not undone.

(The way to undo a save is to retrieve the backup copy, as described later in this chapter. The way to undo a plot is to fold up the paper and then throw it in the recycling bin.)

UNDO and U have no effect on commands and system variables that open, save, export, and close files, change the arrangement of windows (as opposed to viewports), and redraw or regenerate (such as **REGEN**) the drawing.

The two undo commands also have no effect on commands that took place before the drawing was opened. For example, you work on a drawing, save it, and then close the drawing. When you open the drawing again, AutoCAD cannot undo the commands of the previous editing session.

UNDOING COMMANDS: ADDITIONAL METHODS

AutoCAD has several additional methods to undo the effects of commands:

- **UNDO** command controls the undo process at the command line.

- **Standard** toolbar lists undo-able actions.

- **Undo** and **Previous** options undo actions in commands; the **Cancel** button in dialog boxes.

Let's look at each.

Undo

The UNDO command provides fine control over the undo process.

> Command: **undo**
>
> Enter the number of operations to undo or [Auto/Control/BEgin/End/Mark/Back] <1>:
>   *(Enter a number or an option.)*

By default, the UNDO command operates like the U command: press **ENTER** at the "Enter the number of operations to undo" prompt, and AutoCAD undoes the last command (if possible). Enter a number, such as **4**, to undo the last four commands. This is the same as entering U four times.

Auto

The **Auto** option groups all the actions of a single command into a single undo. The **Auto** option is not available when the **Control** option is turned off.

> Enter UNDO Auto mode [ON/OFF] <On>: *(Enter **ON** or **OFF**.)*

Control

The **Control** option limits the effect of the UNDO command:

> Enter an UNDO control option [All/None/One] <All>: *(Enter an option.)*

The **All** option turns on the UNDO and U commands, and allows them to undo all the way to the beginning of the drawing session.

The **None** option turns off the UNDO and U commands, grays out the **Undo** button on the Standard toolbar, and discards the undo history. Use this option only if your computer is low on disk space.

The **One** option restricts the UNDO command to a single undo, and the **Auto**, **Begin**, and **Mark** options are unavailable.

BEgin/End

The **BEgin** option groups several commands into a *set*.

> Command: **undo**
>
> Enter the number of operations to undo or [Auto/Control/BEgin/End/Mark/Back] <1>: **be**

After you enter the **BEgin** option, the UNDO command ends, but AutoCAD starts recording all the commands you enter, until you enter the **End** option:

> Command: **undo**
>
> Enter the number of operations to undo or [Auto/Control/BEgin/End/Mark/Back] <1>: **e**

The UNDO, REDO, and U commands now treat the set of commands as a single undo:

> Command: **u**
>
> GROUP

Mark/Back

The **Mark** option places a marker in the undo collection:

Command: **undo**

Enter the number of operations to undo or [Auto/Control/BEgin/End/Mark/Back] <1>: **m**

The **Back** option undoes all commands back to the marker:

Enter the number of operations to undo or [Auto/Control/BEgin/End/Mark/Back] <1>: **b**

Mark encountered

You may place as many marks as you require; the **Back** option moves back through the undo collection one mark at a time, removing each mark.

Note: It is dangerous to use the **UNDO** command's **Back** option without marks, because it undoes every action in the drawing:

Command: **undo**

Enter the number of operations to undo or [Auto/Control/BEgin/End/Mark/Back] <1>: **b**

This will undo everything. OK? <Y> *(Enter **Y**.)*

Everything has been undone

Standard Toolbar

As an alternative to the undo commands, you can use the **Undo** drop list on the Standard and Standard Annotation toolbars. Click the arrow next to **Undo**, and you see a list of commands.

Select the commands you wish to undo. You cannot, however, selectively choose noncontiguous commands.

Combined Zooms and Pans

When you perform several zooms and pans in a row, undoing them could be annoying if AutoCAD were to redraw each view change. Fortunately, AutoCAD groups all sequential view changes into a single undo through the **Combine zoom and pan commands** option. You find it in the Undo/Redo section of the User Preferences tab of the Options dialog box.

Undo, Previous Options and Cancel Buttons

Several commands include methods of undoing operations within them.

Undo Option

Commands that execute more than one action often include the **Undo** option. This allows you to reverse the effect of the last action without leaving the command. For example, the LINE command's **Undo** option erases the last-drawn segment. Other commands that have an Undo option include **PLINE**, **PEDIT**, and **3DPOLY**.

Previous Option

Instead of an undo option, other commands have an equivalent option called "Previous." The ZOOM command's **Previous** option, for example, displays the previous view, whether created by the ZOOM, PAN, or -SHADEMODE command. This option remembers up to ten previous views.

Other commands with a **Previous** option include SELECT and UCS. One command is itself a "previous" command: LAYERP restores previous layer states.

Cancel Button

Many dialog boxes have a button labeled **Cancel**. Click this button after making changes to the dialog box, if you don't want to keep the changes.

 ## REDO AND MREDO

The REDO and MREDO (short for "multiple redo") commands are the antidotes to the U and UNDO commands: they reverse the undoing.

- **REDO** — redoes the last undo; this is the quick command to use when just one or two changes need to be reversed.

- **MREDO** — provides a couple of options for redoing the effects of undo; this advanced command is also available as a droplist on the toolbar.

The REDO command must be used immediately after one of the undo commands.

TUTORIAL: REDOING UNDOES

1. To undo an undo command, start the **REDO** command with one of these methods:
    - From the **Edit** menu, choose **Redo**.

    - At the keyboard, press **CTRL+Y**.

    - At the 'Command:' prompt, enter the **redo** command.

        Command: **redo** *(Press* ENTER.*)*

2. AutoCAD redoes the previous undo, and then displays the name of the redone command:
    ARC

    When you redo back to the first undo, AutoCAD reports:
    Everything has been redone

REDOING UNDOES: ADDITIONAL METHODS

AutoCAD has several additional methods to reverse the effect of undo commands:

- **MREDO** command controls the undo process at the command line.

- **Standard** toolbar lists redo-able actions.

Let's look at each.

MRedo

The MREDO command provides additional control over the redo process.

Command: **mredo**

Enter number of actions or [All/Last]: *(Enter a number or an option.)*

Enter the number of undone commands you want reversed.

All

The **All** option reverses all previous undo actions.

Last

The **Last** option reverses the last undo action only; this is like the **REDO** command.

Standard Toolbar

As an alternative to the redo commands, use the **Redo** droplist on the Standard and Standard Annotative toolbars. Click the arrow next to **Redo**, and you see a list of commands.

Select the commands you wish to redo. You cannot pick noncontiguous commands. When the **MRedo** button is gray, there are no actions to redo.

RENAME

The **RENAME** command changes the names of linetypes, layers, and so on. Very often, the commands that create those entities, such as **LINETYPE** and **LAYER**, include an option for renaming them. The Rename dialog box, moreover, is a handy way for changing the names of many items in one place.

If you are new to AutoCAD, you might not be familiar with the concept of *named objects*. Named objects are anything in the drawing that you name. These are:

Blocks	Text styles	Table style	Dimension styles	Plot styles	Materials
Layers	Named views	Linetypes	Named viewports	User coordinate systems	

Certain names cannot be changed, such as layer 0 and the Global material; these do not appear in the Rename dialog box. As well, plot styles do not appear when they are not enabled for the drawing.

If we were to apply the technical definition of named objects, being "any object appearing in the Tables section of the drawing file," then this command is not all-encompassing. Missing are named objects, such as visual styles and multiline styles; you need to use the associated commands (**VISUALSTYLES** and **MLSTYLE** command) to change their names.

As an alternative to the **RENAME** commands, many dialog boxes have a provision to rename items, such as layers, text styles, and dimension styles.

TUTORIAL: RENAMING NAMED OBJECTS

1. To rename named objects, start the **RENAME** command with one of these methods:
   - From the **Format** menu, choose **Rename**.
   - At the 'Command:' prompt, enter the **rename** command:

     Command: **rename** *(Press ENTER.)*
   - Alternatively, enter the **ren** alias.

2. In all cases, AutoCAD displays the Rename dialog box.
3. Under the **Named Objects** list, select a table name, such as "Linetypes."
   Notice that AutoCAD displays a list of linetype names; if none in the drawing can be renamed, the list remains blank.

4. Under **Items**, select a linetype name.

Notice it appears in the **Old Name** text box.

5. In the **Rename To** text box, enter a new name, and then click **Rename To**.

Notice that the name changes in the list under **Items**.

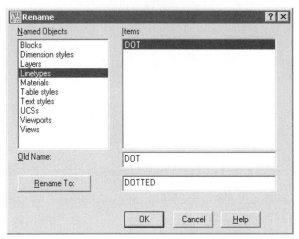

6. Click **OK** to accept the changes.

If you change your mind about the new name, you can use the **U** command.

RENAMING NAMED OBJECTS: ADDITIONAL METHODS

AutoCAD has one other method for renaming objects:

- **-RENAME** changes the names of named objects at the command line.

Let's look at it.

-Rename

The **-RENAME** command renames named objects at the command line:

Command: **-rename**

Enter object type to rename

[Block/Dimstyle/LAyer/LType/Material/Style/Tablestyle/Ucs/VIew/VPort]: *(Enter an option, such as* **la***.)*

Enter old layer name: *(Enter a valid layer name.)*

Enter new layer name: *(Enter a new name for the layer.)*

Rename Option	Comment
Dimstyle	Renames dimension styles.
LType	Renames linetypes.
UCS	Renames user-defined coordinate systems.
VPort	Renames viewports.

You have to know the names of the items to change ahead of time; when you enter the incorrect "old" name, AutoCAD complains:

Cannot find layer "name".

When you enter names that cannot be changed, such as layer 0, AutoCAD complains:

Cannot rename layer "0".

Invalid

PURGE

The **PURGE** command removes unused layers, blocks, and so on from drawings.

Why would you want to remove these objects? The primary reason is that they can clutter up drawings. In the course of creating a drawing, you sometimes create objects you never use, such as text styles, blocks, or dimension styles. For example, I find it easier to load all linetypes, and then later purge the unused ones — this is a lot faster than loading linetype definitions one at a time as I need them.

In the old days of computing, when disk space was very limited, drafters purged drawings to reduce their file size. Today, with gigabytes of disk space being cheap like borsch, we no longer use that excuse. Instead, the clutter occurs in other ways, as in long lists of layer and linetype names.

You needn't worry that the **PURGE** command might erase items of importance. Its built-in safety guard prevents it from touching named objects in use. "In use" means, for example, layers with objects on them, or text styles used in the drawing.

In addition, **PURGE** does not touch named objects set as "current" and those found in a new, empty drawing. The "untouchables" are:

Table Object	Names Untouched
Dimension Styles	Standard
Layers	0, Defpoints
Linetypes	Continuous, Bylayer, Byblock
Materials	ByBlock, ByLayer, *GLOBAL*
Multilines	Standard
Multileader	Standard
Blocks	*Nested blocks, xrefs*
Plot Styles	Normal
Table Styles	Standard
Text Styles	Standard
Visual Styles	2D Wireframe, 3D Wireframe, 3D Hidden, Conceptual, and Realistic

As an alternative to the **PURGE** command, you can remove unused named objects with the **Delete** button in certain dialog boxes. These include:

> Layers through the Layer Properties Manager dialog box (**LAYER** command).
> Linetypes through the Linetype Manager dialog box (**LINETYPE** command).
> Materials through the Materials palette (**MATERIALS** command).
> Multileaders through the Multiline Styles dialog box (**MLSTYLE** command).
> Styles through the Text Style dialog box (**STYLE** command).
> Table styles through the Table Style dialog box (**TABLESTYLE** command).
> Visual styles through the Visual Styles Manager palette (**VISUALSTYLES** command).

The Purge dialog box handles these, and as an added bonus, deletes all unused named objects at once.

One last concept to discuss is the *nested purge*. This action is required for nested objects, such as nested blocks. That's where one block contains other blocks. The Purge dialog box removes nested items automatically, when that option is turned on.

Notes: You can use the **U** command to reverse the effect of the **PURGE** command, and return purged objects to the drawing.

Some items may appear to be unused but can't be purged. A layer may appear to be empty, but is used in a block definition. A linetype override may be applied to an object.

TUTORIAL: PURGING UNUSED OBJECTS

1. To purge a drawing of unused objects, start the **PURGE** command:
    *   From the **File** menu, choose **Drawing Utilities**, and then **Purge**.
    *   At the 'Command:' prompt, enter the **purge** command:

        Command: **pu** *(Press* ENTER.*)*
    *   Alternatively, enter the **pu** alias.

    In all cases, AutoCAD displays the Purge dialog box.

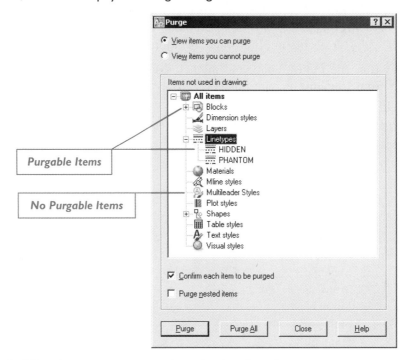

2. Ensure the **View items you can purge** option is selected.
3. Look at the list under **Items not used in drawing**.

    Notice that some items have a + (plus sign) next to them. This indicates there are purgable objects for that item. (If there is no plus sign next to an item, there is nothing to purge.)
4. Click the **+** sign. Notice that the list expands to show the purgable items.
5. Click **Purge All**.

    When the **Confirm each item to be purged** option is turned on, AutoCAD displays a dialog box asking your permission to purge:

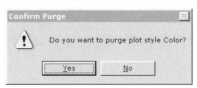

    Click **Yes**.
6. When done, click **Close**.

The dialog box's other radio button (the round buttons near the top of the dialog box) is labeled **View items you cannot purge**. It displays the inverse of the purge list: named objects that cannot be purged, along with an explanation at the bottom of the dialog box. (You can right-click the explanation area, and copy the text to the Clipboard.)

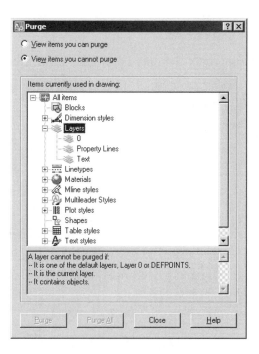

PURGING UNUSED OBJECTS: ADDITIONAL METHODS

AutoCAD has a couple of other methods for purging unused objects:

- **-PURGE** command executes the purge at the command line.

- **WBLOCK** command purges drawings when saved to disk.

-Purge

The **-PURGE** command provides AutoCAD's purge services at the command line. The fastest way to complete the purge process is to answer the prompts in this manner:

> Command: **-purge**
>
> Enter type of unused objects to purge
>
> [Blocks/Dimstyles/LAyers/LTypes/MAterials/Plotstyles/SHapes/textSTyles/Mlinestyles/ Tablestyles/Visualstyles/Regapps/All]:: **a**
>
> Enter name(s) to purge <*>: *(Press* ENTER.*)*
>
> Verify each name to be purged? [Yes/No] <Y>: **n**

As an alternative, you can purge each object type and item individually. But if that's the case, it's better to use the dialog box.

The **Regapps** option allows unused ObjectARx applications to be purged from the drawing. Missing from this command are the "mleader styles" (new to AutoCAD 2008) found in the dialog box version of this command.

WBlock

With versions of PURGE found in AutoCAD 2000 and earlier, you often needed to repeat the command, because AutoCAD purged only one level of nesting at a time. Indeed, this was such a nuisance that power users employed a hidden feature of the WBLOCK command to cleanse the entire drawing. (The **Purge nested items** option in the Purge dialog box cleans nested objects automatically.)

To make WBLOCK purge drawings, you must use the **Entire drawing** option, and then save it to disk.

RECOVER, AUDIT, AND RECOVERALL

The **RECOVER** command attempts to fix corrupted drawings, while **AUDIT** checks drawings for errors.

In perfect worlds, everything works as planned. In our world, however, we encounter difficulties. One day you will see a message from AutoCAD that reads:

INTERNAL ERROR

(followed by a host of numbers)

or

FATAL ERROR.

This means that AutoCAD has encountered a problem severe enough that it cannot continue. You are usually given a choice of whether or not you wish to save the changes you have made since the last time you saved your work. The following message is displayed:

AutoCAD cannot continue, but any changes to your drawing made up to the start of the last command can be saved.

Do you want to save your changes? <Y>: **y**

If you enter "Y," AutoCAD attempts to write the changes to disk. If it is successful, AutoCAD displays the following message:

DRAWING FILE SUCCESSFULLY SAVED

If the save is unsuccessful, the "INTERNAL ERROR" or "FATAL ERROR" messages are displayed again. When you see these ominous messages, you can wave good-bye to unsaved changes. In some severe cases, AutoCAD cannot recover the drawing, and reports:

Unable to recover this drawing.

Of course, as a good CAD operator, you save your work regularly, right? Recall the **SAVETIME** system variable discussed earlier in this book. With AutoCAD 2004, Autodesk has improved the drawing file format to better prevent corruption, as well as reducing the default time between automatic saves from an ineffective two hours down to a more reasonable ten minutes.

Other times, you open a drawing but see the following dialog box:

In either case, you can attempt to recover the damaged drawing files with the **RECOVER** command.

TUTORIAL: RECOVERING CORRUPT DRAWINGS

1. To attempt to recover a damaged drawing, start the **RECOVER** command with one of these methods:
   * From the **File** menu, choose **Drawing Utilities**, and then **Recover**.
   * At the 'Command:' prompt, enter the **recover** command:

     Command: **recover** *(Press ENTER.)*

2. AutoCAD displays the Select File dialog box.
   Select the drawing, and then click **Open**.

3. AutoCAD mutters as it works its way through the drawing, looking for errors:

    Drawing recovery.
    Drawing recovery log.
    Scanning completed.
    Validating objects in the handle table.
    Valid objects 519   Invalid objects 1
    Validating objects completed.
    Creating new ACAD_LAYOUT dictionary
    Creating new ACAD_MATERIAL dictionary
    Creating new ACAD_COLOR dictionary16 error opening *Model_Space's layout.
    Setting layout id to null.
    16 error opening *Paper_Space's layout.
    Setting layout id to null.
    Used contingency data.
        Salvaged database from drawing.
    Removed 2 unread objects from entity lists.
    Opening a Release 13 format file.
     0   Blocks audited
    Pass 1 491   objects audited
    Pass 2 491   objects audited
    Pass 3 500   objects audited
    Total errors found 0 fixed 0
    Regenerating model.

When AutoCAD detects a damaged drawing file when opening it with the **OPEN** command, it performs the audit automatically. If the recovery is successful, the drawing is loaded; if not, the drawing is typically unrecoverable.

If the recovery is successful, and you save the drawing, you can load it normally the next time. When you exit the drawing without saving, the "repair" performed by AutoCAD is discarded.

RECOVERING CORRUPT DRAWINGS: ADDITIONAL METHODS

AutoCAD has one other method for checking drawing integrity:

* **AUDIT** command executes the purge at the command line.

* **RECOVERALL** command recovers and updates drawings and related xrefs.

Audit

The **AUDIT** command looks for damage in drawings already loaded into AutoCAD. Here's the difference: **RECOVER** checks drawings as they open, while **AUDIT** checks them after they are open. This diagnostic tool also corrects damage to a file.

    Command: **audit**

    Fix any errors detected? [Yes/No] <N>: *(Enter* **Y** *or* **N.***)*

If you answer "N" AutoCAD displays a report, but does not fix the errors.

When you answer "Y," AutoCAD displays the report and fixes errors.

In addition to providing the screen report, the **AUDIT** command can save its report to a text file. This is controlled by the **AUDITCTL** system variable, where 0 means no file is produced (the default), and 1 means AutoCAD writes the *.adt* file. The file has the same name as the drawing, and is located in the same folder. You can read and print the audit file with any word processor.

 **RecoverAll**

The **RECOVERALL** command recovers and updates drawings and attached xrefs.

The command works through these steps:

1. AutoCAD opens the selected drawing and its related xrefs.
2. The files are recovered; as well, custom objects are updated if their object enablers are available.
3. The files are saved in AutoCAD 2008 format; the original files are saved as *.bak* files.
4. The files are closed.

The **RECOVERALL** command works like this:

From the **File** menu, choose **Drawing Utilities**, and then **Recover Drawing and Xrefs**. Or, at the command prompt, enter:

Command: **recoverall**

Notice the informative dialog box that's mislabeled "Warning"; click **Continue**.

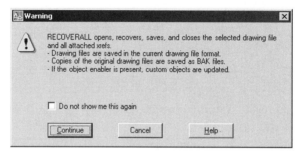

In the Select File dialog box, select a single drawing file (such as *8th floor.dwg* from the Companion CD), and then click **Open**. As the command performs its work, it reports its progress in the Drawing Recovery Log and Recover Progress dialog boxes.

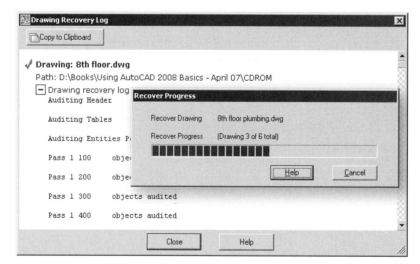

When the recover process is done, read over the log report, and then click **Close** to close the dialog box. AutoCAD automatically closes the drawing and its xrefs.

DRAWINGRECOVERY

The **DRAWINGRECOVERY** command displays a palette listing the names of damaged drawing files. (The **DRAWINGRECOVERYHIDE** command closes the palette.)

Each time you use the **QSAVE** command, AutoCAD renames the current *.dwg* file with the *.bak* extension, and then saves the current state of the drawing as a new *.dwg* file. When AutoCAD crashes, it tries to rename the current backup file so that it doesn't replace the previous backup file. AutoCAD renames it using the file extension of *.bk1*. If a file with such an extension already exists, it uses *.bk2* on through *.bkz*.

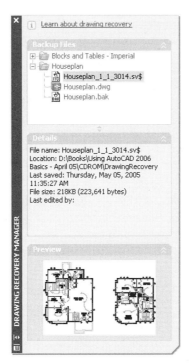

Whenever AutoCAD performs one of its automatic backups, a copy of the drawing is saved as a *.sv$* file. AutoCAD stores the *.sv$* files in the folder specified by **Options | Files | Automatic Save File Location**. Whenever AutoCAD performs an automated save, it reports the path and file name:

> Automatic save to C:\Documents and Settings*username*\Local
> Settings\Temp\Houseplan_1_1_6960.sv$

By default, AutoCAD automatically saves drawings every ten minutes — provided changes have occurred in the drawing. When you go off for a half-hour-long coffee (tea) break, AutoCAD does not keep saving the drawing every ten minutes, but also takes a break.

The Drawing Recovery Manager displays the names of all drawing and backup files that were open when AutoCAD or Windows crashed. It lets you choose which files to save as the *.dwg* file; it does not perform recovery services, like the **RECOVER** command.

You rarely enter the **DRAWINGRECOVERY** command, because AutoCAD displays the Drawing Recovery Manager palette automatically when *.sv$* files exist from previous editing sessions.

Every drawing that was open in AutoCAD at the time of the crash is listed in the window. With each drawing name are up to four file names:

Extension	Meaning
.dwg or *.dws*	Original drawing or standards files.
.dwg or *.dws*	Recovered drawing or standards files.
.sv$	Automatic backup file (saved by AutoCAD).
.bak or *.bk*n	Backup files (saved by users).

To open a drawing, right-click any of the listed drawing files, and then select **Open**. AutoCAD opens the file as a drawing, giving it the extension of *.$sv.dwg* or *.bak.dwg*. You can inspect the backup to see if you want to keep it.

RecoveryMode

The **RECOVERYMODE** system variable controls the Drawing Recovery Manager (DRM):

RecoveryMode	Meaning
0	Recovery information not recorded and DRM not displayed; recovery information removed from the system registry.
1	Recovery information recorded; DRM not displayed.
2 (*default*)	Recovery information recorded; DRM displayed automatically.

File Manager

An alternative to **DRAWINGRECOVERY** is the File Manager (or Windows Explorer) included with Windows. It also allows you to retrieve drawing backups. Should you need to access the previous version of the drawing, you can use Explorer to change the extension from *.bak* to *.dwg*.

To rename a file, select it in Explorer. Right-click, and then select **Rename**.

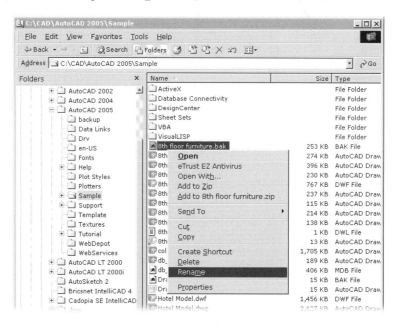

QUIT AND CLOSE

If all else fails, quit the drawing, and AutoCAD discards all changes you've made.

The **QUIT** command discards editing changes, and exits AutoCAD. The **CLOSE** command does the same, but without exiting AutoCAD.

TUTORIAL: DISCARDING CHANGES TO DRAWINGS

1. To discard changes you've made to a drawing, start the **QUIT** command with one of these methods:

   • From the **File** menu, choose **Exit**.

   • At the keyboard, enter the **CTRL+Q** shortcut.

   • At the 'Command:' prompt, enter the **quit** command:

     Command: **quit** (*Press* ENTER.)

   • Alternatively, enter the **exit** alias.

2. In all cases, AutoCAD displays this dialog box.

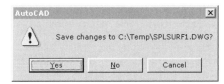

3. Click **No** to discard the changes, and exit AutoCAD.

Click **Yes** if you've had a change of heart, and want to save the changes after all, and then exit AutoCAD.

Click **Cancel** to *not* save changes *and* remain in AutoCAD.

To use the CLOSE command instead of QUIT, from the **Window** menu, select **Close**.

Notes: To change a drawing, yet preserve it in its original form, use the **SAVEAS** command to save the changed drawing by another name, and then use **CLOSE** or **QUIT** to exit the drawing without saving changes.

EXERCISES

1. In this exercise, you work with the **UNDO** and **REDO** commands.
   a. Start AutoCAD with a new drawing.
   b. Choose the **LINE** command, and then draw a line segment. (Press **ESC** to end the command.)
   c. Enter **U** at the keyboard, and then press **ENTER**. Did the line segment disappear?

   d. Use the **LINE** command to draw several line segments.
   e. While the **LINE** command is still active, use its **Undo** option to remove one segment.
   f. End the **LINE** command, and then restart it.
   g. Draw more line segments.
   h. Use the **U** command to undo the lines drawn with the last **LINE** command. Which segments were undone?
   i. Press spacebar to repeat the **U** command. What happened?

   j. Use the **LINE** command to draw two line segments.
   k. Before entering the last point, press function key **F8** to turn on ortho mode. Notice the ORTHO button on the status line.
   l. Now use the **U** command to undo the sequence. Is the ortho mode on now?
   m. Press **F8** to turn on ortho mode.
   n. Now use the **LINE** command to draw a line segment.
   o. Next, use the **U** command to undo the sequence. Is ortho mode still turned on? What is the difference between this and the last sequence you performed?
   p. Use the **LINE** command to draw several line segments.
   q. Use the **U** command to undo the lines.

   r. Now use the **REDO** command. Did the lines reappear?
   s. Enter the **REDO** command again. What does the prompt line say? Why?

2. Use the **RECOVER** command to open the *splsurf1.dwg* drawing file (found on the Companion CD).
   Read the report. How many invalid objects did AutoCAD find?

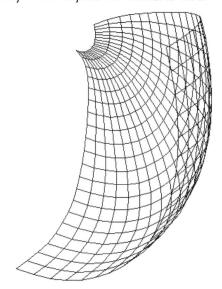

3. From the Companion CD, open the *airport.dwg* file.

   Use the **PURGE** command to remove unused objects. Which items were removed?

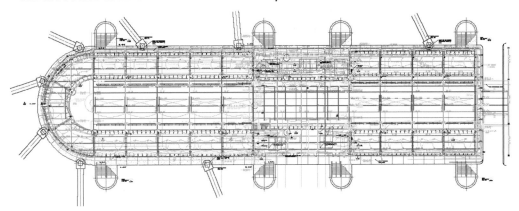

4. From the Companion CD, copy the *forkift.bak* file to your computer's hard drive.

   Use Windows Explorer to rename the file to *forklift.dwg*.

   Open the drawing in AutoCAD. Did the drawing open normally?

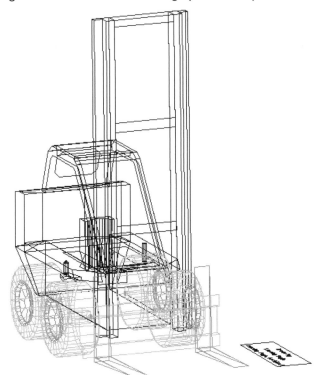

CHAPTER REVIEW

1. Can the **U** command be used to undo a sequence of commands?

2. Can the **U** command be entered from the menu bar as well as from the keyboard?

3. If you use the **Undo** option while in the **LINE** command, do all the segments disappear or is each segment stepped-through backwards?

4. What is the **AUDIT** command used for?

   How does it differ from the **RECOVER** command?

5. What happens to the drawing if you use the **UNDO** command's **Back** option without setting any marks?

6. What is the keyboard shortcut for the following commands:

   UNDO

   REDO

   QUIT

7. What effect does the **UNDO** command have on the drawing after you use the **SAVE** command?

8. What happens to a block if you use the **U** command immediately after creating the block?

9. How far back can you use the **UNDO** command?

10. Name a limitation to using the **REDO** command.

11. How does the **MREDO** command differ from the **REDO** command?

12. Describe the purpose of the **RENAME** command.

13. Can you rename layer 0?

14. Under what conditions are you unable to rename plot styles?

15. List two ways in which the **PURGE** command is useful:

    a.

    b.

16. When would a *nested purge* be required?

17. Name two alternatives to the **PURGE** command:

    a.

    b.

18. What does the message "Internal error" indicate?

19. Which command closes individual drawings?

20. Describe how **RECOVER** differs from **DRAWINGRECOVERY**.

CHAPTER 9

Direct Editing of Objects

The earliest releases of AutoCAD used the command line for entering the names of commands and their options. That was the way computer software operated twenty-five years ago when CAD software was first developed. In the following decades, programmers worked to make the user interface interactive: the first innovation was the menu, followed by icon-bearing toolbars; more recently, programmers have concentrated on getting CAD software to allow direct manipulation of objects on the screen.

In Chapter 7, "Changing Object Properties," you gained experience in changing the properties of objects through the Properties toolbar. In this chapter, you learn to use AutoCAD's direct editing and drawing methods. These are:

Shortcut Menus provide direct access to appropriate commands.

Grips allow direct editing of objects.

Stretch, **Move**, **Rotate**, **Scale**, **Copy**, and **Mirror** comprise the grips editing options.

Arcs and **Dynamic Blocks** have task-specific grips.

Right-click Move/Copy/Insert moves, copies, and inserts objects.

Direct Distance Entry allows "pen down" movement during commands.

TRACKING and **OTRACK** allow "pen up" movement during drawing and editing commands.

M2P finds the midpoint between two picked points.

In Chapter 19, "Editing 3D Objects," you learn about interactive editing with 3D models.

FINDING THE SHORTCUT MENUS

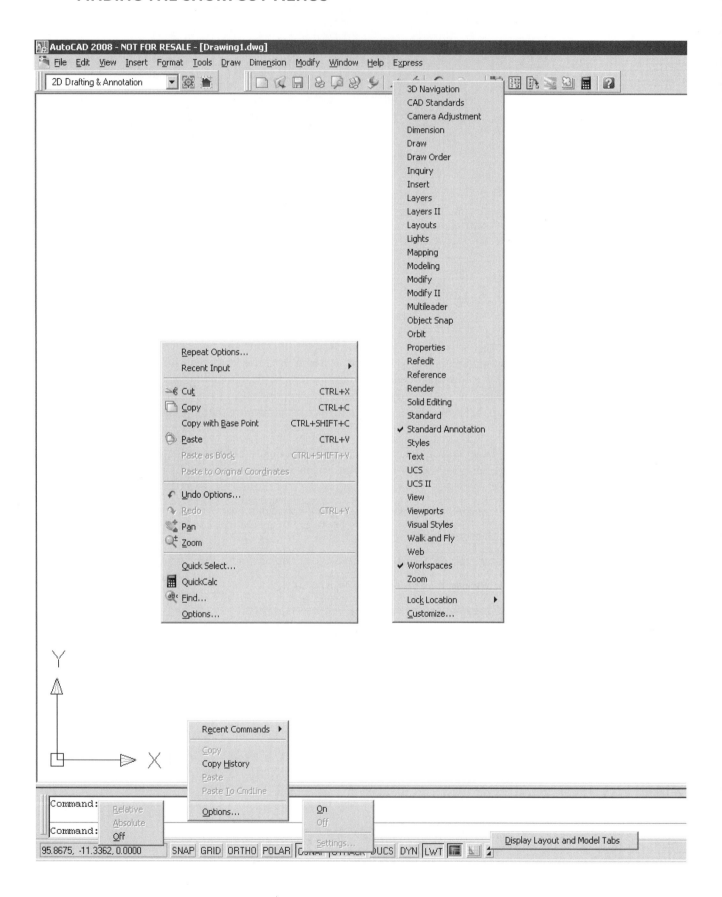

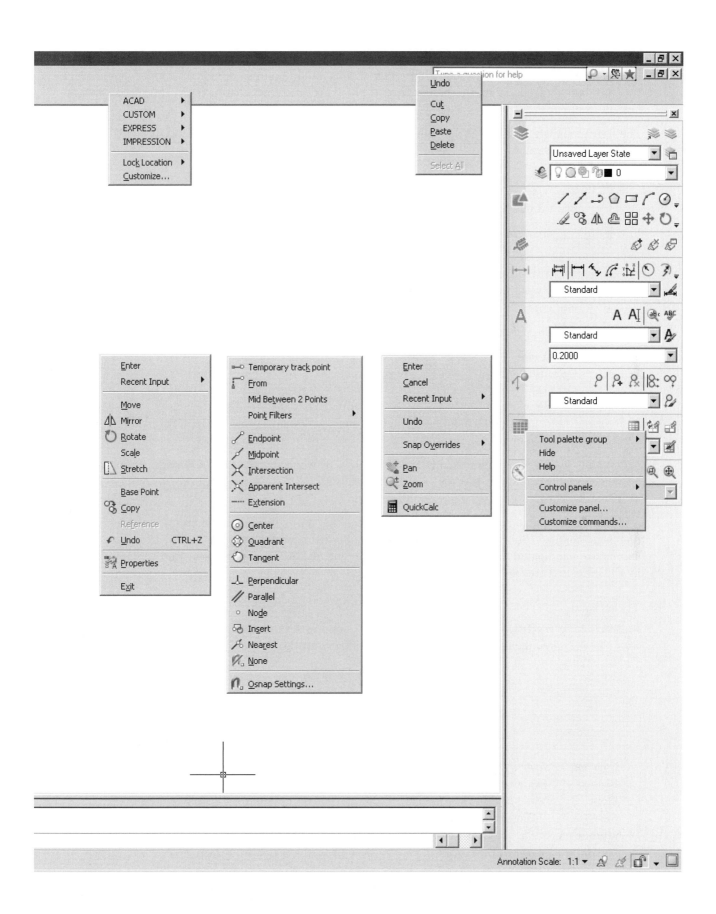

SHORTCUT MENUS

Shortcut menus provide direct access to commands appropriate to the context.

AutoCAD displays a shortcut menu almost every time you press the mouse's right button; these menus are also sometimes called "right-click menus" and "context menus." *Context* means that the commands displayed by the menu depend on where in AutoCAD you press the mouse's right button.

The figure opposite illustrates the shortcut menus displayed in different areas of AutoCAD. On the top part of the window, there are shortcut menus for:

- Title bar
- Toolbars
- Blank toolbar area

In the drawing area, there are shortcut menus for:

- Blank drawing area
- **SHIFT**+blank drawing area
- Active commands
- Hot grips

And in the bottom part of AutoCAD, there are shortcut menus for:

- Layout tabs
- Command line
- Drag bar
- Status bar

In addition, AutoCAD's text window and palettes, such as Properties and DesignCenter, have context-sensitive shortcut menus.

Commands in shortcut menu usually are in black text; when in gray, the command is unavailable. Some commands work with shortcut keystrokes; these are shown to the right of the command, such as: **Close ALT+F4**.

A check mark next to a command means it is a toggle, and is turned on; when off, an empty box is displayed instead. Some shortcut menus have *submenus*, with further options; the presence of submenus is indicated by the arrowhead. Some commands display dialog boxes, and this is indicated by the ellipsis (three dots ...).

TITLE BAR

Right-clicking the title bar displays the shortcut menu with commands for changing the AutoCAD window (the same menu found with all other Windows software).

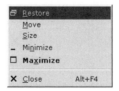

Command	Comment
Restore	Restores the AutoCAD window from a maximized or minimized state.
Move	Moves the AutoCAD window within the computer monitor's screen.
Size	Changes the size of the AutoCAD window.
Minimize	Drops the AutoCAD window down to the Windows taskbar.
Maximize	Maximizes the AutoCAD window so that it fills the entire screen.
Close	Exits AutoCAD, after asking if you want to save changes for drawings.

TOOLBARS

Right-clicking any toolbar displays the shortcut menu with the names of toolbars available in AutoCAD. (This works only if at least one toolbar is displayed.) The check marks indicate the toolbars that are currently displayed. Select a toolbar name to toggle its display, on or off.

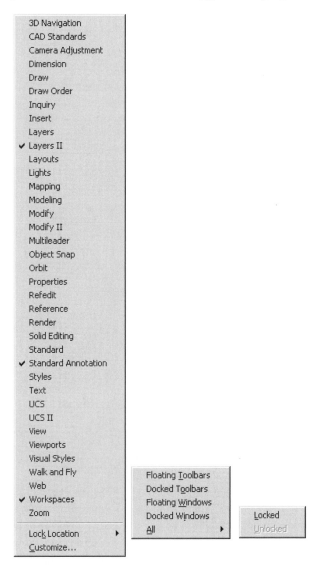

Lock Location locks the positions of toolbars and windows, docked and floating. **Customize** displays the Customize User Interface dialog box.

When all toolbars are turned off, you can access them with the undocumented **-TOOLBARS** command. Use the **All** option to turn them all on.

BLANK TOOLBAR AREA

Right-clicking a blank part of the toolbar area at the top of the AutoCAD window displays the shortcut menu with the names of partial *.cui* files currently loaded. Each *partial menu* holds one or more toolbars, as well as items for the menu bar.

The menu above shows several partial menus, with names like Custom and Express. "Acad" is the standard menu provided with AutoCAD. "Custom" is a blank menu that is available for customizing. You can write custom menus; see *Using AutoCAD: Advanced*. "Impression" is the name of the menu associated with Autodesk's Impression software. If the optional Express Tools package is installed, then it also shows up here as "Express."

Selecting the name of a partial menu displays a submenu of toolbar names.

DRAWING AREA

Right-clicking anywhere in the drawing area (while no command is active) displays the shortcut menu with the names of commands related to viewing, using the Clipboard, and common utilities.

Command	Comment
Repeat	Repeats the last command.
Recent Input	Displays the 20 last commands entered at the 'Command:' prompt.
Cut	Cuts objects from the drawing, and sends them to the Clipboard.
Copy	Copies objects from the drawing, and sends them to the Clipboard.
Copy with Base Point	Copies objects to the Clipboard, after prompting for a base point.
Paste	Pastes objects from the Clipboard into the drawing.
Paste as Block	Pastes objects from the Clipboard in the drawing as a block; available only after using the **Copy with Base Point** menu item.
Paste to Original Coordinates	Pastes objects from the Clipboard into another drawing.
Undo	Reverses the last command.
Redo	Reverses the most-recent undo; available only when the previous command was u.
Pan	Enters real-time pan mode.
Zoom	Enters real-time zoom mode.
Quick Select	Displays the Quick Select dialog box; selects objects by properties.
QuickCalc	Displays the QuickCalc palette for performing calculations.
Find	Displays the Find and Replace dialog box for searching for text.
Options	Displays the Drafting tab of the Options dialog box.

SHIFT+BLANK DRAWING AREA

Holding down the **SHIFT** key (or the **CTRL** key — it matters not) and then right-clicking anywhere in the drawing area displays the shortcut menu with the names of object snap modes and other drawing aids.

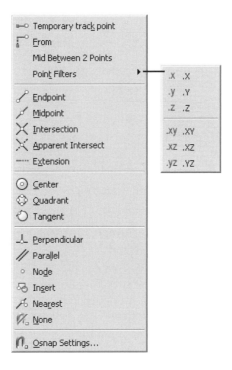

Command	Comment
Temporary track point	Moves the cursor during drawing and editing commands.
From	Locates an offset point during drawing and editing commands.
Mid Between 2 Points	M2P: finds midpoint between two points.
Point Filters	Submenu of point filters; see *Using AutoCAD: Advanced.*
Object snap modes	See Chapter 5, "Drawing with Precision."
Osnap Settings	Displays the Object Snap tab of the Drafting Settings dialog box.

The object snaps listed in this shortcut menu are described fully in Chapter 5, "Drawing with Precision."

ACTIVE COMMANDS

Right-clicking during an active command displays a shortcut menu specific to the command, along with all its options. The shortcut menu is different for every command, because each command has a different set of options.

In addition, the shortcut menu can vary during a command sequence, because different options apply at different times during the command. As an example, the figures below illustrate the two shortcut menus that appear during the **LINE** command:

Command: **line**

Specify first point: *(No shortcut menu available; right-clicking attempts to continue from the previous line or arc.)*

Specify next point or [Undo]: *(Shortcut menu displays the command's **Undo** option.)*

Specify next point or [Close/Undo]: *(Shortcut menu adds the command's **Close** option.)*

Common to all active-command shortcut menus are these commands:

Command	Comment
Enter	Accepts default values, or ends the option or command, depending on the context; equivalent to pressing the ENTER key.
Cancel	Cancels the command; equivalent to pressing the ESC key.
Recent Input	Displays the last ten items entered at the 'Command:' prompt.
Pan	Enters real-time pan mode; see Chapter 5, "Drawing with Precision."
Zoom	Enters real-time zoom modes; see Chapter 5.
QuickCalc	Displays the QuickCalc palette for performing calculations.

HOT GRIPS

Right-clicking a hot (red) grip displays a shortcut menu with the names of commands available during grip editing: **Move**, **Mirror**, and so on, through to **Undo**.

These commands are explained more fully later in this chapter.

LAYOUT TABS

Right-clicking any of the layout tabs displays the shortcut menu with the commands related to layouts and their tabs.

These commands are explained more fully in *Using AutoCAD: Advanced*.

COMMAND LINE

Right-clicking the command line displays the shortcut menu with actions that affect the command line, as well as a submenu listing recently-used commands.

Command	Comment
Recent Commands	Displays a submenu listing recently use commands. Select a command to re-execute it.
Copy	Copies selected text from the command line; this command is available only when text has been selected with the cursor.
Copy History	Copies all the text from the command line to the Clipboard.
Paste	Takes text from the Clipboard and places it into the command line. (Command is available only when the Clipboard contains text.) *Warning*: AutoCAD may react unexpectedly to text pasted from another source.
Paste to CmdLine	Places text in the command line, provided the Clipboard contains appropriate data.
Options	Displays the Display tab of the Options dialog box.

DRAG BAR

Right-clicking the drag bar at the end of the command-line palette displays the shortest shortcut menu. The check mark indicates that the palette will dock at the edge of AutoCAD, when dragged there.

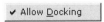

STATUS BAR

Right-clicking any button on the status bar displays shortcut menus specific to the button. Shown below is a typical menu:

Command	Comment
On	Turns on the status bar button.
Off	Turns off the button.
Settings	Displays the appropriate dialog box.

TUTORIAL: CONTROLLING RIGHT-CLICK BEHAVIOR

1. To control the behavior of the right-click, display the Right Click Customization dialog box:

    *   From the **Tools** menu, choose **Options**.

    *   At the 'Command:' prompt, enter the **option** command:

        Command: **options** (*Press* ENTER.)

    *   Alternatively, enter the **op** alias at the 'Command:' prompt.

2. In all cases, AutoCAD displays the Options dialog box.
    Choose **User Preferences** tab, and then the **Right-click Customization** button.
    Notice the Right-click Customization dialog box.

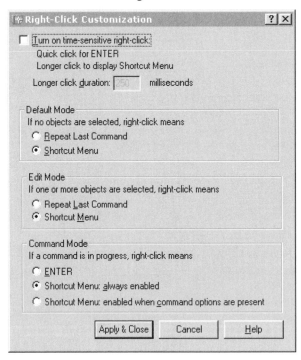

3. Change the settings, and then click the **Apply & Close** button.
    Click **OK** to exit the Options dialog box.

The Right-Click Customization dialog box has the following options:

Turn on Time-Sensitive Right-Click

This option determines how AutoCAD reacts to right-click behavior; this option is normally turned off. When on, a short click acts like pressing the ENTER key. A long click displays the appropriate shortcut menu.

The difference between the short and long click is specified in milliseconds. The default value of 250 milliseconds is a quarter-second.

Default Mode

These two options determine the action when you right-click in the drawing area when no objects are selected and no commands are in progress.

Repeat Last Command repeats the last command.

Shortcut Menu displays the default shortcut menu (default).

Edit Mode

These two options determine the action when you right-click selected objects in grips mode (with no command active):

Repeat Last Command repeats the last command.

Shortcut Menu displays the edit (grips) shortcut menu (default).

Command Mode

These three options determine the action when you right-click in the drawing area while a command is active:

ENTER is the same as pressing ENTER; shortcut menus are disabled.

Shortcut Menu: Always Enabled displays shortcut menus applicable to the command in progress.

Shortcut Menu: Enabled When Command Options Are Present displays shortcut menus only when options are shown (enclosed in square brackets) on the command line. When the command displays no options as available, a right-click acts like pressing ENTER (default).

SHORTCUT MENUS: ADDITIONAL METHOD

The **SHORTCUTMENU** system variable determines whether the default, edit, and command-related shortcut menus are displayed in the drawing area. To turn on two or more options, enter their sum. For example, the default is 11 (1 + 2 + 8):

> Command: **shortcutmenu**
>
> Enter new value for SHORTCUTMENU <11>: *(Enter a value, such as **1**.)*

ShortcutMenu	Meaning
0	Disables the default, edit, and command shortcut menus. (This makes newer AutoCADs act like AutoCAD Release 14.)
1	Displays shortcut menus when no command is active.
2	Displays shortcut menu available during grips editing.
4	Displays shortcut menus during commands.
8	Displays shortcut menus during commands, but only when options are available at the command line.
16	Displays shortcut menus when the right button is held down longer.

GRIPS

Grips allow direct editing of objects.

In traditional AutoCAD usage, drafters enter an editing command, and then select the objects to edit (verb-noun).

But the reverse procedure is also possible in AutoCAD: first select the objects, and then edit them (noun-verb). With no command active, you pick one or more objects. As feedback, AutoCAD highlights the selected objects by showing them with dashed lines, and displays *grips* (called "handles" in other software applications). You then edit the objects by manipulating their grips — until you press ESC to exit direct editing mode.

Grips are small, colored squares that indicate where the object can be edited. For example, lines have grips at both ends and at the midpoint; circles have grips at the center point and at the four quadrant points.

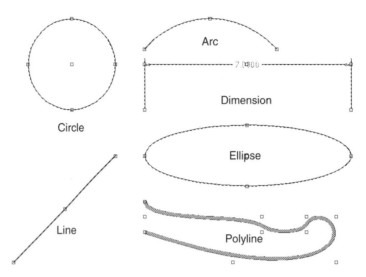

Blocks are a special case. Normally, AutoCAD displays a single grip at the block's insertion point. If you wish to display grips on the objects within the block, select the **Enable grips within blocks** check box in the Options dialog box's **Selection** tab.

 Note: For grips editing to work, the following criteria must be met:

Grips editing must be enabled. You can tell by the small selection box (called the *pickbox*) displayed at the intersection of the crosshairs when no command is active.

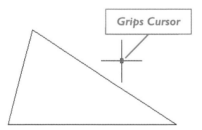

If not enabled, go to the Selection tab of the Options dialog box, and then turn on these two options: **Enable Grips** and **Noun/Verb Selection**.

No commands can be active. A blank area after the 'Command:' prompt indicates no command active.

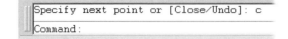

If necessary, press ESC until the command area is blank.

TUTORIAL: DIRECT OBJECT SELECTION

1. Open a drawing in AutoCAD. Or, in a new drawing, place some objects.
2. To select objects in the drawing, pick them using one of the direct selection modes:
   - Selecting one object at a time:

     Command: *(Pick an object.)*

   Notice that AutoCAD highlights the object you selected, and displays one or more blue grips.

   To select additional objects, continue picking:

     Command: *(Pick another object.)*

   To "unselect" objects, hold down the **SHIFT** key while picking them. (If **PICKADD** mode is turned on, you need to hold down **SHIFT** to *add* objects.)

   - Selecting all objects in the drawing:

     Command: *(Press **CTRL+A**.)*

   AutoCAD highlights all objects in the drawing, with the exception of those on layers that are frozen or locked.

   - Selecting more than one object at a time:

     Command: *(Pick in a blank area of the drawing.)*

     Specify opposite corner: *(Move the cursor to form a selection window.)*

   Moving the cursor to the left forms a crossing selection window (selects all objects within and crossing the window), as illustrated below.

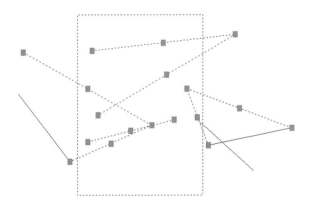

   Moving to the right forms a window selection window (selects only objects fully within the window).

3. You can now edit the selected objects, as described in the sections following.
4. To unselect the objects, press **ESC** once or twice (until the highlighting and grips disappear).

GRIP COLOR AND SIZE

As a visual aid, standard grips are assigned colors and names:

Grip Color	Name	Meaning
Blue	Cold	Grip is not selected.
Green	Hover	Cursor is positioned over grip.
Red	Hot	Grip is selected, and object can be edited.

You can change the color and size of grips in the Selection tab of the Options dialog box. Any color can be chosen, but it makes sense to keep the ones AutoCAD assigned. The grip box size can range from 1 to 255 pixels.

Task-specific grips have other colors and shapes, as described later.

Arcs

Arcs have additional grips that perform specific editing tasks (as do dynamic blocks and 3D surfaces and solids). The task-specific grips are shown as triangles, instead of squares.

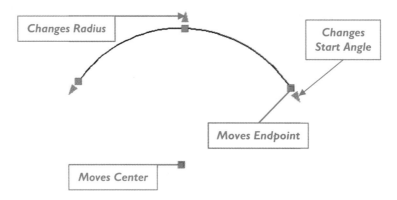

The square grips move the arc and its endpoints; the triangular grips change the start and end angles, as well as the radius.

GRIP EDITING OPTIONS

Stretch, **Move**, **Rotate**, **Scale**, **Copy**, and **Mirror** are the primary grips editing options.

After objects are selected, these options appear on the command line and the shortcut menu. In addition to these six, you can use other AutoCAD commands with grips. Examples include ERASE, COPYCLIP, PROPERTIES, MOVE, SCALE and STRETCH. Other editing commands, such as TRIM and OFFSET, cannot be used with grips.

TUTORIAL: EDITING WITH GRIPS

1. Select any object in the drawing. The object is highlighted.

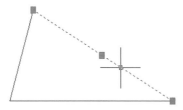

Notice that one or more blue grips are placed on the object (sometimes called "cold"). You can select as many objects as you wish by selecting more than once.

If you cannot select more than one object, hold down the SHIFT key on the keyboard when selecting the grips.

2. Move the cursor over a blue grip.

As the cursor gets close, notice that it jumps to the grip, and that the grip changes color to green (sometimes called "hovers" or "warm").

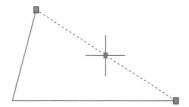

3. Click the green grip.

   Notice that the grip box changes color to red to denote selection (sometimes called "hot"). The hot grip is the base point from which some editing actions, such as stretch and rotate, take place.

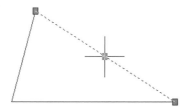

4. At the command line, AutoCAD prompts:

   ** STRETCH **

   Specify stretch point or [Base point/Copy/Undo/eXit]:

   Edit the grip as described below; right-click a red grip to see the editing options.
   Or, press the spacebar to see the other editing options: Move, Rotate, Scale, and Mirror.

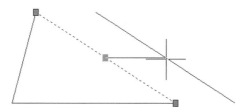

5. Press **ESC** once to clear the hot grip and selected objects.

GRIP EDITING MODES

When you select a single grip (so that it turns red), AutoCAD displays editing options on the command line. The selections cycle as you press **SPACEBAR** or **ENTER**. (Right-click to see a cursor menu listing the same options.)

> ** STRETCH **
>
> Specify stretch point or [Base point/Copy/Undo/ eXit]: *(Press spacebar.)*
>
> ** MOVE **
>
> Specify move point or [Base point/Copy/Undo/ eXit]: *(Press spacebar.)*
>
> ** ROTATE **
>
> Specify rotation angle or [Base point/Copy/Undo/Reference/eXit]: *(Press spacebar.)*
>
> ** SCALE **
>
> Specify scale factor or [Base point/Copy/Undo/Reference/eXit]: *(Press spacebar.)*
>
> ** MIRROR **
>
> Specify second point or [Base point/Copy/Undo/ eXit]: *(Press spacebar.)*

Press the spacebar until the desired edit option is listed, and then proceed with the command. For example, to resize the object, press the spacebar until **SCALE** appears, and then enter a scale factor at the "Specify scale factor:" prompt.

The selected grip point becomes the base point for editing. Alternatively, select one of the options listed on the command line. The commands and their options are covered later in this section.

GRIP EDITING OPTIONS

Each grip editing option has a set of suboptions in common. These are:

Base point

Enter **B** at the command line to specify a base point other than the hot grip. AutoCAD prompts you:

> Specify base point: *(Pick a point.)*

The point you pick becomes the new base point.

Copy

Enter **C** to copy the selected object, leaving the original intact. The prompt changes to:

> ** STRETCH (multiple) **
>
> Specify stretch point or [Base point/Copy/Undo/eXit]: *(Pick a point.)*

The selected object is copied to the point you pick. The prompt repeats itself so that you can make multiple copies, as indicated by the word "(multiple)." Press **ENTER** or **ESC** to exit the command.

Undo

Enter **U** to undo the last operation.

Reference

Enter **R** to provide a reference for the **Scale** and **Rotate** modes. For Scale mode, AutoCAD prompts:

> Specify reference length <1.0000>: *(Enter a length, or pick two points.)*

Sometimes the reference option can be difficult to understand. Here is an example. Suppose you have an object that is 6 units in length, but you wish to resize it to 24 units in length. Select the **Reference** option, and then enter original length, as follows:

> Specify scale factor or [Base point/Copy/Undo/Reference/eXit]: **r**
>
> Reference length <1.0000>: **6**

AutoCAD then prompts for the new length.

> <New length>/Base point/Copy/Undo/Reference/eXit: **24**

AutoCAD calculates the scale factor, and then resizes the object.

For **Rotate** mode, AutoCAD prompts:

> Specify reference angle <0>: *(Enter an angle, or pick two points.)*

The reference angle is the angle at which the object is currently rotated. AutoCAD next prompts for the "New angle." The new angle is the angle you want to rotate the object.

eXit

Enter **X** or press **ESC** to exit grips editing.

Note: It is possible to work with two hot grips at a time, notes the technical editor. For instance, draw two lines, and then select both. While holding down the **SHIFT** key, select two (or more) grips. Release the **SHIFT** key, and then drag one of the hot grips. As you do, the other line also moves along.

GRIP EDITING OPTIONS

Let's look at each grip editing option.

Stretch

Stretch functions like the **STRETCH** command, except that AutoCAD uses the hot grip to determine the stretch results. Because only one object can have a hot grip, only one object at a time can be stretched or moved — unless two lines share a hot grip, as illustrated on the next page.

> ** STRETCH **
>
> Specify stretch point or [Base point/Copy/Undo/eXit]: *(Pick a point, or enter an option.)*

When you select the end grip of a line or arc, or quad grip of a circle, the object is *stretched*. Lines and arcs "stretch" longer or shorter, while circles change their diameter. The figure illustrates stretching these objects. (For clarity, the hot grips are shown as black squares, and the cold grips as white.)

One vertex of the triangle is being moved, changing its size, and the radius of the circle is being changed. The triangle is made of lines, so AutoCAD is stretching two lines at one, because of their common endpoints. The circle is not being stretched into an ellipse, as you might think from the word "stretch." Instead, its diameter is being changed.

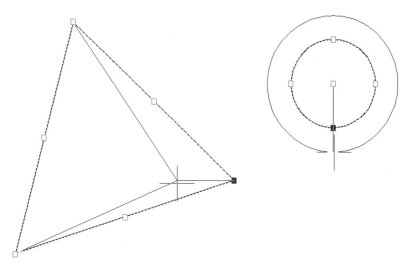

When you select the midpoint grip of a line, or arc, or the center of a circle, the object is *moved*, but not stretched. The figure illustrates **Stretch** mode moving the selected objects. (Again, for clarity, the hot grips are shown as black squares, and the cold grips as white.) One line of the triangle is being moved, because the grip is at the center of the line. The entire circle is being moved.

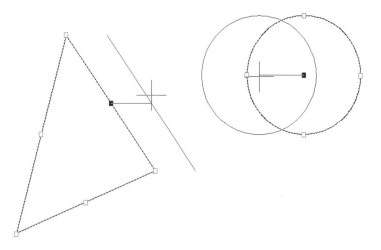

Move

Move moves the selected objects. All selected objects are moved, relative to the hot grip.

> ** MOVE **
>
> Specify move point or [Base point/Copy/Undo/eXit]: *(Pick a point, or enter an option.)*

Objects move by the distance from the base point to the point you pick. Recall that the base point is either the hot grip, or another point specified by the **Base point** option. You can either pick a point, or enter absolute or relative coordinates to specify the distance and direction of the move. For example, enter relative coordinates that specify a distance of 50 units at an angle of 45 degrees from the x-axis:

> Specify move point or [Base point/Copy/Undo/eXit]: **@50<45**

Rotate Mode

The **Rotate** option rotates the selected objects around the base point. All selected objects are rotated, relative to the hot grip.

> ** ROTATE **
>
> Specify rotation angle or [Base point/Copy/Undo/ Reference/eXit]: *(Pick a point, or enter an option.)*

Unless you use the **Base point** option to position the base point at a location other than a grip, the rotation occurs around the hot grip.

When rotating objects, you can dynamically set the rotation angle by moving the crosshairs, or by specifying a rotation in degrees.

Scale Mode

If you wish to resize the selected objects, cycle through the mode list until **Scale** is listed on the command line. All selected objects are resized, relative to the hot grip.

> .** SCALE **
>
> Specify scale factor or [Base point/Copy/Undo/ Reference/eXit]: *(Pick a point, or enter an option.)*

Scale mode resizes the selected object(s). The hot grip serves as the base point for resizing. Dynamically resize the objects by moving the crosshairs away from the base grip.

You can also resize the selected objects by entering a scale factor. A scale factor greater than one enlarges the objects by that multiple. For example, a scale factor of 2.0 makes objects twice the size. Entering a decimal scale factor makes objects smaller than the original. For example, entering a scale factor of 0.5 shrinks the objects to one-half the original size.

Mirror Mode

Mirror mode mirrors the selected object(s). All selected objects are mirrored relative to the hot grip. Objects are moved (not copied) about a *mirror* line created by the cursor. The first point of the mirror line is established by the hot grip (unless you change it with the **Base point** option). AutoCAD prompts you for the second point of the mirror line:

> ** MIRROR **
>
> Specify second point or [Base point/Copy/Undo/ eXit]: *(Pick a point, or enter an option.)*

The figure illustrates the mirror line stretching between the hot grip and the crosshair cursor.

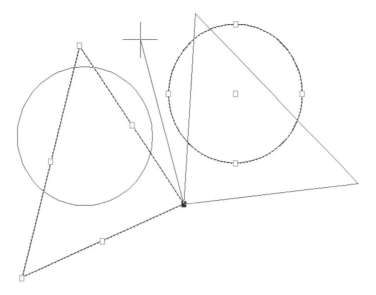

DYNAMIC BLOCKS

Until AutoCAD 2006, blocks were static. They could be *redefined* (replaced by different blocks) or edited with the awkward **REFEDIT** command. After Autodesk dropped its Actrix technical diagraming software several years ago, the company hinted that the ActiveShapes technology would migrate to AutoCAD; they now appear in AutoCAD as *dynamic blocks*.

Dynamic blocks behave in ways unavailable to static blocks. For instance, they can align with other objects, stretch logically (lengthening the table with added chairs, for example), and rotate at predefined angles. The behaviors are indicated by the shape of the related grips:

Grip	Name	Comment
☐	Standard	Moves in any direction, like standard grips.
◁	Linear	Moves back and forth in a line
○	Rotation	Rotates about an axis.
⇨	Flip	Flips (mirrors) about an axis.
△	Alignment	Aligns with other objects.
▽	Lookup	Displays lists of options.

For example, the figure illustrates a door symbol with four dynamic grips.

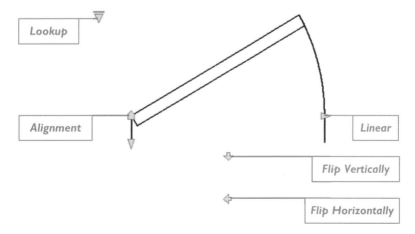

There are, unfortunately, no tooltips to provide hints about the purpose of each dynamic grip; you just need to click them, and see what happens.

TUTORIAL: EDITING DYNAMIC BLOCKS

1. From the Comnpanion CD, open *dynamicblock.dwg*, a drawing of a toilet symbol.

2. Select the block to see the dynamic grips.

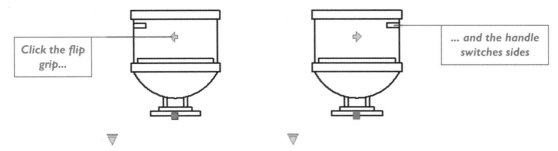

3. Click the *flip* grip (looks like a horizontal arrow). Notice that the handle switches from the left side to the right. This is like using the **MIRROR** command without needing to specifying the mirror line.

4. Click the *lookup* grip. Notice the shortcut menu of alternate toilet symbols.

5. Select **Elongated (Plan)**. Notice that the front view changes to the plan (top) view. Also notice the two new dynamic grips: rotate and align.

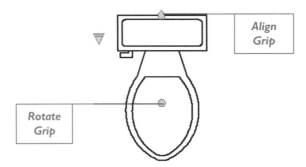

6. Select the *rotate* grip, and then drag it. Notice that the toilet rotates, and that you do not need to specify a base point, as with the **ROTATE** command. (The base point was predefined by the designer of the block.)

Turn on ortho mode if you want to restrict rotation to 90 degrees.

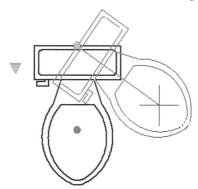

7. For the next step, draw a diagonal line near (but not on) the block.

8. Select the toilet, and then select the *align* grip.

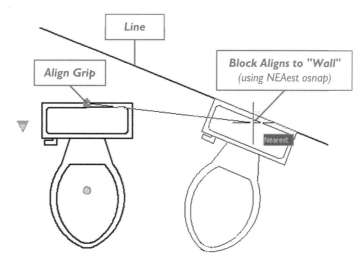

9. Drag the block to the line. Notice that the block aligns itself with the line. (AutoCAD automatically turns on NEArest osnap mode during the dragging process.)

 Notes: When dynamic blocks get really mixed up, you can reset them to their original state with the **RESETBLOCK** command. AutoCAD includes a collection of dynamic blocks in its \sample\dynamic blocks folder. You create dynamic blocks using the **BEDIT** command, as described in *Using AutoCAD: Advanced*.

OTHER GRIPS COMMANDS

You can also use these commands when objects are highlighted with grips.

ERASE deletes the gripped objects from the drawing. As an alternative, press the **DEL** key. *Caution:* selecting all objects and then pressing **DEL** erases the entire drawing! Use **U** to recover.

COPYCLIP copies the gripped objects to the Clipboard; **CUTCLIP** cuts the objects. As alternatives, you can press **CTLR+C** and **CTRL+X**, respectively.

PROPERTIES displays the Properties palette for the gripped objects.

MATCHPROP copies the properties of one gripped object; if more than one object is gripped, AutoCAD complains, "Only one entity can be selected as source object."

ARRAY creates rectangular and polar arrays of the gripped objects.

BLOCK creates a block of gripped objects.

EXPLODE explodes gripped objects; if they cannot be exploded, AutoCAD complains, "1 was not able to be exploded."

RIGHT-CLICK MOVE/COPY/INSERT

A simpler alternative to grips editing is *right-click move/copy/insert*. It is meant as a faster method to copy and move objects, as well as insert them as blocks. (This action is not documented by Autodesk, so I don't know its official name.)

It works like this:

1. Select one or more objects.

2. Holding down the right mouse button, drag the object.
   Important! Do not right-click the grip.

3. Let go of the mouse button. AutoCAD displays a shortcut menu.

   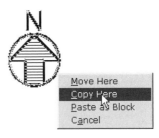

4. Select an action from the shortcut menu:

Menu	Comment
Move Here	Moves the object to the new location.
Copy Here	Copies the object to the new location.
Paste as Block	Inserts the object as a block at the new location.
Cancel	Cancels the move/copy operation.

The **Paste as Block** option quickly coverts a group of objects into a block — much faster than using the BLOCK command. AutoCAD gives the block an "anonymous" name, like A$C6FF64EEA. Later, you can use the RENAME command to change the name of the block to something meaningful.

DIRECT DISTANCE ENTRY

Direct distance entry allows "pen up" movement during drawing and editing commands. (*Pen up* means that the cursor changes its position without drawing or editing objects.)

Direct distance entry is an alternative to entering polar or relative coordinates that rely on a distance and an angle. To show the angle, you move the mouse; then you type the distance. With ortho or polar modes turned on, direct distance entry is an efficient way to draw lines.

Direct distance entry can be used any time a command prompts you to specify a point, such as "Specify next point:".

TUTORIAL: DIRECT DISTANCE ENTRY

1. To draw a line 10 units long using direct distance entry, enter a drawing or editing command:
    Command: **line**

2. Before you can use direct distance entry, you must pick an initial point:
    Specify first point: *(Pick a point.)*

3. At the next prompt, move the mouse to indicate the angle, and then enter the distance at the keyboard:
    Specify next point: *(Move the mouse, and you see a rubber band line. At the keyboard, type **10** and press* ENTER.*)*

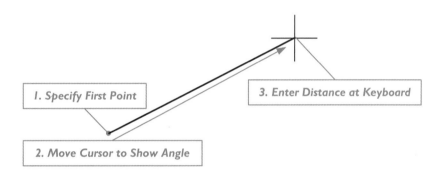

4. End the command:
    Specify next point: (Press ENTER)

AutoCAD draws a line segment ten units long in the direction you move the mouse. You can use direct distance entry for drawing polylines, arcs, multilines, and most other objects. Direct distance entry does not make sense with some drawing commands, like CIRCLE and DONUT.

Additionally, you can use direct distance entry with editing commands, such as MOVE, STRETCH, and COPY.

TRACKING

The **Tracking** modifier allows "pen up" movement during drawing and editing commands.

Whereas direct distance entry lets you draw relative distances, tracking lets you move relative distances within a command. Tracking is not a command, but a command option; you can use tracking only within another command.

Upon entering tracking mode, AutoCAD automatically switches to ortho mode, and then prompts for the "First tracking point:". (If polar mode is on, AutoCAD switches it off and turns on ortho mode.) Once you exit tracking mode, AutoCAD changes ortho and polar back to their original states.

Some commands keep prompting you for additional points, such as the **LINE** and **PLINE** commands. With these, you can go in and out of tracking mode as often as you like.

TUTORIAL: TRACKING

1. To employ tracking, enter a drawing or editing command:
   Command: **line**

2. At a "Specify first point:" or "Specify next point:" prompt, enter **tracking**, **track**, or **tk**.
   Specify first point: **tk**

3. AutoCAD enters tracking mode. Notice that ortho mode is turned on. (Look at the ORTHO button on the status bar.)
   Move the mouse a distance, and then click.
   First tracking point: *(Move the mouse and click.)*

4. This time, move the mouse, and then enter a distance.
   Next point (Press Enter to end tracking): *(Move the mouse in another direction, and then enter a distance, such as 2.)*

5. Press **ENTER** to exit tracking mode, and resume drawing with the **LINE** command.
   If ortho mode was off when you started the line command, AutoCAD switches it off automatically upon exiting tracking mode.
   Next point (Press Enter to end tracking): *(Press ENTER.)*
   To point: **2,3**
   To point: **10,5**

6. Switch back to tracking mode, and enter distances as absolute x,y coordinates:
   To point: **tk**
   First tracking point: **5,10**

7. Exit tracking mode, draw another line, and exit the **LINE** command.
   Next point (Press Enter to end tracking): *(Press ENTER.)*
   To point: **3,2**
   To point: *(Press ENTER.)*

If you want tracking to move in angles other than 90 degrees, use the **Rotate** option of the **SNAP** command to change the angle. Since the **SNAP** command is transparent, you can change the tracking angle in the middle of tracking, as follows:

   Command: **line**
   Specify first point: *(Pick.)*
   Specify next point: **tk**

First tracking point: *(Move cursor and pick a point.)*

Next point (Press Enter to end tracking): **'snap**

>>Specify snap spacing or [ON/OFF/Aspect/Rotate/Style/Type] <0.5000>: **r**

>>Specify base point <0.0000,0.0000>: *(Press* **ENTER** *to accept default.)*

>>Specify rotation angle <0>: **45**

Resuming LINE command.

Next point (Press Enter to end tracking): *(Move cursor and pick a point.)*

Next point (Press Enter to end tracking): *(Press* **ENTER** *to end tracking.)*

Specify next point: *(Pick a point.)*

Specify next point: *(Press* **ENTER** *to exit command)*

 Note: Tracking normally assumes you want to switch direction each time you use it. For example, if you first move north (or south), AutoCAD assumes you next want to move east (or west).

It is not easy, however, to back up or track forward in the same direction. For example, if you track north and then want to track north some more, you find the cursor wanting to move east or west, but not north or south. To make the tracking cursor continue in the same direction, take it back to its most recent starting point. Then move in the direction you want.

OSNAP WITH OBJECT TRACKING

Technical editor Bill Fane points out that it is possible to employ tracking without using the TK option. AutoCAD can acquire coordinates from other objects through object tracking. Follow these steps:

1. Turn on object tracking: click the OTRACK button on the status bar.

2. Turn on ENDpoint object snap: right-click the OSNAP button on the status bar, and then choose Settings. In the dialog box, ensure endpoint is on, and then exit the dialog box.

3. Draw a line at an angle, similar to the one illustrated below.

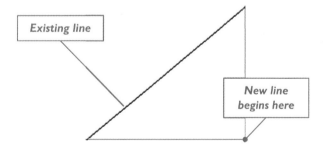

4. You want to draw a second line that starts at the point illustrated above. Enter the **LINE** command:

    Command: **line**

5. At the prompt, move the cursor over the lower endpoint of the existing line. Wait until the tag "Endpoint" appears

    Specify first point: *(Move cursor over endpoint.)*

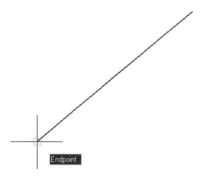

AutoCAD has acquired the coordinates of the endpoint.

6. Move the cursor to the existing line's other endpoint, and pause. Again, the "Endpoint" tag appears after a moment.

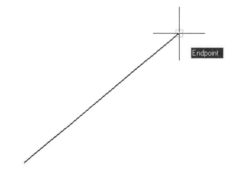

AutoCAD has acquired the coordintes of the other endpoint.

Notice the barely perceptible + marker at the first endpoint. It indicates that AutoCAD is still remembering its coordinates.

7. Move the cursor to where the new line is to begin, at the "intersection" of the two endpoints. You'll know you've arrived when the object tracking lines appear, as illustrated below.

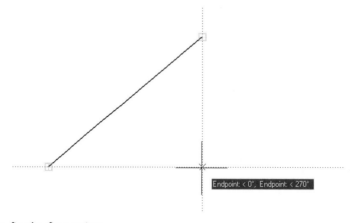

8. Now click to fix the first point:

    Specify first point: *(Pick the point.)*

9. You can continue using this technique to find other points through object tracking, such as the "intersection" of the midpoint and endpoint, as illustrated below.

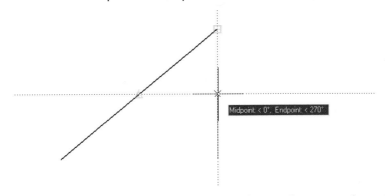

This technique is not limited to lines, but can be used with any objects, such as placing a circle in the center of a hexagon.

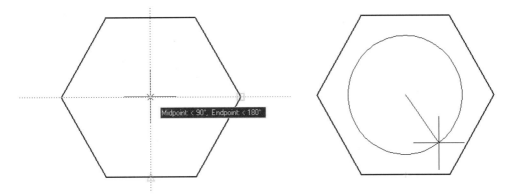

M2P

The **M2P** modifier finds the midpoint between two picked points — hence "m2p." (Alternatively, you can enter "mtp.")

M2P sounds like an object snap, but is not. It is entered during any prompt that asks you to specify a point. You can employ object snaps during the M2P process.

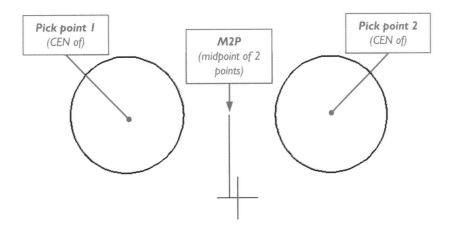

TUTORIAL: M2P

1. To benefit from **M2P**, enter a drawing or editing command:

   Command: **line**

2. At a "Specify first point:" or "Specify next point:" prompt, enter **m2p** or **mtp**.

   Specify first point: **m2p**

3. AutoCAD prompts you for the first "mid" point. I recommend using object snaps to help place the point accurately. In this example, the CENter object snap is specified:

   First point of mid: **cen**

   of *(Pick point 1.)*

4. AutoCAD prompts you for the second point:

   Second point of mid: **cen**

   of *(Pick point 2.)*

5. The **LINE** command carries on. AutoCAD draws the line starting at the midpoint of the two pick points.

   Specify next point: *(Continue on with the command.)*

Technical editor Bill Fane reports that he once wrote an AutoLISP routine that acted just like this keyboard modifier. He finds it useful for placing a circle between two circles, or to create symmetrical (not mirrored) copies.

EXERCISES

1. In the following exercise, you practice using shortcut menus to control AutoCAD and execute commands.

    Right-click a blank part of the status bar, and then turn off the **PAPER/MODEL** button. Does the button disappear? Repeat to return the button.

    Right-click the **ORTHO** button, and then select **Settings**. Which dialog box appears? Click **OK** to dismiss the dialog box.

    Right-click the **LWT** button, and then select **Settings**. Does a different dialog box appear? Click **OK** to dismiss the dialog box.

    Draw a circle. When you are done, right-click the drawing area. Do you see **Repeat CIRCLE** at the top of the shortcut menu?

    Right-click again, and select **Undo**. Does the circle disappear?

2. In this exercise, use grips to edit objects in the drawing.
    From the CD, open the *17_35.dwg* file, a drawing of an angle.

    Select a circle, and then click on the center grip.
    Drag the grip to another place in the drawing. Does the circle move?

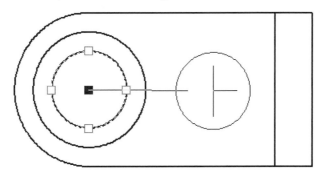

    Now click one of the four quadrant grips.
    Drag the grip. Does the circle become larger?

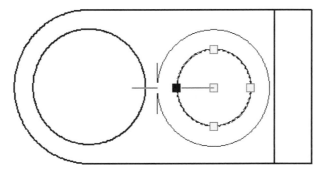

    Press **ESC** to remove the grips.

Select the entire drawing, and then click on any grip.
Press the spacebar to see ** MOVE ** on the command line.
Drag the grip. Do all parts of the drawing move?

Press the spacebar again to see ** ROTATE ** on the command line.
Drag the grip. Does the drawing rotate around the grip?

Enter **b** to access the base point option:
  Specify base point: *(Pick another point.)*
Move the cursor. Does the drawing now rotate about the new point?

Press the spacebar again to see ** SCALE ** on the command line.
Drag the grip. Does the drawing change its size?

Press the spacebar again to see ** MIRROR ** on the command line.
Drag the grip. Do you see a mirrored copy?

3. Use the **LINE** command with direct distance entry to draw the object defined by the
   following relative coordinates:

   Point 1: 0,0
   Point 2: @3,0
   Point 3: @0,1
   Point 4: @−2,0
   Point 5: @0,2
   Point 6: @−1,0
   Point 7: 0,0

4. Use the **PLINE** command with direct distance entry to draw the object defined by the
   following relative coordinates:

   Point 1: 0,0
   Point 2: @4<0
   Point 3: @4<90
   Point 4: @4<180
   Point 4: @4<270

5. Use direct distance entry and tracking to draw the following fuse link. Each square
   represents one unit.

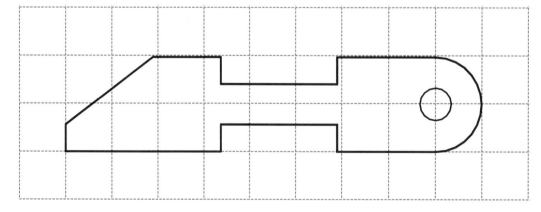

6. Use direct distance entry and tracking to draw the following cylinder. Each square represents 2 units.

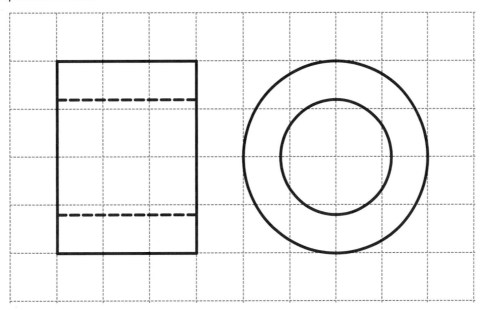

7. Use direct distance entry and tracking to draw the following baseplate. Each square represents a half-unit.

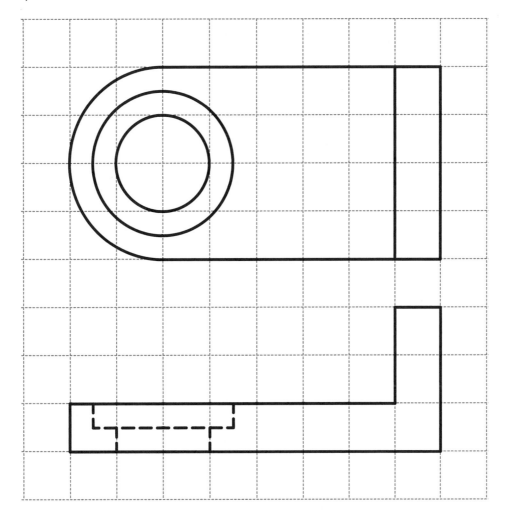

8. Use direct distance entry and tracking to draw the following wrench. Each square represents one unit.

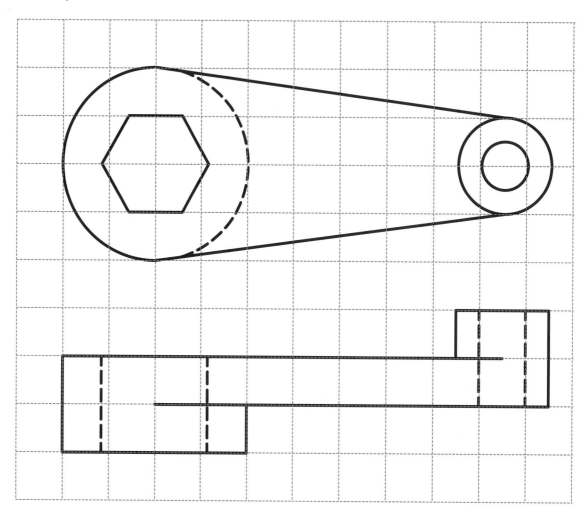

CHAPTER REVIEW

1. What is the difference between a *hot* and a *cold* grip?
2. What are the default colors for the following grips:

   Hot:

   Cold:

   Hover:
3. Name three editing operations you can perform with grips.
4. How do you deselect gripped objects?
5. What is the primary difference between *tracking* and *direct distance entry*?
6. How do you access shortcut menus?
7. Which shortcut menu do you see when holding down the **SHIFT** key in a blank part of the drawing?
8. Is the same shortcut menu shown during all commands?
9. What is the primary purpose of grips editing?
10. Name two conditions under which grips editing can take place:

    a.

    b.
11. Where are grips located on lines?

    On circles?

    On blocks?
12. Can more than one object be selected to perform simultaneous grips editing?
13. Using grips, can more than one object be stretched at a time?
14. Describe the purpose of the **Base point** option in grips editing.
15. What happens when you drag a circle by its center point grip?

    What happens when you drag a circle by its quad grip?
16. Can direct distance entry be used with editing commands?
17. What is the alias for tracking?
18. Name the mode that AutoCAD automatically turns on when you enter tracking mode?
19. Other than the standard six grips editing commands, name three other editing commands that work with gripped objects:

    a.

    b.

    c.
20. Can you use the **BLOCK** command with gripped objects?
21. How would you find the midpoint between two other points?
22. Describe three actions that dynamic blocks are capable of:

    a.

    b.

    c.
23. What is the purpose of the *lookup* grip?
24. Explain how the *align* grip works.
25. When rotating a dynamic block, do you need to specify the base point?
26. Which command resets a dynamic block to its original state?

27. From the CD, open *dyn-arrow.dwg*, the drawing of an arrow. Experiment by manipulating the arrow's dynamic grips.

Describe the actions performed by the grips:

a.

b.

c.

d.

CHAPTER 10

Constructing Objects

Drawing the same objects over and over was a tedious part of hand drafting. In Chapter 7 "Drawing with Efficiency," you learned to insert blocks to place parts quickly in drawings. In some circumstances, however, blocks are not the best method.

This chapter introduces other means of making copies — mirrored copies, parallel offsets, and arrays of copies. After completing this chapter, you will be able to use the following AutoCAD's commands for constructing objects from existing objects:

COPY makes one or more identical copies of objects.

MIRROR makes mirrored copies.

MIRRTEXT determines whether text is mirrored.

OFFSET makes parallel copies.

OFFSETDIST and **OFFSETGAPTYPE** preset parameters for offsetting objects.

MEASURE and **DIVIDE** place copies of points and blocks along objects.

ARRAY and **-ARRAY** construct linear, rectangular, and polar copies.

FILLET and **CHAMFER** create rounded and angled corners.

JOIN joins similar objects into one.

REVCLOUD creates revision clouds.

MARKUP inserts marked-up *.dwf* files from Design Review (DWF Composer).

NEW TO AUTOCAD 2008 IN THIS CHAPTER

- The **COPYMODE** system variable determines whether the **COPY** command repeats automatically.

FINDING THE COMMANDS

On the Dashboard's **2D Draw** panel:

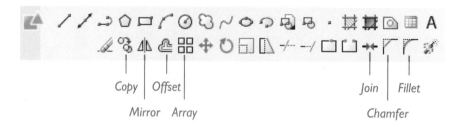

On the **Modify** and **Draw** toolbars:

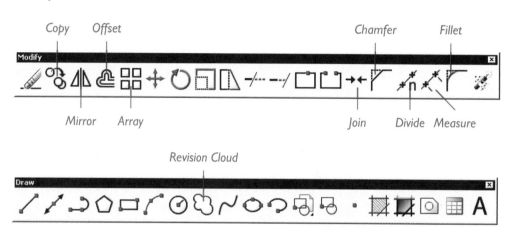

COPY

The COPY command makes one or more copies of objects.

The copies are identical to the original; all that changes is their location in the drawing. This command works within drawings; to copy objects between drawings, use the COPYCLIP command.

To use the COPY command, you need to tell AutoCAD three things: (1) the objects to be copied, (2) the point from which the copying takes place, and (3) the location to place the copies. The most recent displacement is the default value the next time the COPY command is used — but only if the **Displacement** option is used. The COPY command repeats until you press ESC.

TUTORIAL: MAKING COPIES

1. To copy one or more objects, start the **COPY** command:
   * From the **Modify** menu, choose **Copy**.

   * From the Modify toolbar, choose the **Copy** button.

   * In the Dashboard's 2D Draw panel, choose the **Copy** button.

   * At the 'Command:' prompt, enter the **copy** command:

     Command: **copy** *(Press ENTER.)*
   * Alternatively, enter the aliases **co** or **cp** at the 'Command:' prompt.

2. In all cases, AutoCAD prompts you to select the objects you want copied:
   Select objects: *(Pick one or more objects.)*

   Select objects: *(Press ENTER to end object selection.)*

3. Identify the point from which the displacement is measured:
   Specify base point or [Displacement] <Displacement>: *(Pick point 1.)*

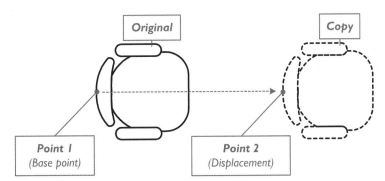

4. Identify the location for the copied object:
   Specify second point of displacement or <use first point as displacement>: *(Pick point 2.)*

5. The command repeats the last prompt so that you can place additional copies. Pick additional points, or press **ESC** to end the command.
   Specify second point or [Exit/Undo] <Exit>: *(Press ESC.)*

Students often have difficulty understanding the concept of *displacement*. This is the distance from the original object (known as the *base point*) to the location of the copied object, and is discussed in greater depth below.

You can choose any point you like as the base point, but some points make more sense than others. Examples include the lower-right corner of rectangles, the center of circles, and the 0,0 origin of drawings. In addition, it is often handy to use object snaps, such as INTersection or INSertion, to pick base points precisely.

To copy objects vertically or horizontally, turn on ortho mode.

AutoCAD has several ways to determine the displacement, which you learn about in the next tutorial.

MAKING COPIES: ADDITIONAL METHODS

This command also has the following options:

- **Displacement** specifies the displacement distance.
- **COPYMODE** system variable determines whether the command repeats.

Displacement

The **Displacement** option displays the displacement from the prior use of the COPY command:

> Specify base point or [Displacement] <Displacement>: **d**
>
> Specify displacement <1.0000, 2.0000, 0.0000>: *(Press ENTER, or enter a new displacement.)*

Whether you press ENTER or enter coordinates, AutoCAD displaces the selected objects, and then exits the COPY command.

More on Displacements

When AutoCAD prompts, 'Specify base point or [Displacement]:' this is a hint of what is to come.

Specify base point means you can enter x, y coordinates (such as 4,5) *or* pick a point in the drawing. Both methods can be used as the base point, but the x, y coordinates could be interpreted differently by AutoCAD. *Displacement* is a hint that the prompt to follow will interpret either method, depending on your next actions.

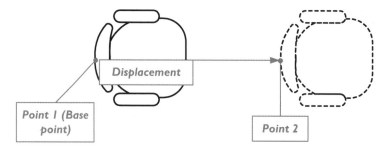

Specify second point of displacement means that you should enter another x, y coordinate or pick another point. Both actions place the copy at a distance that AutoCAD calculates from the two sets of x, y coordinates or two pick points. You can, of course, mix and match coordinate entry and pick points.

It's not clear from the prompt, but AutoCAD wants you to press ENTER at *<use first point as displacement>*. AutoCAD interprets the coordinates you entered at the earlier prompt (such as 4,5) as relative distances. In this case, there is no need to use the @ prefix. AutoCAD places the copy 4 units right and 5 units up from the original.

This option can, unfortunately, have an unexpected result, because, as this book's technical editor Bill Fane once remarked, "Pressing ENTER sometimes makes the copied objects end up near Hawaii — not a problem," he adds, "if you live in Maui." This problem occurs when you use the mouse to pick the first point, and then just press ENTER for the second point; AutoCAD interprets the second point as 0,0. He recommends using the **ZOOM Extents** command to find the "missing" copy.

1. Start the COPY command, and then select objects to copy:

    Command: **copy**

    Select objects: *(Select one or more objects.)*

    Select objects: *(Press ENTER.)*

2. Enter coordinates for the base point:

    Specify base point or [Displacement] <Displacement>: **4,5**

3. Press **ENTER** to place the copy by a relative distance and end the command:

    Specify second point of displacement or <use first point as displacement>: *(Press **ENTER** to interpret the first point as relative coordinates.)*

    Specify second point or [Exit/Undo] <Exit>: *(Press **ENTER** to exit the command.)*

Note: You may enter x, y coordinates for 2D displacement, or x, y, z coordinates for 3D displacement. In addition, you can use direct distance entry to specify the displacement.

NEW IN
2008   **CopyMode**

The **COPYMODE** system variable determines whether this command repeats.

Command: **copymode**

Enter new value for COPYMODE <0>: **1**

0 — The COPY command repeats automatically.

1 — The COPY command makes one copy, and then ends.

Notes: As an alternatives to the **COPY** command, you can use the right-click copy/move/insert technique described in Chapter 9.

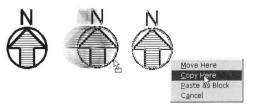

Or, use the **Copy** option of grips editing:

** STRETCH **

Specify stretch point or [Base point/Copy/Undo/eXit]: **c**

Be familiar with AutoCAD's many object selection modes, because they are crucial for working efficiently with the commands in this (and the next) chapter. You may wish to review the **SELECT** command in Chapter 6, "Drawing with Efficiency."

 MIRROR

The **MIRROR** command makes mirrored copies.

The command saves time when you are drawing symmetrical objects. Draw a half, or one-quarter, of objects, and then construct the other parts by mirroring. You have the option to retain or delete the original objects, as well as to decide whether text should be mirrored or not.

Other drawing programs use the phrases "flip horizontal" and "flip vertical" in place of mirror. As alternatives to this command, you can use the **Mirror** option found in grips editing, or insert blocks with a negative x- or y-scale factor (which mirrors the block).

To use the MIRROR command, you need to tell AutoCAD three things: (1) the objects to be mirrored, (2) the line about which the mirroring takes place, and (3) whether the source objects should be erased.

TUTORIAL: MAKING MIRRORED COPIES

1. To make mirrored copies of one or more objects, start the MIRROR command:
   * From the **Modify** menu, choose **Mirror**.

   * From the Modify toolbar, choose the **Mirror** button.

   * In the Dashboard's 2D Draw panel, choose the **Mirror** button.

   * Or, at the 'Command:' prompt, enter the **mirror** command:

     Command: **mirror** *(Press ENTER.)*

   * Alternatively, enter the **mi** alias at the 'Command:' prompt.

2. In all cases, AutoCAD prompts you to select the objects you want mirrored:
   Select objects: *(Pick one or more objects.)*
   Select objects: *(Press ENTER to end object selection.)*

3. Identify the points that define the mirror "line":
   Specify first point of mirror line: *(Pick point 1.)*
   Specify second point of mirror line: *(Pick point 2.)*

   The *mirror line* is the line about which the objects are mirrored. It need not be an actual line; two points will do.

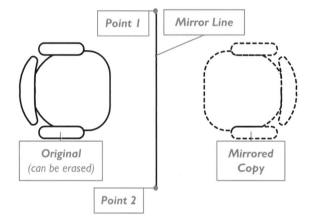

4. Decide whether you want the original objects(s) erased:
   Delete source objects? [Yes/No] <N>: *(Enter Y or N.)*

Notes: To make the mirror line absolutely horizontal or vertical, turn on ortho mode.

Polar mode is handy for making mirrored copies at specific angles.

MAKING MIRRORED COPIES: ADDITIONAL METHOD

Whether text is mirrored is determined by a system variable.

* **MIRRORTEXT** system variable decides whether mirrored text is mirrored.

MirrText

Sometimes objects contain text. The dilemma is whether to mirror the text, which makes it read backwards — or not mirror, which makes it read normally.

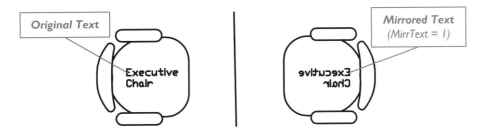

The **MIRRTEXT** system variable determines this, as follows:

> Command: **mirrtext**
>
> New value for MIRRTEXT <0>: *(Enter* **1** *or* **0**.*)*

The 0 and 1 have the following meaning:

MirrorText	Comment
0	Text is not mirrored.
1	Text is mirrored (default in AutoCAD 2004 and earlier).

Entering **0** produces non-mirrored text, which is the default setting. Enter **1** if you need the text to be mirrored.

 **OFFSET**

The **OFFSET** command makes parallel copies.

This command constructs copies parallel to objects; AutoCAD limits you to making offset copies of one object at a time. When the objects have curves or are closed, the copies become larger or smaller, depending on whether they are on the inside or outside.

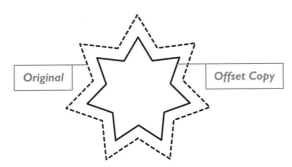

To create offset copies, AutoCAD needs to know three pieces of information, in this order: (1) the offset distance, (2) the objects to offset, and (3) the side on which to place the offset copies.

It may seem counter-intuitive *first* to specify the distance and *then* select the object, but "that's the way the Mercedes bends," as the driver said after his automobile accident. (Pun credit: technical editor.)

AutoCAD offsets lines, arcs, circles, ellipses, elliptical arcs, polylines (2D only), splines, rays, and xlines. Sometimes, unexpected results happen, as illustrated below. The dashed lines are copied offset from the original polyline (shown as the heavy line).

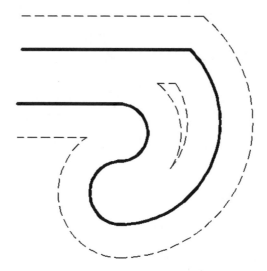

The thick line is the original polyline; the dashed lines indicate offset copies.

As an alternative to this command, use the MLINE command to place parallel lines.

TUTORIAL: MAKING OFFSET COPIES

1. To make offset copies of an object, start the **OFFSET** command:
    *   From the **Modify** menu, choose **Offset**.

    *   From the Modify toolbar, choose the **Offset** button.

    *   In the Dashboard's 2D Draw panel, choose the **Offset** button.

    *   Or, at the 'Command:' prompt, enter the **offset** command:

        Command: **offset** *(Press ENTER.)*
    *   Alternatively, enter the **o** alias at the 'Command:' prompt.

2. In all cases, AutoCAD displays the current settings:
       Current settings: Erase source=No   Layer=Source   OFFSETGAPTYPE=0

    And then prompts you for the offset distance:
       Specify offset distance or [Through/Erase/Layer] <Through>: *(Enter a distance.)*

3. Select the object to offset:
       Select object to offset or [Exit/Undo] <Exit>: *(Select one object.)*

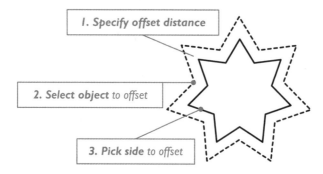

4. Pick the side on which the offset should be placed:
       Specify point on side to offset or [Exit/Multiple/Undo] <Exit>: *(Pick point 3.)*

5. Press **ENTER** to exit the command:
       Specify point on side to offset or [Exit/Multiple/Undo] <Exit>: *(Press ENTER.)*

The "Select object to offset:" prompt repeats to allow you to offset as often as you wish, but just one object at a time. Press ENTER to terminate the command.

For the offset distance, you can pick two points on the screen, or enter a number representing the distance. If you get the message, "That object is not parallel with the UCS," this means that the direction of the object's Z axis was not parallel to the current user coordinate system.

MAKING OFFSET COPIES: ADDITIONAL METHODS

The OFFSET command's **Through** option lets you specify a point through which the copy is offset. In addition, two system variables let you preset parameters.

- **Through** option combines the distance and side options.
- **Erase** option erases the source object.
- **Layer** option specifies the destination layer.
- **OFFSETDIST** system variable presets the offset distance.
- **OFFSETGAPTYPE** system variable determines how polyline gaps are handled .

Let's look at each.

Tutorial: Through

The **Through** option constructs the offset copy "through" a point. In effect, it combines the "Specify offset distance" and "Side to offset" options.

1. Draw a circle.
2. Start the OFFSET command, and then enter the **Through** option.

    Command: **offset**

    Specify offset distance or [Through]: **t**

3. Select the circle, and then pick the point through which the copy should be offset:

    Select object to offset or <exit>: *(Pick the circle.)*

    Specify through point: *(Pick point 1.)*

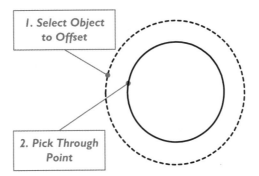

4. Continue making offset copies, or press ENTER to end the command.

    Select object to offset or <exit>: *(Press ENTER.)*

Erase

The **Erase** option erases the source object:

Erase source object after offsetting? [Yes/No] <No>: *(Type **Y** or **N**.)*

"Yes" erases the source object, while "No" retains it. AutoCAD remembers this setting until you change it. (There is no system variable associated with it.)

Layer

The **Layer** option specifies the destination layer:

> Enter layer option for offset objects [Current/Source] <Source>: *(Type **C** or **S**.)*

"Current" places offset objects on the current layer, while "Source" places them on the same layer as the source object.

OffsetDist

With the **OFFSETDIST** system variable, you preset the offset distance. As well, you can use it to preselect the **Through** option. Then, during the **OFFSET** command, you need only press ENTER at the "Specify offset distance" prompt.

> Command: **offsetdist**

> Enter new value for OFFSETDIST <1.00>: *(Enter a distance, positive or negative, such as **2.54** or **-1**.)*

Enter a distance, and that becomes the default offset distance next time you use the **OFFSET** command. The prompt looks like this (with the changes emphasized in blue):

> Specify offset distance or [Through] <**2.5400**>: *(Press ENTER.)*

If, on the other hand, you enter a negative number, such as **-1**, then the **OFFSET** command sets the **Through** option as the default:

> Specify offset distance or [Through] <**Through**>: *(Press ENTER.)*

OffsetGapType

Offsetting polylines can be tricky. For this reason, AutoCAD includes the **OFFSETGAPTYPE** system variable to help you decide how potential gaps between polyline segments should be handled. The choices are extending lines, creating fillets (arcs), or creating bevels (chamfers).

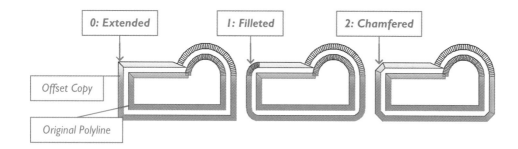

> Command: **offsetgaptype**

> Enter new value for OFFSETGAPTYPE <0>: *(Enter a number between 0 and 2.)*

OffsetGapType	Comment
0	Gaps filled with extended line segments (default).
1	Gaps filled with filleted segments, creating arcs.
2	Gaps filled with chamfered line segments, creating beveled edges.

 **DIVIDE** AND **MEASURE**

The **DIVIDE** and **MEASURE** commands place copies of points and blocks along objects.

DIVIDE divides an object into an equal number of parts, while **MEASURE** works with a specific distance — *number* of segments versus *length* of segment. Notice that the toolbar icons show the letter "n" for **DIVIDE**, and a pair of extension lines for **MEASURE**. Both commands place either blocks or points along the object.

The **MEASURE** command has nothing to do with measuring distances or lengths; for those, use the **DIST** and **LIST** commands.

Like the **OFFSET** command, dividing and measuring works with just one object at a time. Use the cursor to select a single object, because you cannot use Window, Crossing, or Last selection modes. In addition, AutoCAD is limited to working with lines, arcs, circles, splines, and polylines; picking a different object results in the complaint:

> Cannot divide that object. * Invalid*

AutoCAD does not place a point or block at the start or end of open objects.

 Note: After using the **DIVIDE** and **MEASURE** commands to place points on an object, you can snap to the points with the **NODe** object snap.

To use the **DIVIDE** and **MEASURE** commands, you need to tell AutoCAD three things: (1) the object to be marked, (2) the number of markers, and (3) whether the markers are points or blocks.

TUTORIAL: DIVIDING OBJECTS

1. To better see the effect of the **DIVIDE** command, first change visibility of points.
   Command: **pdmode**

   Enter new value for PDMODE <0>: **4**

2. To divide an object into equal parts, start the **DIVIDE** command:
   - From the **Draw** menu, choose **Point**, and then **Divide**.

   - At the 'Command:' prompt, enter the **divide** command:

   Command: **divide** *(Press* ENTER.*)*

   - Alternatively, enter the **div** alias at the 'Command:' prompt.

3. In all cases, AutoCAD prompts you to select the single object to divide:
   Select object to divide: *(Select one object.)*

4. Specify the number of divisions:
   Enter the number of segments or [Block]: *(Enter a number between 2 and 32767.)*

The **DIVIDE** command results in evenly spaced points, and with one fewer points than you would expect: divide a line by *six*, and AutoCAD places *five* points, creating *six* divisions. (Count the number of points and divisions on the line illustrated below.) The points and blocks are independent of the line; you can move and erase them at will.

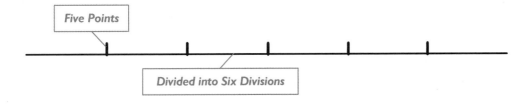

Five Points

Divided into Six Divisions

The MEASURE command operates similarly to the DIVIDE command, the difference being that you specify the length of segment along which to space the points.

TUTORIAL: "MEASURING" OBJECTS

1. To place points along an object at specific distances, start the MEASURE command:
   * From the **Draw** menu, choose **Point**, and then **Measure**.
   * At the 'Command:' prompt, enter the **measure** command.

      Command: **measure** *(Press* ENTER.*)*
   * Alternatively, enter the **me** alias at the 'Command:' prompt

2. In all cases, AutoCAD prompts you to select the single object to divide:

   Select object to measure: *(Select one object.)*

3. Specify the number of divisions:

   Specify the length of segment or [Block]: *(Enter a number between 2 and 32767.)*

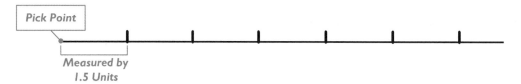

This may sound arcane, but the MEASURE command needs a *starting point*, which depends on the object. As the figure above illustrates, measurement does evenly distribute points as does division; there is usually a section left over at the end.

For open objects — lines, arcs, splines and open polylines — the starting point is the endpoint closest to your pick point; for closed polylines, it's the point where you began drawing the polyline. For circles, it is at the current snap angle, which is usually 0 (at the circle's 3 o'clock point); the measurement is made in the counterclockwise direction.

DIVIDING AND MEASURING: ADDITIONAL METHODS

Both commands' **Block** option lets you specify a block to place along the object, in place of a point. And, since points tend to be invisible, the PDMODE and PDSIZE system variables (accessed by DDPTYPE) are useful.

* **Block** option places blocks along the object.
* **DDPTYPE** command changes the look of the points.

Let's look at each.

Block

The **Block** option places blocks (symbols) along the object. The block must already exist in the drawing. The option operates identically for both commands:

1. Start either command. Enter the **b** option, and then the name of a block:

   Enter the number of segments or [Block]: **b**

   Enter name of block to insert: *(Type name.)*

2. Decide whether you want the block aligned with the object (Y) or at its own orientation (N):

   Align block with object? [Yes/No] <Y>: *(Enter **Y** or **N**.)*

3. Continue with the command.

"Yes" means the inserted block turns with the divided object, such as arc, circle, or spline. "No" means the block is always oriented in the same direction. If you are unsure of the block's name or even of its existence in the drawing, use the DesignCenter to help you.

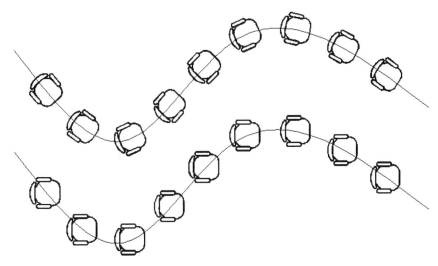

Blocks aligned with a spline (top) and unaligned (bottom).

DdPType

Recall from Chapter 4, "Drawing with Basic Objects," that the **DDPTYPE** command changes the look and size of points. Points are normally invisible (for all intents and purposes), so it may be useful to change their size. The figure below illustrates the before-and-after difference.

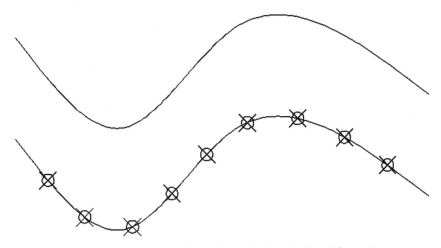

Spline divided by points of normal type (top), and enlarged (bottom).

⊡ ARRAY

The **ARRAY** command makes evenly-spaced copies in linear, rectangular, and round patterns.

There are times when you want to place multiple copies of an object in a pattern. Consider the rows of seats in a movie theater, or the columns of parking spaces at a shopping center. If you were using traditional drafting techniques, you would draw each one separately.

Earlier in the chapter, you saw how the **COPY** command (with its **Multiple** option) and the **DIVIDE** and **MEASURE** commands can place many copies of objects in the drawing. The **ARRAY** command, however, proves superior for making copies in precise rows, columns, matrices, circles, and semicircles. After you array an object, each copy can be edited separately — unlike the similar **MINSERT** command. For placing many copies in random places, **COPY** is better.

The largest number of rows and columns you can enter is 32,767; the smallest number is 1. Because a 32767 x 32767-array creates a billion elements, AutoCAD limits the total number to 100,000 — otherwise your computer system would overload. (The seemingly arbitrary value of 32,767 comes from 2^{15}).

You can change the upper limit to another value between 100 and 10,000,000 with the **MAXARRAY** system registry variable. At the command prompt, enter "MaxArray" exactly as shown:

Command: **(setenv "MaxArray" "1000")**

"1000"

To create arrays, AutoCAD needs to know (1) the type of array, rectangular or polar, (2) the object(s) you plan to array, which must already exist in the drawing, and (3) the parameters of the array.

Note: Sometimes it's difficult to distinguish between *rows* and *columns*. Rows go side to side, while columns go up and down. Still puzzled? Look at the preview window, or click the **Preview** button for a sneak peak.

TUTORIAL: LINEAR AND RECTANGULAR ARRAYS

A linear array copies objects in the horizontal direction (in a row or the x direction) or the vertical direction (in a column or the y direction). A rectangular array copies the objects in a rectangular pattern made up of rows and columns.

1. To make an array of copies, start the **ARRAY** command:
   - From the **Modify** menu, choose **Array**.
   - From the Modify toolbar, choose **Array**.
   - In the Dashboard's 2D Draw panel, choose the **Array** button.
   - At the 'Command:' prompt, enter the **array** command:

     Command: **array** (Press ENTER.)
   - Alternatively, enter the **ar** alias at the 'Command:' prompt.

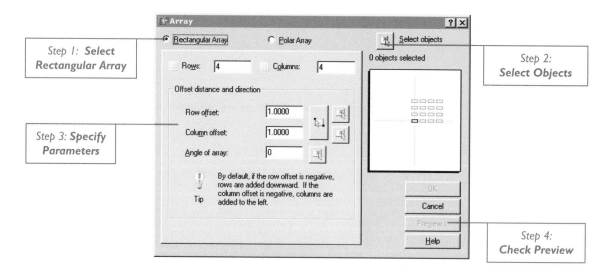

2. In all cases, AutoCAD displays the Array dialog box:
   Notice the dialog box has two radio buttons (the round buttons at the top of the dialog box) that determine the type of array:
   - **Rectangular** displays options for creating linear and rectangular arrays.
   - **Polar** displays options for creating polar (circular and semicircular) arrays.

3. Choose **Select Objects** to select the object(s) to array. The dialog box disappears, and this prompt appears:

> Select objects: *(Select one or more objects.)*
>
> Select objects: *(Press* ENTER *to return to the dialog box.)*

Note: You can avoid this step by selecting the objects before starting the **ARRAY** command.

4. In the **Rows** and **Columns** text boxes, specify the number of copies to make in rows and columns. When you enter a 1 for both, AutoCAD complains,

> Only one element; nothing to do.

because a 1x1 array consists of the original object only.

5. The **Row Offset** and **Column Offset** options measure the distance between the elements of the array. You can enter a specific distance, or click the adjacent buttons to select the distances in the drawing:

The **Pick Both Offsets** button clears the dialog box, and AutoCAD prompts you at the command line:

> Specify unit cell: *(Pick a point.)*
>
> Other corner: *(Pick another point.)*

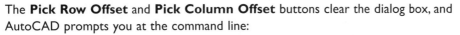

Pick two points, creating a rectangle that specifies the row and column distance between elements in the array. After you pick the second point, the dialog box returns.

The **Pick Row Offset** and **Pick Column Offset** buttons clear the dialog box, and AutoCAD prompts you at the command line:

> Specify the distance between rows (or columns): *(Specify a distance.)*

Enter a distance, or pick two points that specify the distance between row (or column) elements in the array. After picking the second point, the dialog box returns.

Note: AutoCAD normally creates rows to the right, and columns upwards. To have AutoCAD draw the array elements in the other direction, enter negative values for the row and column offsets.
If you use the mouse to pick two points, and the second point is below or to the left of the first, then AutoCAD automatically sets negative values.

6. **Angle of array** tilts the rectangular array at an angle; note that the object itself does not tilt, but the elements of the array are staggered by the angle. Entering an angle of 180 degrees draws the array downward. (To select the angle in the drawing, click the **Pick Angle of Array** button.)

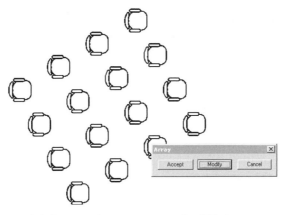

A 4x4 rectangular array at an angle of 30 degrees.

7. Click **Preview** to see what the array will look like. Notice the dialog box with its three buttons:

    Accept keeps the array, and exits the **ARRAY** command.

    Modify returns to the Array dialog box.

    Cancel doesn't keep the array (leaves things the way they were), and exits the **ARRAY** command.

8. Don't like the look of the preview? Click **Modify**, change parameters, and then click **Preview** until you're satisfied.

    Finally, click **Accept**.

TUTORIAL: POLAR AND SEMICIRCULAR ARRAYS

Polar arrays arrange objects in circular patterns. To construct a polar array, you must define the angle between the items (from center to center, not actually between) and either the number of items or degrees to fill. You have the option of rotating (or not rotating) each object as it is arrayed.

1. Start the **ARRAY** command.

2. Click the **Polar Array** radio button. Now let's take a look at the options for creating a round array.

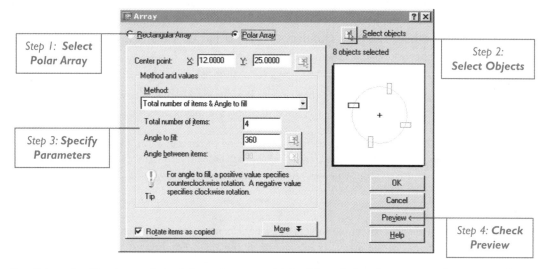

3. Select the **Center point** of the polar array; think of it as being the same as the center of a circle. AutoCAD automatically picks a point for itself using a method I have never figured out. (The technical editor agrees: "It's totally random!") But you can change it: enter new x, y-values or click the **Pick Center Point** button.

4. To create a polar array, you must provide data for any two of the following options. The **Method** droplist lets you pick the pair of options from the three possible choices:

    • **Total Number of Items** specifies the number of elements in the array. AutoCAD draws them to fit the polar route.

    • **Angle to Fill** changes the angle. By default, AutoCAD creates a 360-degree polar array, but you can create an arc array, instead. For example, specifying 270 degrees gives you three-quarters of a polar array.

    • **Angle Between Items** specifies the angle between the elements of the array.

5. Decide whether you want AutoCAD to **Rotate Items as Copied**. When turned on (the check mark appears), AutoCAD makes sure the items "face the center."

6. Click the **More** button to see more options for polar arrays:

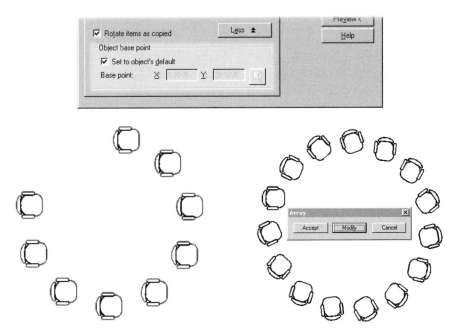

Left: 270-degree polar array with unrotated objects.
Right: 360-degree polar array with rotated objects.

The **Object base point** option is tricky to grok, but essentially you get to pick where AutoCAD measures the distance from the object to the center of the polar circle. AutoCAD uses the following default values, which change depending on the object being arrayed:

Object	Default Base Point
Lines, polylines, donuts, 3D polylines, rays, splines	Starting point.
Arcs, circles, ellipses	Center point.
Polygons, rectangles	First corner drawn.
Xlines	Midpoint.
Blocks, mtext, text	Insertion point.
Regions	Grip point.

7. Click **Select objects** to pick one or more objects to array. Click **Preview** to see what the array will look like.

CONSTRUCTING ARRAYS: ADDITIONAL METHODS

The **-ARRAY** command creates arrays at the command line, while **3DARRAY** creates them in 3D space.

- **-ARRAY** command creates arrays at the command line.
- **3DARRAY** command creates three-dimensional arrays.

Let's look at **-ARRAY** here; the **3dARRAY** command is discussed in *Using AutoCAD: Advanced*.

-Array

The **-ARRAY** command produces arrays after you enter options at the command line. To produce a rectangular array, use the **R** option, as follows:

Command: **-array**

Select objects: *(Select one or more objects.)*

Select objects: *(Press ENTER to end object selection.)*

Enter the type of array [Rectangular/Polar] <P>: **r**

Enter the number of rows (---) <1>: *(Enter the number of rows, such as **4**.)*

Enter the number of columns (| | |) <1> *(Enter the number of columns, such as **3**.)*

Enter the distance between rows or specify unit cell (---): *(Enter the distance between rows, such as **2.5**.)*

Specify the distance between columns (| | |): *(Enter the distance between columns, such as **3.3**.)*

To rotate the rectangular array, change the angle of the SNAPANG system variable before starting the -ARRAY command:

Command: **snapang**

Enter new value for SNAPANG <0>: *(Enter a new angle, such as **30**.)*

To produce a polar array, use the **P** option, as follows:

Command: **-array**

Select objects: *(Select one or more objects.)*

Select objects: *(Press ENTER to end object selection.)*

Enter the type of array [Rectangular/Polar] <P>: **p**

Specify center point of array or [Base]: *(Pick a point.)*

Enter the number of items in the array: *(Enter a number, such as **16**.)*

Specify the angle to fill (+=ccw, -=cw) <360>: *(Press ENTER, or enter an angle.)*

Rotate arrayed objects? [Yes/No] <Y>: *(Enter **Y** or **N**.)*

The options are similar to those available in the dialog box. Entering a negative angle draws the polar array clockwise; do not enter an angle of 0 degrees.

 FILLET AND CHAMFER

The **FILLET** and **CHAMFER** commands create rounded and angled corners, respectively.

The **FILLET** command connects two lines or polylines with a perfect intersection, or with an arc of specified radius. Fillets can also connect two circles, two arcs, a line and a circle, a line and an arc, or a circle and an arc.

The two objects need not need touch to be filleted — this includes parallel lines. This allows you to intersect two non-touching lines; in the case of parallel lines, a 180-degree arc is drawn between their ends.

To use the **CHAMFER** and **FILLET** commands, you need to tell AutoCAD two things: (1) the size of chamfer or fillet, and (2) the two objects to be chamfered or filleted.

Holding down the **SHIFT** key temporarily changes the fillet and chamfer distances to zero.

TUTORIAL: FILLETING OBJECTS

1. To fillet a pair of objects, start the **FILLET** command:

   * From the **Modify** menu, choose **Fillet**.

   * From the Modify toolbar, choose the **Fillet** button.

   * In the Dashboard's 2D Draw panel, choose the **Fillet** button.

   * At the 'Command:' prompt, enter the **fillet** command:

     Command: **fillet** *(Press* ENTER.*)*

   * Alternatively, enter the **f** alias at the 'Command:' prompt.

2. In all cases, AutoCAD first displays the current fillet settings, and then asks you to select two objects:

   Current settings: Mode = TRIM, Radius = 0.0000

   Select first object or [Undo/Polyline/Radius/Trim/Multiple]: *(Pick object 1.)*

   Select second object or shift-select to apply corner: *(Pick object 2.)*

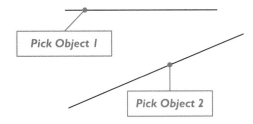

3. And the command is done! Because the radius was set to 0 (the fresh-off-the-distribution-CD default value), AutoCAD creates a clean intersection, without the arc you might have been expecting. (Notice that AutoCAD extended the two lines so that they intersect.)

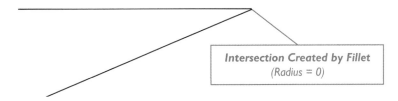

DIFFERENT FILLET RESULTS FOR DIFFERENT OBJECTS

Depending on the objects involved, the fillet differs.

Lines

Two lines are trimmed back (or extended, as necessary), so that an arc fits between them. A zero-radius fillet connects two lines with a perfect intersection. The picked segments are trimmed off.

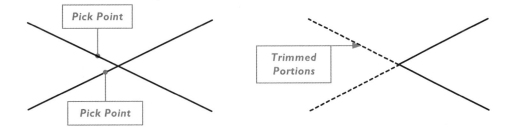

Two parallel lines fillet with a radius equal to their offset distance.

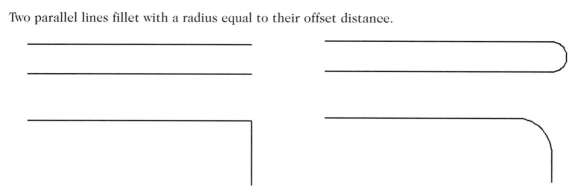

Two pairs of lines before and after being filleted with radius = 1.

Polylines

You can fillet an entire polyline in one operation when you select the **P** option (short for "polyline"). The fillet radius is placed at all vertices of the polyline. If arcs exist at any intersections, they are changed to the new fillet radius. Note that the fillet is applied to one continuous polyline.

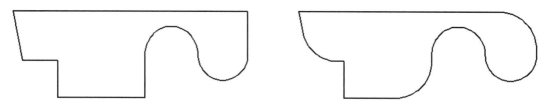

Left: *Original polyline.*
Right: *Filleted with the Polyline option;*
notice that two vertices were too short to be filleted with radius = 1.

If the fillet radius is too large for a line or polyline segment, AutoCAD does not apply the fillet:

> Select 2D polyline: *(Select a single polyline.)*
>
> 4 lines were filleted
>
> 3 were too short

When a line and a polyline are filleted, all three objects (the line, the fillet arc, and the polyline) are converted to a single polyline.

Arcs and Circles

Lines, arcs, and circles can also be filleted. When you fillet such objects, however, there are often several possible fillet combinations. You specify the type of fillet by the points picked when you select the objects. AutoCAD attempts to fillet the endpoint closest to the selection point.

The figures illustrate several combinations between a line and arc. Observe the placement of the points used to pick the objects in the middle row, and the resulting fillet (shown in the bottom row).

Notes: When you select two objects for filleting, and get an undesirable result, use the **Undo** option to undo the fillet. Try to respecify points closer to the endpoints you want to fillet.

If you have several filleted corners to draw, construct your intersections at right angles, and then fillet each later. This allows you to continue the **LINE** command without interruption, and requires fewer commands.

Changing an arc radius by fillet is cleaner and easier than erasing the old arc and cutting in a new one. Let AutoCAD do the work for you!

You "clean up" line intersections by setting the fillet radius to zero and filleting the intersections.

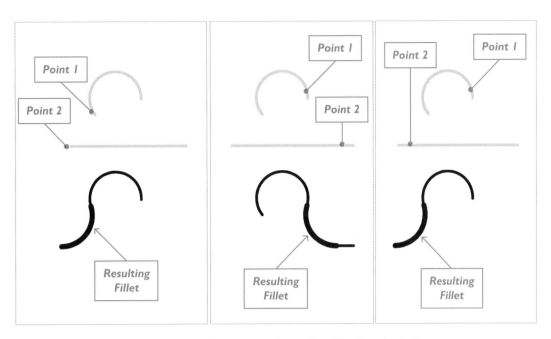

The original line and arc shown in gray; the resulting fillet shown by the heavy arcs.

As with lines and arcs, the result of filleting two circles depends on the location of the two points you use to select the circles. The figure illustrates three possible combinations, each using different selection points.

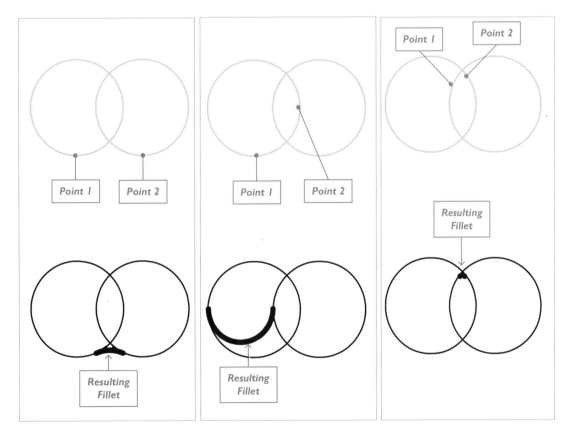

The original pre-fillet circles shown in gray; the resulting fillet shown by the heavy arcs.

CONSTRUCTING FILLETS: ADDITIONAL METHODS

The command has several options for special cases:

- **Undo** undoes the last fillet operation.

- **Polyline** treats polylines differently.

- **Radius** changes the radius of the three-dimensional arrays.

- **Trim** determines what happens to the leftover bits.

- **Multiple** continues the command to fillet additional objects.

Let's look at each.

Undo

The **Undo** option undoes the last fillet operation, without requiring you to exit the command to access the U or UNDO commands.

Polyline

The **Polyline** option fillets all vertices of a single polyline. When you fillet a polyline without using this option, AutoCAD expects you to place a fillet between two adjacent segments. At the "Select first object" prompt, select a single polyline.

> Select 2D polyline: *(Select a single polyline.)*

Radius

The **Radius** option determines the radius of the fillet arc. When set to 0, the command ensures that the two lines match precisely; an arc is not created.

> Specify fillet radius <1.0000>: *(Enter a radius.)*

If the fillet radius is too large for a line, AutoCAD complains:

> Radius is too large *Invalid*

In that case, use a radius smaller than the shortest line.

Press **SHIFT** to temporarily override the radius to 0.0.

Trim

The **Trim** option determines what happens to the trimmed bits. When on (the default), the command trims away the selected edges, up to the fillet arc's endpoint. When off, no trim occurs.

> Enter Trim mode option [Trim/No trim] <Trim>: *(Enter **T** or **N**.)*

Multiple

The **Multiple** option (formerly **mUltiple**) repeats the command until you press ESC:

> Select first object or [Undo/Polyline/Radius/Trim/Multiple]: *(Press **ESC** to exit command.)*

CHAMFER

The **CHAMFER** command trims segments from the ends of two lines or polylines, and then draws a straight line or polyline segment between them. The distance to be trimmed from each segment can be different or the same. The two objects do not have to intersect, but they must be capable of intersecting if there were extended. Unlike fillets, parallel lines cannot be chamfered.

Chamfering only works with line segments, such as lines, 2D polylines, and traces. It does not work with arc segments, such as arcs, circles, and ellipses.

TUTORIAL: CHAMFERING OBJECTS

1. To chamfer a pair of objects, start the **CHAMFER** command:
   - From the **Modify** menu, choose **Chamfer**.

   - From the Modify toolbar, choose the **Chamfer** button.

   - In the Dashboard's 2D Draw panel, choose the **Chamfer** button.

   - At the 'Command:' prompt, enter the **chamfer** command:

     Command: **chamfer** *(Press* ENTER.*)*

   - Alternatively, enter the **cha** alias at the 'Command:' prompt.

2. In all cases, AutoCAD first displays the current chamfer settings:
   (TRIM mode) Current chamfer Dist1 = 0.0000, Dist2 = 0.0000

3. Before selecting objects, change the chamfer distance from the current setting of 0:
   Select first line or [Undo/Polyline/Distance/Angle/Trim/mEthod/Multiple]: **d**
   Specify first chamfer distance <0.0000>: *(Enter a value, such as* **.5**.*)*

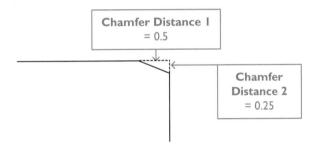

4. Chamfers work with two distances, which can be different. Enter the second distance:
   Specify second chamfer distance <0.5000>: *(Enter another value, such as* **.25**.*)*

5. Now pick the two lines to chamfer:
   Select first line or [Undo/Polyline/Distance/Angle/Trim/mEthod/Multiple]: *(Pick line 1.)*
   Select second line or shift-select to apply corner: *(Pick line 2.)*

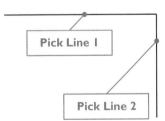

AutoCAD creates a chamfer at the intersection.

You can specify the amount to be trimmed by entering either a numerical value, or by showing AutoCAD the distance using two points on the screen.

CONSTRUCTING CHAMFERS: ADDITIONAL METHODS

The command has several other options for special cases:

- **Undo** undoes the last chamfer operation.

- **Polyline** chamfers a single polyline.

- **Angle** specifies the chamfer angle.

- **Trim** determines whether end pieces are saved.

- **mEthod** specifies the chamfer method.
- **Multiple** continues the command to fillet additional objects.

Let's look at each.

Undo

The **Undo** option undoes the last chamfer operation, without requiring you to exit the command to access the **U** or **UNDO** commands.

Polyline

Like filleting, chamfering a polyline is different from chamfering a pair of lines. If you had used the **PLINE** command's **Close** option to finish the polyline, AutoCAD chamfers *all* corners of the polyline; if not, the final vertex is not chamfered.

Select 2D polyline: *(Pick a polyline.)*

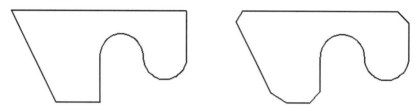

Left: *Polyline before being chamfered.*
Right: *Chamfered with the* **Polyline** *option; notice that arc vertices are not chamfered.*

If the polyline contains arcs, they are not chamfered. If some parts of the polyline do not chamfer, it could be that the segments are too short or parallel to each other. In that case, AutoCAD warns: "*n* were too short."

Angle

As an alternative to specifying a chamfer by two distances, you can specify a distance and an angle.

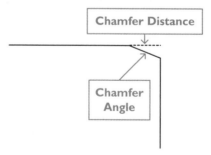

Specify chamfer length on the first line <1.0000>: *(Enter the distance.)*
Specify chamfer angle from the first line <0>: *(Enter the angle.)*

Trim

Normally, the **CHAMFER** command erases the line segments not needed after the chamfer. The **Trim** option, however, determines whether you keep the excess lines.

Enter Trim mode option [Trim/No trim] <Trim>: *(Enter **T** or **N**.)*

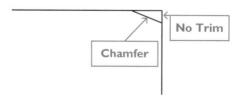

mEthod

The default method specifies two distances. If you prefer the distance-angle method, the **mEthod** option (formerly **Method**) lets you change the default:

> Enter trim method [Distance/Angle] <Angle>: *(Type **D** or **A**.)*

Multiple

The **Multiple** option (formerly **mUltiple**) repeats the command until you press ESC:

> Select first object or [Polyline/Radius/Trim/mUltiple]: *(Press* ESC *to exit command.)*

 JOIN

The JOIN command joins similar objects into one.

The following objects can be joined, subject to these conditions:

- **Colinear lines** joined into a single line, including those with overlaps and gaps between them. (*Colinear* means the lines are lined up in a row like cars in a train.)

- **Arcs** in the same imaginary circle joined into a single arc counterclockwise from the source arc; they can have overlaps and gaps between them. This command also closes arcs into circles. When arcs are dimensioned with **DIMARC**, the converted circles are disassociated from the dimensions.

- **Elliptical arcs** in the same imaginary ellipse joined into a single elliptical arc counterclockwise from the source arc; they can have gaps between them. This command also closes elliptical arcs into ellipses.

- **Polylines**, lines, and arcs joined into a single polyline; they *cannot* have gaps between them. When the first object selected is a polyline, then this works like the **PEDIT Join** command.

- **Splines** joined into a single spline; they *cannot* have gaps between them.

- The source object determines the properties of the joined objects.

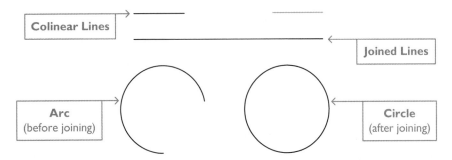

To use the JOIN command, you need to tell AutoCAD two things: (1) source object, and (2) the other objects to join; when arcs and elliptical arcs are selected, there is also the choice of whether you want the open objects closed.

TUTORIAL: JOINING OBJECTS

1. To join two or more objects, start the JOIN command:
   - From the **Modify** menu, choose **Join**.
   - From the Modify toolbar choose the **Join** button.
   - In the Dashboard's 2D Draw panel, choose the **Join** button.

- At the 'Command:' prompt, enter the **join** command:

    Command: **join** *(Press* ENTER.*)*

- Alternatively, enter the **j** alias.

2. AutoCAD prompts you to select the source object, which can be a line, arc, elliptical arc, polyline, or spline:

    Select source object: *(Select one object.)*

3. AutoCAD prompts you to select the objects to join. The prompt varies, depending on the source object selected; the following prompt is for arcs:

    Select arcs to join to source or [cLose]: *(Select one or more objects.)*

4. AutoCAD repeats the prompt until you press ENTER to exit the command:

    Select arcs to join to source or [cLose]: *(Press* ENTER.*)*

JOINING OBJECTS: ADDITIONAL METHODS

The JOIN command has one option:

- **cLose** closes arcs and elliptical arcs.

Let's look at it.

cLose

The **cLose** option closes arcs to make circles, and elliptical arcs to make ellipses.

    Command: **join** *(Press* ENTER.*)*

    Select source object: *(Select one arc or ellipse.)*

    Select arcs to join to source or [cLose]: **l**

    Arc converted to a circle.

Note: Use the BREAK command to convert circles into arcs; use the JOIN command to convert arcs into circles.

 REVCLOUD

The REVCLOUD command creates revision clouds.

Revision clouds are often used to highlight areas in drawings that require attention, such as a revision or a potential error. Revision clouds are sometimes called "markups" or "redlines," because they were often drawn with red pencils to stand out in drawings.

AutoCAD's REVCLOUD command creates revision clouds, or converts other objects into revision clouds — specifically circles, ellipses, closed polylines, and closed splines. Because you cannot invoke transparent zooms and pans (other than through the mouse wheel) during the command, ensure you can see the entire area before starting.

Note: Before starting REVCLOUD, switch to a layer set to red. That makes it easier to turn on and off the display of the revision cloud, and makes it the traditional color of red to boot.

To use the REVCLOUD command, you need to tell AutoCAD two things: (1) the starting point, and (2) the cloud path.

TUTORIAL: REDLINING OBJECTS

1. To markup drawings, start the **REVCLOUD** command:
    *   From the **Draw** menu, choose **Revision Cloud**.
    *   From the Draw toolbar choose the **Revcloud** button.
    *   At the 'Command:' prompt, enter the **revcloud** command.

    Command: **revcloud** *(Press* **ENTER.***)*

2. AutoCAD displays the current settings, and then prompts you to start drawing the revision cloud:

    Minimum arc length: 0.5000   Maximum arc length: 0.5000   Style: Normal

    Specify start point or [Arc length/Object/Style] <Object>: *(Pick a point.)*

3. Move your cursor. Notice that AutoCAD automatically creates the cloud pattern. As an alternative, you can define the size of arcs by picking points.

    Guide crosshairs along cloud path...

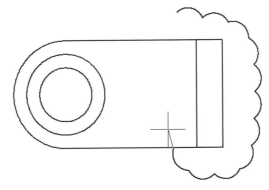

4. When the cursor is close to the start point, AutoCAD automatically closes the cloud:

    Revision cloud finished.

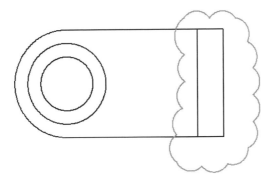

After placing the revision cloud, you can edit it like other objects in the drawing — through copying, grips editing, erasing, and so on.

REDLINING OBJECTS: ADDITIONAL METHODS

To help you in redlining, AutoCAD has these options and commands:

*   **Arc length** option defines the size of the arcs making up the clouds.
*   **Object** option converts existing objects into revision clouds.
*   **Style** option switches between simple and calligraphic arcs.
*   **MARKUP** command imports marked-up *.dwf* files from DWF Composer.

Let's look at each.

Arc length

The **Arc length** option determines the size of the arcs making up the clouds, which are scale-dependent, like linetypes and hatch patterns. **REVCLOUD** saves the arc length as a factor of the **DIMSCALE** system variable, so that clouds drawn after the dimension scale changes still look the right size when the scale factor changes.

When you specify different minimum and maximum lengths, AutoCAD draws clouds with random-size arcs. AutoCAD limits the maximum arc length to three times the minimum length.

Specify minimum length of arc <0.5000>: *(Enter a value.)*

Specify maximum length of arc <0.5000>: *(Enter a value.)*

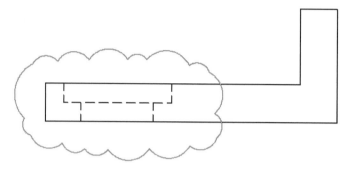

A revision cloud with arcs of varying sizes.

Object

The **Object** option lets you convert existing objects into revision clouds.

Select object: *(Select one object.)*

Reverse direction [Yes/No] <No>: *(Enter **Y** or **N**.)*

Revision cloud finished.

Top: *The original polyline object.*
Middle: *Object converted to revision cloud.*
Bottom: *Revision cloud with reversed arcs.*

You can convert closed polylines, circles, ellipses, and closed splines to revision clouds. After applying this option, what happens to the original object? That depends on the setting of the **DELOBJ** system variable. If set to 1 (the default), the original is erased; if set to 0, the original object stays in the drawing.

> Command: **delobj**
>
> Enter new value for DELOBJ <1>: *(Enter **1** or **0**.)*

Style

The **Style** option switches between simple and calligraphic arcs. Calligraphic arcs, added in AutoCAD 2005, have a variable width, as illustrated below.

Left: *Normal arcs.*
Right: *Calligraphic arcs.*

To switch between normal and calligraphic arcs, use the **Style** option, as follows:

> Specify start point or [Arc length/Object/Style] <Object>: **s**
>
> Select arc style [Normal/Calligraphy] <Normal>: **c**
>
> Arc style = Calligraphy

To convert calligraphic arcs to normal arcs, use the **PEDIT** command's **Width** option.

 ## MARKUP

The **MARKUP** command imports marked-up *.dwf* files from Design Review (formerly DWF Composer; DWF is short for "design Web format). It displays the Markup Set Manager, which places and edits redline DWF markups in drawings.

Despite its name, **MARKUP** does not mark up drawings; its purpose is to manage marked up drawings; the markup process takes place in Design Review. This command works only with *.dwf* files containing markup data; it will not open any other kind of markup file, nor does it "import" non-marked-up DWF data, or work with drawings marked up with the **REDLINE** or **RMLIN** commands.

(*History*: The **MARKUP** command replaces the **RMLIN** command (short for "Red Markup Line In"), which imported XML-format *.rml* markup files created by Autodesk's old Volo View drawing viewer software. **RMLIN** was removed from AutoCAD 2006.)

Design Review

To use the **MARKUP** command, you must have access to Design Review, the free viewing, redlining, and printing software from Autodesk. You can download a copy from www.autodesk.com/designreview, following registration.

Design Review acts as a publisher, merging together files from many sources — such as raster images, word processing documents, spreadsheets, and drawings from other CAD packages — into a single, multi-page document. To convert non-AutoCAD documents into DWF format, download and install the free DWF Writer printer driver. More information is available at www.autodesk.com/dwf.

The markup cycle works like this:

In AutoCAD:

1. Create and edit the drawings.
2. Export the drawing from AutoCAD in DWF format using the **PUBLISH** or **AUTOPUBLISH** commands. (See Chapter 18 for more details.)

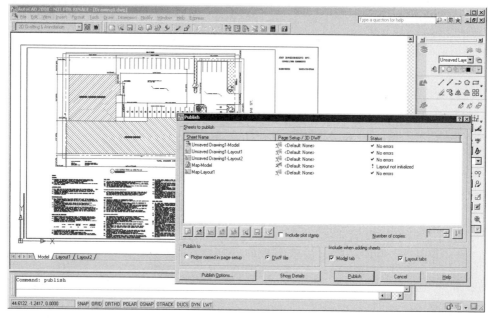

In Design Review:

3. Open the *.dwf* files in Design Review.
4. Mark up the drawings using the markup tools found on the toolbar.

5. Save the markups with the **File | Save** command.

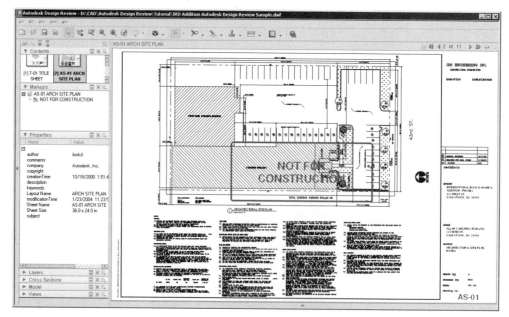

(If Design Review is not available, you can open the sample *.dwf* files found in AutoCAD's *\sample* folder.)

Back in AutoCAD:

6. Open the marked-up *.dwf* files with the **MARKUP** command.

7. To view the markups, double-click the markup name in the Markup Set Manager window, "Markup1" in the figure below.

Markup Options

Many of the **MARKUP** command's options are "hidden" in shortcut menus. Depending on where you right-click in the Markup Set Manager palette, you get menus with differing sets of commands. The most-commonly used shortcut menu is found by right-clicking a markup sheet, as indicated by the red revision cloud icon.

The menu's options are:

Command	*Comment*
Open Markup	Finds and opens the original *.dwg* drawing file, and makes active the layout associated with the markup; the *.dwf* file is also opened, but is hidden until you press **ALT+4**.
Markup Status	Allows you to change the status of markups: **<None>** has no status. **Question** indicates additional information is needed. **For Review** indicates changes should be reviewed by another. **Done** indicates the markup is implemented.
Restore Initial Markup View	Restores the original markup view.
Republish Markup DWF	Overwrites the previous *.dwf* file with changes made to the drawing and the markup status.

Marked-up *.dwf* files can also be opened in AutoCAD with the **OPENDWFMARKUP** command. The **MARKUPCLOSE** command closes the Markup Set Manager window.

EXERCISES

1. From the Companion CD, open the *edit3.dwg* drawing file, an architectural elevation.

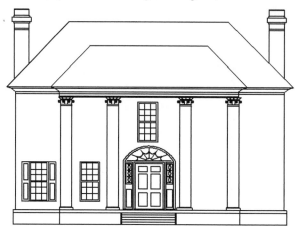

   Use the **COPY** command to copy the windows from the left side to the right side. Then, copy all the windows on the lower level (including those you just copied) to the upper level. When copying, use the object selection options you think will work best.

2. From the Companion CD, open the *edit4.dwg* drawing file, a landscaping plan.

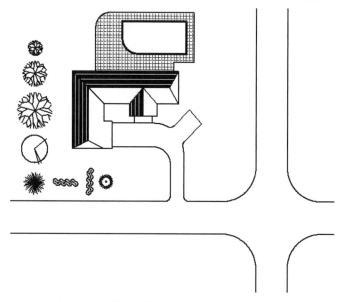

   Use the **COPY** command to copy the landscaping blocks (trees, shrubbery, and so on) and to create a landscape scheme.

3. Connect two lines with fillets of varying radius.
   First, draw lines similar to those in the illustration.

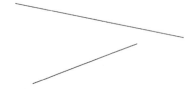

   a. Connect the two lines with a fillet of radius of 0.15.
   b. Use the **SHIFT** key to set the fillet radius to zero, and then apply the fillet to the two lines again. Do the two lines now connect in a perfect intersection?

4. From the Companion CD, open the *edit6.dwg* drawing file, a practice drawing.
    Fillet the objects to achieve the results shown on right side of the figure.

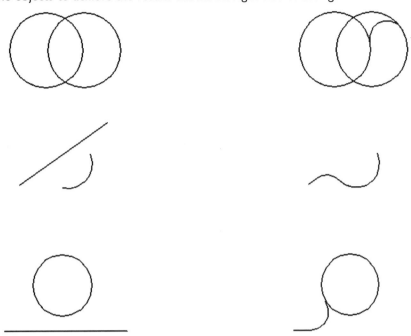

5. Suppose that you have to add another circle to your drawing that is identical to an existing circle.
    Draw a circle, and then use the **COPY** command to place a second circle.
    Is the second circle an exact copy of the first?

6. In this exercise, you create a rectangular array.
    Start a new drawing.
    As a visual aid, turn on the grid with a value of one.
    Draw a circle with the center point located at 1,1 and a radius of one.
    Select the circle, and then start the **ARRAY** command.
    In the dialog box, enter the following options:

Type of array	**Rectangular**
Rows	3
Columns	5
Row offset	2
Column offset	2

    Your array should look like the following illustration:

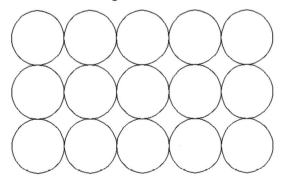

    Save the drawing with the name *array.dwg*.

7. Start another drawing to create a polar array.

Draw a square with sides of one unit each.

Start the **ARRAY** command, and enter these options:

Select objects	*Select the square*
Type of array	**Polar**
Center point	**2,2**
Total number of items	**4**
Angle to fill	**270**
Rotate items as copied	**off**

The polar array should look like the one in the figure.

Save the drawing with the name *parray.dwg*.

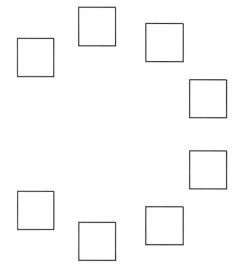

8. From the Companion CD, open the *mirror1.dwg* drawing file, a floor plan.

Suppose that you are designing a house, and you want to reverse the layout of the bathroom.

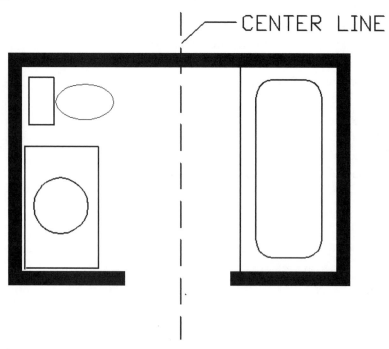

Reverse the room using the **MIRROR** command.

Hint: Turn on ortho mode (**F8**), and delete the source objects.

9. In this exercise, the **PDMODE** system variable allows you to see points placed by the **DIVIDE** command.

    Start a new drawing.

    Draw a circle: center the circle in the viewport, and use a radius of 3.

    Start the **DIVIDE** command, and use 8 for the number of segments.

    Do you see any difference to the circle?

    Set the **PDMODE** system variable to 34.

    If necessary, use the **REGEN** command to make the new point style visible.

10. Draw a symbol similar to the one in the figure.

    Which two commands make it easier to draw the symbol?

    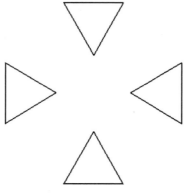

    Turn the symbol into a block, and name it "Symbol."

    a. Draw a circle on the screen, and then start the **DIVIDE** command.

        Use 8 segments, along with the **Block** option, with blocks aligned with the object.

    b. Repeat the exercise, but this time don't rotate the block.

11. Let's "measure" an object.

    Draw a horizontal line 6 units in length.

    Make the point mode setting of 34 to make the points visible.

    With the **MEASURE** command, specify a segment length of 1.0.

12. In this exercise, you use the **ELLIPSE** command to draw a can.

    Draw one ellipse with the **ELLIPSE** command. Don't worry about the size.

    Next, use the **COPY** command to copy the ellipse to create the top of the can.

    Hint: Use ortho mode to align the copy perfectly with the original ellipse.

    Now use the **LINE** command to draw lines between the two ellipses, as shown in the figure.

    Hint: Use QUADrant object snap to capture the outer quadrants of the ellipses.

    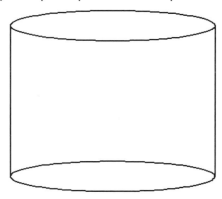

13. In this exercise, you use the **OFFSET** command to help draw a city scape plan. From the Companion CD, open the *offset.dwg* drawing file.

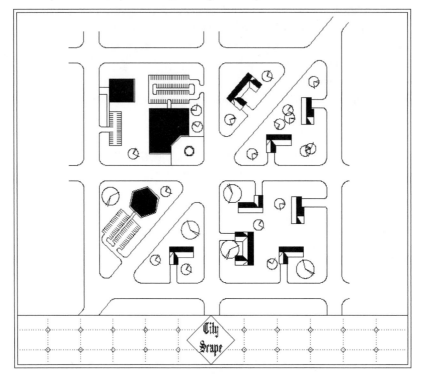

The edges of the streets are drawn with polylines.

Use the **OFFSET** command's **Through** option to offset the curbs by 6".

14. Draw the object shown in the figure below.

Use the **CHAMFER** command to bevel all its corners.

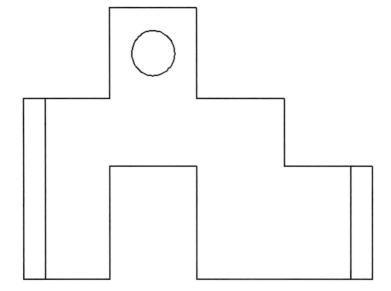

15. From the Companion CD, open the *edit5.dwg* drawing, a piping diagram.

    Use the **REVCLOUD** command to place revision clouds around the two unfinished areas, as illustrated by the figure.

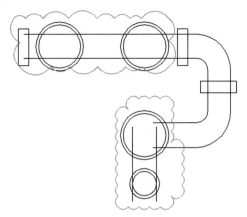

16. Draw the profiles of the aluminum extrusions diagrammed and dimensioned below. Apply fillets, where required, and hatch with a solid fill. All dimensions are in inches.

    a. Angle.

A	B	C	D
4.000	1.000	0.155	0.125

    b. Square tube.

A	T
4.000	0.145

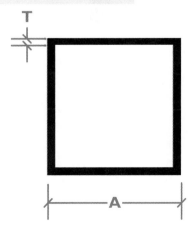

c. Rectangular tube with rounded corners.

A	B	T	R1	R2
2.250	1.750	0.125	0.125	0.125

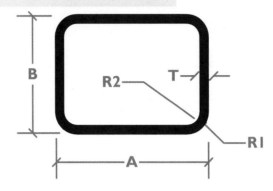

d. Rectangular bar with rounded corners.

A	B	R
3.250	0.375	0.030

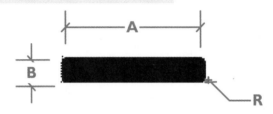

e. Tee.

A	B	T1	T2	R
2.000	2.000	0.125	0.125	0.015

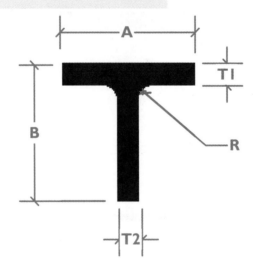

f. Channel with rounded corners.

A	B	C	D	R
3.000	1.500	0.375	0.375	0.375

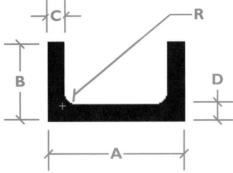

17. Draw the outline and fold lines of the diecut cardboard packaging that holds a tube of toothpaste. Cutlines are shown in black; fold lines in white. All dimensions are in mm. Print the drawing, and then cut and fold along the lines to recreate the cardboard box.

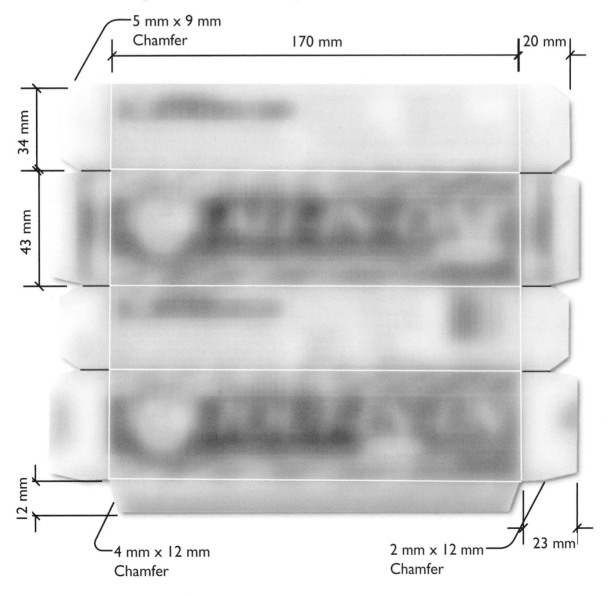

18. From the Companion CD, open the *join.dwg* drawing.

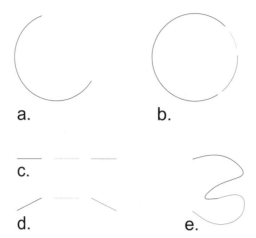

a.   b.

c.

d.   e.

Use the JOIN command to perform the following tasks:

a. Turn the arc into a circle.

b. Join the three arcs into one circle.

c. Join the three lines into one line.

d. Join the three lines into one line.

e. Join the two splines into one spline.

Which one of the above tasks did not work? Why?

What color did each joined entity become? Why?

CHAPTER REVIEW

1. What two changes can a fillet make to an intersection?

2. When making multiple copies with the **COPY** command, is each copy relative to the first base point entered, or to the point of the last copy made?

3. Can objects of different types be filleted (such as a line and an arc), or must they be alike?

4. What are the three types of arrays?

5. Describe the function of the **MIRROR** command.

6. Which system variable controls whether text is mirrored?

7. What is the *mirror line*?

8. What objects can you use with the **MEASURE** command?

9. How many objects can be offset at one time?

10. Describe the purpose of the **Through** option of the **OFFSET** command?

11. Draw an example of using the **OFFSET** command's **Through** option.

12. What is the purpose of the **CHAMFER** command?

13. What is the procedure for setting **CHAMFER** distances?

14. Describe the purpose of the **ARRAY** command.

15. Explain how the **-ARRAY** command differs from the **ARRAY** command.

16. Can the **ARRAY** command draw arrays at an angle?

17. Under what condition does the **ARRAY** dialog box's **OK** button stay grayed (and unavailable)?

18. To construct an array in the -x direction, you enter a negative value in which option?

19. What could happen to your computer if you construct too large an array?

20. Define *displacement*, as used by the **COPY** command.

21. Which mode helps you copy objects precisely horizontally and vertically?

22. What are the aliases for the following commands:

    COPY

    MIRROR

    OFFSET

    FILLET

23. What problem can occur when the **OFFSET** command is applied to curved objects?

24. How does the **OFFSETGAPTYPE** system variable affect polylines?

25. Does the **MEASURE** command measure the lengths of objects?

26. What two kinds of objects does the **DIVIDE** command place along objects?

27. When you specify 5 segments for the **DIVIDE** command, how many objects does it place?

28. Can parallel lines be filleted?

29. Describe the purpose of the **FILLET** and **CHAMFER** commands' **Trim** option.

30. What is the difference between a fillet and a chamfer?

31. Can all vertices of a polyline be chamfered at once?
    If so, how?

32. Describe the purpose of the **REVCLOUD** command.

33. What is another name for "redlines"?

34. How does the **DELOBJ** system variable affect the **Object** option of the **REVCLOUD** command?

35. Explain the purpose of the **MARKUP** command.

36. Describe the role of the Design Review (DWF Composer) software.

37. How are DWF files created?

38. What command converts arcs into circles?
39. Which key do you hold down to make a zero-radius fillet? A zero-distance chamfer?
40. What is the purpose of the **Undo** option in the COPY command?
41. What are *colinear* lines?

Additional Editing Options

In previous chapters, you learned numerous methods of creating and changing objects. In Chapter 9, "Direct Editing of Objects," for instance, you changed objects directly through grips editing. In this chapter, you learn AutoCAD's other editing commands, including some that may already be familiar to you from grips editing.

This chapter summarizes the remaining editing commands used on basic objects in two-dimensional drafting:

ERASE deletes objects from drawings.

OOPS brings them back.

BREAK removes portions of objects.

TRIM, EXTEND, and **LENGTHEN** change the length of open objects.

STRETCH makes portions of objects larger and smaller.

MOVE moves objects in the drawing.

ROTATE rotates objects.

SCALE changes the size of objects.

CHANGE changes the size, properties, and other characteristics of objects.

EXPLODE and **XPLODE** reduce complex objects into their simplest forms.

PEDIT edits polylines.

SPLFRAME, SPLINETYPE, and **SPLINESEGS** control the look of polyline splines.

NEW TO AUTOCAD 2008 IN THIS CHAPTER

- Selections made during the **STRETCH** command are now cumulative.

FINDING THE COMMANDS

On the Dashboard's **2D Draw** panel:

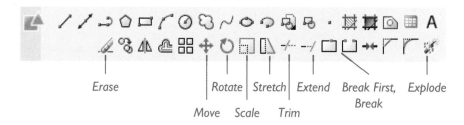

On the **Modify** and **Modify II** toolbars:

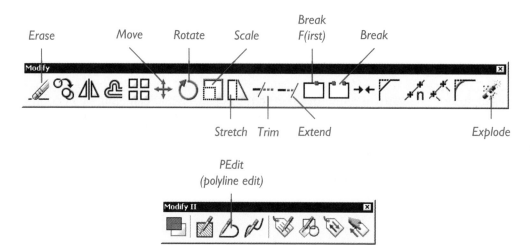

ERASE

The **ERASE** command deletes objects from drawings. As an alternative to this command, you can select one or more objects, and press the **DEL** (delete) key.

TUTORIAL: ERASING OBJECTS

1. From the CD, open the *erase.dwg* drawing file.
2. To erase one or more objects, start the **ERASE** command:
   - From the **Modify** menu, choose **Erase**.
   - From the Modify toolbar, choose the **Erase** button.
   - In the Dashboard's 2D Draw panel, pick the **Erase** button.
   - At the 'Command:' prompt, enter the **erase** command:

     Command: **erase** *(Press* **ENTER.**)
   - Alternatively, enter the **e** alias at the 'Command:' prompt.
3. In all cases, AutoCAD prompts you to select the objects you want erased:

   Select objects: *(Pick one or more objects.)*

   Select objects: *(Press* **ENTER** *to end object selection.)*

   Notice that the selected objects disappear from the drawing.

At the "Select objects:" prompt, the crosshair cursor changes to the tiny, square pickbox for selecting one object at a time (the *point* method of selection).

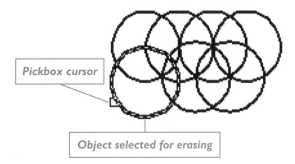

Pickbox cursor

Object selected for erasing

If you wish to use another object selection option, such as Window or Crossing, enter the option name at the "Select objects:" prompt. Following erasure, it may be necessary to use the REDRAW command to refresh the screen.

RECOVERING ERASED OBJECTS

AutoCAD has two commands for restoring erased objects:

- **OOPS** returns the last erased objects.
- **U** undoes the last operation, including erasure.

Let's look at each.

Oops

The **OOPS** command restores the objects that were last erased from the drawing.

   Command: **oops**

Notice that the erased object(s) are returned to the drawing.

This command also works in conjunction with the **BLOCK** and **-BLOCK** commands, when the original objects were erased in making the block.

U

If you find yourself unable to use **OOPS** to return erased objects, remember that you can also reverse the deletion with the **U** command.

Notes: To erase the object you just drew, enter **ERASE** with the **Last** option. A peculiarity is that AutoCAD only erases the last-drawn object if it is visible on the screen.

Similarly, to erase the previous selection set, use **ERASE** **Previous**. You can erase the entire drawing with the **ERASE** **All** command. To do the same using keyboard shortcuts, press **CTRL+A** and then **DEL**.

BREAK

The **BREAK** command partially and fully erases objects, as well as cracking them.

To "break" an object, select two points on it. The portion of the object between the two points is erased. This command removes a portion of almost any object, and shortens open objects — depending on the place you pick. The following objects can be broken: arcs, circles, ellipses, elliptical arcs, lines, polylines, rays, splines, and xlines.

BREAK can remove an object entirely, like the **ERASE** command. Here's how: for the first point, snap to one end of a line, and then snap to the other endpoint for the second point. The entire line disappears from the drawing. Use **U** to recover, if necessary.

BREAK is useful for "cracking" objects: click the same point twice, and **BREAK** creates two segments that touch. **BREAK** operates on just one object at a time.

TUTORIAL: BREAKING OBJECTS

1. To remove a portion of an object, start the **BREAK** command:
   * From the **Modify** menu, choose **Break**.
   * From the Modify toolbar, choose the **Break** button.
   * In the Dashboard's 2D Draw panel, pick the **Erase** button.
   * At the 'Command:' prompt, enter the **break** command:

       Command: **break** *(Press ENTER.)*
   * Alternatively, enter the **b** alias at the 'Command:' prompt.

2. In all cases, AutoCAD prompts you to select the single object you want broken:
       Select object: *(Pick one object at point 1.)*

   The point you pick is crucial, because it becomes the first point of the break.
3. Pick a second point, the other end of the break:
       Specify second break point or [First point]: *(Pick point 2.)*

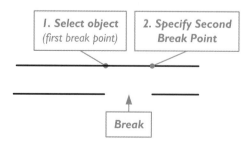

Note: AutoCAD displays a padlock icon when you attempt to select objects on locked layers. Objects on locked layers can be seen but not edited.

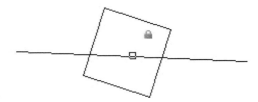

To edit these objects, you must first unlock their layer(s).

BREAKING OBJECTS: ADDITIONAL METHOD

By default, the first point you select on the object becomes the *first* break point. To redefine the first break point, enter F in response to the prompt. Select another point to be the first break point.

Redefining the first point is useful when the drawing is crowded, or when the break occurs at an intersection where pointing to the object at the first break point might result in the wrong object being selected.

• **First** defines the first pick point.

Let's look at this option.

 First

The **First** option allows you to redefine the first pick point, providing greater control over the segment being broken out.

Command: **break**

Select object: *(Pick object.)*

Specify second break point or [First point]: **f**

Specify first break point: *(Pick point 1.)*

Specify second break point: *(Pick point 2.)*

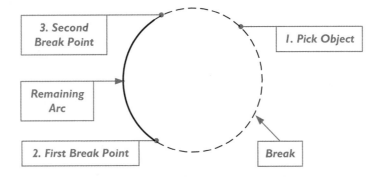

 Note: The first tip I ever learned from technical editor Bill Fane was that sometimes it is easier to create an arc by drawing a circle, and then using the **BREAK** command to remove the unwanted portion.

The **BREAK** command affects objects in different ways:

Object	Break Action
Arcs, Lines	Removes portion of line or arc between the pick points. When one point is on the line, but the other point is off the end of the line, the line is "trimmed back" to the first break point.
Circles	Breaks into an arc. Unwanted piece is determined by going counterclockwise from the first point to the second point.
Ellipses	Same manner as circles, except that it results in an elliptical arc.
Splines	Same manner as lines and arcs.
Traces	Same manner as lines, except that the new endpoints are trimmed square.
Polylines	Cuts wide polylines squarely, as with traces. Breaking closed polylines creates open polylines.
Viewports	Cannot be broken.
Xlines	Become rays.

TRIM, EXTEND, AND LENGTHEN

The **TRIM**, **EXTEND**, and **LENGTHEN** commands change the length of open objects.

The **TRIM** command shortens objects by defining other objects as *cutting edges*; any portion of the object beyond the cutting line is "cut off."

AutoCAD needs to know two pieces of information for trimming: (1) the object(s) to be used as the cutting edges, and (2) the portion of the object to be removed.

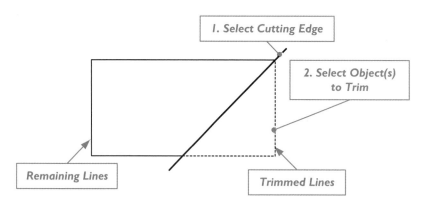

Cutting edges are any combination of arcs, circles, ellipses, lines, viewports, rays, regions, splines, text, xlines, hatches, 2D and 3D polylines. Polylines with nonzero width trim to the centerline of the polyline.

A quick method is to select *all* objects as cutting edges: at the 'Select objects <select all>:' prompt, press **ENTER**, and AutoCAD selects all objects in the drawing, including those not in the current view. A previously undocumented method of quickly selecting many objects to trim is the **Fence** selection mode. The part that is picked is trimmed.

TUTORIAL: SHORTENING OBJECTS

1. To shorten one or more objects, start the **TRIM** command:
   - From the **Modify** menu, choose **Trim**.
   - From the Modify toolbar, choose the **Trim** button.
   - In the Dashboard's 2D Draw panel, pick the **Trim** button.

- At the 'Command:' prompt, enter the **trim** command:

    Command: **trim** *(Press* ENTER.*)*

- Alternatively, enter the **tr** alias at the 'Command:' prompt.

2. In all cases, AutoCAD reports the current trim settings, and then prompts you to select cutting edges:

    Current settings: Projection=UCS Edge=None

    Select cutting edges ...

    Select objects or <select all>: *(Pick one or more objects as cutting edges, or press* ENTER *to select all objects in the drawing.)*

    Select objects: *(Press* ENTER *to end object selection.)*

3. Select the object to be trimmed. *Warning!* AutoCAD trims the portion you select.

    Select object to trim or shift-select to extend or

    [Fence/Crossing/Project/Edge/eRase/Undo]: *(Pick objects to trim, or enter an option.)*

4. AutoCAD repeatedly prompts you to select additional objects to trim. When done trimming, press **ENTER**:

    Select object to trim or shift-select to extend or

    [Fence/Crossing/Project/Edge/eRase/Undo]: *(Press* ENTER *to exit.)*

When you select objects that cannot serve as cutting edges, such as blocks and dimensions, AutoCAD displays the message:

    No edges selected.

If an object cannot be trimmed, AutoCAD displays:

    Cannot TRIM this object

If the object to be trimmed does not intersect a cutting edge, AutoCAD complains:

    Object does not intersect an edge.

Trimming Closed Objects

To trim closed objects, such as circles and polygons, they must be intersected at least twice by the cutting edge, such as by a line drawn through two points on a circle's circumference. If only one cutting edge intersects the closed object, AutoCAD complains:

    Circle must intersect twice.

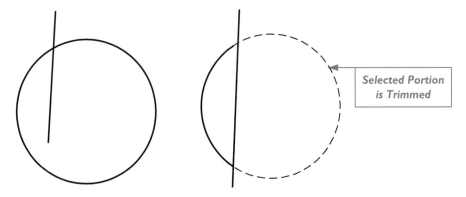

Left: Circle cannot be trimmed, because the line intersects it at only one location.
Right: This circle can be trimmed, because the line intersects it twice.
You can use two cutting edges (two lines) that cross the circle.

Trimming Polylines

Polylines are trimmed at the intersection of the center line of the polyline and the cutting edge. The trim is a square edge. Therefore, if the cutting edge intersects a polyline of nonzero width at an angle, the square-edged end may protrude beyond the cutting edge.

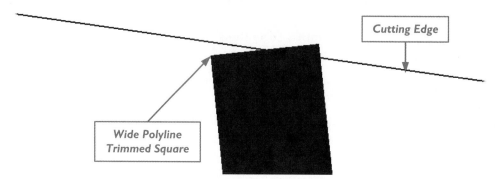

Cutting Edge

Wide Polyline Trimmed Square

Notes: The **TRIM** and **EXTEND** commands can switch roles:

To *extend* an object during the **TRIM** command, hold down the **SHIFT** key while selecting the object. Similarly, to *trim* an object during the **EXTEND** command, hold down the **SHIFT** key while selecting the object.

Hatch patterns can be cutting edges or the objects being trimmed. In other words, the **TRIM** command can trim hatch and fill patterns, and use them to trim other objects. (The **EXTEND** and **LENGTHEN** commands have no effect on hatches.)

TRIMMING OBJECTS: ADDITIONAL METHODS

The **TRIM** command has several options:

- **Select all** selects all objects as cutting edges.

- **Fence** and **Crossing** select objects to be trimmed.

- **Project** determines trimming in 3D space.

- **Edge** toggles actual and implied cutting edges in 3D space.

- **eRase** erases objects.

- **Undo** undoes the last trim.

- **SHIFT-select** extends the object to the cutting line, instead of trimming.

Let's look at each option.

Select All

The **Select all** option selects all objects in the drawing as cutting edges — at least those not on frozen layers. When you press **ENTER** at the 'Select objects or <select all>:' prompt, AutoCAD selects all objects, and then stops repeating the 'Select objects:' prompt.

Fence and Crossing

The **Fence** and **Crossing** options select objects to be trimmed. Their inclusion in the prompt may be puzzling to you: can't any selection mode be used? No. For some peculiar reason, the original **TRIM** command allowed you only to pick objects singly for trimming.

The **Fence** option prompts:

Specify first fence point: *(Pick a point.)*

Specify next fence point or [Undo]: *(Pick another point.)*

And continues repeating the prompt until you press ENTER. You can back up along the fence path by entering the **Undo** option.

Using the **Crossing** option can be ambiguous; Autodesk notes that AutoCAD follows the rectangular crossing window clockwise from the first pick point. The **Crossing** option prompts:

> Specify first corner: *(Pick a point.)*

> Specify opposite corner: *(Pick another point.)*

The crossing window must cross objects; otherwise nothing is trimmed. When you pick an object at the 'Specify first corner:' prompt, AutoCAD abandons **Crossing** mode, and selects the object instead.

Project

The **Project** option is used in 3D drafting to specify the projection AutoCAD uses when it trims objects.

> Enter a projection option [None/Ucs/View]: *(Enter an option.)*

The sub-options are:

None specifies no projection; AutoCAD trims only objects that actually intersect with cutting edges.

UCS projects the objects onto the x, y-plane of the current UCS (user-defined coordinate system). AutoCAD trims objects, even if they do not intersect with the cutting edge in 3D space—in effect, flattening 3D objects onto the 2D plane.

View specifies a projection along the current view direction. AutoCAD trims objects that look as if they should be trimmed from your viewpoint, even if they don't physically intersect.

Edge

The **Edge** option is also used in 3D drafting, and determines whether objects are trimmed at an actual cutting edge (objects cross physically), or at an *implied* cutting edge (objects appear to cross):

> Enter an implied edge extension mode [Extend/No extend]: *(Enter an option.)*

The options are:

Extend projects the cutting edge so that it intersects the object(s).

No extend trims objects at cutting edges that physically intersect; cutting edges are not extended.

eRase

The **eRase** option erases objects, just like using the ERASE command itself. It is a convenient way of removing whole objects that don't require trimming.

Undo

The **Undo** option undoes the last trim. This allows you to undo trim errors without leaving the command.

Extend (SHIFT-select)

When you hold down the SHIFT key at the "Select object to trim or shift-select to extend" prompt, AutoCAD extends the object to the cutting line, instead of trimming — in effect, reversing the TRIM command. The complementary option is available in the EXTEND command.

EXTEND

The EXTEND command extends objects in a drawing to meet a boundary object. It functions very much like the TRIM command. Instead of a cutting edge, this command uses *boundary* objects — the boundary to which selected objects are extended.

AutoCAD needs to know two things: (1) the object(s) to be used as boundary objects, and (2) the object to extend.

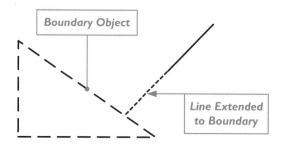

Boundary objects can be lines, arcs, circles, polylines, ellipses, splines, rays, xlines, regions, blocks, and viewports (in paper space). If several boundary edges are selected, the objects are extended to the first boundary encountered. If none can be reached, AutoCAD complains:

> No edges selected.

Generally, closed objects cannot be extended, specifically text, splines, 3D solids, xlines, mutlilines, ellipses, donuts, hatches, revclouds, regions, 3D surfaces, and points. (Dimensions can be extended.) If the object cannot be extended, AutoCAD displays:

> Cannot EXTEND this object.

Take care! You must point close to the end of the object from which to extend.

TUTORIAL: EXTENDING OBJECTS

1. To extend one or more objects, start the **EXTEND** command:
   * From the **Modify** menu, choose **Extend**.
   * From the Modify toolbar, choose the **Extend** button.
   * In the Dashboard's 2D Draw panel, pick the **Extend** button.
   * At the 'Command:' prompt, enter the **extend** command:

     Command: **extend** *(Press ENTER.)*

   * Alternatively, enter the **ex** alias at the 'Command:' prompt.

2. In all cases, AutoCAD reports the current extend settings, and then prompts you to select boundary edges:

   Current settings: Projection=UCS Edge=None

   Select boundary edges ...

   Select objects or <select all>: *(Pick one or more objects as boundaries.)*

   Select objects: *(Press ENTER to end object selection.)*

3. Select the object to be extended. AutoCAD extends the end that you select.

   Select object to extend or shift-select to trim or [Fence/Crossing/Project/Edge/Undo]: *(Pick objects to be extended.)*

4. AutoCAD repeatedly prompts you to select additional objects to extend. When done extending, press **ENTER**:

   Select object to extend or shift-select to trim or [Fence/Crossing/Project/Edge/Undo]: *(Press ENTER to exit command.)*

Extending Polylines

Polylines of nonzero width are extended until the centerline meets the boundary object; similarly, objects extend to the centerline of the polylines. Only open polylines can be extended. If you attempt to extend a closed polyline, AutoCAD complains:

> Cannot extend a closed polyline.

When wide polylines and the boundary intersect at an angle, a portion of the square end of the polyline may protrude over the boundary. Extending tapered polylines adjusts the length of the segments that taper — the taper extends over the longer length.

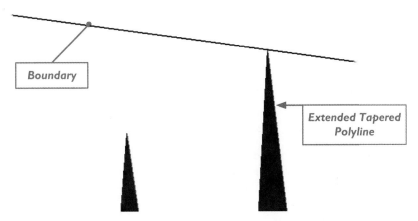

A tapered polyline extended to a boundary, its taper extended as well.

EXTEND OPTIONS

The **EXTEND** command's options are identical to those of **TRIM**'s.

LENGTHEN

The **LENGTHEN** command is a faster version of **TRIM** and **EXTEND**: it changes the length of open objects, making them longer or shorter; it does not work with closed objects. **LENGTHEN** is faster, because you do not need cutting edges or boundaries: you just point anywhere in the drawing for the change to occur, or else specify a percentage change numerically.

This command works with lines, arcs, open polylines, elliptical arcs, and open splines.

TUTORIAL: LENGTHENING OBJECTS

1. To change the length of an open object, start the **LENGTHEN** command:
   - From the **Modify** menu, choose **Lengthen**.
   - At the 'Command:' prompt, enter the **lengthen** command:

     Command: **lengthen** *(Press ENTER.)*
   - Alternatively, enter the **len** alias at the 'Command:' prompt.

2. In all cases, AutoCAD prompts you to select an object. AutoCAD reports the length of the object:

   Select an object or [DElta/Percent/Total/DYnamic]: *(Select one open object.)*

   Current length: 24.6278

3. AutoCAD repeats the prompt. It wants you to select an option:

   Select an object or [DElta/Percent/Total/DYnamic]: *(Enter an option, such as **dy**.)*

4. Curiously, AutoCAD has forgotten which object you selected; you need to select it a second time — or, you can select a different object:

   Select an object to change or [Undo]: *(Select one open object.)*

5. As you move the cursor, notice that AutoCAD ghosts in the lengthened line. When satisfied with the new length, click.

   Specify new end point: *(Move the cursor, and then click.)*

6. AutoCAD repeats the prompt. When done lengthening, press **ENTER**:

   Select an object to change or [Undo]: *(Press ENTER to exit the command.)*

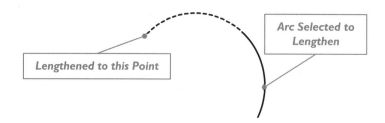

Lengthened to this Point

Arc Selected to Lengthen

Note: When you select an object at the initial "Select an object:" prompt, AutoCAD reports on its length, as follows: *Current length: 10.2580.* For an arc, AutoCAD also reports its angle:

Current length: 8.4353, included angle: 192.

When you select a closed object, such as a circle or closed polyline, AutoCAD complains, "This object has no length definition." You cannot lengthen dimensions, hatch patterns, and splines.

LENGTHENING OBJECTS: ADDITIONAL METHODS

The LENGTHEN command changes the length of open objects by four methods:

Delta — changes by an incremental amount.

Percent — changes by a percentage.

Total — changes to the total amount.

DYnamic — changes by cursor movement.

Let's look at how each option.

Delta

The **DElta** option changes the length by adding the indicated amount to the object. Enter a negative number to shorten the object. As an alternative to typing a value, you can indicate the amount by picking two points anywhere in the drawing.

> Command: **lengthen**
>
> Select an object or [DElta/Percent/Total/DYnamic]: **de**
>
> Enter delta length or [Angle] <0.0000>: *(Enter a number, or pick two points on screen.)*
>
> Select an object to change or [Undo]: *(Select object.)*
>
> Select an object to change or [Undo]: *(Press ESC to exit the command.)*

Note that this command lengthens the open object at the end you select.

The **Angle** option changes the angle of a selected arc or polyarc. The command lengthens the end you select:

> Enter delta angle <0>: *(Type an angle, or pick two points.)*
>
> Select an object to change or [Undo]: *(Select the arc.)*
>
> Select an object to change or [Undo]: *(Press ESC to exit the command.)*

The arc is lengthened by the angle you specify. When you select two points, AutoCAD uses the angle of the rubber-band line, not the length of the line.

The angle you specify must be small enough for the arc not to total 360 degrees. If the angle is too large, AutoCAD curtly informs you: "Invalid angle."

Percent

The **Percent** option changes the length by a percentage of the object. For example, entering **25** shortens a line to 25 percent of its original length, while entering **200** doubles its length.

Select an object or [DElta/Percent/Total/DYnamic]: **p**

Enter percentage length <100.0000>: *(Enter a percentage, such as* **25**.*)*

Select an object to change or [Undo]: *(Select the object.)*

Select an object to change or [Undo]: *(Press* **ESC** *to exit the command.)*

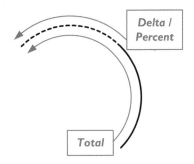

Total

The **Total** option changes the length of a line by an absolute length. For example, a value of **5** changes the line to a length of 5.0 units, no matter its existing length.

Select an object or [DElta/Percent/Total/DYnamic]: **t**

Specify total length or [Angle] <1.0000>: *(Enter a value, such as* **5**.*)*

Select an object to change or [Undo]: *(Select the object.)*

Select an object to change or [Undo]: *(Press* **ESC** *to exit the command.)*

This option changes arcs to an absolute length. The **Angle** option changes the length of an arc to the included specified angle.

Dynamic

The **DYnamic** option visually changes the length of open objects. Notice that this option reverses the order: (1) first you select the object to lengthen, and (2) then you specify the length.

Select an object or [DElta/Percent/Total/DYnamic]: **dy**

Select an object to change or [Undo]: *(Select the object.)*

Specify new end point: *(Pick a point.)*

Select an object to change or [Undo]: *(Press* **ESC** *to exit the command.)*

After you select the object, the length of the object changes as you move the cursor. You can use object snap modes to make the dynamic lengthening more accurate.

 ## STRETCH

The **STRETCH** command makes portions of objects larger and smaller, *and* moves selected objects, while retaining their connections to other objects. As you will see, it is one of the most useful editing commands in AutoCAD — a distant second only to the **UNDO** command. The command works with lines, arcs, elliptical arcs, solids, traces, rays, splines, and polylines.

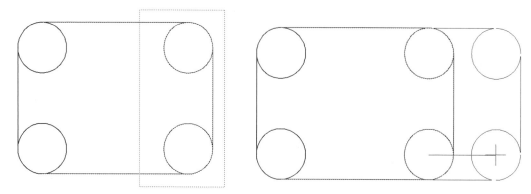

Objects within the selection window being moved; those crossing the window being stretched.

STRETCHING — THE RULES

You must understand the rules associated with the **STRETCH** command to execute it properly:

Rule 1: Crossing Selection. You can use any object selection mode, but at least one must be **Crossing** or **CPolygon**. If you do not use a *windowed* selection, AutoCAD complains, "You must select a crossing or polygon window to stretch." The command is no longer restricted to using the last window specified; instead, **STRETCH** now remembers all the objects selected by multiple **Crossing** selection rectangles. The window must include at least one vertex or endpoint of the object.

Rule 2: Move vs. Stretch. If the object is completely inside the selection, it will be moved rather than stretched. Objects — specifically arcs, elliptical arcs, lines, polyline segments, 2D solids, rays, traces, and splines — entirely within the selection window are moved as with the **MOVE** command. The endpoints within the selection window are moved, while those outside remain fixed.

Arcs are stretched like lines, except that the arc's center, start, and endpoints are adjusted so the distance from the midpoint of the chord to the arc is constant. You can stretch either end of the polyline, and move its vertices. Spline-curve and fit-curve polylines can be stretched if the crossing window includes the original vertex (even though it is invisible). When tangent-arc polyline segments are stretched, they lose tangency.

This command does not stretch 3D solids or text.

Rule 3: Definition Points. Some objects, such as circles and blocks, cannot be stretched; they are either moved or left alone by the **STRETCH** command, depending on their *definition point*. If this point lies inside the selection window, the object is moved; if outside, it is not affected.

Object	Definition Point
Point	Center of the point.
Circle	Center point of the circle.
Block	Insertion point.
Text	Insertion point of the text line.

TUTORIAL: STRETCHING OBJECTS

1. To stretch one or more objects, start the **STRETCH** command:
   - From the **Modify** menu, choose **Stretch**.
   - From the Modify toolbar, choose the **Stretch** button.
   - In the Dashboard's 2D Draw panel, pick the **Stretch** button.
   - At the 'Command:' prompt, enter the **stretch** command:

     Command: **stretch**

- Alternatively, enter the **s** alias at the 'Command:' prompt.

2. In all cases, AutoCAD prompts you to select the objects using a crossing selection mode. If you do not enter a selection option, such as **Crossing**, AutoCAD defaults to **AUtomatic**.

   Select objects to stretch by crossing-window or crossing-polygon...

   Select objects: **c**

   Specify first corner: *(Pick a point.)*

   Specify opposite corner: *(Pick another point.)*

   Select objects: *(Press ENTER to end object selection.)*

3. Identify the points that measure the displacement:

   Specify base point or Displacement <Displacement>: *(Pick a point, using an object snap mode if necessary.)*

   Specify second point of displacement: *(Pick the destination point.)*

For the **Displacement** option, you can enter Cartesian, polar, cylindrical, or spherical coordinates — without the @ prefix, because AutoCAD assumes relative coordinates. The next time you use the STRETCH command, enter the **Displacement** option to recall the previous displacement distance.

An alternative to the STRETCH command is grips editing, as described in Chapter 9.

 MOVE

The MOVE command moves objects in the drawing.

It is similar to the COPY command in that AutoCAD needs to know two things: (1) the objects to move, and (2) the displacement — the distance to move (see Chapter 10, "Constructing Objects").

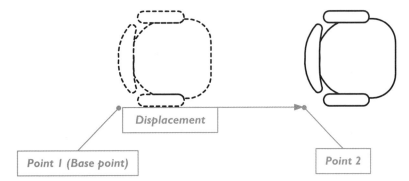

Displacement

Point 1 (Base point)

Point 2

TUTORIAL: MOVING OBJECTS

1. To move one or more objects, start the MOVE command:
   - From the **Modify** menu, choose **Move**.
   - From the Modify toolbar, choose the **Move** button.
   - In the Dashboard's 2D Draw panel, pick the **Move** button.
   - At the 'Command:' prompt, enter the **move** command:

     Command: **move** *(Press ENTER.)*
   - Alternatively, enter the **m** alias at the 'Command:' prompt.

2. In all cases, AutoCAD prompts you to select the objects to move:

   Select objects: *(Select one or more objects.)*

   Select objects: *(Press ENTER to end object selection.)*

3. Identify the point from which the displacement is measured:

> Specify base point or Displacement <Displacement>: *(Pick point 1.)*

4. And pick the second point:

> Specify second point of displacement or <use first point as displacement>: *(Pick point 1.)*

The first point need not be on the object to be moved; using a corner point or another convenient point of reference on the object makes the displacement easier to visualize. I find that object snap and ortho mode help make the move more precise.

The next time you use the MOVE command, use the **Displacement** option to recall the previous displacement distance.

An alternative to the MOVE command is grips editing, as described in Chapter 9. The GTDEFAULT system variable determines whether the MOVE or MOVE3D command is activated when entering "move" in 3D views; the same holds for ROTATE and ROTATE3D.

Command-less Moving

You can move objects in AutoCAD without using any commands. Here's how:

1. Select an object. Notice the blue grips.
2. Grab the object away from the grips by holding down the left mouse button.
3. Wait.

4. After a second or two, the move cursor appears. You can now drag the object in the drawing, effectively moving it.

 **ROTATE**

The ROTATE command rotates objects about a base point. (The *base point* is the point about which selected objects rotate.)

You can specify the rotation angle three ways: by entering an angle, by dragging the angle, or by choosing a reference angle. (The similar sounding REVOLVE command creates 3D solid models.)

TUTORIAL: ROTATING OBJECTS

1. To turn one or more objects, start the ROTATE command:
   - From the **Modify** menu, choose **Rotate** .
   - From the Modify toolbar, choose the **Rotate** button.
   - In the Dashboard's 2D Draw panel, pick the **Rotate** button.
   - At the 'Command:' prompt, enter the **rotate** command:

     > Command: **rotate** *(Press ENTER.)*

   - Alternatively, enter the **ro** alias at the 'Command:' prompt.

2. In all cases, AutoCAD indicates the current rotation settings, and then prompts you to select the objects you want rotated:

   > Current positive angle in UCS: ANGDIR=counterclockwise ANGBASE=0

   > Select objects: *(Pick one or more objects.)*

   > Select objects: *(Press ENTER to end object selection.)*

3. Identify the point about which the objects are rotated:

   Specify base point: *(Pick a point, such as the lower left corner.)*

   The base point need not be on the object, but can be anywhere in the drawing.

4. Specify the angle of rotation:

   Specify rotation angle or [Copy/Reference] <90>: *(Enter an angle, such as **45**.)*

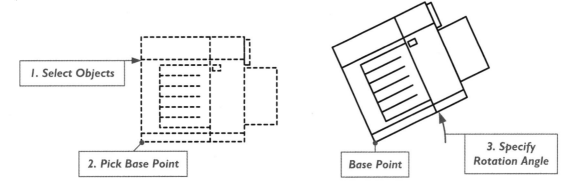

1. Select Objects

2. Pick Base Point

Base Point

3. Specify Rotation Angle

You can specify a simple angle, or change one angle to another. For example, if an object is currently oriented at 58 degrees, and you wish to rotate the object to 26 degrees, rotate the object by the difference of –32 degrees. Positive angles rotate counterclockwise; negative angles, clockwise.

ROTATING OBJECTS: ADDITIONAL METHODS

The **ROTATE** command has alternative options for specifying the angle. In addition, rotation is affected by two system variables.

- **Reference** option rotates relative to another angle.

- **Copy** option rotates a copy of the object.

- **Dragging** option rotates the object in real-time.

- **ANGDIR** system variable specifies the direction of positive angles.

- **ANGBASE** system variable specifies the direction of 0 degrees.

Let's look at each.

Reference

The **Reference** option aligns the object to another. You don't need to know the angle of the other object, but object snaps are most useful. In this example, we want Desk B to be at the same slant as Desk A.

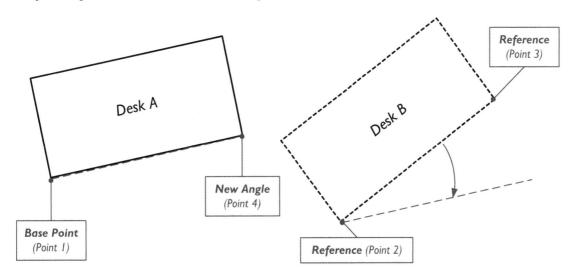

Desk A

Desk B

Reference (Point 3)

New Angle (Point 4)

Base Point (Point 1)

Reference (Point 2)

Specify base point: **end**

of *(Pick point 1.)*

Specify rotation angle or [Reference]: **r**

Specify the reference angle <0>: **end**

of *(Pick point 2.)*

Second point: : **end**

of *(Pick point 3.)*

Specify the new angle: **end**

of *(Pick point 4.)*

Copy

The **Copy** option rotates a copy of the object, leaving the original in place. When you enter **c** at the 'Specify rotation angle or [Copy/Reference] <90>:' prompt, AutoCAD reports:

Rotating a copy of the selected objects.

The option is temporary. The next time you use the **ROTATE** command, you have to reenter the **Copy** option.

Dragging

To rotate the objects in real-time, move the cursor in response to the "Specify rotation angle or [Reference]:" prompt. Keep an eye on the status line as it reports the x, y, z coordinates and angle. Click to fix the object at the location shown on the screen.

AngBase

The **ANGBASE** system variable sets the direction of 0 degrees. In AutoCAD, 0 degrees can point in any direction; the default is at 3 o'clock or east, in the direction of the positive x-axis. To change the position of zero degrees, enter a new angle, as follows:

Command: **angbase**

Enter new value for ANGBASE <0>: *(Enter an angle, such as **90**.)*

Entering 90, for example, rotates 0 degrees by 90 degrees, so that it now points to 12 o'clock, or north.

AngDir

The **ANGDIR** system variable specifies the direction of positive angles. It takes on two values: 0 means positive angles are measured counterclockwise, while 1 means they are measured clockwise.

Command: **angdir**

Enter new value for ANGDIR <0>: *(Enter **0** or **1**.)*

Note: Both **ANGBASE** and **ANGDIR** are usually not changed, but is it is helpful to know about them in case your drafting discipline requires that angles be represented differently, or AutoCAD isn't responding to angle input as you expect. For example, in surveying, 0 degrees points North, and angles are measured clockwise.

An alternative to the **ROTATE** command is grips editing, as described in Chapter 9.

 **SCALE**

The SCALE command changes the size of objects. Contrary to what the name suggests, this command does not change the *scale* of objects in the drawing — but their *size*. The x, y, and z directions of the objects are changed equally.

AutoCAD needs to know three things: (1) the objects to be resized, (2) the base point from which the objects are resized, and (3) the amount to resize.

TUTORIAL: CHANGING SIZE

1. To resize one or more objects, start the SCALE command:
    - From the **Modify** menu, choose **Scale**.
    - From the Modify toolbar, choose the **Scale** button.
    - In the Dashboard's 2D Draw panel, pick the **Scale** button.
    - At the 'Command:' prompt, enter the **scale** command:

      Command: **scale** *(Press ENTER.)*
    - Alternatively, enter the **sc** alias at the 'Command:' prompt.

2. In all cases, AutoCAD prompts you to select the objects you want resized:
    Select objects: *(Pick one or more objects.)*
    Select objects: *(Press ENTER to end object selection.)*

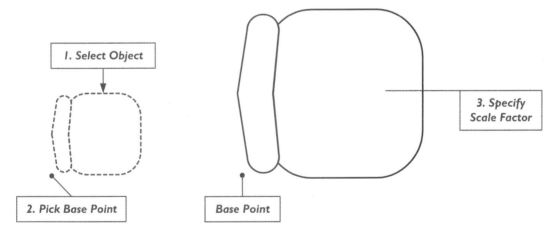

3. Identify the point about which the objects are rotated:
    Specify base point: *(Pick a point, such as the lower left corner.)*

    The base point can be located anywhere in the drawing. This point remains stationary, and the object is resized from that point.
4. Specify the resize factor:
    Specify scale factor or [Copy/Reference] <1.0000>: *(Enter a factor, such as **2** or **0.5**.)*

Entering a decimal factor makes objects smaller: a factor of 0.25 makes an object 25 percent of the original size. Factors larger than 1.0 increase the size of the objects. For example, entering 2.0 makes the object twice the original size in the direction of all three axes.

Instead of entering the scale factor, you can specify the size by dragging the cursor or by referencing a known length, and then entering a new length.

CHANGING SIZE: ADDITIONAL METHODS

The SCALE command has two options:

- **Copy** rotates a copy of the objects.

- **Reference** changes the size relative to other distances.

Let's look at them.

Copy

The **Copy** option scales a copy of the objects, leaving the original in place. When you enter **c** at the 'Specify scale factor or [Copy/Reference] <1.0>:' prompt, AutoCAD reports:

Scaling a copy of the selected objects.

The option is temporary. The next time you use the SCALE command, you have to reenter the **Copy** option.

Reference

The **Reference** option adjusts objects to a "correct" size. This is particularly useful when you scan a drawing, and then bring the raster image into AutoCAD with the IMAGE command. To trace over the scanned image, you first have to change the image to the right size: you achieve this with the **Reference** option. All you need is the true length of just one line in the scanned image.

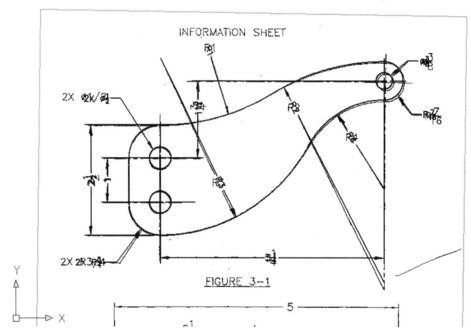

The **Reference** option of the SCALE and ROTATE commands
correcting the size and skew of scanned images.

1. At the "Select objects:" prompt, select the raster image:
   Command: **scale**

   Select objects: *(Pick the edge of the raster image.)*

   Select objects: *(Press ENTER.)*

2. For the base point, select the lower left corner of the raster image.
   Specify base point: *(Pick.)*

3. Specify the **Reference** option:
   Specify scale factor or [Copy/Reference] <1.0>: **r**

4. Select the end of the known line in the raster image.

     Specify reference length <1>: *(Pick one end of the raster line.)*

5. Select the other end of the raster line; this allows AutoCAD to measure the distance of the line:

     Specify second point: *(Pick other end of raster line.)*

6. For the new length, enter the value of the known dimension.

      Specify new length: *(Enter value.)*

AutoCAD resizes the scanned image, making it the correct size. Use the **ROTATE** command to correct a skewed image. (If the scanned image is warped, you need to correct it with "rubber sheeting," a feature not available in AutoCAD.) To change the aspect ratio of an image, convert it to a block, and then insert with differing x and y scale factors; explode the block, and the image maintains its new aspect ratio.

The **SCALE** command is also useful for converting an Imperial drawing to metric, and vice versa. An alternative to the **SCALE** command is grips editing, as described in Chapter 9.

CHANGE

The **CHANGE** command changes the size, properties, and other characteristic of objects. It is similar to, but more advanced than, **CHPROP**, but **PROPERTIES** is the easiest and most versatile of all. (See Chapter 7, "Changing Object Properties.")

TUTORIAL: CHANGING OBJECTS

1. To change one or more objects, start the **CHANGE** command:
   - At the 'Command:' prompt, enter the **change** command:

     Command: **change** *(Press ENTER.)*

   - Alternatively, enter the **-ch** alias at the 'Command:' prompt.

2. In all cases, AutoCAD prompts you to select the complex objects you want to change:

     Select objects: *(Pick one or more objects.)*

     Select objects: *(Press ENTER to end object selection.)*

3. The **Properties** options are somewhat different from the **CHPROP** command:

     Specify change point or [Properties]: *(Pick a point.)*

The **CHANGE** command's **Elev** option changes the elevation of objects, an option missing from **CHPROP**.

Although the Properties window is generally handier than this command-line command, **CHANGE** offers advantages: the **Change Point** option reacts differently, depending on the object selected, as described next.

CHANGING OBJECTS: ADDITIONAL METHODS

Here's how the **CHANGE** command's **Change Point** option affects different objects:

- **Lines** change their endpoints.
- **Circles** change their radius.
- **Blocks** change their insertion point and rotation angle.
- **Text** changes its properties.

- **Attribute Text** changes its properties.

Let's look at each.

Circles

The **Change Point** option changes the circle's radius. When more than one circle is selected, AutoCAD repeats the prompt for each circle:

> Specify new circle radius <no change>: *(Enter a new radius, or press* ENTER.*)*

Press ENTER to keep the circle's radius.

Lines

The **Change Point** option moves the endpoints of lines closest to the pick point.

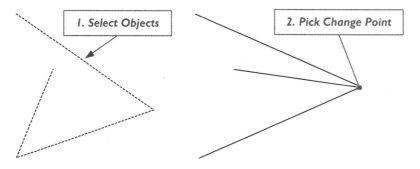

When ortho mode is on, AutoCAD makes the lines parallel to either the x or y axis, depending on which is closest. This is a quick way to straighten out a bunch of crooked lines, and is not available in any other AutoCAD command.

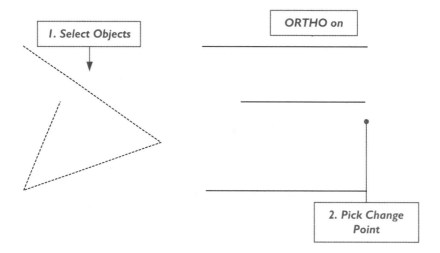

Blocks

The **Change Point** option changes the location and rotation of blocks.

> Specify new block insertion point: *(Pick a new insertion point, or press* ENTER.*)*
>
> Specify new block rotation angle <current>: *(Enter a new angle, or press* ENTER.*)*

Press ENTER to keep each option in place.

Text

The **Change Point** option changes the position of the text, as well as the text's other properties. (Until the PROPERTIES command's predecessor, DDMODIFY, came along, this was the only way to change text properties in a drawing.)

Specify new text insertion point <no change>: *(Pick a new insertion point, or press* ENTER.*)*

Enter new text style <current>: *(Enter a different text style, or press* ENTER.*)*

Specify new height <current>: *(Enter a new height, or press* ENTER.*)*

Specify new rotation angle <current>: *(Enter a new angle, or press* ENTER.*)*

Enter new text <current>: *(Enter a new line of text, or press* ENTER.*)*

Press ENTER to keep each option as is. If you select more than one line of text, AutoCAD repeats the prompts for the next one.

Attribute Definitions

The **Change Point** option changes the text and properties of attributes that are not part of a block.

Specify new text insertion point: *(Pick a new insertion point, or press* ENTER.*)*

Enter new text style <current>: *(Enter a different text style, or press* ENTER.*)*

Specify new height <current>: *(Enter a new height, or press* ENTER.*)*

Specify new rotation angle <current>: *(Enter a new angle, or press* ENTER.*)*

Enter new text <current>: *(Enter new text, or press* ENTER.*)*

Enter new tag <current>: *(Enter a new tag, or press* ENTER.*)*

Enter new prompt <current>: *(Enter a new prompt, or press* ENTER.*)*

Enter new default value <current>: *(Enter a new default value, or press* ENTER.*)*

Press ENTER to keep each option in place.

EXPLODE AND XPLODE

The **EXPLODE** and **XPLODE** commands reduce compound objects to their simplest forms.

The commands "break down" blocks, polylines, and other compound objects into basic lines, arcs, and text. Some objects must be exploded several times before they are finally reduced to vector primitives.

TUTORIAL: EXPLODING COMPOUND OBJECTS

1. To explode one or more compound objects, start the **EXPLODE** command:
   - From the **Modify** menu, choose **Explode**.
   - From the Modify toolbar, choose the **Explode** button.
   - In the Dashboard's 2D Draw panel, pick the **Explode** button.
   - At the 'Command:' prompt, enter the **explode** command:

     Command: **explode** *(Press* ENTER.*)*
   - Alternatively, enter the **x** alias at the 'Command:' prompt.

2. In all cases, AutoCAD prompts you to select the compound objects you want to explode:

   Select objects: *(Pick one or more objects.)*

   Select objects: *(Press* ENTER *to end object selection.)*

The objects are exploded, but you might not see any difference. Different compound objects react differently to being exploded.

Arcs

Arcs do not explode, unless they are part of a nonuniformly scaled block. When the block is exploded, arcs become elliptical arcs.

Note: When a block is inserted with different scale factors in the x, y, and/or z directions, it is called a "nonuniformly scaled block." When exploded, the results may be different from what you expect. Arcs, for example, are converted to elliptical arcs to maintain their distortion.

Sometimes, nonuniformly scaled blocks contain objects that cannot be exploded, such as bodies, 3D solids, and regions. In that case, AutoCAD places them in an *anonymous* block with the ***E** prefix in the block's name.

Circles

Circles do not explode, unless they are part of a nonuniformly scaled block. When the block is exploded, circles becomes ellipses.

Blocks

Blocks explode into the lines and other objects originally used to define them. Nested blocks must be exploded a second time. Attribute text is deleted; attribute definitions are redisplayed.

Xrefs and blocks inserted with the **MINSERT** command cannot be exploded.

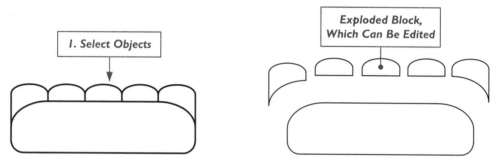

Left: Block before being exploded.
Right: After exploding, with parts pulled apart for clarity.

Polylines

Polylines explode into lines and arcs. Width and tangent information is discarded; the resulting lines and arcs follow the center line of the old polyline. If the exploded polyline has segments of width, AutoCAD displays the message:

Exploding this polyline has lost width information.

The UNDO command will restore it.

The new lines and arcs are placed on the same layer as the polyline, and inherit the same color.

Left: Polyline with variable width before being exploded.
Right: Polyline after being exploded, with lines and arcs pulled apart for clarity.

3D Polylines

Three-dimensional polylines explode into line segments. Linetypes are retained.

Dimensions

Associative dimensions explode into basic objects (lines, polyline arrowheads, and text), which are placed on the same layer as the original dimensions, and inherit that layer's properties.

Leaders

Leaders explode into lines and multiline text; the arrowheads explode into 2D solids. Depending on how the leader was constructed, resulting objects could also include splines, block inserts of arrowheads and annotation blocks, and tolerance objects. More than one explode may be required.

Multiline Text

Multiline text explodes into single-line text.

Multilines

Multilines explode into lines and arcs.

Polyface Meshes

Single-vertex polyface meshes explode into point objects. Two-vertex meshes explode into lines. Three-vertex meshes, into 3D faces.

Tables

Tables explode into lines and mtext.

3D Solids and Surfaces

3D solids explode into bodies, which can be further exploded into regions. Regions, in turn, explode into lines, arcs, and circles. Surfaces explode into lines, arcs, circles, and polylines.

XPLODE

The XPLODE command controls what happens when compound objects are exploded.

TUTORIAL: CONTROLLED EXPLOSIONS

1. To explode compound objects with control, start the **XPLODE** command:
   - At the 'Command:' prompt, enter the **xplode** command.

     Command: **xplode** *(Press* ENTER.*)*

2. In all cases, AutoCAD prompts you to select the compound objects you want to explode:

   Select objects to XPlode.

   Select objects: *(Pick one or more objects.)*

   Select objects: *(Press* ENTER *to end object selection.)*

3. Enter an option:

   Enter an option

   [All/Color/LAyer/LType/LWeight/Inherit from parent block/Explode] <Explode>: *(Enter an option.)*

4. If you had selected more than one object, an additional prompt appears:

   Enter an option [Individually/Globally] <Globally>: *(Enter* **I** *or* **G**.*)*

The **Global** option applies the previous options (color, layer, and so on) to all selected objects. The **Individually** option applies the previous options to each object, one at a time.

All

The **All** option prompts you to specify the color, linetype, lineweight, and layer on which the exploded objects should land. The prompts are the same as those for the four following options.

Color

The **Color** option sets the color of exploded objects:

> [Red/Yellow/Green/Cyan/Blue/Magenta/White/BYLayer/BYBlock/Truecolor/Colorbook]
> <BYLAYER>: *(Enter a color name or number.)*

Layer

The **Layer** option sets the layer of the exploded objects:

> Enter new layer name for exploded objects <current>: *(Enter the name of an existing layer.)*

LType

The **LType** option sets the linetype of the exploded objects:

> Enter new linetype name for exploded objects <BYLAYER>: *(Enter the name of a linetype.)*

Inherit from Parent Block

The **Inherit from parent block** option sets the color, linetype, lineweight, and layer of the exploded objects to that of originating block.

Explode

The **Explode** option reduces compound objects in the same manner as the **EXPLODE** command.

 PEDIT

The **PEDIT** command (short for "polyline edit") edits polylines.

This command also edits 3D polylines drawn with the **3DPOLY** command, and 3D polyfaces drawn with AutoCAD's 3D surfacing commands.

TUTORIAL: EDITING POLYLINES

1. To edit one or more polylines, start the **PEDIT** command:
   - From the **Modify** menu, choose **Objects**, and then **Polyline**.
   - From the Modify II toolbar, choose the **Edit Polyline** button.
   - At the 'Command:' prompt, enter the **pedit** command:

     Command: **pedit** *(Press ENTER.)*
   - Alternatively, enter the **pe** alias at the 'Command:' prompt.

2. In all cases, AutoCAD prompts you to select the polyline you want edited:
   Select polyline or [Multiple]: *(Pick a polyline.)*

3. Many options are listed:
   Enter an option [Close/Join/Width/Edit vertex/Fit/Spline/Decurve/Ltype gen/Undo]: *(Enter an option, or press ESC to exit the command.)*

As a reminder, here are the parts of a polyline:

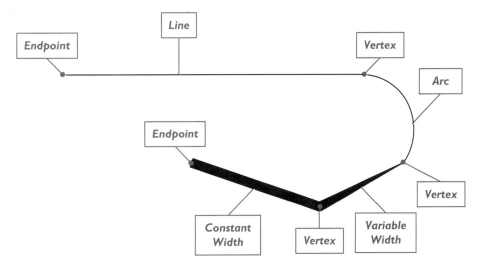

Close/Open

The **Close** option closes open polylines. It connects the last point to the first point of the polyline. This is like the **PLINE** command's **Close** option, except it allows you to close the polyline after exiting **PLINE**.

The **Open** option replaces the **Close** option when the polyline is closed. It opens closed polylines by removing the last segment created by the **Close** option. If the polyline looks closed, but was not closed with **Close**, the **Open** option has no effect.

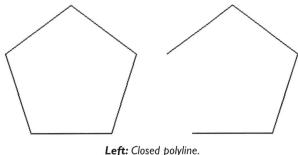

Left: *Closed polyline.*
Right: *Open polyline.*

Join

The **Join** option converts and connects polylines and non-polyline objects (lines and arcs only) to the original polyline. AutoCAD prompts:

Select objects: *(Select one or more lines, arcs, and polylines.)*

The objects you select become part of the original polyline; if the polyline was curve-fitted, it is first decurved.

AutoCAD determines which arcs and lines share common endpoints with the original polyline, and then merges them into that polyline. To join successfully, the objects must touch, or be within a *fuzz factor* distance, as described later in this chapter. If endpoints are too far away to join, use the **FILLET** or **CHANGE** command to extend them to a perfect match. For alternatives, see the **Multiple** and **Jointype** options discussed later in this chapter.

Width

The **Width** option changes the width of the entire polyline. Unlike the **PLINE** command's **Width** option, you can't have tapers or different widths for each segment; the width is applied uniformly to all segments of the polyline. AutoCAD prompts:

Specify new width for all segments: *(Enter a value.)*

You can type the width at the keyboard, or else pick two points in the drawing. The figure illustrates the effect of this option:

Left: A polyline with varying widths.
Right: The polyline after applying the **Width** option.

Edit vertex

The **Edit vertex** options edits the vertices of a polyline. (Recall that the vertex is the connection between polyline segments, as well as between the two endpoints of open polylines.) The name, "edit vertex," is somewhat misleading, because it also edits the segments between pairs of vertices.

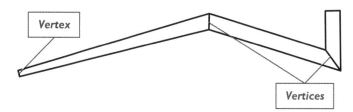

To keep track of which vertex is being edited, AutoCAD displays a marker in the shape of an X. Press N and P to move the marker to the next and previous vertex.

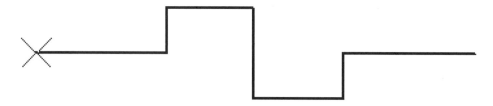

The **X** which marks the vertex.
*(Pressing **N** and **P** moves the marker from vertex to vertex.)*

A second marker, an arrow, is displayed when you work with the **Tangent** option.

Command: **pedit**

Select polyline or [Multiple]: *(Pick a polyline.)*

Enter an option [Close/Join/Width/Edit vertex/Fit/Spline/Decurve/Ltype gen/Undo]: **e**

[Next/Previous/Break/Insert/Move/Regen/Straighten/Tangent/Width/eXit]: *(Enter an option.)*

Note: Quite frankly, the **Edit vertex** options are a pain to use. In many cases, it much easier to edit the polyline using grips. Use these options only to insert and delete vertices. Use the Properties palette to set the widths of individual segments.

The many sub-options include the following:

Next/Previous

The **Next** option moves the x-marker to the next vertex, while the **Previous** option moves it to the previous vertex.

Break

The **Break** option removes one or more segments between two vertices. AutoCAD places the x-mark at the first vertex on one side of the break. You mark the other vertex, on the other side of the break, using the following options:

> Enter an option [Next/Previous/Go/eXit]: *(Enter an option.)*

Press **N** and **P** to move the x-marker to another vertex, and then press **G** to effect the break. Press **X** to exit this mode without breaking the polyline.

The entire segment is removed; to remove a portion of a segment, use the **BREAK** command. This option does not work under two conditions: when the same vertex is marked twice, so that no segment is selected for removal; or when the polyline's two endpoints are marked, in effect removing the entire polyline.

Insert

The **Insert** option adds a vertex to the polyline at your pick point. AutoCAD prompts:

> Specify location for new vertex: *(Pick a point anywhere in the drawing.)*

AutoCAD redraws the polyline so that it includes the new vertex.

Move

The **Move** option moves the marked vertex. AutoCAD prompts:

> Specify new location for marked vertex: *(Pick a point anywhere in the drawing.)*

Regen

The **Regen** option regenerates the polyline, without requiring you to exit the command. This option is required after applying the **Width** option.

Straighten

The **Straighten** option removes vertices, line segments, and arcs between two marked vertices, and replaces them with a single segment. It works like the **Break** option: AutoCAD remembers the x-marked vertex before you enter this option, and then you move the x-marker to another vertex.

> Enter an option [Next/Previous/Go/eXit]: *(Enter an option.)*

Press **N** and **P** to move the x-marker to another vertex, and the press **G** to straighten. Press **X** to exit this mode without changing the polyline.

Tangent

The **Tangent** option attaches a tangent direction to the marked vertex for use later with the **Fit** and **Spline** options. AutoCAD prompts:

> Specify direction of vertex tangent: *(Pick a point, or enter an angle.)*

AutoCAD displays an arrow pointing in the direction of tangency.

Width

The **Width** option changes the starting and ending widths for the segment immediately following the marked vertex. AutoCAD prompts:

> Specify starting width for next segment: *(Enter a width, or pick two points.)*

> Specify ending width for next segment: *(Enter a width, or pick two points.)*

Use the **Regen** option to see the new width.

Exit

The **Exit** option exits **Edit vertex** mode, and returns to PEDIT's original prompt line.

Fit

The **Fit** option constructs smooth curves from the vertices in the polyline. The curve consists of arcs joined at vertices, as illustrated by the figure. Notice how each segment has at least one arc associated with it. AutoCAD inserts extra vertices and arcs, where necessary; in the figure, two segments have two arcs each.

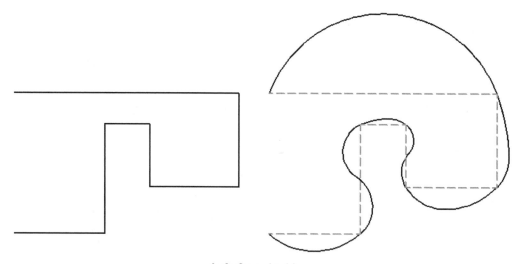

Left: *Original polyline.*
Right: *Polyline after applying the* **Fit** *curve option; the original polyline shown overlaid by gray dashed lines.*

Spline

The **Spline** option creates a spline curve from the polyline, using the vertices as the control points to approximate a B-spline curve. This is not a true B-spline, as constructed by the SPLINE command, but only approximates it.

When the original polyline also contains arcs, they are first converted to straight segments. When the original polyline also has widths, they are tapered from the first vertex to the last.

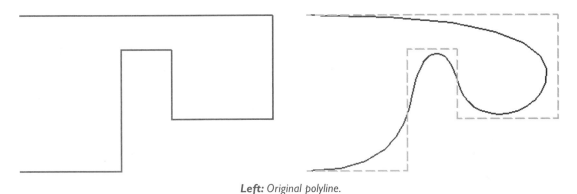

Left: *Original polyline.*
Right: *Polyline after applying the Spline option; the original polyline shown overlaid by gray dashed lines.*

Decurve

The **Decurve** option negates the effect of the two curve options — **Fit curve** and **Spline** — by straightening all segments of the polyline, as well as removing extra vertices that may have been added.

Ltype gen

The **Ltype gen** option (short for "Linetype generation") controls the generation of linetypes through the vertices of the polyline. Normally, AutoCAD generates the dashes, dots, and gaps for each segment of the polyline; the pattern starts and stops at each vertex.

Enter polyline linetype generation option [ON/OFF] <Off>: *(Enter* **ON** *or* **OFF**.*)*

When the option is turned on, the polyline is treated as a single segment, resulting in a uniform linetype pattern along the entire length of the polyline. In most cases, you probably want this option turned on.

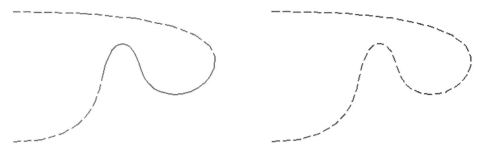

Left: *Polyline with **Ltype gen** turned off, part of the polyline lacking the line pattern.*
Right: *Polyline after turning on the option, all of the polyline patterned..*

 Note: Linetype generation does not work on polylines with tapered segments.

Undo

The **Undo** option reverses the last PEDIT operation without requiring you to exit the command to access the U command.

EDITING POLYLINES: ADDITIONAL METHODS

Of the **PEDIT** command's many options, some are not obvious, and some are accessed through system variables:

- **Select** option selects a line or arc.
- **Multiple** option selects more than one polyline.
- **Jointype** option determines how multiple objects are joined into a single polyline.
- **SPLFRAME** system variable toggles the spline frame.
- **SPLINETYPE** and **SPLINESEGS** system variables determine the spline's type and quality.

Let's look at each.

Select

The "Select polyline" prompt is misleading, because you can also select a non-polyline object — specifically a line or an arc. In that case, AutoCAD notices:

Command: **pedit**

Select polyline or [Multiple]: *(Pick a line or arc.)*

Object selected is not a polyline.

Do you want to turn it into one? <Y>: *(Enter **Y** or **N**.)*

When you respond with "Y", the object is converted to a polyline. If you had selected an object other than a line or arc, the conversion fails, and AutoCAD complains:

Object selected is not a polyline.

Mutiple

You can select more than one polyline to edit, but only by entering the **Multiple** option, as follows:

Command: **pedit**

Select polyline or [Multiple]: **m**

Select objects: *(Pick one or more polylines.)*

Select objects: *(Press **ENTER** to end polyline selection.)*

Jointype

The **Join** option merges lines, arcs, and polylines into a single polyline, but they all need to be touching. When you use the **Multiple** option at the start of the **PEDIT** command, however, then the objects *need not touch*! AutoCAD instead displays the following prompt:

Join Type = Extend

Enter fuzz distance or [Jointype]<0.0000>: *(Enter a distance, or type **J**.)*

The *fuzz distance* determines how far apart the objects can be, and still be joined to the original polyline. The **Jointype** option specifies how distant objects should join:

Enter a vertex editing option

Enter join type [Extend/Add/Both] <Extend>: *(Enter an option.)*

The **Extend** option joins the selected objects by extending and trimming them to fit.

The **Add** option joins the selected objects by bridging them with a straight segment.

The **Both** option joins the selected objects by extending or trimming; if that is not possible, AutoCAD adds the straight segment.

When you attempt to join two non-touching polylines, AutoCAD complains:

0 segments added to polyline

No matter which option you pick, fit-curved and splined polylines lose their curvature, and are converted back to the original frame.

SplFrame

The **SPLFRAME** system variable toggles the visibility of the *spline frame* — the original polyline from which the spline curve was created with the **PEDIT** command's **Spline** option.

The frame is normally turned off, because there isn't much need to see it; it just clutters the drawing. To see the frame, turn on the system variable, and follow that with a drawing regeneration:

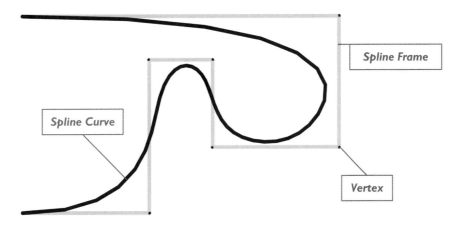

In black: *A splined polyline.*
In gray: *The original polyline, now its frame.*

Command: **splframe**

Enter new value for SPLFRAME <0>: **1**

Command: **regen**

Editing commands act differently with the spline and its frame:

Editing Command	Comment
COPY, ERASE, MIRROR, MOVE, ROTATE, SCALE	Affects both spline curve and frame, whether frame is visible or not.
EXTEND	Adds a vertex to the frame where it intersects the boundary.
BREAK, EXPLODE, TRIM	Deletes the frame, and generates a new spline.
OFFSET	Copies the spline.
STRETCH	Refits spline after spline and frame are stretched.
DIVIDE, MEASURE, HATCH, FILLET, CHAMFER, and AREA's **Object** option	Affects the spline curve only, not the frame.

SplineType

The **SPLINETYPE** system variable gives you the choice of quadratic and cubic spline-fit polylines: a value of 5 approximates quadratic B-splines, while a value of 6 approximates cubic B-splines (values of 1 through 4 having no meaning). Technically, a *B-spline* is a generalization of the Bézier curve; the quadratic B-spline produces a tighter curve than does the cubic B-spline.

Command: **splinetype**

Enter new value for SPLINETYPE <6>: *(Enter 5 or 6.)*

The setting is not retroactive; it comes into effect the next time you convert a polyline into a spline.

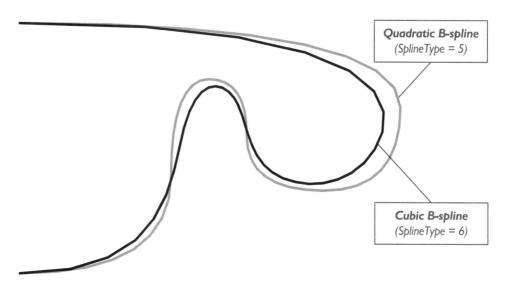

In black: Cubic B-spline polyline.
In gray: Quadratic B-spline.

SplineSegs

The **SPLINESEGS** system variable (short for "spline segments") controls the *apparent* smoothness of spline and fit-curves, apparent because AutoCAD approximates the spline curve with straight line segments. The default value is 8 segments between vertices; higher values construct more segments.

Command: **splinesegs**

Enter new value for SPLINESEGS <8>: *(Enter a value.)*

Setting **SPLINESEGS** to a negative value forces AutoCAD to change the spline to a fit-curve — the same as using the **PEDIT** command's **Fit** option.

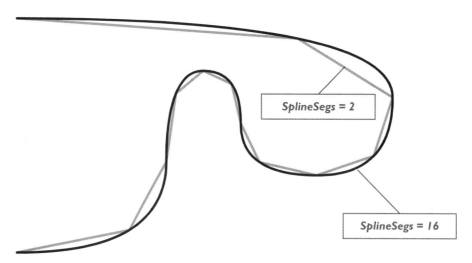

In black: 16 segments per vertex.
In gray: 2 segments per vertex.

 Note: The **SPLFRAME** system variable applies to all polylines in the drawing; AutoCAD cannot display the frames of selected polylines.

In contrast, the **SPLINETYPE** and **SPLINESEGS** system variables apply to polylines individually.

EXERCISES

1. In this exercise, you erase objects from a drawing, and then bring them back.

    From the CD, open the *edit1.dwg* drawing file, a valve housing.

    Use the **ERASE** command to delete some of the objects.

    Next, issue the **OOPS** command. Did the objects return?

    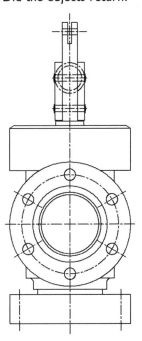

    Repeat the **ERASE** command with the **All** option. Did everything disappear?

    This time, use the **U** command to bring back the drawing.

2. In this exercise, you assemble a jigsaw puzzle.

    From the CD, open the *edit2.dwg* file, a drawing of puzzle pieces.

    Use the **MOVE** command to move the pieces into position, leaving a small space between them. The most effective method is to move the pieces roughly into position, then zoom in, and finely position the pieces using object snap.

    You may also want to use grips editing to move pieces into place.

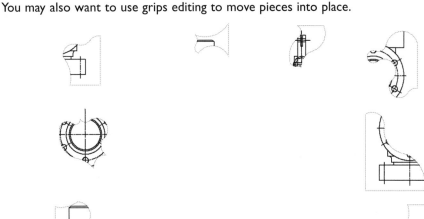

3. In this exercise, you practice breaking and trimming lines.

    From the CD, open the *edit5.dwg* drawing file, a piping diagram.

    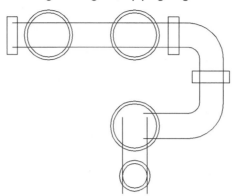

    Use the **BREAK** command to break each of the objects in the drawing, achieving the result shown.

    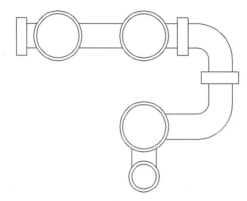

    Quit the drawing — don't save your work!

    Open *edit5.dwg* again; this time use the **TRIM** command to produce the same result as shown above.

    Here's a timesaving tip: at the "Select cutting edges" prompt, just press **ENTER** to select all objects in the drawing.

    You may need to use the **ERASE** command to clean up. Which command did you find easier — **BREAK** or **TRIM**?

4. In this exercise, you work with several editing commands.

    From the Companion CD, open the *edit7.dwg* drawing file, a piece of sheet metal.

    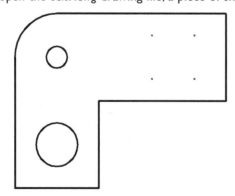

    Suppose that you are instructed to remove the lower circle from the drawing.

    From the **Modify** menu, use the **Erase** command.

    Now, erase the four points on the object by using a window selection.

Your boss changes his mind.

Put back the four points you just erased.

After erasing the lower circle, your boss decides the remaining circle should be moved down by 2.0 units.

Use CENter object snap, ortho mode, and direct distance entry to move it accurately.

After reviewing the drawing, your boss's boss feels the sheet metal needs two holes.

Add another circle, identical to the remaining circle. Use grips editing to copy the circle.

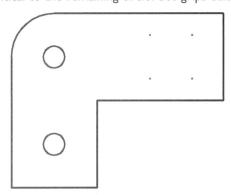

You learn from the product design department that the sheet metal is too large. A notch needs to be taken out.

Turn on snap mode, and set the snap spacing to 0.25. Use the **BREAK** and **LINE** commands to draw the notch.

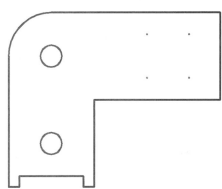

The product safety department reviews the drawing, and feels the sheet metal may be hazardous to children under three years of age. You are asked to round off two corners.

Add a radius of 0.5 to each of the two right corners.

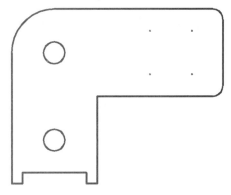

At last, all departments are satisfied. While the scenario in this exercise may seem silly, you will probably experience such changes to your drawings in the workplace.

Save the drawing.

5. In this exercise, you trim objects. Draw four intersecting lines about 6 units long and 2 units apart, as illustrated by the figure. Which command quickly makes the second, parallel line?

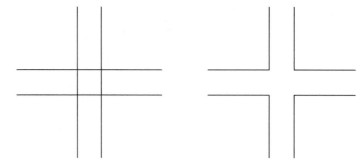

Left: *Four lines before trimming.*
Right: *After trimming.*

Use the **TRIM** command to trim the intersections. Remember to use the trick of responding **all** to the "Select cutting edges" prompt.

6. In this exercise, you extend objects. Draw two vertical parallel lines 6 units long and 6 units apart. Between them, draw one horizontal line 4 units long, as illustrated by the figure.

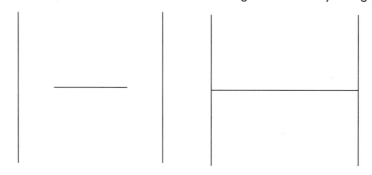

Left: *Line before extending.*
Right: *After extending both ends.*

Extend the horizontal line to the vertical lines with the **EXTEND** command.
To extend the other end (remember, choose both vertical lines as boundaries), respond to the repeating "Select object to extend:" prompt by selecting a point at the other end of the horizontal line.

7. In this exercise, you rotate objects. Draw the arrow symbol illustrated by the figure. Using the **ROTATE** command, turn the arrow by 45 degrees about the center of the arrow's base. Which object snap mode helps you find the midpoint of a line? Using grips editing, turn the arrow a further 90 degrees.

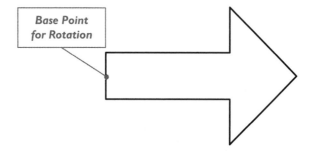

Base Point for Rotation

Draw a cam similar to the one illustrated by the figure.

Use Window object selection to rotate the entire cam by -45 degrees.

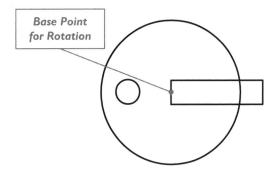

8. In this exercise, you resize objects.

With the **POLYGON** command, draw the triangle illustrated by the figure. The edge is 2 units long.

Use the **SCALE** command to double the size of the triangle.

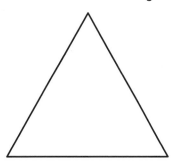

Repeat the command, but this time use the **Reference** option further to double the size of the triangle.

> Specify scale factor or [Reference]: **r**
>
> Specify reference length <1>: **2**
>
> Specify second point: **4**

Did AutoCAD resize all the lines making up the triangle?

9. In this exercise, you "stretch" a window to move it within the wall — without editing the wall.

From the CD, open the *stretch1.dwg* file, a drawing of a wall cross section.

Use the **STRETCH** command to move the window along the wall.

Which selection mode must you use?

Did the wall stretch?

10. From the CD, open the *stretch2.dwg* file, a drawing of a pencil.

Use the **STRETCH** command to make the pencil shorter.

11. In this exercise, you lengthen and shorten objects with a different command.

Draw an arc, using these parameters:

Start point	**2,5**
Second point	**0,3**
Endpoint	**2,1.5**

Start the **LENGTHEN** command, and select the arc. What is its length and included angle?

Use the **DElta** option to add 2 units to the length.

Use the **Total** option to change the arc to 5 inches. Did it become longer or shorter?

Use the **Percent** option to change the arc to 100%. Did the arc change?

12. In this exercise, you change objects.

Ensure ortho mode is turned off.

Draw four lines, randomly, as illustrated by the figure.

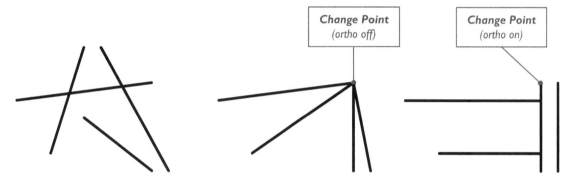

Use the **CHANGE** command to select the lines, and then use the **Change Point** option to give the lines a common endpoint.

Now use the **CHANGE** command's **Properties** option to change the lines to red.

13. In this exercise, you explode objects.

From the CD, open the *edit2.dwg* drawing file, the drawing of the puzzle.

Select one of the puzzle pieces. What does the single grip tell you?

With the piece still selected, use the **EXPLODE** command. Does the piece look different?

Select the piece again. Has the number of grips changed? What does this tell you?

Use the **EXPLODE** command a second time. What message does AutoCAD give you?

Select a different, unexploded piece.

Start the **XPLODE** command, and specify a color of red. What happens to the puzzle piece?

14. In this exercise, you edit polylines.

From the CD, open the *offset.dwg* drawing file, the drawing of the small town.

Select any of the polylines that define the edges of streets.

Use the **PEDIT** command's **Width** option to change the width to 0.1 units.

Exit the command, and start the **PROPERTIES** command.

Select *all* the polylines in the drawing by clicking the **Quick Select** button (funnel with the lightening strike) and entering these parameters:

Apply to	**Entire drawing**
Object type	**Polyline**
Properties	**Layer**
Operator	**=**
Value	**0**

After you click **OK**, AutoCAD highlights all polylines.

In the **Geometry** section of the **Properties** window, change *VARIES* to 0.1 units.

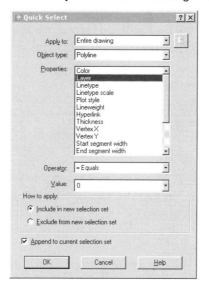

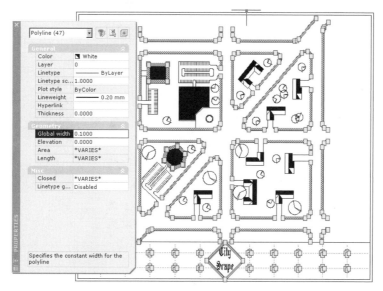

Press **ESC** to remove the grips and highlighting. The drawing should look like the figure below.

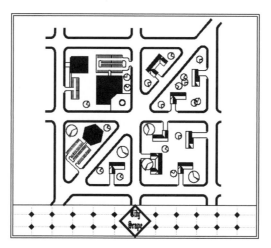

CHAPTER REVIEW

1. What is the difference between hot and cold grips?
2. When moving objects, must the base point be on the selected objects?
3. What objects are affected by the **BREAK** command?
4. In breaking objects, what happens if you do not enter **F** for selection of the first point?
5. Which commands restore objects just erased?
6. What happens when you explode blocks?
7. Describe what happens when you explode polylines.
8. What are *cutting edges*?
9. At what point are polylines trimmed?
10. What is the special requirement before circles can be trimmed?
11. Name the border line used by the **EXTEND** command.
12. Describe three ways to rotate objects.
    a.
    b.
    c.
13. Can objects be resized differently in the x and y axes with the **SCALE** command?
14. What value changes objects to half their original size?
15. Describe the purpose of the **Copy** option of the **ROTATE** and **SCALE** commands.
16. Name the command that explicitly edits polylines?
17. Which parts of the polyline are the *vertices*?
18. How do you edit polyline vertices?
19. What types of vertex editing can you perform?
20. List the aliases for the following commands:

    TRIM

    BREAK

    EXTEND

    PEDIT
21. When can the **OOPS** command be used?
22. Is it possible to erase all objects in the drawing?
23. Can the **BREAK** command shorten objects?
24. What do circles become after applying the **BREAK** command?
    An ellipse?
    Xlines?
25. Describe how the **LENGTHEN** command differs from the **TRIM** and **EXTEND** commands.
26. Can the **TRIM** command extend open objects?
    If so, how?
27. Can the **EXTEND** command extend closed objects?
    If so, how?
28. Do blocks and dimensions work as cutting edges?
    Do hatches?
29. In what kind of drafting are the **TRIM** command's **Project** and **Edges** options used?
30. List the two things AutoCAD needs to know before extending objects:
    a.
    b.
31. Do blocks work as boundary edges?
    Do hatches?

32. Can circles be extended?

33. Is it acceptable to select all objects as cutting edges?

34. Can arcs be lengthened with the **LENGTHEN** command?
    Shortened?

35. Describe how the **LENGTHEN** command's **Total** and **Delta** options differ.

36. When does the **STRETCH** command only move objects?

37. Can objects be moved using grips editing?

38. What does the **ROTATE** command rotate objects about?

39. Where does 0 degrees point, by default?
    Can the direction be changed?
    If so, how?

40. In which direction does AutoCAD measure negative angles, by default?

41. What three pieces of information does AutoCAD need to know before resizing objects?
    a.
    b.
    c.

42. Describe how the **CHANGE** command's **Change Point** option changes lines:
    Ortho mode turned off:
    Ortho mode turned on:

43. What is the difference between the **EXPLODE** and **XPLODE** commands?

44. How are arcs in nonuniformly scaled blocks exploded?

45. Under what condition must blocks be exploded more than once?

46. What are polylines exploded into?

47. Can you specify the color of exploded objects?
    If so, how?

48. Describe how AutoCAD reacts when you select lines and arcs with the **PEDIT** command?

49. Can lines and arcs that don't touch be turned into a single polyline?
    If so, how?

50. During polyline vertex editing, which keystrokes move the x marker from vertex to vertex?

51. How does the **Straighten** option affect a polyline?

52. What is the difference between the **Fit** and **Spline** options of the **PEDIT** command?

53. Are the splines created by **PEDIT** true splines?

54. Explain why the **Ltype gen** option is important?

55. What is the purpose of the **SPLFRAME** system variable?

56. Can the type of polyline spline be changed?
    If so, how?

57. What does the padlock icon indicate?

UNIT IV

Text and Dimensions

CHAPTER 12

Placing and Editing Text

Text is used in many areas of drawings: title blocks that identify drawings, callouts that identify parts, bills of material, and paragraphs of text that warn and explain. Traditionally, placing text by hand in a paper drawing was laborious; drafters drew each letter individually. In contrast, placing text in an AutoCAD drawing requires little effort. This chapter shows how to place text in drawings, and then how to change the text and its properties. The commands are:

TEXT and **-TEXT** place text in drawings, one line at a time.

MTEXT and **-MTEXT** place paragraphs of formatted text.

TEXTTOFRONT displays text in front of other objects.

QLEADER, **LEADER**, and **MLEADER** (new to AutoCAD 2008) place callouts in drawings.

DDEDIT edits text.

STYLE and **-STYLE** define named text styles based on fonts.

MLEADERSTYLE defines named multiline leader styles (new to AutoCAD 2008).

SPELL checks the drawing for unfamiliar words.

FIND searches and optionally replaces text.

QTEXT displays lines of text as rectangles.

SCALETEXT and **JUSTIFYTEXT** change the size and justification of text.

Annotation property displays correctly-scaled text in model space (new to AutoCAD 2008).

COMPILE converts PostScript font files for use in drawings.

NEW TO AUTOCAD 2008 IN THIS CHAPTER

- All text can have the annotative scaling property, including mtext, single-line text, and leaders.
- The **MTEXT** command supports multiple columns, forced justification, new paste special options; and sports a redesigned paragraph dialog box.
- The **SPELL** command begins spell checking immediately, and zooms into unrecognized words.
- The **MLEADER** command allows multiple leader lines and multiple blocks per leader.

FINDING THE COMMANDS

On the Dashboard's **Text** panel:

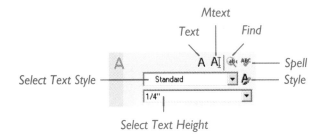

On the Dashboard's **Mleader** panel:

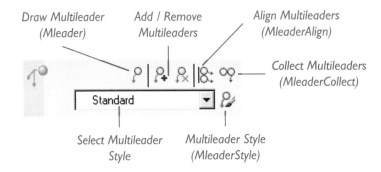

On the **Text** toolbar:

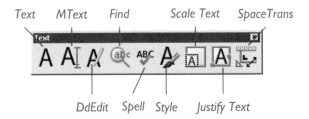

On the **Styles** toolbar:

On the **Mleader** toolbar:

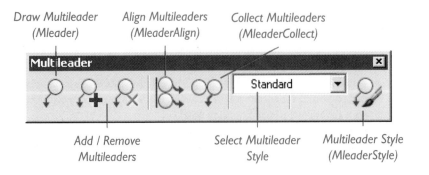

On the **Dimension** toolbar:

*Select / Apply
Text Styles*

*Create / Edit
Text Styles*

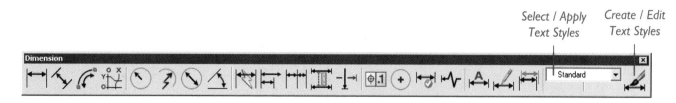

On the status bar:

*Lock Viewport
Scale Factor*

*Automatically Add
Scale Factors*

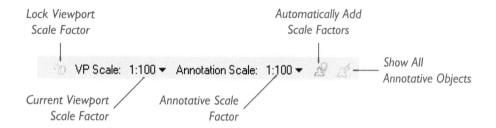

*Show All
Annotative Objects*

*Current Viewport
Scale Factor*

*Annotative Scale
Factor*

TEXT IN DRAWINGS

The figure illustrates many uses of text in drawings. Can you spot the notes, general notes, title block (shown enlarged on the next page), section view numbers, leaders, and dimensions?

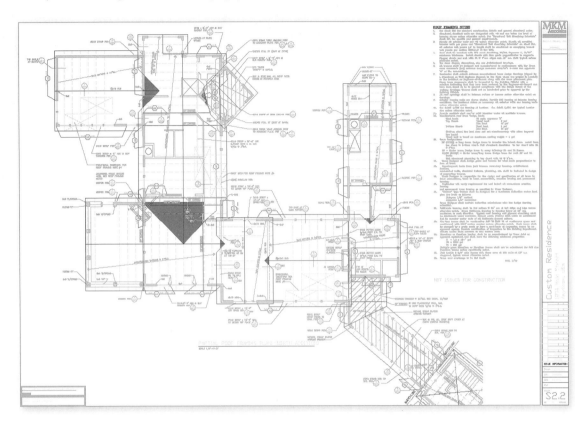

The use of text in drafting is usually governed by standards. Many North American industries use the American National Standards Institute (ANSI) style of letters and numbers. European companies use ISO lettering (International Organization for Standardization). Other companies and countries set standards of their own.

The most important aspect of lettering is that it be clear and concise. In the days of hand drafting, neophyte drafters first learned how to print neatly, and only then how to draw lines. Today, AutoCAD does the printing, but the same rule applies: clear text is crucial to understanding drawings. Here are some guidelines for placing text:

- **Headings** are 3/16-inch high.
- **Note text** is 1/8-inch high.
- **Text** is left-justified (each line of text aligned at its left edge).

You can calculate the size of the text manually, or let AutoCAD do it automatically through the annotative scale property (new to AutoCAD 2008).

FONTS

AutoCAD is versatile in displaying fonts. Many (but not all) can be stretched, compressed, obliqued (slanted), mirrored (reversed), or drawn in a vertical stack. You can apply colors, lineweights, and plot styles to text, but not linetypes.

The default font in every new AutoCAD drawing is the rather ugly **Txt** font. The clear font often used by drafters is called **Simplex** or **RomanS** by AutoCAD.

Txt font   Simplex

Left: *AutoCAD's default Txt font.*
Right: *AutoCAD's Simplex font preferred by many drafters.*

Fonts are the "design" of text letters. Caslon, for example, is the name of the font used for text in this paragraph, while the **GillSans** font is used for headings and tutorial text. Arial, TimesRoman, and `Courier` are among the most commonly-used fonts.

AutoCAD provides a large collection of fonts in its own format known as "SHX" (*.shx* files), but it is now more common to use the TrueType fonts (*.ttf* files) found on your computer. After conversion, PFB PostScript fonts (*.pfb* files) can also be used.

A͟I̤ TEXT

The **TEXT** command places lines of text in drawings; the related **MTEXT** command places paragraphs of text, as detailed later in this chapter. **TEXT** is like the minimalist Notepad text editor, while **MTEXT** is more akin to the full-featured Word word processing application.

TUTORIAL: PLACING LINES OF TEXT

1. To place lines of text in drawings, start the **TEXT** command:
   - From the **Draw** menu, choose **Text**, and then **Single Line Text**.
   - From the Text toolbar, choose the **Single Line Text** button.
   - In the Dashboard's Text panel, pick the **Single Line Text** button.
   - At the 'Command:' prompt, enter the **text** command:

     Command: **text** *(Press* ENTER.*)*

   - Alternatively, enter the **dt** or **dtext** (the old name for this command) aliases at the 'Command:' prompt.

2. In all cases, AutoCAD reports the current text style, height, and annotation scale settings, and then prompts you for the location of the start point:

   Current text style: "Standard" Text height: 0.2000 Annotative: No

   Specify start point of text or [Justify/Style]: *(Pick a point, or specify x,y coordinates.)*

3. Specify the height and rotation angle:

   Specify height <0.2000>: **0.75**

   Specify rotation angle of text <0>: *(Press* ENTER *to accept the default.)*

4. Enter a line of text, and then press **ENTER**.

   Using AutoCAD *(Press* ENTER.*)*

5. After you press **ENTER**, the cursor jumps to the start of the next line. When you enter additional lines of text, AutoCAD places them under the previous line.

   Basics *(Press* ENTER.*)*

(Title block along left margin:)

MKM
Associates

Custom Residence
1234 Main Street
Anytown, USA

S2.2

6. To exit the command, press **ENTER** a second time:

Enter text: *(Press ENTER to exit the command.)*

If you make a mistake, press **BACKSPACE** to erase one character at a time; alternatively, highlight a group of characters to change or erase them.

As you type text, AutoCAD displays the *I-beam cursor*, called that because it looks like the letter I. The cursor has a pair of short horizontal lines that represent the top of uppercase letters and the bottom of descenders. (A *descender* is the part of lowercase letters that hangs below the baseline — g, j, p, q, and y have descenders.)

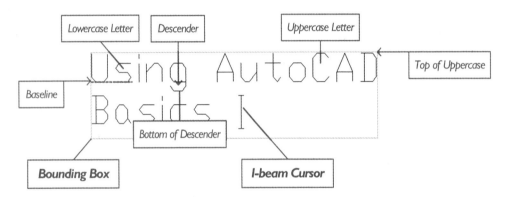

The text you enter is surrounded by a gray *bounding box*, which shows the extent of the text.

While the **TEXT** command is active, menu selections, command options, and other functions are "locked out." Only keyboard entry is permitted while placing text. If you cancel the **TEXT** command by pressing **ESC**, the current line of text is erased; so, remember to press **ENTER** instead.

The next time you use the **TEXT** command, the last text string is *highlighted* (shown in dashed lines).

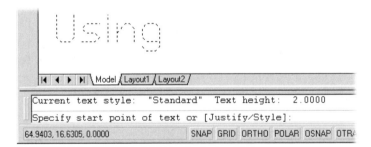

AutoCAD does this for a good reason. Press **ENTER** at the "Specify start point of text:" prompt, and the the I-beam cursor is placed on the next line after that last text string — as though you had not exited. This is useful for placing additional lines of text.

Command: *(Press ENTER to repeat the TEXT command.)*

TEXT Current text style: "Standard" Text height: 2.0000

Specify start point of text or [Justify/Style]: *(Press ENTER to continue text below the last line.)*

AutoCAD skips the height and rotation prompts, because it assumes you want them to be the same as before.

PLACING TEXT ALL OVER DRAWINGS

The **TEXT** command is useful for placing text quickly in many different places on drawings. It is more efficient than **MTEXT**, because you simply click the cursor in another location and then continue typing. (AutoCAD calls these "blocks" of text.) This trick is handy for filling out title sheets and other text intensive jobs.

During the **TEXT** command, you can move between blocks of text by pressing the **TAB** key. Press **SHIFT+TAB** key to return to the previous block; press **TAB** to jump ahead to the next block.

To split a line of text in two, use the cursor arrow keys (or the mouse) to move the cursor, and then press **ENTER**.

PLACING TEXT: ADDITIONAL METHODS

The **TEXT** command allows you to control text justification and style, as listed by the options below. In addition, the **-TEXT** command handles non-text characters in a unique manner.

- **Justify** specifies a text justification mode.
- **Style** selects a predefined text style.
- **Height** specifies the height of the text.
- **Rotation angle** specifies the angle at which the line of text is rotated.
- **Control Codes** add formatting and special characters.
- **-TEXT** and **TEXTEVAL** evaluate AutoLISP expressions during text placement.
- **DTEXTED** system variable controls the look and feel of the command.

Let's look at each.

Justify

When you respond to the prompt "Specify start point of text" by picking a point or typing x,y coordinates, this becomes the starting point for the text string. The **Justify** option specifies different text alignments relative to the starting point (also known as the "insertion point").

The default is *left-justified*, which means that lines of text start at a common left edge. Other commonly used justifications include middle-aligned for title blocks and right-justified for text to the left of leader lines.

To see a list of all text justification options, respond with a **J** (short for "justify") at the 'Specify start point of text or [Justify/Style]:' prompt. AutoCAD prompts you to enter an alignment option:

> Specify start point of text or [Justify/Style]: **j**
>
> Enter an option [Align/Fit/Center/Middle/Right/ TL/TC/TR/ML/MC/MR/BL/BC/BR]:

There are many modes, because text can be justified (aligned) in both directions: vertically and horizontally. If you know the alignment you want, it is not necessary to enter the **J** option; simply enter the one- or two-letter abbreviation for the alignment. For example, to right-justify text:

> Specify start point of text or [Justify/Style]: **r**

Two justification modes operate differently from the others, because they require two pick points: **Align** and **Fit**.

AutoCAD uses one- and two-letter abbreviations to designate each of its alignment options. (*History*: The one-letter options were present in the original AutoCAD 25 years ago; the two-letter options were added more recently.)

When it comes to two-letter options, the first letter describes the vertical alignment; the second, the horizontal alignment. For example, **TL** is top-left. The figures illustrate the alignment modes listed in the tables.

Justify	Meaning
Start point	Default; equivalent to Left and BL.
Center	Equivalent to BC.
Middle	Equivalent to MC.
Right	Equivalent to BR.
Align	Fits text between two points, and adjusts height appropriately.
Fit	Fits text between two point at a specific height.

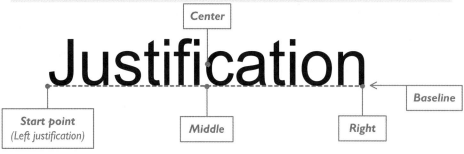

Justify	Meaning
TL	Top left.
TC	Top center.
TR	Top right.
ML	Middle left.
MC	Middle center.
MR	Middle right.
BL	Bottom left.
BC	Bottom center.
BR	Bottom right.

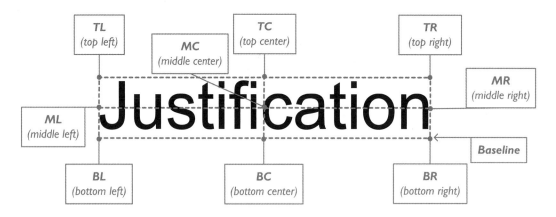

Align

The **Align** option requires you to select two points that the text will fit between. AutoCAD adjusts the text height so that the baseline of the text fits perfectly between the two points. Note that the two points can be placed at any angle in relation to each other.

Command: **text**

Current text style: "Standard" Text height: 0.2000

Specify start point of text or [Justify/Style]: **a**

Specify first endpoint of text baseline: *(Pick point 1.)*

Specify second endpoint of text baseline: *(Pick point 2.)*

Enter text: *(Type text.)*

Enter text: *(Press* ENTER *to exit command.)*

There are no "height" and "rotation angle" prompts, because the pick points determine the height and angle of the text. AutoCAD automatically determines the height.

When you pick the two points in reverse order (right to left), AutoCAD draws the text upside down, as illustrated below.

Fit

The **Fit** option is similar to Aligned: it prompts you for two points between which to place the text. The difference is that it also prompts for the text height. AutoCAD forces the text to fit between the two points, but draws the text at the height you specify.

Specify start point of text or [Justify/Style]: **f**

Specify first endpoint of text baseline: *(Pick point 1.)*

Specify second endpoint of text baseline: *(Pick point 2.)*

Specify height <0.2>: *(Enter a height.)*

Enter text: *(Type text.)*

Enter text: *(Press* ENTER *to exit command.)*

The text is squashed or stretched horizontally to fit between the two points, as illustrated below. Fitted text is often used in constrained areas, for example when labeling small closets. ("Or," suggests, the technical editor, "You could come out of the closet and put the label in the hall.")

Examples of text fitted between two points, but assigned different heights.

As with Aligned, when you pick the second point to the left of the first, AutoCAD draws the text upside down.

Style

The **Style** option selects a predefined text style. *You must first create a text style before you can use it.* The exceptions are the Standard and Annotative styles, which are included in every new drawing.

Styles (covered in detail later in this chapter) define text parameters, such as the font name and whether it is boldface or has a predefined height.

Note: The **Style** option of the **TEXT** command and the **STYLE** command are not the same. The **Style** option *selects* a style, while the **STYLE** command *creates* styles.

Height

The **Height** option determines the height of the text, which can be any distance — as tall as the orbit of Pluto, if need be. (The technical editor corrects the author by noting that text can in fact be taller than the orbit of Pluto, as tall as 9.49×10^{94} units — considerably taller than the known universe.) The default is 0.2 units.

> Specify height <0.2000>: *(Enter a height such as **17.5**, or press ENTER.)*

Once the text height is set, it remains the default until changed.

There are two conditions under which the 'Specify height' prompt does not appear. One is when the height is predefined in the style; another is during the Align option.

Determining Text Height

Like hatch patterns and linetypes, text must be scaled appropriately for the drawing. When you sketch a picture of your house on a piece of paper, for instance, you draw the house small enough to fit the paper. (There might be a sheet of paper big enough to draw the house full size, but it would be hard to find!) Because *scale* is so important for text, a review may be in order.

In the sketch of my 50-foot house, I drew it 4 1/4" long. The drawing of the house is 140 times smaller than its actual size. Converted to inches, 50' is 600", so the scale works out to be:

$$\frac{4.25"}{600"} = 1:140 \text{ (roughly)}$$

The 1:140 is the scale factor. The text I wrote on the paper, however, is *full size*, which happens to measure 1/8" tall in the sketch.

In AutoCAD, scaling is done in reverse order. Instead of drawing the house smaller, I draw it full size: the 50-foot house is drawn 50 feet long in AutoCAD, at a scale of 1:1.

The text cannot be drawn at 1:1 — it must to be 140 times larger. The standard in drafting is to draw text $^1/_8$" tall for "normal-sized" text. When I place the text in my house drawing, I specify a height of 17.5" tall ($^1/_8$" x 140). This may seem to you much too tall (nearly one-and-a half feet tall!) but trust me: when plotted, it looks exactly right.

The figure below illustrates text placed at several heights — some too small, some just right, and some too large. Whether it is right or wrong depends on the plotted size.

Annotative Scaling

There are two approaches to solving the text height problem. One is to work out the text size, as describe above, and then enter it at the prompt:

> Specify height <0.2000>: **17.5**

The second solution is to use *annotative scaling*. AutoCAD determines the text height automatically: you tell it the plotted height (a.k.a. paper height), and AutoCAD sizes it based on the viewport scale factor.

Annotative scaling is available only when the property is turned on in the style with the **STYLE** or **PROPERTY** commands. (All new drawings contain the Annotative text style, although the Standard continues to be the default style.) The **TEXT** command displays the height prompt with the word "paper" added:

Specify <u>paper</u> height <0.2000>: **0.125**

You enter the height at which you wish the text to appear plotted on <u>paper</u>. Thus, text $1/_8$" (0.125 units) tall is plotted $1/_8$" tall, no matter the scale of the viewport.

(There are some other nuances involved in annotative scaling, as detailed near the end of this chapter.)

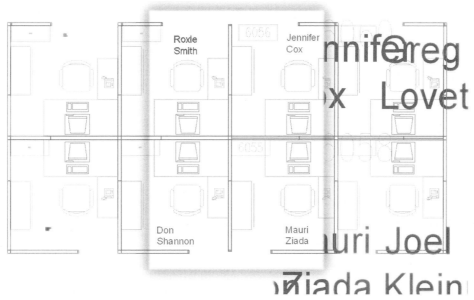

Left: *Text height of 1" being too small.*
Center: *Text height of 6" being just right.*
Right: *Text height of 2' being too tall.*

Rotation Angle

Lines of text can be placed at any angle in drawings. You specify the angle at which the text is drawn by designating the angle when you are prompted:

Specify rotation angle of text <0>: *(Enter an angle, or pick two points.)*

Press **ENTER** to accept the default angle, 0 degrees in this case. The text is rotated about the *insertion point*, which varies depending on the justification mode.

Once the text angle is set, the angle remains the default until it is changed. The figure illustrates text placed at several angles.

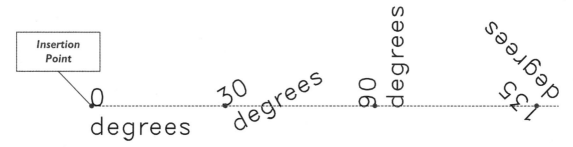

The angle is affected by the **ANGBASE** and **ANGDIR** system variables, as are other angles in AutoCAD.

Control Codes

AutoCAD allows you to add *metacharacters* to the text known as "control codes." Metacharacters mean something other than themselves. (As a professor of English literature, the copy editor notes that *all* text means something other than itself. We will sidestep the philosophical discussion over what is meta and what isn't.) Control codes are used to specify underlined text and to include symbols, such as the degree and plus/minus symbols.

In the TEXT command, this is accomplished by typing codes that start with two percent characters: %%. The following table lists control codes and their functions:

Control Code	Meaning	Sample
%%c	Diameter symbol	Ø
%%d	Degree symbol	o
%%o	Overline	
%%%	Percent symbol	%
%%p	Plus-minus symbol	±
%%u	Underline	
%%nnn	ASCII character nnn	

Let's look at an example with the following text string:

> Command: text
>
> Current text style: "Standard" Text height: .02 Annotative: No
>
> Specify start point of text or [Justify/Style]: *(Pick a point.)*
>
> Specify height <0.02>: *(Press ENTER.)*
>
> Specify rotation angle of text <0>: *(Press ENTER.)*
>
> Enter text: **If the piece is fired at 400%%dF for %%utwenty%%u hours,**
>
> Enter text: **it will achieve %%p95%%% strength.**
>
> Enter text: *(Press ENTER.)*

Codes are initially displayed instead of the effect; once you begin typing the next text, the code disappears, and the formatting appears. AutoCAD converts the metacharacters and draws the text strings like this:

> If the piece is fired at 400°F for <u>twenty</u> hours,
> it will achieve ±95% strength.

Notice that you turn on and off the underscore (a.k.a. underlining) by typing %%u twice — once to turn on, and once to turn off; if you don't turn off underscoring, AutoCAD automatically turns it off at the end of the line.

You can "overlap" symbols, for instance, using the degree symbol between the underscore symbols, like this:

> **%%uFire at 400%%dF%%u.**

Which results in:

> **<u>Fire at 400°F.</u>**

ASCII Characters

ASCII characters refer to the ASCII (short for "American Standard Code for Information Interchange") character set, which assigns number codes to symbols.

To use a symbol in the text, enter two percent signs (%%) followed by the ASCII character code. For example, to place a tilde (~) in your text, enter %%126.

-Text and TextEval

The **-TEXT** command is like **TEXT**, except that it obeys the setting of the **TEXTEVAL** system variable:

TextEval	Meaning
0	Text strings are read literally: text is text, even if it looks like AutoLISP.
1	Text prefixed with (or ! is evaluated as an AutoLISP expression.

By changing the value of **TEXTEVAL** to 1...

> Command: **texteval**
>
> Enter new value for TEXTEVAL <0>: *(Enter **1**.)*

...AutoLISP expressions can be entered and evaluated. Start **-TEXT**, and then enter an AutoLISP expression:

> Command: **-text**
>
> Current text style: "STANDARD" Text height: 0.2"
>
> Specify start point of text or [Justify/Style]: *(Pick a point.)*
>
> Specify height <0'-6">: *(Press **ENTER**.)*
>
> Specify rotation angle of text <0.00>: *(Press **ENTER**.)*
>
> Enter text: **(+ 2 3)**
>
> 5

AutoCAD evaluates the AutoLISP expression, (+ 2 3), and places the result as text (5) in the drawing.

DTextEd

The **TEXT** command has changed over the years, and the **DTEXTED** system variable controls the changes:

0 — displays the Inplace Text Editor for writing and editing text; you cannot click to continue text elsewhere. This simulates the action of **TEXT** in AutoCAD 2006.

1 — displays the 'Enter text:' prompt at the command line and the Edit Text dialog box (**DDEDIT** command) for editing; you can click elsewhere in the drawing to start a new text string. This simulates the action of **TEXT** in AutoCAD 2005 and earlier.

2 — displays the Inplace Text Editor for writing and editing text; you can click to continue text elsewhere in the drawing.

I recommend you leave this system variable set to 2.

TEXT SHORTCUT MENU

The **TEXT** command sports a shortcut menu with several useful options. When **DTEXTED** is set to 0 or 2, right-click during the command to see the menu. (When **DTEXTED** is set to 1, there is no shortcut menu.)

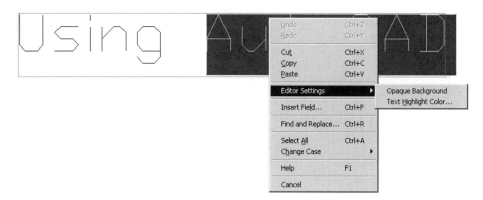

The significant options are described next.

Opaque Background

The **Opaque Background** option fills the bounding box with gray. The color appears only while editing; when you exit the command, the color disappears.

Left: *Opaque background.*
Right: *No background.*

Text Highlight Color

The **Text Highlight Color** option specifies the color of highlighted text, dark blue by default. "AutoCAD" is shown with the highlight color in the figure below.

Insert Field

The **Insert Field** option displays the Field dialog box. This allows you to add automatic text, described more fully in Chapter 16, "Tables and Fields." Examples of field text include the current date and time, the file name of the drawing, and the plot scale.

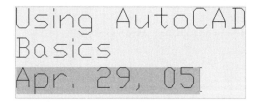

Field text is identified by the dark gray background. Some fields appear as "----" until the drawing is saved or plotted for the first time, or as "####" if values have not yet been assigned.

After the field is placed in the text, the shortcut menu shows two additional options:

Update Field

The **Update Field** option forces AutoCAD to update the value of fields. Normally, AutoCAD waits to update fields until certain events occur, such as the file when is opened, saved, plotted, regenerated, or when the **ETRANSMIT** command is used. Use this option to update fields immediately.

Convert Field to Text

The **Convert Field to Text** option does exactly that: the field is converted to static text. This is useful when you no longer want the field to update itself.

Find and Replace

The **Find and Replace** option displays a simple dialog box for finding (and optionally replacing) text.

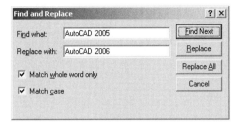

I am not sure of the utility of this option, because the **TEXT** command is meant for small amounts of text, and so finding a specific word is not a problem. A better method is to use the **FIND** command, which searches for any kind of text in the entire drawing.

Select All

The **Select All** option selects all the text in the bounding box.

Change Case

The **Change Case** option changes text to all UPPERCASE or all lowercase. The change applies to selected text only; it does not apply to field text.

Note: The **TEXT** command is efficient for quickly placing strings of text in the drawing. (*Strings* is another way of saying "lines of text.") **MTEXT** is best for paragraphs of text and text that contains formatting, such as **boldface** and **color**.

 **MTEXT**

The **MTEXT** command places paragraphs of text in drawings, and provides many formatting options.

While **TEXT** places text in the drawing one line at a time, the **MTEXT** command (short for "multiline text") creates multiple paragraphs of text. It fits the text into an invisible boundary that you define. (The boundary is not printed or plotted.) It permits more text enhancements than **TEXT**, such as varying heights and colors, and stacking fractions.

The figure below illustrates paragraph text.

ROOF FRAMING NOTES

1. See sheet SD1 for standard construction details and general structural notes.
2. Structural sheathed walls are designated with ~G and are below the level of framing shown unless otherwise noted. See "Structural Wall Sheathing Schedule" sheet SD1 for specific and general requirements.
3. Sheath all exterior walls per ~R unless otherwise noted. Sheath all specified interior walls per plans and "Structural Wall Sheathing Schedule" on sheet SD1. All exterior wall panels ≥4' in length shall be considered as complying braced wall panels per section 2320.11.3 of the UBC.
4. Roof shall be sheathed with APA rated sheathing, 32/16, Exposure 1, 15/32" minimum thickness. Install sheets with face grain perpendicular to supports. Stagger sheets and nail with 8d @ 6"o.c. edges and 12" o.c. field typical unless otherwise noted.
5. For truss shapes, dimensions, etc. see Architectural drawings.
6. All trusses shall be designed and manufactured in conformance with the Truss

TUTORIAL: PLACING PARAGRAPH TEXT

1. To place paragraphs of text in drawings, start the **MTEXT** command:

   • From the **Draw** menu, choose **Text**, and then **Multiline Text**.

   • From the Text toolbar, choose the **Multiline Text** button.

   • In the Dashboard's Text panel, pick the **Multiline Text** button.

   • At the 'Command:' prompt, enter the **mtext** command:

   Command: **mtext** *(Press* ENTER.*)*

   • Alternatively, enter the aliases **t** or **mt** at the 'Command:' prompt.

2. In all cases, AutoCAD reports the current settings, and then prompts you to pick the two corners of the bounding box :

   Current text style: "Standard" Text height: 0.2 Annotative: No

   Specify first corner: *(Pick point 1.)*

   Specify opposite corner or [Height/Justify/Line

   spacing/Rotation/Style/Width/Columns]: *(Pick point 2.)*

   The bounding box lets you specify the space in which the text should fit. Don't worry about the box being exactly the right size; you can always make it bigger or smaller later. In any case, AutoCAD automatically makes it longer as text is added; the direction that the bounding box is expanded is indicated by the arrow icon that you see in the box.

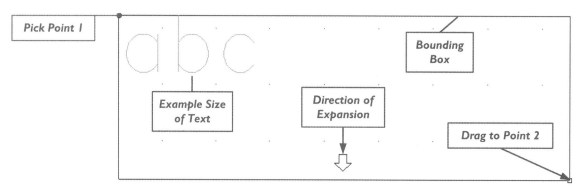

   (You can use the **MTJIGSTRING** system variable to change the preview text, shown by "abc" above.)

3. The text editor appears. It consists of two parts: a Text Formatting toolbar, and a text entry area topped by a tab bar.

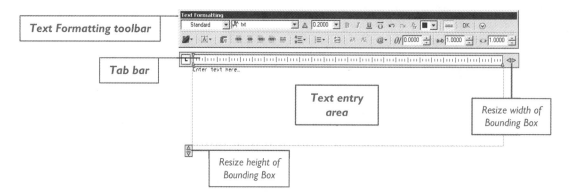

4. Type text into the text entry area.

   To format the text, you need first to select it by *highlighting* it: drag the cursor over the text you want formatted. Then, from the Text Formatting toolbar, select a format option. (Additional options are available by clicking the **More** button.)

5. When done, click **OK**.

   Notice that AutoCAD places the text in the drawing. To exit the mtext editor without saving changes, press **ESC**.

 Note: To insert the Euro symbol, turn on **NUMLOCK**, hold down the **ALT** key, and then type **0128** on your keyboard's numeric keypad. The € (Euro currency) symbol is included with all AutoCAD fonts since Release 2000.

TEXT FORMATTING TOOLBAR - TOP HALF

The Text Formatting toolbar allows you to apply styles to text; more important, it lets you override styles with other properties, such as font and color. We first look at the top half of the toolbar, followed by the bottom half. (The toolbar was changed in AutoCAD 2008.)

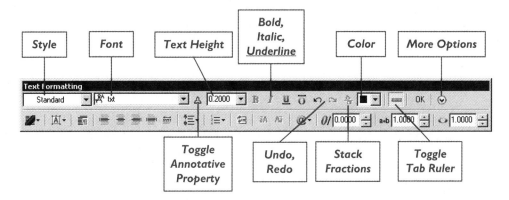

Style

Select a style name from the **Style** drop list. Only text styles defined in the drawing are listed here; you create additional styles with the **STYLE** command.

The style applies to all text in the bounding box; you cannot selectively apply styles. You can, however, override styles with font, text height, and other properties provided in the Text Formatting toolbar.

Font

Select a font name from the **Font** droplist. The fonts listed are those installed on your computer. The list probably includes all TrueType fonts included with Windows, as well as fonts installed by AutoCAD and other applications. TrueType fonts have a tiny **T** logo in front of them; AutoCAD fonts are prefixed by a tiny Autodesk ⬆ logo. (Printer fonts, PostScript fonts, and other fonts are not listed, because AutoCAD does not support them directly.)

You can use one or more fonts in one block of mtext, but remember to highlight the text before selecting the font name. To highlight text, click and drag; to select all text, press **CTRL+A**. Alternatively, right-click the text and choose **Select All** from the shortcut menu.

Annotative Property

Click the annotative button to change the font scaling to annotative. You learn more about this option near the end of this chapter.

Text Height

If you want some (or all) of the text at different heights, highlight the text, and then select a height from the **Text Height** dropbox.

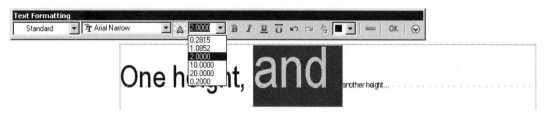

If a height you need isn't on the list, simply type the measurement into the droplist, and then press **ENTER**. AutoCAD adds the height to the list.

Bold, Italic, and Underline

To **boldface** text, highlight the text, and then choose the **B** button. Choose the button a second time to "unboldface" the text. Pressing **CTRL+B** performs the same function.

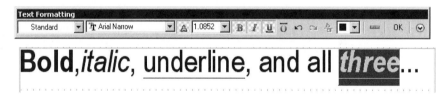

When the **B** on the button is gray, the current font cannot be boldfaced — usually a problem with certain AutoCAD fonts.

To *italicize* text, highlight the text, and then choose the **I** button. When the **I** button is gray, the text cannot be italicized. As an alternative, press **CTRL+I**.

To underline text, highlight the text, and then choose the **U** button. All fonts can be underlined; **CTRL+U** is the keystroke shortcut.

Undo and Redo

To reverse the last operation, choose the **Undo** button. Or press **CTRL+Z**.

Choose the **Redo** button to "undo" the undo. Or press **CTRL+Y.**

Stack

The **a/b** button stacks fractions by changing side-by-side text, like 11/32, to show the 11 over the 32. Using this feature takes these steps:

1. Type the fraction, such as 11/32.
2. Highlight the entire fraction.
3. Choose the **a/b** button.

AutoCAD replaces the numbers with a stacked fraction. Repeat steps 2 and 3 to "unstack" stacked fractions.

This button is not limited to stacking numbers: it stacks any combination of text, numbers, and symbols — whatever is in front of the slash goes on top; whatever is behind goes underneath.

Also, you are not limited to working with slash symbols. The carat (^) and the hash mark (#) indicate different kinds of stacking:

Slash (/) stacks with a horizontal bar. 15/16 $\frac{15}{16}$

Carat (^) stacks without a bar (tolerance style). 12^12 $\frac{12}{12}$

Pound (#) stacks with a diagonal bar. 12#12 $^{12}/_{12}$

When you right-click stacked text, AutoCAD adds an item to the shortcut menu. Select **Stack Properties** to display the following dialog box for controlling the "autostack" feature:

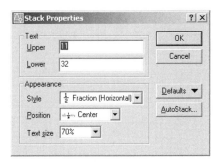

Stack Properties

The **Upper** and **Lower** fields specify the fraction text. This lets you change the values.

The **Appearance Style** droplist chooses the three slash styles noted above: horizontal, diagonal, and none.

The **Position** droplist chooses the position of the fraction relative to the baseline of text on either side of the fraction: top, center, or bottom.

The **Text size** droplist specifies the size of fraction text relative to regular text. The range is from 50% (half as tall) to 100% (same size); the default is 70%.

The **Defaults** button saves the current settings as the default, or restores previously saved defaults.

The **Stack** button displays the AutoStack Options dialog box.

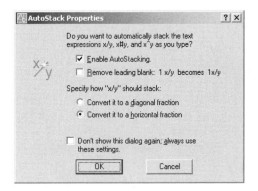

AutoStack Properties

The **Enable AutoStacking** option toggles the automatic stacking of numbers on either side of the /, #, and ^ characters.

Remove Leading Blank removes blanks between whole numbers and stacked fractions. This option is available only when AutoStacking is turned on.

Convert It to a Diagonal Fraction creates stacked numbers with the diagonal slash, regardless of the stack character (/, #, and ^).

Convert It to a Horizontal Fraction creates stacked numbers with the diagonal slash, regardless of the stack character.

Note: When drawings with diagonal stacked fractions are opened in AutoCAD Release 14 or earlier, they are converted to horizontal fractions, but are restored when opened in AutoCAD 2000 or later.

Back to the MTEXT command's Text Formatting toolbar...

Color

Highlight a portion of the text, and then select a color from the **Color** list box. You can choose Bylayer, Byblock, the seven basic AutoCAD colors, and the other 16.7 million colors and color books.

Toggle Tab Ruler

Click the **Toggle Tab Ruler** button to hide and display the tab ruler, located atop the boundary box.

Options and OK

Clicking the ⊙ Options button displays a menu of additional options, described later.

Clicking the **OK** button exits the MTEXT command; you can edit the mtext later by double-clicking it.

TEXT FORMATTING TOOLBAR - BOTTOM HALF

The bottom half of the Text Formatting toolbar shows the following options.

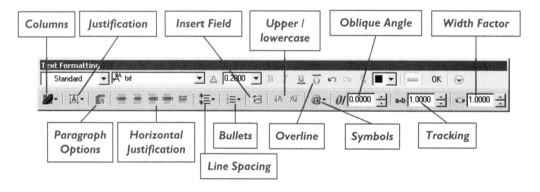

Columns

The Columns button sets the number of columns. It splits the mtext block into two or more columns. This is useful when you need to fit the text into an area that is wider than it is deep.

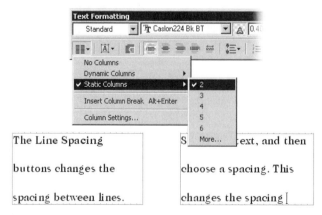

You can choose from dynamic or static columns, or else force the start of a new column with ALT+ENTER.

Dynamic — AutoCAD creates columns automatically as text is added to the bounding box.

Static — you tell AutoCAD the number of columns you want, 2 or more.

The Columns Settings dialog box lets you define all of the options in one place.

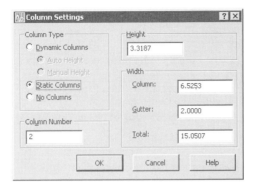

Justification

The **Justification** button changes the justification of the entire mtext block relative to the bounding box. You can choose from left, center, and right; and from top, middle, and bottom.

Vertical justification works only when the bounding box is noticeably longer than the paragraphs of text. You would select **Bottom**, for example, when you want the text block to rest at the bottom of the bounding box.

⊙ 2008 Horizontal Justification

The **Horizontal Justification** buttons change the justification of the text relative to the sides of bounding box. You can choose from Left, Center, Right, Justify, and Distribute (new to AutoCAD 2008).

Both **Justify** and **Distribute** force both the left and right margins to be even. The difference between the two is how the last line in a paragraph is handled: **Justify** left justifies it, while **Distribute** forces the last line to fit between the left and right margins.

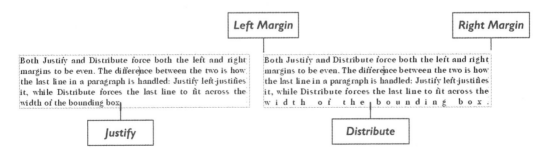

2008 Line Spacing

The **Line Spacing** buttons change the spacing between lines. Select the text, and then choose a spacing. This increases the spacing, as illustrated below.

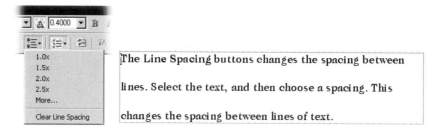

Bullets

The **Bullets** button specifies the format of bullets: letters, numbers, symbols, or none.

AutoCAD automatically numbers (or letters) and indents the text for you. Each paragraph (text that ends with a hard return) is bulleted separately.

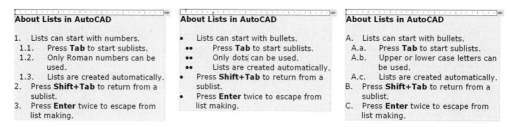

You can use the following characters as punctuation after letter and number bullets: periods (.), commas (,), close parentheses ( )), close angle brackets (>), close square brackets ( ]), and close curly braces (}).

The technical editor advises us how to get this to work: ensure that all bullet options are turned off. Enter a letter or number, the desired character, and then press **TAB**. Enter the text, and then press **ENTER**. The automatic lettering or numbering turns on.

To create sub-bullets, press **TAB** one or more times at the start of the next line. To outdent, press **Shift+ TAB**.

To change the bullet style, click the **Options** button (looks like a downward pointing arrow), and then select **Bullets and Lists** from the menu.

Off — removes bulleting from the mtext. This is the same as turning off the three bullet buttons on the Text Formatting toolbar.

Lettered — applies letter bullets to paragraphs of text. Once it reaches Z, it continues with AA, BB, and so on. Letters can be all uppercase or all lowercase. This is the same as toggling the **Letters** bullet button on the Text Formatting toolbar.

Numbered — applies number bullets to paragraphs of text; numbers are Roman numerals only. This is the same as toggling the **Numbering** bullet button on the Text Formatting toolbar.

Bulleted — makes bullets round. This is the same as toggling the **Bullets** button on the Text Formatting toolbar.

Restart — restarts letter bullets from A; number bullets restart from 1.

Continue — adds selected paragraphs to the previous list.

Allow Auto-list — applies bullet formatting as you type.

Use Tab Deliminter Only — restricts automatic bullet creation to when the TAB key is pressed.

Allows Bullets and Lists — applies bullets to all text objects that look like lists (namely those that begin with one or more letters or numbers or a symbol, followed by punctuation, a space created by TAB, and end with ENTER or SHIFT+ENTER). When off, this option removes bullet formatting and turns off all other options, except this option.

Insert Field

The **Insert Field** button displays the Field dialog box; see Chapter 16, "Tables and Fields."

UPPER and lower Case

The **Uppercase** and **Lowercase** buttons change selected text to all UPPERCASE or all lowercase.

Overline

The **Overline** button places an overline (opposite of an underline) over selected text.

Insert Symbols

The **Insert Symbols** button displays a menu of symbols commonly used in drawings. Click **Other** to display the Windows Character Map dialog box, which provides access to all symbols. More details follow in this chapter.

Oblique Angle, Tracking, and Width Factor

The **Oblique Angle** option changes the slant of selected text, almost like applying italics, except that text can be slanted forward and backwards by a specific amount (up to 85 degrees).

The **Tracking** option changes the distance between characters. By reducing the tracking, you can squeeze more text into a line.

The **Width Factor** option makes selected characters wider and narrower. Again, by making characters narrower, you can squeeze more text into a line. More details follow.

MTEXT TEXT ENTRY AREA

The text entry area is rectangle that represents the mtext bounding box. The box is topped by a tab bar that sets tabs, indents, and margins. A different indent and tab setting can be applied to each paragraph.

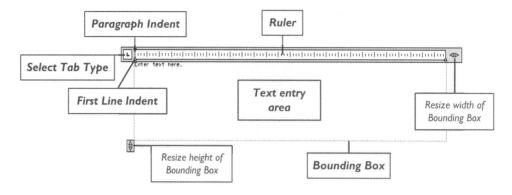

First Line Indent

The *first line indent* shows how far in the first line of each paragraph is indented. You move its arrowhead along the tab bar by dragging it back and forth with the cursor.

You can also create *hanging indents*: move this indent farther left than the paragraph indent. Hanging indents are created automatically when you apply bullets and lists.

Paragraph Indent

The *paragraph indent* shows how far the entire paragraph should be indented. Only the left indent can be modified; to change the right indent, drag the boundary box margin.

Tabs

Click the tab bar to set tabs; existing tabs can be dragged back and forth along the tab bar. AutoCAD supports the commonly used left tabs (illustrated below), as well as right, center, and tabs for lining up text by decimals.

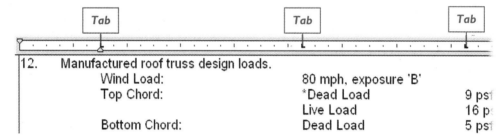

Change the tab type by clicking the **L** icon at the left end of the tab bar. Right-clicking the tab bar reveals a shortcut menu:

 Paragraph

The **Paragraph** option (named **Indents and Tabs** in earlier releases of AutoCAD) displays a dialog box that controls all aspects of paragraphs.

The **Tab** section lets you enter exact tab stop positions. *Tab stops* line up columns of data.

The **Left Indent** and **Right Indent** sections allow you to enter precise distances to indent the first line and the paragraph (i.e. every line).

The **Paragraph Alignment** section selects the type of paragraph justification.

The **Paragraph Spacing** section sets the distance between paragraphs, defined as a distance before and after the paragraph.

The **Paragraph Line Spacing** section sets the spacing between *lines*, despite the the word "paragraph" in the name.

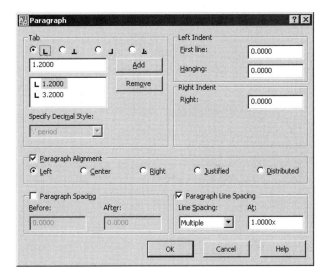

Set Mtext Width and  Set Mtext Height

The **Set Mtext Width** and Set Mtext Height options display a dialog box that allows you to enter a precise width or height for the bounding box.

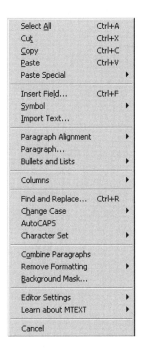

As an alternative, you can drag the right edge of the bounding box, making it wider and narrower.

MTEXT SHORTCUT MENU

Right-clicking the text entry area displays a shortcut menu that reveals many additional options. The content of the menu varies according to the text selected. Some options are new to AutoCAD 2008.

The first four options on the shortcut menu are standard: **Select All** (CTRL+Z), **Cut** (CTRL+X), **Copy** (CTRL+C), and **Paste** (CTRL+V).

 Paste Special

The **Paste Special** item provides ways to paste text without losing formatting. Its options can be confusing to sort out, not least because Autodesk's documentation doesn't explain them.

Paste — pastes text with all formatting intact. For instance, if the text contains several fonts and colors, they are preserved when pasted as mtext.

Paste without Character Formatting — pastes the text according to the current AutoCAD text style.

Paste without Paragraph Formatting — pastes text without right or centered justification.

Paste without Any Formatting — pastes text with the current style.

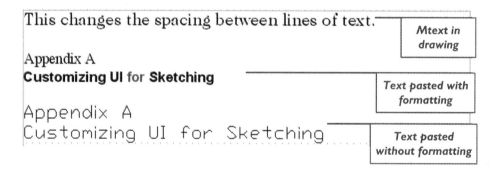

Insert Field

The **Insert Field** option displays the Field dialog box. See Chapter 16, "Tables and Fields."

Symbol

The **Symbol** option displays a submenu listing common drafting and other symbols.

Degrees	%%d
Plus/Minus	%%p
Diameter	%%c
Almost Equal	\U+2248
Angle	\U+2220
Boundary Line	\U+E100
Center Line	\U+2104
Delta	\U+0394
Electrical Phase	\U+0278
Flow Line	\U+E101
Identity	\U+2261
Initial Length	\U+E200
Monument Line	\U+E102
Not Equal	\U+2260
Ohm	\U+2126
Omega	\U+03A9
Property Line	\U+214A
Subscript 2	\U+2082
Squared	\U+00B2
Cubed	\U+00B3
Non-breaking Space	Ctrl+Shift+Space
Other...	

Behind the symbol names are Unicode numbers for each character, such as such as \U+2248. Some of the symbols can be used for Greek characters; for example, **Ohm** is the same as **omega**.

One symbol is invisible: the "non-breaking space." When you place it between two words, it prevents AutoCAD from using that space to wrap the sentence. For instance, if your text includes the phrase "one inch," you might want a non-breaking space between the "one" and "inch" to keep the two words together.

Other opens the Windows Character Map dialog box, which contains all the characters available for the font. Inserting a symbol in this way requires the rather awkward process of (1) selecting the symbol, (2) clicking **Select,** (3) clicking **Copy,** (4) clicking **x** to close the dialog box, (5) right-clicking in the mtext editor, and (6) selecting **Paste**.

Import Text

The **Import Text** option displays a dialog box for selecting documents saved in plain text (ASCII format) or RTF (rich text format). You cannot import text saved in a word processing format, such as Write (WRI), Word (DOC), or WordPerfect (WP) files. Attempting to import these formats results in this warning:

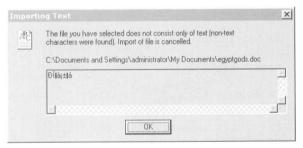

AutoCAD can import files up to 32KB in size, although a warning dialog box incorrectly states the limit is 16KB. Excel spreadsheets imported through RTF format are truncated at 72 rows, unless created in Office 2002 (with service pack 2 installed) or later. Text color is set to the current color; some (but not all) other formatting is preserved, such as font names and sizes, though not justification or columns.

As a better alternative, you can copy text from any document to the Clipboard, and then paste it into the mtext editor. The text formatting is retained.

Paragraph Alignment, Paragraph, and Bullets and Lists

The **Paragraph Alignment** item displays a submenu listing the justification modes described earlier.

The **Paragraph** option (formerly **Indents and Tabs**) displays the dialog box described earlier.

The **Bullets and Lists** option displays a submenu with options for creating lists and prefixing them with a variety of bullets. These options were described in detail earlier.

Columns

The **Columns** item displays a submenu for creating static or dynamic columns, as described earlier.

Find and Replace

The **Find and Replace** option displays the same dialog box for searching for and optionally replacing text as displayed by the TEXT command.

Change Case

The **Change Case** option changes selected text to all UPPERCASE or all lowercase.

UPPERCASE Ctrl+Shift+U
lowercase Ctrl+Shift+L

There are a couple of keyboard shortcuts that bypass the menu. Select text, and the press the following keys:

CTRL+SHIFT+U converts selected text to all uppercase.

CTRL+SHIFT+L converts selected text to all lowercase;.

AutoCAPS

The **AutoCAPS** toggle forces the computer's CapsLock mode, so that all typed text and imported text are displayed in uppercase characters. If the CapsLock light won't go off on your keyboard, it's because this option is still turned on.

("Or," suggests the copy editor, "you spilled Gatorade on your keyboard.")

Character Set

The **Character Set** option changes the character set to match local language requirements. Not all fonts support all character sets.

Combine Paragraphs

The **Combine Paragraphs** option groups selected paragraphs into a single paragraph. You need to select at least two paragraphs for this option to operate.

Remove Formatting

The **Remove Formatting** option removes bold, italic, and underline formatting from the selected text.

Remove Character Formatting Ctrl+Space
Remove Paragraph Formatting
Remove All Formatting

Remove Character Formatting — removes formatting from characters, such as boldface; paragraph formatting remains. You can use the **CTRL+SPACE** shortcut to remove formatting from selected text.

Remove Paragraph Formatting — removes formatting from paragraphs, such as justification.

Remove All Formatting — removes all format overrides. This option returns the font to the current style.

Background Mask

The **Background Mask** option places a colored rectangle behind paragraphs of text. The size of the background mask is the same as the mtext window, which may be larger or smaller than the area of the text. You can resize the mask with grips editing.

To create the background mask, select **Background Mask** from the shortcut menu. In the dialog box, turn on the **Use Background Mask** option, and then select the color.

If you turn on the **Use Drawing Background Color** option, the background color is (usually white), masking out the other drawing entities behind it. You may need to use the TEXTTOFRONT command to display text on top of other objects.

 Note: To create white text on a colored background, as illustrated above, you need to select a True Color as follows:
1. In the Text Formatting toolbar's **Color** droplist, select **Select Color**.
2. In the Select Color dialog box, click the **True Color** tab.
3. Select any color.
4. Drag the **Luminance** slider to the top, turning the text color to white. Click **OK**.

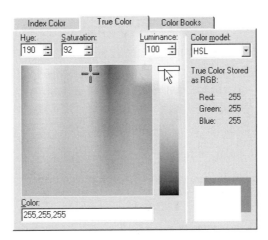

Editor Settings

The **Editor Settings** item controls the display of elements of the mtext editor.

Show Toolbar — toggles the display of the entire Text Formatting toolbar.

Show Options — toggles the lower half of the Text Formatting toolbar.

Show Ruler — toggles the display of the tab ruler.

Opaque Background — toggles the gray background of the text entry area.

 Text Highlight Color — selects a color for highlighted text.

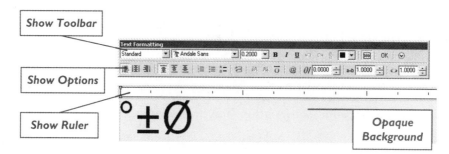

Show Toolbar

Show Options

Show Ruler

Opaque Background

With all options turned off, you have a bare text editor:

All options turned off.

Learn About MText

This option displays online help of the New Features Workshop with information on the "inplace editor."

Cancel

Exits the MTEXT command.

PLACING PARAGRAPH TEXT: ADDITIONAL METHODS

The MTEXT command has many options. As well, there is the command-line version of the command, and a system variable that selects the text editor to use. Each time you respond to an option (except the **Width** option), AutoCAD redisplays the MTEXT prompt until you pick the opposite corner:

Specify opposite corner or [Height/Justify/Line spacing/Rotation/Style/Width/Columns]:

- **Height** specifies the height of text.

- **Justify** specifies the justification of text in the bounding box.

- **Line spacing** specifies the spacing between lines of text.

- **Rotation** rotates the bounding box.

- **Style** selects the text style.

- **Width** specifies the width of the bounding box.

- **Columns** specifies the number of columns.

- **MTEXTED** system variable selects the mtext editor.

- **-MTEXT** command accepts text entry at the command-line.

- **TEXTTOFRONT** command brings text in front of other objects.

Let's look at each.

Height

The **Height** option changes the height of the text used by the mtext editor. It prompts you:

Specify height <0.2000>: *(Type a number or indicate a height.)*

Justify

The **Justify** option selects the justification and positioning of text within the bounding box. AutoCAD prompts:

Enter justification [TL/TC/TR/ML/MC/MR/BL/BC/BR] <TL>: *(Enter an option, or press* ENTER.*)*

The justification options are the same as for the MTEXT command, except that in this case, there are two areas where justification applies: (1) text and (2) flow.

Text justification is left, center, or right, relative to the left and right boundaries of the rectangle. *Flow justification* positions the block of text top, middle, and bottom relative to the top and bottom boundaries of the rectangle.

Linespacing

The **Linespacing** option changes the spacing between lines of text, sometimes called the "interline spacing" or "leading." The option has two methods for specifying the spacing: **At least** and **Exactly**. The **At least** option specifies the minimum distance between lines, while **Exactly** specifies the precise distance between lines of text.

Enter line spacing type [At least/Exactly] <At least>: **a**

Enter line spacing factor or distance <1x>: *(Enter a value, including the* **x** *suffix.)*

The value you enter is a multiplier of the standard line spacing distance, which is defined in the font.

Rotation

The **Rotation** option specifies the rotation angle of the bounding box.

Specify rotation angle <default>: *(Enter an angle, or pick two points.)*

This option rotates the entire block of text. For example, specify 90 degrees to place text sideways at the edge of a drawing.

Note: When you use the mouse to show the rotation angle, AutoCAD calculates the angle as follows:

- The *start* of the angle is the X-axis.
- The *end* of the angle is the line anchored by the "Specify first corner:" pick; the other corner is now in line with the indicated angle.

Style

The **Style** option selects a text style to use for the multiline text.

Enter style name (or ?) <Standard>: *(Enter a name, or press* **?** *for a list of style names.)*

When you enter **?** in response to this prompt, AutoCAD displays the names of text styles defined in the drawing. You create new text styles with the STYLE command (described later in this chapter) or with a style borrowed from another drawing using the **DesignCenter**.

Width

The **Width** option specifies the width of the bounding box.

Specify width: *(Enter a value, or pick a point to show the width.)*

A width of 0 (zero) has special meaning. AutoCAD draws the multiline text as one long line (no word wrap), as if there were no bounding box at all.

Columns

The **Columns** option specifies the number of columns.

MTextEd

The MTEXTED system variable defines the text editor used to edit mtext.

> Command: **mtexted**
>
> Enter new value for MTEXTED, or . for none <"Internal">: *(Enter a name.)*

MTextEd	Meaning
. *(period)*	Uses the default mtext editor; does not disable mtext editor.
Internal	Uses the default mtext editor.
oldeditor	Uses the old AutoCAD 2005 mtext editor.
:lisped	Uses an AutoLISP-based editor.
notepad.exe	Uses Notepad as the editor.

As an alternative, you can specify any word processor, but you must include the full path and executable name. For example, to use the Atlantis word processor:

> Command: **mtexted**
>
> Enter new value for MTEXTED, or . for none <"Internal">: **"c:\program files\atlantis\atlantis.exe"**

External text editors and word processors do not understand AutoCAD's mtext format codes, and thus display them literally, which can be an advantage if you wish to examine the code.

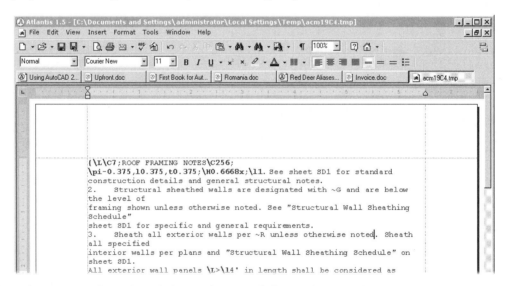

The **:lisped** option works only with lines of text with fewer than 80 characters.

-MText

The -MTEXT command operates identically to the MTEXT command, except that it prompts you to enter text at the command line.

> Command: **-mtext**
>
> Current text style: "ROMAND" Text height: 3/16"
>
> Specify first corner: *(Pick a point.)*
>
> Specify opposite corner or [Height/Justify/Line spacing/Rotation/Style/Width]: *(Pick another point, or enter an option.)*
>
> MText: *(Enter text.)*
>
> MText: *(Press ENTER to exit the command.)*

At the "MText:" prompt, you can type lines of text; as an alternative, you can right-click and select **Paste** so that text in the Clipboard is pasted to the command line and placed in the drawing. The drawback is that the text cannot contain blank lines, which AutoCAD misinterprets as its cue to exit the **-MTEXT** command.

TextToFront

The **TEXTTOFRONT** command brings text to the front of the display order. It works with text and dimensions. (Use the **HATCH** command's **Draw Order** option for hatch patterns, and the **DRAWORDER** command for all other objects.)

You can force AutoCAD to display all text over top all other objects in the drawing, like this:

> Command: **texttofront**
>
> Bring to front [Text/Dimensions/Both] <Both>: **t**
>
> *n* object(s) brought to front.

This command is much more efficient than the **DRAWORDER** command, because it selects text only, instead of any object.

Left: Text behind another object.
Right: Text brought to the front.

LEADER AND 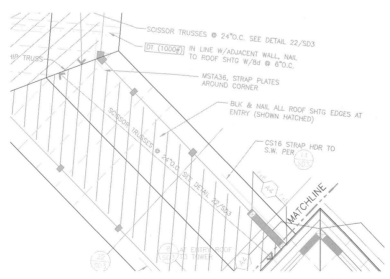 QLEADER

The **LEADER** and **QLEADER** commands place *callouts* (a.k.a leaders) in drawings. The figure illustrates examples of leaders in a drawing.

AutoCAD has three commands for creating leaders. The differences among them are, as follows:

- **LEADER** — all options are entered at the command line. This is the original method of creating leaders, dating back to AutoCAD Release 13.

- **QLEADER** (short for "quick leader") — includes a Settings dialog box for fashioning the leader's properties. Introduced in AutoCAD Release 14.

- 🔵 **MLEADER** (short for "multiline leader") — groups multiple leader lines and block annotation together. Mleaders can have preset styles through the **MLEADERSTYLE** command. Introduced with AutoCAD 2008. Unlike leaders and qleaders, mleaders are limited to mtext and blocks; they cannot include tolerances or selected objects.

TUTORIAL: PLACING LEADERS

1. To place leaders in drawings, start the **QLEADER** command:
   - From the **Dimension** menu, choose **Leader**.
   - From the Dimension toolbar, choose the **Quick Leader** button.
   - At the 'Command:' prompt, enter the **qleader** command:

     Command: **qleader** *(Press ENTER.)*
   - Alternatively, enter the **le** alias at the 'Command:' prompt.

2. In all cases, AutoCAD prompts you for the first point, which is where the tip of the arrowhead is placed:

   Specify first leader point, or [Settings] <Settings>: *(Pick point 1.)*

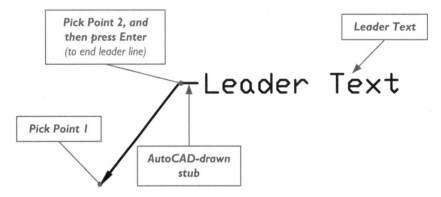

3. Continue picking points. Each pick indicates the location of the next vertex in the leader line. Pressing **ENTER** stops drawing the leader line. Later, AutoCAD draws the short horizontal stub automatically.

   Specify next point: *(Pick point 2 , and then press ENTER to end the leader line.)*

4. If the text needs to be constrained in width, enter the width here; otherwise, enter 0 for unconstrained width:

   Specify text width <0.0000>: *(Press ENTER.)*

5. Enter the text, and then press **ENTER**.

   Enter first line of annotation text <Mtext>: *(Enter text.)*

6. Press **ENTER** twice in a row to exit the command.

   Enter next line of annotation text: *(Press ENTER to exit command.)*

If the leader's start point touches an object using object snap, it is *associated* with the object. This means that when you move the object, the arrowhead moves with the object, although the leader text stays in place. (Associativity is available when associative dimensioning is turned on through the **DIMASSOC** system variable — which is the case in most drawings.)

Many of the leader's properties, such as line color and arrowhead scale, are determined by system variables related to dimensioning, all of which start with **DIM**. To modify the properties, use the

DIMSTYLE command, as described in the next chapter.

You can use grips editing to modify the leader's position. The figures illustrates some possibilities.

Left: *Moving the arrowhead.*
Right: *Moving a vertex.*

To edit the leader text, double-click the text (not the leader line), and AutoCAD displays the familiar mtext editor.

PLACING LEADERS: ADDITIONAL METHODS

There are these alternatives to creating leaders:

- **Settings** specifies leader characteristics through a dialog box of the **QLEADER** command.

- **LEADER** command enters all options at the command line.

Let's look at them.

Settings

The **Settings** option displays a dialog box for specifying the look and operation of "q"-leaders.

> Command: **qleader** *(Press* ENTER.*)*
>
> Specify first leader point, or [Settings] <Settings>: *(Type* **S**.*)*

Notice the dialog box:

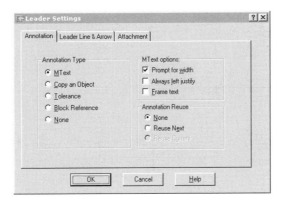

Annotation Type

The options you select under **Annotation Type** affect the prompt displayed by the QLEADER command.

MText (default) — displays prompts for creating leader text from mtext:

> Specify text width <0.0000>: *(Enter a width, or press* ENTER.*)*
>
> Enter first line of annotation text <Mtext>: *(Enter text.)*
>
> Enter next line of annotation text: *(Press* ENTER *to end the command.)*

Copy an Object — prompts you to select another line of text already in the drawing. This can be multiline text (including other leader text), single-line text, a tolerance object, or a block.

> Select an object to copy: *(Select one object.)*

Tolerance — displays the Tolerance dialog box when the command requires annotation; see Chapter 15, "Geometric Dimensioning and Tolerancing."

Block Reference — prompts you to insert a block in place of the leader text:

> Enter block name or [?]: *(Enter the name of a block)*
>
> Specify insertion point or [Scale/X/Y/Z/Rotate/PScale/PX/PY/PZ/PRotate]: *(Pick a point.)*
>
> Enter X scale factor, specify opposite corner, or [Corner/XYZ] <1>: *(Enter a scale factor.)*
>
> Enter Y scale factor <use X scale factor>: *(Press ENTER.)*
>
> Specify rotation angle <0>: *(Enter an angle.)*

None — removes extra prompts; draws the standard arrowhead and leader line.

MText Options

The **MText Options** are available only when the **MText** option was selected (see above).

Prompt for Width (default = on) — toggles the display of prompts for the width of the mtext leader text:

> Specify text width <0.0000>: *(Enter a width, or press ENTER.)*

Always Left Justify (default = off) — toggles left-justification of mtext, regardless of leader location. AutoCAD normally adjusts the text justification based on the leader line's orientation. When this option is turned on, multiple lines of text are left-justified when the leader angles left, as illustrated by the figure below. When this option is turned off, multiple lines of text are right-justified.

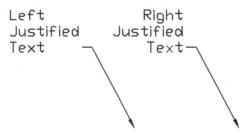

Frame Text (default = off) — toggles the addition of a rectangular frame around the mtext annotation:

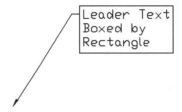

Annotation Reuse

The options under **Annotation Reuse** determine whether the leader text or other annotation is reused by subsequent **QLEADER** commands.

None (default) — causes AutoCAD to prompt you for an annotation, as described above.

Reuse Next — uses the next annotation you create for subsequent leaders. There is no prompt for specifying mtext, blocks, tolerances, or copying objects.

Reuse Current — reuses the current annotation for future leaders.

Leader Line

Straight — draws the leader line using straight line segments.

Spline — draws the leader from a spline object, resulting in a curved leader.

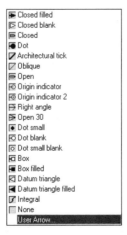

Number of Points

No Limit — command prompts you to "Specify next point:" until you get tired and press ENTER.

Maximum — command stops prompting you after a fixed number, such as 3. You can always stop earlier by pressing ENTER. You must set the number to one more than the number of leader segments you want to create: 3 means that you are prompted for two segments; recall that AutoCAD draws the third segment automatically. The minimum maximum is 2; maximum is 999!

Arrowhead

Select one of the arrowheads shown by the droplist:

```
Closed filled
Closed blank
Closed
Dot
Architectural tick
Oblique
Open
Origin indicator
Origin indicator 2
Right angle
Open 30
Dot small
Dot blank
Dot small blank
Box
Box filled
Datum triangle
Datum triangle filled
Integral
None
User Arrow...
```

User Arrow — lists blocks in the drawing, of which one can be selected for use as the leader's arrowhead.

Angle Constraints

First Segment — specifies the angle of constraint for the first segment. "Any angle" means the segment is not constrained; otherwise, select from Horizontal (0 degrees), 90, 45, 30, or 15 degrees.

Second Segment — specifies the angle-of-constraint for the second leader segment.

Multiline Text Attachment

The **Multiline Text Attachment** options are available only when the **Mtext** option is selected on the **Annotation** tab. These options align the text with the leader line, and can be set differently for left- and right-justified text.

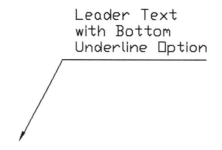

Top of Top Line — aligns the leader line with the top of the top mtext line.

Middle of Top Line — aligns the leader line with the middle of the top mtext line.

Middle of Multiline Text — aligns the leader line with the middle of the mtext.

Middle of Bottom Line — aligns the leader line with the middle of the bottom mtext line.

Bottom of Bottom Line — aligns the leader line with the bottom of the bottom mtext line.

If none of the above options works for you, you may have selected the wrong side: select an option for the correct side.

Underline Bottom Line — attaches the leader line to the bottom of the mtext, and underlines the last mtext line.

Leader Command

To place leaders in drawings with the LEADER command:

> Command: **leader** *(Press ENTER.)*
>
> Specify leader start point: *(Pick a point.)*
>
> Specify next point: *(Pick another point.)*
>
> Specify next point or [Annotation/Format/Undo] <Annotation>: *(Press ENTER to specify text.)*

The **Annotation** and **Format** options are similar to those of QLEADER, but prompt you at the command line. For example the **Annotation** option prompts:

> Enter first line of annotation text or <options>: *(Press enter for options; don't enter "O"!)*
>
> Enter an annotation option [Tolerance/Copy/Block/None/Mtext] : *(Enter an option.)*

And the **Format** option prompts:

> Enter leader format option [Spline/STraight/Arrow/None] <Exit>: *(Enter an option.)*

The **Undo** option undoes the last action.

MLEADER

The MLEADER command draws leaders with multiple lines, among other effects.

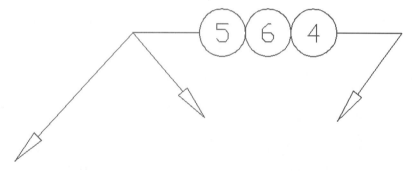

The initial mleader can be drawn with one or more leader lines; the MLEADEREDIT command adds and removes leader lines, as do the undocumented AIMLEADEREDITADD and AIMLEADEREDITREMOVE commands.

Multiline leaders can be attached to text or blocks, like the LEADER and QLEADER commands, but not tolerance symbols. The MLEADERSTYLE command determines the default look of mleaders; the CMLEADERSTYLE system variable reports the current mleader style name.

By default, the blocks are bubbles, as illustrated above, but can also be any user-defined block. In addition, mleaders can have multiple blocks per leader line; the MLEADERCOLLECT command handles that task.

When the drawing contains a number of mleaders near each other, the MLEADERALIGN command makes them look neat and tidy by lining them up. Mleaders support annotative scaling, so that they appear at the correct size in model space.

Mleaders have triangular grips not found on regular leaders; these adjust the shoulder length (landing).

TUTORIAL: DRAWING MULTILINE LEADERS WITH TEXT (MLEADER)

1. To draw leaders with multiple leader lines, start the **MLEADER** command:
   - From the **Dimension** menu, choose **Multileader**.
   - From the Multileader toolbar, choose the **Multileader** button.
   - In the Dashboard's Multileader panel, choose the **Multileader** button.
   - At the 'Command:' prompt, enter the **multileader** command:

     Command: **mleader** *(Press ENTER.)*
   - Alternatively, enter the **mld** alias at the 'Command:' prompt.

2. In all cases, AutoCAD prompts you to specify the start point of the leader:

   Specify leader arrowhead location or [leader Landing first/Content first/Options] <Options>: *(Pick point 1.)*

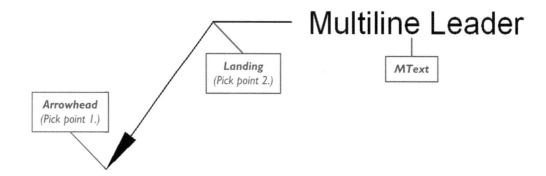

3. The prompts that follow vary, depending on the mleader style. The default style, "Standard," next prompts you to pick the end of the leader line, which also happens to be the start of the landing line:

   Specify leader landing location: *(Pick point 2.)*

4. As soon as you pick the landing location, the mtext editor opens.

   Enter text for the leader, and then click the mtext editor's **OK** button to exit the command.

This command has the following options:

   Specify leader arrowhead location or [leader Landing first/Content first/Options] <Options>:

Specify leader arrowhead location — specifies the endpoint of the arrowhead; the reset of the leader is drawn from there.

leader Landing first — draws the landing first, followed by the arrowhead location.

Content first — draws the "content" (mtext or blocks) first, followed by the arrowhead location.

Options — leads you to a further set of options. Quite frankly, it is easier to set these options with the MLEADERSTYLE command, but for the sake of completeness I include them here:

> Enter an option [Leader type/leader lAnding/Content type/Maxpoints/First angle/Second angle/eXit options] <eXit options>:

Leader type — draws the leader from straight lines, a spline, or none at all.

leader lAnding — specifies the length of the landing, or no landing at all.

Content type — specifies mtext, a block, or no content. If a block, then its name. After the MLEADER command has been used at least once in a drawing, then these bubble blocks become available:

_DetailCallout	_TagBox	_TagCircle
_TagSlot	_TagTriangle	_TagHexagon

Maxpoints — specifies the maximum number of leader vertices; default and minimum = 2.

First angle — constrains the angle of the first leader segment.

Second angle — constrains the angle of the second leader segment; default = 0 degrees (horizontal).

TUTORIAL: ADDING LEADERS TO MLEADERS (MLEADEREDIT)

Once one multiline leader is in a drawing, you can add additional leader lines to it with the MLEADEREDIT command. This command adds and removes leader lines; it works with one mleader object at a time.

1. To add leaders to mleaders, start the MLEADEREDIT command:
   - From the Multileader toolbar, choose the **Multileader Edit** button.
   - In the Dashboard's Multileader panel, choose the **Multileader Edit** button.
   - At the 'Command:' prompt, enter the **multileaderedit** command:

     Command: **mleaderedit** *(Press ENTER.)*
   - Alternatively, enter the **mle** alias at the 'Command:' prompt.

2. In all cases, AutoCAD prompts you to select an mleader:
   Select a multileader: *(Pick an mleader.)*

3. Specify whether you want to add or remove leader lines; for this tutorial, press ENTER to take the default, Add:
   Select an option [Add leader/Remove leader] <Add leader>: *(Press ENTER.)*

4. Indicate where you want the arrowhead located; AutoCAD then positions the leader line:
   Specify leader arrowhead location: *(Pick a point.)*

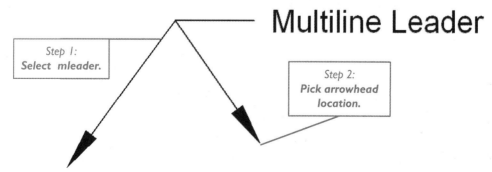

 Note: The second leader line can end up on the other side of the text, depending on how you position the cursor.

5. Add additional leader lines, or press **ENTER** to exit the command:

    Specify leader arrowhead location: *(Press ENTER.)*

To remove leader lines, restart this command and then select the **Remove** option. Pick the leaders to be removed, as follows:

    Command: **mleaderedit**

    Select a multileader: *(Pick a multileader.)*

    Select an option [Add leader/Remove leader] <Add leader>: **r**

    Specify leaders to remove: *(Pick one or more leader lines.)*

    Specify leaders to remove: *(Press ENTER to exit the command.)*

 Notes: The **Remove** option allows you to remove all leaders from mleaders, leaving just the mtext or block. Since it is still seen as an mleader object, you can add leader lines to it again.

The Multileader toolbar and panel don't use the **MLEADEREDIT** command to add and remove leaders. Instead, they use a pair of commands undocumented by Autodesk:

> **AIMLEADEREDITADD** — adds leaders to mleaders.
> **AIMLEADEREDITREMOVE** — removes leaders from mleaders.

The advantage of this pair of commands is that they do their work more quickly, because they exclude the "Select an option [Add leader/Remove leader]" prompt.

TUTORIAL: DRAWING MLEADERS WITH BUBBLES (MLEADERSTYLE)

Multiline leaders can show bubbles in place of text. To do so, you first need to set up a new mleader style. This is done with the MLEADERSTYLE command.

1. To create a new mleader style, start the **MLEADERSTYLE** command:
    - From the **Format** menu, choose **Multileader Style**.

    - From the Multileader toolbar, choose the **Multileader Style** button.

    - In the Dashboard's Multileader panel, choose the **Multileader Style** button.

    - At the 'Command:' prompt, enter the **mleaderstyle** command:

      Command: **mleaderstyle** *(Press ENTER.)*

    - Alternatively, enter the **mls** alias at the 'Command:' prompt.

2. In all cases, AutoCAD displays the Multiline Leader Style dialog box. To create the new style, click **New**.

3. Enter a name for the style. For this tutorial, enter "Round Bubble," and then click **Continue**.

4. Since we want to change the content of the mleader, choose the **Content** tab.

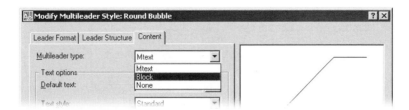

5. From the Multileader Type droplist, select **Block**.
6. Notice that the dialog box changes to show options related to blocks.

   From the Source Block droplist, select **Circle**.

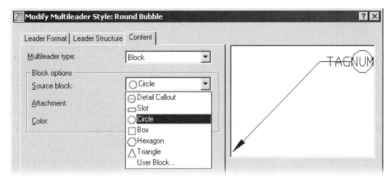

7. Click **OK** to exit the first dialog box.

   Click **Set Current** to make this new style current.

   Click **Close** to exit the second dialog box.
8. Use the **MLEADER** command to place multiline leaders with the new style.

   Command: **mleader**

   Specify leader arrowhead location or [leader Landing first/Content first/Options] <Options>: *(Pick a point.)*

   Specify leader landing location: *(Pick another point.)*

   Enter attribute values

   Enter tag number <TAGNUMBER>: **1**

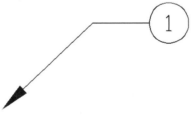

 Notes: The other style options are very similar to those of the Settings dialog box found in the **QLEADER** command. The **CMLEADERSTYLE** system variable stores the name of the current mleader style name.

You can easily create styles from existing mleaders that you've modified:
1. Select the mleader, and then right click.
2. From the shortcut menu, select **Multileader Style**, and then **Save As New Multileader Style**.
3. Notice the dialog box. Enter a name, and then click **OK**.

To use the new style, select it from the droplist in the Multiline toolbar or panel. Styles created this way remember the mtext, and will prompt you:

Overwrite default text [Yes/No] <No>: *(Type Y or N.)*

TUTORIAL: COLLECTING BUBBLES INTO ONE MLEADER (MLEADERCOLLECT)

Multiline leaders can have two or more bubbles each. To do so, you first create a number of mleaders with a single bubble each, and then use the **MLEADERCOLLECT** command to combine them. This is commonly done when the leaders all point to the same feature, as illustrated below.

1. Draw two or more bubble mleaders following the steps described above.

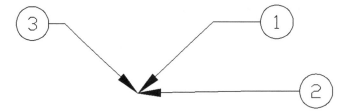

2. To collect multiple multiline leaders into one, start the **MLEADERCOLLECT** command:
   * From the Multileader toolbar, choose the **Collect Multileaders** button.
   * In the Dashboard's Multileader panel, choose the **Collect Multileaders** button.
   * At the 'Command:' prompt, enter the **mleadercollect** command:

     Command: **mleadercollect** *(Press ENTER.)*
   * Alternatively, enter the **mlc** alias at the 'Command:' prompt.

2. In all cases, AutoCAD prompts you to select two or more mleaders:

   Select multileaders: **all**

   Select multileaders: *(Press ENTER to exit object selection.)*

3. AutoCAD asks where you would like to position the collected mleader:

   Specify collected multileader location or [Vertical/Horizontal/Wrap] <Horizontal>: *(Pick a point, or enter an option.)*

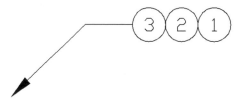

The **MLEADERCOLLECT**'s options include:

Vertical — stacks the bubbles vertically.

Horizontal — aligns the bubbles horizontally, as illustrated above.

Wrap — wraps the bubbles; AutoCAD asks you how many or how far to wrap them:

   Specify wrap width or [Number]: *(Enter a distance, or the number of bubbles.)*

You can always do the stacking or wrapping yourself, because the added bubbles (#2 and #1, in the figure above) are not part of the leader with #3.

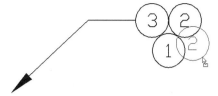

You can drag the bubbles to new locations, as required, and as illustrated above.

 Notes: AutoCAD calls numbers inside the bubbles "tags," but they actually are attributes. To edit them, double-click the bubble (which is a block). In the Edit Attributes dialog box, you can change the tag number to any other number or text.

You can change the bubble type through the **PROPERTIES** command. In the Block section, click the Source block droplist, and then choose another style of bubble, as illustrated below.

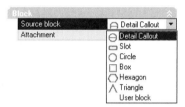

The predefined blocks (a.k.a bubbles) are illustrated below.

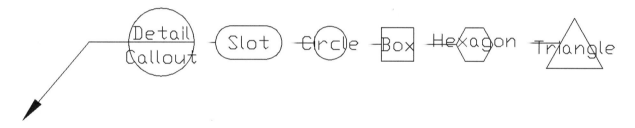

TUTORIAL: ALIGNING MLEADERS (MLEADERALIGN)

When numerous mleaders are in the same area of the drawing, they can look messy to some eyes. The **MLEADERALIGN** command lines them up nicely.

1. Draw two or more bubble mleaders following the steps described earlier.

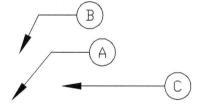

2. To align multiple multiline leaders into a row, start the **MLEADERALIGN** command:
   * From the Multileader toolbar, choose the **Align Multileaders** button.

   * In the Dashboard's Multileader panel, choose the **Align Multileaders** button.

   * At the 'Command:' prompt, enter the **mleaderalign** command:

     Command: **mleaderalign** *(Press* ENTER.*)*
   * Alternatively, enter the **mla** alias at the 'Command:' prompt.

2. In all cases, AutoCAD prompts you to select two or more mleaders:

    Select multileaders: **all**

    Select multileaders: *(Press ENTER to exit object selection.)*

3. AutoCAD asks you to pick one mleader; the others will align with it.

    Current mode: Use current spacing

    Select multileader to align to or [Options]: *(Pick one multiline leader. In this tutorial, I've picked C.)*

4. Pick a point to define the alignment. This is like drawing an xline:

    Specify direction: *(Pick a point.)*

    The screen gets messy looking as AutoCAD shows the mleaders in their original position and their current position. Notice that as you move the cursor, the bubbles of the other mleaders align themselves between the cursor and the base mleader.

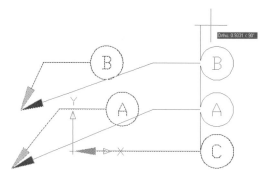

Turn on ortho mode to line mleaders vertically or horizontally.

This command has an **Options** option:

> Enter an option [Distribute/make leader segments Parallel/specify Spacing/Use current spacing] <Use current spacing>:

Distribute — distributes the mleader content (bubbles or mtext) between two points. The distance and angle between the two points determine the spacing and angle of the content.

> Specify first point or [Options]: *(Pick a point.)*
>
> Specify second point: *(Pick another point.)*

make leader segments Parallel — makes the last leader segment of each mleader parallel. AutoCAD prompts you to select the one mleader against which the other leader lines are made parallel.

specify Spacing — allows you to specify the spacing between each leader.

> Specify spacing <0.000000>: *(Enter a number, such as 0.5.)*
>
> Select multileader to align to or [Options]: *(Pick one mleader.)*
>
> Specify direction: *(Pick a point to indicate the angle.)*

Use current spacing — keeps the spacing of the leaders (default).

 **DDEDIT**

The **DDEDIT** command edits text. (**DDEDIT** is short for "dynamic dialog editor," *dynamic dialog* being an old reference to dialog boxes.)

This command handles all types of AutoCAD text: single-line text, mtext, dimensions, fields, and attributes. To do so, it displays a different user interface for each type of text.

TUTORIAL: EDITING TEXT

1. To edit text in drawings, start the **DDEDIT** command:
    * From the **Modify** menu, choose **Object, Text**, and then **Edit**.

    * From the Edit toolbar, choose the **Edit Text** button.

    * At the 'Command:' prompt, enter the **ddedit** command:

        Command: **ddedit** *(Press* ENTER.*)*

    * Alternatively, enter the **ed** alias at the 'Command:' prompt.

2. In all cases, AutoCAD prompts you to select the text:

    Select an annotation object or [Undo]: *(Select another line of text, or press* ESC *to exit the command.)*

    AutoCAD reacts differently, depending on the text you selected:
    * Inplace editor for single-line text placed with the **TEXT** command:

    * Text Formatting toolbar for multi-line text placed with the **MTEXT** command, as well as text in fields, tables, leaders, and dimensions.

    * Edit Attribute Definition dialog box for attribute definitions placed with the **ATTDEF** command, as well as text in mleader bubbles.

 Notes: As an alternative to entering the **DDEDIT** command, you can double-click single-line and multiline text, and attribute definitions. AutoCAD automatically displays the appropriate editor.

Double-clicking leaders and dimension text displays the Properties window, which can also be used to edit text.

Attributes placed with the **INSERT** command are edited with a separate command, **ATTEDIT**.

Single-line Text Editor

When you select single-line text placed by TEXT, AutoCAD displays the inplace editor. Each line of text is edited independently of the others. See the TEXT command at the beginning of this chapter.

You can edit a group of text blocks by holding down the ALT key, and then selecting each text block.

When done editing, press ENTER. If you want to discard the editing changes, press ESC.

Multiline and Leader Text Editor

When you select leaders, dimensions, or text created by the MTEXT command, AutoCAD displays the Text Formatting bar — identical to that displayed by MTEXT.

Make your changes, and choose the **OK** button. AutoCAD returns to the prompt:

> Select an annotation object or [Undo]: *(Select another paragraph of text, or press* ESC *to exit the command.)*

Attribute Text Editor

When you select attribute definitions created by the **ATTDEF** command, AutoCAD displays the Edit Attribute Definition dialog box. (Attributes are discussed in *Using AutoCAD: Advanced.*)

Make your changes to the tag, prompt, and default. Choose the **OK** button.

Enhanced Attribute Editor

When you select a block containing attributes, AutoCAD displays the Enhanced Attribute Editor dialog box. This is the same dialog box displayed by the **EATTEDIT** command, as described in *Using AutoCAD: Advanced.*

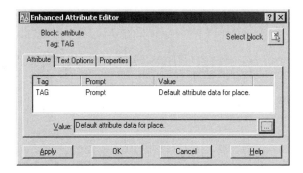

Make your changes to the attributes, and then choose the **OK** button

EDITING TEXT: ADDITIONAL METHODS

An alternative to the **DDEDIT** command is the **PROPERTIES** command. It displays a palette that allows you to edit text and change many properties — color, layer, insertion point, toggling the background mask, and so on. The only property that cannot be changed is the font. which must be changed with the **STYLE** command.

To access this palette, select the text, and then right-click. From the shortcut menu, select **Properties**. AutoCAD displays the Properties palette.

The **Contents** area displays the text of the text object. When you see strange characters — such as {\L and \P — these are mtext formatting codes.

If the text is paragraph text, choose the [...] button, which causes AutoCAD to open the familiar mtext editor.

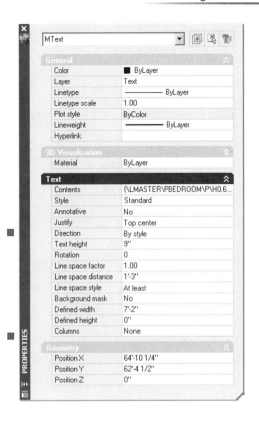

 STYLE

The **STYLE** command defines named text styles based on fonts.

A *style* is a collection of properties applied to the font of your choice, such as slanted text, or perhaps condensed text. Styles are often used in word processing and desktop publishing, and work the same way in AutoCAD. The idea is to preset most text properties so that you don't need to do it each time with every line of text. (Some properties, however, cannot be set by styles, for example color and layer.)

Styles are used by the **TEXT** and **MTEXT** commands, as well as by attributes, fields, leaders, and dimensions. Styles are stored and accessed by a name of your choice.

Text styles are composed of the following information, although not all fonts support all properties:

- **Style name** describes the style — can be up to 255 characters in length.

- **Font file** associates a *.ttf* (TrueType) or *.shx* (AutoCAD) file with the style. PostScript *.pfb* font files can be used after conversion to *.shx* with the **COMPILE** command.

- **Font style** selects regular, boldface, or italicized text.

- **Height** specifies the height of the text. Normally, this is set to 0, which means that the **TEXT** command prompts you for the height. Entering a height here means the command doesn't pester you later.

- **Annotative** toggles annotative scaling (new to AutoCAD 2008).

- **Orientation** makes text is independent of drawing and layout rotation (new to AutoCAD 2008).

- **Width factor** specifies a multiplier making the text wider or narrower. A width of 1 is standard; a width factor of 0.5 produces text at one half the width.

- **Obliquing angle** determines the slant of the text. A positive angle produces a forward slant; a negative angle, a backward slant.

- **Backwards** mirrors the text.

- **Upside down** places the text upside down.

- **Vertical** places the text vertically.

When you change a style, all text assigned to that style also changes.

TUTORIAL: DEFINING TEXT STYLES

1. To define one or more text styles, start the **STYLE** command:
   - From the **Format** menu, choose **Text Style**.

   - From the Text toolbar, choose the **Text Style** button.

   - From the Styles toolbar, choose the **Text Style Manager** button.

   - In the Dashboard's Text panel, pick the **Text Style** button.

   - At the 'Command:' prompt, enter the **style** command.

     Command: **style** *(Press ENTER.)*

   - Alternatively, enter the aliases **st** or **ddstyle** (the old name) at the 'Command:' prompt.

2. In all cases, AutoCAD displays the dialog box (redesigned in AutoCAD 2008):

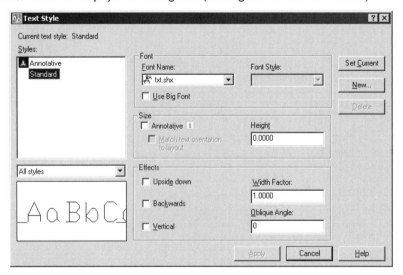

   The **Styles** list provides the names of styles already in the drawing. Use this list to change the properties of existing styles. (Options that do not apply to a font or style are grayed out.)

3. To create a new style, click **New**. Notice that AutoCAD displays the New Text Style dialog box. This is where you name the style.

   Change the default name "style1" to anything else — up to 255 characters long — and then choose **OK**. Notice that AutoCAD adds the name to the **Styles** list.

Notes: To rename a style, right-click its name, and then choose **Rename** from the shortcut menu. You can rename all styles except Standard.

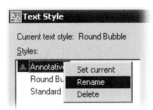

The **Delete** option erases selected styles from drawings — but only when they are unused. If used by text, AutoCAD complains, "Style is in use, can't be deleted." The Standard style can never be deleted.

4. From the **Font Name** droplist, select a font.

There are many font files available for AutoCAD, which comes with an large selection: specialty fonts for map making (symbols instead of letters), cursive writing, and so on.

 Note: The font file should be appropriate to the application. Most mechanical applications use the ANSI type lettering, often called "Leroy." The Simplex font (also known as "RomanS" — roman, single stroke) closely approximates this style.

Architectural drawings typically use "hand-lettered" fonts, such as City Blueprint. Engineering applications often use the Simplex font, since it closely resembles a traditional font constructed with a lettering template.

Multistroke fonts use a code, with "S" denoting single stroke, "D" duplex or double stroke, "C" complex (also double stroke), and "T" triple stroke.

5. Select any of the following options, if available (not grayed out):

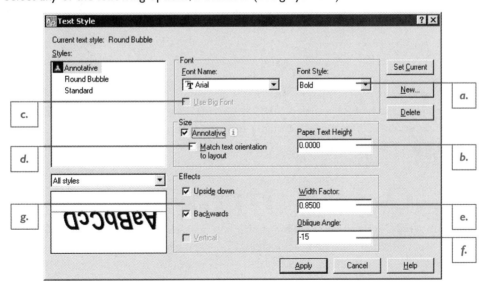

a. The **Font Style** drop list selects Regular, **Boldface**, *Italic*, or ***Bold Italic***. Not all font styles are available, because some fonts cannot handle these effects.

b. The **Height** text box specifies the height of the text, which is measured from the baseline to the top of uppercase letters. There is three ways to specify the height:

- Determine the scale of the drawing before deciding the text height, as described earlier in this chapter..
- Enter a value of 0. This gives you the flexibility to change the height as the text is being placed. This applies particularly if you are not sure of the scale until the end of the drawing process.

Turn on the **Annotative** option, and then specify the true (paper) height.

c. The **Use Big Font** option refers to a class of AutoCAD fonts that handle more than 256 characters, primarily for Chinese and other languages with thousands of characters.

d. **Match Text Orientation to Layout** ensures the text is always displayed "correctly," no matter how the rest of the drawing is rotated (new to AutoCAD 2008). The name of this option is somewhat misleading, because it works even when the layout is not rotated. In the figure below, the entire drawing (including the text) is rotated 90 degrees. The text keeps its orientation, which makes it easier to read.

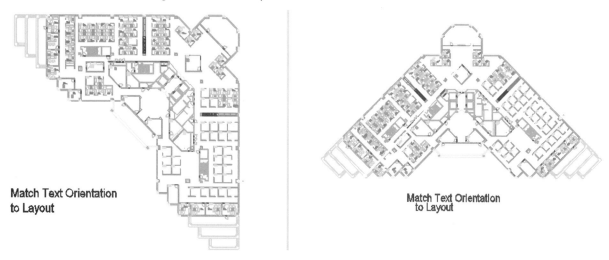

Match Text Orientation to Layout

Match Text Orientation to Layout

e. The **Width Factor** determines the width of characters. A width factor of 1.0 is "standard." A decimal value, such as 0.85, draws text narrower by 15%, which is useful for condensed text. Values greater than 1 expand text.

0.5 Width Factor

2.0 Width Factor

f. **Oblique Angle** determines the slant of characters. A zero obliquing angle draws text that is "straight up." Positive angles, such as 15 degrees, draw text that slants forward, while negative angles draw backward slants. Valid values are between −85 and 85 degrees.

15° Oblique

-15° Oblique

g. **Upside Down** draws text upside down.

∩pside down Text

Backwards means the text is drawn in reverse. Backwards text is useful for plotting drawings on the back side of clear media, or for mold drawings used by the casting and injection-molding industries.

Reverse Text

Vertical draws text vertically, which is *not* the same as rotating text by 90 degrees. Text is drawn so that each letter is vertical and the text string itself is vertical. TrueType and some AutoCAD fonts cannot be drawn vertically.

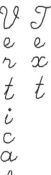

 Note: Obliquing, underscoring, or overscoring should not be used with vertical text, since the result will look incorrect.

6. Click **Set Current** to make the style current; this means any text you draw now takes on this style.
7. Click **OK** to exit the dialog box.

USING TEXT STYLES

The difference between fonts and styles is sometimes confusing. You cannot use fonts directly in AutoCAD; fonts can only be used through styles. (The exception is in the mtext editor, but even there the font only overrides a style.) Styles modify the properties of a font, such as its height, width, and slant. One font can be used by many styles, but each style can specify only one font.

1. To use a font, you must create a style with the **STYLE** command.
2. Then, in the **TEXT** and **MTEXT** commands, you specify the style to use for the text you place.

Each line or paragraph of text is assigned a style; you may change the style with the **PROPERTIES** command. When the style changes, the look of the text also changes. To change text globally, change the style with the **STYLE** command.

Often, drafting offices create a standard selection of styles, each of which is used for a particular type of text in drawing. For example, one style is used for notes, another for titles, and others for the different parts of the title block. Using styles ensures consistency across drawings, no matter the drafter.

Styles Panel

An alternative method to applying styles is using the Dashboard's Styles panel.

1. Select one or more lines of text.
2. Pick a style name from the style droplist:

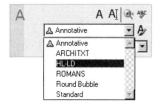

Notice that AutoCAD changes the look of the text to match the style. In the panel, clicking the **Text Style** button displays the Text Style dialog box.

DEFINING STYLES: ADDITIONAL METHODS

The **-STYLE** command defines styles at the command prompt. In addition, several system variables affect the look and quality of TrueType fonts.

* **-STYLE** command defines styles at the command prompt.

* **FONTALT** system variable specifies font substitution.

* **TEXTFILL** system variable toggles the fill of TrueType fonts for plots.

* **TEXTQLTY** system variable adjusts the quality of TrueType fonts for plots.

Let's look at each.

-Style

The **-STYLE** command prompts you to create text styles at the command line:

> Command: **-style**
>
> Enter name of text style or [?]: *(Enter a name, type **?**, or press ENTER.)*

Press **ENTER** to redefine an existing font. AutoCAD prompts:

> Existing style. Full font name = current:
>
> Specify full font name or font file name <TTF or SHX>: *(Enter the name of a .ttf or .shx font file.)*

To specify the name of a Big Font, enter a *.shx* file name, a comma, and the Big Font file name. Enter ~ to display the Select Font File dialog box. Enter **?** to list the names of text styles in the drawing.

The remaining prompts are similar to the options in the Style dialog box:

> Specify height of text or [Annotative]: *(Enter a height, type A, or enter **0**.)*
>
> Specify width factor: *(Enter a factor.)*
>
> Specify obliquing angle: *(Enter an angle between -85 and 85 degrees.)*
>
> Display text backwards? [Yes/No] <N>: *(Enter **Y** or **N**.)*
>
> Display text upside-down? [Yes/No] <N>: *(Enter **Y** or **N**.)*
>
> Vertical? <N>: *(Enter **Y** or **N**.)*

Annotative — displays two more prompts:

> Create annotative text style [Yes/No] <Yes>: *(Enter **Y** or **N**.)*
>
> Match text orientation to layout? [Yes/No] <Yes>: *(Enter **Y** or **N**.)*

FontAlt

The **FONTALT** system variable specifies the name of a font to use when other fonts cannot be found.

Often, different computers have different collections of installed fonts. When you receive a drawing from another AutoCAD drafter, the drawing may use fonts not found on your computer. So that text doesn't go missing, AutoCAD substitutes the *simplex.shx* font for missing fonts. (If the Simplex font cannot be found, AutoCAD displays the Alternate Font dialog box, so that you can select another font file.)

> Command: **fontalt**
>
> Enter new value for FONTALT, or . for none <"simplex">: *(Enter the name of a TrueType or AutoCAD font.)*

For missing TrueType fonts, AutoCAD can substitute another font without warning you, which may be annoying. On my computer, AutoCAD substitutes the Absalom font, which makes drawing texts hard to read, like this:

TextFill

The **TEXTFILL** system variable toggles the fill of TrueType fonts for plots. When on (the default), fonts are filled; when off, only the outline of the font is plotted. This system variable affects only the plotting of fonts; it does not affect the display of fonts in drawings.

> Command: **textfill**
>
> Enter new value for TEXTFILL <1>: **0**

TextQlty

The **TEXTQLTY** system variable specifies the resolution of TrueType font outlines (technically, the tessellation fineness) when plotted: 0 means the text is not smoothed; 100 means maximum smoothness. The difference is seen only when drawings are plotted; this system variable does not affect the display.

> Command: **textqlty**
>
> Enter new value for TEXTQLTYL <50>: **100**

SPELL

The **SPELL** command checks the drawing for words unfamiliar to AutoCAD.

TUTORIAL: CHECKING SPELLING

1. To check the spelling of text, start the **SPELL** command:
   • From the **Tools** menu, choose **Spell**.

   • In the Dashboard's Text panel, pick the **Spell Check** button.

   • At the 'Command:' prompt, enter the **spell** command:

   Command: **spell** (*Press* ENTER.)

   • Alternatively, enter the **sp** alias at the 'Command:' prompt.

   In all cases, AutoCAD displays the Check Spelling dialog box.

2. Click **Start** to begin checking the spelling of words.
3. Confirm the spelling of words AutoCAD does not recognize.
   When done, AutoCAD reports: "Spelling check complete."

This command works with words placed or imported by the **TEXT**, **-TEXT**, **MTEXT**, **-MTEXT**, **LEADER**, **QLEADER**, **MLEADER**, and **ATTDEF** commands. When it comes to attributes, values are spell checked, but not tags. Because 'SPELL can be run transparently, you can use it while other commands are active.

 FIND

The FIND command searches for, and optionally, replaces text.

TUTORIAL: FINDING TEXT

1. To find and replace text, start the FIND command:
   - From the **Edit** menu, choose **Find**.

   - From the Text toolbar, choose the **Find and Replace** button.

   - In the Dashboard's Text panel, pick the **Find** button.

   - At the 'Command:' prompt, enter the **find** command.

   Command: **find** (Press ENTER.)

2. In all cases, AutoCAD displays the Find and Replace dialog box.

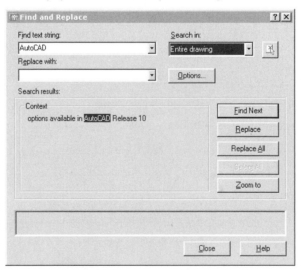

3. Enter a word or phrase to search for in the drawing.
4. To narrow the search, click **Options**.

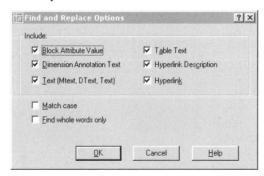

This dialog box lets you narrow the search to specific kinds of words, such as those found only in dimensions or in table text. Field text is considered a form of mtext.

5. Optionally, enter a replacement word or string.

6. Click **Find Next** to find the first instance of the phrase.

   If necessary, click **Replace** or **Replace All**.

7. When done, click **Close**.

The U command reverses changes made by this command.

QTEXT

The **QTEXT** command displays lines of text as rectangles, which speeds up the display of drawings containing a lot of text (short for "quick text").

The rectangular boxes represent the height and length of the text. The change from text to boxes does not take place until the next regeneration; therefore you should follow with the **REGEN** command. The figure illustrates the effects of applying **QTEXT/REGEN** to a drawing.

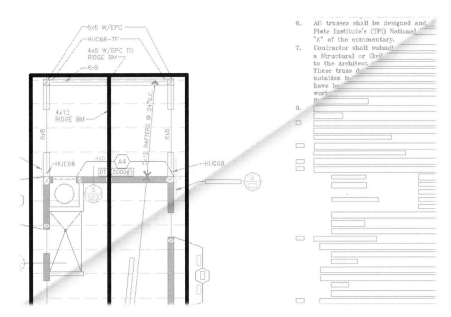

Left: *Normal display of text —* **QTEXT** *off.*
Right: *Text represented by rectangles —* **QTEXT** *on.*

Notes: When **QTEXT** is on, it displays and plots text as rectangles. Being a transparent command, **'QTEXT** can be used during other commands.

History: This command was useful in the early days of AutoCAD, when computers were very slow. Back then, turning off text helped speed up the redraw and regeneration times. Today, visual styles (not text) typically slow display speed.

TUTORIAL: QUICKENING TEXT

1. To display text as rectangles, start the **QTEXT** command:
   • At the 'Command:' prompt, enter the **qtext** command.

   Command: **qtext** *(Press* ENTER.*)*

2. AutoCAD displays the tersest prompt of all commands :
    ON/OFF <OFF>: **on**

3. To see the rectangles, follow with the **REGEN** command:
    Command: **regen**

Even after a regeneration, some text may still appear normal. That's because it lies in another space, either model or paper space. Switch to model space or to a layout to force the regeneration. The figure below illustrates the problem:

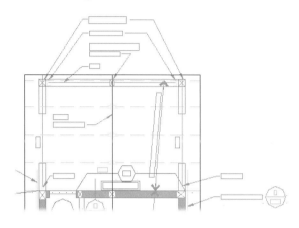

 SCALETEXT AND **JUSTIFYTEXT**

The **SCALETEXT** and **JUSTIFYTEXT** commands change the size and justification of text.

After placing text in drawings, you may find it necessary to change the height of the text, or its justification.

AutoCAD provides the **SCALETEXT** command to change the height (scale) of text. It makes selected text larger or smaller. The advantage of this command over **PROPERTIES** or **SCALE** is that it resizes multiple text objects without changing their location; the **PROPERTIES** command may not do it accurately, while the **SCALE** command scales everything, including location, relative to the base point.

The **JUSTIFYTEXT** command changes the justification of text. It can, for example, change left-justified text to right-justified. The style of justification of mtext can be changed with the mtext editor through the **MTEDIT** command.

TUTORIAL: RESIZING TEXT

1. To resize text, start the **SCALETEXT** command:
    - From the **Modify** menu, choose **Object**, **Text**, and then **Scale**.
    - From the Text toolbar, choose the **Scale Text** button.
    - At the 'Command:' prompt, enter the **scaletext** command.

        Command: **scaletext** *(Press ENTER.)*

2. In all cases, AutoCAD prompts you to select text.
    Select objects: *(Pick one or more lines of text.)*
    Select objects *(Press ENTER to end object selection.)*

3. Pick a base point, or select a justification point:
    Enter a base point option for scaling [Existing/Align/Fit/Center/Middle/Right/TL/TC/TR/
    ML/MC/MR/BL/BC/BR] <Existing>: *(Enter an option, or press ENTER.)*

4. Specify the new height:

> Specify new height or [Match object/Scale factor] <3/16">: *(Enter a height, or an option.)*

When AutoCAD prompts you to select objects, you can choose a mix of text and non-text objects; AutoCAD filters out the non-text objects automatically.

The SCALETEXT command changes the height of selected text relative to a *base point*. The base point can be the existing insertion point or one of the many text justification points. By default, the base point is the existing insertion point.

The **Match Object** option matches the height to that of another text object:

> Select a text object with the desired height: *(Pick another text object.)*
>
> Height=3/16"

The **Scale Factor** option scales the text by a factor. A factor larger than 1 enlarges the text, while a value under 1 reduces it.

> Specify scale factor or [Reference]: *(Enter a factor, such as **2**.)*

The **Reference** option is identical to that of the SCALE command: the text is scaled relative to another size.

TUTORIAL: REJUSTIFYING TEXT

1. To change the justification of text, start the JUSTIFYTEXT command:
   * From the **Modify** menu, choose **Object**, then **Text**, and then **Justify**.
   * From the Text toolbar, choose the **Justify Text** button.
   * At the 'Command:' prompt, enter the **justifytext** command.

   > Command: **justifytext** *(Press ENTER.)*

2. In all cases, AutoCAD prompts you to select text.

   > Select objects: *(Pick one or more lines of text.)*
   >
   > Select objects *(Press ENTER to end object selection.)*

3. AutoCAD prompts you to select a justification option; the option you select will override the justification of the existing text.

   > Enter a justification option [Existing/Align/Fit/Center/Middle/Right/ TL/TC/TR/ML/MC/MR/ BL/BC/BR] <Existing>: *(Press ENTER.)*

The same rules apply to the JUSTIFYTEXT and SCALETEXT commands regarding object selection. The U command can undo any damage.

ANNOTATION SCALING

Annotation scaling ensures only correctly-scaled objects appear in Model tab and model space viewports.

"Annotation" refers to objects that annotate drawings. (*Annotation* means "explanatory text.") In AutoCAD, this includes the following objects:

Single-line text	Mtext	Field text
Attribute text	Dimensions	Tolerances
Multiline leaders	Hatch patterns	Linetypes

As described earlier in this chapter, text needs to be scaled larger to appear at the correct size when the drawing is plotted smaller to fit the paper. Annotation scaling is a system devised by Autodesk to solve the problem semi-automatically. It works like this:

A layout viewport has a scale factor of 1:100. Draw text with an annotative style. AutoCAD automatically assigns the text the inverse scale factor of 100:1.

There is no "Annotation" command; instead, annotation is a property integrated into some commands, styles, and user interface elements. Let's work through a tutorial, and then examine the additional options.

TUTORIAL: AUTOMATICALLY SCALING TEXT

1. In AutoCAD, open the *Annotative.dwg* drawing file from the Companion CD.
2. Switch to layout mode by clicking the Layout1 tab.
3. Enter model space by double-clicking inside the viewport order.

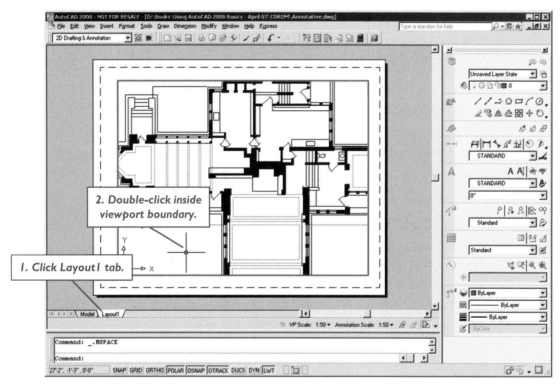

Notice the tools for working with annotative scaling. These can appear in two places, depending on how AutoCAD is set up; one possible location is on the drawing status bar above the command bar, as illustrated below.

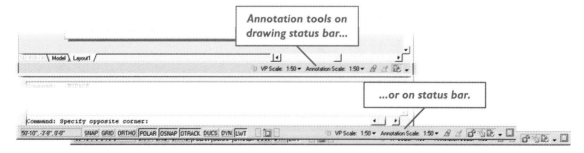

The other location is on the status bar below the command bar, as illustrated above.

4. The drawing is not correctly scaled in the viewport; it should be scaled to meet two criteria: (1) the entire drawing fits the viewport, and (2) the drawing being at a standard scale factor.

    Here's how to select an appropriate scale factor:

    a. Use the **ZOOM Extents** command to fit the drawing to the viewport.

    b. Notice the scale factor next to **VP Scale**: it reads 0.010998 or similar.

    c. Click **VP Scale**, and then select the nearest standard scale factor of **1:100**. (See figure at right.) Notice that the drawing reacts by zooming slightly smaller, and that the Annotation Scale factor matches.

    Notes: The **VP Scale** button displays a long list of scale factors. You can edit this list with the **SCALELISTEDIT** command, removing and adding scale factors. This single command affects the scale factors available to viewport scaling, plot scaling, and annotation scaling.

    The annotation tools can appear on the drawing status bar or the status bar. The location is determined by the **OPTIONS** command: in the Display tab, toggle the **Display Drawing Status Bar** option.

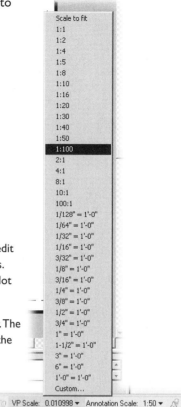

5. To test the annotation feature, place some text with the **TEXT** command. Follow these steps:

    a. Switch to Model tab.

The technical editor explains why: "If you zoom and pan while entering text in a layout tab, you mess up the viewport's scale. When you re-enter model space, the annotations are 'missing,' because their scale no longer matches the viewport scale. For this reason, it is best to enter annotative text in Model tab."

    b. From the Dashboard's Text panel or the Styles toolbar, select the Annotative text style.

    c. Start the **TEXT** command.

    Command: **text**

    d. Notice that the prompt reports that the annotative property is turned on:

    Current text style: "Annotative" Text height: 0'-0" Annotative: Yes

    Specify start point of text or [Justify/Style]: *(Pick a point in the viewport.)*

    e. Notice that the prompt asks for "paper height": this is the height of the text as you want it to appear when plotted. In this case, enter a height of 0.2", which is close to the ideal of 3/16":

    Specify paper height <0'-0">: **0.2"**

    Specify rotation angle of text <0>: *(Press ENTER.)*

    Annotative Text *(Press ENTER twice.)*

Notice that the text appears at the correct size in the viewport. As illustrated below, the floorplan is 70 feet wide, yet the text is legible despite it being just 0.0017 feet tall (0.2").

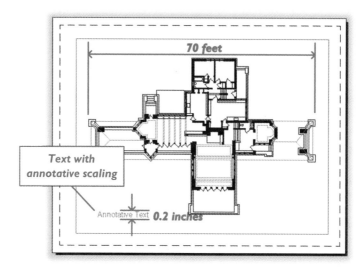

6. To see why this is possible, use the **LIST** command, and then select the text. I've highlighted the important parts below:

TEXT	Layer: "0"
Space:	Model space
Handle =	5d5
Style =	"Annotative"
Annotative:	Yes
Annotative scale:	**1:100**
Typeface =	Arial
start point,	X= 9'-0" Y= -14'-6" Z= 0'-0"
paper text height	**0'-0"**
model text height	**1'-8"**
text	Annotative Text

The text contains the annotative scale property of 1:100, and then applies it to text in model space ("model text height"). We entered a height of 0.2". Behind the scenes, AutoCAD automatically inverted the scale to draw the text at 0.2" x 100:1 = 20" (same as 1'-8" in feet-inches).

The **paper text height** is set to 0'0", even though you entered 0.2". That's because AutoCAD reverse-calculates the height from the annotative scale factor multiplied by the model text height. The paper text height changes automatically when the associated annotative scale changes.

Note: AutoCAD identifies objects with annotative scaling through the triangular icon ⧄. As you pass the cursor over objects, they highlight, and those with annotative scaling display the icon near the cursor.

(*History:* The icon represents the end view of a triangular scale ruler used by architects and engineers. Being triangular in cross-section, it provided 12 scale factors, two along each of the six sides.)

7. To see the effect of annotative scaling, change the scale of the viewport. From the status bar, click **VP Scale** and select **1:50**. Notice that the text disappears. (If necessary, pan the drawing to where the text is/was located.)

8. One of the status bar buttons lets you see annotative-scaled objects, no matter what the viewport scaling is. Click the ⧉ **Annotation Visibility** button. (This button toggles the **ANNOALLVISIBLE** system variable.) Notice that the text reappears and is twice as large.

9. When you work with several viewports scales, it can be a pain to assign annotative scales to a large numbers of objects. To have AutoCAD automatically assign the scale factors, follow these steps"

   a. Click the **Automatically Add Scales** button. (This button toggles the **ANNOAUTOSCALE** system variable.)

   b. Change **VP Scale** to something else, such as **1:40**. Notice that the text does not become larger (as you might expect from step 7), but returns to its former size.

   c. Change **VP Scale** to **1:50**. Again, the text is the "correct" size. No matter which viewport scale you now select, the annotatively-scaled text appears at constant size.

Notes: When objects have more than one annotative scale, the selection cursor shows a double triangle icon, as illustrated below

Autodesk warns against objects having too many annotative scale factors, saying that this could slow down your computer. Curiously, the **LIST** and **PROPERTY** commands do not list multiple scale factors; instead, you need to use the **OBJECTSCALE** command to view, add, and remove scale factors from objects.

WHERE ANNOTATIVE SCALING HAPPENS

Annotative scaling is scattered about AutoCAD in many nooks and crannies. It is a property that can be set in styles and commands, and changed after the fact with these commands and system variables. Annotative styles are indicated in droplists by the triangle icon, such as in the toolbar illustrated below.

Style Commands
DIMSTYLE command — affects dimensions and tolerances.
MSTYLELEADER command — affects multiline leaders.
STYLE command — affects single-line text, mtext, and fields.

Drawing Commands
ATTDEF and **BATTMAN** commands — create and edit attributes.
INSERT command — reports annotatively-scaled blocks and dynamic blocks.
HATCH and **HATCHEDIT** commands — draw and edit hatches.
TEXT, **MTEXT**, and **DDEDIT** commands — draw and edit text, mtext, and fields.
MSLTSCALE system variable — scales all linetypes annotatively.

Properties Command
The **PROPERTIES** command changes the annotative property for single-line text, mtext, fields, dimensions, hatches, attributes, multileaders, leaders, qleaders, and tolerances. Typical annotation-related properties are shown below for text.

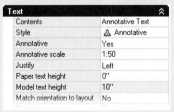

ANNOTATION BAR

There is no command for controlling annotation scaling, because it is a property. Instead, it is controlled through the "annotation bar" (illustrated below), system variables, properties, some editing and style commands, and the Properties palette.

In Model tab, these annotation controls are displayed on the status bar:

In Layout tab, only two controls are displayed when in paper space:

When in a layout's model space, the following controls are shown:

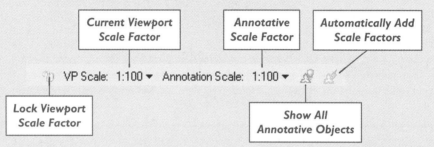

Lock Viewport Scale Factor — toggles locking of the viewport scale; also locks the annotation scale. This option is also available through the viewport's properties.

Current Viewport Scale Factor — reports the scale of the current viewport; click the arrow to select another scale factor.

Annotative Scale Factor — reports the annotation scale in the current viewport; click the arrow to select another scale factor.

Show All Annotative Objects — toggles the display of annotative objects:
    On — all objects are displayed, regardless of their annotative scale.
    Off — only objects whose annotative scale matches the viewport scale are displayed.

Automatically Add Scale Factors — toggles addition of annotative scales to objects:
    On — when the viewport scale changes, the new scale factor is added to annotative objects automatically.
    Off — objects keep their annotative scale factors fixed.

ANNOTATION SCALING: ADDITIONAL METHODS

While no single command controls annotative scales, these system variables and commands fine tune the system.

- **OBJECTSCALE** and **-OBJECTSCALE** commands edit annotative scales associated with objects.

- **SELECTIONANNODISPLAY** and **XFADECTL** system variables dim the display of annotative objects whose scale differs from the viewport scale.

- **ANNOUPDATE** command updates objects to match a recently-changed style.

- **ANNORESET** command forces the scale and visibility to match the current settings.

- **CANNOSCALE** and **CANNOSCALEVALUE** system variables report the current annotation scales.

- **ANNOTATIVEDWG, DIMANNO**, and **MSLTSCALE** system variables control the annotative properties of drawings inserted as blocks, dimensions, and linetypes, respectively.

- **SAVEFIDELITY** system variable determines whether drawings are saved with visual fidelity to earlier releases of AutoCAD.

Let's look at each.

ObjectScale

The **OBJECTSCALE** command edits annotative scales associated with objects through a dialog box.

Command: **objectscale**

Select annotative objects: *(Pick one or more objects.)*

Select annotative objects: *(Press ENTER to end object selection.)*

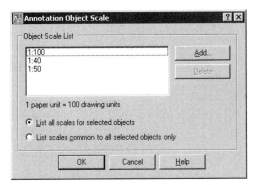

The dialog box lists the scale factor(s) assigned to the selected object(s).

Add — adds scale factors to the objects, in a roundabout manner; displays the following dialog box:

This dialog box copies the list maintained by the **SCALELISTEDIT** command, which also handles scale factors for viewport and plot scaling; if you don't see the scale factor you need, then you need to use the **SCALELISTEDIT** command to add it, as detailed in Chapter 22.

Delete — deletes scale factors. You can select one or more for deletion, but you cannot delete all of them, because at least one annotative scale must remain. (To remove the last remaining annotative scale, use the **PROPERTIES** command to turn off the object's Annotative property.)

The following options determine which scale factors are listed by the Annotation Scale Factor dialog box:

⊙ **List All Scales For Selected Objects** — displays all scales factors from all the objects you selected.

○ **List Scales Common to All Selected Objects Only** — displays only the scale factors that the selected objects share.

-ObjectScale

The **-OBJECTSCALE** command edits annotative scales at the command line using a numerical code.

>Command: **-objectscale**
>
>Enter an option [Add/Delete/?] <Add>:

Add — adds scale factors to the objects; displays the following prompt:

>Enter named scale to add or [?] <1:100>: *(Enter a scale factor identifier, such as 6.)*

If you enter an actual scale factor, such as the 1:100 suggested by the prompt, the command retorts:

>0 objects scale added.

Instead, you need to enter a number assigned to the scale factor. Enter **?** to see the list.

? — lists the scale factors and their identifiers, such as the following:

Scale Name	Paper Units	Drawing Units	Effective Scale
1: 1:1	1"	1"	1"
2: 1:2	1"	2"	1"
3: 1:4	1"	4"	0"
4: 1:5	1"	5"	0"
5: 1:8	1"	8"	0"
6: 1:10	1"	10"	0"
7: 1:16	1"	1'-4"	0"
8: 1:20	1"	1'-8"	0"
9: 1:30	1"	2'-6"	0"
10: 1:40	1"	3'-4"	0"
etc.			

Delete — deletes scale factors; displays the following prompt:

>Enter named scale to delete or [?] <1:100>: *(Enter a scale factor identifier, such as 6.)*

SelectionAnnoDisplay and XFadeCtl

The **SELECTIONANNODISPLAY** system variable toggles the display of annotative objects whose scale differs from the viewport scale.

>Command: **selectionannodisplay**
>
>Enter new value for SELECTIONANNODISPLAY <1>: *(Type 1 or 0.)*

0 — Only the annotative objects that match the viewport scale are displayed.

1 — All annotative objects are displayed, with non-current ones faded (default).

At left in the figure below, **SELECTIONANNODISPLAY** is on. The annotative text is selected and displays its multiple-scale representations; those not at the current viewport scale are faded (light gray).

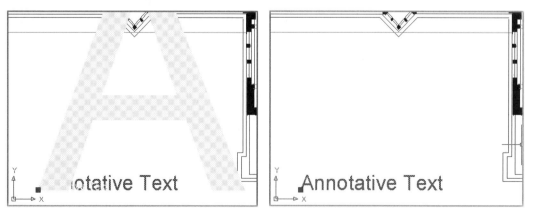

Left: SELECTIONANNODISPLAY = 1, XFADECTL = 90.
Right: SELECTIONANNODISPLAY = 0. (Value of XFADECTL doesn't matter.)

At right, **SELECTIONANNODISPLAY** is off, and so only the representation that matches the viewport scale is displayed.

The level of dimming is controlled by a second system variable, **XFADECTL** (short for "xref fade control"). The level is a percentage between 0 and 90:

> Command: **xfadectl**
>
> Enter new value for XFADECTL <50>: *(Enter a value between 0 and 90.)*

0 — Off, no dimming.

90 — Maximum dimming.

Notes: When annotative text has multiple representations, you can move each independently of the others, as illustrated below. To do this, set a viewport scale that matches one representation, and then move it; select the next viewport scale, and so on.

While you can move multiple representations independently of each other, you cannot change properties, such as color, independently.

The **DRAWORDER** command appears to have no effect on changing the display order when multiple representations overlap.

AnnoUpdate

The **ANNOUPDATE** command updates objects to match a recently-changed style.

> Command: **annoupdate**
>
> *n* found *n* were updated

You use this command to update annotative objects when the related styles are changed, such as text and dimension styles.

AnnoReset

The **ANNORESET** command forces the location of scale representations to match the current settings.

> Command: **annoreset**
>
> Reset alternate scale representations to current position
>
> Select objects: *(Pick one or more objects.)*
>
> Select objects: *(Press* ENTER *to end object selection.)*

The technical editor notes that this command does not reset annotatively-scaled objects to their original locations, but instead moves them to match the current location of the one used to trigger the change.

CAnnoScale

The **CANNOSCALE** system variable reports the current annotation scales in the current viewport (short for "current annotation scale"). Different viewports may report different scale factors.

> Command: **cannoscale**
>
> Enter new value for CANNOSCALE, or . for none <"1:50">: *(Enter an annotation scale name, such as "1:100".)*

The "name" of the annotation scale must be selected from the list found in the **SCALELISTEDIT** dialog box, illustrated below. See Chapter 22, "Plotting Drawings."

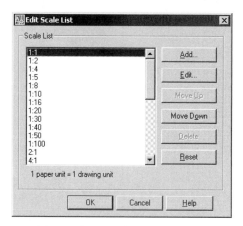

Macros would use this system variable to change the annotation scale.

CAnnoScaleValue

The **CANNOSCALEVALUE** system variable reports the current annotation scales (read-only).

> Command: **cannoscalevalue**
>
> CANNOSCALEVALUE = 0.020000000 (read only)]

The value of 0.02 is equivalent to a scale factor of 1:50.

AnnotativeDwg

The **ANNOTATIVEDWG** system variable controls the annotative property of drawings inserted as blocks into other drawings. It is read-only when the drawing contains at least one annotative object; otherwise, you can change the value between 0 and 1.

> Command: **annotativedwg**
>
> ANNOTATIVEDWG = 0 (read only)

0 — Behaves nonannotatively.

1 — Behaves annotatively.

DimAnno

The **DIMANNO** system variable reports whether the current dimension style has the annotative property.

> Command: **dimanno**
>
> DIMANNO = 1 (read only)

0 — dimension style is not annotative.

1 — dimension style is annotative; use the **DIMSTYLE** command to change the annotative property, which affects the size of dimension text and arrows. See Chapter 14, "Editing Dimensions."

MsLtScale

The MSLTSCALE system variable toggles the annotative property of linetypes in model space (short for "model space linetype scale").

> Command: **msltscale**
>
> Enter new value for MSLTSCALE <0>: *(Type 1 or 0.)*

0 — linetypes are not scaled by the annotation scale; the default for drawings imported from AutoCAD 2007 and earlier.

1 — linetypes are scaled by the annotation scale; the default for drawings created in AutoCAD 2008 and later.

This system variable does not come into effect until after the next drawing regeneration. (Use the REGENALL command, if necessary.) To use this system variable effectively, all other linetype scale-related system variables should also be set to 1:

> LtScale = 1 or a scale correct for plotting.
> CeLtScale = 1
> PsLtScale = 1

SaveFidelity

The SAVEFIDELITY system variable determines whether drawings with multiple annotatively-scaled objects are saved with *visual fidelity*. Earlier releases of AutoCAD cannot handle these objects, and so visual fidelity ensures that they are saved to separate layers

> Command: **savefidelity**
>
> Enter new value for SAVEFIDELITY <1>: *Type 1 or 0.)*

0 — Drawings save only annotative objects at the current scale, and discard the others.

1 — Drawings save each scaled representation of annotative objects on separate layers (default).

The problem with multiple representations of annotatively-scaled text is that they display correctly only in AutoCAD 2008 and later. This system variable determines what to do when drawings are opened in earlier releases of AutoCAD.

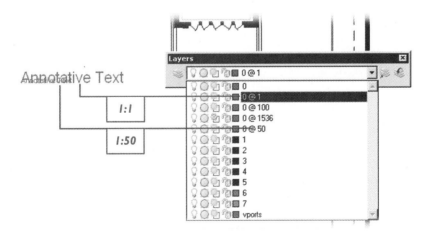

When set to 1 (the default), drawings contain extra layers whose names are marked with the scale factor, such as "0 @ 50," as illustrated above. This layer contains all annotative objects found on layer 0 bearing the scale factor of 1:50. The layer is frozen when first opened in AutoCAD 2007 (or earlier), and can be thawed to view the text.

COMPILE

The **COMPILE** command converts *.shp* and *.pfb* files into *.shx* files.

Many kinds of PostScript fonts are defined by *.pfb* files. Some older versions of AutoCAD directly supported PostScript fonts, which are the *de facto* standard in desktop publishing and the popular Acrobat Reader software — not surprisingly, since all three were defined by Adobe. When Microsoft licensed the competitive TrueType font technology from Apple Computer, TrueType quickly became popular with Windows users, because it was free. Autodesk followed suit, replacing support for PostScript with TrueType in AutoCAD; PostScript files can still be used in drawings, but only after conversion with the **COMPILE** command.

AutoCAD uses *.shx* files to define its own font format, as well as shapes, an older but more efficient version of blocks. (Technically, *.shp* files are the source code for compiled *.shx* files.)

TrueType fonts cannot be converted, nor is there a need to.

TUTORIAL: COMPILING POSTSCRIPT FONTS

1. To use PostScript fonts in drawings, start the **COMPILE** command:
   * At the 'Command:' prompt, enter the **compile** command.

     Command: **compile** *(Press ENTER.)*

2. AutoCAD displays the Select Shape or Font File dialog box.
   In **Files of Type,** select "PostScript Fonts (*.pfb)".
   Select a *.pfb* file, and then click **Open**.

3. AutoCAD converts the font file:

   Compiling shape/font description file

   Compilation successful. Output file

   C:\Adobe\Acrobat\Distillr\Data\Fonts\coo_____.shx contains 40891 bytes.

4. Use Widows Explorer to copy the .shx files from the folder in which they were converted, to the \autocad 2008\fonts folder — otherwise AutoCAD cannot find them.

5. Exit AutoCAD, and then restart the software. This forces AutoCAD to update its list of fonts.

You can now use the converted PostScript fonts in the drawing. As illustrated by the figure, the fonts appear unfilled; they are outlined only. This can be an advantage or a disadvantage, depending on your needs. A definite advantage, however, is that converted PostScript fonts display and plot faster than equivalent TrueType fonts.

PostScript Font

Note: Most fonts are *proportionally spaced*, which means the letter **i** takes up less width than the letter **w**. This makes it easier to read the text.

AutoCAD includes a font meant especially for columns of text, where it is important that the text line up vertically. The *monotxt.shx* font is designed so that every letter takes up the same width. A sample is shown below:

3. Install sheets with face grain perpendicular to suports.
   Stagger sheets and nail with 8d @ 6'o.c. edges and 12' o.c. field typicl
   unless otherwise noted.

4. For truss shapes, dimensions, and so on, see the Architectural drawings.

EXERCISES

1. In this exercise, you practice placing text.

    Start AutoCAD with a new drawing. With the **TEXT** command, place following lines of text:

    THIS IS THE FIRST LINE OF TEXT

    THIS IS THE SECOND & CENTERED LINE OF TEXT

2. In this tutorial, you place text centered on a point.

    The text "Part A" is to be centered both vertically and horizontally on a selected point. Use the **TEXT** command to place the text. Which justification mode did you use?

3. In this exercise, you place text with a variety of alignments.

    The figure illustrates several text strings. The placement point is marked with a solid dot. Use the **TEXT** command to place the text as shown.

    Kitchen   Radius

    AutoCAD

    Drill-thru   Title

    Isometric View

    Capacitor   Section A-A

4. In this exercise, you place the following line of text at a variety of angles:

    This text is rotated

    Using the **TEXT** command, place the following line of text at angles of:

    a. 0 degrees.

    b. 45 degrees.

    c. 90 degrees.

    d. 135 degrees.

    e. 180 degrees.

    Use the same base point for each line.

5. In this exercise, use control codes to construct the following text string in the drawing.

    The story entitled <u>MY LIFE</u> is ±50% true.

6. In this exercise, you practice using mtext. Start a new drawing, and use the **MTEXT** command to place several lines of text in the drawing, such as your name and address.

    Your First and Last Name

    1234 First Avenue

    Anytown, BC

    V8C 1T2 Canada

    Click the **OK** button to exit the mtext editor.

7. Continuing from the previous exercise, double-click the mtext. Does the mtext editor reappear?

   Make the following changes to parts of the text:

   Font: **Times New Roman**

   Color: **Red**

   Click the **OK** button to exit the mtext editor. Did the changes come into effect?

8. Continuing from the previous exercise, turn on the **QTEXT** command.

   Did the text change its look?

   If no, which command did you forget to use?

9. In this exercise, you import text from another file.

   If necessary, create a brief text file in Notepad, the text editor included with every version of Windows. Save the file in *.txt* format.

   Import the text file into your drawing with the **MTEXT** command.

10. Create five different styles that represent very different text appearances, such as wide text, slanted, and so on.

    Which command creates styles?

    Place examples of each style in the drawing.

11. Using only the **MTEXT** command, design your own business card with at least two fonts, three font variations, and two colors.

    Do not design a logo — just position the text. Include at least your name, address, and Internet information.

    Standard business cards are rectangles that measure 3.5" wide by 2" tall.

    Which command is useful for drawing rectangles?

    How many business cards can you fit onto a vertical A-size sheet?

    (The technical editor quotes Lynn Allen: you know you've been using AutoCAD too long when you use it to design wedding invitations.)

12. Use a combination of text and drawing commands to design a logo for an engineering office. The logos often include variations on the owner's names.

13. In this and the next exercise, you practice drawing leaders.

    Draw a rectangle.

    Attach a leader to each of the four corners.

    Use the following text for each leader:

    North East Corner

    North West Corner

    South East Corner

    South West Corner

14. Continuing from the previous exercise, place a splined leader pointing to the center of the rectangle. Change the arrowhead to a dot.

    Use the following annotation:

    Parcel of Land.

15. Continuing from the previous exercise, double-click the annotation of the splined leader, and change the text to:

    Disputed Parcel of Land.

    Not to be Subdivided.

16. In this and the following exercise, you correct and change text in drawings.

    From the Companion CD, open the *spelling.dwg* drawing file, a drawing that contains text with spelling errors.

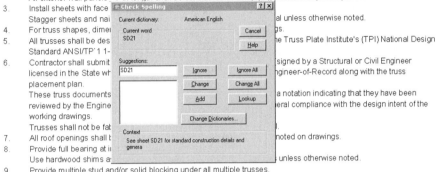

Correct the spelling.

How many mistakes does AutoCAD's **SPELL** command find?

How many *real* spelling mistakes are there?

17. Continuing with the same drawing, find and replace the following phrases:

Find	Replace With
sheet SD1	sheet AA-01
Building Official	local Building Official

Which command did you use to find and replace words in drawings?

18. From the Companion CD, open the *insert.dwg* drawing file, a house plan.

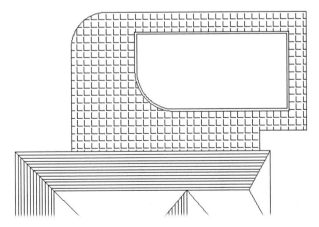

Use the **MTEXT** command to place the following note over a white rectangle on the patio area at the rear of the house. Use the background mask feature.

    Interlocking Belgium Bricks

19. Continuing with the same drawing, use the **SCALETEXT** command to reduce the size of the text you added in the previous exercise.

20. Continuing with the previous exercise, use the **JUSTIFYTEXT** command to change the justification to **Center**. How does the text change?

21. In this exercise, we see the effect of using the following buttons on annotative text:

Annotation Scale: 1:1 ▼

From left to right, these are:

- Annotation Scale (**CANNOSCALE** system variable) — changes the annotation scale.

- Annotation Visibility (**ANNOALLVISIBLE** system variable) — toggles visibility of text at all annotation scales.

- Add Annotation Scales Automatically (**ANNOAUTOSCALE** system variable) — toggles addition of scales to objects as annotation scale changes.

a. Start a new drawing, and then turn off all of the settings, as listed below. You can use the controls on the status bar, or change the values of the system variables.

Annotation Scale	1:1	(CANNOSCALE = 1:1).
Annotation Visibility	off	(ANNOALLVISIBLE = 0)
Add Annotation Scales Automatically	off	(ANNOAUTOSCALE = -4)

Creating an Annotative Text Style

b. Create a new text style named "Anno" with the following properties:

Annotative	☑ (on)
Paper Text Height	**0.25**
Font Name	**Arial**

Set the style current, and then close the dialog box.

c. With the **TEXT** command, place the following text in the drawing in model space:

Annotative Text = 0.25

d. With the **RECTANGLE** command, draw a rectangle that encompasses the extent of the text, as illustrated below. This rectangle will help you see the size and location of the original text as the annotation scale changes.

Annotative Text = 0.25

e. Change **Annotation Scale** to **1:2**.
What happens to the text?
Why did the text change in that way?

Controlling Annotation Visibility

f. Click the **Annotation Visibility** button (or set **ANNOALLVISIBLE** = 1).
Why does the text appear?

g. Change **Annotation Scale** to **1:4**,
Click the **Add Annotation Scales Automatically** button (or set **ANNOAUTOSCALE** = 4)
Why does the text change its look?

h.   Change **Annotation Scale** back to **1:2**, and then select the text.

What do you see on the screen?

Resetting and Updating Annotative Text

i.   With the text still selected, use its grip to move it a few inches away.

What happens to the other representations of the text?

j.   With the text still selected, use the **ANNORESET** command.

To see the effect of the command, select the text.

What happens to the other representations of the text?

k.   Return to the **STYLE** command, and change the **Paper text height** to **0.5**.

Click **Apply** and then **Close**.

What happens to the text?

l.   Select the text, and then enter the **ANNOUPDATE** command.

What happens to the text?

m.   Select the text.

What happens to the other representations?

(Keep this drawing open for the following exercise.)

22.   In this exercise, you create *oriented* text. Continue with the drawing from the previous exercise.

a.   Start the **STYLE** command, and then turn on the **Match Text Orientation to Layout** option for the "Anno" text style.

Click **Apply** and then **Close**.

b.   Use the **ANNOUPDATE** command to update the style of the text in the drawing.

c.   Use the **ROTATE** command to rotate all objects in the drawing by 90 degrees.

What happens to the annotative text?

23.   Use the **MLEADER** command to create the following leaders:

a.   b.

b.   Draw the following multi-bubble mleader. Which command did you use to create it?

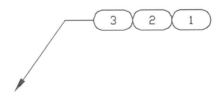

d.   Open the *mleader.dwg* file from the Companion CD. Add one more leader to make it look like the following figure. Which command did you use to add the leader?

CHAPTER REVIEW

1. Name three ways you can see a list of the text styles stored in the drawing.

    a.

    b.

    c.

2. What are *fonts*?

    How do styles differ from fonts?

3. What four text properties are altered by the **STYLE** command?

    a.

    b.

    c.

    d.

4. There are several methods of placing text. List three of them, with a brief explanation of each:

    a.

    b.

    c.

5. When an underscore or overscore is added to text, is the entire string altered, or can words be treated separately?

    Explain.

6. When AutoCAD asks for a text height, is a numerical entry required?

    Explain.

7. List five alignment modes available through the **TEXT** command.

    a.

    b.

    c.

    d.

    e.

8. Can text be rotated at any angle?

    How does the oblique angle differ from the text angle?

9. How do the **TEXT** and **-TEXT** commands differ?

10. How can the text height be altered each time you enter text without redefining the style?

11. How can you compress and expand the text width?

12. Which formats of text can AutoCAD import into drawings?

13. What height should note text be in drawings?

14. Name four areas where text is used in drawings:

    a.

    b.

    c.

    d.

15. Can AutoCAD use *any* TrueType font found on your computer?

16. What happens when you open a drawing that contains a font not found on your computer?

17. Which command is better for placing lines of text at many locations over the drawing: **TEXT** or **MTEXT**?

18. List the command associated with each alias:

    dt

    t

    ed

    st

19. Can you access the menu while entering text?

20. What do the three horizontal lines of the I-beam cursor represent?

    a.

    b.

    c.

21. Which one of the following letters has a descender?

    a

    B

    d

    j

    L

22. Explain the meaning of the following justification codes:

    TL

    MC

    BR

    TR

23. Is BR justification the same as the **Right** justification?

    Explain.

24. Describe the one difference between **Align** and **Fit** justification modes.

25. What height should text be for a drawing scaled at:

    1:100

    1" = 50'

    1:1

    1" = 6"

    Show your work for each calculation.

26. Why is it wrong to use 1/8"-high text in an A-size drawing of a house?

27. When might you place rotated text in drawings?

28. Explain the meaning of the following control codes:

    %%u

    %%%

    %%d

    %%c

29. If you fail to turn off underlining, what happens on the next line of text?

30. What does *string* mean?

31. If the Euro symbol is not on your keyboard, how would you enter it?

32. Can you use more than one font with the TEXT command?

    With the MTEXT command?

33. Explain the meaning of the following mtext editor keyboard shortcuts:

    CTRL+B

    CTRL+I

    CTRL+A

34. How does AutoCAD stack text on either side of these characters?

    /

    ^

    #

35. Can different colors be applied to text placed with the **TEXT** command?
    With the **MTEXT** command?

36. Explain the meaning of the following zeros:
    Text Height = 0
    Mtext Width = 0

37. Can only numbers be used for stacked text?

38. Is it possible to use a text editor or word processor other than AutoCAD's built-in mtext editor?
    If so, how?

39. List the three primary parts of a leader:
    a.
    b.
    c.

40. What is the quickest way to edit the text of a leader?
    To edit the position of leader line?

41. Can leaders use only one kind of arrowhead?

42. Name three kinds of leader annotation:
    a.
    b.
    c.

43. What does a splined leader line look like?

44. Can leaders be attached to objects?

45. List three ways to edit text:
    a.
    b.
    c.

46. Is a different command needed to edit text created by the **TEXT** and **MTEXT** commands?

47. Does a different editor display for text created by the **TEXT** and **MTEXT** commands?

48. Can you have more than one text style in drawings?

49. What happens when you apply a different style to a line of text?

50. What happens to text when you make changes to a style?

51. Which font most closely approximates the Leroy lettering guides?

52. Can TrueType fonts have the vertical style?

53. Name one advantage to using styles in drawings.

54. What is the purpose of the **TEXTFILL** and **TEXTQLTY** system variables?
    When do they come into effect?

55. Can the **SPELL** command be used for text placed by commands other than **TEXT** and **MTEXT**?

56. Explain a benefit of the **MTEXT** command's background mask.

57. Which command turns lines of text into rectangles?

58. Describe the purpose of the **SCALETEXT** command.
    And the **JUSTIFYTEXT** command.

59 How would you change the size of text in a drawing being converted from imperial to metric units?

60. Can PostScript fonts be used in AutoCAD drawings?

61. How does the **MLEADER** command differ from the **QLEADER** command?

62. Define "content" of multiline leaders.

63. Label the parts of the multi-line leader illustrated below:

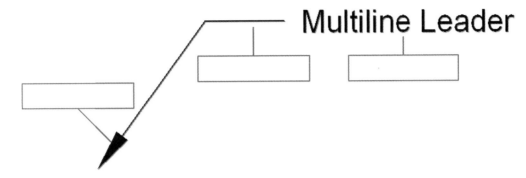

64. Describe how annotative text differs from normal text.

65. When would you use the **OBJECTSCALE** command?

66. How is annotative scaling applied to text?

67. What happens to orientated text when it is rotated?

68. Does annotative scaling apply to hatch patterns and dimensions?

CHAPTER 13

Placing Dimensions

Dimensions play an important role in defining the size, location, angle, and other attributes of objects in drawings. Dimensions show the length and width of objects, the diameter of holes, and the angles of sloped parts.

This chapter describes how to place dimensions in the drawing using these commands:

DIMHORIZONTAL, DIMVERTICAL, and **DIMROTATED** place horizontal, vertical, and rotated dimensions.

DIMLINEAR places horizontal, vertical, and rotated dimensions.

DIMALIGNED places dimensions aligned with objects.

DIMBASELINE and **DIMCONTINUE** add baseline and continuous dimensions.

QDIM generates a variety of continuous dimensions.

DIMRADIUS and **DIMDIAMETER** place radial and diameter dimensions.

DIMJOGGED draws radial or linear dimensions with jogged leader or dimension lines.

DIMARC measures the lengths of arcs.

DIMCENTER places center marks on arcs and circles.

DIMANGULAR places angular dimensions.

NEW TO AUTOCAD 2008 IN THIS CHAPTER

- The **DIMANGULAR** command locks angular dimensions within specific quadrants.
- The **DIMDIAMETER**, **DIMRADIUS**, and **DIMJOGGED** commands have a new **Extension** option.
- Dimension support annotative scaling.

FINDING THE COMMANDS

On the Dashboard's **Dimension** panel:

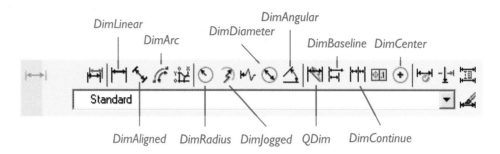

On the **Dimension** toolbar:

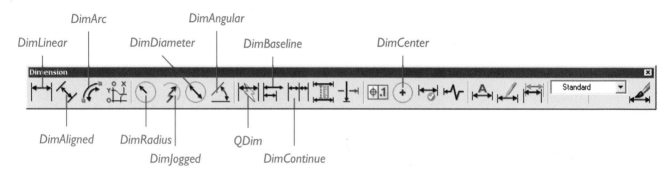

INTRODUCTION TO DIMENSIONS

Dimensions in drawings eliminate the need to use scale rulers and protractors to work out measurements and angles, and are far more accurate than measuring drawings with a ruler.

In some disciplines, such as architecture, the most important parts of the drawing are dimensioned, not every detail. In contrast, mechanical parts are completely dimensioned. Either way, dimensions should not interfere with the drawing.

The figure below illustrates metric dimensions in a sample Chinese drawing provided with AutoCAD.

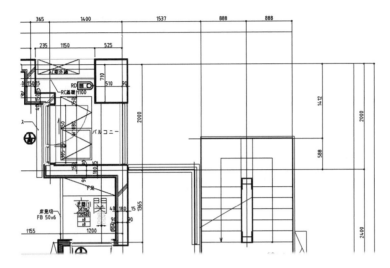

THE PARTS OF DIMENSIONS

Dimensions consist typically of these parts: a dimension line with an arrowhead at each end, a pair of extension lines, and dimension text indicating the measured distance.

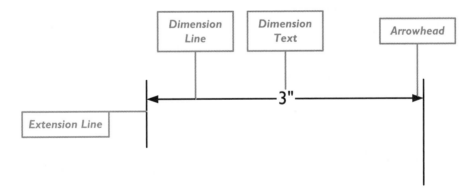

Dimension Lines

The *dimension line* indicates the distance or angle being measured. It is the line with the arrows or "ticks" at each end. When there is too little room between the extension lines, the dimension line is placed outside the extension lines or, in some cases, not drawn.

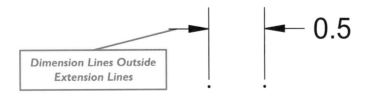

When you measure angles, the dimension line is an arc, instead of a straight line.

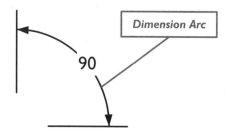

Dimension lines normally take on the properties of their layer. AutoCAD allows you to override the layer setting by specifying the lineweight, color, and linetype for the dimension and extension lines separately.

To report the distance or angle, text is placed on or near the dimension line. AutoCAD automatically creates a gap in the dimension line to make room for the text.

Extension Lines

Nearly all dimensioning commands begin by prompting you to pick two points: these become the start points for *extension lines*. Extension lines (sometimes called "witness lines") indicate the geometric features being dimensioned. Extension lines are usually perpendicular to the dimension line, but need not be.

There are usually two extension lines, one at either end of the dimension line. Sometimes, the first or second extension line is not drawn, and in rare cases, neither is drawn.

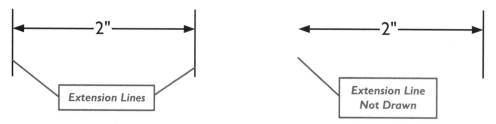

AutoCAD *offsets* the extension lines from the pick points by a distance of 0.0625 units to mimic standard drafting practice. The offset distance is named "offset from origin," and can be changed in the Dimension Style Manager dialog box.

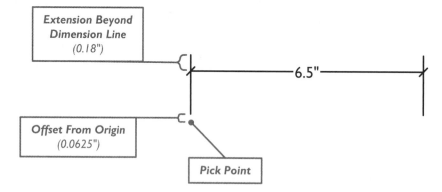

In addition, AutoCAD draws the extension lines a short distance beyond the dimension line. Again, this is standard drafting practice. The default distance is 0.18 units, but can be changed in the Dimension Styles Manager dialog box.

AutoCAD normally draws the extension lines long enough to span the distance from the origin to just beyond the dimension line — but AutoCAD can also allow the length of extension lines to be fixed.

Arrowheads

Arrowheads are placed at either end of the dimension line, pointing to the extension lines. If there is too little room between the extension lines, the arrowheads are placed outside. The Dimension Style Manager dialog box lets you specify whether arrowheads or text should be moved outside first, or forced to fit inside the extension lines.

The arrowheads can be replaced with tick marks and other symbols. AutoCAD includes many different arrowheads, and you can define your own. The figure at right illustrates the arrowhead styles included with AutoCAD; to create custom arrowheads, you first define them as blocks.

➤ Closed filled
▷ Closed blank
⇒ Closed
● Dot
✒ Architectural tick
╱ Oblique
⇒ Open
-○ Origin indicator
-◎ Origin indicator 2
→ Right angle
⇾ Open 30
● Dot small
-○ Dot blank
○ Dot small blank
◁ Box
◀ Box filled
◁ Datum triangle
◀ Datum triangle filled
𝑓 Integral
None

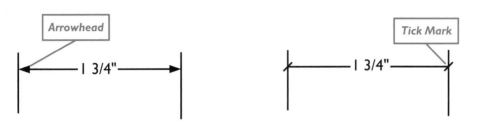

Arrowheads are usually drawn with a 1:3 aspect ratio: the arrow is three times longer than it is wide. Other arrowheads, such as tick marks and circles, are 1:1.

The length of the arrowhead varies, depending on the scale of the drawing. Arrowheads that are too small or large prove difficult to read and can distract. Generally, arrowheads are 1/8 inch in length when used in small drawings, and 3/16 inch in larger drawings.

Dimension Text

The *dimension text* reports the distance or angle between the extension lines. Sometimes, the text reports other information, such as tolerances or alternate units of measurement. Typically you let AutoCAD measure the distance and determine the text; you can, however, override it and enter your own text with the mtext editor.

Dimension text appears in or near the dimension line, depending on whether there is enough room and on the drafting standard. In architectural and structural drawings, the dimension is placed on top of the dimension line. In mechanical drawings, it is usually placed within the dimension line. When there is too little room between the extension lines, the text is placed outside, or even some distance away, and in some cases referenced with a leader line.

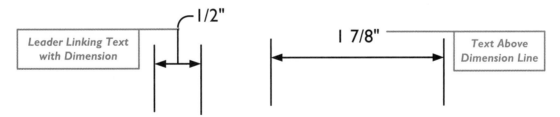

When the dimension text is horizontal and centered, the text is said by AutoCAD to be in the "home" position. Dimension text can be centered, left justified, or right justified on the dimension line. The text can be horizontal, vertical, or rotated.

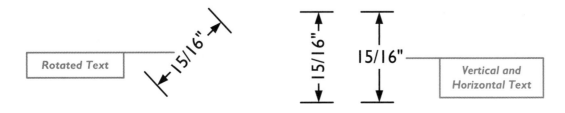

Feet and inches are separated by a dash, as in **5'-4"**. If there are no inches, a zero is usually included: **6'-0"**. AutoCAD allows you to choose whether you want 0 feet and/or 0 inches displayed. If dimensions are stipulated strictly in inches, such as **72"**, use the inch mark to avoid confusion with feet or other units; in mechanical design, inch marks are not used.

Fractions are given either as common fractions, such as **1/2**, **3/4**, **9/16**, and so on, or as decimal fractions, such as **0.50** and **0.75**. In CAD, some text fonts lack "stacked" fractions, such as $^1/_2$ and $^3/_4$. For this reason, the dash is used to separate inches from fractions and to avoid confusion, such as **3-1/2"**.

Dimension text indirectly follows the text style set by the **STYLE** command and the numerical format and precision set by the **UNITS** command. The style, units, and precision of dimensions are determined by the **DIMSTYLE** command. You can specify the style name, color of text, fill (background) color, and fixed height, as well as decide whether to surround the text with a box.

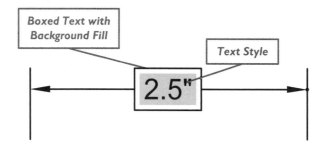

Annotative Scaling

Like text and leaders, dimension can be annotatively scaled. This ensures that they appear at the correct size at specific model space scale factors, and eliminates the need to calculate a scale factor to make dimensions plot correctly.

Tolerance Text

Dimension *tolerances* are plus and minus amounts appended to the dimension text. AutoCAD can add the tolerances automatically, but typically you specify the plus and minus amounts, which can be equal ("symmetrical") or unequal ("deviation").

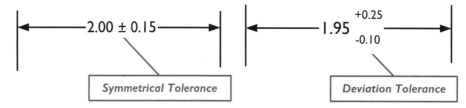

If tolerances are symmetrical, they are drawn with the plus/minus symbol. If deviation, they are drawn one above the other: above for the plus amount, below for the minus.

Limits Text

Instead of showing dimension tolerances, the tolerance can be applied to the text itself — added to and subtracted from the dimension text. This is called "limits." The example below is a measurement of 4.00 units, with a tolerance of ±0.15.

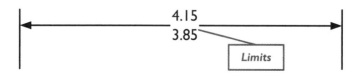

Alternate Units

"Alternate units" show two forms of measurement, such as English and metric, on the same dimension line. The second set is usually shown in square brackets.

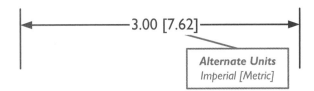

When alternate units are turned on in the dimension style, AutoCAD adds them automatically. When turned off, you can have AutoCAD calculate and add alternate units by entering a pair of square brackets (**[ ]**) when editing the dimension text.

DIMENSIONING OBJECTS

AutoCAD places dimensions between any two points in drawings. In addition, AutoCAD has a *direct dimensioning* mode, where it dimensions specific objects automatically: lines, polyline segments, arcs, polyline arcs, circles, vertices, and single points.

Lines and Polyline Segments

AutoCAD can dimension the lengths of lines and polyline segments with a single pick. The dimension is drawn completely, consisting of the dimension and extension lines, arrowheads, and text. This is accomplished typically with the DIMLINEAR and DIMALIGNED commands, as detailed later.

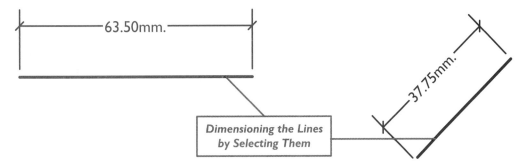

Arcs, Circles, and Polyarcs

Arcs are dimensioned by their radius or length. When dimensioning the radius, place the leader line at an angle to avoid horizontal and vertical placements. The dimension line for the length of an arc follows the curve of the arc. The symbol "∩" designates arc length dimensions.

The letter "R," designating radius, prefixes the dimension text, such as **R2.125**. If a circle is dimensioned by its diameter, the dimension text is prefixed by the diameter symbol, such as Ø4.25.

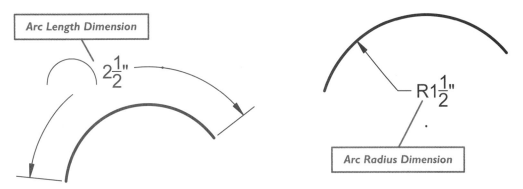

Objects constructed of several arcs are dimensioned in two stages: (1) locate the center of the arcs with horizontal or vertical dimensions, and (2) show their radii with radius dimensioning.

AutoCAD dimensions arcs and polyline arcs with a single pick. The resulting dimension depends on the command you use:

The **DIMLINEAR** and **DIMALIGNED** commands place dimension and extension lines, arrowheads, and text. The **DIMRADIUS** and **DIMDIAMETER** commands place dimension lines or leaders, arrowheads, and text.

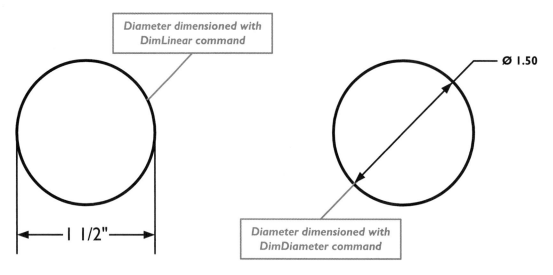

Arcs are usually dimensioned with radii, while circles are usually dimensioned with diameters.

Wedges and Cylinders

Wedges are dimensioned in two views using three distances: length, width, and height.

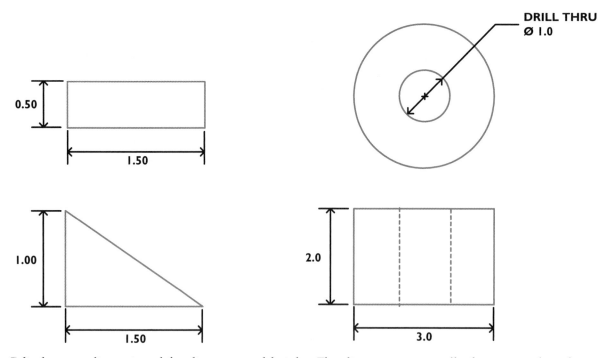

Cylinders are dimensioned for diameter and height. The diameter is typically dimensioned in the non-circular view. If a drill-through is dimensioned, it is described by a diameter leader.

Cones and Pyramids

Cones are dimensioned at the diameter and the height. Some conical shapes, such as truncated cones, require two diameter dimensions. Pyramids are dimensioned like cones.

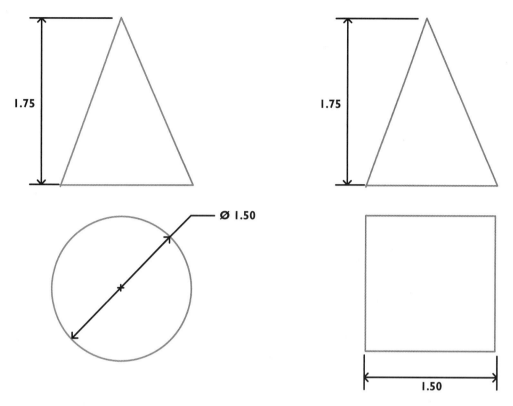

Holes

Holes are created by drilling, reaming, boring, punching, or coring materials. Dimension holes in drawings with notes that give the diameter, operation, and the number of holes, if more than one. The operation describes such techniques as counter-bored, reamed, and countersunk.

Whenever possible, point the dimension leader to the hole in the circular view. Holes made up of several diameters can be dimensioned in their section.

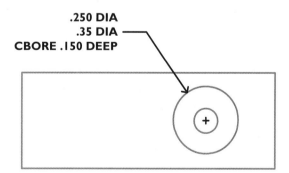

Standards dictate that drill sizes be designated as decimal fractions.

Vertices

AutoCAD measures the angle of lines, arcs, and polyline arcs with two picks using the **DIMANGULAR** command.

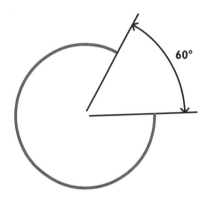

Single Points

To "dimension" a single point, you typically use the **LEADER**, **QLEADER**, **MLEADER**, or **DIMORDINATE** commands. Leaders are typically pieces of text that point at a spot in the drawing through lines. Ordinate dimensions measure the x and y coordinates of objects relative to a base point.

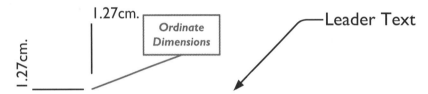

For drawing leaders, see Chapter 12, "Placing and Editing Text"; for ordinates, see Chapter 15, "Geometric Dimensioning and Tolerancing."

Chamfers and Tapers

A chamfer is an angled surface applied to an edge. Use a leader to dimension chamfers of 45 degrees, with the leader text designating the angle and one (or two) linear distances.

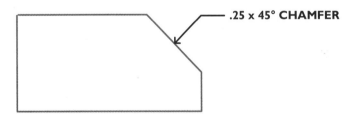

If the chamfer is not 45 degrees, dimensions showing the angle and the linear distances describe the part.

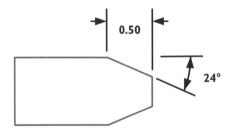

A taper can be described as the surface of a *cone frustum* (cone with the top sliced off). Dimension tapers by giving any three of these parameters: start diameter, end diameter, rate of taper, and/or the length.

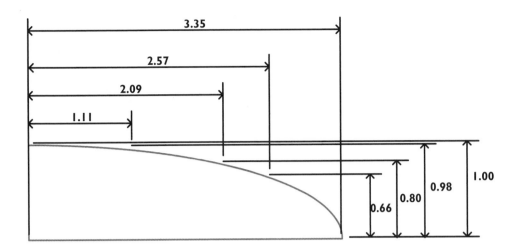

DIMENSION MODES

AutoCAD constructs dimensioning *semiautomatically*. All it needs from you is some basic information, such as what to dimension, or where to start and end the dimensions. Then it constructs the dimension for you, measures the distances and angles, and then draws all the elements of the dimension.

AutoCAD's dimensions are *associative*. That means they automatically update when associated objects are moved and stretched. This is a very powerful feature and a great time-saver.

Dimensions, however, do not need to be associative; AutoCAD can works with several levels of associativity:

Fully associative is the standard mode, where AutoCAD attaches the ends of the extension lines to geometry of the object. Move the object, and the dimension moves with it; stretch the object, and the dimension text updates automatically. This is the preferred method.

Partly associative was Autodesk's first attempt at associativity. Extension lines are attached to *dimension points* placed on layer Defpoints (short for "definition points"). When stretching or moving objects, you must include the defpoints in the selection set; otherwise, the dimension does not update correctly.

Non-associative means the dimension is not attached to objects in any way.

Trans-spatial means that dimension created in paper space are attached to objects in model space; changing the model updates the dimensions in paper space.

Dimensions are placed as if they were blocks: the lines, arrows, and text act as a single object. You can use the **EXPLODE** command to break apart dimensions.

Note: When AutoCAD reports "Non-associative dimension created," this means that it was not able to attach a dimension to the object.

Dimension Variables and Styles

Dimension variables determine how the dimensions are drawn; they are system variables specific to dimensions, sometimes called "dimvars" for short. Some dimvars store values, such as the color of the dimension line and the look of the arrowhead; others are toggles that turn values on and off, such as whether the first or second extension line should be displayed.

Just as styles determine the look of text, *dimension styles* determine the look of dimensions. Dimension styles ("dimstyles" for short) are created and modified with the **DIMSTYLE** command, which collects all the settings of dimension variables. Dimvars and dimstyles are discussed in Chapter 14, "Editing Dimensions."

Dimension Standards

There are many ways to draw dimensions, because different industries and countries define standards differently. AutoCAD includes the ISO-25 dimensioning standard with its *acadiso.dwt* template drawing. (Other standards which were included in previous releases of AutoCAD, such as DIN, JIS, and Gb, are not part of AutoCAD 2008.)

Most standards organizations make the dimension standards available to members at a cost (for example at www.nationalcadstandard.org, — that's .org, not .com), while some are freely available at Web sites (for example at www.cadinfo.net/editorial/archdim1.htm).

Object Snaps

Object snap modes are very helpful in placing dimensions accurately. For example, the ENDpoint and INTersection osnaps are useful for capturing the ends and intersections of lines, while the CENter and TANgent osnaps help with dimensions of arcs and circles.

DIMHORIZONTAL

The **DIMHORIZONTAL** command places horizontal dimensions.

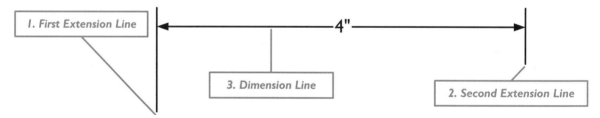

This command restricts the dimension line to the horizontal. To locate extension lines, you select an object (line, arc, circle, or polyline) or pick two points (1 and 2, in the figure above). Another pick point (3) is required to locate the dimension line. The text and arrowheads are located automatically by AutoCAD.

While the **DIMLINEAR** command is more flexible (combining horizontal, vertical, and rotated dimensioning), the **DIMHORIZONTAL** command may be easier for users new to AutoCAD.

PLACING DIMENSIONS HORIZONTALLY

1. To place a horizontal dimension, start the **DIMHORIZONTAL** command:
   - At the 'Command:' prompt, enter the **dimhorizontal** command.

     Command: **dimhorizontal** *(Press ENTER.)*
2. AutoCAD prompts you to select the object you want dimensioned:
     Specify first extension line origin or <select object>: *(Press ENTER.)*
     Select object to dimension: *(Pick one object.)*
3. Pick a point to place the dimension line:
     Specify dimension line location or [Mtext/Text/Angle]: *(Pick a point to place the dimension line.)*

4. Notice that AutoCAD reports the length measurement at the command prompt, and draws the horizontal dimension.

    Dimension text = 4"

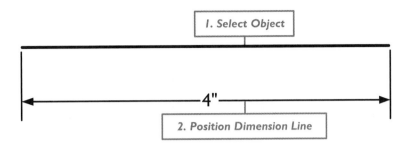

HORIZONTAL DIMENSIONS: ADDITIONAL METHODS

The **DIMHORIZONTAL** command has options to change the dimension text.

*   **MText** displays the mtext editor.

*   **Text** prompts you to edit the dimension text.

*   **Angle** rotates the dimension text.

Let's look at each option.

MText

The **MText** option displays the mtext editor, allowing you to edit the dimension text. The editor should be familiar to you from Chapter 12, "Placing and Editing Text."

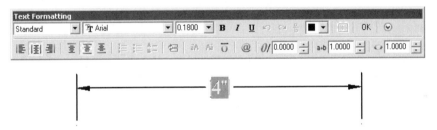

The mtext editor highlights the default dimension text. (Prior to AutoCAD 2006, the double angle bracket <> was used as a shorthand notation for the default dimension text.)

Here are some examples of how you can work with the default text:

*   Replace default text with other text, and AutoCAD shows the new text on the dimension line.

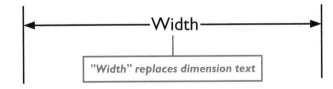

*   Add text on either side of the default text, and AutoCAD shows the added text.

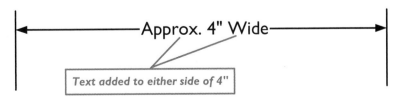

- Replace the default text with a space character, and AutoCAD erases the dimension text.

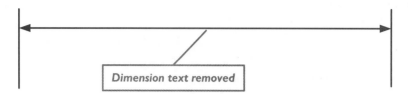

Dimension text removed

Special Symbols

To add special characters, use the mtext editor's **Symbols** submenu.

Alternatively, you can enter metacharacters to show these common symbols:

° (degrees) with %%d

± (plus or minus) with %%p

Ø (diameter) with %%c

Text

The **Text** option prompts you to edit the dimension text at the command prompt.

Enter dimension text <4.0000>: **The distance is <> inches**

Include the double angle brackets (<>) to specify the location of AutoCAD's measured distance with the text you add. The result is: **The distance is 4.0000 inches**.

Angle

The **Angle** option rotates the dimension text about its center point.

Specify angle of dimension text: *(Enter an angle, such as **45**, or use the mouse to show the angle.)*

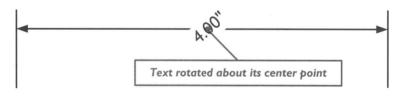

Text rotated about its center point

DIMVERTICAL AND DIMROTATED

The **DIMVERTICAL** command places vertical dimensions, while the **DIMROTATED** command places dimensions at an angle.

Vertical Dimension

Rotated Dimension

These two commands operate identically to **DIMHORIZONTAL**, except that **DIMVERTICAL** forces the dimension line to be vertical, while **DIMROTATED** asks you for a rotation angle, and then draws the dimension line at that angle.

PLACING DIMENSIONS VERTICALLY

1. To place a vertical dimension, start the **DIMVERTICAL** command:
   - At the 'Command:' prompt, enter the **dimvertical** command.

   Command: **dimvertical** *(Press* ENTER.*)*

2. AutoCAD prompts you to select the object you want dimensioned:

   Specify first extension line origin or <select object>: *(Press* ENTER.*)*

   Select object to dimension: *(Pick one object.)*

3. Pick a point to place the dimension line:

   Specify dimension line location or [Mtext/Text/Angle]: *(Pick a point to place the dimension line.)*

   Dimension text = 10

PLACING DIMENSIONS AT AN ANGLE

1. To place a rotated dimension, start the **DIMROTATED** command:
   - At the 'Command:' prompt, enter the **dimrotated** command.

   Command: **dimrotated** *(Press* ENTER.*)*

2. AutoCAD asks for the dimension's angle:

   Specify angle of dimension line <0>: *(Enter an angle, such as **45**, or show the angle by picking two points in the drawing.)*

3. AutoCAD prompts you to pick the endpoints of the extension lines:

   Specify first extension line origin or <select object>: *(Pick a point.)*

   Specify second extension line origin: *(Pick another point.)*

4. Pick a point to place the dimension line:

   Specify dimension line location or [Mtext/Text/Angle]: *(Pick a point to place the dimension line.)*

   Dimension text = 1"

DIMLINEAR

The **DIMLINEAR** command places horizontal, vertical, and rotated dimensions.

After you specify the location of the extension lines, AutoCAD automatically determines whether to draw a horizontal or vertical dimension — depending on where you place the dimension line:

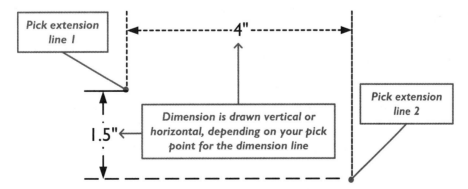

This command also applies dimensioning directly to objects, and is more flexible than the dimension commands described earlier.

PLACING LINEAR DIMENSIONS

1. To place a linear dimension, start the DIMLINEAR command:

   - From the **Dimension** menu, choose **Linear**.

   - From the Dimension toolbar, choose the **Linear Dimension** button.

   - In the Dashboard's Dimension panel, choose the **Linear** button.

   - At the 'Command:' prompt, enter the **dimlinear** command.

     Command: **dimlinear** *(Press ENTER.)*

   - Alternatively, enter the aliases **dli** or **dimlin** at the 'Command:' prompt.

2. In all cases, AutoCAD prompts you to select the object you want dimensioned:

   Specify first extension line origin or <select object>: *(Press ENTER.)*

   Select object to dimension: *(Pick one object.)*

3. Pick a point to place the dimension line:

   Specify dimension line location or

   [Mtext/Text/Angle/Horizontal/Vertical/Rotated]: *(Pick a point to place the dimension line horizontally or vertically.)*

   Dimension text = 4"

LINEAR DIMENSIONS: ADDITIONAL METHODS

The DIMLINEAR command has the same options as DIMHORIZONTAL, as described earlier in this chapter; see that command for more information on the four options listed below:

- **MText** displays the mtext editor.

- **Text** prompts you to edit the dimension text.

- **Angle** rotates the dimension text.

In addition, the command has these options:

- **Horizontal** forces the dimension line horizontally.

- **Vertical** forces the dimension line vertically.

- **Rotated** rotates the dimension line.

Let's look at the three new options.

Horizontal and Vertical

The **Horizontal** option forces the dimension line the horizontal, no matter where you locate the dimension line. This makes it operate like DIMHORIZONTAL. Similarly, the **Vertical** option forces the dimension line to the vertical, just like the DIMVERTICAL command.

Rotated

The **Rotated** option rotates the dimension line, just like the DIMROTATED command. When you enter **r** at the prompt, AutoCAD asks for the angle:

Specify dimension line location or [Mtext/Text/Angle/Horizontal/Vertical/Rotated]: **r**

Specify angle of dimension line <0>: **34**

You can enter different forms of angle measurement by including their units designation. For example, to enter the angle in grads, include the **g** suffix:

Specify angle of dimension line <0>: **38g**

To enter the angle in radians, include the **r** suffix:

> Specify angle of dimension line <0>: **0.59r**

Or pick two points to show the angle:

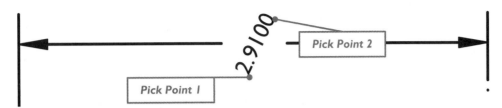

> Specify angle of dimension text: *(Pick point 1.)*
>
> Specify second point: *(Pick point 1.)*

 ## DIMALIGNED

The **DIMALIGNED** command places dimensions aligned to two points.

This command is like the **DIMLINEAR** command, except that the pick points for the extension lines determine the angle of the dimension line.

When you employ direct dimensioning, that is, press **ENTER** and then select an object, **DIMALIGNED** reacts differently, depending on the object you select:

> **Lines, Arcs, and Polylines** — endpoints of lines, arcs, and polylines determine the angle of the dimension line; AutoCAD measures the length of the line and polyline segments, and the chord length of arcs.

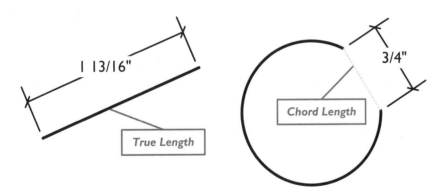

> **Circles** — a point on the circle determines the angle of the dimension line; AutoCAD measures the circle's diameter.

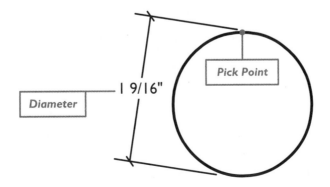

PLACING ALIGNED DIMENSIONS

1. To place a dimension aligned to two points, start the **DIMALIGNED** command:
   - From the **Dimension** menu, choose **Aligned**.
   - From the Dimension toolbar, choose the **Aligned Dimension** button.
   - In the Dashboard's Dimension panel, choose the **Aligned** button.
   - At the 'Command:' prompt, enter the **dimaligned** command.

     Command: **dimaligned** (*Press* ENTER.)
   - Alternatively, enter the aliases **dal** or **dimali** at the 'Command:' prompt.

2. In all cases, AutoCAD prompts you to select the object you want dimensioned:

   Specify first extension line origin or <select object>: (*Press* ENTER.)

   Select object to dimension: (*Pick one object.*)

3. Pick a point to place the dimension line:

   Specify dimension line location or

   [Mtext/Text/Angle]: (*Pick a point to place the dimension line.*)

   Dimension text = 10.0

This command's **Extension line**, **Mtext**, **Text**, and **Angle** options operate identically to those of the **DIMHORIZONTAL** command.

DIMBASELINE AND ⊢⊢⊢ DIMCONTINUE

The **DIMBASELINE** command (also called "parallel dimensioning") continues dimensions from baselines, while the **DIMCONTINUE** command (also called "chain dimensioning") continues dimensions from the previous dimension.

Baseline dimensions tend to stack over one another, while *continuous* dimension are usually located next to each other. The stack distance is 0.38, but can be changed. Continuous dimensions share common extension lines with each other.

Both commands continue from the last placed dimension, which can be linear, angular, or ordinate. (These commands do not work with leaders, radial, or diameter dimensions.) You cannot use either command until at least one other dimension has been placed in the drawing. When no dimensions were created during the current drafting session, AutoCAD prompts you to select a dimension as the base.

PLACING ADDITIONAL DIMENSIONS FROM A BASELINE

1. To place baseline dimensions, start the **DIMBASELINE** command:
   - From the **Dimension** menu, choose **Baseline**.
   - From the Dimension toolbar, choose the **Baseline Dimension** button.
   - In the Dashboard's Dimension panel, choose the **Baseline** button.
   - At the 'Command:' prompt, enter the **dimbaseline** command.

     Command: **dimbaseline** (*Press* ENTER.)
   - Alternatively, enter the aliases **dba** or **dimbase** at the 'Command:' prompt.

2. If necessary, AutoCAD prompts you to select a dimension to use as the base:

   Select base dimension: (*Pick a linear, angular, or ordinate dimension.*)

3. AutoCAD prompts you to pick the endpoint for the next dimension:

Specify a second extension line origin or [Undo/Select] <Select>: *(Pick a point.)*

Dimension text = 10.0

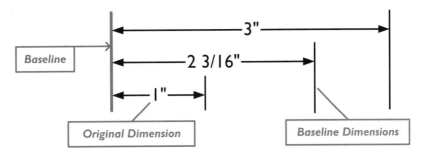

The baseline dimension continues from the first extension line of the previous dimension.

4. AutoCAD repeats the prompt for the next dimension:

    Specify a second extension line origin or [Undo/Select] <Select>: *(Pick another point.)*

5. Press **ESC** to exit the command:

    Specify a second extension line origin or [Undo/Select] <Select>: *(Press ESC.)*

CONTINUING DIMENSIONS

The **DIMCONTINUE** command is identical to **DIMBASELINE**, except that it continues dimensions by placing new ones next to each other.

1. To continue dimensioning, start the **DIMCONTINUE** command:

    • From the **Dimension** menu, choose **Continue**.

    • From the Dimension toolbar, choose the **Continue Dimension** button.

    • In the Dashboard's Dimension panel, choose the **Continue** button.

    • At the 'Command:' prompt, enter the **dimcontinue** command.

        Command: **dimcontinue** *(Press ENTER.)*

    • Alternatively, enter the aliases **dco** or **dimcont** at the 'Command:' prompt.

2. If necessary, AutoCAD prompts you to select a dimension to use as the base:

    Select base dimension: *(Pick a linear, angular, or ordinate dimension.)*

3. AutoCAD prompts you to pick the endpoint for the next dimension:

    Specify a second extension line origin or [Undo/Select] <Select>: *(Pick a point.)*

    Dimension text = 10.0

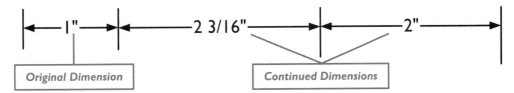

Notice that the dimension continues from the second extension line of the previous dimension.

4. AutoCAD repeats the prompt for the next dimension:

    Specify a second extension line origin or [Undo/Select] <Select>: *(Pick another point.)*

5. Press **ESC** to exit the command:

    Specify a second extension line origin or [Undo/Select] <Select>: *(Press ESC.)*

CONTINUED DIMENSIONS: ADDITIONAL METHODS

The **DIMBASELINE** and **DIMCONTINUE** commands have the following options:

- **Undo** undoes the last dimension placed.

- **Select** selects another dimension to continue from.

Let's look at both.

Undo

The **Undo** option undoes the last placed dimension. Because these two commands repeat automatically, this option is handy for undoing a dimension you did not mean to place.

Select

The **Select** option selects another dimension from which to continue. AutoCAD prompts:

Select continued dimension: *(Pick an extension line.)*

By picking an extension line, you determine the direction in which the dimension continues — to the left or to the right — assuming the dimension is horizontal.

 QDIM

The **QDIM** command places baseline and continuous style dimensions in one fell swoop.

It is a more powerful version of the **DIMCONTINUE** and **DIMBASELINE** commands: it places continued radial dimensions, applies the continued dimensions all at once, and more. Most importantly, it can dimension more than one object at a time. On the negative side, it is more complex than the two previous commands.

The figure below illustrates the continuous radius (at left) and continuous linear dimensions generated by this command. "While this is how AutoCAD dimensions radii," notes the technical editor, "it is poor practice. Dimensions should always be *outside* the parts."

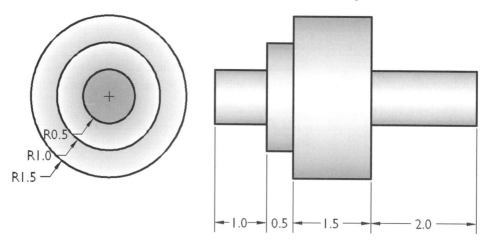

Perhaps the most powerful aspect of **QDIM** is its ability to change the type of dimension. For instance, you might first dimension objects in staggered mode, but some time later want to change to ordinate mode. Making the change is this simple: restart the command, select the same objects, and then specify ordinate mode. AutoCAD changes the staggered dimensions to ordinate dimensions.

QUICK CONTINUOUS DIMENSIONS

1. To place continuous-style dimension quickly, start the **QDIM** command:
   - From the **Dimension** menu, choose **Quick Dimension.**
   - From the Dimension toolbar, choose the **Quick Dimension** button.
   - In the Dashboard's Dimension panel, choose the **Quick Dimension** button.
   - At the 'Command:' prompt, enter the **qdim** command.

     Command: **qdim** *(Press* **ENTER.***)*

2. In all cases, AutoCAD reports the dimensioning priority, and then prompts you to select the geometry to dimension:

   Associative dimension priority = Endpoint

   Select geometry to dimension: *(Select one or more objects.)*

   Select geometry to dimension: *(Press* **ENTER** *to end object selection.)*

   Unlike other dimensioning commands, you may select more than object.

3. Pick a point to define the position of the dimension line:

   Specify dimension line position, or

   [Continuous/Staggered/Baseline/Ordinate/Radius/Diameter/datumPoint/Edit/seTtings]
       <Radius>: *(Pick a point.)*

   Depending on the objects you select, AutoCAD guesses at how to dimension them.

QUICK DIMENSIONS: ADDITIONAL METHODS

The QDIM command boasts a host of options to force the type of continuous dimension and specify settings:

- **Continuous** forces continued dimensions.
- **Staggered** places staggered dimensions.
- **Baseline** places baseline dimensions.
- **Ordinate** places ordinate dimensions.
- **Radius** places radial dimensions.
- **Diameter** places diameter dimensions.
- **datumPoint** specifies the start point for baseline and ordinate dimensions.
- **Edit** edits continuous dimensions.
- **seTtings** selects the object snap priority.

Let's look at all the options.

Continuous

The **Continuous** option forces continued dimensions, like those created by DIMCONTINUE.

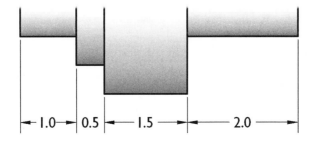

Staggered

The **Staggered** option forces staggered dimensions, which are like stacked continuous dimensions. This option, however, requires an even number of dimensionable objects. As illustrated in the figure, the odd-numbered segment, unfortunately, is left out and is not dimensioned.

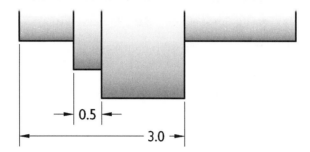

Baseline

The **Baseline** option forces baseline dimensions, like those created by the **DIMBASELINE** command. The figure below shows the dimensions starting at the left end of the flywheel; you can change the start point with the **Datum point** option, as described later.

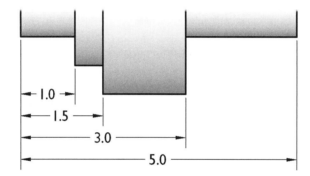

Ordinate

The **Ordinate** option forces ordinate dimensions, like those created by the **DIMORDINATE** command, with two differences. **DIMORDINATE** prevents ordinate dimensions from ending on top of each other, while **QDIM** does not. Also, **DIMORDINATE** measures relative to the UCS, while this command measures relative to a *datum point*.

The figure illustrates x-ordinate dimensions starting with a value of 5.0, which is the distance from the origin (0,0). You can change the origin point with the **Datum point** option.

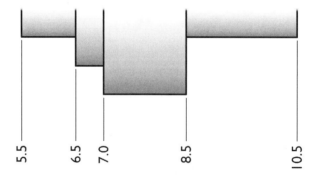

To create y-ordinate dimensions, move the cursor to the left or right side of the object during the "Specify dimension line position" prompt.

Radius

The **Radius** option forces radius dimensions, and works only with circles and arcs. Selecting other circular objects causes AutoCAD to complain, "No arcs or circles are currently selected." The radial dimensions are placed at the point you pick at the "Specify dimension line position" prompt.

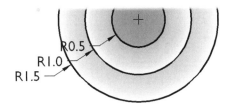

Diameter

The **Diameter** option forces diameter dimensions. It operates identically to the **Radius** option. In both cases, it's a good idea to avoid placing dimensions horizontally, because they tend to bunch up. Because placing dimensions on parts is poor practice, you should manually move the radial and diameter dimensions. This is easily done with grips editing.

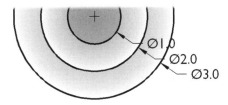

datumPoint

The **datumPoint** option prompts you to select a new base point from which to measure baseline and ordinate dimensions.

> Select new datum point: *(Pick a point.)*

Edit

The **Edit** option enters editing mode, and prompts you to add and remove dimension markers. These markers show up as an **x** at the endpoints of lines, and the center points of circles and arcs.

> Indicate dimension point to remove, or [Add/eXit] <eXit>: *(Enter an option.)*

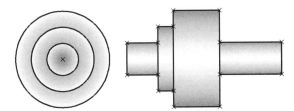

Remove removes markers from objects. To remove, pick a marker with the cursor. AutoCAD reports, "One dimension point removed."

Add adds markers to the drawing. To add, pick a point anywhere in the drawing; it need not be on an object. AutoCAD reports, "One dimension point added."

eXit exits **Edit** mode, and returns to the "Specify dimension line position" prompt.

seTtings

The **seTtings** options doesn't set much, except whether endpoints or intersections have priority, and even then I am not convinced either setting makes any difference. AutoCAD displays the following prompt:

> Associative dimension priority [Endpoint/Intersection]: *(Enter **E** or **I**.)*

DIMRADIUS AND DIMDIAMETER

The **DIMRADIUS** command places dimensions that measure the radius of circular objects, while the **DIMDIAMETER** command places dimensions that measure diameters.

These commands dimension circles, arcs, and *polyarcs* (arcs that are parts of polylines). They do not dimension other curved objects, such as ellipses, splines, and regions.

The dimensions created by these two commands are more like leaders than the two-extension-line dimensions you have seen so far in this chapter. They don't prompt for a point, but immediately ask you to select a circle or an arc.

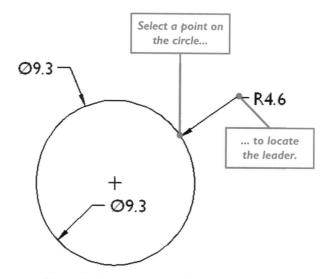

The second prompt asks you to "Specify dimension line location." Although this sounds like a single item (the dimension line), the point you pick determines several things at once, which can be difficult for new users:

- The angle of the leader line,
- and the length of the leader line,
- and the location of the dimension text,
- and the placement of the center mark,
- and the extent of an extension line (new to AutoCAD 2008).

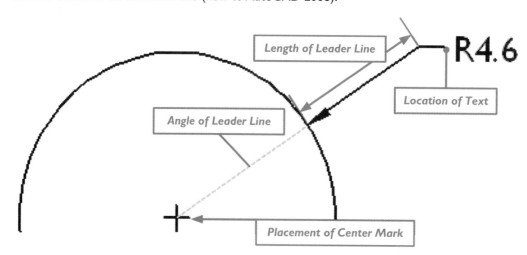

PLACING RADIAL DIMENSIONS

1. To dimension a radius, start the **DIMRADIUS** command:

    • From the **Dimension** menu, choose **Radius**.

    • From the Dimension toolbar, choose the **Radius Dimension** button.

    • In the Dashboard's Dimension panel, choose the **Radius** button.

    • At the 'Command:' prompt, enter the **dimradius** command.

       Command: **dimradius** *(Press* ENTER.*)*

    • Alternatively, enter the aliases **dra** or **dimrad** at the 'Command:' prompt.

2. In all cases, AutoCAD prompts you to select the arc or circle you want dimensioned:

    Select arc or circle: *(Pick one arc, circle, or polyarc.)*

    Dimension text = 10.0

3. Pick a point to place the dimension leader:

    Specify dimension line location or [Mtext/Text/Angle]: *(Pick a point to position the dimension leader.)*

When you position the dimension beyond the end of the arc, AutoCAD adds an extension line automatically, as illustrated below:

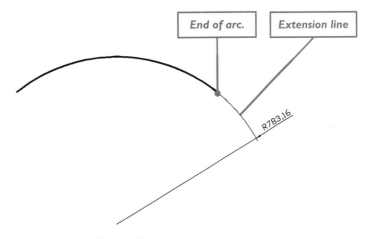

The **DIMDIAMETER** command operates identically:

    Command: **dimdiameter**

    Select arc or circle: *(Pick one arc, circle, or polyarc.)*

    Dimension text = 10.0

    Specify dimension line location or [Mtext/Text/Angle]: *(Pick a point to position the dimension.)*

The **Mtext**, **Text**, and **Angle** options are the same as those described earlier in this chapter under the **DIMHORIZONTAL** command.

⚡ DIMJOGGED

The **DIMJOGGED** command draws radial dimensions with a jogged leader line (also called "foreshortened" radius dimensions).

This command has more options, because AutoCAD also needs to know (1) where to relocate the dimension origin, and (2) where to place the jog.

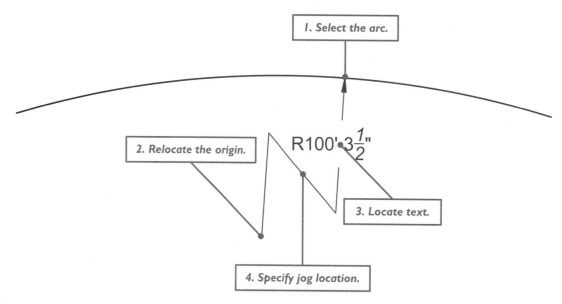

The relocated origin replaces the center point, even though it should always point to the center. The jog location determines the middle point of the jog; the angle of the jog (called the "transverse angle") is determined by the dimension style.

PLACING JOGGED RADIAL DIMENSIONS

1. To dimension a radius with a jogged leader, start the **DIMJOGGED** command:
   - From the **Dimension** menu, choose **Jogged**.
   - From the Dimension toolbar, choose the **Jogged** button.
   - In the Dashboard's Dimension panel, choose the **Jogged** button.
   - At the 'Command:' prompt, enter the **dimjogged** command.

     Command: **dimjogged** *(Press ENTER.)*

   - Alternatively, enter the aliases **jog** or **djo** at the 'Command:' prompt.

2. In all cases, AutoCAD prompts you to select the arc or circle you want dimensioned:
   Select arc or circle: *(Pick one arc, circle, or polyarc.)*

3. Pick a point to relocate the origin point closer to the arc:
   Specify center location override: *(Pick a point.)*
   Dimension text = 101'-3 1/2"

4. Indicate the center of the dimension text:
   Specify dimension line location or [Mtext/Text/Angle]: *(Pick a point or enter an option.)*

5. Locate the jog's center:
   Specify jog location: *(Pick a point.)*

DIMARC

The DIMARC command measures the lengths of arcs.

This command places dimensions along the circumference of arcs. The dimension can be along the full length of the arc, or just a portion.

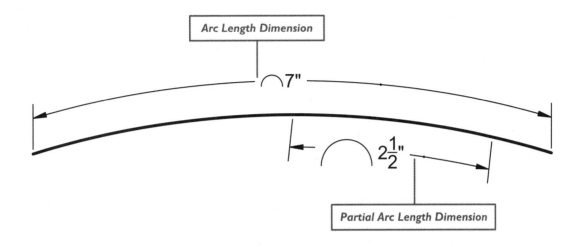

PLACING ARC LENGTH DIMENSIONS

1. To dimension the length of an arc, start the DIMARC command:
   - From the **Dimension** menu, choose **Arc Length**.
   - From the Dimension toolbar, choose the **Arc Length** button.
   - In the Dashboard's Dimension panel, choose the **Arc Length** button.
   - At the 'Command:' prompt, enter the **dimarc** command.

     Command: **dimarc** *(Press ENTER.)*
   - Alternatively, enter the **dar** alias at the 'Command:' prompt.

2. In all cases, AutoCAD prompts you to select an arc or polyline arc:
   Select arc or polyline arc segment: *(Pick one arc or polyarc.)*

3. Indicate the position of the dimension text:
   Specify arc length dimension location, or [Mtext/Text/Angle/Partial/Leader]: *(Pick a point or enter an option.)*

ARC LENGTH DIMENSIONS: ADDITIONAL METHODS

The DIMARC command has three options you are already familiar with — **Mtext**, **Text**, and **Angle**. Options unique to arc lengths are:

   - **Partial** measures parts of arc lengths.
   - **Leader** toggles the addition of leaders.

The figure below shows the effect of the partial and leader options.

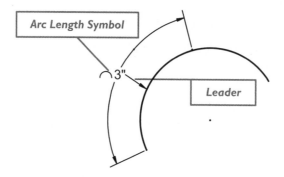

Partial

The **Partial** option dimensions just part of the length of the arc. AutoCAD prompts you to pick two points along the arc between which the dimension is placed:

> Specify first point for arc length dimension: *(Pick a point on the arc.)*
>
> Specify second point for arc length dimension: *(Pick another point.)*

If ENDpoint osnap is on, this command will snap to the endpoints of the arc, negating the attempt at partial dimensioning.

The results differ, depending on where you place the partial dimension. If the dimension is between extension lines, the extensions lines are parallel to each other; when the dimension is outside, the extension lines are radial to the arc's center point.

Leader

The **Leader** option toggles (turns on and off) the addition of a leader to arcs greater than 90 degrees. (This option does not appear when the arc is less is 90 degrees.) The leader points from the dimension text to the arc.

After entering the **Leader** option, the prompt changes to **No leader**:

> Specify arc length dimension location, or [Mtext/Text/Angle/Partial/Leader]: l
>
> Specify arc length dimension location, or [Mtext/Text/Angle/Partial/No leader]:

⊕ DIMCENTER

The **DIMCENTER** command places center marks and lines on arcs and circles.

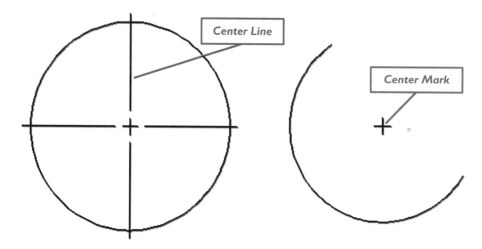

TUTORIAL: PLACING CENTER MARKS

1. To mark the center of arcs and circles, start the **DIMCENTER** command:
   - From the **Dimension** menu, choose **Center Mark**.

   - From the Dimension toolbar, choose the **Center Mark** button.

   - In the Dashboard's Dimension panel, choose the **Center Mark** button.

   - At the 'Command:' prompt, enter the **dimcenter** command.

      Command: **dimcenter** *(Press* ENTER.*)*

   - Alternatively, enter the **dce** alias at the 'Command:' prompt.

2. In all cases, AutoCAD prompts you to select the arc or circle you want dimensioned:

      Select arc or circle: *(Pick one arc, circle, or polyarc.)*

   Notice that AutoCAD places the center mark.

DimCen

The **DIMCEN** system variable defines the size and look of the center mark. This variable affects center marks created by **DIMRADIUS** and **DIMDIAMETER**, although their center marks are drawn only when the dimension is placed outside the circle or arc. The default value is 0.09, which means that **DIMCENTER** draws the center mark lines 0.09 units long.

DimCen	Meaning
0	Draws neither center lines nor center marks.
<0	Draws center lines and center marks.
>0	Draws center marks only.

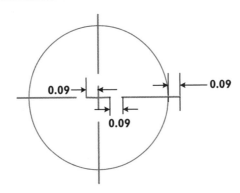

When the value is negative, such as -0.09, AutoCAD leaves a gap of 0.09 units between the mark and the line, as well as extends the line 0.09 units beyond the circumference of the circle or arc.

To draw center lines as well, change the value of **DIMCEN** to -0.09 (the negative value), as follows:

      Command: **dimcen**
      Enter new value for DIMCEN <0.0900>: **-0.09**

Normally, this value is set in the dimension style using the **DIMSTYLE** command.

DIMANGULAR

The **DIMANGULAR** command dimensions angles, arcs, circles, and lines.

Unlike the other dimensioning commands in this chapter, this dimensions angles only — not lengths or diameters. The command requires a *vertex* and some means of determining the angle. The result depends on the object selected:

Two Lines — the vertex is the real or apparent intersection of the two lines; the angle is measured between the two lines.

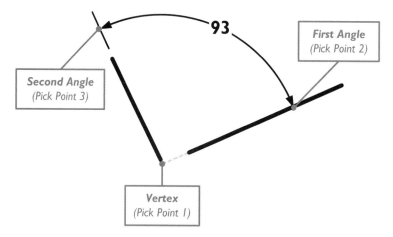

Arc — the vertex is the arc's center point; the angle is measured between two endpoints.

Circle — vertex is the circle's center point; the angle is measured between the "Select circle" pick and the next pick point.

Three Points — the vertex is at the first pick point; the angle is measured between the next two pick points.

For intersecting lines, **DIMANGULAR** can draw four possible angles, each a supplementary pair adding up to 180 degrees. In the figure below, the supplementary pairs are shown in the same color.

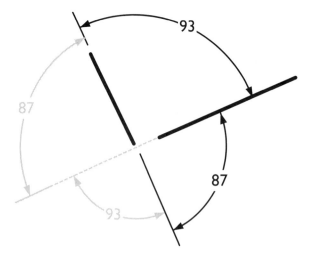

In the case of arcs and circles, two angles are possible: the *minor arc* (shown in black, below) and the *major arc* (shown in gray). Together they add up to 360 degrees.

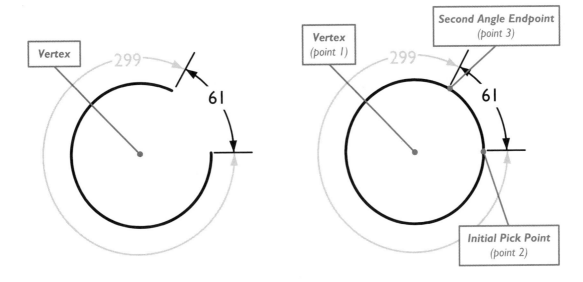

TUTORIAL: PLACING ANGULAR DIMENSIONS

1. To dimension an angle, start the **DIMANGULAR** command:
   - From the **Dimension** menu, choose **Angular**.
   - From the Dimension toolbar, choose the **Angular Dimension** button.
   - In the Dashboard's Dimension panel, choose the **Angular** button.
   - At the 'Command:' prompt, enter the **dimangular** command.

     Command: **dimangular** *(Press ENTER.)*
   - Alternatively, enter the aliases **dan** or **dimang** at the 'Command:' prompt.

2. In all cases, AutoCAD prompts you to select the object you want dimensioned:

   Select arc, circle, line, or <specify vertex>: *(Press ENTER for the vertex option)*

3. Pick three points to specify the angle:

   Specify angle vertex: *(Pick a vertex at point 1.)*

   Specify first angle endpoint: *(Pick point 2.)*

   Specify second angle endpoint: *(Pick point 3.)*

   Non-associative dimension created.

4. Pick a point to locate the dimension:

   Specify dimension arc line location or [Mtext/Text/Angle/Quadrant]: *(Pick a point to position the dimension.)*

   Dimension text = 90

The **Mtext**, **Text**, and **Angle** options are the same as those described earlier in this chapter under the **DIMHORIZONTAL** command. The unique options is as follows:

   - **Quadrant** locks the dimension to one of four sectors (new to AutoCAD 2008).

Quadrant

The **Quadrant** option forces the angular dimension to reside in one of the four quadrants:

   Specify quadrant: *(Drag the dimension around the arc.)*

Pick a quadrant for the dimension. When you drag the dimension beyond the end of the arc, AutoCAD adds an extension, just as with the **DIMRADIUS** command.

TUTORIAL: DIMENSIONING

In the following exercise, you dimension the fuse link below with some of the commands learned in this chapter. As you work through the tutorial, refer to the numbered points in the figure below.

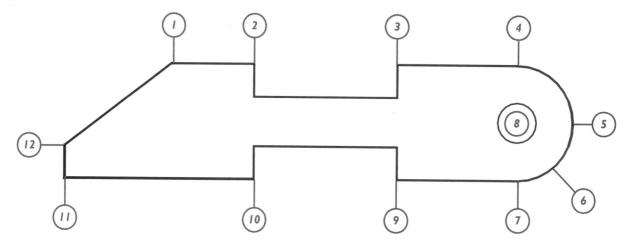

1. From the companion CD, open the *dimen.dwg* file, a drawing of a fuse link.
   This drawing has layers, dimension styles, and object snaps preset for you.
   Check the Layers toolbar to ensure layer "Dimensions" is current.

2. Start by placing a linear dimension along the top of the fuse link.
   From the **Dimension** menu, select **Linear**.
   Press **ENTER** to use the **Select Object** option, and then select the line between points 1 and 2.
   Locate the dimension line roughly 0.5 units above.

3. Continue dimensioning with the **DIMCONTINUE** command.
   Use INTersection object snap at points 3 and 4.
   Complete the dimension at point 5; do not dimension the arc at this time.
   Remember to press **ESC** to exit the command.
   Do the new dimension lines continue at the same height as the first dimensions?

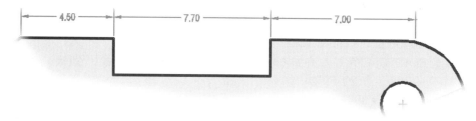

4. Now start a linear dimension along the bottom of the fuse link with the **DIMLINEAR** command.
   Place extension lines at points 11 and 10.
   Does INTersection object snap help you?

5. Switch to baseline dimensioning. From the **Dimension** menu, choose **Baseline**.
   Pick points 9 and 7 as the second extension line origins.
   Complete the command by picking point 7; do not dimension the arc yet.

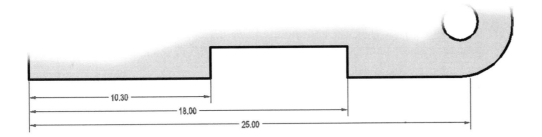

6. Place a vertical dimension by entering the **DIMLIN** alias.
   Choose point 11 as the first extension line origin, and point 12 as the second.
   If you make a mistake, use the **UNDO** command to eliminate the last dimension.

7. Dimension the angle at the left side with the **DIMALIGNED** command.
   Use the **Select Object** option to select the angled line between points 12 and 1.

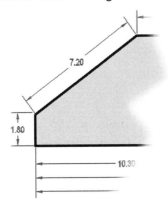

8. Select **Diameter** from the **Dimension** menu.
   Dimension the circle at point 8.
   Make sure the leader line extends outside the fuse link.

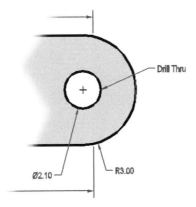

9. Dimension the arc with **DIMRADIUS**.
   Pick point 6, and then place the dimension leader outside the fuse link.

10. Construct a leader by selecting **Leader** from the **Dimension** menu. (See Chapter 12, "Placing and Editing Text," for more about leaders.)

Start the leader at the circle (point 8), and end it outside the object.

For the leader text, enter "Drill Thru."

Remember to press **ENTER** to exit the **QLEADER** command.

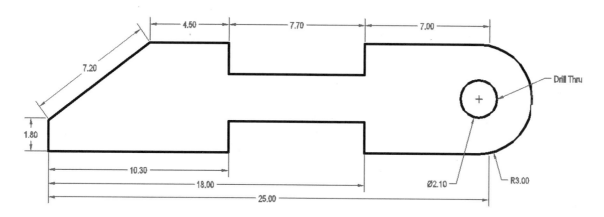

11. Save the completed drawing as *dimen.dwg*.

TUTORIAL: ANNOTATIVE DIMENSIONING

Your boss has emailed you *Anno-Dim.dwg*, requesting that you dimension it. In his note, he remarks that the drawing will be needed in detail sheets with two different scale factors, one at 1:8 scale and the other at 1:16.

In this tutorial, you use annotative scaling to dimension the drawing.

1. From Companion CD, open the *Anno-Dim.dwg* file.

2. On the status bar, change the **Annotation Scale** to 1:8, the first value your boss wants it dimensioned at.

3. You start the **DIMLINEAR** command:

    Command: **dimlinear**

    Specify first extension line origin or <select object>: *(Press ENTER.)*

    Select object to dimension: *(Select a line.)*

    Specify dimension line location or [Mtext/Text/Angle/Horizontal/Vertical/Rotated]: *(Pick a point.)*

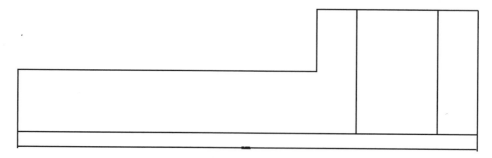

... but find that the dimension elements are too small. Of course, you should be using annotative scaling.

4. You note that the drawing file lacks an annotative dimension style, and so you create one.
   a. In the Dashboard's Dimensions panel, click Dimension Style button
   b. Click New, and then name it "Annotative."

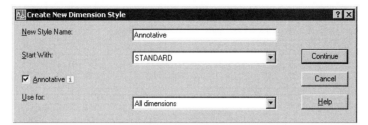

   c. Ensure that the **Annotative** option is turned on, and then click Continue.
   d. You don't need to change anything else, so click **OK**, **Set Current**, and **Close**.
      The new annotative dimension style is now the current one.

5. Reuse the **DIMLINEAR** command. That's better!

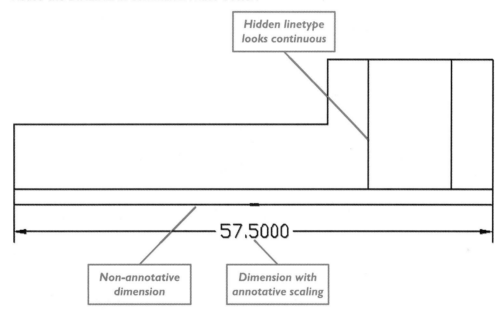

   Notice that annotative scaling takes over the duty of the **DIMSCALE** system variable.

6. Erase the first linear dimension, the one that's scaled too small.

7. The two vertical lines should be showing hidden lines, but they look continuous. The Properties palette confirms to you that they have the Hidden linetype, but are scaled wrong.
   Change the value of msltscale to 1, and then use the regen command to update the drawing.

   Command: **msltscale**

   Enter new value for MSLTSCALE <0>: **1**

   Command: **regen**

Notice how annotative scaling takes care of scaling linetypes correctly.

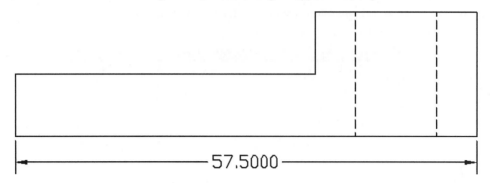

8. Add the other dimensions required by the drawing.

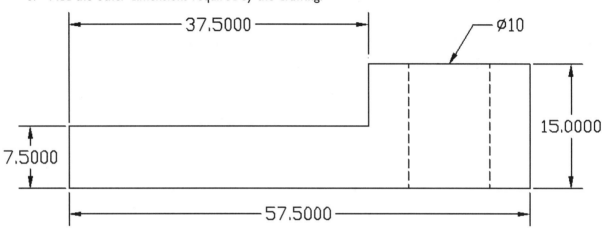

9. The other scale your boss wanted was 1:16. Change **Annotation Scale** to 1:16.

10. Re-run the **dimlinear** command.

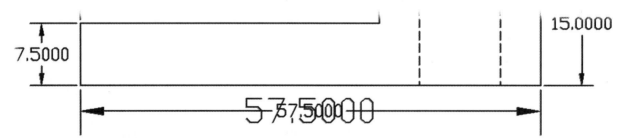

Notice that the dimension text and arrowheads are twice as large — annotative scaling takes care of that.

11. But, oh no! You can see both scaled versions of the linear dimension. The quick way to deal with that is to turn off **ANNOALLSIBIBLE**: click the **Annotation Visibility** button on the toolbar.

That's better.

12. Finish dimensioning the drawing, and then save it.

An alternative approach is to apply one set of dimensions at one annotative scale. Once applied, select all, and then use the Properties palette to all additional annotative scale factors.

EXERCISES

1. Start a new drawing.

   With the **POLYGON** command, draw a square approximately one-third the size of the screen.

   From the **Dimension** menu, select **Linear**. Use object snap INTersection to capture the lower left and lower right corners of the square.

   Place the dimension line below the bottom line of the square.

2. Continuing from the previous exercise, repeat the command for the upper line of the square, with a change. Instead of picking the corners, dimension the line by selecting it.

   Is the dimension text the same for the lower and upper lines?

3. Continuing, select **Linear** from the **Dimension** menu, and then dimension one of the vertical sides of the box.

   Is the vertical dimension the same as the horizontal dimension?

4. Next, dimension the remaining vertical line.

   At the "Specify dimension line location or [Mtext/Text/Angle/Horizontal/Vertical/ Rotated]" prompt, enter **m**.

   Does the mtext editor dialog box appear?

   Type your name.

   Choose **OK** when finished.

   What appears along the dimension line?

   Save the drawing as *square1-4.dwg*.

5. Start a new drawing.

   Draw a scalene triangle.

   Use the **DIMALIGNED** command to dimension the three angled sides.

   Does AutoCAD draw the dimension lines parallel to the sides?

6. Draw two lines that are not parallel, similar to those shown below.

   Place an angular dimension on the two lines.

   Repeat the command to draw the supplemental angle.

7. Draw an arc and a circle.

   Dimension the minor and major arcs with the **DIMANGULAR** command.

   Save the drawing as *dimen5-7.dwg*.

8. Start a new drawing.

   Draw a diagonal line, similar to the ones in exercise #6.

   Use the following commands to dimension the line three times: horizontally, vertically, and rotated.
   a. **DIMHORIZONTAL**
   b. **DIMVERTICAL**
   c. **DIMROTATED**

9. Undo the three dimensions you drew in the previous exercise.

   Use the **DIMLINEAR** command to draw the same three orientations of dimension.

   Which commands did you find easier to use?

   Save the drawing as *linear8-9.dwg*.

10. Start a new drawing.

    Draw an arc and a circle on the screen, and then dimension each with the **DIMDIAMETER** command.

    Practice placing diameter dimensions inside and outside the circle, and at different locations around the arc.

11. Repeat exercise #9 with the **DIMRAD** command.

    Save the drawing as *diarad10-11.dwg*.

12. From the companion CD, open the *qdim.dwg* drawing of a flywheel.

    Dimension the two views with the **QDIM** command.

    When complete, save the drawing as *flywheel12.dwg*.

13. Draw the following baseplates, and then place the dimensions at the locations shown.
    a. Save the drawing as *baseplate13-a.dwg*.

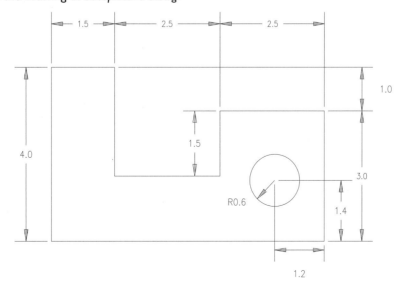

b. Save the drawing as *baseplate13-b.dwg*.

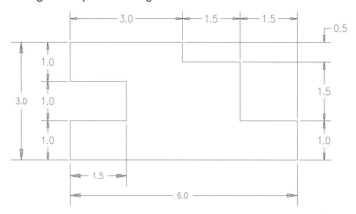

c. Save the drawing as *baseplate13-c.dwg*.

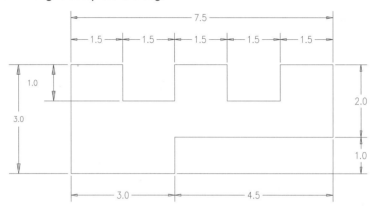

14. Draw the two views of the clamp, and then use dimension commands to place the vertical dimensions at the locations shown. Save the drawing as *clamp14.dwg*.

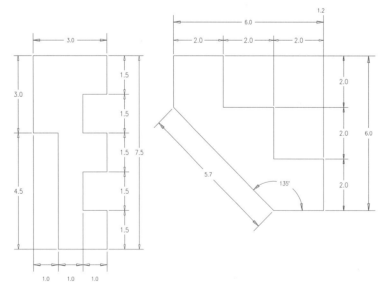

15. Construct the drawing of the corner shelf, and then place all the dimensions as shown. Save the drawing as *shelf15.dwg*.

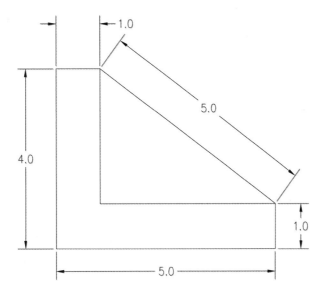

16. Draw the following 5.5" x 3.5" object, and then use radius dimensioning to specify the two 1.5" fillets. Save the drawing as *radial16.dwg*.

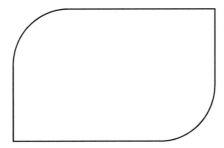

17. Construct the drawing of the corner shelf, and then place all the dimensions as shown. Save the drawing as *shelf17.dwg*.

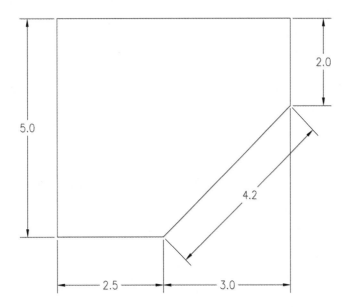

18. Construct the drawing of the jig, and then place the baseline dimensions. Save the drawing as *jig18.dwg*.

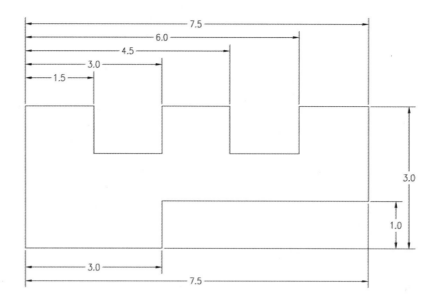

19. From the companion CD, open the *15-47.dwg* of a gasket. Dimension the drawing with the **DIMRAD** and **DIMDIA** commands. When complete, save the drawing by the same name.

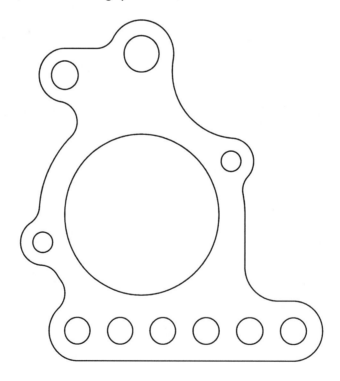

20. From the companion CD, open the following drawings, and then dimension the gaskets.

   a. Save as *5-43.dwg*.

   b. Save as *15-44.dwg*.

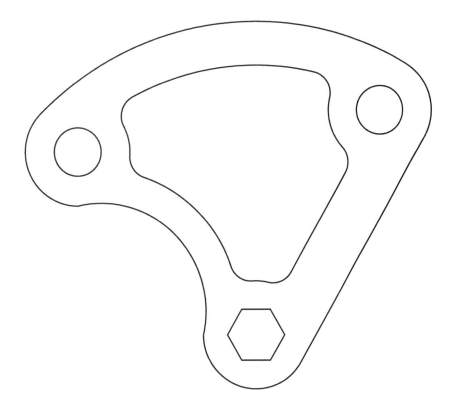

c. Save as *15-45.dwg.*

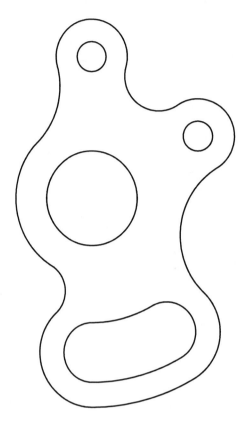

d. Save as *15-46.dwg.*

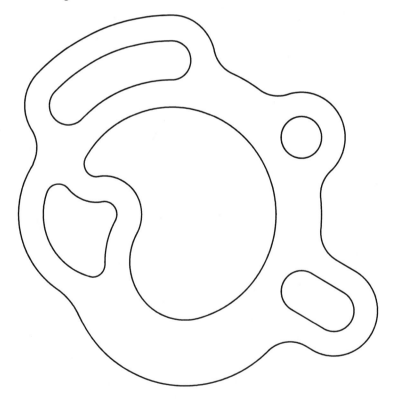

21. From the Companion CD, open the following drawings. Dimension each drawing, and then save your work.

   a. *17_19.dwg*: Attachment for a three-point hitch.

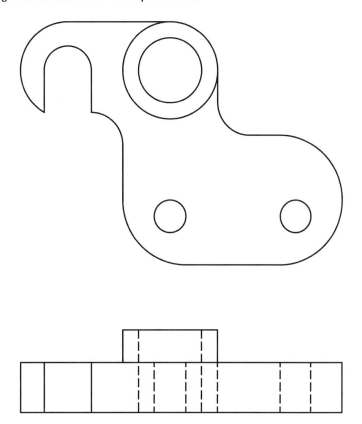

   b. *17_20.dwg*: Flag pole stabilizer.

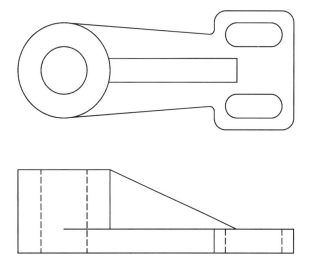

c. *17_21.dwg*: Reciprocating linkage.

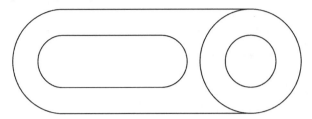

d. *17_22.dwg*: Socket linkage.

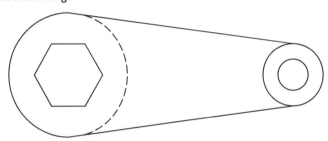

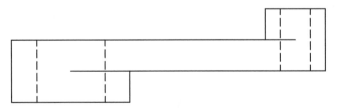

e. *17_23.dwg*: Attachment plate.

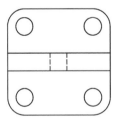

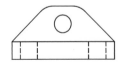

f. *17_24.dwg:* Hitch yoke.

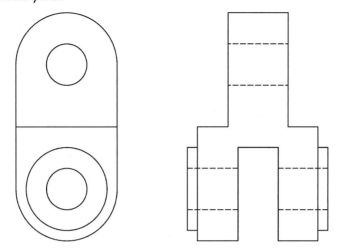

g. *17_25.dwg:* Hitch linkage.

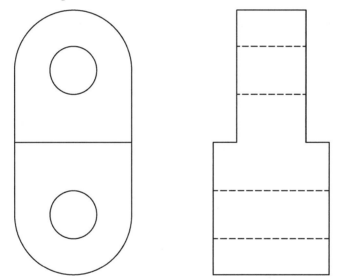

h. *17_26.dwg:* Axle support.

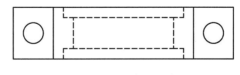

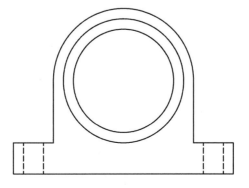

i. *17_27.dwg*: Cone of silence. (Technical editor comments: "Yes! Maxwell Smart lives!" Copy editor comments: "Sorry about that, Chief.")

j. *17_28.dwg*: Bushing.

k. *17_29.dwg*: Wedge.

l. *17_30.dwg:* Concrete footing.

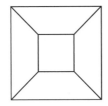

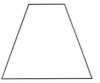

m. *17_31.dwg:* Heavy-duty axle support.

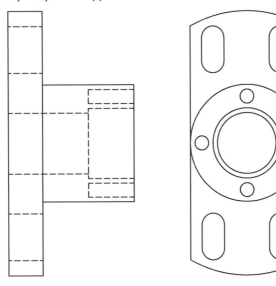

n. *14_32.dwg:* Paintbrush handle.

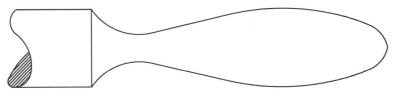

o. *17_33.dwg:* Spacer.

p. *17_34.dwg*: Baseplate.

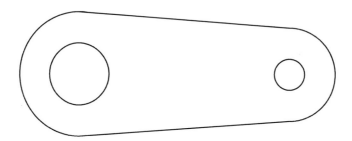

q. *17_35.dwg*: Angled baseplate. Use the **QDIM** command.

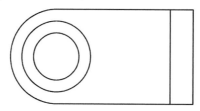

CHAPTER REVIEW

1. A dimension measures _____.
2. Should extension lines touch the object they reference?
3. Where should dimensions be placed, whenever possible?
4. Leaders indicate _____.
5. In which menu do you find the dimensioning commands?
6. Should you attempt to dimension every detail?
7. Under what conditions are dimension lines not drawn?
8. What is another name for "witness lines"?
9. How are the endpoints of extension lines determined?
10. Must a dimension always have two extension lines?
11. Sketch and label the parts of a measurement dimension. Include the following parts:
    Dimension line
    Arrowheads
    Extension lines
    Text
12. Sketch and label the parts of a radial dimension. Include the following parts:
    Leader line
    Arrowhead
    Text
    Radius Symbol
13. Sketch and label the parts of an extension line:
    Extension line
    Offset
    Extension beyond dimension line
14. List the types of linear dimensions that the **DIMLINEAR** command draws.
    a.
    b.
    c.
15. Explain the function of the **DIMDIAMETER** command?
16. What is the purpose of the **QDIM** command?
17. Draw an example of a baseline dimension:
18. Must arrowheads always be arrows?
19. What is the aspect ratio of arrowhead?
20. List three places where dimension text can be placed.
21. Must dimension text always be horizontal?
22. Write out an example of each of the following terms:
    Deviation Tolerance
    Symmetrical Tolerance
    Limits
    Alternate Units
23. Explain why *direct dimensioning* is beneficial.
    Can any object be dimensioned directly?
24. Which dimensioning commands do you use most often with arcs and circles?
25. What is the purpose of the **DIMANGULAR** command?
26. Describe the meaning of "fully associative dimensioning."

27. Are all dimensions associative?

    Give an example.

28. How is the look of dimensions controlled?

29. Where in AutoCAD might you find dimension standards?

30. Name three object snaps useful for dimensioning:

    a.

    b.

    c.

31. Write out the names of commands that can draw dimensions horizontally.

32. Explain the meaning of the following options available in many dimensioning commands:

    Angle

    Mtext

    Text

33. What is the **<>** symbol meant for?

34. A dimension reads 2.5. Write out the text string that changes the dimension to read:

    a. "Adjust to 2.5m"

    b. "Turn by 2.5° increments."

    c. "Temperature <u>range</u> is 2.5°±0.5°"

35. Which part of an arc does the **DIMALIGNED** command measure?

    Of a circle?

    Of a line?

36. How do these commands differ?

    DIMBASELINE

    DIMCONTINUE

37. Can you apply the **DIMCONTINUE** command to a diameter dimension?

38. Describe the steps to create y-ordinate dimensions with **QDIM**.

39. Can you dimension ellipses and splines with the **DIMRADIUS** command?

40. The **DIMJOGGED** command is best suited for which kinds of arcs and circles?

41. When would you use the **DIMARC** command?

42. Write out the command for each alias:

    dra

    dco

    dli

    dimali

43. Explain the purpose of the center mark.

    Describe the difference between a center mark and a center line?

    How do you instruct AutoCAD to draw one or the other?

44. What angle does the **DIMANGULAR** command measure for the following objects:

    Circle

    Two Lines

    Arc

45. How many possible angles can **DIMANGULAR** draw on:

    Circles?

    Pair of lines?

CHAPTER 14

Editing Dimensions

You often need to change existing dimensions. This chapter describes how to edit the position and properties of dimensions. After completing this chapter, you will have an understanding of these commands:

Grips edits dimensions and leaders directly.

AIDIMFLIPARROW flips the direction of arrowheads.

DIMJOGLINE adds jogs to dimension lines (new to AutoCAD 2008).

DIMBREAK adds breaks to dimension and extension lines (new to AutoCAD 2008).

DIMSPACE spaces groups of dimensions evenly (new to AutoCAD 2008).

DIMEDIT slants extension lines.

DIMTEDIT relocates dimension text.

DDEDIT changes dimension text with the mtext editor.

DIMINSPECT adds and removes inspection text (new to AutoCAD 2008).

AIDIM and **AI_DIM** prefix a group of undocumented dimension editing commands.

Dimension Variables specify the look and position of dimension elements.

DIMSTYLE creates and modifies dimension styles.

DIMOVERRIDE applies changed dimvars to selected dimensions.

TEXTTOFRONT changes the display order of dimensions.

NEW TO AUTOCAD 2008 IN THIS CHAPTER

- The **DIMBREAK, DIMJOGLINE, DIMINSPECT,** and **DIMSPACE** commands are new.
- The **DIMSTYLE** dialog box has new options.
- The **DIMANNO** system variable reports whether the dimension style is annotative.

FINDING THE COMMANDS

On the Dashboard's **Dimension** panel:

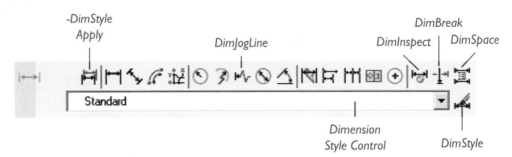

On the **Dimension** toolbar:

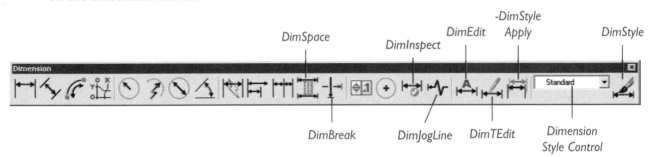

On the **Styles** toolbar:

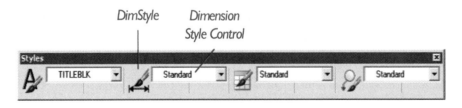

GRIPS EDITING

Grips allow you to edit dimensions and leaders directly — without needing to enter the names and options of dimension editing commands.

The grips on associative dimensions change the location of the dimension line, text, extension lines, and so on. The tricky part is knowing which grip performs what action; the grips themselves provide no clues, unfortunately. Thus, this section describes their actions. (Grips that move and rotate entire dimensions are described in Chapter 9, "Direct Editing of Objects.")

TUTORIAL: EDITING DIMENSIONS THROUGH OBJECTS

1. Draw a line, and then dimension it using the **Select Object** option, as follows:

   Command: **dimlinear**

   Specify first extension line origin or <select object>: *(Press* ENTER.*)*

   Select object to dimension: *(Pick the line.)*

   Specify dimension line location or [Mtext...]: *(Pick a point to position the dimension line.)*

   Dimension text = 2.00

2. Select the line — not the dimension! Notice the blue grips.
3. Select one of the end grips; it turns red.
4. Stretch the *object*, longer or shorter. Notice that the associated dimension updates itself, as illustrated by the figure.

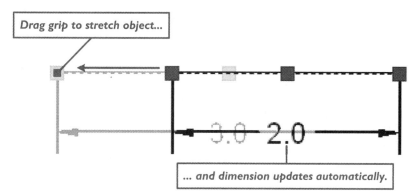

Drag grip to stretch object...

... and dimension updates automatically.

5. Select middle grip to move the line. Notice that the dimension moves along.
6. Press ESC to end grips editing.

GRIPS EDITING OPTIONS

Grips perform six editing operations: stretch, copy, move, rotate, scale, and mirror. Here is how each one affects dimensions:

Stretch changes the position of dimension parts, such as extension lines or text, while keeping the remainder of the dimension in place. The dimension remains partially associated with its object. When stretching the dimension line, the dimension text is updated; other parts of the dimension react differently to stretching, as detailed later.

Copy copies the dimension, subject to the peculiarities of the selected grip.

Move moves the entire dimension, but disassociates it from its object.

Rotate rotates the entire dimension about the selected grip, but disassociates it from its object.

Scale resizes the dimension, using the selected grip as the base point. The dimension remains associated with its object; dimension text is updated to reflect the new size. As with copying, scaling is subject to the peculiarities of the selected grip.

Mirror does not mirror dimensions; instead, this option rotates them about the selected grip. Text is not mirrored, ignoring the setting of the MIRRTEXT system variable. The "mirrored" dimension is disassociated from its object.

EDITING LEADERS WITH GRIPS

The LEADER and QLEADER commands draw identical-looking leaders; the only difference is the location of the grip on the text, which is immaterial for grips editing.

Leaders have at least three grips, one each for changing the position of the arrowhead end of the leader, the vertices, and the leader endpoint. The leader text is placed independently, and thus is edited separately from the leader line.

The MLEADER command draws multiline leaders with additional grips.

Stretch is detailed below. **Copy** copies the leader line, independently of the text. **Move** moves the leader line. **Rotate** rotates the leader line about the selected grip. **Scale** resizes the leader along the axis of the line. **Mirror** mirrors the leader line, and erases the original.

Stretching the Arrowhead

The arrowhead end of the leader line is relocated by dragging its grip:

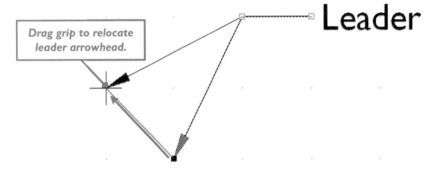

Drag grip to relocate leader arrowhead.

Stretching the Vertex

There are grips located at each vertex along the leader line. The vertex is relocated by dragging its grip:

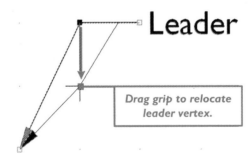

Drag grip to relocate leader vertex.

Stretching the Endpoint

The endpoint of the leader line is also relocated by moving its grip. (Moving the leader line does not affect the text; it has its own grip.)

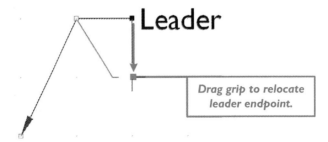

Drag grip to relocate leader endpoint.

Stretching the Text

The text is independent of the leader line; you must select it separately. The text is relocated by dragging its grip. (Moving the text brings the leader line along with it.)

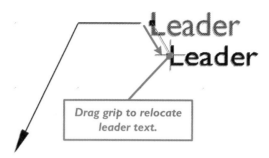

Drag grip to relocate leader text.

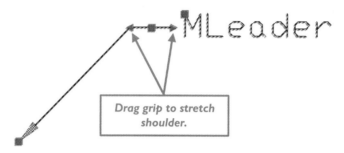 MLeader Grips

NEW IN 2008

Multiline leaders have two more grips. The triangular grips resize the shoulder (the horizontal stub in front of the leader text). They are triangular in shape to indicate that they are object-specific; in this case, they only move horizontally.

Drag grip to stretch shoulder.

GRIPS EDITING LINEAR AND ALIGNED DIMENSIONS

Linear dimensions are drawn by many dimensioning commands such as **DIMLINEAR** and **DIMROTATED**. Editing these dimensions with grips is identical for each one — horizontal, vertical, or rotated. Linear dimensions tend to have five grips: one at either end of the dimension line, one at the end of each extension line, and one on the text.

Stretch is detailed below. **Copy** copies the dimension, subject to the grip selected:

- Extension line grips constrain copying in the direction of the dimension line.

- Dimension line grips constrain copying in the direction of the extension lines.

- Text grip constrains copying in the direction of the extension lines, but also moves the text.

Move moves the entire dimension freely, but disassociates it from its object. **Rotate** and **Mirror** rotate the dimension about the selected grip. **Scale** resizes the dimension, subject to the grip selected:

- Extension line grips resize the dimension relative to the selected grip.

- Dimension line grips constrain resizing in the direction of the dimension line, relative to the selected grip.

- Text grip constrains resizing in the direction of the dimension line, but resizes symmetrically. (Both extension lines move in mirrored fashion.)

Stretching the Dimension Line

There is a grip at each end of the dimension line. The two grips perform the identical task: moving the dimension line and text closer to, and further away from, the object. Movement is constrained along the extension line.

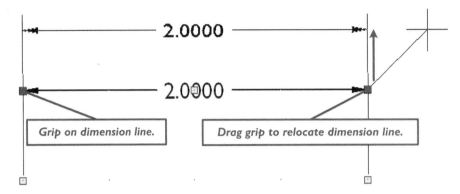

Grip on dimension line.

Drag grip to relocate dimension line.

Stretching the Extension Lines

Each extension line has a grip at its endpoint that has two functions: changing the length of the extension line and the width of the dimension line, which, in turn updates the dimension text. It is helpful to turn on ortho or polar mode to constrain movement to the horizontal and vertical. *Warning!* Stretching extension lines causes dimensions to lose their associativity.

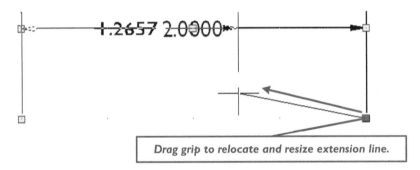

Drag grip to relocate and resize extension line.

Stretching the Text

Text has a single grip at the center that performs two functions similar to extension line grips: changing the position of the text line along the dimension line, and moving the dimension line.

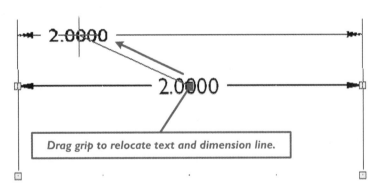

Drag grip to relocate text and dimension line.

GRIPS EDITING ANGULAR DIMENSIONS

Angular dimensions are drawn by the **DIMANGULAR** command, and have five grips: one at the vertex, one on the dimension arc, one at the end of each extension line, and one on the text.

Stretch is detailed below. **Copy** copies the dimension, subject to the grip selected:

- Extension line grips constrain copying in the direction of the dimension line.
- Dimension line grips constrain copying in the direction of the extension lines.
- Text grip constrains copying in the direction of the extension lines, but also moves the text.

Move moves the entire dimension freely, no matter which grip is selected. The dimension is disassociated from its object. **Rotate** rotates the dimension about the selected grip. **Scale** resizes the dimension, using the selected grip as the base point. **Mirror** mirrors the dimension about the selected grip; the original is erased. Text is not mirrored, ignoring the MIRRTEXT system variable.

Stretching the Dimension Arc

Stretching the dimension arc grip moves it and the text towards, and away from, the vertex.

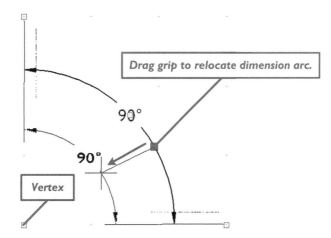

Drag grip to relocate dimension arc.

Vertex

Stretching the Text

Stretching the text grip performs two functions simultaneously: it moves the text along the dimension arc and moves the dimension arc itself (as described above).

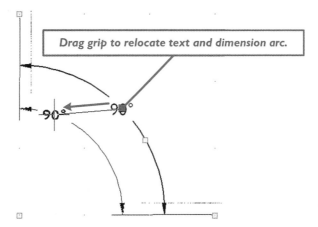

Drag grip to relocate text and dimension arc.

Stretching the Extension Lines

Stretching the grips at the ends of the extension lines also performs two functions simultaneously: it changes the length of the extension line, and moves the extension line. Moving the extension line changes the angle, and so the text also updates.

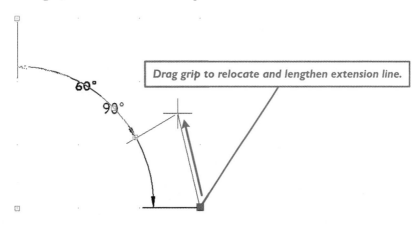

Drag grip to relocate and lengthen extension line.

Stretching the Vertex

Stretching the grip at a vertex is mind bending. The text grip and all other non-grip parts of the angular dimension change — or rather, bend: as you move the vertex grip, the extension lines and dimension arc change their lengths and angles, but are constrained by the extension line and dimension arc grips that remain fixed. As the angle changes, the text is also updated.

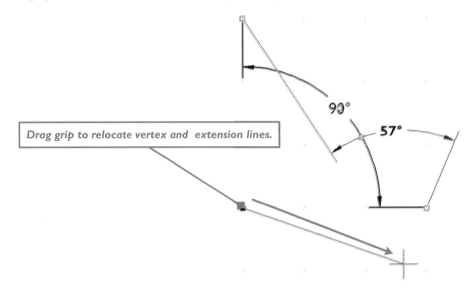

Drag grip to relocate vertex and extension lines.

EDITING DIAMETER AND RADIUS DIMENSIONS

The radial dimensions are drawn by the **DIMDIAMETER** and **DIMRADIUS** commands. These commands dimension circles, arcs, and polyarcs.

Each dimension has three grips: one where the dimension attaches to the circle or arc, one at the diameter or radius point, and one in the center of the text.

Stretch is detailed below. **Copy** copies the dimension, subject to the grip selected:

- Extension line grips constrain copying in the direction of the dimension line.
- Dimension line grips constrain copying in the direction of the extension lines.
- Text grip constrains copying in the direction of the extension lines, but also moves the text.

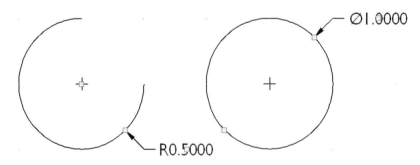

Left: *The grips found on radius dimensions.*
Right: *The grips found on diameter dimensions.*

Move moves the entire dimension freely, no matter which grip is selected. The dimension is disassociated from its object. **Rotate** and **Mirror** rotate the dimension about the leader and text grips. When the center point grip is selected, the radius dimension rotates about the circle. The text and leader stubs maintain their horizontal orientation. **Scale** resizes the dimensions, using the selected grip as the base point.

Stretching the Circumference

Radius dimensions have a grip located on the circumference of circles and arcs, while diameter dimensions have two. Drag the grip to relocate the dimension around the circle or arc; the text maintains its orientation, and the leader, its length.

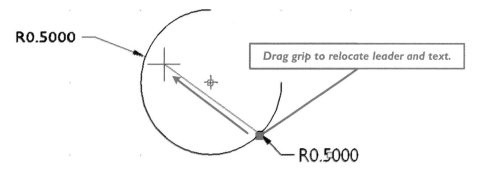

Stretching the Center Mark

Only radius dimensions have a grip at the center mark. Drag the grip to relocate the leader away from the arc or circle; the dimension text stays in place, roughly.

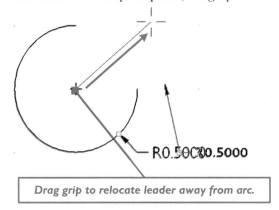

Stretching the Text

Both radius and diameter dimensions have a grip located in the center of the dimension text. Drag the grip to relocate the dimension around the circle or arc; the text maintains its orientation, but the leader changes its length.

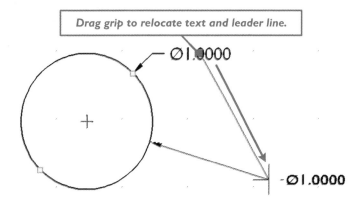

EDITING JOGGED DIMENSIONS

Recall that the **DIMJOGGED** command can dimension the radius of arcs with a jogged radius line. This command is meant for circle, arcs, and polyarcs with a very large radii.

Each dimension has four grips: one where the leader's arrowhead meets the arc, another at the arc's center point, one to change the location of the jog, and one to relocate the text.

Stretch is detailed below. **Copy** copies the dimension, subject to the grip selected:

- Arrowhead grip moves a copy of the dimension, sliding the arrowhead end along the arc, and staying anchored to the dimension's center point.

- Center point grip moves a copy of the dimension away from the dimension's center point.

- Jog grip moves a copy of the dimension, but keeps it anchored; only the jogged portion moves.

- Text grip moves a copy of the dimension, but keeps it anchored to the dimension's center point.

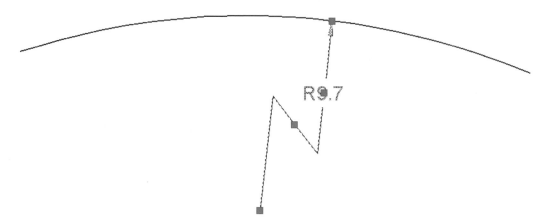

Move moves the entire dimension freely, no matter which grip is selected. The dimension is disassociated from its object. **Rotate** and **Mirror** rotate the dimension about selected grip. When the center point grip is selected, the radius dimension rotates about the circle. The text maintains its horizontal orientation. **Scale** resizes the dimensions, using the selected grip as the base point. The dimension text updates to reflect the change in size.

Stretching the Arrowhead and Text

The jogged radial dimension has a grip at the arrowhead, where the dimension touches the arc. Dragging this grip changes the location of the jogged leader; movement is constrained along the curve of the arc. The identical movement occurs when the text grip is moved.

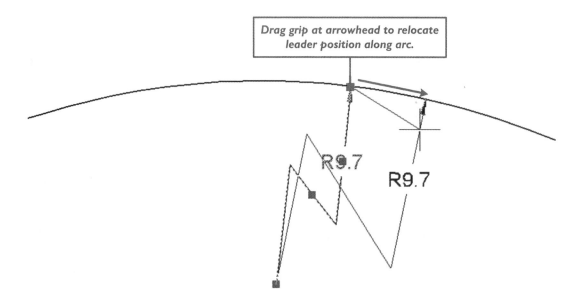

Drag grip at arrowhead to relocate leader position along arc.

Stretching the Jog

When the grip located at the center of the jog (the N-shaped portion of the leader) is dragged, the location of the jog changes. Movement is constrained radially, between the two endpoints of the leader. The position of the text is unaffected.

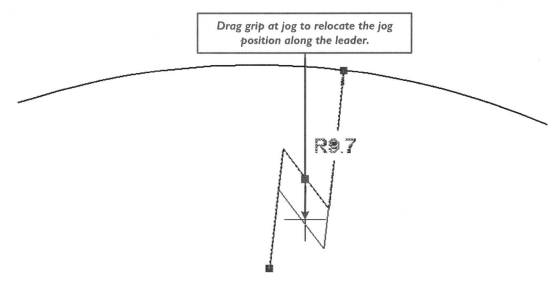

Drag grip at jog to relocate the jog position along the leader.

Stretching the Endpoint

When the grip at the end of the leader is dragged, the location of the endpoint changes; it is no longer at the original "center location override" specified by the **DIMJOGGED** command. The position of the text is unaffected, as is the arrowhead half of the leader.

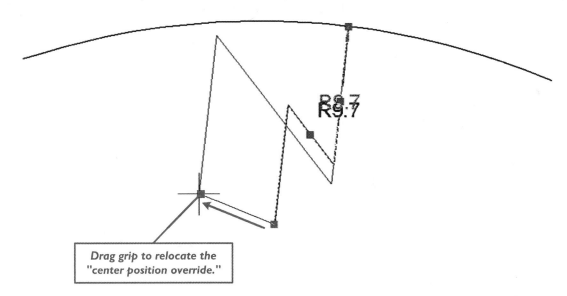

Drag grip to relocate the "center position override."

EDITING ARC LENGTH DIMENSIONS

The **DIMARC** command dimensions the length of arcs and polyarcs. The dimension has four grips: two where the extension lines meet the arc, one to change the diameter of the dimension, and one to relocate the text.

Stretch is detailed below. **Copy** copies the dimension, subject to the grip selected:

- Extension line grips do not copy, but move the entire dimension.

- Dimension line grip constrains copying in the radial direction of the arc's center point.

- Text grip copies the dimension and relocates the text.

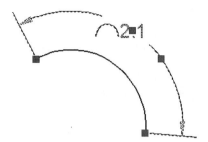

Move moves the entire dimension freely, no matter which grip is selected. The dimension is disassociated from its object. **Rotate** and **Mirror** rotate the dimension about the selected grip. When the center point grip is selected, the radius dimension rotates about the circle. The text maintains its horizontal orientation. **Scale** resizes the dimensions, using the selected grip as the base point. The dimension text updates to reflect the change in size.

Stretching the Extension Lines

The arc length dimension has one a grip on each extension line. Dragging them changes the location of the extension line and updates the text. This is equivalent to using the **Partial** option of the DIMARC command.

Stretching the Dimension Line

The dimension line is an arc parallel to the arc being measured. Stretching it moves the dimension line radially: it moves closer to, and further away from, the arc's center point.

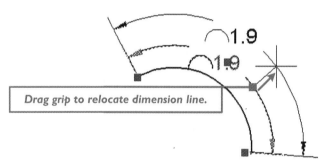

Move the grip past the arc's center point, and the dimension reverses, measuring the open part of the arc.

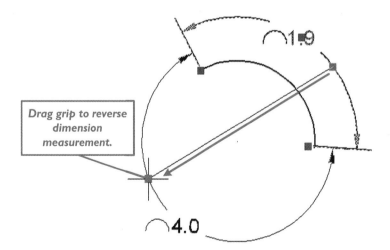

Stretching the Text

Dragging the grip attached to the text moves both the text and the dimension line; the value of the text remains unchanged.

AIDIMFLIPARROW

The AIDIMFLIPARROW command flips the direction of arrowheads.

This command flips the direction of selected arrowheads by 180 degrees. This action is useful for touching up dimensions when arrowheads would look better placed outside the extension lines, or vice versa. It is not documented by Autodesk, but appears in the shortcut menu when a dimension is selected.

TUTORIAL: FLIPPING ARROWHEADS

1. Draw a linear dimension with the DIMLINEAR command.
2. To change the direction of one or both arrowheads, start the AIDIMFLIPARROW command:
   - Select the dimension, and then select **Flip Arrow** from the shortcut menu.
   - At the 'Command:' prompt, enter the **aidimfliparrow** command.

     Command: **aidimfliparrow** *(Press ENTER.)*
3. AutoCAD prompts you to select one or more objects, but it really wants you to select one or more arrowheads to flip:

     Select objects: *(Select one or more arrowheads.)*

     Select objects: *(Press ENTER to end object selection.)*

   AutoCAD flips the arrowheads, and draws a short dimension line.

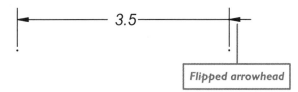

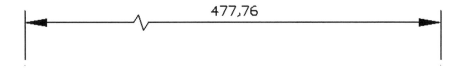

Flipped arrowhead

If you select dimensions instead of arrowheads, AutoCAD flips the arrowhead nearest to your pick point.

 **DIMJOGLINE**

The DIMJOGLINE command adds jogs to linear dimension lines.

Very wide drawings, such as road cross-sections, use jogged dimension lines to show overall distance. This command adds and removes the jogs; you use the PROPERTIES command to specify the size of the jog, and can relocated the jog using grips editing.

477,76

The DIMJOGLINE command works only with linear dimensions. You cannot give a dimension line more than one jog; to add jogs to radial extension lines, use the DIMJOGGED command, as described in the previous chapter.

TUTORIAL: JOGGING DIMENSION LINES

1. To add a jog to dimension lines, start the DIMJOGLINE command:
   - From the **Dimension** menu, choose **Jogged Linear**.
   - From the Dimension toolbar, choose the **Jogged Linear** button.
   - In the Dimension panel, choose the **Dimjogline** button.

- At the 'Command:' prompt, enter the **dimjogline** command.

  Command: **dimjogline** *(Press* ENTER.*)*

- Alternatively, enter the **djl** alias at the 'Command:' prompt.

2. In all cases, AutoCAD prompts you to select a dimension:

   Select dimension to add jog or [Remove]: *(Pick one dimension.)*

3. Specify the location of the jog:

   Specify jog location (or press ENTER): *(Pick a point.)*

When you press ENTER at the last prompt, AutoCAD places the jog halfway between the first extension line and the dimension text.

To change the size of the jog, use the PROPERTIES command: in the Lines & Arrows section, change the value of **Jog Height Factor**.

To move the jog, use grips editing (as illustrated below), or else reuse this command and pick another location for the jog. Since dimension lines can have only one jog, specifying a "second" one relocates the jog.

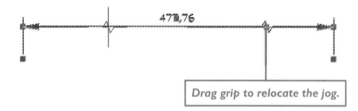

Drag grip to relocate the jog.

The **Remove** option removes the jog from the dimension line.

 **DIMBREAK**

The **DIMBREAK** command breaks extension and dimensions lines where they cross other objects.

Breaking lines can make dimensions easier to read. AutoCAD remembers the breaks, so that when you move a dimension, the break moves appropriately.

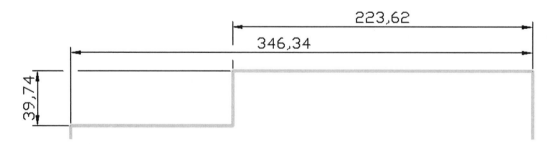

TUTORIAL: BREAKING DIMENSION AND EXTENSION LINES

1. To break lines in dimensions, start the DIMBREAK command:

   - From the **Dimension** menu, choose **Dimension Break**.

   - From the Dimension toolbar, choose the **Dimension Break** button.

   - In the Dimension panel, choose the **Dimbreak** button.

   - At the 'Command:' prompt, enter the **dimbreak** command.

     Command: **dimbreak** *(Press* ENTER.*)*

2. In all cases, AutoCAD prompts you to select one or more dimensions:

   Select a dimension or [Multiple]: *(Select a dimension.)*

3. Choose an option; I recommend the **Auto** option, which lets AutoCAD handle the breaking job:

   Select object to break dimension or [Auto/Restore/Manual] <Auto>: **a**

   Notice that dimension and/or extension lines are broken.

To place breaks manually, enter "M" for the **Manual** option, which operates like the BREAK command:

Specify first break point: *(Pick a point.)*

Specify second break point: *(Pick another point.)*

This option allows you to place the break anywhere, and not just where lines are overlapping. This options works better when osnaps are turned off.

To break all dimensions in the drawing at once, use the **Multiple** option, as follows:

Select a dimension or [Multiple]: **m**

Select dimensions: **all**

Select dimensions: *(Press ENTER to exit object selection.)*

Enter an option [Break/Restore] <Break>: *(Press ENTER to accept the default.)*

When you stretch or move dimensions, the breaks follow. Should the dimensions no longer cross, the breaks heal, but reappear when you move the dimensions back.

To remove breaks, use the **Restore** option. You can restore dimensions one by one, or all of them at once.

 **DIMSPACE**

The **DIMSPACE** command evenly spaces groups of dimensions.

The purpose of this command is to give a neater look to drawings with many dimensions; it evenly spaces them. When you stretch or move dimensions, the break follows.

TUTORIAL: EVENLY SPACING DIMENSIONS

1. To learn how to space dimensions evenly, draw several linear dimensions, as illustrated below.

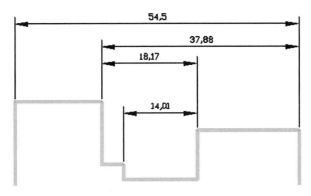

2. Start the **DIMSPACE** command:

   - From the **Dimension** menu, choose **Dimension Space**.

   - From the Dimension toolbar, choose the **Dimension Space** button.

   - In the Dimension panel, choose the **Dimspace** button.

   - At the 'Command:' prompt, enter the **dimspace** command.

     Command: **dimspace** *(Press ENTER.)*

3. In all cases, AutoCAD prompts you to select the base dimension. This is the one from which the others will be spaced:

   Select base dimension: *(Pick one dimension.)*

4. Select the other dimensions to be spaced:

   Select dimensions to space: *(Pick one or more dimensions.)*

   Select dimensions to space: *(Press ENTER to exit object selection.)*

5. Enter a number for the size of the spacing between dimensions; or, do it the easy way, and press ENTER to let AutoCAD work out the spacing:

   Enter value or [Auto] <Auto>: *(Press ENTER.)*

   Notice that the dimensions have the same spacing between them.

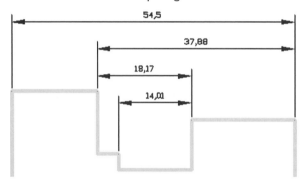

The **Auto** option determines the spacing distance by multiplying the dimension text height; Autodesk says the factor is 2x, but in the figure above, the factor works out to 5x. This system sometimes does not work well when DIMSCALE is set to extreme values.

You can enter a value of 0, which reduces the spacing to zero, making the dimensions look like overlapping continuous dimensions, as illustrated below.

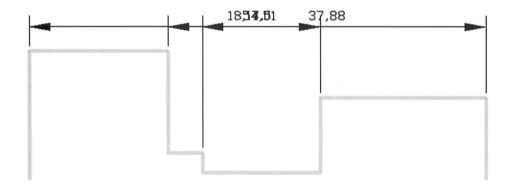

DIMEDIT, DIMTEDIT AND DDEDIT

The **DIMEDIT** and **DIMTEDIT** commands edit extension lines and dimension text.

Extension lines can be *obliqued* with the **DIMEDIT** command, so that they are no longer at right angles to the dimension line. Obliqued (slanted) extension lines are required by isometric drawings, and are also useful in difficult dimensioning situations, such as dimensioning along shallow curves. Obliquing is not applied to leaders, such as diameter, radius, and ordinates, because their angles can be changed with grips editing.

The **DIMTEDIT** command changes the location of dimension text relative to the dimension line by rotating, relocating, and justifying (left, center, or right). After these changes, either command can be used to "home" the text by sending it back to its default location: centered on the dimension line.

One command's name is short for "dimension edit," while the other is short for "dimension text edit," yet both edit dimension text. The toolbar icons are unhelpful, **DIMTEDIT** showing a pencil, implying the dimension itself can be edited when it cannot.

The two commands share text editing tasks, making the difference between them confusing. The table below compares and contrasts the abilities of each command:

Editing Operation	DimEdit	DimTEdit
Extension Lines:		
Apply obliquing angle	☑	...
Dimension Text:		
Rotate	☑	☑
Move to default location (home)	☑	☑
Edit text	☑	...
Drag text to new location	...	☑
Center text on dimension line	...	☑
Right- and left-justify text on dimline	...	☑

It would make more sense if **DIMTEDIT** handled *only* the editing of dimension text, while **DIMEDIT** handled the non-text parts of dimensions, such as extension lines, as well as functions ignored by AutoCAD, such as relocating and removing arrowheads. But they don't, and so we carry on. Here's how to keep them straight:

DimEdit obliques extension lines (ignoring its other capabilities).

DimTEdit changes the position of dimension text.

DdEdit edits dimension text with the mtext editor.

TUTORIAL: OBLIQUING EXTENSION LINES

1. To change the angle of extension lines, start the **DIMEDIT** command:
   - From the **Dimension** menu, choose **Oblique**.
   - From the Dimension toolbar, choose the **Dimension Edit** button.
   - At the 'Command:' prompt, enter the **dimedit** command.

      Command: **dimedit** *(Press ENTER.)*
   - Alternatively, enter the aliases **ded** or **dimed** at the 'Command:' prompt.

2. In all cases, AutoCAD lists the options available. Enter **o** for the oblique option:

    Enter type of dimension editing [Home/New/Rotate/Oblique] <Home>: **o**

3. Select the dimensions whose extensions lines to slant.

    You can enter **all**, and AutoCAD filters out non-dimension objects.

    Select objects: *(Pick one or more dimensions.)*

    Select objects: *(Press* **ENTER** *to end object selection.)*

4. Specify the angle of slant between 0 and 360 degrees:

    • To slant extension lines for isometric drawings, enter **30**.

      Enter obliquing angle (press ENTER for none): **30**

    • To slant in the other direction, enter negative angles, such as **-30**.

    • To keep the extension lines at their current angle (no change), press **ENTER**.

    • To return obliqued extension lines to perpendicular, enter **0** or **90**.

    • Or, pick two points to show AutoCAD the angle.

    AutoCAD obliques the extension lines of selected dimensions.

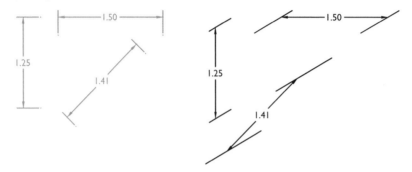

Left: *Dimensions with normal, perpendicular extension lines.*
Right: *Dimensions with extension lines obliqued by 30 degrees.*

 Note: Be careful with obliquing angles that are very close to the angle of the dimension (1.0 degrees for horizontal dimensions, 89.0 degrees for vertical dimensions, and so on). They stretch the dimension into an unrecognizable jagged line. You can always use the **U** command to reverse undesirable obliquing.

REPOSITIONING DIMENSION TEXT

1. To change the position of dimension text, start the **DIMTEDIT** command:

    • From the **Dimension** menu, choose the **Align Text** flyout.

    • From the Dimension toolbar, choose the **Dimension Text Edit** button.

    • At the 'Command:' prompt, enter the **dimtedit** command.

      Command: **dimtedit** *(Press* **ENTER.**)*

    • Alternatively, enter the **dimted** alias at the 'Command:' prompt.

2. AutoCAD prompts you to select one dimension:

    Select dimension: *(Select one dimension.)*

3. Enter an option to relocate the text:

    Specify new location for dimension text or [Left/Right/Center/Home/Angle]: *(Move cursor, or enter an option.)*

    • **Specify new location** relocates the text along the dimension line dynamically , and at the same time moves the dimension line perpendicular to the extension lines.

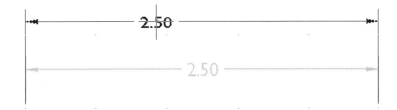

- **Left** moves the text to the left end of the dimension line.
- **Right** moves the text to the right end of the dimension line.
- **Center** centers the text on the dimension line.
- **Home** returns the text to its original position.
- **Angle** rotates text about its center point. AutoCAD prompts you:

  Specify angle for dimension text: *(Enter an angle, such as **30**.)*

Angled text is used with isometric dimensions.

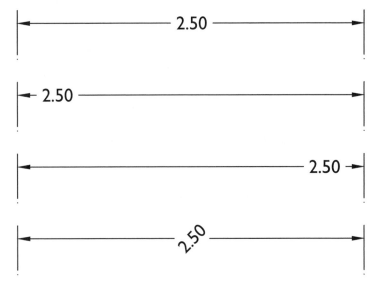

Top: *Dimension text in home and center positions.*
Above: *Text in left, right, and rotated positions.*

EDITING DIMENSION TEXT

1. To edit dimension text, start the **DDEDIT** command:
    - From the **Modify** menu, choose **Object**, then **Text**, and then **Edit**.

    - From the Text toolbar, choose the **Edit Text** button.

    - At the 'Command:' prompt, enter the **ddedit** command.

      Command: **ddedit** *(Press **ENTER**.)*

    - Alternatively, enter the **ed** alias at the 'Command:' prompt.

2. In all cases, AutoCAD prompts you to select the text to edit.
    Select a single dimension:

    Select an annotation object or [Undo]: *(Select a dimension or leader text.)*

3. AutoCAD opens the mtext editor with the dimension text highlighted in blue.

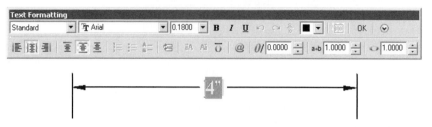

Recall from the previous chapter that you can edit the dimension text (with the blue background):

- **Erase** the text from the dimension.
- **Change** the dimension text.
- **Add** text before and after to create prefixes and suffixes.

4. To apply a background mask to the dimension text, right-click and select **Background Mask**.

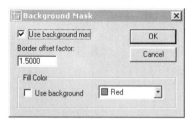

In the dialog box, turn on the **Use background mask** option, and then select a color. Click **OK** to exit the dialog box.

(If this dialog box does not work, use the **Fill** option of the Properties window instead.)

5. When done editing the text, click **OK** on the Text Formatting bar.

6. AutoCAD repeats the prompt.

Select another dimension, or any other text object.

When done editing, press **ENTER**.

Select an annotation object or [Undo]: *(Press ENTER to exit command.)*

 **DIMINSPECT**

The **DIMINSPECT** command adds and edits inspection-style text to dimensions.

Inspection dimensions are used in quality assurance (QA) testing to ensure that the manufactured part has precisely the correct dimension. For example, if the percentage is 80%, then 80% of the manufactured parts must meet this dimension to pass QA. This command modifies existing dimension text to make three parts:

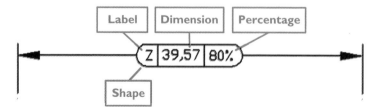

Label — identifies the dimension in tables and notes. You can enter any text for the label.

Dimension — specifies the original dimension text. Like other dimensions, inspection dimensions can contain prefixes, suffixes, and tolerances.

Percentage — specifies the inspection rate. You can enter any value, such as 23/74 or "Quite a few"; it need not be a percentage.

Shape — surrounds the dimension to make it stand out. AutoCAD allows ovals, diamonds, or no shape at all.

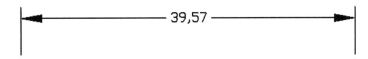

You select the type and value of inspection through a dialaog box; the -DIMINSPECT command performs the same function at the command line.

TUTORIAL: ADDING INSPECTION TEXT TO DIMENSIONS

1. To learn how to add inspection text to dimensions, draw a linear dimension similar to the one illustrated below.

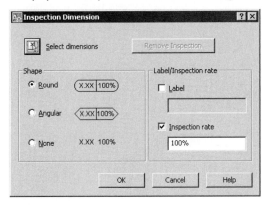

2. To add inspection text, start the **DIMINSPECT** command:
   * From the **Dimension** menu, choose **Inspection**.

   * From the Dimension toolbar, choose the **Inspection** button.

   * In the Dashboard's Dimension panel, choose the **Diminspect** button.

   * At the 'Command:' prompt, enter the **diminspect** command.

      Command: **diminspect** *(Press ENTER.)*

3. In all cases, AutoCAD displays the Inspection Dimension dialog box:

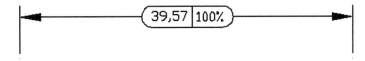

4. Click the **Select Dimensions** button.
5. Notice that AutoCAD prompts you at the command line:
      Select dimensions: *(Select one or more dimensions.)*

      Select dimensions: *(Press ENTER to return to the dialog box.)*

6. Choose inspection options, and then click **OK**. The default options change the dimension to this:

To change the values and/or shape of inspection dimensions, rerun the DIMINSPECT command, and then select the dimension. Alternatively, you can use the Misc section of the Properties palette to edit inspection dimensions.

NOTE: You can use this command to interactively create dimensions with prefixes and suffixes, such as the one illustrated below. Set **Inspection Label** to "Length = " and **Inspection Rate** to " Inches." Alternatively, you can use the Properties palette to add and edit prefixes and suffixes.

AIDIM AND AI_DIM

AutoCAD has a group of undocumented commands that edit dimension text. These commands are meant for use in menu and toolbar macros, but are also handy for making quick fixes to mucked-up dimensions. The commands are shortcuts to options within other commands and dimension variables:

(The "AI" prefix indicates custom commands written by Autodesk Incorporated. Why do some commands start with **AI** and some with **AI_**? It seems that those starting with **AI** have multiple options, while those with **AI_** perform a single operation.)

Command	Shortcut For	Comment
AIDIMPREC	DIMDEC	Changes precision of fractions and decimals selectively.
AIDIMSTYLE	DIMSTYLE	Saves and applies up to six dimension styles.
AIDIMTEXTMOVE	DIMTMOVE	Relocates text with optional leader.
AI_DIM_TEXTABOVE	DIMTAD 3	Moves text above dimension line for JIS compliance.
AI_DIM_TEXTCENTER	DIMTEDIT C	Centers dimension text.
AI_DIM_TEXTHOME	DIMTEDIT H	Returns text to home position.

You find these commands in a shortcut menu: select a dimension, and then right-click.

Here is how the commands relate to the shortcut menu:

Shortcut Menu	Submenu	Related Command	Option
Dim Text position	▸ Move text alone	AIDIMTEXTMOVE	0
	▸ Move with leader	AIDDIMTEXTMOVE	1
	▸ Move with dim line	AIDIMTEXTMOVE	2
	▸ Above dim line	AI_DIM_TEXTABOVE	...
	▸ Centered	AI_DIM_TEXTCENTER	...
	▸ Home text	AI_DIM_TEXTHOME	...
Precision	...	AIDIMPREC	0 – 6
Dim Style	▸ Save as New Style	AIDIMSTYLE	Save
	▸ Standard, etc.	AIDIMSTYLE	1 – 6
	▸ Other	AIDIMSTYLE	Other
Flip Arrow	...	AIDIMFLIPARROW	

AIDIMPREC

The **AIDIMPREC** command selectively changes the precision of fractional and decimal dimension text and angles of all dimensions type, except leaders. To affect a permanent change, change the dimstyle.

This command is more convenient than using its official alternative: changing the **DIMDEC** system variable, followed by the **-DIMSTYLE** command's **Apply** option. Decimal text and angles are rounded to the nearest decimal place, while fractional text is rounded to the nearest fraction.

AiDimPrec	Decimal Units	Fractional Units
0	0	1"
1	0.0	1/2"
2	0.00	1/4"
3	0.000	1/8"
4	0.0000	1/16"
5	0.00000	1/32"
6	0.000000	1/64"

CHANGING DISPLAY PRECISION

1. To change the precision of dimension text, start the **AIDIMPREC** command:
   Command: **aidimprec** *(Press ENTER.)*

2. AutoCAD prompts you to specify a precision:
   Enter option [0/1/2/3/4/5/6] <4>: *(Enter the number of decimal places.)*

3. Select one or more dimension objects.
   Select objects: *(Enter all or pick one or more dimensions.)*

   Select objects: *(Press ENTER to end object selection.)*

   Entering **all** selects everything in the drawing; AutoCAD filters out non-dimension objects.
   This command does not change the precision of leader objects.
   Notice that AutoCAD changes the display precision of the selected dimensions.

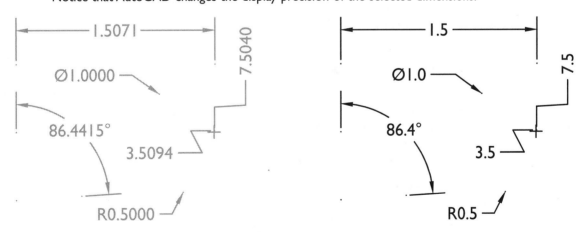

Left: Dimension text displayed to four decimal places.
Right: Precision reduced to one decimal place.

Notes: The change in precision is not permanent; AutoCAD remembers the actual measurements and angles — which is why it is called "display precision." You can reapply the command to increase and decrease the precision displayed by the dimensions.

Because the **AIDIMPREC** command rounds off dimension text, it can create false measurements. For instance, if a dimension measures 3.4375", setting **AIDIMPREC** to 0 rounds down to 3". Similarly, a measurement of 2.5" rounds up to 3".

AIDIMTEXTMOVE

The **AIDIMTEXTMOVE** command relocates text, either with or without moving the dimension line, and optionally adds a leader.

This command is more convenient than using its official alternative, changing the **DIMTMOVE** system variable, followed by the **DIMOVERRIDE** command. **AIDIMTEXTMOVE** has the same options as **DIMTMOVE**:

AiDimTextMove	DimTMove	Comment
0	0	Moves text within dimension line; moves dimension line.
1	1	Adds a leader to the moved text.
2 *(Default)*	2	Moves text without leader line; moves the dimension line.

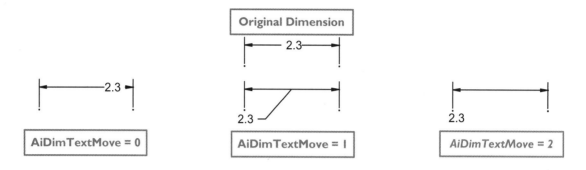

MOVING DIMENSION TEXT

1. To move the dimension text, start the **AIDIMTEXTMOVE** command:
   Command: **aidimtextmove** *(Press ENTER.)*

2. AutoCAD prompts you to specify a precision:
   Enter option [0/1/2] <2>: *(Enter an option.)*

3. Select one dimension.
   Although the command allows you to select more than one dimension, it operates on the first-selected dimension only.
   Select objects: *(Pick one dimension.)*
   Select objects: *(Press ENTER to end object selection.)*

4. Move the cursor to relocate the dimension text.

The following trio of commands quickly relocates dimension text relative to the dimension line: above, centered, or in the dimension line. These commands operate on just one dimension at a time, unfortunately. Using this command is more convenient than changing the **DIMTAD** system variable, followed by the **-DIMSTYLE** command's **Apply** option.

Command	DimTAD	Comment
AI_DIM_TEXTCENTER	0	Centers text vertically.
AI_DIM_TEXTHOME	2	Centers text horizontally and vertically.
AI_DIM_TEXTABOVE	3	Moves text above dimension line.

(When **DIMTAD** is 1, AutoCAD moves text above the dimension line when the dimension line is horizontal, and text inside the extension lines isn't forced horizontally.)

Ai_Dim_TextCenter

The **AI_DIM_TEXTCENTER** command centers text along the dimension line. If the text is above or below the dimension line, AutoCAD keeps the vertical position, but centers the text horizontally.

Command: **ai_dim_textcenter**

Select objects: *(Pick one dimension.)*

Select objects: *(Press* ENTER *to end object selection.)*

Ai_Dim_TextHome

The AI_DIM_TEXTHOME command centers the dimension text on the side of the dimension line farthest from the defining points.

Command: **ai_dim_texthome**

Select objects: *(Pick one dimension.)*

Select objects: *(Press* ENTER *to end object selection.)*

Ai_Dim_TextAbove

The AI_DIM_TEXTABOVE command moves text above the dimension line, which is helpful for making drawings JIS compliant.

Command: **ai_dim_textabove**

Select objects: *(Pick one dimension.)*

Select objects: *(Press* ENTER *to end object selection.)*

DIMENSION VARIABLES AND STYLES

How dimensions look and act depends on the settings of *dimension variables* (called "dimvars" for short). The variables control many properties of dimensions, such as whether dimension text is placed within or above the dimension line, the use of arrowheads or tick marks, the color of the extension lines, and the overall scale. (Technically, dimvars are no different from system variables.)

The names of all dimvars begin with "dim," which makes them easy to identify, such as DIMSCALE for dimension scale and DIMTXT for the text style. Other names, however, can seem strange and are hard to decode. Who could guess that DIMTOFL means "draw the dimension line between the extension lines, even when the Text is Forced Outside due to Lack of space." I suppose "tofl" could be short for "text outside forced line." Or how about DIMTAD — "Text Adjusted (relative to the) Dimension (line)," perhaps? The technical editor suggests a new dimension variable, DIMWIT for "Weird Indecipherable Terminology."

Like system variables, dimvars can be toggled (turned on and off), hold numeric values, or report on the status of dimensions. The advantage to dimvars is that they control almost every aspect of dimensions; the disadvantage is that there so many of them, nearly 80, that it is difficult to remember them all. Two-thirds are dedicated to formatting dimension text.

That's why AutoCAD has dimensions styles. Like text styles, these remember dimvars and control the look of dimensions. When you create or change a dimensions style with the DIMSTYLE command's Dimension Style dialog box, you are changing the values of dimvars.

Creating dimension styles ("dimstyles" for short) is similar to creating text styles. In both, you specify options, and then save them by name. New AutoCAD drawings based on the *acad.dwt* template drawing have dimension styles named "Standard" and "Annotative."

CONTROLLING DIMVARS

AutoCAD's dimension variables can be set through several methods, such as using the Dimension Style Manager dialog box or the Properties palette, or by entering dimvar names at the 'Command:' prompt.

For example, to set an overall scale factor for dimensions, it can be faster to enter the **DIMSCALE** at the command prompt than to hunt it down in the multi-tabbed Dimension Style Manager dialog box:

Command: **dimscale**

New value for DIMSCALE <1.0>: *(Enter a scale factor, such as* **2.5***.)*

The scale is not retroactive, however, and only applies to dimensions drawn subsequently. To change the scale of existing dimensions, you need to use the **DIMSTYLE** dialog box's **Override** option.

There are two dimvars are not set by the Dimension Style Manager dialog box: **DIMSHO** (toggles whether dimensions are updated while dragged) and **DIMASO** (determines whether dimensions are associative or non-associative). Both have a value of 1, which means they are turned on. You can turn them off, but there isn't any good reason to do so.

History: Both dimvars exist only for compatibility with old versions of AutoCAD. When computers were slow, generating highlighted images of dragged dimensions slowed the computer even more; this is no longer an issue.

SOURCES OF DIMSTYLES

When you create new drawings based on template drawings, they contain specific dimstyles. AutoCAD includes a *.dwt* template file that meets the international metric standard: *acad_iso.dwg* contains a dimstyle named "ISO," which conforms to the dimensioning standards of the International Organization for Standardization.

Comparing Dimvars

To compare the values of dimvars in different dimension styles, use the **DIMSTYLE** command's **Compare** button. It opens a dialog box that lists the values of two dimension styles.

The **Copy** button copies the data to the Clipboard in tab-delimited ASCII format. You can then paste the data into a spreadsheet or other document.

Dimstyles from Other Sources

To copy dimension styles from other drawings, use DesignCenter. In the tree view of DesignCenter, open the **Dimstyles** item in the drawing from which you wish to copy the dimstyles. Right-click the dimstyle name, and then select **Add Dimstyle(s)** from the cursor menu.

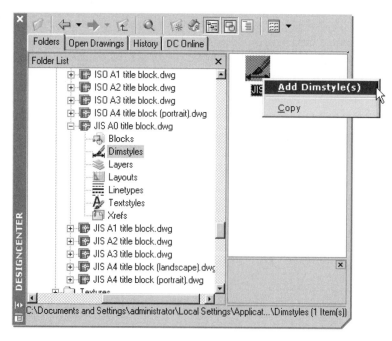

To check that the dimstyle was added to the drawing, click the **Dimstyle Control** list box on the Styles toolbar or Dashboard's Dimension panel. The name should appear on the list.

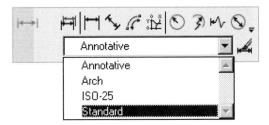

AutoCAD has several additional methods of sharing dimension standards between drawings:

Express Tools includes the **DIMEX** and **DIMIM** commands for exporting and importing files that contain dimension styles. From the **Express** menu, select **Dimension**, and then **DimStyle Export** or **DimStyle Import**. AutoCAD displays the dialog boxes shown below.

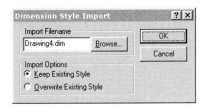

When working with externally-referenced drawings, you can use the **XBIND** command to *bind* (add in) dimension styles found in attached drawings. You cannot, however, set an externally-referenced dimstyle as current.

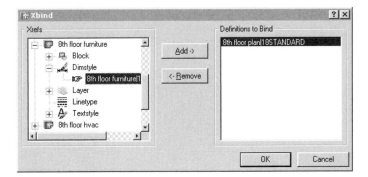

SUMMARY OF DIMENSION VARIABLES

The following sections summarize the meaning of dimension variables, grouped by purpose.

General

DimAnno (<u>anno</u>tative) reports whether the current dimension style uses annotative scaling; read-only. When on (set to 1), **DIMSCALE** is set to 0.

DimAso (<u>asso</u>ciative) toggles dimensions between associative and non-associative.

DimAssoc (<u>assoc</u>iative) determines how dimensions are created: exploded, attached to defpoints, or attached to objects.

DimScale specifies the overall scale factor for dimensions. When set to zero, AutoCAD calculates the scale factor for dimensions by two methods:

- When **DIMANNO** = 1, AutoCAD divides the scale factor of the current model space viewport by that of paper space.

- When **DIMANNO** = 0, AutoCAD uses the current viewport scale factor.

DimSho (show) updates dimensions while dragging.

DimStyle specifies the current dimension style as set by the **DIMSTYLE** command. (Note that the system variable and the command share the same name, so use the **SETVAR** command to access the dimvar.)

Dimension Lines

Extension and Offset

DimDLE (dimension line extension) specifies how far the dimension line extends past the extension lines.

DimLwD (lineweight dimension) specifies the lineweight of dimension lines.

DimDLI (dimension line increment) specifies the offset for baseline dimensions.

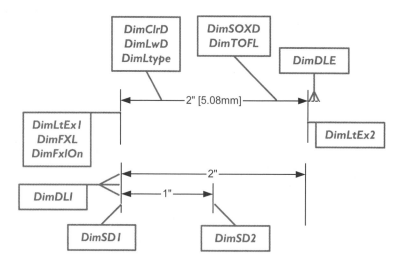

Suppression and Position of Dimension Lines

DimSD1 (suppress dimension 1) suppresses the first dimension line.

DimSD2 (suppress dimension 2) suppresses the second dimension line.

DimSOXD (suppress outside extension dimension) suppresses dimension lines outside extension lines.

DimTOFL (text outside forced line) forces dimension lines inside extension lines when text is outside the extension lines.

Color and Lineweight

DimClrD (color dimension) specifies the color of dimension lines.

DimLwD (lineweight dimension) specifies the lineweight for dimension lines.

Linetype

DimLtype (linetype) specifies the linetype for dimension lines.

Extension Lines

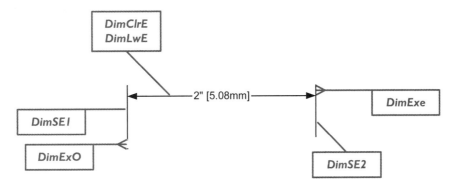

Suppression and Linetype

DimFXL (<u>f</u>ixed-length e<u>x</u>tension <u>l</u>ines) specifies the default length of fixed-length extension lines.

DimFxlOn (<u>f</u>ixed-length e<u>x</u>tension <u>l</u>ines <u>on</u>) toggles the use of fixed-length extension lines.

DimLtEx1 (<u>l</u>inetype for e<u>x</u>tension line <u>1</u>) specifies the linetype for the first extension line.

DimLtEx2 (<u>l</u>inetype for e<u>x</u>tension line <u>2</u>) specifies the linetype for the second extension line.

Color and Lineweight

DimClrE (<u>c</u>ol<u>or</u> <u>e</u>xtension) specifies the color of extension lines.

DimLwE (<u>l</u>ine<u>w</u>eight <u>e</u>xtension) specifies the lineweight of extension lines.

Extension and Offset

DimExE (<u>ex</u>tension <u>e</u>xtend) extends the extension lines above the dimension line.

DimExO (<u>ex</u>tension <u>o</u>ffset) offsets the extension line from the origin.

Suppression and Position

DimSE1 (<u>s</u>uppress <u>e</u>xtension <u>1</u>) suppresses the first extension line.

DimSE2 (<u>s</u>uppress <u>e</u>xtension <u>2</u>) suppresses the second extension line.

Arrowheads

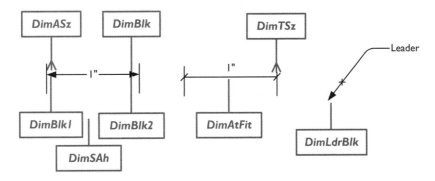

Size and Fit

DimASz (<u>a</u>rrow <u>si</u>ze) determines the length of arrowhead blocks.

DimTSz (<u>t</u>ick <u>si</u>ze) specifies the size of tick strokes.

DimATFit (<u>a</u>rrow <u>t</u>ext <u>fit</u>) determines under which conditions arrowheads and text are fitted between extension lines; see Alphabetical Listing of Dimvars later in this chapter.

Names of Blocks

DimBlk (<u>block</u>) names the arrowhead block.

DimBlk1 (<u>block 1</u>) names the first arrowhead block.

DimBlk2 (<u>block 2</u>) names the second arrowhead block.

DimLdrBlk (<u>leader block</u>) names the leader arrowhead.

DimSAh (<u>s</u>eparate <u>ar</u>row<u>h</u>eads) determines whether separate arrowhead blocks are used.

Center Marks

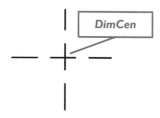

DimCen (<u>cen</u>ter) specifies the mark size and line.

Arc Length Dimensions

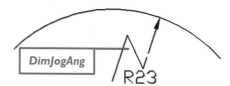

DimArcSym (<u>arc</u> <u>sym</u>bol) specifies the location of the arc length symbol.

Jogged Dimensions

DimJogAng (<u>jog</u> <u>ang</u>le) specifies the default angle of jogged radial dimension lines.

Text

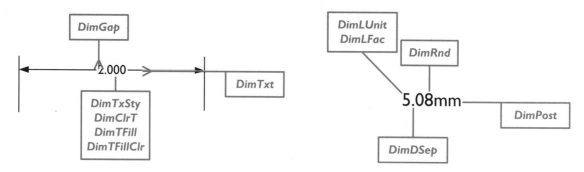

Color and Format of Text

DimClrT (<u>c</u>olor <u>t</u>ext) specifies the color of the text.

DimGap determines the gap between the dimension line and text.

DimTxSty (text style) specifies the text style.

DimTxt (text) stores the text height.

Background Fill

DimTFill (text fill) switches the background fill of dimension text among these: none, same as drawing background color, or as specified by DIMTFILLCLR.

DimTFillClr (text fill color) specifies the color of the background fill.

Units, Scale, and Precision

DimLUnit (linear unit) specifies the format of linear units.

DimPost (postfix) specifies prefixes and suffixes of dimension text.

DimLFac (linear factor) specifies the linear unit scale factor.

DimDSep (decimal separator) specifies the decimal separator.

DimRnd (round) rounds distances.

DimFrac (fractions) determines how fractions are stacked: horizontal, diagonal, or not stacked.

DimZIn (zero inches) suppresses zeroes in feet-inches units.

DimADec (angular decimal) specifies the precision of angular dimensions.

DimAUnit (angular unit) specifies the format for angular dimensions.

DimAZIn (angular zero inches) controls how zeros are suppressed in angular dimensions.

Justification

DimJust (justify) specifies alignment (justification) of horizontal text: left, center, right.

DimTAD (text adjust dimension) positions dimension text vertically: top, center, outside, JIS.

DimTVP (text vertical position) specifies the vertical text position when DIMTAD is off.

DimTIH (text inside horizontal) determines whether text inside extensions is horizontal.

DimTIX (text in extension) forces text between extension lines.

DimTMove (text move) determines how dimension text is relocated: with dimension line, an added leader, or without a leader.

DimTOH (text outside horizontal) determines whether text outside extension lines is horizontal.

DimUPT (user position text) toggles whether user positions dimension line and/or text.

Alternate Text

DimAlt (alternate) toggles alternate units.

DimAltU (alternate units) determines the format of alternate units.

DimAltD (<u>al</u>ternate <u>d</u>ecimals) specifies decimal places of alternate units.

DimAltF (<u>al</u>ternate <u>f</u>actor) specifies the scale factor of alternate units.

DimAltRnd (<u>al</u>ternate <u>ro</u>u<u>nd</u>ing) rounds off alternate units.

DimAltZ (<u>al</u>ternate <u>z</u>ero) suppresses zeros in alternate units.

DimAPost (<u>al</u>ternate <u>post</u>fix) determines the prefixes and suffixes for alternate text.

Limits

DimLim (<u>lim</u>its) toggles dimension limits.

Tolerance Text

DimTol (<u>tol</u>erance) toggles whether tolerances are drawn.

DimTZIn (<u>t</u>olerance <u>z</u>ero <u>in</u>ches) suppresses zeros in tolerances.

DimTFac (<u>t</u>olerance <u>fac</u>tor) scales the tolerance text height.

DimTolJ (<u>tol</u>erance <u>j</u>ustify) justifies tolerance text vertically.

DimTP (<u>t</u>olerance <u>p</u>lus) specifies the plus tolerance value.

DimTM (<u>t</u>olerance <u>m</u>inus) specifies the minus tolerance value.

Primary Tolerance

DimDec (<u>dec</u>imals) specifies the decimal places for the primary tolerance.

DimTDec (<u>t</u>olerance <u>dec</u>imals) specifies the decimal places for primary tolerance units.

Alternate Tolerance

DimAltTD (<u>al</u>ternate <u>t</u>olerance <u>dec</u>imals) specifies decimal places for tolerance alternate units.

DimAltTZ (<u>al</u>ternate <u>t</u>olerance <u>z</u>ero) suppresses zeros in tolerance alternate units.

ALPHABETICAL LISTING OF DIMVARS

The table summarizes dimension variables, default values in the *acad.dwt* template, and optional values.

DimVar	Default	Settings and Options

A

DimVar	Default	Settings and Options
DimADec	0	Angular dimension precision:
		-1 Use **DimDec** setting (default).
		0 Zero decimal places (minimum).
		8 Eight decimal places (maximum).
DimAlt	Off	Alternate units:
		On Enabled.
		Off Disabled.
DimAltD	2	Alternate unit decimal places.
DimAltF	25.4000	Alternate unit scale factor.
DimAltRnd	0.0000	Rounding factor of alternate units.
DimAltTD	2	Tolerance alternate unit decimal places.
DimAltTZ	0	Alternate tolerance units zeros:
		0 Zeros not suppressed.
		1 Suppresses all zeros.
		2 Includes 0 feet, but suppresses 0 inches.
		3 Includes 0 inches, but suppresses 0 feet.
		4 Suppresses leading zeros.
		8 Suppresses trailing zeros.
DimAltU	2	Alternate units:
		1 Scientific.
		2 Decimal.
		3 Engineering.
		4 Architectural; stacked.
		5 Fractional; stacked.
		6 Architectural.
		7 Fractional.
		8 Windows desktop units setting.
DimAltZ	0	Zero suppression for alternate units:
		0 Suppresses 0 ft and 0 in.
		1 Includes 0 ft and 0 in.
		2 Includes 0 ft; suppress 0 in.
		3 Suppresses 0 ft; include 0 in.
		4 Suppresses leading 0 in decimal dims.
		8 Suppresses trailing 0 in decimal dims.
		12 Suppresses leading and trailing zeroes.
(2008) DimAnno	0	Reports whether the current dimension style is annotative.
DimAPost	""	Prefix and suffix for alternate text.
DimArcSym	""	Position of the arc length symbol:
		0 Before the dimension text.
		1 Above dimension text.
		2 Not displayed.
DimAso	On	Toggle associative dimensions:
		On Dimensions are associative.
		Off Dimensions are not associative.
DimAssoc	2	Controls creation of dimensions:
		0 Dimension elements are exploded.
		1 Single dimension object, attached to defpoints.
		2 Single dimension object, attached to geometric objects.
DimASz	0.1800	Arrowhead length.

DimVar	Default	Settings and Options
DimAtFit	3	When insufficient space between extension lines:
		0 Text and arrows outside extension lines.
		1 Arrows first outside, then text.
		2 Text first outside, then arrows.
		3 Either text or arrows, whichever fits better.
DimAUnit	0	Angular dimension format:
		0 Decimal degrees.
		1 Degrees.Minutes.Seconds.
		2 Grad.
		3 Radian.
		4 Surveyor units.
DimAZin	0	Suppress zeros in angular dimensions:
		0 Display all leading and trailing zeros.
		1 Suppress 0 in front of decimal.
		2 Suppress trailing zeros behind decimal.
		3 Suppress zeros in front and behind the decimal.

B

DimVar	Default	Settings and Options
DimBlk	""	Arrowhead block name:
		Architectural tick: "Archtick"
		Box filled: "Boxfilled"
		Box: "Boxblank"
		Closed blank: "Closedblank"
		Closed filled: "" *(default)*
		Closed: "Closed"
		Datum triangle filled: "Datumfilled"
		Datum triangle: "Datumblank"
		Dot blanked: "Dotblank"
		Dot small: "Dotsmall"
		Dot: "Dot"
		Integral: "Integral"
		None: "None"
		Oblique: "Oblique"
		Open 30: "Open30"
		Open: "Open"
		Origin indication: "Origin"
		Rightangle: "Open90"
DimBlk1	""	Name of first arrowhead's block; uses same names as DimBlk.
		. Return to default arrowhead.
DimBlk2	""	Name of second arrowhead's block; uses same names as DimBlk.
		. Return to default arrowhead.

C

DimVar	Default	Settings and Options
DimCen	0.0900	Center mark size:
		-n Draws center lines.
		0 No center mark or lines drawn.
		+n Draws center marks of length *n*.

D

DimVar	Default	Settings and Options
DimClrD	0	Dimension line color:
		0 BYBLOCK (default).
		1 Red.
		...
		255 Dark gray.
		256 BYLAYER.
DimClrE	0	Extension line and leader color.
DimClrT	0	Dimension text color.
DimDec	4	Primary tolerance decimal places.

DimVar	Default	Settings and Options
DimDLE	0.0000	Dimension line extension.
DimDLI	0.3800	Dimension line continuation increment.
DimDSep	"."	Decimal separator (must be a single character).

E

DimExe	0.1800	Extension above dimension line.
DimExO	0.0625	Extension line origin offset.

F

DimFrac	0	Fraction format when **DimLUnit** is set to 4 or 5:

    0   Horizontal.
    1   Diagonal.
    2   Not stacked.

DimFXL	1.0000	Specifies the default length of fixed-length extension lines.
DimFxlOn	Off	Toggles fixed-length extension lines:

    On   Extension lines are displayed at a fixed length.
    Off   Extension lines are displayed at the user-drawn length

G

DimGap	0.0900	Gap from dimension line to text.

J

DimJogAng	45	Default angle for jogged radial dimension lines.
DimJust	0	Horizontal text positioning:

    0   Center justify.
    1   Next to first extension line.
    2   Next to second extension line.
    3   Above first extension line.
    4   Above second extension line.

L

DimLdrBlk	""	Block name for leader arrowhead; uses same name as **DimBlock**.

    .   Return to default.

DimLFac	1.0000	Linear unit scale factor.
DimLim	Off	Generate dimension limits.
DimLtEx1	""	Linetype used for the first extension line.
DimLtEx2	""	Linetype used for the first extension line.
DimLtype	""	Linetype used for dimension lines.
DimLUnit	2	Dimension units (except angular); replaces **DimUnit**:

    1   Scientific.
    2   Decimal.
    3   Engineering.
    4   Architectural.
    5   Fractional.
    6   Windows desktop.

DimLwD	-2	Dimension line lineweight; valid values are BYLAYER, BYBLOCK, or an integer multiple of 0.01 mm.
DimLwE	-2	Extension lineweight; valid values are BYLAYER, BYBLOCK, or an integer multiple of 0.01 mm.

P

DimPost	""	Default prefix or suffix for dimension text (maximum 13 characters):

    " "   No suffix.
    <>mm   Millimeter suffix.
    <>Å   Angstrom suffix.

DimVar	Default	Settings and Options
R		
DimRnd	0.0000	Rounding value for dimension distances.
S		
DimSAh	Off	Separate arrowhead blocks:
		Off Use arrowhead defined by DimBlk.
		On Use arrowheads defined by DimBlk1 and DimBlk2.
DimScale	1.0000	Overall scale factor for dimensions:
		0 Value is computed from the scale between current model space viewport and paper space.
		>0 Scales text and arrowheads.
DimSD1	Off	Suppress first dimension line:
		On First dimension line is suppressed.
		Off Not suppressed.
DimSD2	Off	Suppress second dimension line:
		On Second dimension line is suppressed.
		Off Not suppressed.
DimSE1	Off	Suppress the first extension line:
		On First extension line is suppressed.
		Off Not suppressed.
DimSE2	Off	Suppress the second extension line:
		On Second extension line is suppressed.
		Off Not suppressed.
DimSho	On	Update dimensions while dragging:
		On Dimensions are updated during drag.
		Off Dimensions are updated after drag.
DimSOXD	Off	Suppress dimension lines outside extension lines:
		On Dimension lines not drawn outside extension lines.
		Off Are drawn outside extension lines.
DimStyle	"STANDARD"	Name of the current dimension style.
T		
DimTAD	0	Vertical position of dimension text:
		0 Centered between extension lines.
		1 Above dimension line, except when dimension line not horizontal and **DimTIH** = 1.
		2 On side of dimension line farthest from the defining points.
		3 Conforms to JIS.
DimTDec	4	Primary tolerance decimal places.
DimTFac	1.0000	Tolerance text height scaling factor.
DimTFill	0	Toggles the display of background fill:
		0 None.
		1 Background color of drawing.
		2 Color specified by DimTFillClr.
DimTFillClr	0	Specifies the color used for the background of dimension text.
DimTIH	On	Text inside extensions is horizontal:
		Off Text aligned with dimension line.
		On Text is horizontal.
DimTIX	Off	Place text inside extensions:
		Off Place text inside extension lines, if room.
		On Force text between the extension lines.
DimTM	0.0000	Minus tolerance.
DimTMove	0	Determines how dimension text is moved:
		0 Dimension line moves with text.
		1 Adds a leader when text is moved.
		2 Text moves anywhere; no leader.

DimVar	Default	Settings and Options
DimTOFL	Off	Force line inside extension lines: **Off** Dimension lines not drawn when arrowheads are outside. **On** Dimension lines drawn, even when arrowheads are outside.
DimTOH	On	Text outside extension lines: **Off** Text aligned with dimension line. **On** Text is horizontal.
DimTol	Off	Generate dimension tolerances: **Off** Tolerances not drawn. **On** Tolerances are drawn.
DimTolJ	1	Tolerance vertical justification: **0** Bottom. **1** Middle. **2** Top.
DimTP	0.0000	Plus tolerance.
DimTSz	0.0000	Size of oblique tick strokes: **0** Arrowheads. **>0** Oblique strokes.
DimTVP	0.0000	Text vertical position when **DimTAD**=0: **1** Turns on **DimTAD**. **>-0.7 or <0.7** Dimension line is split for text.
DimTxSty	"STANDARD"	Dimension text style.
DimTxt	0.1800	Text height.
DimTZin	0	Tolerance zero suppression: **0** Suppress 0 ft and 0 in. **1** Include 0 ft and 0 in. **2** Include 0 ft; suppress 0 in. **3** Suppress 0 ft; include 0 in. **4** Suppress leading 0 in decimal dim. **8** Suppress trailing 0 in decimal dim. **12** Suppress leading and trailing zeroes.

U

DimUPT	Off	User-positioned text: **Off** Cursor positions dimension line. **On** Cursor also positions text.

Z

DimZIN	0	Suppression of 0 in feet-inches units: **0** Suppress 0 ft and 0 in. **1** Include 0 ft and 0 in. **2** Include 0 ft; suppress 0 in. **3** Suppress 0 ft; include 0 in. **4** Suppress leading 0 in decimal dim. **8** Suppress trailing 0 in decimal dim. **12** Suppress leading and trailing zeroes.

Two dimvars are considered obsolete by Autodesk, but remain in AutoCAD for compatibility reasons:

Obsolete DimVar	Default	Comments
DimFit	3	Use DimATfit and DimTMove instead.
DimUnit	2	Replaced by DimLUnit and DimFrac.

 **DIMSTYLE**

The **DIMSTYLE** command displays a dialog box for creating and modifying named dimension styles.

The tabbed dialog boxes of the Dimension Style Manager control almost all dimension variables, which in turn affect the look of dimensions in your drawings. A preview window provides a graphical overview of the current dimstyle.

New drawings contains at least two dimension styles, either "Standard" in Imperial drawings or "ISO-25" in metric drawings, and "Annotative" in all drawings. You can create as many dimstyles as you require, each with its own name — one for every drawing, if need be.

The dialog box is called a "manager," because it manages dimension styles in the drawing. With it, you create new styles, override and modify existing styles, compare differences between two styles, and delete styles.

WORKING WITH DIMENSION STYLES

To create and change dimension styles, start the **DIMSTYLE** command:

- From the **Dimension** menu, choose **Dimension Style**.

- From the Dimension or Styles toolbars, choose the **Dimension Style** button.

- In the Dashboard's Dimension panel, choose the **Dimension Style** button.

- At the 'Command:' prompt, enter the **dimstyle** command.

    Command: **dimstyle** *(Press* ENTER.*)*

- Alternatively, enter the aliases **d**, **dst**, **dimsty**, or **ddim** (the command's old name) at the 'Command:' prompt.

In all cases, AutoCAD displays the Dimension Style Manager dialog box.

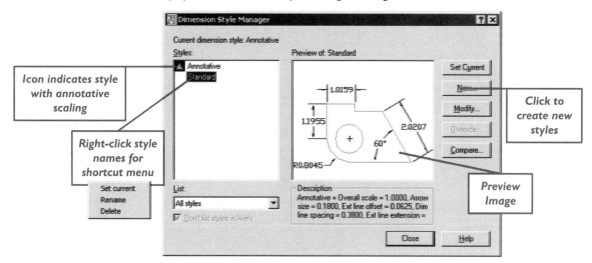

The five buttons on the right perform the following important functions:

- **Set Current** — sets the selected dimstyle current. After clicking this button, you usually click **OK** to exit the dialog box. After you exit, new dimensions follow this style.

- **New** — creates new dimstyles by naming them; usually works by renaming and modifying an existing style, such as Standard or Annotative.

- **Modify** — modifies retroactively, and globally updates, all existing dimensions created using the style. You tend to use this button more often than **Override**.

- **Override** — overrides the settings of the current dimstyle. Overrides change only dimensions created after the override is applied. Externally-referenced dimstyles cannot be modified or overridden.
- **Compare** — displays a dialog box for comparing the differences between two dimension styles.

All of these are discussed in detail in the following sections.

NEW — CREATING NEW DIMSTYLES

1. To create a new dimension style, click the **New** button, which leads to the Create New Dimension Style dialog box.
2. Making a new style is as easy as typing a name of up to 255 characters long.

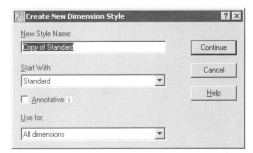

3. Choose **Continue**. Notice that AutoCAD displays the New Dimension Style dialog box, which is identical to the Modify version of the dialog box detailed below.

Notes: The standard method of creating a new style is to copy an existing one, make changes, and then save the result. For this reason, AutoCAD displays "Copy of Standard" (or whatever the current style name is) in the **New Style Name** text field.

Start With text field lets you select which dimstyle to copy — provided the drawing contains two or more dimstyles. This allows you to start with an existing dimstyle (such as the DIM, JIS, Gb, and ISO styles stored in the template drawings) and modify it according to your needs.

Annotative determines whether the style uses annotative scaling.

Use for droplist lets you apply changes to all, or a limited group of, dimensions:

- All dimensions
- Angular dimensions
- Radius dimensions
- Leaders and tolerance
- Linear dimensions
- Ordinate dimensions
- Diameter dimensions

MODIFY — MODIFYING DIMSTYLES

The **Modify** button leads to the Modify Dimension Style dialog box, which contains most of the options that affect dimensions. Here you modify the values to create and change dimstyles. Because there are so-o-o many options affecting dimensions, the dialog box segregates them into a series of tabs. Here is an overview of what each tab contains:

- **Lines** — properties of dimension and extension lines.
- **Symbols and Arrows** — properties of arrowheads, sizes of center marks and arc length symbols, and angle of jogs.
- **Text** — format, placement, and alignment of dimension text.
- **Fit** — placement of dimension text, overall scaling of dimensions, and toggling of annotation.

- **Primary Units** — formats of units for primary linear and angular dimensions, and zero suppression.

- **Alternate Units** — formats of alternate (secondary) units.

- **Tolerances** — format of dimension tolerance text.

LINES TAB

The Lines tab specifies the properties of dimension lines and extension lines.

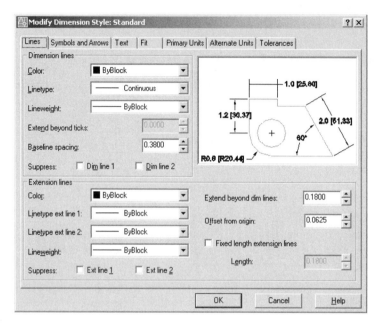

Dimension Lines

The Dimension Lines section specifies the properties of dimension lines.

Color

The **Color** droplist specifies the color of dimension lines. The default color is **Byblock**, because dimensions are considered blocks. "ByBlock" means that dimension lines take on the color of the entire dimension; in contrast, the dimension's default color is Bylayer, meaning the dimension takes on the color defined by its layer.

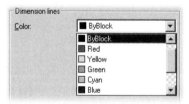

The color of the dimension line is stored in the dimension variable named DIMCLRD (short for "DIMension CoLoR Dimension line").

 Note: Although you can choose another color, it is best to place dimensions on layers that specify the color of the dimension.

You can see the effect of changing variables through the Preview box: select the color blue from the **Color** droplist. Notice in the Preview box that the dimension line color changes from black to blue.

Linetype

To assign linetypes to dimension lines, click the **Linetype** droplist (stored in DIMLTYPE). The preview window updates to show the new linetype.

You cannot change the linetype if *any* dimstyle override is active. As with colors, it is better to specify the lineweight via the layer, rather than to change the lineweight here.

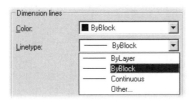 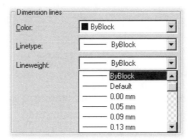

Lineweight

To change the lineweight of the dimension line, choose the droplist next to **Lineweight** and select a predefined width. The dimension line lineweight is stored in dimension variable DIMLWD.

As with colors and linetypes, it is better to specify the lineweight via the layer, rather than to change the lineweight here.

Extend Beyond Ticks

The **Extend beyond ticks** option sets the distance that extension lines should extend beyond the dimension line (stored in DIMDLE). The default is 0.0.

This option is available only when the "Architectural Tick" is selected for arrowheads, an option that's set in the Symbols and Arrows tab.

Baseline Spacing

The **Baseline spacing** option determines the distance between dimension lines when they are automatically stacked by the DIMBASELINE and QDIM commands.

The distance is stored in dimvar DIMDLI.

Suppress

AutoCAD can suppress the display of either or both dimension lines. To do so, select the **Suppress** check boxes to suppress **Dim Line 1** (stored in dimvar DIMSD1) or **Dim Line 2** (dimvar DIMSD2). These options also suppress the display of associated arrowheads.

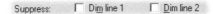

Dim Line 1 refers to neither the left nor right part of the dimension line. Instead, it is the first dimension line drawn, as determined by the initial pick points that place the extension lines.

When the text does not cut the dimension line in two, these options do not affect the dimension line.

Extension Lines

The **Extension Lines** section specifies the properties of extension lines.

Color, Linetype, and Lineweight

The **Color**, **Linetype**, and **Lineweight** options are similar to those of dimension lines; you can specify the color (stored in dimvar **DIMCLRE**) and lineweight (**DIMLWE**) of extension lines. In the case of linetypes, you can specify a different one for each extension line (**DIMLTEX1** and **DIMLTEX2**).

Suppress

The **Suppress** option hides the first (**DIMSE1**) or second (**DIMSE2**) or both extension lines. "Ext line 1" is the first extension line drawn, based on the points picked during the dimensioning command.

(*History*: this was an issue for pen plotters. When multiple extension lines were drawn on top of each other with pens, the paper eventually wore through.)

Additional Options

The **Extend beyond dim lines** option specifies the distance the extension line protrudes beyond the dimension line (stored in **DIMEXE**).

The **Offset from origin** option sets the distance the extension line begins away from the pick point of the object being dimensioned (**DIMEXO**).

The **Fixed length extension lines** option specifies a fixed length for all extension lines (**DIMFXLON**). The default value is 0.18 units (**DIMFXL**).

SYMBOLS AND ARROWS

The Symbols and Arrows tab specifies the properties of extension lines.

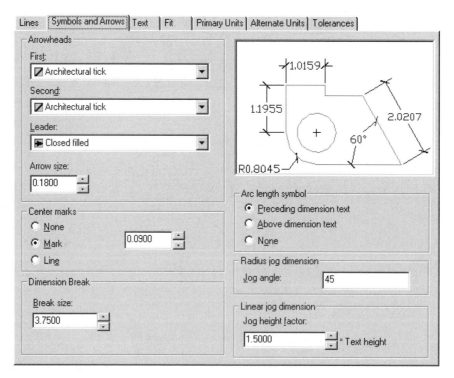

Arrowheads

The Arrowheads section defines the type and size of arrowheads used for dimensions and leaders.

First, Second, and Leader

The **First** droplist lets you choose from 20 predefined arrowhead styles (stored in **DIMBLK**). In addition to the standard arrowhead, AutoCAD includes arrowheads, such as the tick, the dot, open arrowhead, open dot, right angle arrowhead, or no head at all; see figure at the side.

When you select an arrowhead from the **First** droplist, AutoCAD changes the **Second** and **Leader** droplists to the same arrowhead.

Alternatively, you can select **User Arrow** to choose a custom-made arrowhead. The customized arrowhead is defined by the creation of an object scaled to unit size, and then saved as a named block with the **BLOCK** command. The arrowhead name is stored in **DIMBLK**. (See the tutorial on the following pages.)

Second and Leader

You can specify a different arrowhead at each end of dimension lines (**DIMBLK1** and **DIMBLK2**) and for the leader dimensions (**DIMLDRBLK**). When you specify a different one for each end, AutoCAD changes dimvar **DIMSAH** to 1.

Although radius and diameter dimensions consist of leaders, they use the arrowhead defined by **Second**; the arrowhead defined by **Leader** affects only the leaders placed by the **LEADER** and **QLEADER** commands. Ordinate dimensions do not use arrowheads.

Arrow Size

The **Arrow size** option specifies the distance from the left to right end; for custom arrowheads, AutoCAD scales the unit block to size (0.18 units, by default). The value is stored in **DIMASZ**.

- ◂ Closed filled
- ▷ Closed blank
- ⇨ Closed
- ● Dot
- ╱ Architectural tick
- ╱ Oblique
- ⇒ Open
- ⦵ Origin indicator
- ⊛ Origin indicator 2
- → Right angle
- ⇉ Open 30
- ◆ Dot small
- ○ Dot blank
- ○ Dot small blank
- ⊐ Box
- ◼ Box filled
- ◁ Datum triangle
- ◀ Datum triangle filled
- ∫ Integral
- None

TUTORIAL: CREATING CUSTOM ARROWHEADS

In this tutorial, you draw a custom arrowhead, and then use it with dimensions.

1. Start AutoCAD with a new drawing.
2. Set the grid to 1.0, or use xlines to demarcate a one inch square, because the arrowhead must fit within a 1"-square.

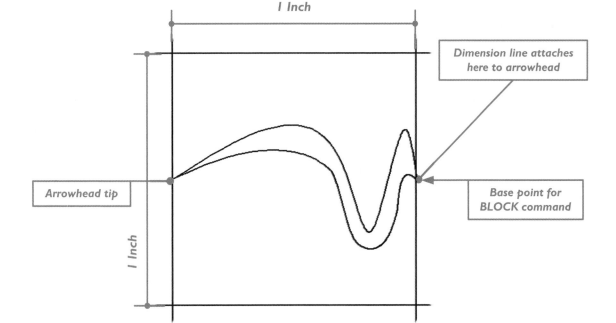

1 Inch

Dimension line attaches here to arrowhead

Arrowhead tip

1 Inch

Base point for BLOCK command

3. Draw the arrowhead with the tip pointing West.
    (The arrowhead illustrated below was drawn with the **SPLINE** and **OFFSET** commands.)
    The arrowhead attaches to the dimension line on the East (see figure below).

4. Use the **BLOCK** command to turn the arrowhead into a block, and then give the block a meaningful name, such as "spiral."
    Important! For the **Base point**, pick the East end of the arrowhead, where it attaches to the dimension line, as illustrated below.

5. To use the new arrowhead with dimensions:
    a. Start the **DIMSTYLE** command, and then click **Modify**.
    b. Select the Arrows and Symbols tab.
    c. Under Arrowheads, in the **First** droplist, select **User Arrow**. Notice the Select Custom Arrow Block dialog box, which lists the names of all blocks defined in the drawing. (You can use any block found in the drawing as an arrowhead. Because they were not designed for use as arrowheads, however, they would probably not work well.)

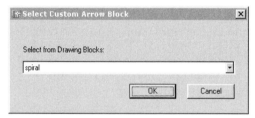

6. Select the name of the block you created earlier, and then click **OK**.
    Notice that AutoCAD automatically fills in the **Second** and **Leader** droplists with the same name. The preview image shows the custom arrowheads.

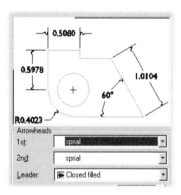

7. Click **OK** to exit the **Symbols and Arrows** tab.
    Choose the **Set Current** button, and then **Close** to close the dialog box.

8. Place some dimensions with the custom arrowhead.

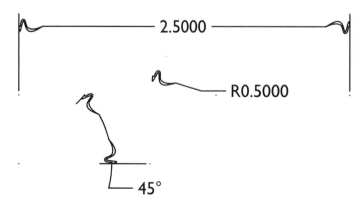

If the arrowhead is too small, increase its size with the **DIMSTYLE** command's **Arrow size** option.

Center Marks for Circles

When you mark the centers of circles and arcs with the DIMCENTER command, AutoCAD determines the type of center mark by the value stored in the DIMCEN dimvar: center mark, center mark with extending lines, or no mark at all. Center marks are also placed by the DIMRADIUS and DIMDIAMETER commands.

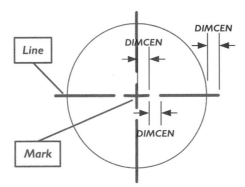

The value of DIMCEN determines the length of the marks, the width of the gap, and the distance the lines extend beyond the circle. The default value is 0.09 units. When DIMCEN is:

-0.09   Center lines and center marks are drawn.

0   No center mark or lines drawn.

0.09   Center marks drawn.

Select a radio button:

- **None** places no center marks in circles and arcs.
- **Mark** places the center mark.
- **Line** places the center mark and lines.

Size specifies the size of the mark, gap, and line extension.

Dimension Break

Break Size specifies the size of the gap created by the DIMBREAK command.

Arc Length Symbol

The **Arc Length Symbol** option specifies the location of the arc length symbol (DIMARCSYM). It looks like an upside down "U"; the symbol cannot be customized:

- **Preceding Dimension Text** places the symbol before the dimension text.

$$\text{———}\ \cap 14.3\ \text{———}$$

$$\text{———}\ \overset{\frown}{14.3}\ \text{———}$$

- **Above Dimension Text** places the symbol above the dimension text.
- **None** suppresses the display of the symbol.

Radius Dimension Jog

The **Radius Dimension Jog** section specifies the angle used by jogged dimensions (DIMJOGANG). The default is 45 degrees.

Use the **Jog Angle** option to change the angle. Autodesk does not document the limits, but the angle must be between 5 and 90 degrees. Enter an angle outside those limits, and AutoCAD does not allow you to exit the tab — until you enter a correct value.

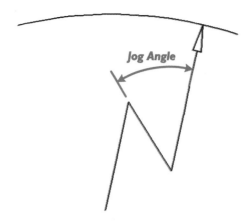

Jog Angle

Linear Jog Dimension

Jog Height Factor specifies the size of the jog created by the **DIMJOGLINE** command.

TEXT

The **Text** tab specifies the format, placement, and alignment of dimension text. Additional settings affecting the display of dimension text are found in the Primary Units tab.

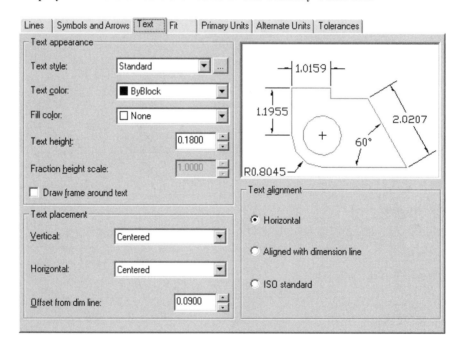

Text Appearance

Dimension text is treated independently of other text in the drawing. For example, dimension text uses the style defined by this dialog box, rather than the style defined by the **STYLE** command.

Text Style

The **Text Style** droplist selects a style previously defined by the **STYLE** command. The default is "Standard." Choose the **...** button to display the **Text Style** dialog box, which lets you create and modify text styles. The dimension text style name is stored in dimvar **DIMTXSTY**.

Text Color

The **Text Color** droplist selects the color of the dimension text (**DIMCLRT**). To choose a color not listed, pick **Select Color** at the bottom of the droplist.

Fill Color

The **Fill Color** option determines the color of a rectangle placed behind the dimension text (DIMTFILL and DIMTFILLCLR).

One of the colors is called "Background," which takes on the color defined by the OPTIONS command. (From the **Tools** menu, select **Options**, **Display**, and then **Colors**. The color for the model and layout backgrounds can be chosen separately.)

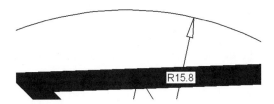

The TEXTTOFRONT command can make dimension text stand out from other objects.

Text Height

The **Text Height** option operates like the height setting in text styles: 0 means use the same height as defined in the text style. Any other height entered here overrides the height defined by the style (DIMTXT).

Fraction Height Scale

The **Fraction Height Scale** option determines the size of fractions and tolerances in dimension text, as a fraction of the dimension text (DIMTFAC). When the scale is 1, fractions and tolerances are the same height as the dimension text; when set to 0.5, fractions and tolerances are half the size of dimension text.

Draw Frame Around Text

The **Draw frame around text** check box adds a box around the dimension text (DIMGAP < 0). The distance between the box and the dimension text is specified later in the tab with the **Offset from dim line** option.

Text Placement

The Text Placement section specifies how dimension text is placed relative to the dimension line.

Vertical

The **Vertical** droplist specifies the vertical placement of dimension text (DIMTAD):

- **Centered** centers dimension text between extension lines and on the dimension line (default).
- **Above** places text above the dimension line.
- **Outside** places text on the side farthest away from the extension line pick points.
- **JIS** places text according to the Japanese Industrial Standards for dimensions, which places text above the dimension line.

Horizontal

The **Horizontal** droplist specifies the horizontal placement of dimension text (DIMJUST).

- **Centered** centers text between the extension lines and on the dimension line (the default).
- **At Ext Line 1** left-justifies text against first extension line.
- **At Ext Line 2** right-justifies text against the second extension line.
- **Over Ext Line 1** positions text vertically over the first extension line.
- **Over Ext Line 2** positions text vertically over the second extension line.

Offset From Dim Line

The **Offset from dim line** option specifies the gap between dimension text and the dimension line, as well as the space between the text and the frame (DIMGAP).

Text Alignment

The **Text Alignment** option determines the orientation of dimension text when inside and outside the extension lines (DIMTIH and DIMTOH).

- **Horizontal** forces text to be horizontal, whether or not it fits inside the extension lines.

- **Aligned with dimension line** forces text to align with the dimension line.

- **ISO Standard** aligns text with the dimension line when it fits inside extension lines; when outside extension lines, it is drawn horizontally.

FIT

The **Fit** tab determines the placement of dimension text, arrowheads, leader lines, and dimension lines.

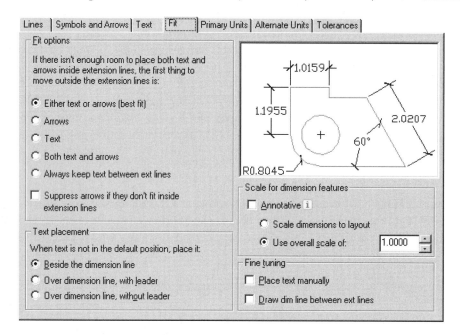

Fit Options

The **Fit Options** section controls where text and arrowheads are placed when the distance between extension lines is too narrow (DIMATFIT).

- **Either the text or the arrows, whichever fits best** places the elements where there is room.

- **Arrows** fits only the arrowheads between extension lines, when space is available.

- **Text** fits text between extension lines, while arrowheads are outside when space is lacking.

- **Both text and arrows** forces both text and arrows outside the extension lines.

- **Always keep text between ext lines** always keeps text between the extension lines (**DIMTIX**).

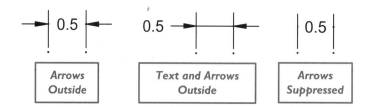

- **Suppress arrows if they don't fit inside the extension lines** draws no arrowheads in that case (**DIMSOXD**).

Text Placement

When AutoCAD cannot place the dimension text normally, you have these options (**DIMTMOVE**):

- **Beside the dimension line** places text beside the dimension line.

- **Over the dimension line, with a leader** draws a leader line between the dimension line and the text when there isn't enough room for the text.

- **Over the dimension line, without a leader** draws no leader.

Scale for Dimension Features

The **Scale for Dimension Features** section controls the size of text and arrowheads independently of the rest of the drawing, and is usually associated with the plot scale. The length of the dimension and extension lines is unaffected. (To change the weight of these lines, use the **Lineweight** option in the Lines and Arrows tab.)

Annotative toggles use of the annotative scaling property (**DIMANNO**). When on, the following two options become unavailable:

- **Use overall scale of** sets the scale factor for text and arrowheads (**DMSCALE**). Entering a value of 2, for example, doubles the size of arrowheads and text.

- **Scale dimensions to layout** scales dimensions to a factor based on the scale between model space and layouts, as follows:

    AutoCAD sets **DIMSCALE** = I in model tab.
    AutoCAD sets **DIMSCALE** = 0 in layout tab.
    AutoCAD then computes **DIMSCALE** in model space viewports of layout tabs from the viewport's scale factor; or from the current value of the **ZOOM XP** command when dimensions are placed in paper space layouts.

Fine Tuning

The **Fine Tuning** section presents a couple of miscellaneous items:

Place text manually when dimensioning determines whether dimension commands prompt you for the position of the dimension text, allowing you to change the text position, if necessary (**DIMUPT**). AutoCAD ignores the horizontal justification settings.

Always draw dim line between ext lines forces AutoCAD always to draw the dimension lines inside the extension lines (**DIMTOFL**). The arrowheads and text are placed outside, if there isn't enough room. This option also forces leaders to be drawn to the circle's center point, even when the leader would not normally be drawn.

PRIMARY UNITS

The **Primary Units** tab sets the format of primary dimension units, as well as the prefix and suffix of dimension text. "Primary" refers to ordinary dimension text, as opposed to alternate units (shown in square brackets) and tolerance text.

Linear Dimensions

The **Linear Dimensions** section specifies the dimensioning units, and whether the dimension text has a prefix or suffix. This section has features similar to those of the **UNITS** command. The Drawing Units dialog box controls the settings of units for all numbers, angles, and dimensions in the drawing; the Primary Units tab overrides those settings.

Units Format

The **Units format** droplist selects the format: scientific, decimal (default), architectural, engineering, fractional, and the units specified by the Windows operating system (**DIMLUIT**). The Windows units are changed by the Numbers tab of the Regional Options dialog box (found in the Control Panel).

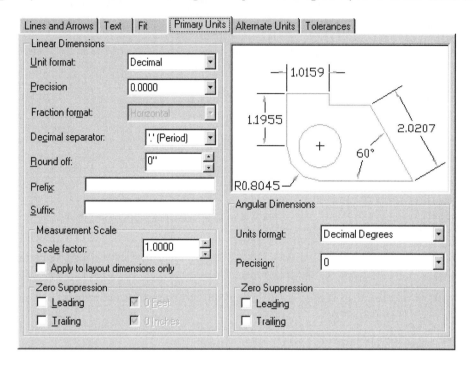

Precision

The **Precision** droplist selects the number of decimal places or fractional divisions to be displayed. The range is from 0 to 8 decimal places, and from 0" to 1/256". The measured values are unaffected by this setting, which changes only the display.

Fraction Format

The **Fraction format** droplist specifies how fractions are stacked: horizontally, diagonally, or not at all (**DIMFRAC**). This option is available only when you select a units format with fractions, such as Architectural; otherwise, it is grayed out.

Decimal Separator

The **Decimal separator** droplist specifies the decimal separator: period (.), comma (,), or space ( ) (**DIMDSEP**). This option is available only when you select **Decimal** as the units format. This is typically for countries that use the comma (,) to separate units from decimals.

Round Off

The **Round off** option specifies how to round off decimals and fractions of numbers; it does not apply to angles (**DIMRND**). For example, setting it to 0.33 rounds the dimension text to the nearest 0.33; the measured values are unaffected by this setting, which changes only the display. This option is different from **Precision**, which *truncates* the display of decimal places.

Prefix and Suffix

The **Prefix** and **Suffix** options provide room for any alphanumeric values, which are then added in front of and behind the dimension text. For example, to prefix every dimension with the word "Verify," enter that in the **Prefix** box. To suffix every dimension with "(TYPICAL)," enter that in the **Suffix** box. The **DIMPOST** dimvar stores both values using this format: "Verify<>(TYPICAL)".

Measurement Scale

The **Measurement Scale** section specifies the scale factor applied to the *value* of the dimension text; to change the size of text and arrows, specify the overall scale factor in the Fit tab.

Scale Factor

The **Scale Factor** option sets the scale factor that multiplies linear dimensions (DIMLFAC). By entering 25.4, for example, imperial dimensions are converted to millimeters (metric units). The factor is not applied to angles; alternate and tolerance values have their own scale factor settings.

Apply to Layout Dimensions Only

The **Apply to layout dimensions only** options applies the linear scale factor only to dimensions created in layouts (paper space). When turned on, the scale factor is stored as a negative value in DIMLFAC.

Zero Suppression

The **Zero Suppression** section determines whether zeros are displayed in dimension text.

Leading and Trailing

The **Leading** and **Trailing** options specify whether leading or trailing zeros, and the zero feet or inches, are suppressed (DIMZIN). When both options are turned on, **0.2500** becomes **.25**, and **1.00** becomes **1**. Note that zero suppression overrides the setting of the **Precision** option.

0 Feet and 0 Inches

The **0 feet** and **0 inches** options determine whether zero feet and zero inches are displayed. When both options are turned on, **0'–1/2"** becomes **1/2"** and **34'–0"** becomes **34'**. These feet and inches options become available only when a fraction unit format is selected, such as Architectural.

Angular Dimensions

The Angular Dimension section specifies the display format for dimensions placed by the DIMANGULAR command. This allows angles to be formatted independently of linear dimensions.

Units Format

The **Units format** droplist specifies the format of angular dimensions (DIMAUNIT) — curiously, the Surveyor's units are missing:

- Decimal degrees (*DdD.dddd*, the default).
- Degrees/ Minutes/Seconds (*DD.MMSSdd*).
- Grads (*DDg*).
- Radians (*DDr*).

Precision

The **Precision** droplist specifies the number of decimal places or fractions (DIMADEC). This affects the display of the angle, and not its measured value.

Zero Suppression

The **Zero Suppression** section suppresses leading and trailing zeros of angles (DIMAZIN). As an example of suppressing leading zeros, the measured angle of **56°07"** is displayed as **56°7"**; and in suppressing trailing zeros, **56°00"** is displayed as **56°**.

ALTERNATE UNITS

The **Alternate Units** tab selects the format of units, angles, dimensions, and scale of alternate measurement units. The options are similar to the Primary Units tab; the following descriptions list only the differences.

AutoCAD allows you to place dimensions with double units: a primary unit plus a second or *alternate* unit. The alternate units appear in square brackets. This is particularly useful for drawings that must show imperial and metric units. Angles cannot have alternate units in AutoCAD.

When you select options in this tab, the values are stored in a different set of dimvars:

DIMALTD specifies the decimal places for alternate units.

DIMALTRND specifies the rounding of alternate units.

DIMALTU specifies the format of alternate units, except for angular dimensions.

DIMALTZ specifies zero suppression for alternate units.

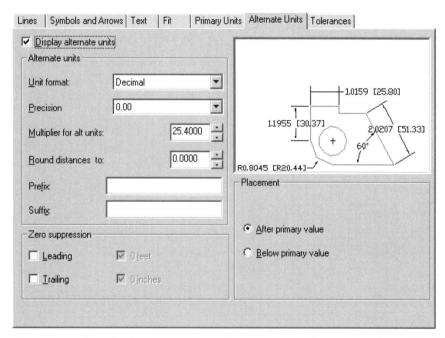

The **Display alternate units** check box turns on alternate units (**DIMALT**); when off, all options are grayed out, meaning they are unavailable.

Alternate Units

The **Alternate Units** section specifies the look of the alternate text; the options are identical to that of the Primary Units tab.

Zero Suppression

The **Zero Suppression** section determines whether leading and trailing zeros are suppressed; the options are identical to that of the Primary Units tab.

Placement

The **Placement** section determines where the alternate units are placed (**DIMAPOST**):

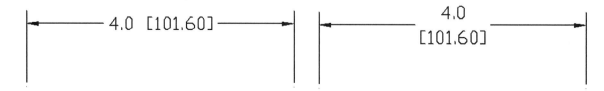

- **After primary value** places the alternate units behind the primary units.

- **Below primary value** places the alternate units below the primary units.

TOLERANCES

The **Tolerances** tab controls the format of dimension text tolerances. It does not format tolerance symbols placed with the TOLERANCE command; they have no format options.

Tolerance Format

The **Tolerance Format** section determines the style and look of tolerances.

Method

The **Method** droplist offers five styles of tolerance text (DIMTOL and DIMLIM):

- **None** suppresses the display of tolerance text (default), and grays out all options in this tab, except **Vertical position**, curiously enough.

- **Symmetrical** adds a single plus/minus tolerance, such as the 0.3 illustrated below.

- **Deviation** adds a plus and a minus tolerance, such as the +0.5 and -0.1 illustrated above.

- **Limits** places two dimensions, such as the 3.0 and 2.4 illustrated below. AutoCAD arrived at these values by adding 0.5 to 2.5 (=3.0), and subtracting 0.1 from 2.5 (=2.4).

- **Basic** draws a box around the dimension text. (The distance between text and box is stored in **DIMGAP**.) No tolerances are displayed, as illustrated above.

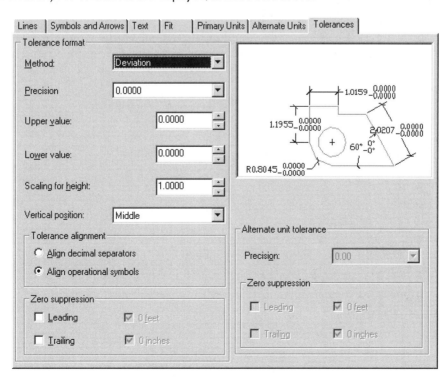

Dimvar **DIMTOL** appends the tolerances to the dimension text; **DIMLIM** displays dimension text as limits. Turning on **DIMLIM** turns off **DIMTOL**.

Precision

The **Precision** droplist determines the number of decimal places displayed by both tolerance values.

Upper Value and Lower Value

The **Upper Value** and **Lower Value** options specify the upper or plus value (**DIMTP**), and the lower or minus tolerance value (**DIMTM**).

Scaling for Height

The **Scaling for height** option determines the size of the tolerance text relative to the main dimension text (**DIMTFAC**). A value of 1 indicates each character of tolerance (or limits) text is the same size as the primary dimension text. Changing the value to **0.5** reduces the tolerance text height by half.

Vertical Position

The **Vertical position** droplist determines the relative placement of tolerance text (**DIMTOLJ**), whether it is aligned to the top, middle, or bottom of the primary dimension text.

Tolerance Alignment

The **Tolerance Alignment** options align the deviation and limits tolerance text by decimal points or prefix symbols, such as + and -.

Zero Suppression

The **Zero Suppression** options are identical to those found in the Primary Units tab.

Alternate Unit Tolerance

The **Alternate Unit Tolerance** specifies the number of decimal places for the tolerance values in the alternate units of a dimension (**DIMALTTD**). This option is available only when alternate units are turned on in the Alternate Units tab. Combining tolerances with alternate units leads to cluttered dimension text.

Setting the Current Dimstyle

To set a dimension style as active, select the dimstyle name from the list under **Styles**, and then choose the **Set Current** button. (The current dimension style name is stored in dimvar **DIMSTYLE**.)

Alternatively, you can use the Styles toolbar, which provides droplists of dimension and text styles in the drawing. To make a dimstyle current, select it from the droplist.

Renaming and Deleting Dimstyles

It is not immediately apparent how to rename or delete dimension styles in the Dimension Style Manager. The commands are "hidden" in shortcut menus: select a dimension style name in the **Styles** list. Right-click to display the cursor menu, and then choose **Rename** or **Delete** .

Not every style can be erased or renamed. If a style is in use, it cannot be erased. The "Standard" dimstyle can be renamed, but not deleted. Externally-referenced dimstyles cannot be renamed or deleted.

Alternatively, you can use the RENAME command to rename dimstyles, and the PURGE command to remove unused ones.

MODIFYING AND OVERRIDING DIMSTYLES

After you create one or more dimension styles for drawings, you might need to change the style. Perhaps the client wants dimension lines thicker than the extension lines, or the units changed to metric, or perhaps some extension lines removed.

Some changes are *global*, which means they apply to every dimension in the drawing, such as the thicker dimension lines. Other changes are *local*, which means they apply to selected dimensions, such as extension lines to be removed.

Some changes are *retroactive*, which means they apply to all dimensions already drawn. Other changes are *proactive*, which means they apply to dimensions not yet drawn.

Here is how AutoCAD handles these four cases.

> **To apply global changes retroactively:** modify the dimension style with the **DIMSTYLE** command's **Modify** button. AutoCAD retroactively changes all dimensions in the drawing.

> **To apply global changes proactively**: modify the dimension style with the **DIMSTYLE** command's **Override** button, which creates "sub-styles." The next dimension you draw reflects the change in dimension style.

> **To apply different dimension styles to selected dimensions (local changes):** the -**DIMSTYLE** command's **Apply** option prompts you to select dimensions, and then applies the dimstyle to them — a local modification.

> **To change dimension variable settings of selected dimensions (local changes):** the **DIMOVERIDE** command asks you for the name of a dimension variable and the new value. It then prompts you to select the dimensions to which the local override should be applied.

 Note: The difference between *modifying* and *overriding* dimension styles:

Modify — modifies retroactively, and globally updates all existing dimensions created using that style.
Override — changes only dimensions created after the override is applied,

In the following tutorial, you learn how to change the style of all dimensions in the drawing at once.

TUTORIAL: GLOBAL RETROACTIVE DIMSTYLE CHANGES

1. Start a new drawing with AutoCAD, and then draw some dimensions.
2. To change the style of all dimensions in the drawing, start the **DIMSTYLE** command:
3. In the dialog box, click **Modify**.

4. In the **Lines** tab, change the dimension line **Color** to red.
   Notice that the preview shows the dimension lines changed to red.
5. Click **OK**, and then **Close**.
   Notice that all dimensions in the drawing now have red dimension lines.

In the next tutorial, you learn how to override the dimension style temporarily, so that new dimensions take on the changed style; existing dimensions remain unaffected.

TUTORIAL: GLOBAL TEMPORARY DIMSTYLE CHANGES

1. Continue with the drawing from the previous tutorial.
2. Start the **DIMSTYLE** command, and then choose the **Override** button.
3. In the **Lines** tab, change the dimension line color to blue.

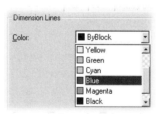

4. Click **OK**.
   Notice that the Standard dimstyle has a sub-style called "<style override>."

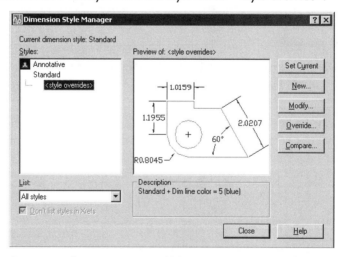

   It is highlighted, meaning that it is current. (After exiting this dialog box, any dimension you draw has blue dimension lines.)

5. In the **Description** area, notice the change to the blue dimension lines: "Standard+Dim line color = 5 (blue)".
6. Click **Close**. Notice that none of the existing dimensions in the drawing changes.
7. Draw a linear dimension with the **DIMLINEAR** command. Notice that its dimension line is blue.

Unfortunately, the names of overrides are not listed by the droplist in the Dashboard panel or Styles toolbar.

 Note: You can turn overrides into permanent dimension styles. Here's how:

To turn overrides into *new* dimstyles, right-click "<style override>" (the AutoCAD-generated name), and then select **Rename** from the shortcut menu. Change the name to something meaningful, like "Blue Dimlines."

To make an override the current dimstyle, right-click "<style override>", and then select **Save to current style** from the shortcut menu. Notice that the override disappears, and that the current style shows the changes, such as blue dimension lines. Click **Close**, and all dimensions change to blue dimlines. (This is equivalent to using the **Modify** button in the first place.)

The dimension style dialog box is lacking in one respect: it cannot apply changed dimension styles to individual dimensions. This responsibility lies elsewhere, with the Dashboard's Dimension panel, the **DIMOVERIDE** command, and the Properties window.

STYLE OVERRIDES, DIMOVERRIDE, AND PROPERTIES

Overriding individual dimensions is called "local" changes.

The Dim Style droplist on the Dashboard's Dimension panel applies dimension styles to selected dimensions, as does the droplist on the Styles toolbar. The **DIMOVERRIDE** command applies changed dimension variables to selected dimensions.

TUTORIAL: LOCAL DIMSTYLE CHANGES

1. For this tutorial, create a drawing with at least two dimension styles and several dimensions.
2. To apply a dimstyle to certain dimensions, select a few dimensions in the drawing. Notice that they are highlighted.
3. From the Dashboard's Dimension panel, click the **Dimension Style** droplist. (You can also do the same with the Dimension Style droplist in the Styles and Dimension toolbars.) Notice the list of dimension style names.

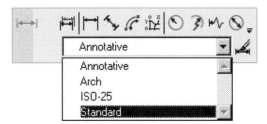

3. Select a different style name. Notice that the selected dimensions change their style.
4. Press **ESC** to remove the highlighting from dimensions.

In the next tutorial, you change the color of one dimension's extension line to yellow.

TUTORIAL: LOCAL DIMVAR OVERRIDES

1. For this tutorial, continue with the drawing from the last tutorial, the one with at least two dimensions styles.

2. To change a dimvar for selected dimensions, start the **DIMOVERRIDE** command.

   • From the **Dimension** menu, choose **Override**.

   • At the 'Command:' prompt, enter the **dimoverride** command.

      Command: **dimoverride** *(Press ENTER.)*

   • Alternatively, enter the aliases **dov** or **dimover** at the 'Command:' prompt.

3. Enter the name of a dimvar, such as **DIMCLRE**, which sets the color of extension lines:

   Enter dimension variable name to override or [Clear overrides]: **dimclre**

4. Enter the new value for the dimvar, such as yellow:

   Enter new value for dimension variable <BYBLOCK>: **yellow**

5. The prompt repeats so that you can override other dimvars. Press **ENTER** to continue and select the dimensions to override:

   Enter dimension variable name to override: *(Press ENTER to continue.)*

6. Select the dimensions to change:

   Select objects: *(Select a dimension.)*

   Select objects: *(Press ENTER to end object selection.)*

   Notice that the extension lines change to yellow.

Note: You can override dimvars *during* dimensioning commands. In the following example of the **DIMALIGNED** command, the dimension line color is changed to green using the **DIMCLRD** dimvar:

   Command: **dimaligned**

   Specify first extension line origin or <select object>: **dimclrd**

   Enter new value for dimension variable <byblock>: **green**

   Specify first extension line origin <select object>: *(Continue with the DIMALIGNED command.)*

New dimension lines are now green, until you override again.

Clearing Local Overrides

If you change your mind, you can clear local overrides. Start the DIMOVERRIDE command, and then enter the **Clear overrides** option:

   Command: **dimoverride**

   Enter dimension variable name to override or [Clear overrides]: **c**

   Select objects: *(Select the dimensions to return to normal.)*

   Select objects: *(Press ENTER to end object selection.)*

The dimension returns to its previous look, because the override is canceled.

PROPERTIES

You can use the Properties palette to change many properties of dimensions locally. Open the palette with the **PROPERTIES** command, and then select the dimensions you wish to change.

DRAWORDER FOR DIMENSIONS

You can force AutoCAD to display dimensions in front of all other objects in the drawing. To do this, use the **TEXTTOFRONT** command, as follows:

> Command: **texttofront**
>
> Bring to front [Text/Dimensions/Both] <Both>: **d**
>
> *n* object(s) brought to front.

This command is more efficient than the **DRAWORDER** command, because it selects all dimensions for you.

Because **TEXTTOFRONT** is a command, and not a setting, you may have to reuse this command as the drawing progresses.

EXERCISES

1. In the following exercises, you modify the Standard dimension style.

a. Start AutoCAD with a new drawing.

b. Draw one each of the following dimensions — linear, aligned, diameter, and leader — similar to the figure below.

Which commands did you need to draw each dimension?

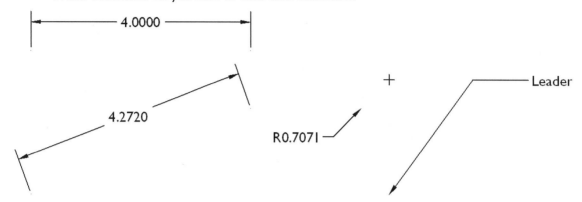

c. Open the Dimension Style Manager dialog box. Which dimstyles are listed there? Modify the dimstyle: change the color of extension lines to red. When you exit the dialog box, what happens?

d. Return to the Dimension Style Manager dialog box. Override the dimstyle: change the color of text to blue. When you exit the dialog box, what happens?

e. Draw another linear dimension. Is it different from the others?

f. Return to the Dimension Style Manager dialog box. Convert the override into a dimstyle called "Blue Text." When you exit the dialog box, what happens?

g. Use the Dashboard's Dimension panel to apply the Blue Text dimstyle to the aligned dimension. Does it change?

Repeat, applying the dimstyle to the leader. Does it change?

2. Continuing with the drawing from above, create a new dimension style with the following settings:

Extension line:

    Color   **8**

Text:

    Font   **Times New Roman**

    Color   **155**

    Vertical Placement   **JIS**

Arrowheads:

    Both   **Architectural Tick**

Hint: change the text style.

Save the style by the name of "JayGray." Apply the style to all dimensions in the drawing. Save the drawing as *jaygray2.dwg*.

3. Continuing with the drawing from the previous exercise, use grips editing on the different grips of each dimension. Do not save the drawing.

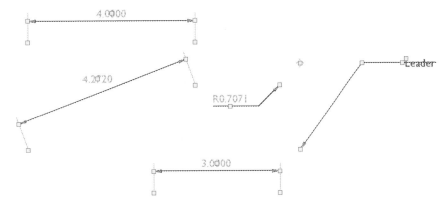

4. Start a new drawing.

   Open the DesignCenter, and then import dimension styles from the following drawing: *Architectural - Annotation Scaling and Multileaders.dwg.*

5. From AutoCAD's *\sample* folder, open the *Architectural - Annotation Scaling and Multileaders.dwg* file, a drawing of a set of stairs.

   Select a dimension, and then edit its text

   Do the same to a leader.

6. Start a new drawing, and then draw a linear dimension.
   a. Add a jog to the dimension line.

      Which command did you use to add the jog?
   b. Draw a second linear dimension that crosses the first one.

      Add a gap to where the two dimensions cross.

      Which command did you use to create the gap?

7. From the Companion CD, open *structure.dwg* file, the drawing of an arbour. Space the linear dimensions evenly.

   Which command did you use to space the dimensions?

CHAPTER REVIEW

1. Which command is used to create, edit, and delete dimension styles?

2. What is the difference between *modifying* and *overriding* dimensions?

3. Can the color and lineweight of dimension lines be set independently of extension lines?

4. Can extension lines be suppressed independently of each other?

5. Is it better to use dimension styles or layers to set the color and lineweight of dimension lines?

6. Can you define your own arrowhead for use by dimensions?

7. When the distance between extension lines is too tight, where does AutoCAD place the dimension text?

8. Describe some situations where you might need to edit dimensions.

9. What are *alternative units*?

10. When you move an object, do associative dimensions move also?

11. Describe how the following grips editing commands affect dimensions:

    Stretch

    Copy

    Move

    Rotate

    Scale

    Mirror

12. What happens to the text when a leader is grips edited?

13. Can the extension lines be grips-edited independently of the dimension line?

14. Explain the meaning of the following abbreviations:

    dimvar

    dimstyle

    dimline

15. What is the "T" in **DIMTEDIT** short for?

16. Describe the function of the **DIMEDIT** command's **Oblique** option.

    When is it useful?

17. How do you edit the wording of the dimension text?

    The properties of dimension text?

    The position of dimension text?

18. Briefly explain the differences among these three similar-sounding dimension-editing commands:

    DDEDIT

    DIMEDIT

    DDIMTEDIT

19. If you make a mistake editing a dimension, which command returns the original dimension?

20. Can dimensions be scaled independently of the drawing?

21. What is the danger in rounding dimension text?

22. When AutoCAD rounds off dimension values, is the change permanent?

23. Round the following numbers to the nearest unit:

    2.5

    1.01

    4.923

    10.00

    6.489

24. Where does the JIS dimension standard expect text to be placed?

25. Describe the purpose of *dimension variables*.

26. Does every drawing have a dimension style?

27. How do you change the scale of dimensions relative to the rest of the drawing?

28. List two sources of dimension styles:

   a.

   b.

29. Why might you want to edit dimensions in the following manner:

   a. Create gaps in extension lines?

   b. Create jogs in the dimension line?

   c. Space dimensions evenly?

30. Briefly describe the purpose of the following dimvars. (For help, look up the tables in this chapter.)

   DIMCLRE

   DIMSE1

   DIMJUST

   DIMCLRT

   DIMAFIT

31. Find the name of the dimvar that does the following task. (For help, look up the tables in this chapter.)

   Specifies the size of the center mark.

   Enables alternate units.

   Specifies the length of the arrowhead.

   Specifies linear dimension units.

   Controls suppression of zeros.

32. Why does the Standard dimension style set **DIMALTF** (alternate scale factor) to 25.4?

   And why does the ISO-25 dimstyle set the same dimvar to 0.04?

33. What is the effect of annotative scaling on dimensions?

34. What is the command that corresponds to the alias?

   ddim

   dimted

   ded

   dov

   d

35. Can you rename dimension styles?

   Purge dimension styles?

36. Can a drawing have more than one dimension style?

   Can a dimension have two different arrowheads?

37. Can you create your own arrowheads for dimensions?

38. Under what condition do the arrowheads and text appear outside the extension lines?

39. What is the purpose of *zero suppression*?

   Why might you want to use zero suppression?

40. When are alternative units used?

   How are alternative units shown?

41. Briefly describe the sequence to change the style of all dimensions in the drawing retroactively.

42. Briefly describe the sequence to change the style of dimensions selectively.

43. Can dimvars be changed during a dimensioning command?

44. If dimensions are hidden by other objects in the drawing, what can you do to see them better?

45. How would you quickly bring all dimensions to the top of the display order?

46. What is the purpose of the **AIDIMFLIPARROW** command?

47. Can extension lines have fixed lengths?

CHAPTER 15

Geometric Dimensions and Tolerances

Ordinate dimensions and GD&T (geometric dimensioning and tolerancing) are used for drawings meant for machined parts, such as metal parts drilled with holes, slots, and contours. These drawings are often used in conjunction with CNC programming (computer numerical control of machinery).

Ordinate dimensions show distances from a common starting point, while tolerancing informs the machine operator about the accuracy of parts being machined.

In this chapter, you learn these commands:

> **DIMORDINATE** and **QDIM** draw ordinate dimensions.
>
> **DIMCONTINUE** draws additional ordinates.
>
> **UCS** relocates the datum, and changes the leader angle.
>
> **TOLERANCE** and **QLEADER** construct GD&T symbols.

NEW TO AUTOCAD 2008 IN THIS CHAPTER

- Dimensions and tolerances can have annotative scaling.

FINDING THE COMMANDS

On the Dashboard's **Dimensions** panel:

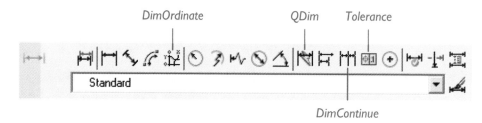

On the **Dimension** toolbar:

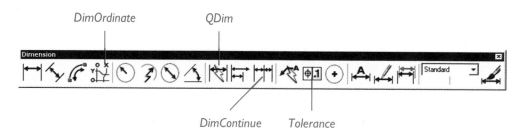

▦ DIMORDINATE

The **DIMORDINATE** command places ordinate dimensions relative to a base point.

Ordinate dimensions (or "datum dimensions") display x and y coordinates at the ends of leaders. They measure perpendicular distances between a common origin (also called the "datum") and features in drawings, commonly corners of parts and centers of holes. When indicating hole centers, apply the Center linetype to the leader line.

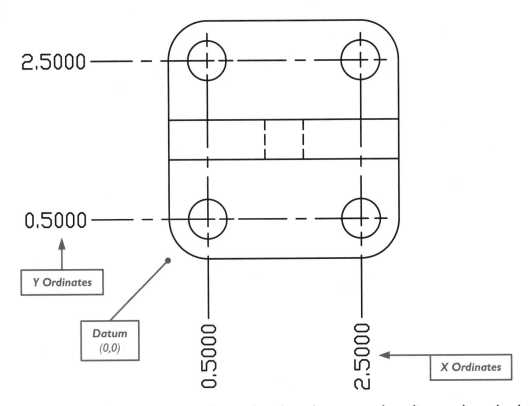

Ordinate dimensions have just one number: either the x distance or the y distance from the datum. They are measured, as follows:

- **X distance** — measured along the x axis, and specified by the *x ordinate*.

- **Y distance** — measured along the y axis, and specified by the *y ordinate*.

AutoCAD's **DIMORDINATE** command has two methods of placing x and y ordinate dimensions. In the first, you specify the **Xdatum** or **Ydatum** options of the **DIMORDINATE** command. In the second, AutoCAD determines the correct ordinate automatically from the cursor position. AutoCAD measures the distance from the cursor location to the origin, and ghosts whichever ordinate is *farthest* from the origin, as illustrated by the figure below.

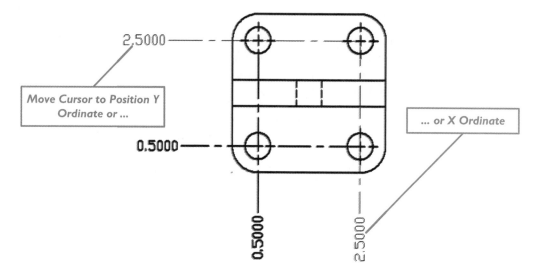

Ordinate dimensions ignore text orientation defined by the dimension style. Instead, the ordinate text aligns vertically or horizontally with the leader — unless you override the text orientation with the **Angle** option. In addition, you can override the calculated ordinate with other text through the **Text** and **MText** options.

Once ordinate dimensions are placed in drawings, the DIMCONTINUE and DIMBASELINE commands quickly place additional dimensions. The two commands operate identically, with the DIMBASELINE command not stacking ordinates, as it normally does for linear and angular dimensions.

Notes: You may find it useful to place ordinate dimensions accurately with object snaps, such as CENter to find the centers of holes.

To avoid jogged leader lines, turn on Polar or Ortho mode.

TUTORIAL: PLACING ORDINATE DIMENSIONS

1. From the CD, open the *ordinate.dwg* drawing file.
2. To place x and y ordinate dimensions, start the **DIMORDINATE** command:
   - From the **Dimension** menu, choose **Ordinate**.
   - From the Dimension toolbar, choose the **Dimension Ordinate** button.
   - In the Dashboard's Dimension panel, pick the **Ordinate** button.
   - At the 'Command:' prompt, enter the **dimordinate** command:

     Command: **dimordinate** (*Press* ENTER.)

   - Alternatively, enter the aliases **dor** or **dimord** at the 'Command:' prompt.
3. In all cases, AutoCAD prompts you to select the feature you want dimensioned:
   Specify feature location: (*Pick point 1.*)
4. Move the cursor to locate the ordinate text:
   Specify leader endpoint or [Xdatum/Ydatum/Mtext/Text/Angle]: (*Move the cursor, and then pick point 2.*)

AutoCAD reports the distance it calculated from the datum (origin).

Dimension text = 1.0000

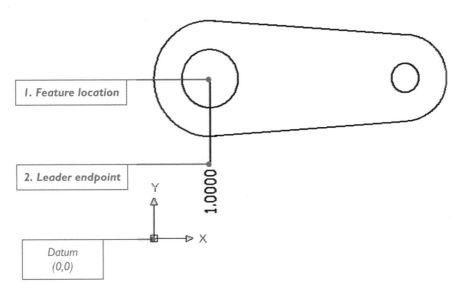

1. Feature location

2. Leader endpoint

1.0000

Y

X

Datum
(0,0)

When you pick an object at the "Specify feature location:" prompt, AutoCAD associates the ordinate dimension with the object. Move the object, and the ordinate dimension updates; move the dimension, and the datum distance also updates automatically.

The dimensions are calculated relative to the *origin of the current UCS*, as described later in this section — not relative to the drawing's origin (0,0). If it seems that ordinates are drawn relative to the drawing's origin, it's because the origin coincides with the base point for the UCS.

The value of the coordinates is always positive, even when you draw ordinates "below" the x or y axis of the UCS. That's because negative distances make no sense in ordinate dimensions.

PLACING ORDINATES: ADDITIONAL METHODS

The **DIMORDINATE** command has the following options:

- **Xdatum** specifies the x ordinate.

- **Ydatum** specifies the y ordinate.

- **Mtext** opens the mtext editor.

- **Text** prompts for text at the command line.

- **Angle** specifies the angle of the text.

Let's look at each option.

Xdatum

The **Xdatum** option forces the ordinate in the x direction. AutoCAD repeats the prompt:

Specify leader endpoint or [Xdatum/Ydatum/Mtext/Text/Angle]: **x**

Specify leader endpoint or [Xdatum/Ydatum/Mtext/Text/Angle]: *(Move the cursor, and then pick a point.)*

Ydatum

The **Ydatum** option forces the ordinate in the y direction in the same way as **Xdatum**.

Mtext

The **Mtext** option displays the mtext editor. You can use the Text Formatting toolbar to apply styles, fonts, and formatting.

In the text edit window, you can add prefix and suffix text to the ordinate, such as the "Approx." and "mm" illustrated below.

To give the ordinate text annotative scaling, click the ▲ (Annotative) icon in the Text Formatting toolbar. (For more on annotation, see earlier chapters on dimensions.)

Click **OK** to exit the mtext editor. AutoCAD repeats the prompt:

> Specify leader endpoint or [Xdatum/Ydatum/Mtext/Text/Angle]: *(Move the cursor, and then pick a point.)*

Text

The **Text** option prompts you to change the ordinate text:

> Enter dimension text <0.7500>: *(Type text, such as **Approx. <> mm**, and then press ENTER.)*

Use the double angle brackets (**<>**) as a placeholder for AutoCAD's calculated ordinate distance. After pressing ENTER, AutoCAD repeats the prompt:

> Specify leader endpoint or [Xdatum/Ydatum/Mtext/Text/Angle]: *(Move the cursor, and then pick a point.)*

Angle

The **Angle** option rotates the ordinate text:

> Specify angle of dimension text: *(Enter an angle, such as **45**.)*

After you press ENTER, AutoCAD repeats the prompt:

> Specify leader endpoint or [Xdatum/Ydatum/Mtext/Text/Angle]: *(Move the cursor, and then pick a point.)*

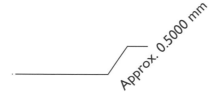

*An ordinate dimension modified by the **Text** and **Angle** options.*

⊢⊢⊣ DIMCONTINUE

The **DIMCONTINUE** command draws additional ordinate dimensions of the same type — helpfully avoiding the pair of prompts displayed by the **DIMORDINATE** command. For example, if you place an x ordinate dimension, **DIMCONTINUE** continues with additional x ordinates.

> Command: **dimcontinue**
>
> Specify feature location or [Undo/Select] <Select>: *(Pick feature.)*
>
> Dimension text = 2.0000
>
> Specify feature location or [Undo/Select] <Select>: *(Pick feature.)*

Dimension text = 3.5000

Specify feature location or [Undo/Select] <Select>: *(Pick feature.)*

Dimension text = 5.0000

Specify feature location or [Undo/Select] <Select>: *(Press ENTER to exit command.)*

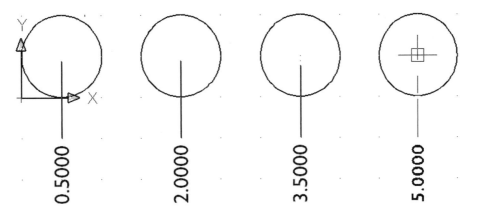

 UCS ORIGIN AND **3POINT**

The **UCS** command defines the base point from which AutoCAD measures ordinate dimensions. Use the command's **Origin** option to move the base point. (This is a shortcut not documented by Autodesk.) You typically move the UCS origin to the lower-left corner of the part, in a procedure called "datum shift." (Ordinate dimensions do not rely on the **BASE** command, which defines the origin of drawings, nor the **SNAPBASE** system variable, which controls the origin of snap, grid markings, and hatch patterns.)

In addition, AutoCAD uses the axes of the UCS to draw the leaders orthogonally. Rotate the axes with the **3point** option, and the ordinate dimensions are drawn at an angle.

To see the effect of the origin on a drawing containing ordinate dimensions, use the **ZOOM Extents** command. Instead of objects filling the screen, as you might expect, they may instead appear in the upper right corner of the drawing. This happens because AutoCAD takes into account the origin of the ordinate dimensions when calculating the extents of the drawing.

TUTORIAL: CHANGING THE ORDINATE BASE POINT

Ensure the UCS icon is turned on for this tutorial.

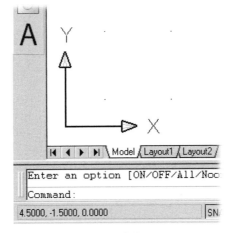

If you do not see the UCS icon, turn it on, as follows:

Command: **uscicon**

Enter an option [ON/OFF/All/Noorigin/ORigin/Properties] <OFF>: **on**

Repeat the command, using the **ORigin** option to force the icon to appear at the origin:

Command: *(Press spacebar.)*

Enter an option [ON/OFF/All/Noorigin/ORigin/Properties] <OFF>: **or**

1. To see the effect of the UCS command on ordinates, first draw an x-ordinate at 2,2:

    Command: **dimordinate**

    Specify feature location: **2,2**

    Specify leader endpoint or [Xdatum/Ydatum/Mtext/Text/Angle]: **x**

    Specify leader endpoint or [Xdatum/Ydatum/Mtext/Text/Angle]: *(Pick a point.)*

    Dimension text = 2.0000

2. Start the **UCS** command, and then specify the **Origin** option:

    Command: **ucs**

    Enter an option [New/Move/orthoGraphic/Prev/Restore/Save/Del/Apply/?/World]

    <World>: **o**

3. Move the UCS origin to 1,1:

    Specify new origin point <0,0,0>: **1,1**

4. Draw another x-ordinate at 2,2 and notice that it appears in a different location.

    Command: **dimord**

    Specify feature location: **2,2**

    Specify leader endpoint or [Xdatum/Ydatum/Mtext/Text/Angle]: **x**

    Specify leader endpoint or [Xdatum/Ydatum/Mtext/Text/Angle]: *(Pick a point.)*

    Dimension text = 2.0000

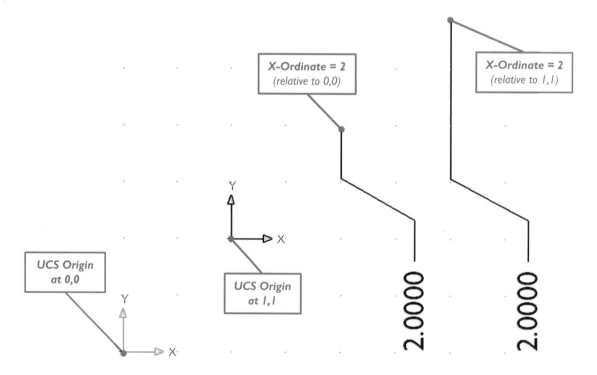

TUTORIAL: ROTATING THE ORDINATE

Ensure the UCS icon is turned on for this tutorial.

1. Start the **UCS** command, and then specify the **3point** option:

    Command: **ucs**

    Enter an option [New/Move/orthoGraphic/Prev/Restore/Save/Del/Apply/?/World]

    <World>: **3point**

2. You can keep the same UCS origin (datum), or specify a new datum:

    Specify new origin point <0,0,0>: *(Press ENTER to keep datum.)*

3. Pick a point to align the x axis at an angle. You can pick a point, or enter coordinates.

    To rotate 45 degrees, for example, enter polar coordinates using the *dist<angle* format.

    Specify point on positive portion of X-axis <1.0,0.0,0.0>: *(Pick a point, or enter polar coordinates, such as* **1.0<45**.*)*

4. Press **ENTER** to keep the y axis perpendicular to the x axis:

    Specify point on positive-Y portion of the UCS XY plane <-0.7,0.7,0.0>: *(Press ENTER.)*

Notice that several things change in the drawing. The UCS icon rotates, but so does the crosshair cursor, as well as the grid dots, snap angle, and the angle for ortho mode (if you have them turned on). This is because the UCS rotates the entire coordinate system.

Coordinates may appear in an unexpected location, as you find out with the final step in this tutorial. (You learn more about UCSs in Volume 2 of this book, *Using AutoCAD 2008: Advanced*, where they are used for three-dimensional design.)

5. Draw another x-ordinate at 2,2 and notice that it appears in a different location.

    Command: **dimord**

    Specify feature location: **2,2**

    Specify leader endpoint or [Xdatum/Ydatum/Mtext/Text/Angle]: **x**

    Specify leader endpoint or [Xdatum/Ydatum/Mtext/Text/Angle]: *(Pick a point.)*

    Dimension text = 2.0000

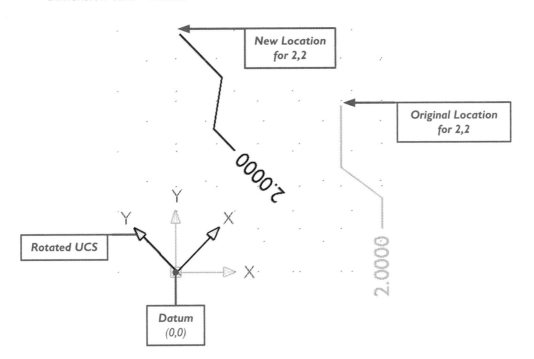

Note: To return the UCS icon back to "normal" — located at the drawing origin and unrotated — use the **UCS World** command:

> Command: **ucs**
>
> Enter an option [New/ ... /?/World] <World>: **w**

 QDIM

The **QDIM** command also places ordinate dimensions.

You may prefer it, because it combines DIMORDINATE, DIMCONTINUE, and UCS Origin into a single command.

ORDINATE DIMENSIONS WITH QDIM

1. Start the **QDIM** command:

   > Command: **qdim**

2. Select the objects to dimension:

   > Associative dimension priority = Endpoint
   >
   > Select geometry to dimension: *(Select one or more objects; press CTRL+A to select all objects.)*
   >
   > Select geometry to dimension: *(Press ENTER to end object selection.)*

3. Position the datum at the lower-left corner of the objects:

   > Specify dimension line position, or
   >
   > [Continuous/Staggered/Baseline/Ordinate/Radius/Diameter/datumPoint/Edit/seTtings] <Continuous>: **p**
   >
   > Select new datum point: *(Pick a point.)*

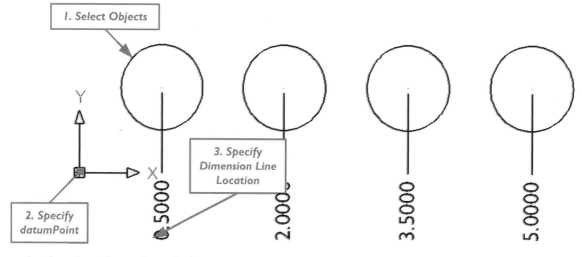

4. Specify ordinate dimensioning:

   > Specify dimension line position, or
   >
   > [Continuous/Staggered/Baseline/Ordinate/Radius/Diameter/datumPoint/Edit/seTtings] <Continuous>: **o**

5. Specify the dimension line location:

   > Specify dimension line position, or
   >
   > [Continuous/Staggered/Baseline/Ordinate/Radius/Diameter/datumPoint/Edit/seTtings] <Continuous>: *(Pick a point.)*

EDITING ORDINATE DIMENSIONS

Once ordinate dimensions are in place, you edit them with grips or with commands that affect dimensions — with some limitations.

Grips Editing

When you select ordinate dimensions, they display several grips: one at either end of the leader line, one on the text, and one at the datum (UCS origin).

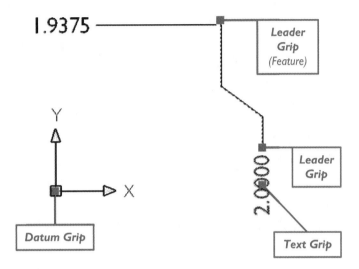

Leader Line

The leader line has grips at the datum, on the text, and at each end of the leader line. When you move the feature-end grip, the other end and the text remain in place; the value of the text updates automatically.

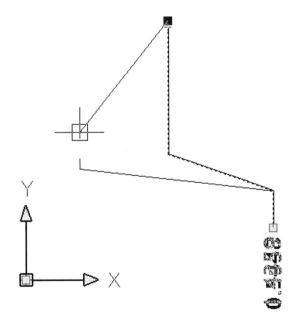

When you move the text-end grip, the text moves with it; the other end remains in place. Unlike regular leaders, ordinate text remains attached to its leader.

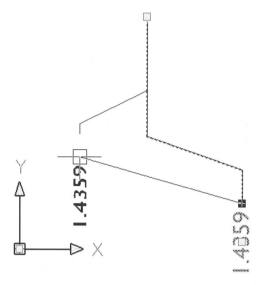

Text

The text grip is located in the center of the ordinate text. It moves the text and leader in exactly the same manner as the leader grip.

Datum

The datum grip is initially located at the UCS origin. When you move it, the datum changes for the selected dimension, but the UCS icon remains in place. The text updates its coordinates.

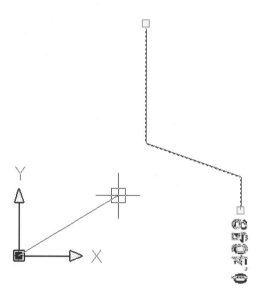

Editing Commands

The **DIMEDIT** command rotates and homes the ordinate dimension text, but does not oblique the leader lines.

The **DIMTEDIT** command relocates the ordinate text, as well as rotates and homes it. The **Left**, **Right**, and **Center** options have no effect on the text.

To edit the properties, double click the ordinate dimension, or enter the **PROPERTIES** command. AutoCAD displays the Properties palette (illustrated at right).

TOLERANCE

The **TOLERANCE** command displays a dialog box for specifying tolerance symbols.

The **TOLERANCE** command creates leaderless frames; use the **QLEADER** command if you need to attach frames to leaders. After placing tolerance frames, you can edit the symbols by double-clicking the frame; AutoCAD displays the Geometric Tolerance dialog box. Geometric tolerances are not associated with geometric objects. The frame can be edited by most editing commands, as well as with grips.

GEOMETRIC DIMENSIONING AND TOLERANCES (GD&T)

Designers use geometric tolerance symbols to show machinists the acceptable deviations from *form*, *profile*, *orientation*, *location*, and *runout* of features. Symbols are used, because they reduce the need for notes in describing complex geometry requirements. The symbols provided with AutoCAD are based on the ASME Y14.5M – 1994 standard.

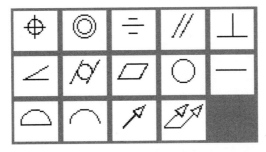

Geometric characteristic symbols.

Left: *Tolerance diameter symbol.*
Right: *Projected tolerance zone symbol.*

Material condition symbols.

Tolerance symbols are defined in the *gdt.shx* shape definition file. Some GD&T symbols are missing from AutoCAD, such as the round datum target.

The symbols are placed in a *feature control frame*. The frame makes it easier to read the symbols, because it separates them into these categories:

1. Geometric characteristics.

2 Tolerance zones and modifiers (MMC, LMC, or RFS).

3. Datum references, and datum reference modifiers (optional).

The parts of a typical frame are shown below:

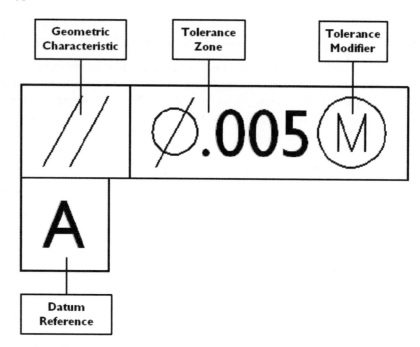

The frame shows that the part **A** needs to be machined to a **parallel** tolerance of **diameter 0.005** inches, **maximum**.

GEOMETRIC CHARACTERISTIC SYMBOLS

The geometric characteristics identify the ideal geometry, such as perpendicular surfaces and cylinders. As with ordinate dimensions, *datums* in tolerances are the origin from which the geometry is established. The *datum target* is a line or area on parts that establishes datums. See www.engineersedge.com/gdt.htm for additional GD&T definitions.

Profile Symbols

⌒ Profile of a Line

Entire length of a feature must lie between two parallel zone lines.

⌓ Profile of a Surface

Entire surface must lie between envelope surfaces separated by the tolerance zone.

Orientation Symbols

∠ Angularity

Surfaces, axes, and center planes must lie between two parallel plates sloped at a specified angle.

⊥ Perpendicularity

Surfaces, axes, median planes, and lines must be exactly 90 degrees to the datum plane or axis.

// Parallelism

All points on the surface or axis must be equidistant from the reference datum.

Location Symbols

⊕ True Position

Center, axis, and center planes must be within the zone of tolerance.

— Symmetry

Features must be symmetrical about the center plane of the datum .

◎ Concentricity

Features must be within the cylindrical tolerance zone.

Runout Symbols

↗ Circular Runout

Circular elements must be within the runout tolerance (full 360-degree rotation about the datum axis).

↗↗ Total Runout

Surface elements across the entire surface must be within runout tolerance.

Form Symbols

—— Straightness

Surface or axis must be a straight line, within the tolerance.

▱ Flatness

Entire 3D surface must be flat, within the tolerance.

◯ Circularity

All points on a surface of revolution — such as cylinders, spheres, and cones — must be equidistant from the axis of the center, within the tolerance.

⌭ Cylindricity

All points on the surface of revolution — such as cylinders — are the same distance from a common axis, within the tolerance.

ADDITIONAL SYMBOLS

Tolerance Symbols

The *tolerance* is the difference between the minium and maximum limits.

No Symbol

No tolerance symbol means the tolerance is linear.

⌀ Tolerance Diameter

This symbol means the tolerance refers to a diameter measurement.

Ⓟ Projected Tolerance Zone

The perpendicularity and mating clearance for holes in which pins, studs, and screws are inserted.

Material Characteristics Symbols

The *material characteristics* indicates the maximum, minimum, or required adherence to the tolerance.

Ⓜ Maximum Material Condition (MMC)

The feature contains the maximum material within the stated limits. Examples include the maximum shaft diameter and the minimum hole diameter.

$\textcircled{L}$ Least Material Condition (LMC)

Feature contains the least amount of material within the stated limits; reverse of MMC. Examples include minimum shaft diameter and maximum hole diameter.

$\textcircled{S}$ Regardless of Feature Size (RFS)

Geometric tolerance or datum reference applies at any increment of size within its tolerance.

SYMBOL USAGE

The following table shows appropriate uses for the symbols:

Geometric Characteristic	Surface	Size	MMC	LMC	P
Form:					
Straightness	☑	☑	☑	☑	...
Flatness	☑	...	...	...	...
Circularity	☑	...	...	...	...
Cylindricity	☑	...	...	...	...
Orientation:					
Perpendicularity	☑	☑	☑	☑	☑
Angularity	☑	☑	☑	☑	☑
Parallelism	☑	☑	☑	☑	☑
Location:					
Positional Tolerance	...	☑	☑	☑	☑
Concentricity	...	☑	...	...	...
Symmetry	...	☑	...	...	...
Runout:					
Circular Runout	☑	☑	...	...	...
Total Runout	☑	☑	...	...	...
Profile:					
Profile of a Line	☑	...	☑	☑	...
Profile of a Surface	☑	...	☑	☑	...

TUTORIAL: PLACING TOLERANCE SYMBOLS

1. To place tolerance frames, start the **TOLERANCE** command:

    • From the **Dimension** bar, choose **Tolerance**.

    • From the Dimension toolbar, choose the **Tolerance** button.

    • In the Dashboard's Dimension panel, pick the **Tolerance** button.

    • At the 'Command:' prompt, enter the **tolerance** command:

      Command: **tolerance** *(Press ENTER.)*

    • Alternatively, enter the **tol** alias at the 'Command:' prompt.

In all cases, AutoCAD displays the Geometric Tolerance dialog box:

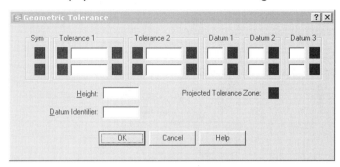

2. Under **Sym**, click the black square. AutoCAD displays the Symbol dialog box.

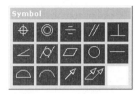

3. Select a Geometric Characteristic symbol. (To select none, click the white square, or press **ESC**.) Notice that the dialog box disappears, and AutoCAD fills in the symbol.

4. If you need the diameter symbol, click the first black square under **Tolerance 1**. To remove the symbol, click the square a second time.

5. Enter the tolerance value in the white text entry rectangle under **Tolerance 1**.

6. To add a material condition symbol, click the second black square under **Tolerance 1**. AutoCAD displays the Material Condition dialog box.

7. Select a material condition symbol. (To select none, click the white square, or press **ESC**.) Notice that the dialog box disappears, and AutoCAD fills in the symbol.

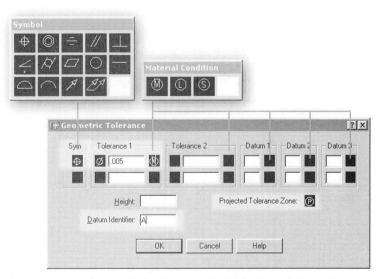

8. If necessary, fill in values and symbols for **Tolerance 2**, **Datum 1**, **2**, and **3**, as well as the height and datum identifier.

9. Click **Projected Tolerance Zone** if you need the **P** symbol.

10. When done, click **OK**.

AutoCAD prompts:

Select tolerance location: *(Pick a point.)*

Pick a point in the drawing to place the tolerance frame.
Using object snap modes can be helpful.

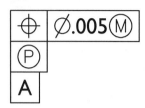

 Note: (New to AutoCAD 2008.) To assign tolerances with annotative scale factor, you can apply an annotative dimension style, or follow these steps:

1. Create the tolerance.
2. Open the Properties palette.
3. Select the tolerance, and then turn on the Annotative property in the palette's Misc section, as illustrated below.

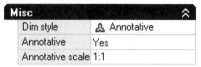

 **QLEADER**

The **QLEADER** command includes an option for placing tolerance marks.

TUTORIAL: PLACING TOLERANCES WITH QLEADER

1. To place tolerance frames at the ends of leaders, start the **QLEADER** command, and then select the **Settings** option:

    Command: **qleader**

    Specify first leader point, or [Settings] <Settings>: **s**

2. In the Settings dialog box, select **Tolerance** in the Annotation tab.

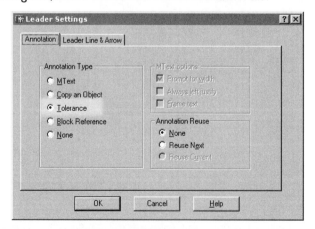

3. Make any other changes you wish in the dialog box, and then click **OK**.

4. AutoCAD prompts you to pick points for the leader's endpoint and vertices:

    Specify first leader point, or [Settings] <Settings>: *(Pick a point.)*

    Specify next point: *(Pick another point.)*

5. Press **ENTER** to end the leader.

   Specify next point: *(Press* **ENTER** *to end leader creation.)*

   AutoCAD displays the Geometric Tolerance dialog box.

6. Enter the tolerance symbols and values, and then click **OK**.

   AutoCAD places the leadered frame in the drawing.

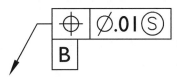

AutoCAD also includes the **LEADER** command, but it is less convenient, because you must specify the tolerances each time you use it:

   Command: **leader**

   Specify leader start point: *(Pick a point.)*

   Specify next point: *(Pick another point.)*

   Specify next point or [Annotation/Format/Undo] <Annotation>: *(Press* **ENTER** *to end leader line.)*

   Enter first line of annotation text or <options>: *(Press* **ENTER** *for options.)*

   Enter an annotation option [Tolerance/Copy/Block/None/Mtext] <Mtext>: **t**

(The **MLEADER** command, new to AutoCAD 2008, does not handle tolerances.)

EXERCISES

1. In the following exercise, you prepare a drawing for ordinate dimensioning.
    From the Companion CD, open the *positioningplate.dwg* file, a drawing of a positioning plate.

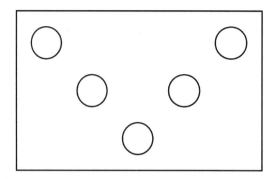

Create a new layer with the following properties:

Name	**Ordinate**
Color	**Blue**
Linetype	**Center**

Make the layer current.

Change the value of **DIMCEN** to -0.09, and then use the **DIMCENTER** command to place center marks at the center of each circle.

Relocate the UCS origin to the lower left corner of the plate.

Dimension each hole with x and y ordinate dimensions. There should be eight dimensions in all.

Save the drawing as *ordinate1.dwg*.

2. From the Companion CD, open the *bar.dwg* file, a drawing of a mounting bar provided as part of a sample drawing included with AutoCAD.

    Dimension each hole with x and y ordinate dimensions.

    How many dimensions did you place?

    Save the drawing.

3. From the Companion CD, open the *brace.dwg* file, a drawing of a locating brace.

    Dimension each hole with x and y ordinate dimensions.

    Save the drawing.

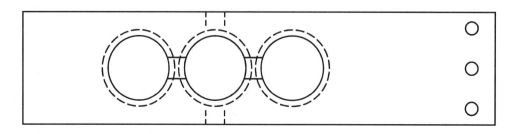

4. From the Companion CD, open the *mount.dwg* file, a drawing of a motor mounting plate. Using the **QDIM** command, dimension the holes with x and y ordinate dimensions. Save the drawing.

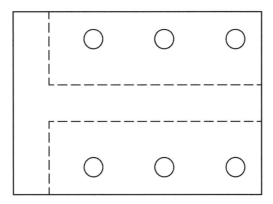

5. In the following exercises, you create tolerance frames. Open a new drawing. Use the **TOLERANCE** command to recreate the following tolerance frames:

    a.

    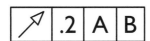

    b.

    c.

    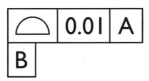

    d.

6. Use the **QLEADER** command to recreate the following tolerance frames:

    a.

    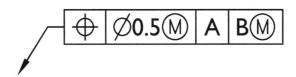

b.

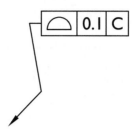

c.

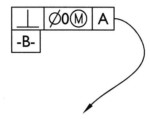

CHAPTER REVIEW

1. On which menu do you find the **TOLERANCE** command?

2. Explain the purpose of the **DIMORDINATE** command.

3. What is another name for **ORDINATE** dimensions?
   For the *origin*?

4. What do ordinate dimensions measure?

5. List two methods for moving the datum:
   a.
   b.

6. Describe how the **DIMCONTINUE** command is useful for placing ordinate dimensions.

7. How many ordinate dimensions are needed for the holes in the following drawing:
   In the x direction?
   In the y direction?
   Sketch the locations of the x and y ordinate leader lines.

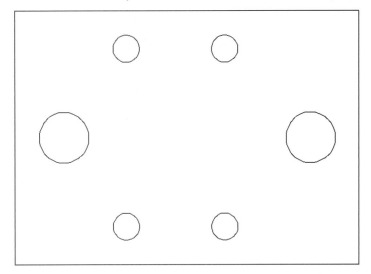

8. Why do you need to move the datum?

9. Two of the following commands relocate the datum for ordinate dimensions.
   Which two are correct?
   SNAPBASE
   BASE
   QDIM datumPoint
   UCS Origin

10. Can the leaders of ordinate dimensions be drawn at an angle?
    If so, how?

11. Can the text of ordinate dimensions be drawn at an angle?
    If so, how?

12. Name the command (and option) that return the datum to "normal."

13. Why is the **QDIM** command a good alternative to the **DIMORDINATE** command?

14. What is *GD&T* short for?

15. Label the parts of the tolerance frame:

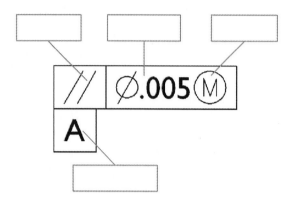

16. Decode the following abbreviations:
    RFS
    LMC
    MMC
    P

CHAPTER 16

Tables and Fields

This chapter shows you how to place tables and field text in drawings, and then change them. Tables present text in orderly rows and columns, much like spreadsheets. Fields automatically update text when conditions change.

The commands are:

TABLE creates the outline structure of tables.

TABLEDIT edits text in table cells.

TABLETOOLBAR toggles display of the table editing toolbar (new in AutoCAD 2008).

TINSERT edits blocks in table cells.

TABLEEXPORT exports table data in CSV format.

TABLESTYLE defines named table styles.

MATCHCELL copies properties from one cell to others.

FIELD inserts updateable text and formulas in drawings.

UPDATEFIELD forces field text to update itself.

NEW TO AUTOCAD 2008 IN THIS CHAPTER

- Tables can be auto filled, created from external data sources (an advanced topic), and split into columns.
- Tables have additional formatting options, including formatting for individual cells.
- The table and Table Style dialog boxes are redesigned.
- The new **TABLETOOLBAR** system variable toggles the display of the table toolbar.
- Table cells can contain more than one piece of content; the order of the content can be changed.

FINDING THE COMMANDS

On the Dashboard's **Tables** panel:

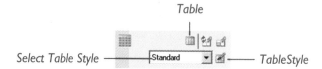

On the **Draw** toolbar:

On the **Styles** toolbar:

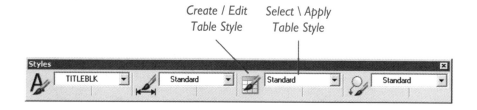

TABLE

The **TABLE** command creates the outline structure of tables, following which you fill cells with text, formulas, and drawings.

Tables consist of *rows* and *columns*. The topmost row is usually reserved for the title of tables, such as "Parts List" in the table illustrated below. AutoCAD calls this the "title row."

Insertion Point

Title Row

Header Row

Data Rows

Parts List				
ITEM	Qty	File Name	Description	Blank
1	1	VW252-02-0203-2	Drive Roller	See Drg VW252-02-203-2
2	2	VW252-02-1000	Brg.Block Slide	See Drg VW252-02-1000
3	2	M10X115LN	AllenBolt M10x115	B.O.
4	2	VW252-02-0304	SS Coil Spring ID-13, Nos of Coil-15, WIre Dia.-3	B.O.
5	1	NYLON_GEAR	Spl Gear	See Drg VW02-210
6	2	M10_NUT	S Hex Nut M10	B.O.

Columns   *Cells*   *Borders*

The second row typically labels the columns, such as "ITEM" and "Qty" (short for "quantity"). AutoCAD calls this the "header row."

The rows and columns define *cells*, just as in spreadsheets. Each cell contains one piece of information, either some text (including numbers and formulas), or the image of a block or drawing. AutoCAD calls these the "data rows."

Tables are not static; you can change them at any time. You can add and remove rows and columns, merge cells to make them longer or taller, format cells to add color or more spacing, and export the table data for use by other software. The title, header, and data rows can be formatted differently from each other, as can individual cells. Tables are changed through grips editing, the Properties palette, and commands found on shortcut menus.

Tables have insertion points, just like text and blocks. The *insertion point* is the anchor from which the table grows and shrinks. The insertion point is typically located at the table's upper-left corner. When you add rows, the table grows downward from its insertion point.

In some drawings, it may make more sense to have the insertion point at the bottom corner, so that the table grows upwards. Below, the title and header rows are at the bottom of the table.

Table Grows Upwards

Insertion Point

6	2	M10_NUT	S Hex Nut M10	B.O.
5	1	NYLON_GEAR	Spl Gear	See Drg VW02-210
4	2	VW252-02-0304	SS Coil Spring ID-13, Nos of Coil-15, WIre Dia.	B.O.
3	2	M10X115LN	Alle *Title and Header Rows*	B.O.
2	2	VW252-02-1000	Brg.Block Slide	See Drg VW252-02-1000
1	1	VW252-02-0203-2	Drive Roller	See Drg VW252-02-203-2
ITEM	Qty	File Name	Description	Blank
Parts List				

BASIC TUTORIAL: PLACING TABLES

1. To place tables in drawings, start the **TABLE** command with one of these methods:
    *   From the **Draw** menu, choose **Table**.
    *   From the Draw toolbar, choose the **Table** button.
    *   In the Dashboard's Tables panel, choose the **Table** button.
    *   At the 'Command:' prompt, enter the **table** command.

        Command: **table** *(Press* ENTER.*)*
    *   Alternatively, enter the **tb** alias at the 'Command:' prompt.

    In all cases, AutoCAD displays the Insert Table dialog box.

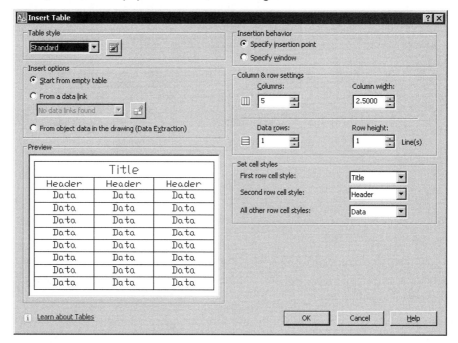

2. Ignore all the options, and click **OK**.
3. AutoCAD prompts you for the insertion point:

    Specify insertion point: *(Pick a point.)*

    AutoCAD displays the Text Formatting bar (identical to that displayed by the **MTEXT** command), and highlights the title row.

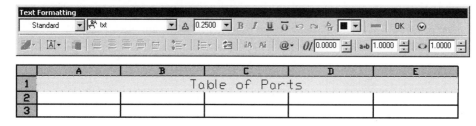

4. Enter a title for the table:

    Table of Parts

5. Press the **TAB** key. Notice that AutoCAD highlights the next cell.
6. Enter a header name for the column:

    Item Number

7. Press the **TAB** key, and the highlight jumps to the next cell. (Press **SHIFT+TAB** to move to the previous cell.)

   You can also use the cursor keys to move to other cells.

   Enter the next header name:

   Quantity *(press TAB.)*

8. AutoCAD auto-fill cells when you drag the cursor across cells in a specific manner (new to AutoCAD 2008). Here's how:

   a. Enter "1" (one) in the cell under Item Number.

   b. Select the cell containing the number "1." (Use a single click.) Notice the cell has five grips:

   > **Square** grips resize the cell.
   >
   > **Diamond** grip performs auto-fill.

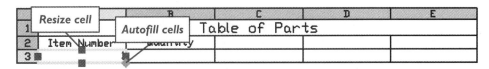

   b. Drag the diamond grip across blank cells. Notice that the tooltip displays the number that will be auto-filled. Numbers are incremented; if you auto-fill with a cell containing a word, the same word is repeated in each cell.

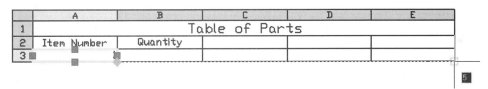

   c. Let go of the mouse button. Notice that the numbers 2 - 5 fill the cells.

	A	B	C	D	E
1	Table of Parts				
2	Item Number	Quantity			
3	1	2	3	4	5

9. When done filling the table's cells, click in a cell to redisplay the Text Formatting bar.

Notes: When you press **TAB** in the last cell, AutoCAD automatically creates another row.

To move around the table, you can also press the four cursor keys on the keypad. Pressing **ENTER** moves the highlight down the table.

To create line breaks (where text is forced to start on the line below), press **ALT+ENTER**. To change the text style of a cell quickly, select a text style from the **Text Style** droplist on the Standard toolbar.

ADVANCED TUTORIALS: INSERTING TABLES

The Insert Table dialog box contains several options for controlling the initial size of the table. The -TABLE command provides similar options at the command line.

Insert Table Dialog Box

The Insert Table dialog box controls the initial look and size of the table.

Location and Size

The Insertion Behavior section of the dialog box provides two basic methods of specifying the overall size of the table:

- **Specify Insertion Point** — you specify where the table must start; AutoCAD determines the size of the table.

- **Specify Window** — you specify the area into which the table must fit; AutoCAD determines how to size the cells to fit.

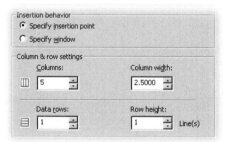

The numbers you specify for the table's rows and column are not critical, because you can always add and remove rows and columns later, as well as change the width and height of cells. Still, the Column & Row Settings section provides two ways changing the size of the cells that make up the rows and columns:

- **Columns** and **Data Rows** — you specify the number columns and rows; AutoCAD determines the size based on column width and row height.

- **Column Width** and **Row Height** — you specify the width of columns and height of rows; AutoCAD determines the number of rows and columns to create.

 Note: The **Row Height** is measured in "lines." A *line* is defined as the text height + top margin + bottom margin. The default value is 0.3 units, which comes from 0.18 (text height) + 0.06 (top margin) + 0.6 (bottom margin).

Table and Cell Styles

When the drawing already has table styles defined, you can select one from the **Table Style Name** droplist. If you want to create or modify a table style, click the  button, which displays the Table Style dialog box. (You can use DesignCenter to borrow table styles from other drawings.)

You can specify style override with the Set Cell Styles options (new to AutoCAD 2008). Although called "cell styles," the styles apply to entire rows: the first row uses the "Title" style by default, the second row the "Header" style, and remaining rows the "Data" style.

These styles can be modified with the **TABLESTYLE** command, as detailed later in this chapter.

 Data Source

New to AutoCAD 2008 is the ability to create tables from data in the drawing or from external data files, such as spreadsheets. This is an advanced topic not covered by this book.

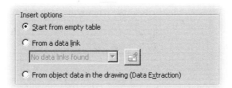

Tables can be inserted from spreadsheet programs: copy the data (using **CTRL+C**), and then paste it (**CTRL+V**) into the AutoCAD drawing. When spreadsheets are copied from Excel XP, they are pasted as table objects. For Excel, use the **PASTESPEC** command, and then select "AutoCAD Entities"; the spreadsheet data are pasted as a table. For other spreadsheets, the data are pasted as mtext.

Preview

The preview image, unfortunately, does not reflect the number of rows, columns, and sizes specified; it reflects only the visual characteristics of the current table style, such as fonts and colors.

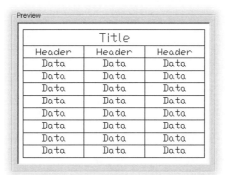

-Table Command

The **-TABLE** command allows you to create the table at the command line; this command is meant for use in scripts and programs.

1. Start the **-TABLE** command. Notice that AutoCAD reports the default values at the command line:

   Command: -table

   Current table style: Standard Cell width: 2.5000 Cell height: 1 line(s)

2. Specify the number of columns and rows.

   Enter number of columns or [Auto/from Style/data Link] <5>: *(Enter a value, or type **A**.)*

   Enter number of rows or [Auto] <1>: *(Enter a value, or type **A**.)*

3. Decide where the table should go:

   Specify insertion point or [Style/Width/Height]: *(Pick a point, or enter an option.)*

   The **Style** option specifies the named table style to use, if any exists in the drawing other than the default style, Standard.

   The **Width** option changes the default width of columns (default = 2.5 units).

   The **Height** option changes the default height of columns (default = 1 line).

Notes: The **Auto** option "automatically" generates the table, based on how you move the cursor; it is similar to the **Specify Window** option of the Insert Table dialog box. After you specify the table's insertion point, AutoCAD ghosts a 3-row by 1-column table, as illustrated below.

As you move the cursor downwards and to the left, AutoCAD adds rows and columns, based on the default cell size of 1 unit by 2.5 units. Remember that tables grow and shrink relative to their insertion points.

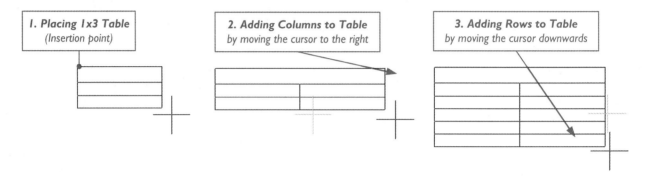

EDITING TABLES

AutoCAD provides many options for editing tables. The methods of editing tables and cells include:

- **TABLEDIT** command edits the text of the selected cell.

- **TABLETOOLBAR** system variable toggles display of the table editing toolbar (new in AutoCAD 2008).

- **TINSERT** command inserts blocks and drawings in cells.

- **PROPERTIES** command changes many properties of tables.

- **TABLEEXPORT** command exports the table data as a comma-delimited text file.

- **Shortcut menus** provide access to most table editing commands.

- **Grips editing** changes the size of tables, rows, and columns.

I find it curious that a command named TABLEDIT edits *cells* (and then just the text in cells) and not the table itself. Another curiosity: the spelling of the TABLEDIT command has one "e," while TABLEEXPORT has two.

Many other table and cell editing and property commands are "hidden" in shortcut menus. For the following tutorials, open the *table.dwg* file from the Companion CD, the drawing of a table.

TablEdit Command

The TABLEDIT command (one "e") edits the text of selected cells. It edits a cell at a time.

1. To edit text in a table cell, start the TABLEDIT command with one of these methods:
   - Double-click the cell you wish to edit.

   - At the 'Command:' prompt, enter the **tabledit** command. AutoCAD prompts you to select a cell:

Command: **tabledit** *(Press* ENTER.*)*

Pick a table cell: *(Select a single cell.)*

- Alternatively, select a cell, and then click the right mouse button. From the shortcut menu, select **Edit Cell Text**.

In all cases, AutoCAD highlights the selected cell in gray, and then displays the Text Formatting bar.

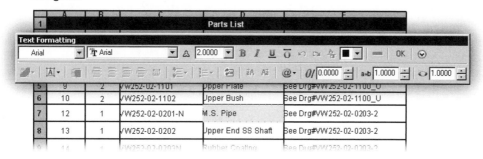

2. Make your editing changes; see the **MTEXT** command in an earlier chapter. You can edit the text, or change properties, for example, make the text boldface, insert a symbol, or change its justification. (These are called "local overrides.")
3. Use the **TAB** key or the cursor keys to edit other cells.
4. When you are finished editing the text, click **OK** or press the **ESC** key.

NEW IN 2008 Table Toolbar System Variable

The **TABLETOOLBAR** system variable toggles the display of the table editing toolbar. This toolbar is not the mtext-like editor you saw above; it is for editing the table itself.

When on (1), this sysvar causes the table editing toolbar to appear (new to AutoCAD 2008); when off, it does not appear, mimicking the behavior of AutoCAD 2007 and earlier.

Command: **tabletoolbar** *(Press* ENTER.*)*

Enter new value for TABLETOOLBAR <1>: *(Type* **1** *or* **0***.)*

The table editing toolbar appears when you select one or more cells, and looks like this:

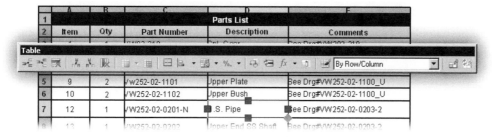

The toolbar's options are very similar to the right-click menu described later in this chapter.

TInsert Command

The **TINSERT** command inserts blocks and drawings in cells (short for "table insert"). Cells can contain blocks or drawings, but just one per cell; text cannot share a cell with them.

1. To insert blocks into cells, start the **TINSERT** command with one of these methods:
   - At the 'Command:' prompt, enter the **tinsert** command. AutoCAD prompts you to select a cell:

   Command: **tinsert** *(Press* ENTER.*)*

   Pick a table cell: *(Select a single cell.)*

- Alternatively, select a cell, and then click the right mouse button. From the shortcut menu, select **Insert Block**.

In both cases, AutoCAD displays the Insert a Block in a Table Cell dialog box (redesigned in AutoCAD 2008).

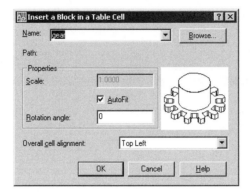

2. Change the dialog box settings, as follows:

Name	**gear**
AutoFit	*(turned on)*
Rotation Angle	**0**
Overall Cell Alignment	**Middle Center**

"Gear" is the name of a block definition provided in the *table.dwg* file.

Browse lets you insert an entire drawing, including *.dwg* and *.dxf* files.

Overall Cell Alignment is like text justification; the **Middle Center** cell alignment centers the block in the cell; you can choose any other alignment.

AutoFit determines the size of the block:

☑ Block scaled automatically to fit the cell.

☐ Cell resized to fit the block; **Scale Factor** option becomes available.

Rotation angle rotates the block; 90 degrees, for example, turns the block on its left side.

3. Click **OK**.

AutoCAD places the block in the cell, replacing the text, if any. When you erase the block, the original text reappears.

Parts List		
mber	**Description**	**Comments**
	Spl. Gear	See Drg#VW202-210
gear	Helical Gear	
01	Upper Plate	See Drg#VW252-02-1100_U
102	Upper Bush	See Drg#VW252-02-1100_U

 Note: To make a cell look as if it holds both text and a block, change the border property so that the line between two cells is invisible.

Description	Comments
Spl. Gear	See Drg#VW202-210
Helical Gear	
Upper Plate	See Drg#VW252-02-1100_U
Upper Bush	See Drg#VW252-02-1100_U

Properties Command

The **PROPERTIES** command changes many (but not all) properties of tables and cells.

The Properties palette has different displays, depending on what you select.

- **Entire table** — the palette displays properties for the entire table.

- **One or more cells** — the palette displays properties appropriate to the cell(s). When a cell contains a block, properties appropriate to blocks are listed, including scale and rotation angle.

Both are illustrated below.

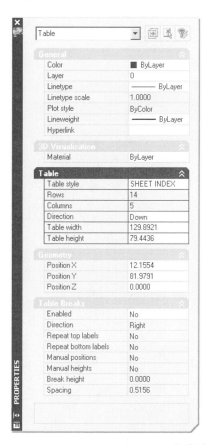

Left: Properties for tables.
Right: Properties for cells.

Properties of tables and cells shown in white can be changed; those in gray cannot. The meaning of cell and table properties are shown below.

Property	Comment
Table properties	
Table Style	Names the table style.
Rows	Indicates the number of rows (read-only).
Columns	Indicates the number of columns (read-only).
Direction	Toggles between Down or Up.
Table Width	Specifies the overall width of table, in current units.
Table Height	Specifies the overall height of table, in current units.
Table Breaks properties	
Enabled	Toggles between Yes and No.
Direction	Specifies right, left, or down.
Repeat Top Labels	Repeats first label rows on each column.

(2008) Repeat Bottom Labels	Repeats last label rows on each column.
(2008) Manual Positions	Allows table columns to be moved freely.
(2008) Manual Heights	Allows independent heights of table columns.
(2008) Break Height	Specifies the height at which tables are broken.
(2008) Spacing	Specifies the distance between table columns.

Cell properties

(2008) Cell Style	Names the cell style.
(2008) Row Style	Names the row style.
(2008) Column Style	Names the column style.
Cell Width	Specifies the width of the cell, in current units.
Cell Height	Specifies the height of the cell, in current units.
Alignment	Specifies the alignment of the text in the cell.
Background Fill	Specifies the color of the cell.
Border Lineweight	Specifies the lineweight of the border lines.
Border Color	Specifies the color of the border lines.
Horizontal Cell Margin	Specifies the distance between text and border.
Vertical Cell Margin	Specifies the distance between text and border.
(2008) Cell Locking	Locks content, format, or both.
(2008) Cell Data Link	Specifies name of the link source.

Content properties

Cell Type	Indicates Text or Block (read-only).
Contents	Specifies the text found in the cell.
Text Style	Names the text style.
Text Height	Specifies the eight of the text, in current units.
Text Rotation	Specifies the rotation angle of the text.
Text Color	Specifies the color of the text.
Data Type	Specifies the type of data, such as angle, date, etc.
Format	Specifies the cell formatting, if not General.
Precision	Indicates the display precision (read-only).
Additional Format	Opens the Table Cell Format dialog box.

Cell Contains a Block

Block Name	Names the block or drawing.
Block Scale	Specifies the scale of the block (cannot be changed when the block inserted using AutoScale).
Block AutoFit	Toggles between Yes or No.
Block Rotation	Specifies the rotation angle of the block.
Block Color	Specifies the color of the block.

Changes made with the Properties palette are known as "local overrides." You can remove them by right-clicking the table, and then selecting **Remove All Property Overrides**.

Grips Editing

Grips editing changes the size of the table, rows, and columns — depending on where you click, and where you grip. Once the cell(s) or table is selected, blue grips appear. You can use the grips to change the size of the table, the rows, and the columns — again, depending on which grips you drag.

When you select:

- **Table border** — entire table is selected. When changing the width of cells, hold down **CTRL** to make the table width change in synchronicity.

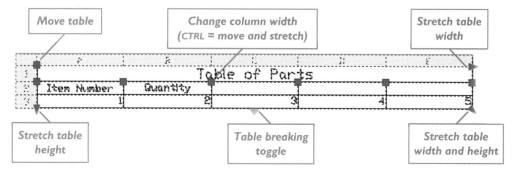

- **Inside a cell** — the cell is selected. When changing the cell's height and width, the entire table also changes its size to match.

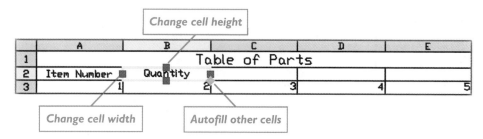

- **Across several cells** — the cells are selected.

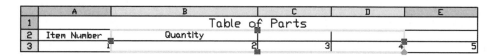

NEW IN 2008 Table Toolbar and Shortcut Menus

The Table toolbar provides access to almost all table editing commands. Indeed, this is where you find commands you can't find elsewhere.

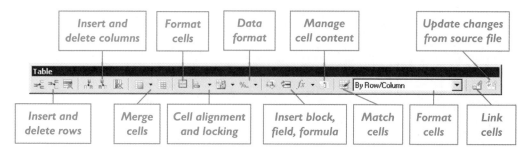

To display the toolbar, select a cell by clicking in the center of the cell. Whether or not the toolbar appears depends on the setting of the **TABLETOOLBAR** system variable, as described earlier.

In addition, shortcut menus provide access to table editing commands. The difference is that the toolbar shows icons, while the shortcut menus use words. Use whichever you prefer. Note that content of the menus varies according to the portion of the table selected.

Cell(s) Selected for Editing

When one or more cells are selected, the shortcut menu has these commands specific to manipulating cells and their contents. (The shortcut menu is substantially changed in AutoCAD 2008.)

Option	Comment
Cell Style	Selects a cell style name.
Alignment	Specifies the alignment of the cell's content relative to its borders: ▶ Top Left ▶ Top Center ▶ Top Right ▶ Middle Left ▶ Middle Center ▶ Middle Right ▶ Bottom Right ▶ Bottom Center ▶ Bottom Left
Borders	Displays the Cell Border Properties dialog box.
Locking	Locks and unlocks the cell's content, format, or both.
Data Format	Displays the Table Cell Format dialog box.
Match Cell	Copies properties from this cell, and applies them to other cells; see the MATCHCELL command.
Remove All Property Overrides	Returns all cells to the properties specified by the table style.
Data Link	Displays the Select Data Link dialog box.
Insert	▶ **Block** — inserts a block or drawing in the cell; see TINSERT. ▶ **Field**— inserts and edits field text; see the FIELD command.
Edit Text	Edits text with the Text Formatting bar; see the TABLEDIT command.
Manage Content	Displays the Manage Cell Content dialog box.
Delete Content	Deletes specific content of the cell.

 Delete All Content	Deletes all content from the cell.	
Columns	▸ **Insert Left** — inserted to the left of the selected cell.	
	▸ **Insert Right** — inserted to the right of the selected cell.	
	▸ **Delete** — deletes the column in which the selected cell resides.	
	▸ **Size Equally** — makes columns of selected cells the same width.	
Rows	▸ **Insert Above** — inserted above the selected cell.	
	▸ **Insert Below** — inserted below selected cell.	
	▸ **Delete** — deletes the row in which the selected cell resides.	
	▸ **Size Equally** — makes rows of selected cells the same height.	
Merge Cells	Merges two or more cells; only the content of the first cell is kept:	
	▸ **All** — merges all selected cells into a single cell.	
	▸ **By Row** — merges rows of cells into a single column.	
	▸ **By Column** — merges columns of cells into a single row.	
Unmerge Cells	Splits previously merged cells (available only for cells which were previously merged).	

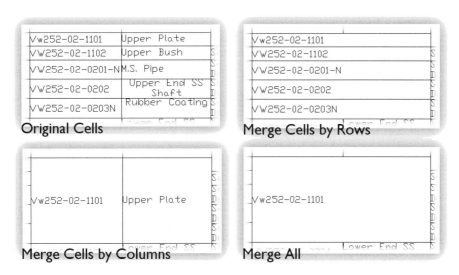

Original Cells   *Merge Cells by Rows*

Merge Cells by Columns   *Merge All*

Warning! When you merge two or more cells with **Merge Cells**, only the text or block of one cell is kept; the content of other cells is erased. AutoCAD warns you, "Only the content of the first cell will be retained when cells are merged." If you mistakenly merge cells, you can use the **U** command to return the cells and their content.

How does AutoCAD define the "first" cell? It is the upper-left of a group of selected cells.

The **Unmerge Cells** option separates merged cells; however, the content of the erased cells is not returned.

Entire Table Selected for Editing

Select the entire table (click on a border line), and then right-click. The shortcut menu has these commands specific to tables:

Command	Comment
Table Style	▸ **Save as New Table Style** — saves table's style by name.
	▸ **Set as Table in Current Table Style** — sets this table's style as the current table style.
	▸ *Style Name* — selects table style name.
Size Columns Equally	Makes columns of the selected cells the same width.
Size Rows Equally	Makes rows of the selected cells the same height.
Remove All Property Overrides	Returns all cells to the properties specified by the table style.
Export	Exports table in CSV format; see the TABLEEXPORT command.
Table Indicator Color	Selects the color for the row and column indicators; stored in the TABLEINDICATOR system variable.
Update Table Data Links	Forces attributes and other data styles to update.
Write Data Links to External Source	Updates table changes to the linked external data file.

 Managing Cell Content

In the past, AutoCAD restricted cells to one piece of content, either text or a block. As of AutoCAD 2008, cells can contain different types of content. A new dialog box controls the display order.

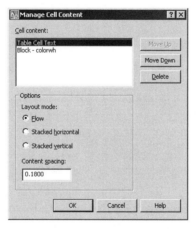

(There appears to be no command for accessing this dialog box. When a cell contains two or more objects, right-click, and then choose **Manage Content** from the shortcut menu.)

The Cell Content list names the contents of the selected cell. Use the **Move Up** and **Move Down** buttons to change the display order. Click **Delete** to erase the selected item.

Flow — determines where to place the cell contents according to the width of the cell; some might be next to each other; others might be stacked on top of each other.

Stacked Horizontal — lays out cell content horizontally.

Stacked Vertical — stacks cell content vertically.

Content Spacing — specifies the width between cell content items.

TableExport Command

The **TABLEEXPORT** command (two "e"s) exports the table data as CSV-format files (short for "comma-separated values").

This means the table data are saved as an ASCII text file, with data separated by commas, one line of data per table row. Here is an example:

"VW252 - Washing Unit Drawing Sheets"
"Sheet Number","Sheet Title"
"00","Cover Sheet"
"01","VW252-Washing-Unit"
"03","Drive Roller Sub Assy"

The *.csv* files can be imported into just about any spreadsheet and database program, as well as some word processors and other software.

AutoCAD exports the table data literally. If the table's direction is up, you may want to use the **PROPERTIES** command to change the table's Direction property to **Down** before exporting the table — otherwise the title and header rows appear at the end of the *.csv* file for tables with an Up direction.

1. To export tables from drawings, start the **TABLEEXPORT** command with one of these methods:
   - At the 'Command:' prompt, enter the **TABLEEXPORT** command. AutoCAD prompts you to select a table:

     Command: **tableexport** *(Press* ENTER.*)*

     Select a table: *(Select a single table.)*

   - Alternatively, select a table, and then click the right mouse button. From the shortcut menu, select **Export**.

   In either case, AutoCAD displays the Export Data dialog box.

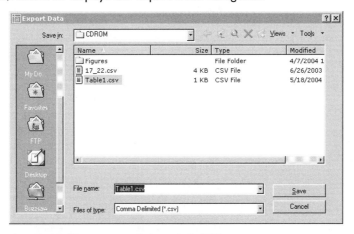

2. Specify the path, enter a file name, and then click **OK**.

   AutoCAD exports the data. You can open the file in a spreadsheet or database program.

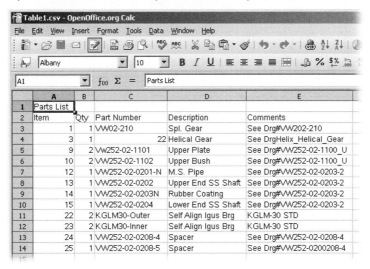

TABLESTYLE

The **TABLESTYLE** command defines named table styles, allowing you to change the look of tables quickly, for example giving them thicker lines, color, and other fonts.

Tables have styles similar to text and dimensions. I find the default style included with AutoCAD, called "Standard," quite ugly; you'll want to change it right away with this command — or use DesignCenter to borrow table styles from other drawings.

As of AutoCAD 2008, table styles include *cell styles*, where you define sub-styles for individual cells. AutoCAD includes predefined cell styles named Data, Header, and Title. You can edit these cell styles, as well as create new ones.

This command allows you to change the properties of cells, header rows, and title rows independently of each other. In addition, you can specify the color and thickness of lines, color of text and color behind cells, margin widths, justification within cells, and more. (The dialog box is substantially changed in AutoCAD 2008.)

BASIC TUTORIAL: CREATING TABLE STYLES

1. From the Companion CD, open the *tablestyle.dwg* file, a drawing of a table with the Standard style.

2. To modify the table style, start the **TABLESTYLE** command with one of these methods:

   • From the **Format** menu, choose **Table Style**.

   • From the Styles toolbar, choose the **Table Styles Manager** button.

   • In the Dashboard's Tables panel, choose the **Table Style** button.

   • At the 'Command:' prompt, enter the **tablestyle** command.

     Command: **tablestyle** *(Press* ENTER.*)*

   • Alternatively, enter the **ts** alias at the 'Command:' prompt.

   In all cases, AutoCAD displays the Table Style dialog box.

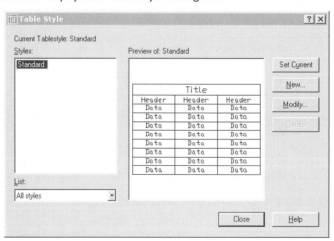

3. To improve the look of the Standard style, click **Modify**. AutoCAD displays the Modify Table Style dialog box.

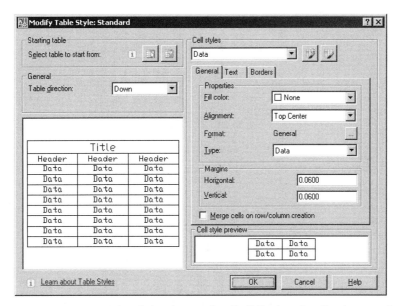

The most important item in this dialog box is the **Cell Styles** droplist, because it displays the names of cell styles — the styles that affect the looks of cells.

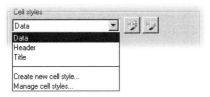

Data — generic style for data cells, all cells that aren't headers or titles.

Header — cell style for header cells, usually in the table's second row.

Title — cell style for title cells, usually the table's first row.

Create New Cell Style 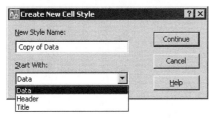 — displays a dialog box for naming new cell styles. Note that new styles are based on existing ones. Clicking **Continue** returns you to the Table Style dialog box.

Manage Cell Styles 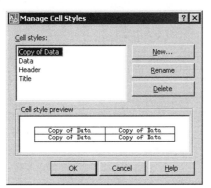 — displays a dialog box for renaming and deleting style names.

4. Let's change the Standard style to make tables more attractive. With the **Data** cell style, make these changes:

   a. Choose the Text tab.

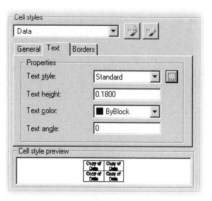

   b. Click the **...** button next to Text Style.
   c. In the Text Style dialog box, select "Arial" from the **Font Name** list.
      (The Upside Down and Backwards options do not work for text in tables.)
   d. Click **Apply**, and then **Close**. Notice that the fonts change in the Cell Style Preview window.

5. From the Cell Styles droplist, choose "Header."

   a. Choose the **General** tab.
   b. From the **Fill Color** droplist, select "Yellow." Note the changes in the preview window.

6. From the **Cell Styles** droplist, choose "Title." Make these changes to the style:

   a. Change the text color to white. (From the **Text Color** droplist, select "Select Color." In the Select Color dialog box, select the **True Color** tab. Change **Luminance** to "100", and then click **OK**. This works only when the AutoCAD screen color is white.)
   b. Change the **Fill Color** to "Black." Note the changes to the preview window.

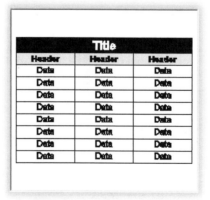

7. Click **OK**, and then **Close** to exit the dialog boxes.
   You have modified the Standard table style. To see what it looks like, use the TABLE command to create a table, and then fill its cells with text.

Note: Changing a table style definition retroactively changes existing tables made with the same style.

ADDITIONAL STYLE STUFF

- **Format** option applies formats to dates, coordinates, and other data.

- **MATCHCELL** command matches cell properties.

- **MATCHPROP** command matches table properties.

- Multiple sections of tables (new to AutoCAD 2008).

Format

The Table Cell Format dialog box formats decimal numbers, text, points (x,y,z coordinates), dates, and angles. (To display it, click the **...** button next to **Format** in the General tab of the Table Style dialog box.)

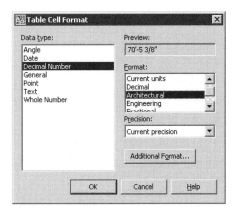

From the dialog box, select (1) a data type, (2) a format, and (3) a precision, if available. There is no formatting for General and Whole Numbers.

Data Type	Available Formats
Angles	None, Current units, Decimal degrees, Deg/min/sec, Grads, Radians, Surveyor's units.
Dates	Days, months, years, time.
Decimal Numbers	None, Current units, Decimal, Architectural, Engineering, Fractional, Scientific.
General	*None.*
Points	None, Current units, Decimal, Architectural, Engineering, Fractional, Scientific.
Text	None, Uppercase, Lowercase, First capital, Title case.
Whole Numbers	*None.*

Click **Additional Format** for more options:

Current Value — displays the value in base drawing units, like a "before" image.

Preview — shows what the value will look like with the changes made in this dialog box.

Conversion Factor — specifies a conversion factor. For example, if you enter 2.54, the current value is multiplied by 2.54, which is handy for inches-to-centimeter conversion. Leave the value set to 1 for no conversion.

Prefix — specifies text to be displayed in front of the value.

Suffix — specifies text to be displayed after the value.

Decimal — selects the separator for the decimal: period (North American), comma (European), or space.

Thousands — selects the separator between thousands: none, comma (North American), decimal (European), or single quote.

Zero Suppression — suppresses leading, trailing, feet, and inch zeros.

MatchCell

The **MATCHCELL** command matches the properties of cells. You select one cell as the template, and then apply its properties to other cells. You typically use this command when you want to override the table style. All properties are copied, except the value of cells (text, blocks, and formulas).

1. To match cell properties, start the **MATCHCELL** command. At the 'Command:' prompt, enter the **MATCHCELL** command.

    Command: **matchcell** *(Press ENTER.)*

2. AutoCAD prompts you to select a cell:

    Select source cell: *(Select a single cell.)*

3. Now start picking — one at a time — the cells that will take on the copied properties:

    Select destination cell: *(Pick a cell.)*

    Select destination cell: *(Pick another cell.)*

4. Press **ESC** to exit the command:

    Select destination cell: *(Press ESC.)*

Change your mind? Use the U command to reverse the changes, or else right-click the table, and then select **Remove All Property Overrides**.

MatchProp

The **MATCHPROP** command matches the properties of the entire table. You select one table as the template, and then apply its properties to other tables. This command even works between two drawings open at the same time.

See the **MATCHPROP** command in Chapter 7, "Changing Object Properties."

 Multiple Sections

Tables can be split into multiple sections. This is useful for tables that are too long to fit the sheet.

You split tables into sections by using two grips found on tables: one grip toggles sectioning; the other stretches the table to other sections. It works like this:

1. Open the *table. dwg* drawing from the Companion CD.
2. Select the table (click on one of the outer boundary lines).

12	23	2	OLM 30-Inner	Self Aligni gus Brg	K9LM-30 STD
13	24	1	VW252-02-0208-4	Spacer	See Drg#VW252-02-0208-4
14	25	1	VW252-02-0208-5	Spacer	See Drg#VW252-0200203-4

Table breaking inactive.
Click and drag to set break height.

3. Notice the pale blue triangular grips. Its purpose is to toggle sectioning.

    Pause the cursor over the grip to view a tooltip that briefly explains the grip's purpose.

4. Drag the grip upwards. As you do, notice that AutoCAD ghosts the size of the split table.

Parts List				
Item	Qty	Part Number	Description	Comments
1	1	VW02-210	Spl. Gear	See Drg#VW202-210
3	1	Helix_Helical gear	Helical Gear	See DrgHelix_Helical_Gear
9	2	Vw252-02-1101	Upper Plate	See Drg#VW252-02-1100_U
10	2	VW252-02-1102	Upper Bush	See Drg#VW252-02-1100_U
12	1	VW252-02-0201-N	M.S. Pipe	See Drg#VW252-02-0203-2
13	1	VW252-02-0202	Upper End SS Shaft	See Drg#VW252-02-0203-2
14	1	VW252-02-0203N	Rubber Coating	See Drg#VW252-02-0203-2
15	1	VW252-02-0204	Lower End SS Shaft	See Drg#VW252-02-0203-2
22	2	KGLM30-Outer	Self Align Igus Brg	KGLM-30 STD
23	2	KGLM30-Inner	Self Align Igus Brg	KGLM-30 STD
24	1	VW252-02-0208-4	Spacer	See Drg#VW252-02-0208-4
25	1	VW252-02-0208-5	Spacer	See Drg#VW252-0200208-4

5. Click the mouse button. Notice that the table shortens by splitting into two parts. Press **ESC** to clear the selection.

Parts List				
Item	Qty	Part Number	Description	Comments
1	1	VW02-210	Spl. Gear	See Drg#VW202-210
3	1	Helix_Helical gear	Helical Gear	See DrgHelix_Helical_Gear
9	2	Vw252-02-1101	Upper Plate	See Drg#VW252-02-1100_U
10	2	VW252-02-1102	Upper Bush	See Drg#VW252-02-1100_U
12	1	VW252-02-0201-N	M.S. Pipe	See Drg#VW252-02-0203-2
13	1	VW252-02-0202	Upper End SS Shaft	See Drg#VW252-02-0203-2
14	1	VW252-02-0203N	Rubber Coating	See Drg#VW252-02-0203-2

15	1	VW252-02-0204	Lower End SS Shaft	See Drg#VW252-02-0203-2
22	2	KGLM 30-Outer	Self Align Igus Brg	KGLM-30 STD
23	2	KGLM 30-Inner	Self Align Igus Brg	KGLM-30 STD
24	1	VW252-02-0208-4	Spacer	See Drg#VW252-02-0208-4
25	1	VW252-02-0208-5	Spacer	See Drg#VW252-0200208-4

Modifying Split Tables

The table is split in two, but doesn't look clear, because there is no header on the second part, and the two halves are stubbornly stuck together. To fix these, open the Properties palette, and then select the table. Look for the section named Table Breaks.

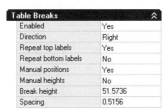

Table Breaks	⌃
Enabled	Yes
Direction	Right
Repeat top labels	Yes
Repeat bottom labels	No
Manual positions	Yes
Manual heights	No
Break height	51.5736
Spacing	0.5156

Change the following properties:

Repeat Top Labels — change to **Yes**. This repeats the title and header rows on the additional sections.

Manual Positions — change to **Yes**. This lets you separate the two tables.

Spacing — change to **1** or other value. This ensures all sections are separated by the same amount, for a clean look.

With these changes made, the split table looks like this:

Parts List				
Item	Qty	Part Number	Description	Comments
1	1	VW02-210	Spl. Gear	See Drg#VW202-210
3	1	Helix_Helical gear	Helical Gear	See DrgHelix_Helical_Gear
9	2	Vw252-02-1101	Upper Plate	See Drg#VW252-02-1100_U
10	2	VW252-02-1102	Upper Bush	See Drg#VW252-02-1100_U
12	1	VW252-02-0201-N	M.S. Pipe	See Drg#VW252-02-0203-2
13	1	VW252-02-0202	Upper End SS Shaft	See Drg#VW252-02-0203-2
14	1	VW252-02-0203N	Rubber Coating	See Drg#VW252-02-0203-2

Parts List				
Item	Qty	Part Number	Description	Comments
15	1	VW252-02-0204	Lower End SS Shaft	See Drg#VW252-02-0203-2
22	2	KGLM30-Outer	Self Align Igus Brg	KGLM-30 STD
23	2	KGLM30-Inner	Self Align Igus Brg	KGLM-30 STD
24	1	VW252-02-0208-4	Spacer	See Drg#VW252-02-0208-4
25	1	VW252-02-0208-5	Spacer	See Drg#VW252-0200208-4

FIELD

The **FIELD** command places "automatic text" in drawings.

Field text can be dates and time, the properties of objects, or formulas. The text is updated automatically under certain conditions, such as when the drawing is opened or plotted, or manually with the **UPDATEFIELD** command. Automatic updating is controlled with the **FIELDEVAL** system variable. Field text has an entire syntax of its own, which AutoCAD handles for you, fortunately.

Fields are useful for labeling drawings, creating tables of content, and numbering legends. Field text is placed just like other text with the mtext editor. It can appear by itself in drawings, or in tables, dimensions, attributes, and so on.

Field text is edited with the **DDEDIT** command. The **FIND** command searches for field text. Field text can be readily identified in drawings, because it has a gray background, as illustrated below. Turn off the gray background with the **FIELDDISPLAY** system variable.

VW252 - Washing Unit Drawing Sheets	
Sheet Number	Sheet Title
00	Cover Sheet
01	VW252-Washing-Unit

BASIC TUTORIAL: PLACING FIELD TEXT

In this tutorial, you practice using field text by placing a drawing's path and file name in the drawing as fields.

1. From the Companion CD, open *field.dwg* file, the drawing of a title block.
2. To place field text, start the **FIELD** command with one of these methods:
   - From the **Insert** menu, choose **Field**.
   - At the 'Command:' prompt, enter the **field** command.

     Command: **field** *(Press ENTER.)*

   In all cases, AutoCAD displays the Field dialog box. You create field text by selecting one item from column A, another from column B, and so on.

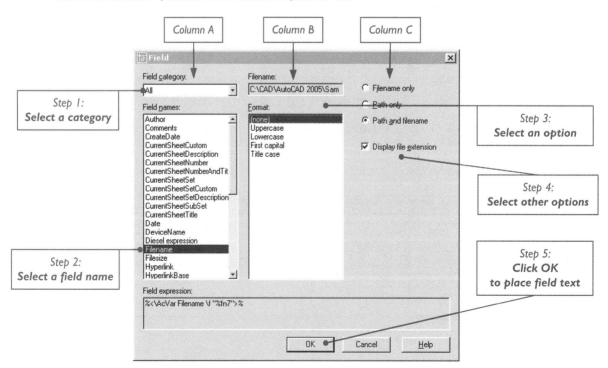

3. <u>Step 1</u>: Select a category from the **Field Category** droplist.

    For this tutorial, select "Document."

Categories are useful when you are not sure of the field name, and want to narrow down the choices. The default category, "All," can be confusing because of the long list of field names. Categories create the following groups:

Category	Comment
All	Lists the names of all fields.
Date and Time	Reports the date and/or time drawing was created, plotted, or saved.
Objects	Reports information about any object in the drawing.
Plot	Provides statistics of the most-recent plot.
Sheetsets	Reports data about sheets, sheet sets, and views.
Document	Reports data stored in the Drawing Properties dialog box.
Linked	Specifies hyperlink addresses.
Other	Evaluates system variables and Diesel expressions.

4. <u>Step 2</u>: Select a field name. The complete list of field names is provided on the facing page. For this tutorial, select **Filename**.

    After you select the field name, columns B and C change to show all available options and suboptions. For the Filename field, the options involve formatting. For other fields, the options vary widely, as illustrated below.

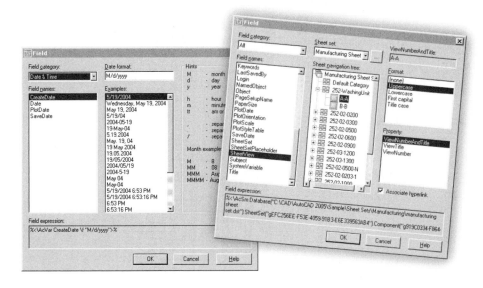

5. <u>Step 3</u>: Select an option from column "B." The Filename's formatting options are:

    (none) — does not format the text.

    Uppercase — changes all characters to UPPERCASE.

    Lowercase — changes all characters to lowercase.

    First capital — sets only the first character of the entire filename to uppercase.

    Title case — sets the first character of every word to uppercase.

    As you select an option, notice that its effect is previewed under **Filename** (top center of the dialog box). For this tutorial, select **Title case**, which capitalizes the first letter of every word.

Note: The complete list of field categories and names is shown below.

Category	Field Names	Comments
Date & Time	Create Date	Date drawing created.
	Date	Current date and time.
	Plot Date	Date last plotted.
	Save Date	Date last saved.
Document	Author	*These items are specified in*
	Comments	*the Summary tab of the*
	Filename	*Drawing Properties dialog*
	Filesize	*box, which is accessed*
	HyperlinkBase	*by the* DWGPROP *command.*
	Keywords	
	LastSavedBy	
	Subject	
	Title	
Linked	Hyperlink	Hyperlink and alternate text.
Objects	NamedObject	Block, dimstyle, layer, linetype, tablestyle, textstyle, and view.
	Object	Object selected from drawing.
	Formula	Average, Sum, Count, Cell, and Formula; for use in tables only.
Other	Diesel Expression	Macros written in Diesel.
	LISP Variable	AutoLISP and VLISP variables.
	System Variable	Any of 400+ system variables.
Plot	DeviceName	Name of the plotter.
	Login	Windows login name of the user.
	PageSetupName	Name of the page setup.
	PaperSize	Size of the media.
	PlotDate	Date the drawing was plotted.
	PlotOrientation	Landscape or Portrait.
	PlotScale	Plot scale.
	PlotStyleTable	Name of the plot style table.
Sheetset	SheetSet	Information about sheet sets.
	SheetSetPlaceholder	
	SheetView	
	CurrentSheetSetSubset	
	CurrentSheetNumber	
	CurrentSheetSet	
	CurrentSheetNumberandTitle	
	CurrentSheetSubset	
	CurrentSheetTitle	
	CurrentSheetCategory	
	CurrentSheetIssuePurpose	
	CurrentSheetRevisionDate	

Category	Field Names	Comments
Sheetsets (*continued*)	CurrentSheetRevisionNumber	Information about sheet sets.
	CurrentSheetCategory	
	CurrentSheetCustom	
	CurrentSheetDescription	
	Current SheetSetCustom	
	CurrentSheetSetDescription	
	CurrentSheetProjectMilestone	
	CurrentSheetSetProjectNumber	
	CurrentSheetProjectName	
	CurrentSheetSetProjectPhase	

The following fields were removed from AutoCAD 2007: CurrentSheetProjectMilestone, CurrentSheetProjectName, CurrentSheetProjectNumber, and CurrentSheetProjectPhase.

6. <u>Step 4</u>: Select a sub-option from column "C." For this tutorial, select **Filename only**.

   The Filename field's formatting suboptions are:

   Filename only — displays the file name and (optionally) extension, such as *filename.ext*.

   Path only — displays only the path to the file name, such as *c:\folder*.

   Path and filename — displays full path, such as *c:\folder\filename.ext*.

   First capital — sets only the first character of the entire filename to uppercase.

   Display the extension — toggles display of the extension.

   As you select options, notice that AutoCAD is generating the field text codes at the bottom of the dialog box, looking like this:

   %<\AcVar Filename \f "%tc4%fn6">%

7. Click **OK**.

8. AutoCAD prompts you to pick the point where the text should be located.

   MTEXT Current text style: "Standard" Text height: 0.0500

   Specify start point or [Height/Justify]: *(Pick a point, or enter an option.)*

   If you wish, enter **H** to change the height and **J** to change the justification (alignment). Notice that AutoCAD places the filename as field text against its distinctive gray background.

	Field Text Tutorial			
	SIZE	FSCM NO.	DWG NO. Field.dwg	REV
	SCALE		SHEET	

Notes: The complete list of date and time codes is shown below. This list is more complete than that provided by Autodesk.

Letters not used for codes are treated literally, such as **g** and **N**. You can use any characters as separators, such as **/ - +** and so on.

Notice that the number of characters often determines how the date or time is displayed. One or two **M**s display the month as a number, while three or four display it as the word. One or two **d**s display the date of the month, while three or four display the day of the week.

Some codes are case-sensitive. For example, uppercase **M** means "month," while lowercase **m** means "minute."

The System Time refers to the format of the date and time as specified by Windows.

Format	Comment	Example
Months (*must be uppercase M*)		
M	Number of month.	February = 2
MM	Number with zero prefix.	02
MMM	Three-letter abbreviation.	Feb
MMMM	Full month name.	February
Dates and Days		
d	Date of the month.	5
dd	Date, with zero prefix.	05
ddd	Abbreviated day of the week.	Thu
dddd	Full day name.	Thursday
Years (*must be lowercase y*)		
y	Single digit year.	2004 = 4
yy	Two-digit year.	04
yyy *or* yyyy	Four-digit year.	2004
Hours		
h	12-hour clock.	4
hh	Hour with zero prefix.	04
t	Single-character AM or PM.	P
tt	Placeholder for AM or PM.	PM
H	24-hour clock.	16
HH	24-hour with zero prefix.	07
Minutes (*must be lowercase m*)		
m	Minutes.	9
mm	Minutes with zero prefix.	09
Seconds		
s	Seconds.	7
ss	Seconds with zero prefix.	07
System Time (*case sensitive*)		
%c	Date and time in short format.	5/20/04 3:17:05 PM
%#c	Date and time.	Thursday, May 20, 2004 3:17:05 PM
%X	Time.	3:17:05 PM
%x	Date in short format.	5/20/04
%#x	Date in long format.	Thursday, May 20, 2004

ADVANCED TUTORIAL: ADDING FIELDS TO MTEXT

In this tutorial, you combine text and field text with the MTEXT command. The text is "Last revised," and the field text is the date-and-time of the last drawing save:

> Date Revised:
>
> May 20/08 3:48 PM

Continue with the *field.dwg* drawing from the previous tutorial.

1. Start the MTEXT command (**Draw | Text | Multiline Text**).

   Command: **mtext** *(Press ENTER.)*

   Current text style: "Standard" Text height: 0.2000 Annotative = No

   Specify first corner: *(Pick point 1.)*

   Specify opposite corner or [Height/Justify/Line spacing/Rotation/Style/Width/Columns]: *(Pick point 2.)*

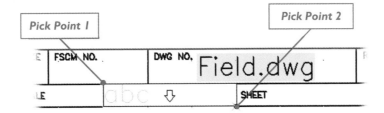

2. From the Text Formatting bar, select the **Roman** style.

3. Enter the following text:

   DATE REVISED: *(Press ENTER to move cursor to the next line.)*

4. To add the field text, use one of these methods:

   • Click **Field**.

   • Press **CTRL+F**.

   • Right-click, and then from the shortcut menu select **Insert Field**.

   In all cases, AutoCAD displays the Field dialog box.

5. In the dialog box, select these options to construct the formatted field text:\

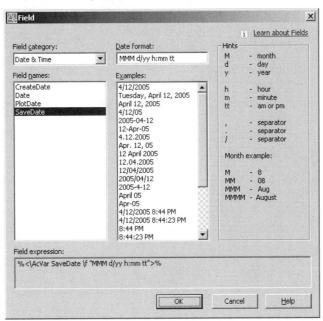

Category: **Date & Time**

Field names: **SaveDate**

Date format: **MMM d/yy h:mm:tt**

The date format means something like the following:

May 20/08 3:48 PM

In the Date Format field, you can enter any format of date and time you want; those listed in the dialog box under **Examples** are mere samples created by Autodesk. You can even create crazy dates, like:

M-H (YYY)

...which would display as **5 - 12 (2004)**.

The meaning of all the date and time codes is shown on another page.

6. Click **OK**. AutoCAD adds the field text with its gray background.

7. Click **OK** to exit the mtext editor.

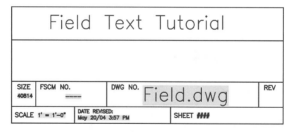

8. Repeat the **FIELD** command to place other fields in the title block, such as the plot scale (**Plot > PlotScale**) and file size (**Document > Filesize**).

Sometimes the field text initially consists of dashes, ---- , as illustrated below under FSCM NO. This means that the field has no value yet. Examples include the PlotDate, if the drawing has never been plotted.

Other times you may see a series of hashes, ####, which means the value is invalid for some reason. (The reasons are not explained by AutoCAD; you have to figure it out yourself.)

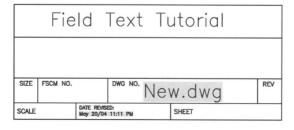

9. Let's watch the automatic updating capabilities of field text in action. Use the **QSAVE** command to save the drawing.

Notice that the time changes (and perhaps the date).

10. Use the **SAVEAS** command to save the drawing by the name *new.dwg*.

Notice that the file name changes (and perhaps the date and time), if they are different.

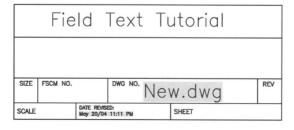

ADVANCED TUTORIAL: CALCULATING FORMULAS IN TABLES

In this tutorial, you combine tables and field text to place formulas in the drawing. You use the Count formula to count the number of items, and the Sum formula to add up the quantity.

Open the *table.dwg* drawing from the Companion CD.

1. Select the last row of the table.
2. Right-click, and from the shortcut menu select **Insert Rows | Below**.
   Notice that AutoCAD adds a blank row to the bottom of the table.

24	1	VW252-02-0208-4	Spacer	See Drg#VW252-02-0208-4
25	1	VW252-02-0208-5	Spacer	See Drg#VW252-0200208-4

3. Double-click the first cell of the new row. Notice the mtext editor.
   (If necessary, change the text height to 2.0).
4. Add the following text:

   > Total Items =

   Use the alignment buttons on the Text Formatting toolbar to center and top align the text.
5. Right-click the cell, and then select **Insert Field** from the shortcut menu. (Alternatively, press **CTRL+F**). Notice the Field dialog box.

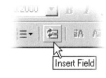

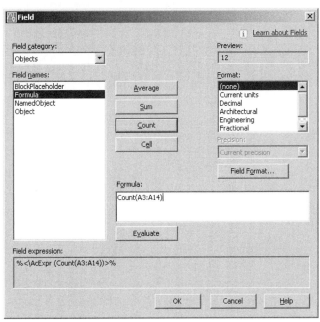

6. In the Field dialog box, select **Objects**, **Formula**, and then click **Count**.
   AutoCAD removes the dialog box from view, and then prompts you to pick the first cell of the range. The *range* is from the first to last rows with data in them:

   > Select first corner of table cell range: *(Pick cell A3.)*

   > Select second corner of table cell range: *(Pick cell A14.)*

Note: AutoCAD uses the same format of formulas as do spreadsheets. For example, to count the number of rows, use the following formula:

=count(a3:a14)

Where:

=	indicates that a formula follows.
count	is the name of the formula.
:	indicates a range of cells (applies the formula to all cells in the range).
a3	is the first cell of the range.
a14	is the last cell of the range.

Item	Qty	Part Num
1	1	VW02-210
3	1	Helix_Helical g
9	2	Vw252-02-110
10	2	VW252-02-110
12	1	VW252-02-020
13	1	VW252-02-020
14	1	VW252-02-020
15	1	VW252-02-020
22	2	KGLM30-Outer
23	2	KGLM30-Inner
24	1	VW252-02-020
25	1	VW252-02-020
Total Items=12		

Cell A3

Cell A14

Formula
=count(a3:a14)

7. Click **OK** to exit the Field dialog box. Notice that the cell displays the number of items (rows): 12.

8. It turns out that you don't need to use the Field dialog box for formulas. You can enter the formulas directly in the mtext editor. The drawback, however, is that you cannot enter any other text in the cell.

   Click the next blank cell. (If you had exited the mtext editor, double-click the cell instead.) Enter the following text:

   =sum(b3:b14)

   Click **OK** to exit the mtext editor. Notice that AutoCAD displays the total quantity: 16.

25	1	VW252-02-0
Total Items=12	16	

9. To see how the total changes, delete one or two items from the table.

FIELDDISPLAY

The **FIELDDISPLAY** system variable toggles the display of the gray background for field text. You might want to turn off the gray rectangles to make your drawings look cleaner.

FieldDisplay	Comment
0	Field text has no background color.
1	Field text has gray background (*default*).

(There is no need to turn off the gray for plotting, because the fills are not plotted.)

TUTORIAL: EDITING FIELD TEXT

AutoCAD treats field text as a variation of mtext. Thus, when you want to edit it, AutoCAD brings up the mtext editor and its Text Formatting toolbar. In this tutorial, you edit the field text placed in the previous tutorials. You convert one piece of field text to "normal" text, and change the formatting of another.

Follow these steps to convert field text to text. This is something you would do when you want fields *not* to update automatically.

1. Double-click the text "New.dwg." AutoCAD opens the mtext editor.
2. Right click the text, and from the shortcut menu select **Convert Field Text to Text.**
3. Click **OK** to exit the mtext editor.
   Notice that the gray background disappears.

To change the format of field text:

1. Double-click the date and time. AutoCAD opens the mtext editor.
2. Right click the text, and from the shortcut menu, select **Edit Field.**
   AutoCAD displays the Field dialog box, with the date format inserted for you.
3. Change the format from MMM d/yy h:mm tt to:
   d MMMM, yyyy at HH:mm
4. Click **OK** to dismiss the dialog box.
   The mtext editor shows the result of the new date and time format:
   20 May, 2004 23:11
5. Click **OK** to exit the mtext editor.

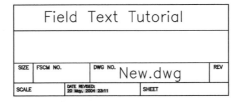

 **UPDATEFIELD**

AutoCAD usually updates fields automatically, but only when certain actions occur. These actions are when drawings are opened, regenerated, plotted, e-transmitted, or saved.

Some fields change their value based on their location in the drawing. For instance, the **PaperSize** field changes when the scale setting changes in different layouts.

Some fields do not change automatically; this is for compatibility between drawings. These fields, listed below, are stored in blocks and external references (xrefs). They are called "contextual fields."

These fields are not updated when you insert the blocks and xrefs in drawings; they continue to display their last value. To force contextual fields to update upon insertion, create them as attributes.

The contextual fields are:

PlotOrientation	CurrentSheetTitle	CurrentSheetSetCustom
DeviceName	CurrentSheetNumber	CurrentSheetDescription
PaperSize	PageSetupName	CurrentSheetNumberAndTitle
PlotDate	PlotStyleTable	CurrentSheetSetDescription
PlotScale	CurrentSheetSubSet	CurrentSheetCustom
CurrentSheetSet		

The settings of fields in xrefs are updated based on the settings of the current drawing, not the externally-referenced source drawing. That means the same field may have two different settings: one when viewed in the original, and another when viewed as an xref.

In some cases, you might want fields to update immediately. Update manually with the **UPDATEFIELD** command. In other cases, you may not want fields to update at all. You can turn off automatic field updating by setting the **FIELDEVAL** system variable to 0.

Autodesk notes that there are some situations where automatic updating fails. Fields are not updated when opened in releases prior to AutoCAD 2005. Fields are updated when drawings are opened in AutoCAD LT, but the **FIELDEVAL** system variable is not available in LT.

When the **DEMANDLOAD** system variable is set to 2, field text is not updated until you use the **UPDATEFIELD** command. (This system variable specifies whether AutoCAD loads external software applications on demand. When set to 2, applications are only loaded when one of their commands is entered. The default is 3.)

TUTORIAL: UPDATING FIELDS MANUALLY

1. For this tutorial, continue with the *field.dwg* drawing from the previous tutorials.
2. To update field text immediately, start the **UPDATEFIELD** command with one of these methods:
   * From the **Tools** menu, choose **Update Field**.
   * At the 'Command:' prompt, enter the **updatefield** command.

     Command: **updatefield** (Press ENTER.)
3. AutoCAD prompts you to select the fields to update:

   Select objects: (Select one or more objects.)

   Select objects: (Press ENTER TO end object selection.)

   n field(s) found

   n field(s) updated

FIELDEVAL

The **FIELDEVAL** system variable controls automatic updating of field text. Its default value is 31, which means that fields are updated when any of the actions listed below occur. When set to 0, fields are never updated. (This setting must be made for each drawing individually.)

Its value is the sum of any of the following numbers:

FieldEval	Fields are updated...
0	never.
1	when the drawing is opened.
2	when the drawing is saved.
4	when the drawing is plotted.
8	when the ETRANSMIT command is used.
16	when the drawing is regenerated.

Note: The **Date** field is never updated automatically; it must be updated manually. The workaround is to select the **Other > System Variable** field, and then specify the **Date** system variable.

As an alternative to using the FIELDDISPLAY and FIELDEVAL system variables, you can control the settings through the Options dialog box. Here's how.

1. From the **Tools** menu, select **Options**.
2. In the Options dialog box, click the **User Preferences** tab.
3. In the **Fields** section, selecting the **Display background of fields** option is equivalent to toggling the FIELDDISPLAY system variable. Click the option to toggle the gray background of field text.
4. Click the **Field Update Settings** button to display the dialog box, which is equivalent to using the FIELDEVAL system variable.

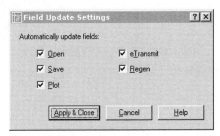

5. Click the command names you wish to change.
   - ☑ When command is executed, all field text is updated automatically.
   - ☐ Command has no effect on field text.

   (The **Save** option includes the SAVE, SAVEAS, and QSAVE commands. The **Regen** option includes the REGEN and REGENALL commands.)
6. Click **Apply & Close** to exit this dialog box, and then click **OK** to close the first dialog box.

EXERCISES

1. Employ AutoCAD's **TABLE** command to create a table with seven rows and five columns.

2. Use the **TABLE** command to recreate the table of reference drawings illustrated below. (A copy of the table is available on the Companion CD in the *refdwg.tif* file.)

B·46858	CONV. C-12 LAYOUT
A·50918	CHUTE AT DISCHARGE END, ARRGT & DET.

REFERENCE DRAWINGS

3. With the **TABLE** command, recreate the table of general notes illustrated below. (A copy of the table is available on the Companion CD in the *generalnotes.tif* file.)

1.	ALL FIELD WELDED CONSTRUCTION (EXCEPT LADDER)
2.	PAINT NOTE: 1 COAT OF RED PRIMER AND 1 COAT OF EQUIPMENT GREY (SHOP PAINT LADDER, FIELD PAINT PLATFORM AFTER INSTALLATION).
3.	ALL WELDING MUST CONFORM TO THE REQUIREMENTS OF THE LATEST ISSUE OF C.S.A. STANDARD W.59 EXCEPT THE REFERENCE TO C.S.A STANDARD W.47 DOES NOT APPLY.

GENERAL NOTES

4. Recreate the bill of materials table illustrated below. (A copy of the table is available on the Companion CD in the *table-scan.tif* file.)
   Use auto-fill to fill the cells under "No. PER UNIT."

BILL OF MATERIAL						
PART No.	No. OF UNITS	No. PER UNIT	DESCRIPTION	MATERIAL	STOCK No.	REMARKS
C-9248-1	2		WASTE OIL DISPOSAL UNIT			
-2	2	2	3/16" PLATE x 1'-1½" ∅	STEEL	2502001	
-3		2	3/16" x 3" FLAT BAR x 3'-6½"	"	2500205	
-4		2	3" x 3" BUTT HINGE	"	1970323	
-5		2	3/8" x 1½" FLAT BAR x 3"	"	2500503	
-6		2	12" STD. BLACK PIPE x 1'-6"	"	3010003	
-7		2	24 MESH SCREEN, 11-3/4" ∅	S.S.	2040201	
-8		2	3/8" ROUND x 3'-0" LG.	STEEL	2501703	

5. From the Companion CD, open the *chapter16.dwg* file, a drawing of a parts list table.

   a. Edit the table to add the values in the **Qty** column (short for "quantity"). What approach did you take to enter the values?

Parts List			
Item	**Qty**	**Part Number**	**Description**
1	5	PC-15	Paper clips
2	4	P-24	Pencils
3	3	PP-33	Paper Pads
4	2	R-42	Rulers
5	1	CB-51	Clipboard

   b. Split the table into two sections.

6. Create the parts list table illustrated below. In the **Description** column, insert the appropriate blocks. Describe the approach you took to insert the blocks.

Parts List			
Item	**Qty**	**Part**	**Illustration**
1	5	Air Injector (Dynamic Pump)	
2	4	Evaporator - Circular Fin	
3	3	Heater Feed with Air Outlet	
4	2	Pressure Gauge	
5	1	Thermometer - Type 2	

7. From the Companion CD, open the *table.dwg* file, and then use the **TABLEEXPORT** command to create a *.csv* file.

   If you have access to a spreadsheet program, open the *.csv* file and view its contents.

   Is the data in the spreadsheet the same as the table in AutoCAD?

8. From the Companion CD, open the *tablestyle.dwg* file.

   Use the **TABLESTYLE** command to make a new style based on the Standard style:

   a. Change font to Times New Roman.

   b. Change fill color of data rows to light gray.

   c. Change border line color to white.

   d. Name the style.

9. From the Companion CD, open the *field-16.dwg* file, a drawing of a title block.

    Fill in each block of information using fields. (The field text shown below is a sample of the result; your result will differ.)

Last Saved By ----	Login Name Administrator	Drawing Filesize 31401		Today's Date May. 21, 04	
		Drawing Filename C:\Books\Using AutoCAD 2005 Basics - April 04\CDROM\field-16.dwg			
		Date Last Plotted ----		Plot Scale 4 41/64" = 1'-0"	Paper Size ANSI A (8.50 x 11.00 Inches)

10. Continue with the *field-16.dwg* file from the previous exercise. Save the drawing.

    Did any fields change?

    If not, what might be the reason?

CHAPTER REVIEW

1. Examine the bill of material below, and then answer the following questions. (Also available as *bofm.tif* on the Companion CD.)

PART No.	No. OF UNITS	No. PER UNIT	DESCRIPTION	MATERIAL	REQ. No.	REMARKS
C-10102-1	1		LUB DRIP TRAY ASSY.			
-?		1	16 GA. x 1-8½"x 7-3" SH MET. TRAY	M.S.	2503101	
-3		1	2"ØSTD. PIPE NIPPLE, 4" LG.	"	3252405	
-4	1		¼" N.C. Hx. HD. BOLT, 1"LG	"	DE-CONTR	TACK-WELD TO TRAY
-5	2		5/16"Ø FLAT WASHER	"	-"-	~~ ONE AS SHOWN
-6		1	½" N.C. WING NUT	"	2320501	
-8		1	3/8"x 2"-2" L , 2'-0" LG	M.S.	2501300	
-9		1	3/8"x1½"x1½" L , 8'-4" LG	"	2501002	
-10		1	-"- , 10" LG	"	-"-	

a. How many columns does the table have?

b. How many rows?

c. What is the wording of the title row?

d. What does AutoCAD call the row containing the phrase "PART No."?

e. Where is the insertion point located?

2. Examine the bill of material below, and then answer the following questions. (Also available as *revisions.tif* on the Companion CD.)

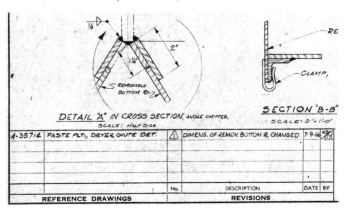

a. How many columns does the table have?

b. How many rows?

c. Is the table's direction up or down?

d. What might be the purpose of the Reference Drawings table?
   Of the Revisions table?

3. Explain the reason behind tables having title, header, and data rows.

4. Can individual cells be formatted differently from each other?

5. Where is the title row located when the table's direction is up?

6. After creating a table, can rows be added?
   Removed?

7. When might you use the **-TABLE** command instead of the **TABLE** command?

8. What is an alternative to creating tables with the **TABLE** command?

9. Describe the purpose of the **TINSERT** command.

10. How many blocks can a cell have?

11. What can you do to ensure the block fits into the cell?

12. When might you use grips editing on tables?

13. Explain the purpose of the **MATCHCELL** command.

14. When would you use the **MATCHPROP** command.

15. Describe a method of copying table styles from other drawings?

16. Can the justification of text in cells be changed?

17. What must you keep in mind when merging cells?

18. How can you bring a table from AutoCAD into a spreadsheet program?

19. When might you use field text instead of mtext?

20. How does AutoCAD show that field text is different from other text?

21. What do the following date and time codes indicate?

    MMMM

    mm

    ddd

    yy

    %c

22. Given **February 25, 1988 at 12:20pm,** write down how the following field codes would represent the date and time:

    MM

    yyyy

    m

    dd

    t

23. Given the following dates, write out the field text codes for each:

    April 8, 05

    09 Oct 2001

    05/07/05 09:07P *(date/month/year)*

    2/8/2003 1:49PM

    2007-Dec-2 Wed 14:08

    System date in long format

24. Can field text be part of mtext?

    Tables?

    Attributes?

    Polylines?

    Formulas?

25. Briefly describe two ways to enter formulas into tables.

26. Explain the meaning of the following formulas:

    =count(a1:z26)

    =sum(b6:b7)

27. Why would you want to convert field text to normal text?

28. How would you prevent field text from updating, and yet remain field text?

29. When field text is part of an xref, which drawing does the field text reflect?

30. What are *contextual fields?*

31. Explain the meaning of the following field displays:

    ####

32. How do table styles differ from cell styles?

33. What is the purpose of the tabletoolbar system variable?

34. List the steps for auto-filling cells.

35. What happens when auto-filling from a cell with a number? A cell with text?

CHAPTER 17

Reporting on Drawings

A benefit of drawing with CAD is that you can obtain information from the drawing. AutoCAD's inquiry commands provide the following kinds of information:

ID reports the x, y, z coordinates of single points.

DIST reports the distance and angle between two points.

AREA reports the area and perimeter of areas.

MASSPROP reports the area, perimeter, centroid, and so on of regions and solids.

LIST, **DBLIST**, and **PROPERTIES** report information about objects.

DWGPROPS reports and stores information about the drawing.

TIME reports on time spent editing the drawing.

STATUS reports on the state of the drawing.

SETVAR reports the settings of system variables.

 Note: AutoCAD always works to 14 decimal places. All reporting commands display accuracy and units according to the current setting of the **UNITS COMMAND**; sometimes, the actual values are not returned. For instance, the value 3.128362563 is stored by AutoCAD, but could be displayed as 3.13, $3^1/_8$, or 3 — depending on the setting of **UNITS**.

NEW TO AUTOCAD 2008 IN THIS CHAPTER

- The **DBLIST** and **LIST** commands report on annotative scale factors, when applicable.

FINDING THE COMMANDS

On the **Standard** and **Standard Annotation** toolbars:

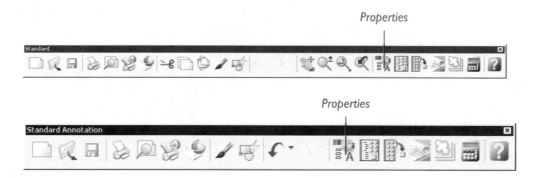

On the **Inquiry** toolbar:

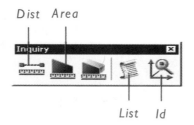

ID

The **ID** command reports the x,y,z coordinates of single points. This command is transparent, so it can be used during other commands.

TUTORIAL: REPORTING POINT COORDINATES

1. To find the coordinates of points, start the **ID** command:
    *   From the **Tools** menu, choose **Inquiry**, and then **ID Point**.
    *   From the Inquiry toolbar, choose the **Locate Point** button.
    *   At the 'Command:' prompt, enter the **id** command.

        Command: **id** *(Press* ENTER.*)*

2. In all cases, AutoCAD prompts you to pick a point in the drawing.
    Specify point: *(Pick a point.)*
    AutoCAD reports the x, y, z coordinates of the point:
    X = 13.3006   Y = 4.4596   Z = 0.0000

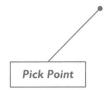

Pick Point

The z coordinate is the same as the elevation. To capture the point of a geometric feature accurately, use an object snap mode, such as ENDpoint or CENter.

LastPoint and @

AutoCAD stores the x,y,z coordinates in the **LASTPOINT** system variable, which you can access in the next command by entering @ at the prompt. Here is an example using the **LINE** command:

> Command: **line**
> Specify first point: **@**
> Specify next point or [Undo]: *(Pick a point.)*

And an example using the **CIRCLE** command:

> Command: **circle**
> Specify center point for circle or [3P/2P/Ttr (tan tan radius)]: **@**
> Specify radius of circle or [Diameter]: *(Pick a point.)*

The coordinates are stored in **LASTPOINT** only until the next 'Specify point' prompt. As soon as you pick another point, or enter coordinates at the keyboard, AutoCAD stores the new coordinates in the **LASTPOINT** system variable.

REPORTING POINT COORDINATES: ADDITIONAL METHODS

In addition to reporting the 3D coordinates of points you pick, the **ID** command can show you coordinates through blip marks.

*   **BLIPMODE** command turns on blips.
*   **Marking points** shows the location of entered coordinates.

Let's look at how they work.

Blipmode

The **BLIPMODE** command toggles the display of blips in the drawing. AutoCAD normally keeps blip mode turned off. When enabled, *blips* are small + markers that appear any time you pick a point in the drawing.

Command: **blipmode**

Enter mode [ON/OFF] <OFF>: **on**

Now when you use the **ID** command (and most other commands), each screen pick leaves behind the blip marker.

Command: **id**

Specify point: *(Pick a point.)*

   X = 7.0   Y = 10.0   Z = 0.0

The size or shape of blip marks cannot be changed. Blip marks are like grid marks: they do not plot. Unlike grid marks, however, blips are not permanent. The next command involving a redraw or regeneration erases them.

Command: **redraw**

Marking Points

The **ID** command can operate in reverse: specify coordinates, and it draws a blip mark at that point, when **BLIPMODE** is on:

Command: **id**

Specify point: **7.9,10.2**

   X = 7.9   Y = 10.2   Z = 0.0

 DIST

The **DIST** command reports the distance and angle between two points.

TUTORIAL: REPORTING DISTANCES AND ANGLES

1. To find the distance and angle between two points, start the **DIST** command:
   - From the **Tools** menu, choose **Inquiry**, and then **Distance**.
   - From the Inquiry toolbar, choose the **Distance** button.
   - At the 'Command:' prompt, enter the **dist** command:

     Command: **dist** *(Press ENTER.)*

   - Alternatively, enter the **di** alias at the 'Command:' prompt.

2. In all cases, AutoCAD prompts you to pick two points in the drawing. (Use an object snap mode to capture geometric points accurately.)

    Specify first point: *(Pick point 1.)*

    Specify second point: *(Pick point 2.)*

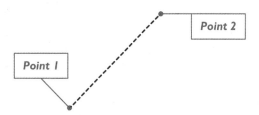

AutoCAD reports the distance and angles between the points several different ways:

Distance = 0.7071, Angle in XY Plane = 45, Angle from XY Plane = 0

Delta X = 0.5000, Delta Y = 0.5000,   Delta Z = 0.0000

For 2D drafting, you need only the x- and y-related information, as illustrated by the figures below:

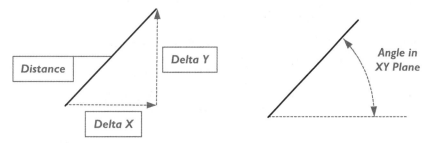

Left: *The distance, delta X and delta Y reported by the DIST command.*
Right: *The angle in the XY plane.*

AutoCAD stores the distance in the **DISTANCE** system variable.

You can use the **DIST** command to show distances by specifying relative or polar coordinates, with blip mode turned on:

    Command: **dist**

    Specify first point: *(Pick first point.)*

    Specify second point: **@10,0**

To find the shortest distance between two objects, use the **PERpendicular** object snap, such as the line and circle illustrated below:

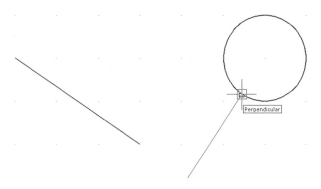

    Command: **dist**

    Specify first point: **per**

    to *(Pick a point on the line.)*

When there are more than two objects in the drawing, AutoCAD does not find the perpendicular on the line until you pick the next point. An ellipsis (...) appears next to the perpendicular icon to indicate this deferment.

> Specify second point: **per**
>
> to *(Pick a point on the circle.)*

AutoCAD measures the shortest distance between the two objects. In the figure above, there is a gap between the line and the point of perpendicularity: AutoCAD "extends" the line segment to find the point of perpendicularity.

AREA

The **AREA** command reports the area and perimeter of areas.

This command measures the area and perimeter of closed objects, such as polygons, circles, and hatches, as well as arbitrary areas of picked points. You can add and subtract areas to arrive at a total.

TUTORIAL: REPORTING AREAS AND PERIMETERS

1. To find the area and perimeters of objects, start the **AREA** command:
   - From the **Tools** menu, choose **Inquiry**, and then **Area**.
   - From the Inquiry toolbar, choose the **Area** button.
   - At the 'Command:' prompt, enter the **area** command.

     Command: **area** *(Press ENTER.)*
   - Alternatively, enter the **aa** alias at the 'Command:' prompt.

2. In all cases, AutoCAD prompts you to pick corner points. These are the vertices of an imaginary polygon that defines the area to be measured.
   (To capture geometric points accurately, use object snap modes.)

     Specify first corner point or [Object/Add/Subtract]: *(Pick point 1.)*

     Specify next corner point or press ENTER for total: *(Pick point 2.)*

     ...

     Specify next corner point or press ENTER for total: *(Pick point 5.)*

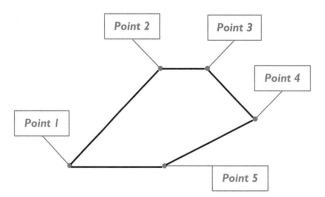

3. Press **ENTER** to end the measurement process. (It is not necessary for your last pick to match the first one; AutoCAD automatically "closes" the polygon for you.)

     Specify next corner point or press ENTER for total: *(Press ENTER to end the command.)*

   AutoCAD reports the area and perimeter:

     Area = 2.0316, Perimeter = 5.9220

AutoCAD stores the area and perimeter measurements in the **AREA** and **PERIMETER** system variables. **AREA** is one of a few system variables with the same name as a related command, and so you must use the **SETVAR** command to access the value stored in the system variable. (Entering **AREA** at the 'Command:' prompt executes the command.)

> Command: **setvar**
>
> Enter variable name or [?]: **area**
>
> AREA = 2.0316 (read only)

REPORTING AREA: ADDITIONAL METHODS

AutoCAD can add and subtract areas from the total area, as well as find the area of circles, ellipses, splines, polylines, polygons, regions, hatches, and 3D solids.

- **Add** option adds another area to the total.
- **Subtract** option subtracts another area from the total.
- **Object** option finds the area of closed objects.

Let's look at each option.

Add

The **Add** option adds areas to the total. You can pick points or select objects to add:

> Command: **area**
>
> Specify first corner point or [Object/Add/Subtract]: **a**
>
> Specify next corner point or press ENTER for total (ADD mode): *(Pick a point.)*
>
> *...and so on.*

The "(ADD mode)" message reminds you that AutoCAD is adding areas.

Subtract

The **Subtract** option removes areas from the total. Again, you can pick points or select objects to remove. AutoCAD can find negative areas by immediately going into **Subtract** mode.

> Command: **area**
>
> Specify first corner point or [Object/Add/Subtract]: **s**
>
> Specify next corner point or press ENTER for total (SUBTRACT mode): *(Pick a point.)*
>
> *...and so on.*

The "(SUBTRACT mode)" message reminds you that AutoCAD is subtracting areas.

Object

The **Object** option finds the area and perimeter of selected objects: circles, ellipses, open and closed splines, open and closed polylines, polygons, regions, hatch patterns, and 3D solids — but not 2D solids.

> Command: **area**
>
> Specify first corner point or [Object/Add/Subtract]: **o**
>
> Select objects: *(Select one object.)*
>
> Area = 0.9085, Perimeter = 3.5000

An open polyline (or spline) has a shorter "perimeter" than a closed polyline of the same area, because only the length of the open polyline is measured as the perimeter.

Even though the plural "objects" of the **Object** option's 'Select objects' prompt suggests you can select more than one object at a time, you are limited to one — unless you first specify the **Add** or **Subtract** option. To add the area of additional objects, use the **Add** option.

When you select open objects (other than polylines), such as arcs and lines, AutoCAD complains, "Selected object does not have an area."

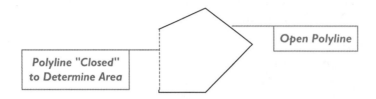

Objects	Measurement
Circles	Area, circumference.
Ellipses, polygons	Area, perimeter.
Regions	Area, perimeter.
Closed polylines and closed splines	Area, perimeter.
Open polylines and open splines	Area, length.
Wide polylines	Area and length/perimeter along centerline.
3D solids	Area of all faces.
Hatch patterns	Area, perimeter.

Note: Here's a quick way to find the area of multiple irregular areas: hatch them with a single pattern, and then look up the area with the Properties palette.

 **MASSPROP**

The **MASSPROP** command reports the area, perimeter, centroid, and so on, of 2D boundaries and 3D solid models.

The boundary must be made from a single region object. You can use the **SUBTRACT** command to remove *islands* from regions. (The application of **MASSPROP** to 3D solid models is described in *Using AutoCAD: Advanced*.)

TUTORIAL: REPORTING INFORMATION ABOUT REGIONS

1. To find out information about regions, start the **MASSPROP** command:
    *   From the **Tools** menu, choose **Inquiry**, and then **Region/Mass Properties**.
    *   From the Inquiry toolbar, choose the **Mass Properties** button.
    *   At the 'Command:' prompt, enter the **massprop** command.

    Command: **massprop** *(Press ENTER.)*

2. In all cases, AutoCAD prompts you to select objects.

    Select objects: *(Select region objects.)*

    Select objects: *(Press ENTER to end object selection.)*

AutoCAD switches to the Text window, and displays a report similar to the following:

```
---------------   REGIONS   ---------------
Area:           1.7211
Perimeter:              14.3193
Bounding box:  X: 1.2616  --  3.7327
                       Y: 0.9440  --  3.0999
Centroid:              X: 2.6010
                       Y: 2.2224
Moments of inertia:    X: 9.0893
                       Y: 12.3840
Product of inertia:    XY: 9.9040
Radii of gyration:     X: 2.2981
                       Y: 2.6825
Principal moments and X-Y directions about centroid:
                       I: 0.5772 along [0.9653 -0.2612]
                       J: 0.7529 along [0.2612 0.9653]
```

3. AutoCAD asks if you wish to save the data to a file.

   Write analysis to a file? [Yes/No] <N>: **y**

4. If you answer **Y,** AutoCAD displays the Create Mass and Area Properties File dialog box. Enter a file name, and then click **Save**. AutoCAD creates the *.mpr* mass properties report file.

5. Press **F2** to return to the drawing window.

The centroid value is useful for finding the geographic center of irregularly shaped objects, like maps, such as that of Greenland illustrated below. The bounding box is a rectangle that completely surrounds the region. The area and perimeter values are the same values reported by the **AREA**, **LIST**, and **PROPERTIES** commands.

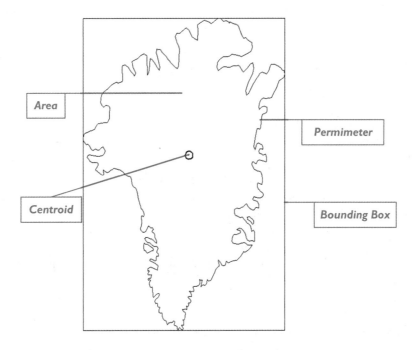

LIST AND DBLIST

The **LIST** command reports information about selected objects, while **DBLIST** produces the same report, but on *every* object in the drawing (short for "database listing").

TUTORIAL: REPORTING INFORMATION ABOUT OBJECTS

1. To find out information about objects, start the **LIST** command:
   - From the **Tools** menu, choose **Inquiry**, and then **List**.
   - From the Inquiry toolbar, choose the **List** button.
   - At the 'Command:' prompt, enter the **list** command:

     Command: **list** *(Press* ENTER.*)*

   - Alternatively, enter the **ls** alias at the 'Command:' prompt.

2. In all cases, AutoCAD prompts you to select objects. While you can select more than one object, it is better to pick just one, because of the amount of information generated.

   Select objects: *(Select one object.)*

   Select objects: *(Press* ENTER *to end object selection.)*

   AutoCAD reports on the object(s) in the Text window.

```
        CIRCLE    Layer: "0"
             Space: Model space
             Handle = 15f22
      center point, X=1231.9752  Y=-1589.5176  Z=   0.0000
             radius  593.9697
      circumference 3732.0214
             area 1108353.7564
```

You can select the text, and then use **CTRL+C** to copy it to the Clipboard for pasting into other documents. Press **F2** to return to the drawing window.

AutoCAD reports information specific to each object, such as for the circle:

Line #	Example	Information Reported
1	CIRCLE	Type of object.
	"0"	Layer on which the object resides.
2	Model space	Space in which the object resides: model or paper space.
3	9D	Handle (unique hexadecimal identifier assigned by AutoCAD.)
4	0.5,7.5,0.0	X, y, z coordinates of the object's location.
Other		Additional geometric information specific to the object:
		Area, perimeter, length.
		Color, linetype, and lineweight, if not set to BYLAYER.
		Thickness, if not zero.

The geometric information varies among objects, as illustrated by the table below. For complex objects, the listing can get very long, because AutoCAD reports on every subpart.

Object	Information Reported
Arc	Center point, Radius, Start angle, End angle, Length.
Box	Position, Length, Width, Height, Rotation.
Circle	Center point, Radius.
Ellipse	Center, Major Axis, Minor Axis, Radius Ratio.
Hatch	Pattern name, Scale, Angle, Associativity, Area, Origin.
Helix	Position, Length, Base radius, Top radius, Turns, Step Dist, and Axis.
Line	From point, To point, Length, Angle in XY Plane, Delta X, Delta Y, Delta Z.
Multileader	Leader Type, Content Type, Landing, Leader Number, Vertex, Annotative scale
PlaneSurface	U-isolines, V-isolines, Bounding Box.
Point	Point.
Polyline	Constant width (or Starting width and Ending width), Area, Perimeter.
Polyarc	Bulge, Center, Radius, Start angle, End angle.
Sphere	Position, Radius.
Spline	Area, Circumference, Order, Properties, Parametric Range, Number of control points, Control Points, Number of fit points, User Data, Start Tangent, End Tangent.
Text	Style, Annotative, Font file, Start point, Height, Text, Rotation angle, Width scale factor, Obliquing angle, Generation
etc.	

Note: The **LIST** command is more useful than **AREA** for individual objects, because it provides more information. For example, **AREA** reports the area and circumference of circles, but **LIST** also reports their radius.

TUTORIAL: REPORTING INFORMATION ABOUT ALL OBJECTS

1. To find out information about all objects in the drawing, start the **DBLIST** command:

   Command: **dblist** *(Press* **ENTER.***)*

2. AutoCAD switches to the Text window, and reports on every object in the drawing. (The report pauses at every screenfull.)

   Press ENTER to continue: *(Press* **ENTER.***)*

3. The report is very long, even for small drawings. To end it, press **ESC**:

   Press ENTER to continue: *(Press* **ESC.***)*

Press F2 to return to the drawing window.

PROPERTIES

The **PROPERTIES** command reports on objects in a format easier to digest and more complete than **LIST** and **DBLIST**. Compare the data for the circle shown at right with that listed on the previous page.

When you select all objects in the drawing, the Property window reports the total number.

A drawback is that the data listed by the Properties palette cannot be copied to the Clipboard. For more information about this command, see Chapter 7, "Changing Object Properties."

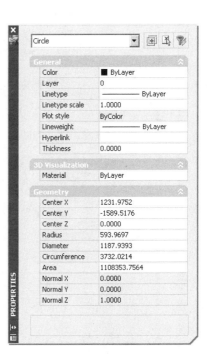

DWGPROPS

The DWGPROPS command reports and stores information about drawing files.

This command displays a dialog box of data related to the file, such as its size and attributes. As well, you can add information about the drawing, such as the name of the drafter and general comments. This data can be searched with AutoCAD's Design Center and the Windows Explorer FIND (or SEARCH) command.

TUTORIAL: REPORTING INFORMATION ABOUT OBJECTS

1. To find out information about objects, start the DWGPROPS command:
   - From the **File** menu, choose **Drawing Properties**.
   - At the 'Command:' prompt, enter the **dwgprops** command.

   Command: **dwgprops** (*Press* ENTER.)

2. In all cases, AutoCAD displays the Properties dialog box.

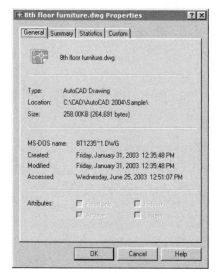

The **General** tab shows information about the drawing file's name, location, size, and dates. (If the drawing has not yet been saved, most fields are blank.) Attributes can be changed only by the Windows Explorer's **Property** option.

3. Click the **Summary** tab.

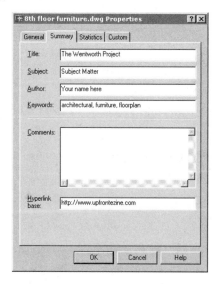

You can add data to the fields, such as your name, the subject of the drawing, and general comments.

- **Title** can be different from the drawing's file name, such as the project name.

- **Keywords** are useful when drawings are searched by database programs.

- **Hyperlink Base** is the base URL address used by relative hyperlinks in the drawing.

4. Click the **Statistic** tab.

- **Created** is the date and time the drawing was first created with the **NEW** command; the date and time are retrieved from the **TDCREATE** system variable.

- **Last Modified** is the date and time the drawing was last edited or had objects added, retrieved from the **TDUPDATE** system variable.

- **Last Saved By** is the user's login name, retrieved from **LOGINNAME** system variable, which in turn is obtained from Windows.

- **Revision Number** is the AutoCAD software revision number.

- **Total Editing Time** is the time the drawing has been open. The name is misleading, because it includes time when the drawing is displayed by AutoCAD without being edited. The value is retrieved from the **TDINDWG** system variable.

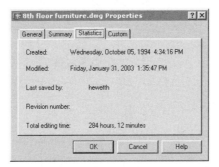

 Note. If the drawing was last saved by a non-Autodesk software product, an additional message appears in this tab:

> This file may have been last saved by a program other than Autodesk software.

In addition, the value of 0 is stored in **DWGCHECK** system variable.

5. Click the **Custom** tab.

You can enter text in up to ten fields, along with a value for each. The custom fields are searchable by the **Find** option of DesignCenter.

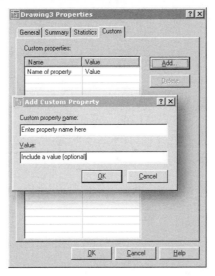

To add a field, click **Add**; enter a name for the field and optionally a value. Click **OK**.

To delete a field, select it, and then click the **Delete** button.

6. Click **OK** to close the dialog box.

'TIME

The **TIME** command reports the time spent editing drawings, and provides timer functions.

This command can be used transparently in the middle of other commands.

TUTORIAL: REPORTING ON TIME

1. To find out time-related information about drawings, start the **TIME** command:

    • From the **Tools** menu, choose **Inquiry**, and then **Time**.

    • At the 'Command:' prompt, enter the **time** command.

      Command: **time** (*Press* ENTER.)

    AutoCAD reports time-related data similar to the following:

      Command: **time**

      Current time: Wednesday, June 25, 2003 at 2:53:15:890 PM

      Times for this drawing:

      Created: Wednesday, June 25, 2003 at 1:04:52:812 AM

      Last updated: Wednesday, June 25, 2003 at 1:04:52:812 AM

      Total editing time: 0 days 13:48:23.203

      Elapsed timer (on):   0 days 13:48:23.047

      Next automatic save in: <no modifications yet>

2. Press ENTER to exit the command:

      Enter option [Display/ON/OFF/Reset]: (*Press* ENTER.)

The accuracy of the time depends on your computer's clock. The meaning of the data reported by this command is as follows:

Time	Meaning
Current Time	Current date and 24-hour time to the nearest millisecond.
Created	Date and time drawing was created.
Last Updated	Latest use of the SAVE and QSAVE commands.
Total Editing Time	Cumulative time drawing has been open in AutoCAD, excluding plotting time, and sessions when drawing was exited without saving.
Elapsed Timer	Stopwatch-like timer.
Next Automatic Save In	Minutes remaining until the next automatic save.

REPORTING ON TIME: ADDITIONAL METHODS

The **TIME** command includes options to redisplay its output and act as a stopwatch.

• **Display** redisplays the time report.

• **ON** starts the timer.

• **OFF** turns off the timer.

• **Reset** resets the timer.

Let's look at each.

Display

The **Display** option redisplays the time report, with updated times.

ON, OFF, and Reset

The **ON** option starts the timer. The **OFF** option stops the time, and the **Reset** option sets the timer back to 0 days and 0 time (0 hours, 0 minutes, and 0 seconds).

STATUS

The **STATUS** command reports on the state of the drawing. (The **DIM:STATUS** command reports the values of dimension variables, while the **STATS** command reports on the most recent rendering.)

Command: **status**

AutoCAD switches to the Text screen, and reports the information similar to that shown below. You can copy the information to the Clipboard by selecting the text, and then pressing **CTRL+C**. To return to the drawing window, press **F2**.

```
134 objects in Drawing.dwg
Model space limits are  X:   0.0000  Y:   0.0000  (Off)
                        X:  12.0000  Y:   9.0000
Model space uses        X:   4.9634  Y:   2.4175
                        X:  10.1843  Y:   7.0376
Display shows           X:   0.1219  Y:  -0.0534
                        X:  11.8781  Y:   9.0534
Insertion base is       X:   0.0000  Y:   0.0000  Z:   0.0000
Snap resolution is      X:   0.5000  Y:   0.5000
Grid spacing is         X:   0.5000  Y:   0.5000

Current space:      Model space
Current layout:     Model
Current layer:      "0"
Current color:      BYLAYER -- 7 (white)
Current linetype:   BYLAYER -- "Continuous"
Current material:   BYLAYER -- "Global"
Current lineweight: BYLAYER
Current elevation:  0.0000  thickness:  0.0000
Fill on  Grid off  Ortho off  Qtext off  Snap off  Tablet off
Object snap modes:    Center, Endpoint, Intersection, Extension,
Free dwg disk (D:) space: 27776.5 MBytes
Free temp disk (C:) space: 26506.5 MBytes
Free physical memory: 285.9 Mbytes (out of 1023.5M).
Free swap file space: 1002.8 Mbytes (out of 1693.4M).
```

Status	Meaning
Model Space limits	X,y coordinates stored in the LIMMIN and LIMMAX system variables.
Paper Space limits	**Off** indicates limits checking is turned off (LIMCHECK).
Model Space uses	X,y coordinates of the lower-left and upper-right extents of objects.
Paper Space uses	**Over** indicates drawing extents exceed the drawing limits.
Display shows	X,y coordinates of lower-left and upper-right of the current display.
Insertion base is	X,y,z coordinates stored in system variable INSBASE
Snap resolution is	Snap and grid settings, as stored in the SNAPUNIT and GRIDUNIT
Grid spacing is	system variables.
Current space	Whether model or paper space is current.
Current layout	Name of the current layout.

Current layer Current color Current linetype Current material Current lineweight Current plot style Current elevation Thickness	Values for the layer name, color, linetype name, elevation, and thickness, as stored in system variables CLAYER, CECOLOR, CELTYPE, CMATERIAL, LWEIGHT, CPLOTSTYLE, ELEVATION, and THICKNESS.
Fill, Grid, Ortho QText, Snap, Tablet	Toggle settings (on or off) for the fill, grid, ortho, qtext, snap, and tablet modes, as stored in the system variables FILLMODE, GRIDMODE, ORTHOMODE, TEXTMODE, SNAPMODE, and TABMODE.
Object Snap modes	Current object snap modes, as stored in system variable OSMODE.
Free dwg disk (C:) space Free temp disk (C:) space Free Physical Memory Free Swap File Space	Hard disk space free on the specified drive. Hard disk space free on the drive specified for AutoCAD's temporary files. Amount of free RAM memory, and total RAM memory. Amount of free swap file space, and total swap file size.

You can also use the **STATUS** command at the 'Dim:' prompt, as follows:

> Command: **dim**
>
> Dim: **status**

AutoCAD flips to the Text window, and displays the value of all dimension variables:

DIMASO	Off	Create dimension objects
DIMSTYLE	Standard	Current dimension style (read-only)
DIMADEC	0	Angular decimal places
DIMALT	Off	Alternate units selected
DIMALTD	2	Alternate unit decimal places
DIMALTF	25.4000	Alternate unit scale factor
DIMALTRND	0.0000	Alternate units rounding value
DIMALTTD	2	Alternate tolerance decimal places
DIMALTTZ	0	Alternate tolerance zero suppression
DIMALTU	2	Alternate units
DIMALTZ	0	Alternate unit zero suppression
DIMAPOST		Prefix and suffix for alternate text
DIMARCSYM	0	Arc length symbol
DIMASZ	0.1800	Arrow size
DIMATFIT	3	Arrow and text fit
et cetera...		

You can copy the information to the Clipboard by selecting the text, and then pressing CTRL+C. To return to the drawing window, press **F2**. Enter **EXIT** to return to the 'Command:' prompt:

> Dim: **exit**
>
> Command:

'SETVAR

The **SETVAR** command reports the settings of system variables (short for "set variables").

System variables store the state of the drawing and control numerous commands. You can change the value of most system variables, thus changing the state of the drawing (snap, grid, and so on) and the actions of commands (mirrored text, current layer, and so on). Some system variables cannot be changed; they are called "read-only."

This command can be used transparently during other commands.

You have used this command many times in this and previous chapters, without realizing it. *History*: System variables could only be changed with the SETVAR command. More recently, Autodesk made it possible to enter system variables at the 'Command:' prompt. Thus, the only time you must use this command is when you need to examine the list of system variables, or when the name of the system variable is the same as the command, such as AREA.

All of AutoCAD's system variables are listed in Appendix C of this book.

TUTORIAL: LISTING SYSTEM VARIABLES

1. To list the values of system variables, start the SETVAR command:
   * From the **Tools** menu, choose **Inquiry**, and then **Set Variable**.

   * From the Modify toolbar, choose the **Copy** button.

   * At the 'Command:' prompt, enter the **setvar** command:

     Command: **setvar** *(Press ENTER.)*

   * Alternatively, enter the **set** alias at the 'Command:' prompt.

2. In all cases, AutoCAD prompts you to enter the name of a system variable. Enter the ? option:

   Enter variable name or [?] <AREA>: **?**

3. AutoCAD asks which variables you wish to see. Use wildcard characters:
   * — means all variables.
   a* — means all variables starting with the letter "a."
   grid* — means all variables starting with the word "grid."
   ? — means a single character.

   Enter variable(s) to list <*>: *(Press ENTER to list all variables.)*

   AutoCAD flips to the Text window, and displays the value of all system variables.

3DDWFPREC	2	
ACADLSPASDOC	0	
ACADPREFIX	"C:\Documents and Settings\Administrator\ApplicationData\Aut..."	(read only)
ACADVER	"17.0s (LMS Tech)"	(read only)
ACISOUTVER	70	
AFLAGS	16	
ANGBASE	0	
ANGDIR	0	
APBOX	0	
APERTURE	10	
AREA	1365453.6670	(read only)
ATTDIA	0	
ATTMODE	1	
ATTREQ	1	
AUDITCTL	0	
AUNITS	0	
AUPREC	0	
AUTOSNAP	39	

   et cetera...

4. With over 400 system variables, the listing is very long. Press ESC to end it early. To return to the drawing window, press F2.

EXERCISES

1. Start AutoCAD with a new drawing. Use this drawing for all the exercises in this chapter. Draw a line with endpoints at 2,2 and 6,6.

    With the **ID** command and ENDpoint object snap, select the first endpoint of the line.

    What coordinates does the command return?

    Repeat for the other end. Are the coordinates what you expect?

2. Draw a line, circle, and arc on the screen.

    Select the line with the **LIST** command.

    What is the line's angle in the XY Plane?

    Select the circle with **LIST** command.

    What is the circle's radius?

    Select the arc with **LIST** command.

    What is the arc's starting angle?

3. Continue with the drawing from exercise #2.

    Select the line with the **PROPERTIES** command.

    Deselect the line, and then select the arc.

    How does the data displayed by the Properties window change?

    Select all three objects. Does the Properties window display a smaller or larger amount of information?

4. Continue with the same drawing, and enter the **DBLIST** command.

    What do you notice about the information listed by AutoCAD?

    Use the **COPY** command's **Multiple** option to copy the objects three times.

    Use the **DBLIST** command again, and notice the length of the listing.

5. Continuing with the same drawing, use the **DIST** command with ENDpoint object snap.

    Select two endpoints of one of the lines on the screen.

    Is the length the same as reported by the **LIST** command?

    Repeat the **DIST** command, but use PERpendicular object snap to find the distance between a line and a circle.

6. Draw two rectangles of different sizes with the **RECTANG** command.

    Use the **AREA** command to determine the area of one rectangle.

    Which option of the **AREA** command did you use?

    Repeat the same procedure on the second rectangle. Add the two areas together, and obtain the total area.

    Use the **AREA** command, but this time use the **Add** option to find the area of both rectangles.

    Does AutoCAD arrive at the same answer as you?

7. From the CD, open the *area.dwg* file, a drawing of geometric shapes.
   Find the area of each object, and then write down the figures.
   Using the **Add** option, find the total area of all objects.

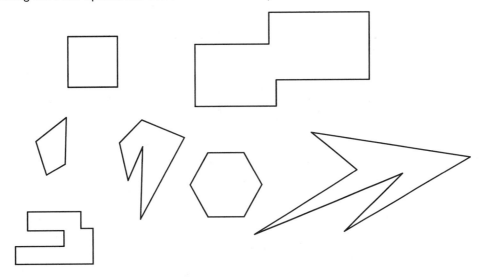

8. Use the **TIME** command to check the current time in the drawing.
   Is the time reported by AutoCAD the same as on your wristwatch?
   The same as shown on your computer?

9. Use the elapsed timer option of the **TIME** command to determine the time needed to complete the following exercise.
   Before starting, write down the time you estimate it will take you to complete exercise #10.

10. Draw the baseplate shown below.
    Use the **AREA** command to find the area.
    Hints: Use the **RECTANGLE** command with its **Fillet** option. Remember to use the **Subtract** option to remove the holes from the total.

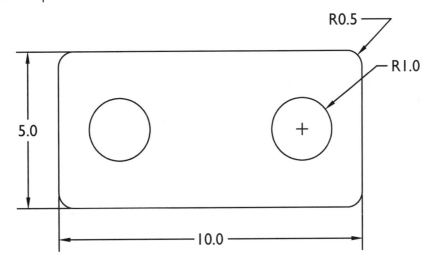

Turn off the timer. How close was your estimate?

11. From the CD, open the *34486.dwg* file, a drawing of the author's 50' x 25' basement floor plan. Each unit in the drawing represents a foot.

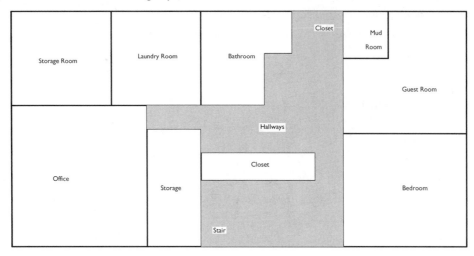

What is the total square footage of the basement?

Create a new layer called "Areas."

Draw rectangles that match each room.

Find the area of each room:

  Storage Room

  Laundry Room

  Bathroom

  Mud Room

  Guest Room

  Bedroom

  Closet

  Storage

  Office

What is the area of the hallways, including stairs?

12. Start a new drawing with *acad.dwt*.

Find the value of the following system variables:

  OFFSETGAPTYPE

  LASTPOINT

  AREA

  MAXSORT

If you do not understand the meaning of a system variable, refer to Appendix C of this book.

13. From the CD, open the *15_44.dwg* file, a drawing of a gasket.

    Ensure the **DELOBJ** system variable is set to 1.

    Start the **BOUNDARY** command, and set object type to **Region**. Click **Pick Points**, and select a point inside the gasket. How many regions is the gasket?

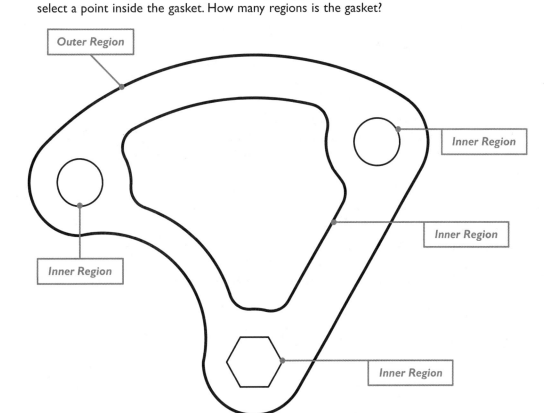

Outer Region

Inner Region

Inner Region

Inner Region

Inner Region

Start the **SUBTRACT** command, and then select the outermost region. When prompted "Select solids and regions to subtract," select the four inner regions (also called "islands").

Apply the **MASSPROP** command to the gasket, and then save the data to an *.mpr* file. Write down the values from the report:

Area: _____

Perimeter: _____

Bounding box: X: _____   Y: _____

Centroid: X: _____   Y: _____

Moments of inertia: X: _____   Y: _____

Product of inertia: XY: _____

Radii of gyration: X: _____   Y: _____

Principal moments and X-Y directions about centroid: I: _____

                                                                                   J: _____

Use the **POINT** command to place a point at the centroid of the gasket.

14. From the CD, open the *15_43.dwg* file, and find the centroid of the gasket.

CHAPTER REVIEW

1. List the two kinds of actions the **ID** command performs:

   a.

   b.

2. Where are the **ID** coordinates stored?

   How do you access them in other drawing and editing commands?

3. What happens when blipmode is turned on?

4. What is the difference among the **PROPERTIES, LIST,** and the **DBLIST** commands?

5. What are the six distances returned by the **DIST** command?

6. What option under **AREA** calculates the area of a circle?

7. After using the **AREA** and **LIST** commands on the circle, which command do you find more useful?

8. How do you determine the last time a drawing was updated?

9. Can the **AREA** command return the length of an open polyline?

10. Why would you want to subtract areas?

11. How do you stop the report generated by **DBLIST**?

12. Describe a purpose of the **DWGPROPS** dialog box.

13. List two reasons why the Total Editing Time might be inaccurate for drawings:

    a.

    b.

14. Can you change the value of all system variables?

UNIT V

Introducing
3D Modeling

CHAPTER 18

Creating and Viewing 3D Models

In 2D drafting, you draw and edit 2D objects — such as lines, arcs, and other basic objects — to construct drawings. In 3D designing, the process is similar: you draw and edit basic 3D objects — boxes, spheres, and other 3D objects — to produce drawings.

In this section of the book, you learn the basics of creating, editing, and rendering 3D models. *These introductory chapters do not present all of AutoCAD's 3D capabilities, only select portions.*

In this chapter, you find the following commands:

PERSPECTIVE switches between parallel and perspective projections.

DASHBOARD provides one-stop access to many 3D commands

SPHERE, BOX, CYLINDER, WEDGE, TORUS, PYRAMID, and **POLYSOLID** create basic 3D solid bodies.

UCS creates new coordinate systems.

3DORBIT changes the 3D viewpoint interactively.

PLAN returns the viewpoint to the plan view.

SWEEP, EXTRUDE and **REVOLVE** create 3D solid bodies from 2D objects.

UNION and **SUBTRACT** create 3D bodies through Boolean operations.

FINDING THE COMMANDS

On the Dashboard's **3D Make** panel:

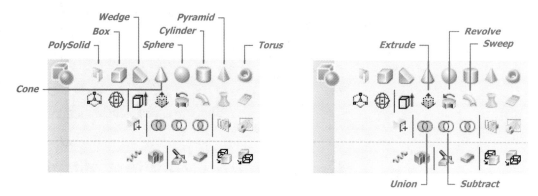

On the Dashboard's **3D Navigate** panel:

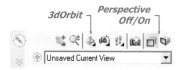

On the **Modeling** toolbar:

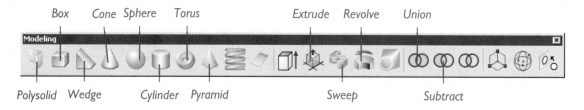

On the **3D Navagation** and **UCS** toolbars:

3D MODELING WORKSPACE

T1he difference between 2D and 3D lies in the nature of the basic objects, the kinds of editing tools available, and the extra dimension (z), which represents elevations and thicknesses. Editing consists of adding and subtracting solids from each other. For example, holes are created by subtracting cylinders from plates.

Solid modeling is used primarily in mechanical engineering, because models of different materials can be easily analyzed. Solid modeling is also sometimes used by architects to represent complex intersections, such as the roofs on buildings. After the models are constructed, they are usually "converted" to standard sets of 2D drawings for construction.

AutoCAD offers a 3D modeling environment: from the Workspace toolbar, select "3D Modeling."

The 3D Modeling workspace usually loads the *acad3d.dwt* template file that creates a 3D-like drawing environment. You may need to load the template manually if you switch during an existing drawing.

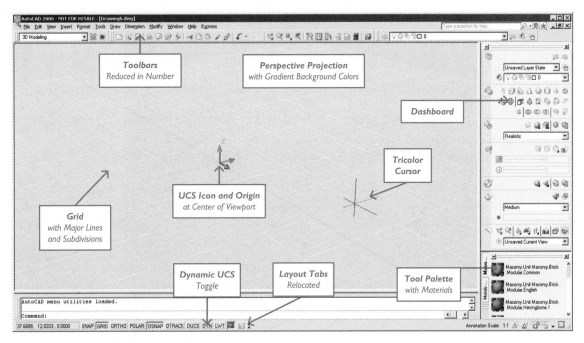

If you don't see the grid, click **GRID** on the status bar.

TOURING THE 3D MODELING WORKSPACE

The 3D Modeling workspace provides a drawing and editing environment suitable for creating 3D models. Autodesk, in fact, has tried to make 3D modeling in AutoCAD similar to working in its more-expensive Inventor mechanical CAD software.

Perspective Projection

The 3D Modeling workspace makes 3D look more realistic through the use of perspective projection. Perspective makes objects, such as the grid, look smaller when they are farther away. (In AutoCAD 2006 and earlier, you could only view models in perspective mode, and not edit them.)

In perspective mode, zooming is limited to real-time, all, extents, window, and previous. One of my favorites, **ZOOM** 0.9, doesn't work in perspective mode. Panning is in real-time only; you cannot pick two points.

You can change the projection between perspective and parallel with the **PERSPECTIVE** system variable. When set to 0, the projection is parallel; when set to 1, perspective.

> Command: **perspective**
>
> Enter new value for PERSPECTIVE <0>: **1**

To help with the illusion of 3D projection, the colors fade from top to bottom to mimic the gradient colors of the sky. These colors do not plot. You can change the colors in the **OPTIONS** command's dialog box: select the **Display** tab, and then click the **Colors** button.

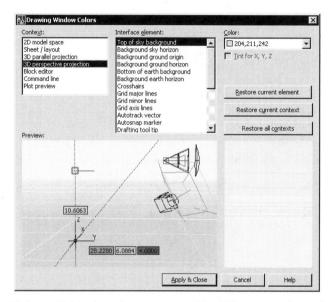

The Drawing Window Colors dialog box lets you specify different color schemes for parallel and perspective projections. For clarity, I prefer to set all background colors to white.

Grid

The grid consists of horizontal and vertical lines, instead of dots. The grid line that matches the x axis is red, and the y axis green. The grid can extend to infinity, or be limited by the setting of the **LIMITS** command. The display of the grid is toggled with the **GRID** button on the status bar.

The grid is initially located at z=0, and can be relocated higher and lower on the z axis with the **ELEV** command. As well, it can temporarily align itself with the dynamic UCS, as described later. Grid lines are unaffected by lineweight and linetype, and they do not plot. (The grid reverts to dots when the visual style is set to "2D wireframe.")

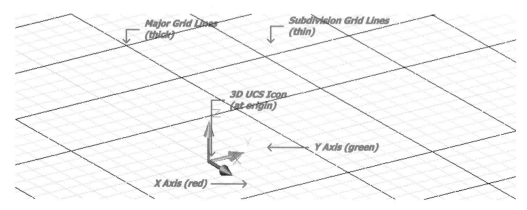

As you zoom in, additional grid lines appear. These are called "subdivisions." Similarly, when you zoom out, fewer grid lines appear. The main grid lines are slightly thicker than the subdivision lines.

You can change the grid settings with **DSETTINGS** and **OPTIONS** commands. In the **Snap and Grid** tab of the Drafting Settings dialog box, you can make these changes:

- **Grid X** and **Y Spacing** — specify the spacing between major grid lines; default = 0.5 units.

- **Major Line Every** — specifies how many subdivision lines appear between each major line; default = 5 (range is 1 to 100).

- **Adaptive Grid** — reduces the number of grid lines when zooming out; default = on.

- **Allow Subdivision Below Grid Spacing** — increases the number of grid lines when zoomed in; default = off.

- **Display Grid Beyond Limits** — increases the number of grid lines when zooming in; default = on.

- **Follow Dynamic UCS** — causes the grid plane to match the dynamic UCS temporarily; default = off.

In the **Display** tab of the Options dialog box, you can make the major and minor grid lines colored or black — as you can the lines where the grid matches the x and y axes. (Click the **Color** button, and then select a grid option. For colored axis lines, select the **Tint for X, Y, Z** option.)

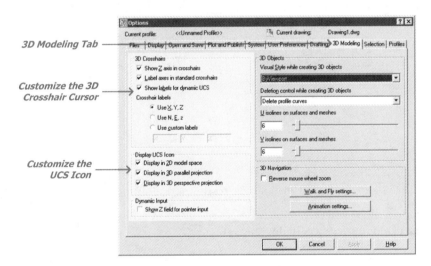

Tricolor Cursor

The 3D cursor shows the three axes, and is color coded: red = x, green = y, and blue = z. The crosshairs of the cursor are asymmetrical: they are longer in the direction of the positive axes than in the negative.

The three cursor axes are labeled X, Y, and Z. You can turn off the labels, or change them to N, E, and z (useful for surveyors), or enter any eight-character words of your choosing. Make these changes in the **3D Modeling** tab of the OPTIONS command's dialog box. In the same tab, you can toggle the display of the z crosshair. The **Display** tab's **Color** button changes the axes between colored and black. (For colored cursor crosshairs, select the **Tint for X, Y, Z** option.)

UCS Icon

The UCS icon appears at the center of the screen, which is initially at the drawing's origin of 0,0,0. Its properties can be changed with the UCSICON command's **Settings** option.

In the **3D Modeling** tab of the Options dialog box, you can specify whether the UCS icon appears in the three model space viewing modes, 2D, 3D parallel, and 3D perspective.

Dynamic UCS

The **DUCS** button on the status bar toggles dynamic UCS (user-defined coordinate system) mode. When on, the UCS automatically aligns itself with the selected face of a 3D model. In the figure below, a circle is easily drawn on the face of the pyramid, because of dynamic UCS.

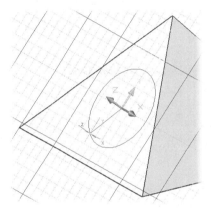

 Notice how the UCS icon's x and y axes are aligned with the face — as is the grid. If snap and/or polar modes are turned on, they also adjust to the new alignment.

To have the grid align with dynamic UCS, turn on the **Follow Dynamic UCS** option in the **Snap and Grid** tab of the DSETTINGS dialog box.

Layout Tabs

The 3D Modeling workspace moves the model and layout tabs to buttons on the status bar. That's because the emphasis in 3D modeling is on model space; paper space is incidental, and is only used to lay out 2D views of the 3D model just before plotting.

The traditional location for layout tabs is next to the scroll bar. You can return the tabs there by right-clicking the **MODEL** or **PAPER** button, and then selecting **Display Layout and Model Tabs** from the shortcut menu.

3D TERMS

Edges outline the faces on non-curved surfaces, such as cubes and the base of cones.

Faces represetn the surface areas between edges.

Vertices locate the corners of edges.

Isolines represent curved surfaces, such as on cones and spheres; the **ISOLINES** system variable changes the number of isolines.

Facets approximate curved surfaces.

Intersection lines appear where solid parts intersect each other.

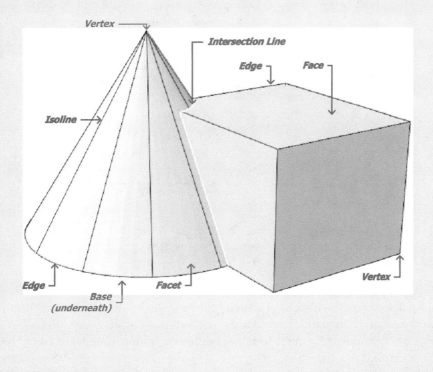

Dashboard Palette

There are fewer toolbars in the 3D Modeling workspace, because the Dashboard palette takes over the function of most of them. If the Dashboard is not displayed, you can open it with the **DASHBOARD** command. It is described in greater detail later.

Tools Palette

The Tools palette is greatly expanded with material and lighting palettes. (We will ignore it until Chapter 20, "Rendering 3D Models.")

Notes: Despite these many 3D user interface elements, AutoCAD still limits you to drawing and editing in the x,y-plane — in 2D. In fact, the z coordinate on the status bar reports the *elevation*, not the z location of the cursor.

The 3D commands are not limited to 3D workspaces; they can also be executed in 2D workspaces.

 **DASHBOARD**

The **DASHBOARD** command displays the Dashboard palette, which provides one-stop access to many of the commands and settings used in 3D modeling, editing, and rendering.

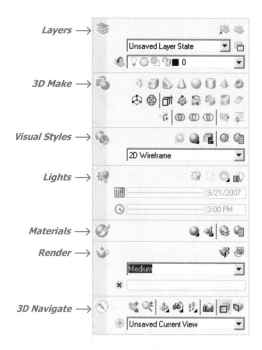

TOURING THE 3D DASHBOARD

The Dashboard palette consists of sections called "control panels." Each panel has buttons, droplists, and sliders that activate commands and change the values of system variables. The 3D Modeling workspace displays the following panels, which are different from those displayed by the 2D Drafting & Annotation workspace, for the most part:

Layers controls layers and layer states; same as in the 2D workspace.

3D Make creates 3D models by combining primitives and converting 2D objects.

Visual Styles specifies visual styles, shadow casting, and sketch effects.

Lights adds point, distant, and spot lights to drawings, and controls the sun light.

Materials attaches materials and textures to models in conjunction with the Tools palette.

Render renders 3D models.

3D Navigate provides 3D viewing.

In this and the following chapters, we use the Dashboard palette to access commands and settings.

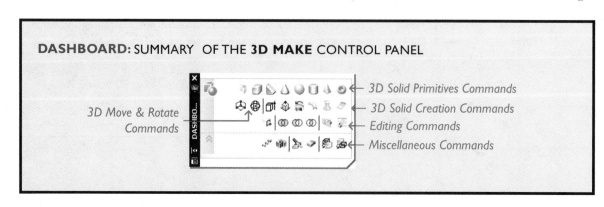

DRAWING 3D PRIMITIVES

One method of creating solid models is with solid primitives, so-called because they are the most basic building blocks — cubes, balls, and so on. In this chapter, we look at these:

Box — draws cubic and rectangular box shapes.

Cylinder — draws cylindrical shapes; often used to make holes.

Cone — draws cone shapes.

PolySolid — draws wall shapes.

Pyramid — draws pyramid shapes.

Sphere — draws ball shapes.

Torus — draws donut shapes.

Wedge — draws wedge shapes.

Before starting the following tutorials, set the visual style to Conceptual:

Command: **vscurrent**

Enter an option [2dwireframe/3dwireframe/3dHidden/Realistic/Conceptual/Other]

<3dwireframe>: **c**

See Chapter 20. "Rendering 3D Models," for more about visual styles.

 SPHERE

The **SPHERE** command creates 3D solid model balls.

This command draws spheres only, and so is the easiest one to use. All AutoCAD needs to know is a location for the center of the sphere and its radius (or diameter).

BASIC TUTORIAL: DRAWING SOLID SPHERES

1. To draw 3D solid spheres, start the **SPHERE** command:
   - From the **Draw** menu, choose **Modeling**, and then **Sphere**.
   - In the Dashboard palette, click the **Sphere** button.
   - At the 'Command:' prompt, enter the **sphere** command.

   Command: **sphere** *(Press* ENTER.*)*

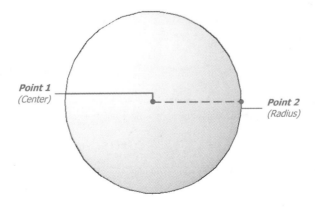

Point 1 (Center)   *Point 2* (Radius)

2. In all cases, AutoCAD prompts you to specify a point for the sphere's center. Pick the point with the cursor, or enter x, y, z coordinates.

    Specify center point or [3P/2P/Ttr]: 0,0 *(Pick point 1, or enter x,y,z coordinates.)*

3. Indicate the size of the sphere by specifying its radius. Enter a value, or show the radius by picking a point with the cursor.

    Specify radius or [Diameter] <32.9>: *(Type a value, or pick point 2.)*

AutoCAD positions the sphere so that its central axis is parallel to the z axis of the current UCS.

DRAWING SPHERES: ADDITIONAL METHODS

The **SPHERE** command has several options that determine the size of the sphere. In addition, a system variable controls the look of all curved solids.

*   **3P** option specifies the diameter by three points.

*   **2P** option specifies the diameter by two points.

*   **Ttr** option specifies the diameter by two points tangent to other objects, plus the radius.

*   **Diameter** option specifies the diameter of the sphere.

*   **ISOLINES** system variable specifies the number of isolines on curved surfaces of solid models.

Let's look at them.

3P

The **3P** option specifies the circumference of the sphere by three points.

> Command: **sphere**
>
> Specify center point or [3P/2P/Ttr]: **3p**
>
> Specify first point: *(Pick point 1, or enter x,y,z coordinates.)*
>
> Specify second point: *(Pick point 2, or enter x,y,z coordinates.)*
>
> Specify third point: *(Pick point 3, or enter x,y,z coordinates.)*

The three points define the sphere's position in space. The z coordinate is equal to the current elevation setting.

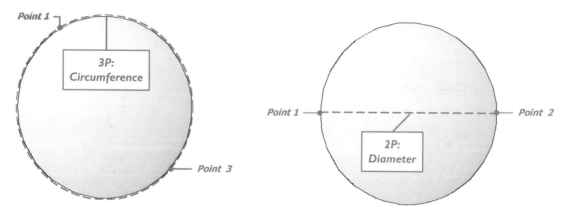

Left: *Defining the sphere by 3 points on its circumference.*
Right: *Defining the sphere by 2 points on its diameter.*

2P

The **2P** option specifies the diameter by two points:

> Command: **sphere**
>
> Specify center point or [3P/2P/Ttr]: **2p**

Specify first end point of diameter: *(Pick point 1, or enter x,y,z coordinates.)*

Specify second end point of diameter: *(Pick point 2, or enter x,y,z coordinates.)*

The two points define the sphere's position in space. The z coordinate of the first endpoint determines the elevation of the sphere.

TTR

The **Ttr** option specifies the diameter by two points tangent to other objects (circles, arc, polyline, or line), plus the radius.

Command: **sphere**

Specify center point or [3P/2P/Ttr]: **t**

Specify point on object for first tangent: *(Pick object 1.)*

Specify point on object for second tangent: *(Pick object 2.)*

Specify radius of circle <15.0>: *(Type a value, or pick two points.)*

You can press **ENTER** to reuse the previous radius or diameter values.

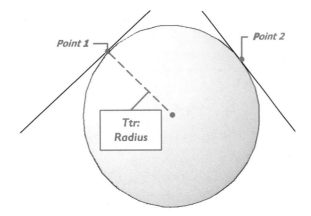

Diameter

The **Diameter** option specifies the diameter of the sphere.

Command: **sphere**

Specify center point or [3P/2P/Ttr]: *(Pick a point.)*

Specify radius or [Diameter] <22.0>: **d**

Specify diameter <44.0>: *(Type a value, or pick a point.)*

The diameter is twice the radius.

Isolines

The **ISOLINES** system variable specifies the number of isolines on curved surfaces of solid models.

Curved surfaces have *isolines*, the vertical and curved lines that show the curvature of solid models and faces. Technically, isolines indicate constant values. An isoline on a solid model connects points of equal value. In the case of spheres, isolines show a constant radius from the center. The latitudinal isolines are parallel to the x,y-plane.

Isolines are a display aid, like grid dots. To turn on the display of isolines, change the value of the **VSEDGES** system variable to 1, as follows:

Command: **vsedges**

Enter new value for VSEDGES <2>: **1**

(A value of 0 turns off the display of isolines, while 2 displays facet edges.) The number of isolines on curved solids can be changed with the **ISOLINES** system variable, followed by the **REGEN** command.

You can enter a value ranging from 0 to 2047, but a value of 12 or 16 looks best.

Command: **isolines**

Enter new value for ISOLINES <4>: **12**

Command: **regen**

Regenerating model.

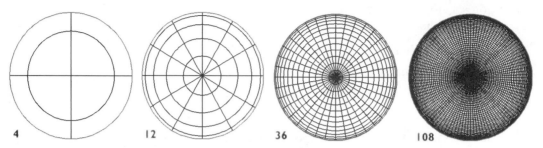

4   12   36   108

Changing the number of isolines from 4 (default) through to 108; shown in plan view.

Note: When you look at drawings straight down the z axis, this is called the "plan view" (as illustrated by the spheres above). To get a better idea of what 3D objects look like, change the viewpoint with the **VIEWPOINT** command, as follows:

Command: **vpoint**

Specify a view point or [Rotate] <display compass and tripod>: **2.5,-2,1**

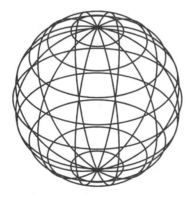

Alternatively, select a viewpoint from the 3D Navigate panel on the Dashboard palette.

BOX

The BOX command creates solid model boxes that are cubic or rectangular in shape.

To draw a box, AutoCAD needs to know three things: (1) a corner point of the base, (2) the opposite corner of the base, and (3) the height. The base is drawn in the current UCS.

BASIC TUTORIAL: DRAWING SOLID BOXES

1. To draw 3D solid boxes, start the BOX command:
   • In the Dashboard palette, click the **Box** button.
   • From the **Draw** menu, choose **Modeling**, and then **Box**.
   • At the 'Command:' prompt, enter the **box** command.

Command: **box** *(Press ENTER.)*

2. In all cases, AutoCAD prompts you for the location of a corner:

    Specify corner or [Center] <0,0,0>: *(Pick point 1, or enter x, y, z coordinates.)*

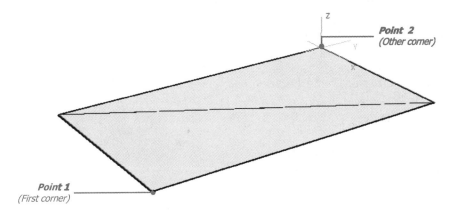

3. Locate the other corner to form the base of the box:

    Specify other corner or [Cube/Length]: *(Pick point 2, or enter x, y, z coordinates.)*

4. Indicate the height:

    Specify height or [2Point] <26.4780>: *(Show the height, or enter a value.)*

    Notice that the box grows as you move the cursor.

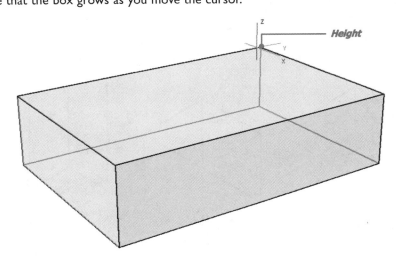

DRAWING BOXES: ADDITIONAL METHODS

The **BOX** command has a number of options that control the size of the box:

- **Center** specifies the center of the box.

- **Cube** draws a cube.

- **Length** specifies the lengths of all three sides.

- **2Point** specifies the height by picking two points.

Let's look at them.

Center

The **Center** option specifies the center of boxes. The "center" is the 3D center, including the z axis:

    Command: **box**

    Specify first corner or [Center]: **c**

Specify center: *(Pick a point, or enter x,y,z coordinates.)*

Specify corner or [Cube/Length]: *(Pick another point, or enter coordinates.)*

Specify height or [2Point] <9.0>: *(Indicate the height.)*

Cube

The **Cube** option draws cubes, where all three sides have the same length:

Command: **box**

Specify first corner or [Center]: *(Pick a point, or enter x,y,z coordinates.)*

Specify other corner or [Cube/Length]: **c**

Specify length: *(Indicate a length.)*

Length

The **Length** option specifies the lengths of all three sides:

Command: **box**

Specify first corner or [Center]: *(Pick a point, or enter x,y,z coordinates.)*

Specify corner or [Cube/Length]: **l**

Specify length: *(Indicate the length of the box's base.)*

Specify width: *(Indicate the width of the base.)*

Specify height or [2Point] <9.0>: *(Indicate the height.)*

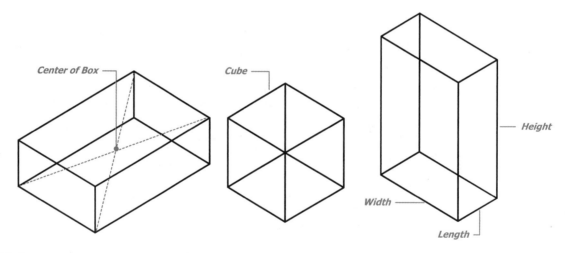

2Point

The **2Point** option specifies the height by picking two points. This allows you to specify the height indirectly, where the box's height matches the size of another object. Note that you need to type "2p" as the abbreviation for this option:

Command: **box**

Specify first corner or [Center]: *(Pick a point, or enter x,y,z coordinates.)*

Specify corner or [Cube/Length]: *(Pick another point.)*

Specify height or [2Point] <9.0>: **2p**

Specify first point: *(Pick point 1.)*

Specify second point: *(Pick point 2.)*

When you pick the points in reverse order from that illustrated below, the box is drawn below the current elevation.

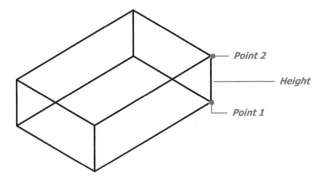

Point 2
Height
Point 1

Notes: AutoCAD draws boxes so that their sides are parallel to the x, y, and z axes of the current UCS. Enter negative values for coordinates and for the height to draw the box downward and/or to the left.

Boxes are displayed by their edges. Isolines do not apply to boxes, because they have no curved surfaces. The edge lines can be turned off with the **No Edges** button in the Dashboard (Visual Styles panel).

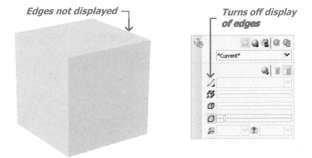

Edges not displayed   *Turns off display of edges*

Sometimes it is easier to work with solids when they are translucent — you can partially see through them. Translucency is turned on with the **Xray Mode** button in the Dashboard (Visual Styles panel).

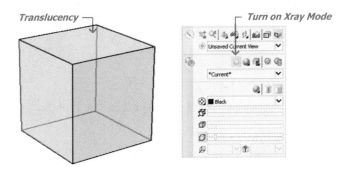

Translucency   *Turn on Xray Mode*

POLYSOLID

The **POLYSOLID** command draws 3D walls.

This command draws 3D walls in a manner similar to drawing polylines. As you pick points, the wall is drawn. Polysolid walls can have straight and arc sections that connect as a single object, just like polylines. The walls are made of the Sweep object.

Alternatively, this command can convert objects to polysolids: lines, 2D polylines, arcs, and circles.

The **PSOLWIDTH** and **PSOLHEIGHT** system variables specify the default width and height of polysolid walls, and you can change the values in the command. The default width is 0.25 units (5 metric); the default height is 4 units (80 metric).

BASIC TUTORIAL: DRAWING POLYSOLIDS

1. To draw 3D walls, start the **POLYSOLID** command:

   - In the Dashboard palette, click the **Polysolid** button.

   - From the **Draw** menu, choose **Modeling**, and then **Polysolid**.

   - At the 'Command:' prompt, enter the **polysolid** command.

   - Alternatively, enter the **psolid** alias.

        Command: **polysolid** *(Press ENTER.)*

2. In all cases, AutoCAD prompts you at the command line:

        Specify start point or [Object/Height/Width/Justify] <Object>: *(Pick a point or enter an option.)*

3. Pick points in the drawing. As you do, AutoCAD draws the wall.

        Specify next point or [Arc/Undo]: *(Pick another point or enter an option.)*

        Specify next point or [Arc/Undo]: *(Pick another point or enter an option.)*

4. Press **ENTER** to exit the command:

        Specify next point or [Arc/Close/Undo]: *(Press ENTER to exit the command.)*

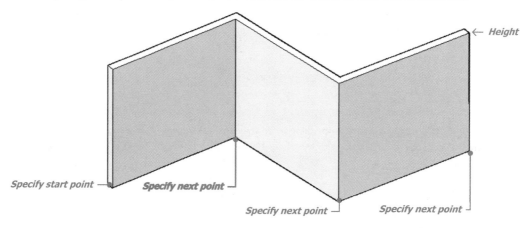

Specify start point — *Specify next point* — *Specify next point* — *Specify next point* — ← *Height*

DRAWING POLYSOLIDS: ADDITIONAL METHODS

The **POLYSOLID** command has a number of options that control the look and size of the 3D walls.

   - **Object** converts 2D objects into 3D walls.

   - **Height** and **Width** specify the width and height of walls.

   - **Justify** determines how walls align with pick points.

   - **Arc** creates curved walls.

   - **Close** draws a segment that closes the walls.

   - **Undo** undraws the last wall segment.

Objects

The **Objects** option creates 3D walls from 2D objects. This command converts just one object at a time.

        Specify start point or [Object/Height/Width/Justify] <Object>: **o**

        Select object: *(Pick a line, circle, arc, or polyline.)*

The figure below shows a 3D wall made from a circle. If you want to use a polyline as the source object, the polyline cannot be self-intersecting. Use the **DELOBJ** system variable to control whether the source objects are erased from drawings.

Height and Width

The **Height** and **Width** options specify the width and height of walls. The values must be greater than 0.

> Specify start point or [Object/Height/Width/Justify] <Object>: **h**
>
> Specify height <4.0>: *(Enter a value.)*
>
> Specify start point or [Object/Height/Width/Justify] <Object>: **w**
>
> Specify width <0.25>: *(Enter a value.)*

The default values are stored in the **PSOLHEIGHT** and **PSOLWIDTH** system variables. For example, you may want to change the default height to 8 feet and width to 4 inches, the standard size for walls in North America.

Justify

The **Justify** option determines how walls align with pick points in a manner similar to multilines.

> Specify start point or [Object/Height/Width/Justify] <Object>: **j**
>
> Enter justification [Left/Center/Right] <Center>: *(Enter an option.)*

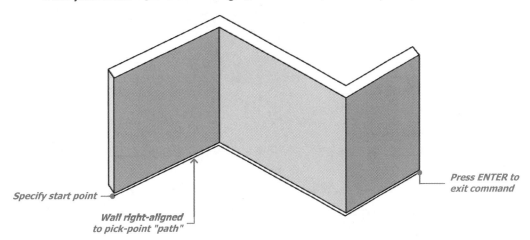

Specify start point

Wall right-aligned to pick-point "path"

Press ENTER to exit command

Left aligns the left edge of the wall to the path marked by your pick points; **Center** centers the wall; and **Right** aligns to the right edge.

Arc

The **Arc** option creates curved walls in a manner similar to drawing polyline arcs. See the **POLYLINE** command earlier in this book for an explanation of options. The arcs are not true curves, but comprised of many short segments.

> Specify next point or [Arc/Undo]: **a**
>
> Specify endpoint of arc or [Close/Direction/Line/Second point/Undo]: *(Pick a point, or enter an option.)*

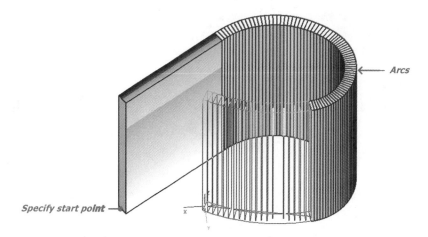

Arcs

Specify start point

 Note: You can apply twist angles along 3D walls to create interesting effects. Here's how: select the wall, right-click, and then select **Properties** from the shortcut menu. Change the **Twist Along Path** angle from 0 degrees. The figure below illustrates a twist angle of 180 degrees applied to a wall made of arcs.

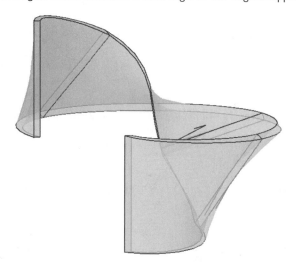

 CYLINDER

The **CYLINDER** command creates solid model cylinders, both round and elliptical, as well as straight and slanted.

To draw a cylinder, AutoCAD needs to know three things: (1) the center point of the base, (2) the radius, and (3) the height of the cylinder. The base is drawn in the current UCS. Cylinders can be used to create holes, by subtracting them from other solids.

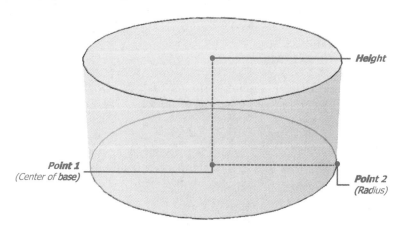

Height

Point 1
(Center of **base**)

Point 2
(Radius)

BASIC TUTORIAL: DRAWING SOLID CYLINDERS

1. To draw 3D solid cylinders, start the **CYLINDER** command:
   * In the Dashboard palette, click the **Cylinder** button.
   * From the **Draw** menu, choose **Modeling**, and then **Cylinder**.
   * At the 'Command:' prompt, enter the **cylinder** command.
   * Alternatively, enter the **cyl** alias.

     Command: **cylinder** *(Press ENTER.)*

2. In all cases, AutoCAD prompts you to specify a point for the cylinder's center. You can pick the point with the cursor, or enter x,y,z coordinates.

   Specify center point of base or [3P/2P/Ttr/Elliptical]: *(Pick point 1, or enter x, y, z coordinates.)*

3. Indicate the radius of the cylinder by entering a value, or show the radius by picking a point with the cursor.

   Specify base radius or [Diameter] <7.5>: *(Type a value, or pick point 2.)*

4. Indicate the height of the cylinder. Enter a negative value to draw the cylinder downwards.

   Specify height or [2Point/Axis endpoint] <15.0>: *(Type a value, or show the height.)*

DRAWING CYLINDERS: ADDITIONAL METHODS

The CYLINDER command has a number of options that control the shape and size of cylinders.

   * **2P**, **3P**, **Ttr**, **Diameter**, and **2Point** operate identically to those of the **SPHERE** and **BOX** commands.
   * **Elliptical** creates a cylinder with an elliptical base.
   * **Axis endpoint** creates a slanted cylinder.

Let's look at the two that are different from earlier commands.

Elliptical

The **Elliptical** option creates cylinders with elliptical bases. Some of the prompts are similar to those of the ELLIPSE command.

   Command: **cylinder**
   Specify center point of base or [3P/2P/Ttr/Elliptical]: **e**
   Specify endpoint of first axis or [Center]: *(Pick a point.)*
   Specify other endpoint of first axis: *(Pick another point.)*
   Specify endpoint of second axis: *(Pick a point.)*
   Specify height or [2Point/Axis endpoint] <12.1>: *(Indicate the height.)*

Axis Endpoint

The **Axis endpoint** option creates slanted cylinders. (This option was formerly **Center of other end**.)

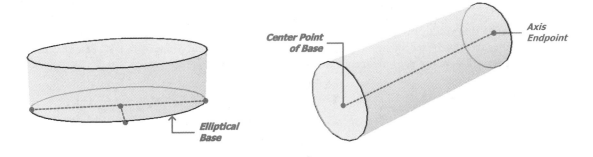

Elliptical Base

Center Point of Base

Axis Endpoint

The cylinder's centerline is drawn between the two points that define the center of the base and the endpoint of the axis:

> Specify height or [2Point/Axis endpoint] <7.5>: **a**
>
> Specify axis endpoint: *(Pick a point, or enter x,y,z-coordinates.)*

 ## WEDGE

The **WEDGE** command creates solid model wedges, with square or rectangular bases.

To draw a wedge, AutoCAD needs to know three things: (1) the location of one corner of the base, (2) the opposite corner, and (3) the height. AutoCAD doesn't allow you to pick the location of the sloping face, so you need to memorize its location: "The sloped face rises from the first corner."

TUTORIAL: DRAWING SOLID WEDGES

1. To draw 3D solid wedges, start the **WEDGE** command:
    *   In the Dashboard palette, click the **Wedge** button.
    *   From the **Draw** menu, choose **Modeling**, and then **Wedge**.
    *   At the 'Command:' prompt, enter the **wedge** command.

        Command: **wedge** *(Press* ENTER.*)*
    *   Alternatively, enter the **we** alias.

2. The prompts are similar to those of the **BOX** command:

    Specify first corner or [Center]: *(Pick point 1, enter x,y,z coordinates, or type* **C**.*)*

3. Locate the other corner to form the base of the box:

    Specify other corner or [Cube/Length]: *(Pick point 2, enter x,y,z coordinates, or enter an option.)*

4. Indicate the height:

    Specify height or [2Point]: *(Show the height, enter a value, or type* **P**.*)*

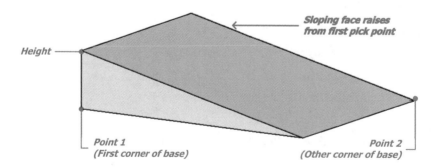

The **CEnter**, **Cube**, **Length**, and **2Point** options are identical to those of the BOX command.

> **Notes:** When you specify a 3D point for the wedge's second corner, AutoCAD extracts the x,y-values and uses them as the second point. The z value is applied to the height. Snapping to a 3D point has the same result. The wedge does not anchor to the snapped point.

CONE

The CONE command creates straight and slanted solid model cones, with round and elliptical bases.

To draw a cone, AutoCAD needs to know three things: (1) the center point of the base, (2) the radius of the base, and (3) the height of the cone. The base is drawn in the UCS.

BASIC TUTORIAL: DRAWING SOLID CONES

1. To draw 3D solid cones, start the **CONE** command:
   - In the Dashboard palette, click the **Cone** button.
   - From the **Draw** menu, choose **Modeling**, and then **Cone**.
   - At the 'Command:' prompt, enter the **cone** command.

     Command: **cone** *(Press* ENTER.*)*

2. The prompts are similar to those of the **CYLINDER** command:
   Specify center point of base or [3P/2P/Ttr/Elliptical]: *(Pick point 1, enter x,y,z coordinates, or enter an option.)*

3. Indicate the radius of the cone's base by entering a value, or show the radius by picking a point with the cursor.
   Specify base radius or [Diameter]: *(Pick point 2, enter a value, or type* **D**.*)*

4. Indicate the height of the cone. Enter a negative value to draw the cone downwards.
   Specify height or [2Point/Axis endpoint/Top radius] <1.0> *(Show the height, enter a value, or enter an option.)*

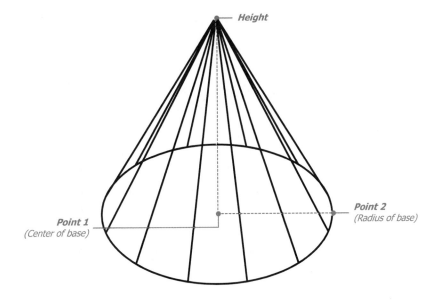

DRAWING CONES: ADDITIONAL METHODS

The CONE command has a number of options that change the shape of the cone. Many of them are identical to that of the CYLINDER command.

- **Elliptical** creates a cone with an elliptical base.
- **Diameter** specifies the diameter of the cone's base.
- **Top radius** creates flat-topped cones.

Elliptical

The **Elliptical** option creates a cone with an elliptical base in a manner identical to the CYLINDER command.

> Specify center point of base or [3P/2P/Ttr/Elliptical]: **e**
>
> Specify endpoint of first axis or [Center]: *(Pick point 1.)*
>
> Specify other endpoint of first axis: *(Pick point 2.)*
>
> Specify endpoint of second axis: *(Pick point 3.)*
>
> Specify height or [2Point/Axis endpoint/Top radius] <2.9>: *(Indicate the height.)*

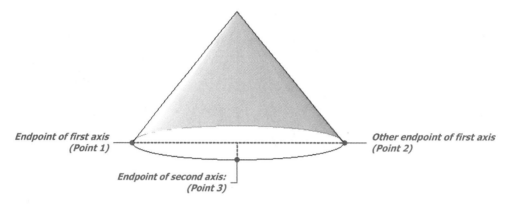

Diameter

The **Diameter** option specifies the diameter of the cone.

> Specify radius for base of cone or [Diameter]: **d**
>
> Specify diameter for base of cone: *(Enter a value, or pick two points.)*

Top radius

The **Top radius** option creates topped cones.

> Specify height or [2Point/Axis endpoint/Top radius] <5.8>: **t**
>
> Specify top radius <0.0>: *(Enter a value, or pick two points.)*

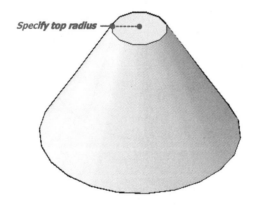

 **PYRAMID**

The PYRAMID command creates 3D solid pyramids with pointy or flat tops, and 3 to 32 sides.

To draw a pyramid, AutoCAD needs to know three things: (1) the center point of the base, (2) the radius of the base, and (3) the height of the pyramid. The base is drawn in the current UCS.

BASIC TUTORIAL: DRAWING SOLID PYRAMIDS

1. To draw 3D solid pyramids, start the PYRAMID command:
   - In the Dashboard palette, click the **Pyramid** button.

   - From the **Draw** menu, choose **Modeling**, and then **Pyramid**.

   - At the 'Command:' prompt, enter the **pyramid** command.

     Command: **pyramid** *(Press* ENTER.*)*

   - Alternatively, enter the **pyr** alias.

2. In all cases, AutoCAD reports the default settings for the pyramid:

   4 sides Circumscribed

   Specify center point of base or [Edge/Sides]: *(Pick point 1, or enter an option.)*

   Specify base radius or [Inscribed] <2.7>: *(Enter a value, pick point 2, or type I.)*

   Specify height or [2Point/Axis endpoint/Top radius] <3.5406>: *(Enter a value, pick point 3, or enter an option.)*

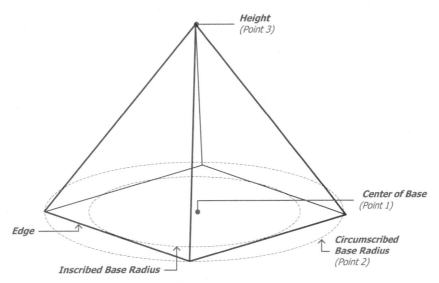

DRAWING PYRAMIDS: ADDITIONAL METHODS

The PYRAMID command has a number of options that change the shape of the pyramid. Many of them are similar to those of the POLYGON command.

   - **Edges** specifies the length of one edge of the pyramid's base.

   - **Sides** specifies the number of sides, ranging from 3 to 32.

   - **Inscribed** fits the pyramid's base outside an imaginary circle.

   - **2Point** and **Axis endpoint** create pyramids at an angle.

   - **Top radius** creates flat-topped pyramids.

 **TORUS**

The **TORUS** command creates solid model donuts.

The torus is a tube wrapped around in a circle. To draw a torus, AutoCAD needs to know three things: (1) the center point, (2) the radius, and (3) the diameter of the tube. The torus is drawn parallel to the plane of the UCS.

BASIC TUTORIAL: DRAWING SOLID TORI

1. To draw 3D solid donuts, start the **TORUS** command:
   * In the Dashboard palette, click the **Torus** button.
   * From the **Draw** menu, choose **Modeling**, and then **Torus**.
   * At the 'Command:' prompt, enter the **torus** command.

     Command: **torus** *(Press* ENTER.*)*
   * Alternatively, enter the **tor** alias.

2. In all cases, AutoCAD asks for the center of the torus:
   Specify center point or [3P/2P/Ttr]: *(Pick a point, enter x,y,z coordinates, or enter an option.)*

3. Indicate the radius of the torus:
   Specify radius or [Diameter] <2.0>: *(Pick a point, enter a value, or type* **D**.*)*

4. And indicate the radius of the tube:
   Specify tube radius or [2Point/Diameter]: *(Pick a point, enter a value, or enter an option.)*

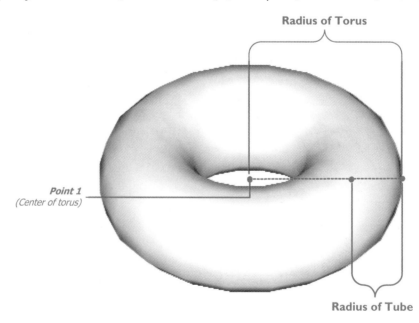

Radius of Torus

Point 1
(Center of torus)

Radius of Tube

The **TORUS** command has additional options for drawing donuts, which are identical to drawing cylinders.

You can specify radii and diameters that "don't make sense," such as when the tube radius exceeds the torus radius. For example, create a torus whose radius is 2, but with tube radius 5. Notice that the tube intersects the center point of the torus.

Another case is giving a negative value to the torus radius. (The tube radius must be a positive number of greater value.) For example, construct a torus with a torus radius of –2 and a tube radius of 5. (The tube radius must be greater than 2 in this case.) The two cases are illustrated below.

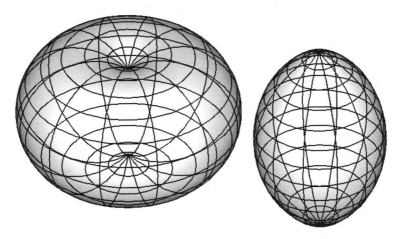

Left: *Pumpkin torus drawn with tube radius greater than the torus radius.*
Right: *Football torus drawn with a negative torus radius.*

VIEWING IN 3D

Computer interfaces are inherently two-dimensional: the screen is flat, the mouse moves on flat surfaces, and the output is on paper — all are 2D. The flat nature of computer interfaces makes it difficult to work with CAD's three-dimensional drawings.

To make it easier to visualize, AutoCAD simulates the third dimension through shading and perspective views. It has adapted its 2D interface so that we can interact with 3D objects.

In this section, you learn about some of AutoCAD's commands for manipulating views in 3D:

> **UCS** creates new coordinate systems.

> **PLAN** returns the viewpoint to the plan view.

> **3DORBIT** changes the 3D viewpoint interactively.

INTRODUCTION TO UCS

To create and edit 3D drawings, you must become comfortable with changing the point of view, and working at any angle in space.

AutoCAD provides a number of tools that create and save 3D viewpoints, such as **UCS** for orienting 2D work planes in 3D space, **3DORBIT** for changing the viewpoint interactively, and **PLAN** for returning to 2D plan view. (*UCS* is short for *user-defined coordinate system.*)

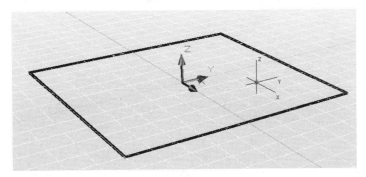

If there are coordinate systems that can be defined by the user, then there must be another, absolute coordinate system to which they refer. It is named the "world coordinate system" (WCS, for short).

When you start AutoCAD, it is in WCS mode: you look down on the x, y-plane, with the z axis coming out of the monitor toward your face. You can change the viewpoint, and it is still the WCS. But when you change the coordinate system, you define a new one — a UCS.

Think about drawing on the surface of a sloped roof: the task is hard if you cannot reorient the drawing plane. The workaround is to create UCSs with the **UCS** command. The UCS icon and the 3D cursor point show you the direction of the x, y, and z axes.

 UCS

The **UCS** command creates and modifies user coordinate systems.

New coordinate systems are created with the **New** option of the **UCS** command. You need to tell AutoCAD two things: (1) the x, y, z coordinates of the origin, and (2) the direction of the x, y, z axes.

BASIC TUTORIAL: CREATING A UCS

Before starting this tutorial, turn on snap and grid. Draw a square with the **RECTANG** command. If necessary, rotate the viewpoint:

> Command: **viewpoint**
>
> Specify a view point or [Rotate] <display compass and tripod>: **60,-40,28**

The square and the UCS icon look rotated:

1. To create new UCSs, start the **UCS** command:
    - From the **Tools** menu, choose **New UCS**.
    - At the 'Command:' prompt, enter the **UCS** command.

    Command: **ucs** *(Press ENTER.)*

2. AutoCAD reports the name of the current UCS, and then displays a long prompt.
    Current ucs name: *NO NAME*

    Specify origin of UCS or [Face/NAmed/OBject/Previous/View/World/X/Y/Z/ZAxis] <World>:

    To align the coordinate system with the current viewpoint, enter **v**. Think of this as aligning the UCS to the current view. (This is probably the easiest option, but perhaps the least useful, because it does not necessarily align with features on 3D objects.)
    Notice that the UCS icon rotates to show just the x and y axes, just as if you were looking at the drawing in plan view. In addition, the grid "faces you."

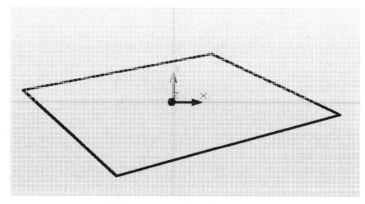

3. Draw another square, this time with the **RECTANG** command. Give the polyline a width of 0.5 to distinguish it from the first square. Notice that you are drawing in the new x,y-working plane.

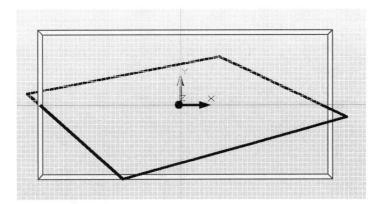

4. To return to the WCS, enter the **World** option:

    Command: **ucs**

    Enter an option [New/Move/orthoGraphic/Prev/Restore/Save/Del/Apply/?/World] <World>: **w**

By changing the coordinate system to match the viewpoint, you now draw just as if you were in plan view.

TUTORIAL: CREATING A WORKING PLANE

In this tutorial, you learn that UCSs make it possible to draw at strange angles.

1. Open AutoCAD with the *Ch18UCS.dwg* drawing file from the CD, a drawing of a simple 3D house with pitched roof.

2. With the **TEXT** command, attempt to place text on one of the roof slopes. You will find it is impossible, because AutoCAD only places text in the x,y-plane of the current coordinate system.

3. The solution is to reorient the coordinate system, so that the roof becomes the new x,y-plane. Create a UCS oriented to one of the roof slopes with the **3point** option, as follows:

    Command: **ucs**

    Current ucs name: *WORLD*

    Specify origin of UCS or [Face/NAmed/OBject/ ... /ZAxis] <World>: **3**

    Notice that I entered **3** (short for "3point") directly.

4. Use object snap modes to make your picks accurate:

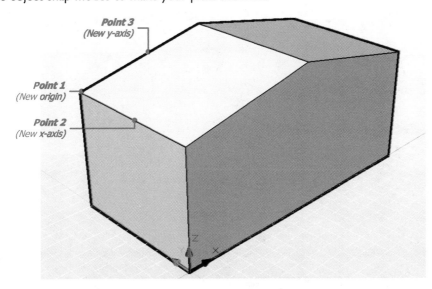

Specify new origin point <0,0,0>: **end**

of *(Pick point 1, the origin for the new UCS.)*

Specify point on positive portion of X-axis <1,25,20>: **nea**

to *(Pick point 2 to show the direction of the new x-axis.)*

Specify point on positive-Y portion of the UCS XY plane <1,25,20>: **nea**

to *(Pick point 3 to pin down the angle of the new x,y-plane.)*

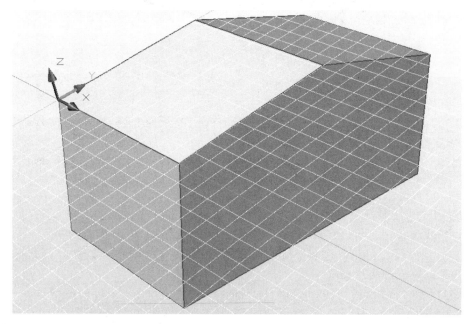

Notice that the UCS icon relocates to the new origin. Its axes show the direction of the x,y-plane (highlighted by me in gray) and z. There is no need to specify the direction of the z-axis, because AutoCAD determines it by the right-hand rule.

5. You created a new coordinate system, a *user*-defined coordinate system (UCS).

That's the first step; the second is to view the new x,y-plane with the **PLAN** command.

Command: **plan**

Enter an option [Current ucs/Ucs/World] <Current>: *(Press ENTER.)*

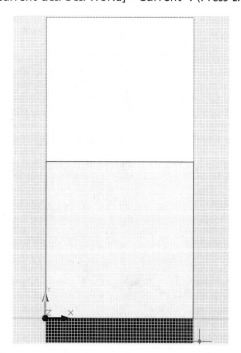

You are now looking straight down on the inclined roof (a.k.a. *true plane*). In the figure, the plane is shaded gray.

6. With the roof oriented in the drawing plane, you can now draw and edit to your heart's content. Try placing text in the new working plane with the **MTEXT** command.

7. After the text is in place, use the **ZOOM Previous** command to see the text-on-the-roof from another viewpoint.

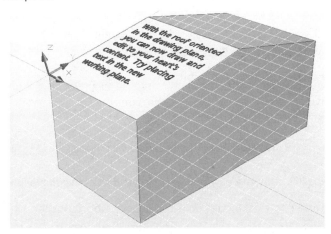

Dynamic UCS

The interactive version of UCS is called "dynamic UCS." When you start to draw on the face of a 3D object, the x,y-plane aligns itself with the face. In addition, the cursor and the grid align with the face.

To turn on dynamic UCS, click the **DUCS** button on the status bar. To determine whether the grid aligns with the dynamic UCS, check the **Follow Dynamic UCS** option in the **Snap and Grid** tab of the **DSETTINGS** command's dialog box.

Dynamic UCS makes it easier to draw on sides and inclined planes. (The dynamic UCS does not align to curves surfaces, such as on spheres and cylinders.) When the drawing command ends, the UCS returns to its original orientation.

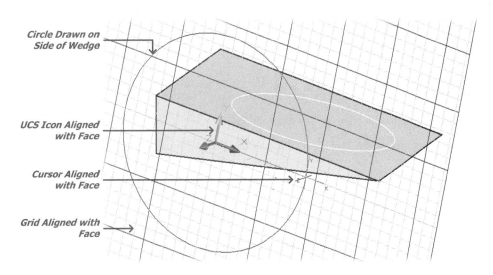

Dynamic UCS also appears during some editing commands, such as grips editing, **3DMOVE**, and **3DROTATE**. It does not appear when erasing, offsetting, breaking, lengthening, and so on.

 **3DORBIT**

The **3DORBIT** command interactively sets 3D viewpoints.

This command tilts and rotates the drawing in real time as you move the cursor. It lets you swing around 3D models, to view them from all sides.

BASIC TUTORIAL: INTERACTIVE 3D VIEWPOINTS

1. To change the 3D viewpoint interactively, start the **3DORBIT** command:
   - From the **Dashboard**, choose **Constrained Orbit**.
   - From the **View** menu, choose **3D Orbit**.
   - At the 'Command:' prompt, enter the **3dorbit** command.

     Command: **3dorbit** *(Press ENTER.)*
   - Alternatively, enter **3do** or **orbit** aliases at the keyboard.

   In all cases, AutoCAD displays a new cursor ⟨⊕⟩ , and the following prompt:

   Press esc or enter to exit, or right-click to display shortcut menu.

2. Drag the cursor. As you do, the viewpoint changes.
3. To exit the command, press **ESC**.

 Note: To enter 3D orbit mode transparently, hold down the **SHIFT** key and then press the middle mouse button. This shortcut may not work with some mouse configurations.

PLAN

The **PLAN** command returns the viewpoint to the plan view — the 2D view.

This command can return to the plan view of the current UCS or of the world coordinate system. (It is equivalent to using the VPOINT 0,0,0 command.)

Command: **plan**
Enter an option [Current ucs/Ucs/World] <Current>: **w**
Regenerating model.

Current UCS — returns to the plan view of the current UCS.

UCS — returns to the plan view of a named UCS; prompts for the name of the UCS.

World — returns to the plan view of the WCS (world coordinate system).

CREATING 3D MODELS FROM 2D OBJECTS

Another method of creating solid models is to turn 2D objects into 3D solids. In this section, we look at these ones:

> **Extrude** — extrudes 2D shapes into 3D solids with optional tapered sides.
>
> **Sweep** — sweeps 2D curves along paths to create 3D solids.
>
> **Revolve** — revolves 2D shapes about an axis to create 3D solids.

Other 2D-to-3D conversion commands include:

> **Loft** — spans two or more 2D boundaries to create 3D solids or surfaces.
>
> **Thicken** — gives thickness to 2D objects, and then converts them to 3D solids.

 SWEEP

The **SWEEP** command sweeps 2D objects along a path to create 3D solid models.

This command is used for giving thickness to 2D objects, and for creating objects like handrails and snaking tubes. AutoCAD needs to know two things: (1) the objects to be swept, and (2) the path along which to sweep. (This command is similar to, but easier to use than, the **EXTRUDE** command, described later.)

In the figure below, the donut profile was swept along the splined path to create the curved solid.

This command sweeps ellipses, closed polylines (including polygons and donuts), traces, closed splines, regions, 2D solids, and 3D faces. (When the profile objects are lines, arcs, elliptical arcs, open polylines, and open splines, the result is a 3D *surface*, instead of a 3D solid.) AutoCAD cannot sweep objects within blocks or self-intersecting objects.

The path can consist of lines, arcs, elliptical arcs, 2D and 3D polylines, 2D and 3D splines, circles, ellipses, helixes, or the edges of surfaces and solids. The extrusion process takes place at right angles to the center line of the path. In the figure below, notice that the profile of the donut-like region is projected at right angles along the path. In short, it matters not where the profile is located, because it gets relocated to the path.

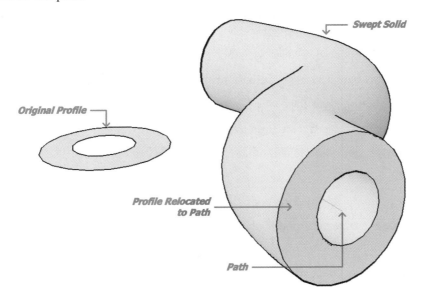

If you plan to extrude objects with holes in them, such as gears with axle holes, then you should convert the objects to a single region. (Use the **SUBTRACT** command to remove holes.)

TUTORIAL: SWEEPING OBJECTS INTO SOLIDS

1. Draw a path and the profile. The objects to be swept (converted into 3D solids) are called "profiles."

2. To sweep the profile object into a 3D solid, start the **SWEEP** command:

    *   In the Dashboard palette, click the **Sweep** button.

    *   From the **Draw** menu, choose **Modeling**, and then **Sweep**.

    *   At the 'Command:' prompt, enter the **sweep** command.

        Command: **sweep** *(Press ENTER.)*

3. In all cases, AutoCAD prompts you select the profile object(s) to sweep:

    Current wire frame density: ISOLINES=12

    Select objects to sweep: *(Select one or more objects.)*

    Select objects to sweep: *(Press ENTER to end object selection.)*

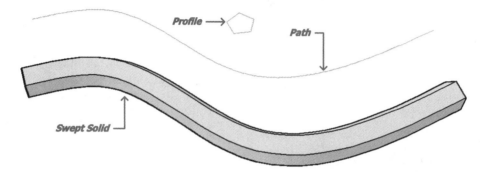

4. Indicate the path along with the object(s) should be swept:

    Select sweep path or [Alignment/Base point/Scale/Twist]: *(Pick a path object.)*

 Note: The **SWEEP** command fails when the radius of the path is small enough to cause kinks on the inside curve.

SWEEPING OBJECTS: ADDITIONAL METHODS

The **SWEEP** command has a number of options that control the look and size of the swept solids, as well as a system variable that affects source objects.

*   **Alignment** option determines whether the profile is aligned with the path.

*   **Base point** option specifies the base point for profiles.

*   **Scale** option applies a uniform scale factor to the profile.

*   **Twist** option applies a twist angle to the swept solid.

*   **DELOBJ** system variable determines the fate of source objects.

Alignment

The **Alignment** option determines whether the profile is aligned with the path; by default, it does.

Align sweep object perpendicular to path before sweep [Yes/No] <Yes>: *(Type **Y** or **N**.)*

Yes — aligns the profile perpendicularly to the path.

No — keeps the profile at its current orientation.

Base Point

The **Base point** option specifies the base point for profiles:

> Specify base point: *(Pick a point.)*

Scale

The **Scale** option applies a uniform scale factor to the profile. This allows the swept solid to be larger or smaller than the profile objects.

> Enter scale factor or [Reference] <1.0>: *(Enter a number, or type **R**).*

The **Reference** option works identically to that of the SCALE command.

Twist

The **Twist** option applies a twist angle to the swept solid. The angle you specify is applied along the entire path.

> Enter twist angle or allow banking for a non-planar sweep path [Bank] <0>: *(Enter a number, or type **B**).*

You can specify an angle between 0 and 359.9 degrees. In the figure below, the twist angle is 180 degrees.

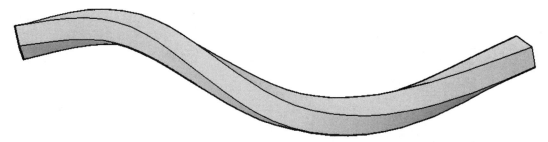

The **Bank** option forces the profile to twist "naturally" along curved paths.

DelObj

After the profile is swept, it is usually erased. If you wish to keep it, change the value of the DELOJB (short for "delete objects") system variable. The default value is 1.

DelObj	Meaning
0	All source objects are retained.
1	Profile curves used for the EXTRUDE, SWEEP, REVOLVE, and LOFT commands are erased.
	Cross sections used for the LOFT command are erased.
2	Paths and guide curves used for the SWEEP and LOFT commands are erased.
-1	Same as 1, but users are prompted, "Erase defining objects?"
-2	Same as 2, but users are prompted, "Erase defining objects?"

 **EXTRUDE**

The **EXTRUDE** command extrudes 2D objects in the z direction or along paths to create 3D solid models.

This command is like SWEEP, but has this advantage: you can specify an angle, which tapers the extrusion, as illustrated below.

This command has a disadvantage compared to SWEEP, however: the extrusion takes place at right angles to the center line of the path — the profile must be perpendicular to the path. If it isn't,

AutoCAD complains, "Profile and path are tangential. Unable to extrude the selected object." The solution is to draw the path in one UCS and the profile in a second UCS at an angle to the first. AutoCAD reorients the profile to make it perpendicular to the path.

Unlike SWEEP, however, EXTRUDE does not need a path, because almost all extrusions are perpendicular to the profile plane. You simply specify the height of extrusion. If you want to extrude along a path, then use SWEEP.

TUTORIAL: EXTRUDING OBJECTS INTO SOLIDS

1. To extrude objects into 3D solids, start the **EXTRUDE** command:
   * In the Dashboard palette, click the **Extrude** button.
   * From the **Draw** menu, choose **Modeling** , and then **Extrude**.
   * At the 'Command:' prompt, enter the **extrude** command.

     Command: **extrude** *(Press* ENTER.*)*
   * Alternatively, enter the alias **ext** at the keyboard.

2. In all cases, AutoCAD prompts you select the object(s) to extrude.

   Select objects: *(Select one or more objects.)*

   Select objects: *(Press* ENTER *to end object selection.)*

3. Indicate the height to which the object(s) should be extruded:

   Specify height of extrusion or [Path]: *(Enter a value.)*

   * **Specify height** — pick two points to indicate the height, or enter a distance.
   * **Path** — pick a line or other open object. The path must be perpendicular to the object being extruded.

4. Optionally, you can taper the extrusion by specifying an angle other than 0. The angle you specify must not allow the extruded solid to intersect itself; thus shallow angles work best. Positive angles taper inward, negative outward — the opposite to most other 3D software.

   Specify angle of taper for extrusion <0>: *(Press* ENTER, *or enter an angle such as* **5**.*)*

 **REVOLVE**

The REVOLVE command rotates 2D objects to create 3D solid models.

To revolve (convert) 2D objects into 3D solids, AutoCAD needs to know two things: (1) the objects to be revolved, and (2) the axis about which to revolve them.

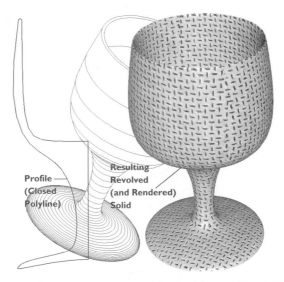

Profile (Closed Polyline)

Resulting Revolved (and Rendered) Solid

The REVOLVE command rotates lines, traces, arcs, elliptical arcs, 2D polylines, 2D splines, 2D solids, circles, ellipses, regions, and flat 3D faces. AutoCAD cannot revolve objects within blocks, or polylines that cross over themselves.

The revolution takes place around the x or y axis, or about an object, such as a line or polyline segment. The length of the axis is unimportant; AutoCAD uses only the axis' orientation in space to determine the location and angle of the revolution. You can specify full 360-degree revolutions, or partial revolutions, with positive or negative angles.

TUTORIAL: REVOLVING OBJECTS INTO SOLIDS

1. Draw a profile.
2. To revolve 2D objects into 3D solids, start the REVOLVE command:
   - In the Dashboard palette, click the **Revolve** button.
   - From the **Draw** menu, choose **Modeling**, and then **Revolve**.
   - At the 'Command:' prompt, enter the **revolve** command.

     Command: **revolve** *(Press* ENTER.*)*
   - Alternatively, enter the alias **rev** at the keyboard.
3. Select one or more objects as the profile. The profile must be entirely on one side of the axis, or part of the axis.

     Current wire frame density: ISOLINES=12

     Select objects to revolve: *(Select one or more objects.)*

     Select objects to revolve: *(Press* ENTER *to end object selection.)*

Notes: To revolve objects about a central axis, draw just one half of the profile, as illustrated on the previous page.

If the profile is an open object, such as a line or arc, this command creates a 3D surface; if a closed objects, such as a circle or region, then a 3D solid.

You can enter a positive or negative angle for the angle of rotation. AutoCAD uses the "right-hand rule" to determine the positive direction of rotation. The positive direction of the axis is from the first (start) point to the second (end) point.

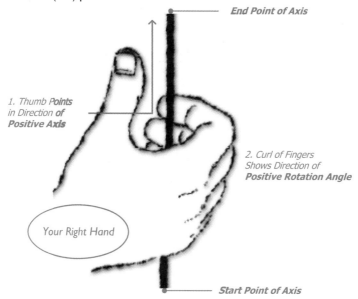

The right hand rule:
1. Your thumb points in the direction of the positive axis.
2. The curl of your fingers shows the direction of the positive rotation angle.

4.   Select the axis about which the object(s) will be revolved.

  Specify axis start point or define axis by [Object/X/Y/Z] <Object> *(Pick two points, or enter an option.)*

You can choose one of the following for the axis:

* **Start point** — two points that indicate the axis.
* **Object** — a line, a polyline segment, or one segment of the profile.
* **X** or **Y** — the x or y axis.

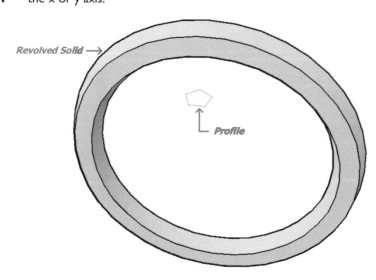

5. Specify the angle of revolution — how far around you want the profile revolved:

Specify angle of revolution or [STart angle] <360>: *(Enter an angle, or type* **ST***.)*

Portion	Angle of Revolution
Full	360 degrees.
3/4	270 degrees.
2/3	240 degrees.
1/2	180 degrees.
1/3	120 degrees.
1/4	90 degrees.

The **STart angle** option allows you to specify an offset for the revolution:

Specify start angle <0>: *(Enter an angle.)*

Specify angle of revolution <360>: *(Enter a second angle.)*

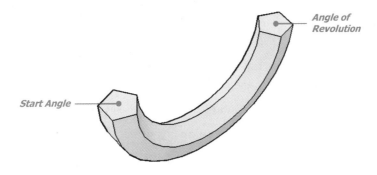

Angle of Revolution

Start Angle

BOOLEAN OPERATIONS

The **UNION**, **INTERSECT**, and **SUBTRACT** commands are collectively known as "Boolean operations," named after George Boole, a 19th-century mathematician who developed Boolean logic.

You may be familiar with using the terms *or* (equivalent to union), *and* (intersect), and *not* (subtract) in narrowing down searches on the Internet. These are based on the work of Mr. Boole. You can learn more about this topic at www.kerryr.net/pioneers/boole.htm.

In this section, we look at how the **UNION** and **SUBTRACT** commands manipulate solid primitives to make complex models.

 UNION

The **UNION** command merges 3D solid models into a single body.

This command also works with 2D regions, but regions and 3D solid models cannot be mixed. The objects do not need to be touching or intersecting to be unioned.

To execute this command, AutoCAD needs to know just one thing: the objects to be joined. In the figure below, three boxes were unioned into a single solid.

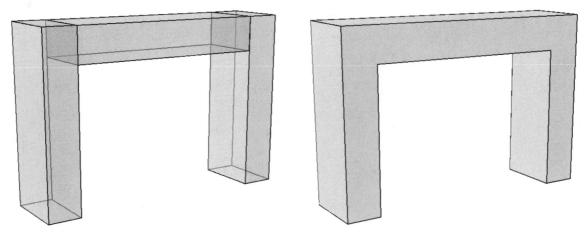

Left: *Three boxes.*
Right: *Unioned into a single shape.*

TUTORIAL: JOINING SOLIDS

1. To join two or more 3D solids, start the **UNION** command:
   - In the Dashboard palette, click the **Union** button.
   - From the **Modify** menu, choose **Solid Editing**, and then **Union**.
   - At the 'Command:' prompt, enter the **union** command.

     Command: **union** *(Press* ENTER*.)*
   - Alternatively, enter the alias **uni** at the keyboard.

2. In all cases, AutoCAD prompts you to select the objects to join. Select at least two:

   Select objects: *(Select one or more objects.)*

   Select objects: *(Press* ENTER *to end object selection.)*

   The selected objects are merged into a single object. You will probably notice some faces or edges missing. This is normal.

⦾ SUBTRACT

The **SUBTRACT** command removes intersecting parts 3D solid models.

The objects must be intersecting to be subtracted. (This command also works with regions, but regions and 3D solid models cannot be subtracted from each other.) To execute this command, AutoCAD needs to know two things from you: (1) the objects from which to subtract, and (2) the objects to subtract. Then you can create Swiss cheese-like objects as illustrated below.

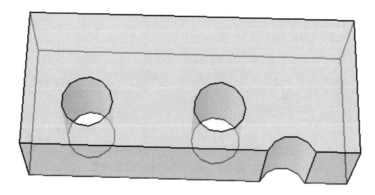

BASIC TUTORIAL: SUBTRACTING SOLIDS

1. To remove 3D solids from each other, start the **SUBTRACT** command:
   * In the Dashboard palette, click the **Subtract** button.

   * From the **Modify** menu, choose **Solids Editing**, and then **Subtract**.

   * At the 'Command:' prompt, enter the **subtract** command.

   * Alternatively, enter the alias **su** at the keyboard.

   Command: **subtract** *(Press ENTER.)*

   In this tutorial, we subtract the cylinders from the box:

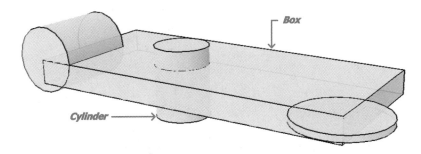

2. In all cases, AutoCAD prompts you to select the objects from which to subtract.

   Select solids and regions to subtract from ...

   Select objects: *(Select one or more objects.)*

   Select objects: *(Press ENTER to end object selection.)*

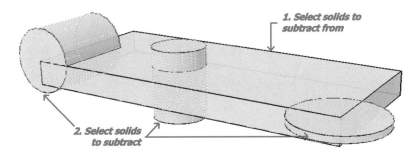

3. Select the objects that will do the subtracting.

   Select solids and regions to subtract ...

   Select objects: *(Select one or more objects.)*

   Select objects: *(Press ENTER to end object selection.)*

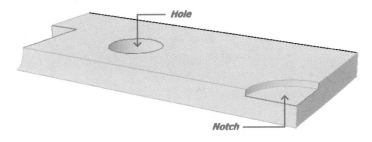

Notes: When two or more objects are selected to "subtract from," they are first joined into a single solid before others are subtracted.

The results may not be what you expect, because the order in which you select objects is important:
1. First, select the objects from which to subtract. These are the objects to be made smaller.
2. Second, select the objects to subtract.

If the result looks wrong, you selected objects in reverse order. Use the **U** command to undo, and try again.

EXERCISES

1. Draw these 3D solid objects using the parameters provided below:

3D Solid	Radius or Length	Width	Height
Sphere	Radius = 2		
Cylinder	Radius = 1.5	Height = 2.5	
Box	Length = 3	Width = 2	Height = 1

2. From the Companion CD, open the *Ch18Glass.dwg* drawing file, a profile of a goblet.
   Use the **REVOLVE** command to create the goblet.

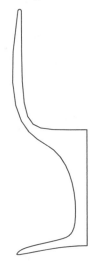

3. Create a profile of your own, and then apply the **REVOLVE** command.

4. From the Companion CD, open the *Ch18Path.dwg* drawing file, a pentagon profile and a path.
   Use the **SWEEP** command to sweep the pentagon along the spline path.

5. From the CD, open the *Ch18Sweep.dwg* drawing file, a profile of a gear.
   Use the **SWEEP** command to make the gear 0.5 units thick.
   What kind of path did you draw?

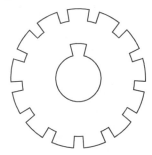

Repeat using the **EXTRUDE** command. Which command did you find easier to use for this exercise?

6. Create a profile of your own, and then apply the **SWEEP** command.

7. Recreate the following profiles in 3D. Make each 2 units long. The result should look similar to the following image:

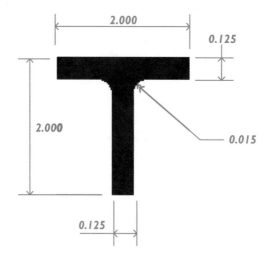

a. Square tube.

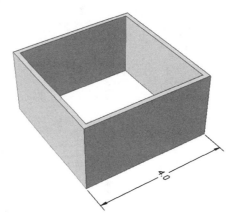

b. T-bar with 0.015 fillets.

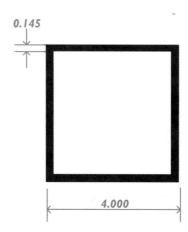

c. 130-pound rail:

> HT = Overall height of rail.
>
> HW = Width of the rail's head.
>
> HD = Depth of the head.
>
> W = Width of the rail's web (at center point).
>
> FD = Fishing (net height of the web).
>
> BD = Depth of the base.
>
> BW = Width of the base.

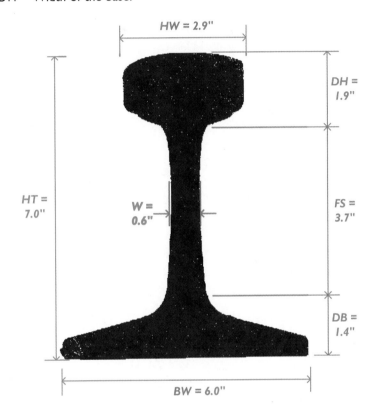

CHAPTER REVIEW

1. The 3D cursor has asymmetrical crosshairs; in which direction do the longer ones point?

2. What is reported by the z coordinate on status bar?

3. What kinds of 3D objects are created with the **REVOLVE** command?

4. What objects are drawn with the following commands?

   SPHERE

   BOX

   CYLINDER

5. Why are the **UNION** and **SUBTRACT** commands called "Booleans"?

6. Describe how to create bolt holes in solid objects.

7. Can the **REVOLVE** command revolve objects less than a full 360 degrees?

8. What is the plan view?

   How do you change to the plan view in AutoCAD?

9. What is the purpose of the **ISOLINES** system variable?

   Do isolines appear on 3D box objects?

10. What does the **3DORBIT** command do?

11. What is the purpose of dynamic UCS?

12. Briefly explain the purpose of the **UCS** command.

13. Which command aligns the view with the current UCS?

14 How does the **SWEEP** command work?

15. What is the purpose of the **DELOBJ** system variable?

CHAPTER 19

Editing 3D Models

AutoCAD provides a number of commands that further shape and change solid models to create more sophisticated designs. In this chapter, you learn about the following commands:

EXPLODE, **CHAMFER**, **FILLET** and other "2D" commands work with 3D solid models.

3DMOVE and **3DROTATE** move and rotate objects in 3D space with new grip tools.

Grips editing allows direct changes to 3D solids.

SHOWHIST records the primitive objects making up complex 3D models.

SECTIONPLANE, **LIVESECTION**, and **JOGSECTION** apply and edit section planes through 3D models.

FINDING THE COMMANDS

On the Dashboard's **3D Model** panel:

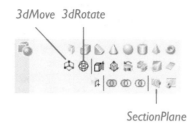

On the **Modify** toolbar:

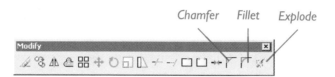

On the **Modeling** toolbar:

"2D" EDITING COMMANDS

AutoCAD has two kinds of commands for editing 3D solids. Some are 2D editing commands that can be used on 3D solids; you are already familiar with these from earlier in this book. Others are commands specific to 3D solids. (I put quotation marks around "2D" because these commands are primarily used for 2D objects, but work equally well with 3D solids.)

In this chapter, we specifically look at how these 2D editing commands work with 3D solids:

- **EXPLODE** converts 3D solid models into 2D regions.
- **CHAMFER** cuts the edges of faces.
- **FILLET** rounds the edges of faces.

Two of these commands have been especially adapted for solid models: **CHAMFER** and **FILLET**. They affect the edges of solid objects, as shown by the figure below.

These two work on *edges* only (the intersection between two faces); they do not work on roundish solid models, like spheres, cylinders, and curves. You can, however, fillet and chamfer edges where they meet curved surfaces.

In addition to these three, you can use **ERASE** to delete solid objects, **SCALE** to change their size and **CHANGE**, **CHPROP**, and **PROPERTIES** to change some properties — specifically color, layer, linetype, linetype scale, lineweight, materials, and embedded hyperlinks. These 2D editing commands work on 3D objects as you might expect. For instance, erasing a 3D solid model is just like erasing a 2D object.

EXPLODE

The **EXPLODE** command converts solid models into groups of *regions* and *bodies*. Flat surfaces of 3D models become 2D regions, while curved surfaces become 3D bodies. The figure below shows a solid model exploded into regions and bodies, and then pulled apart with the **MOVE** command.

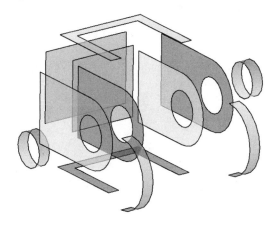

When you apply the command a second time to the resulting regions and bodies, the 2D regions become lines and arcs. Bodies become circles and arcs. (The isolines disappear.)

LINETYPES

Applying linetypes to solid models results in funky effects, such as the Batting linetype applied to the sphere shown below, with isolines set to 6. The linetype effect, however, applies only to the isolines that define the surface and edges of the solid, and not to the solid itself.

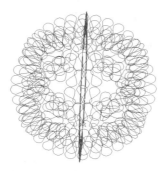

2D COMMANDS THAT "DON'T WORK"

Not all 2D editing commands work with solid models. The **STRETCH** command only moves solid models, but does not stretch them. **EXTEND** and **LENGTHEN** apply only to open objects, which solid models are not; the workaround is to use the **SOLIDEDIT** command, described later in this chapter.

BHATCH, **HATCH**, and **HATCHEDIT** don't work, because solid models cannot be hatched. The closest workaround is to change their color with the **COLOR** command; alternatively, use the **MATERIALS** command to apply a material to the solid model, which is then displayed by the **RENDER** command.

You cannot break solid models with the **BREAK** command. As a workaround, use the **SLICE** command to cut off a portion of the solid model.

DIVIDE and **MEASURE** do not apply to solid models. The **TRIM** command does not work with solid models; the workaround is to use the **INTERSECT** or **SLICE** command. Solid models cannot be offset with the **OFFSET** command; the workaround is to use the **COPY** or **SOLIDEDIT** command.

Most grips editing options work with 3D solids: **Copy**, **Rotate**, **Move**, and **Scale**. The **Stretch** option moves them, while the **Mirror** option rotates.

 Note: AutoCAD LT can view, but not edit, solid models.

SELECTING 3D OBJECTS AND FEATURES

Three-dimensional objects are much more complex than 2D objects. In the 3D world, an object consists of faces, edges, isolines, facets, and vertices. These are called "features." Here is a reminder of solid model parts:

Body is the entire 3D solid model.

Faces are the flat and round sides of bodies.

Edges are the intersections between faces; rectangular faces have four edges.

Vertices are the intersections of edges.

Isolines show curved surfaces.

Facets are used to display curved surfaces more quickly.

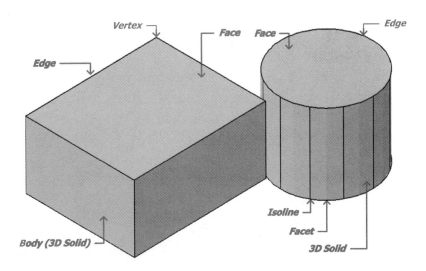

SELECTING 3D SOLIDS

When you pick a 3D solid, the entire solid is selected. Its edges and isolines are highlighted in the drawing. In addition, AutoCAD displays grips for manipulating the solids; these are discussed later in this chapter.

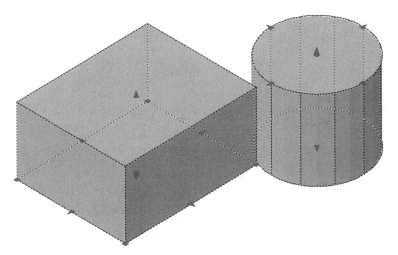

SELECTING FACES, EDGES, AND VERTICES

To select a feature, hold down the **CTRL** key. As you move the cursor over the 3D solid, AutoCAD highlights the edges of the nearest face, as illustrated below.

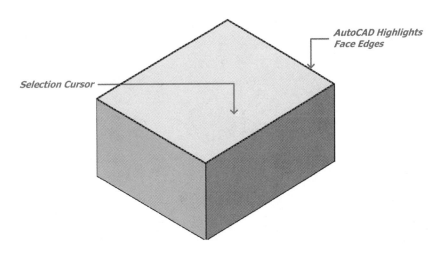

Pick the feature that interests you. AutoCAD highlights the feature, and then displays a grip that identifies the type of feature selected.

- Position the cursor over a face, and click — AutoCAD selects the face, and displays a grip in the shape of a large dot. The edges of the face are highlighted.

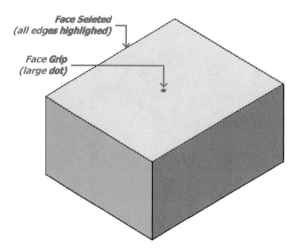

- Position the cursor over an edge, and click — AutoCAD selects the edge, and displays a dash grip.

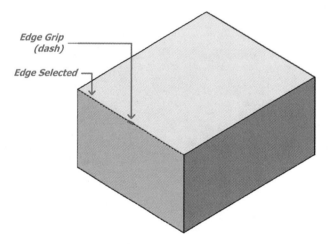

- Position the cursor over a vertex, and click — AutoCAD selects the corner, and displays a small dot grip. No features are highlighted.

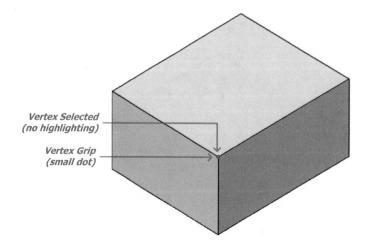

You can select more than one feature at a time. Keep holding down the CTRL key and keep picking features, as illustrated below. You can use the grips to modify the shape of the solid.

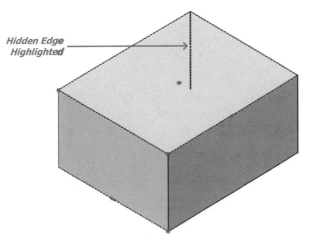

You can even select hidden features. As the cursor passes over them, AutoCAD highlights hidden edges, as illustrated above. To make it easier to select hidden features, turn on Xray mode, as described in the following chapter.

Chamfer and Fillet Selections

The **CHAMFER** and **FILLET** commands ask you to select *chains* (two or more edges) and *loops* (all the edges belonging to one face).

When you pick an edge, AutoCAD highlights the adjacent faces, and reports the number of faces it found. The figure below shows the two faces highlighted: the side and the top.

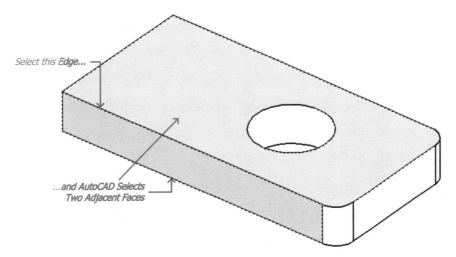

To remove the faces you don't want to work with, use the **Remove** option to pick an *unique* edge (one not shared by the two selected faces).

> Select faces or [Undo/Remove/ALL]: **r**
>
> Remove faces or [Undo/Add/ALL]: *(Pick a unique edge).*
>
> 2 faces found, 1 removed.

Press ENTER when you are finished with the selection process:

> Remove faces or [Undo/Add/ALL]: *(Press* ENTER *to end the selection process.)*

Notice that AutoCAD now highlights the single face (shown by dashed lines).

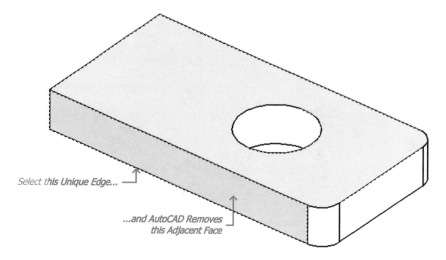

Select this Unique Edge...

...and AutoCAD Removes this Adjacent Face

SELECTING ISOLINES

Curved surfaces have *isolines*, the vertical and curved lines that show the curvature of the face. When you select a sphere or a fillet, you select its entire (single) face.

Isolines are a display aid, like grid dots. You can vary the number of isolines displayed on curved solids with the **ISOLINES** system variable, followed by the **REGEN** command. You can enter a value ranging from 0 to 2047, but a value of 12 or 16 looks best.

Command: **isolines**

Enter new value for ISOLINES <4>: **12**

Command: **regen**

Regenerating model.

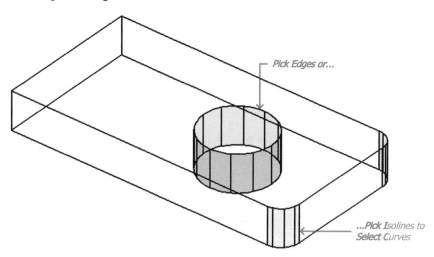

Pick Edges or...

...Pick Isolines to Select Curves

 **CHAMFER**

The **CHAMFER** command chamfers adjacent faces of 3D solid objects.

When you select a solid model, this command automatically switches to solids-editing mode, and presents a set of prompts different from those you saw in Chapter 10, "Constructing Objects." This command operates on a single solid model at a time, but chamfers two or more faces at a time.

BASIC TUTORIAL: CHAMFERING OBJECTS

1. To chamfer the intersection of two faces of a solid model, start the **CHAMFER** command:
   * From the menu bar, choose **Modify**, and then **Chamfer**.

   * From the Modify toolbar, choose the **Chamfer** button.

   * At the 'Command:' prompt, enter the **chamfer** command.

     Command: **chamfer** *(Press* ENTER.*)*

   * Alternatively, enter the **cha** alias at the 'Command:' prompt.

2. In all cases, AutoCAD displays prompts familiar to you from 2D editing:
   (TRIM mode) Current chamfer Dist1 = 0.0000, Dist2 = 0.0000

   Select first line or [Polyline/Distance/Angle/Trim/Method/mUltiple]: *(Pick a solid model.)*

3. In selecting the solid model, you pick an edge (the line between two faces) or an isoline (the lines that define curved surfaces).
   * Select an edge — AutoCAD creates a bevel between two the two adjacent faces.
   * Select an isoline of a hole — AutoCAD creates a countersink.

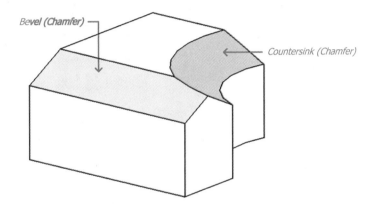

Select an Edge — AutoCAD does two things: (1) highlights one face by outlining its edges with dashed lines, and (2) prompts you to select the *base surface* (a.k.a. the first face).

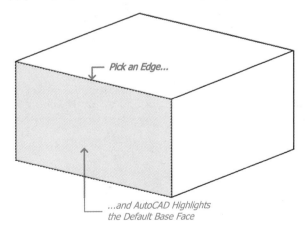

Because the two chamfer distances can be unequal, AutoCAD needs to know which distance applies to which face. The first chamfer distance applies to the base face, which AutoCAD calls the "base surface chamfer distance."

Select an isoline, and AutoCAD reverses the order: first it asks for the chamfer distances, and then for the other face.

Notice that AutoCAD highlights one of the two surfaces adjoining the selected edge. The following prompt allows you to select a base face other than the one highlighted:

Base surface selection...

Enter surface selection option [Next/OK (current)] <OK>: (Type **N** or **OK**.)

- Enter **N** (next) to select the other face as the base surface.

- Enter **OK** when the two chamfer distances are the same, or when AutoCAD correctly guesses the base surface.

4. Provide the chamfer distance(s); you cannot specify an angle, as with 2D chamfering. Notice the reference to the "base surface chamfer distance":

Specify base surface chamfer distance: **0.5**

Specify other surface chamfer distance <0.5000>: (Press **ENTER**, or enter a value.)

If the chamfer distance is too large, AutoCAD complains, "Cannot blend edge with unselected adjacent tangent edge. Finding connected blend set failed. Failure while chamfering," and exits the command. Restart the command, and enter a smaller value.

5. AutoCAD can chamfer just the selected edge, or a *loop*. A "loop" consists of all the edges that touch the base surface. A rectangular surface, as illustrated below, has a loop with four edges.

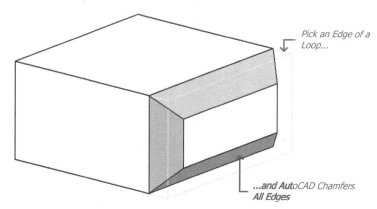

Pick an Edge of a Loop...

...and AutoCAD Chamfers All Edges

Select an edge or [Loop]: (Pick an edge, or type **L**.)

Select an edge or [Loop]: (Press **ESC**.)

6. The command repeats until you press **ESC**.

TUTORIAL: ADDING BEVELS TO SOLIDS' EDGES

Before applying the CHAMFER command, first create a simple solid model of a hole inside a cube.

1. Start AutoCAD with a new drawing.
2. Draw a cube, as follows:

Command: **box**

Specify corner of box or [CEnter] <0,0,0>: (Press **ENTER**.)

Specify corner or [Cube/Length]: **c**

Specify length: **1**

3. Add the cylinder, as follows:

Command: **cylinder**

Specify center point of base or [3P/2P/Ttr/Elliptical]: **0.5,0.5,0**

Specify base radius or [Diameter] <12.0>: **0.125**

Specify height or [2Point/Axis endpoint] <20.0>: **1**

4. To view the cube better, rotate the viewpoint, as follows:

    Command: **vpoint**

    Current view direction: VIEWDIR= 0,0,0

    Specify a view point or [Rotate] <display compass and tripod>: **2.5,-2,1**

    Regenerating model.

(To improve the image further, click the **LWT** button to thicken the lines, set **ISOLINES** to **16**, zoom to the extents, and then **ZOOM 0.5x**.)

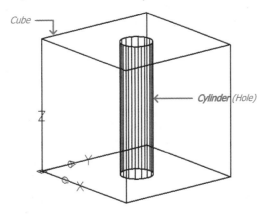

5. Subtract the cylinder from the box to create the hole:

    Command: **subtract**

    Select solids and regions to subtract from ..

    Select objects: *(Pick the cube.)*

    1 found Select objects: *(Press* **ENTER**.*)*

    Select solids and regions to subtract ...

    Select objects: *(Pick the cylinder.)*

    1 found Select objects: *(Press* **ENTER**.*)*

(Following the subtraction operation, the model will look exactly the same.)

6. From the **Modify** menu, select **Chamfer**, and then select an edge of the cube.

    Command: **chamfer**

    (TRIM mode) Current chamfer Dist1 = 0.5000, Dist2 = 0.5000

    Select first line or [Polyline/Distance/Angle/Trim/Method]: *(Pick any edge of the cube.)*

    Base surface selection...

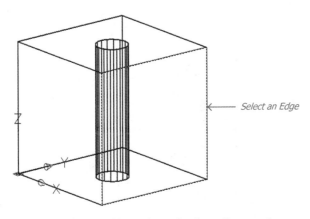

(To highlight a different face, enter **N** — short for "next" — at the prompt; otherwise press **ENTER**.)

Enter surface selection option [Next/OK (current)] <OK>: *(Press ENTER.)*

7. Specify the chamfer distances here. (The **Distance** option earlier in this command is only for 2D entities.)

Specify base surface chamfer distance <0.5000>: **0.2**

Specify other surface chamfer distance <0.5000>: **0.2**

8. Select the edges you want chamfered: pick any two edges, and then press **ENTER** to exit the command.

Select an edge or [Loop]: *(Pick edge 1.)*

Select an edge or [Loop]: *(Pick edge 2.)*

Select an edge or [Loop]: *(Press ENTER.)*

Notice that AutoCAD performs a "double" chamfer: the intersection between the two chamfered edges is another 45-degree angle.

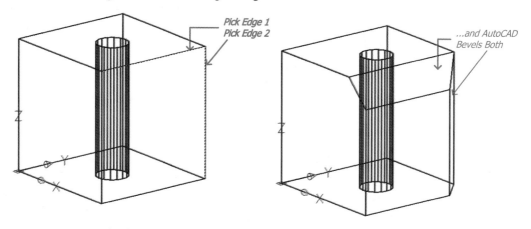

TUTORIAL: ADDING COUNTERSINKS

In this tutorial, you continue from the previous one to create a second type of chamfer, one that countersinks the hole. Continue with the drawing from the previous tutorial.

1. Start the **CHAMFER** command again, and then select a top edge on the cube (adjacent to the hole).

Command: **chamfer**

(TRIM mode) Current chamfer Dist1 = 0.2000, Dist2 = 0.2000

Select first line or [Polyline/Distance/Angle/Trim/Method]: *(Pick an edge on the top face of the cube, but not one of the chamfer edges.)*

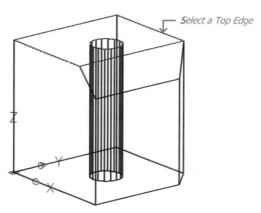

If the "wrong" face is highlighted by AutoCAD, use the **Next** option to pick the top face:

Enter surface selection option [Next/OK (current)] <OK>: **n**

Enter surface selection option [Next/OK (current)] <OK>: *(Press ENTER.)*

2. Specify a chamfer distance of 0.3 units:

    Specify base surface chamfer distance <0.2000>: **0.3**

    Specify other surface chamfer distance <0.2000>: **0.3**

3. Select the top of the cylindrical hole:

    Select an edge or [Loop]: *(Select the top of the cylindrical hole.)*

    Select an edge or [Loop]: *(Press ENTER.)*

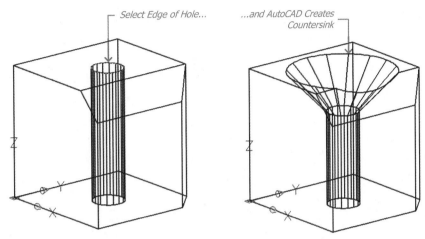

Notice two things: the chamfer created a countersink, and the countersink extends into the first chamfer you created. To view the countersink better, change the visual style to **Conceptual** with the **VSCURRENT** command.

 **FILLET**

Applying the **FILLET** command to solids is similar to using the **CHAMFER** command, except that it creates rounded corners.

When you select a solid model, this command automatically switches to solids-editing mode. While **FILLET** operates on only a single solid model at a time, it allows you to fillet as many edges as you need during a single operation. It cannot fillet between two solid models. AutoCAD is limited to constant radius fillets; it cannot apply a variable radius to fillets. (It does not work with surface models.)

BASIC TUTORIAL: FILLETING OBJECTS

1. To fillet a solid model, start the **FILLET** command:
    *   From the menu bar, choose **Modify,** and then **Fillet.**
    *   From the Modify toolbar, choose the **Fillet** button.
    *   At the 'Command:' prompt, enter the **fillet** command.

        Command: **fillet** *(Press ENTER.)*
    *   Alternatively, enter the **f** alias at the 'Command:' prompt.

2. In all cases, AutoCAD first displays the current fillet settings, and then asks you to select objects:

    Current settings: Mode = TRIM, Radius = 0.0

    Select first object or [Undo/Polyline/Radius/Trim/Multiple]:*(Select a solid model.)*

3. Provide the fillet radius:

    Enter fillet radius: **0.1**

4. Select the edge to fillet:

    Select an edge or [Chain/Radius]: *(Pick an edge, or enter an option.)*

    *   **Radius** option allows you to specify a different radius for each edge. Specify the radius, and then select the edge:

        Select an edge or [Chain/Radius]: **r**

        Enter fillet radius <0.1000>: **.05**

        Select an edge or [Chain/Radius]: *(Pick an edge.)*

        The trio of prompts repeat until you press **ENTER.**

        Select an edge or [Chain/Radius]: *(Press ENTER to end the command.)*

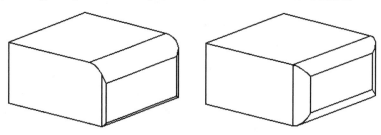

Left: *Radius option that allows different fillet radii on selected edges.*
Right: *Chain option that applies same fillet radius on selected edges.*

    *   **Chain** option applies the same radius of fillet to all the edges that you pick. Select edges, and then press **ENTER** to exit the command:

        Select an edge or [Chain/Radius]: **c**

        Select an edge chain or [Edge/Radius]: *(Pick an edge.)*

        Select an edge chain or [Edge/Radius]: *(Pick another edge.)*

        Select an edge chain or [Edge/Radius]: *(Press ENTER.)*

    You can switch between chain and radius modes at any time during the command.

TUTORIAL: FILLETING SOLID EDGES

1. In this tutorial, you draw a cylinder on top of a rectangular box, and then add fillets. Begin by drawing the box and cylinder, as follows:

    Command: **box**

    Specify corner of box or [CEnter] <0,0,0>: *(Press ENTER.)*

    Specify corner or [Cube/Length]: **l**

Specify length: **9**

Specify width: **4**

Specify height: **1**

Command: **cylinder**

Current wire frame density: ISOLINES=16

Specify center point for base of cylinder or [Elliptical] <0,0,0>: **4.5,2,1**

Specify radius for base of cylinder or [Diameter]: **1**

Specify height of cylinder or [Center of other end]: **3**

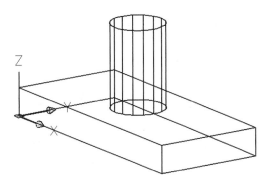

2. Start the **FILLET** command, and fillet the four edges of the end of the box:

    Command: **fillet**

    Current settings: Mode = TRIM, Radius = 0.0

    Select first object or [Undo/Polyline/Radius/Trim/Multiple]: *(Pick an edge.)*

    Enter fillet radius <0.0>: **0.2**

    Select an edge or [Chain/Radius]: *(Pick another edge.)*

    Select an edge or [Chain/Radius]: *(Pick a third edge.)*

    Select an edge or [Chain/Radius]: *(Pick a fourth edge.)*

    Select an edge or [Chain/Radius]: *(Press ENTER.)*

    4 edge(s) selected for fillet.

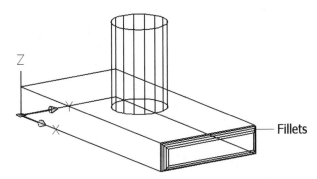

3. Another fillet is to be placed at the base of the cylinder. But the **FILLET** command cannot fillet between two solid objects. The workaround is to use the **UNION** command to join the box and the cylinder into a single object.

    Command: **union**

    Select objects: *(Pick the box.)*

    1 found Select objects: *(Pick the cylinder.)*

1 found, 2 total Select objects: *(Press ENTER.)*

The objects won't look any different.

4. Apply the fillet between the box and the cylinder:

Command: **fillet**

Current settings: Mode = TRIM, Radius = 0.2

Select first object or [Undo/Polyline/Radius/Trim/Multiple]: *(Pick the bottom edge of the cylinder.)*

Enter fillet radius <0.2>: *(Press ENTER to reuse the same radius.)*

Select an edge or [Chain/Radius]: *(Press ENTER to exit the command.)*

1 edge(s) selected for fillet.

Notice the curve at the base of the cylinder.

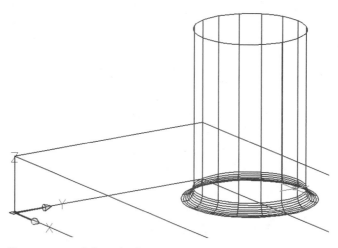

5. Apply one last fillet, on top of the cylinder.

Command: **fillet**

Current settings: Mode = TRIM, Radius = 0.2

Select first object or [Undo/Polyline/Radius/Trim/Multiple]: *(Pick the top edge of the cylinder.)*

Enter fillet radius <0.2>: *(Press ENTER to reuse the radius.)*

Select an edge or [Chain/Radius]: *(Press ENTER.)*

1 edge(s) selected for fillet.

Notice that the top of the cylinder is gracefully curved. Select **View | Shade | Gouraud** to see the shaded model.

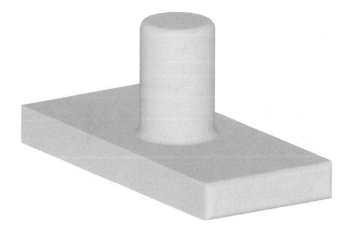

3DMOVE

When you enter 3D coordinates, the MOVE command works in 3D. But when you want to show the distance and direction of the move with the cursor, the command works only in the current UCS. The **3DMOVE** command overcomes this limitation with the *grips tool*, illustrated below.

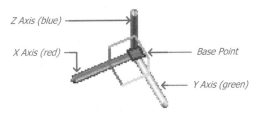

When you start the **3DMOVE** command, the grips tool appears. Select one of the colored bars, x, y, or z. The axis line changes to yellow. You can now move the object along that axis. (When dynamic UCS is turned on, the grip tool aligns itself with the selected face.)

You can drag the base grip (in the middle of the grip tool) to relocate the base point.

BASIC TUTORIAL: MOVING OBJECTS IN 3D

For this tutorial, draw a 3D solid primitive, such as a wedge.

1. To move objects in 3D space, start the **3DMOVE** command:
   - From the **Modify** bar, choose **3D Operations,** and then **3D Move.**
   - From the Modeling toolbar, choose the **3D Move** button.
   - At the 'Command:' prompt, enter the **3dmove** command.

     Command: **3dmove** *(Press ENTER.)*
   - Alternatively, enter the **3m** alias at the 'Command:' prompt.

2. In all cases, AutoCAD prompts you to select the objects to move:

   Select objects: *(Pick a solid model.)*

   Select objects: *(Press ENTER.)*

3. Notice that the grips tool appears. As you move the cursor, the tool moves with it. Move the tool onto the slopped plane of the wedge. Notice how the tool aligns its red-green (x, y) axes with the slope. (Also notice that AutoCAD displays all five of the wedge's faces, so that you can pick any one of them.)

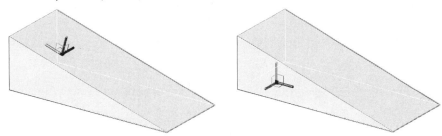

   Move the tool to the wedge's side. Again the tool aligns its red-green axes with the vertical side.

4. At the command prompt, AutoCAD asks for the base point from which the move takes place:

   Specify base point or [Displacement] <Displacement>: *(Pick a point.)*

   Picking a point places the tool.

   Moving the cursor now moves the object in the current UCS.

5. Move the cursor over one of the colored bars, such as the blue (z) bar. Notice that a blue line shoots out from either end of the bar, and that the bar turns yellow.

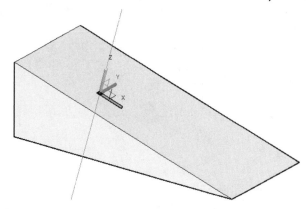

6. Click the yellow bar. *Now cursor movement is constrained along the selected axis (z, in this case).*

7. Pick a point to finish this command:

   Specify second point or <use first point as displacement>: *(Pick a point.)*

In the same way, selecting the red bar constrains movement along the x axis, and the green along the y axis.

RELATED SYSTEM VARIABLES

The GTAUTO system variable toggles the display of grips tools:

0 — grip tools are displayed only during the **3DMOVE** and **3DROTATE** commands.

I — grip tools are also displayed when objects are selected (default).

The GTDEFAULT system variable determines whether the **3DMOVE** or the MOVE command activates automatically in 3D space when you enter "move" at the command prompt. The same goes for the **3DROTATE** command when you enter "rotate."

0 — entering "move" and "rotate" executes the **MOVE** and **ROTATE** commands, respectively (default).

I — entering "move" and "rotate" executes the **3DMOVE** and **3DROTATE** commands.

The GTLOCATION system variable determines whether the grip tool appears initially:

0 — grip tool is placed at and aligned with the UCS icon (default).

I — grip tool is placed at and aligned with the last selected object or face.

3DROTATE

The **3DROTATE** command is similar to **3DMOVE**, except that its grips tool selects the axis of rotation, as illustrated below.

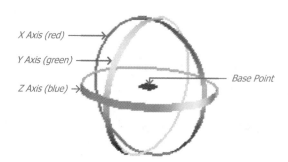

When you start the **3DROTATE** command, the grips tool appears. Select one of the colored circles, x, y, or z. The rotation circle changes to yellow. You can now rotate the object in that plane about the base point. You can drag the base grip (in the middle of the grip tool) to relocate the base point.

According to Autodesk documentation, you can press the spacebar to switch between **3DROTATE** and **3DMOVE**, but neither I nor the technical editor has been able to get it to work.

BASIC TUTORIAL: ROTATING OBJECTS IN 3D

For this tutorial, draw a 3D solid primitive, such as a wedge.

1. To rotate objects in 3D space, start the **3DROTATE** command:
   - From the **Modify** bar, choose **3D Operations**, and then **3D Rotate**.
   - From the Modeling toolbar, choose the **3D Rotate** button.
   - At the 'Command:' prompt, enter the **3drotate** command.
   - Alternatively, enter the **3r** alias at the 'Command:' prompt.

     Command: **3drotate** *(Press* ENTER.*)*

2. In all cases, AutoCAD prompts you to select the objects to rotate:

     Current positive angle in UCS: ANGDIR=counterclockwise ANGBASE=0

     Select objects: *(Pick a solid model.)*

     Select objects: *(Press* ENTER.*)*

3. Notice that the tri-circle grips tool appears. As you move the cursor, the tool moves with it. Move the tool onto the slopped plane of the wedge. Notice how the tool aligns its red-green (x, y) axes with the slope.

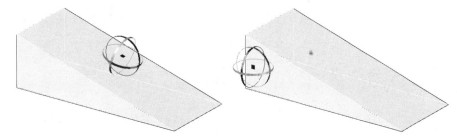

Move the tool to the wedge's side. Again the tool aligns its red-green axes with the vertical side.

4. At the command prompt, AutoCAD asks for the base point about which to rotate:

     Specify base point: *(Pick a point.)*

   Picking a point places the tool.

5. Move the cursor over one of the colored rings, such as the blue (z) ring.

     Pick a rotation axis: *(Pick a rotation axis.)*

   Notice that a blue line that aligns with the axis of the ring, and that the ring turns yellow.

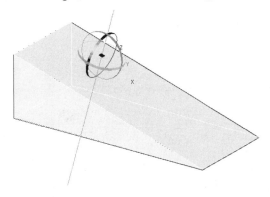

When you now rotate the cursor, movement is constrained about the z axis.

6. Specify the starting and ending angles:

   Specify angle start point: *(Pick a point, or enter an angle.)*

   Specify angle end point: *(Pick another point.)*

In the same way, selecting the red ring constrains rotation about the x axis, and the green about the y axis.

GRIPS EDITING

Like other objects in AutoCAD, 3D solid models can be edited using grips: the square grip moves or stretches objects; triangular ones change the size of objects in specific directions, such as changing the radius of cylinders or the width of boxes.

When the entire model is selected, grips appear with these shapes:

   ■ grips — move or stretch 3D solids.

   ▲ grips — change the radius (diameter) or height (width, length) of 3D solids.

To edit with a specific grip, select it so that it turns red, and then drag the grip to edit. To exit grips editing, press the **ESC** key once or twice.

EDITING SPHERES

Spheres have these editing grips:

One ■ grip in the center of the sphere moves the sphere.

Four ▲ grips along the sphere's equator increase and decrease the radius.

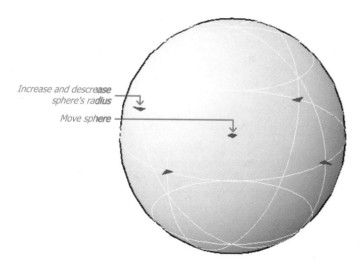

Increase and descrease sphere's radius

Move sphere

EDITING BOXES

Boxes have these editing grips:

One ■ grip in the center of the base moves the box.

Four ■ grips at the corners of the base stretch the corners.

Two ▲ grips, one each at the top and bottom increase and decrease the thickness.

Four ▲ grips at the edges lengthen and shorten the sides.

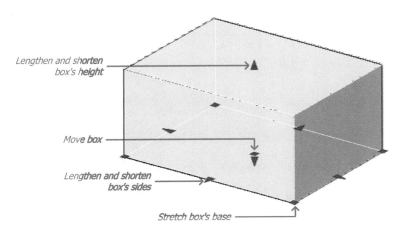

EDITING CYLINDERS

Cylinders have these editing grips:

One ■ grip in the center of the cylinder's base moves the cylinder.

Two ▲ grips, one each at the top and bottom increase and decrease the height.

Four ▲ grips around the base increase and decrease the radius.

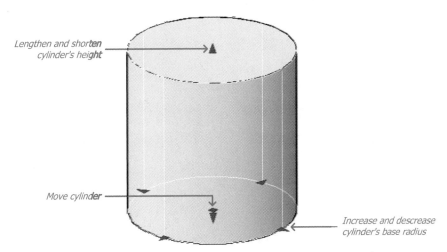

The grips cannot change the fundamental shape of cylinders. For example, you cannot use grips editing to give them elliptical bases or to slant them in space.

COMPLEX BODIES AND FACES

When complex bodies are selected, only the move grip is available, as shown below:

One ■ grip in the center of the body's base moves the body.

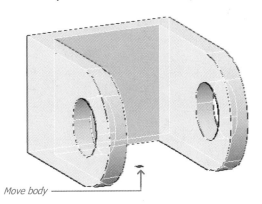

To grip edit a specific face, hold down the CTRL key, then pick the desired face. The stretch and move grips are now available for that face.

One ■ grip in the center of the face's base moves the face.

Four ■ grips at the corners of the base stretch the corners.

Two ▲ grips, one each at the top and bottom increase and decrease the height.

Four ▲ grips around the base increase and decrease the size.

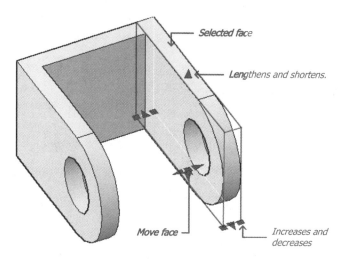

Even holes can be selected; they have the same editing grips as the solids from which they were made. In the figure below, the hole acts like a cylinder:

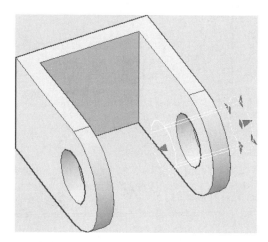

As a result, the body's parts can be extensively changed, as illustrated below.

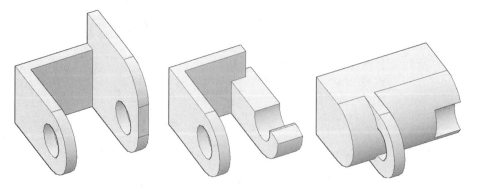

EDITING FACES, EDGES, AND VERTICES

Features of 3D solids have these editing grips:

One large ● grip in the center of the selected face stretches the face and adjacent faces.

One — grip in the center of the selected edge stretches the edge and adjacent edges.

One small • grip at the selected vertex moves or stretches the vertex and adjacent edges.

In the figure below, the edge grip is stretched, changing the shape of the box.

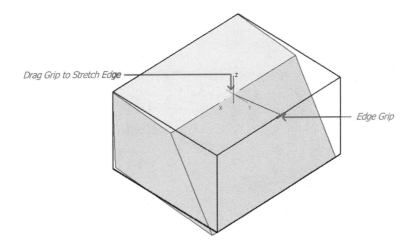

 Note: As you drag the grip, press the **CTRL** key to change the style of stretch. Examples include constraining the stretch, stretching a vertex or all edges connected to the vertex, or twisting the entire model.

SHOWHIST AND SOLIDHIST

After solid models are modified, you can examine the original parts by setting the **SHOWHIST** system variable to **2**. The default value is 1, which means that AutoCAD does not override the history settings of each part. This feature does not work with solid models created in AutoCAD 2006 or earlier, and it does not work when **SOLIDHIST** is set to 0.

For instance, the following model appears to have a large cylinder cut out of a rather flat box.

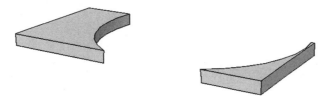

Turning on **SHOWHIST** reveals that at least two primitives appear to have been involved: a box, a cylinder, and perhaps something else.

```
Command: showhist
Enter new value for SHOWHIST <0>: 2
```

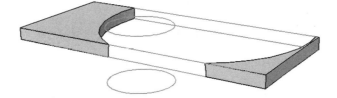

Moving the cursor over the solid shows that the model was in fact made from three primitives:

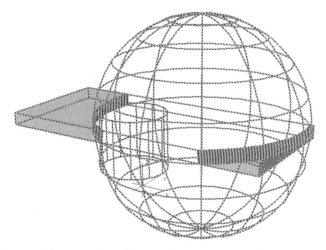

The individual primitives can be moved and resized using grips editing, changing the design of the solid model, as illustrated below:

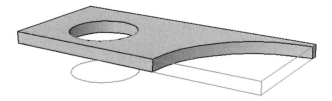

SECTIONPLANE, LIVESECTION, AND JOGSECTION

The **SECTIONPLANE** command creates temporary sections of 3D objects.

This command allows you to "draw" section planes, which reveal the insides of 3D models. It works in conjunction with the **LIVESECTION** command, which activates (makes live) and edits the section plane's properties. The **JOGSECTION** command seems like an afterthought: it adds 90-degree jogs to section planes. In brief, this trio of commands is used in this order:

1. **SECTIONPLANE** creates the initial section plane by placing the cutting plane through the 3D model. (This command should not be confused with the **SECTION** command, which creates permanent sections.) Section planes can have multiple segments, including angles.

2. **LIVESECTION** activates the section by revealing the insides of the 3D model.

3. **JOGSECTION** adds 90-degree jogs to existing section planes, if required.

BASIC TUTORIAL: SECTIONING 3D OBJECTS

For this tutorial, open *Ch19-SectionPlaneEx.dwg* from the CD-ROM, a drawing of a part to be sectioned.

1. To section 3D objects, start the **SECTIONPLANE** command:
   - From the **Draw** menu, choose **Modeling**, and then **Section Plane**.
   - At the 'Command:' prompt, enter the **sectionplane** command.

     Command: **sectionplane** *(Press ENTER.)*

   - Alternatively, enter the **splane** alias at the 'Command:' prompt.

2. In all cases, AutoCAD prompts you at the command line. Type "d" to draw the section plane.

   Select face or any point to locate section line or [Draw section/Orthographic]: **d**

3. Use MIDpoint object snap to draw a section plane clear through the middle of the 3D part:

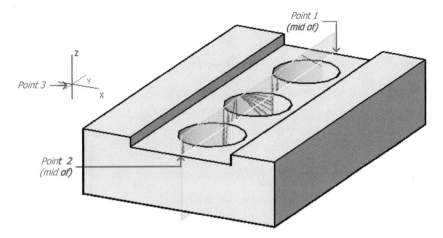

   Specify start point: **mid**

   of *(Pick point 1.)*

   Specify next point: **mid**

   of *(Pick point 2.)*

   Specify next point or ENTER to complete: *(Press* **ENTER** *to stop drawing.)*

4. As the last step, AutoCAD wants to know which side to keep when the section plane is later activated. The side is not critical, because you can change it later with the **LIVESECTION** command.

   Specify point in direction of section view: *(Pick a point behind the model.)*

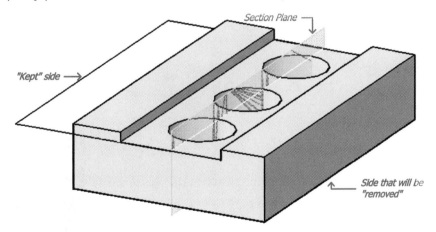

ADVANCED TUTORIAL: EDITING SECTION PLANES

For this tutorial, keep the *Ch19-SectionPlaneEx.dwg* file open.

1. To edit section planes, start the **LIVESECTION** command:
   • At the 'Command:' prompt, enter the **livesection** command.

   Command: **livesection** *(Press* **ENTER.***)*

2. AutoCAD prompts you to select the section plane:

   Select section object: *(Select a section plane.)*

Notice that half of the 3D object disappears, and you can see inside.

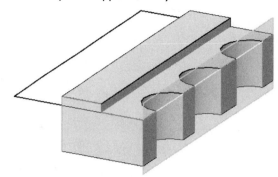

3. Select the section. Notice that several grips appear (shown in plan view):

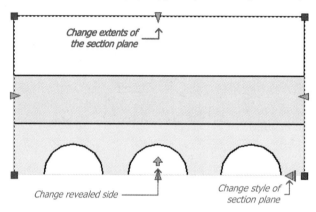

Four ■ grips at the corners stretch the section.

Four ▲ grips at the edges lengthen and shorten the section, as illustrated below.

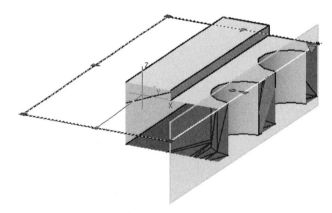

One ⬆ grip at the center of the section plane switches the sectioned side, as shown below.

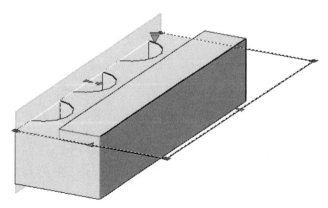

One ▽ grip at the end of the section plane changes the style of the section plane through a shortcut menu:

Section Plane — one-dimensional section that cuts away only in the x direction.

Section Boundary — two-dimensional section that cuts away in the x and y directions. (illustrated above).

Section Volume — three-dimensional section that cuts away in the x, y, and z directions.

4. To see the cut-away portion, right-click the section plane, and then choose **Show Cutaway Geometry** from the shortcut menu. It is shown in transparent red:

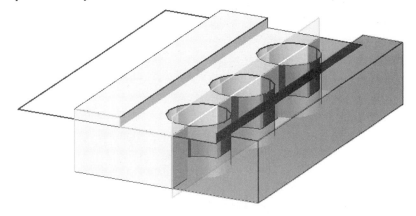

5. To add a jog to the section plane, use the **JOGSECTION** command:

    Command: **jogsection**

    Select section object: *(Pick the section object.)*

    I find it easier to use the NEArest object snap to grab the section plane:

    Specify a point on the section line to add jog: **nea**

    to *(Pick a point on the section plane.)*

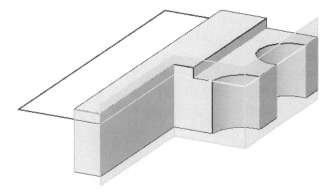

The 90-degree jog is added at your pick point.

Note: The JOGSECTION command applies only 90-degree jogs. If you want a different angle, use the **Draw** option of the SECTIONPLANE command.

6. To remove the section plane, select it, and then press DEL.

CHANGING SECTION PLANE PROPERTIES

To change the properties of section planes, select one, right-click, and then select **Live Section Settings** from the shortcut menu.

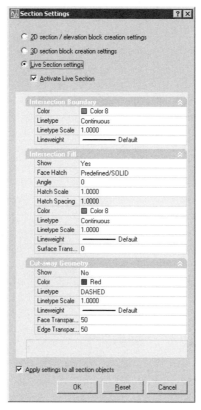

○ **2D Section / Elevation Block Creation Settings** — specifies how 2D sections are displayed when generated from 3D objects.

○ **3D Section Block Creation Settings** — specifies how 3D generated objects are displayed.

⊙ **Live Section Settings** — specifies how live section objects are displayed.

☑ **Activate Live Section** — toggles live sectioning.

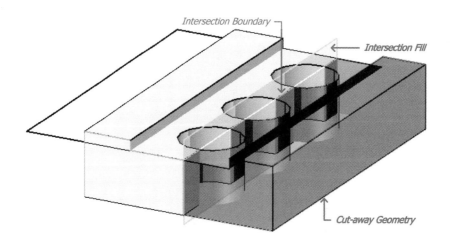

Intersection Boundary

These options specify the properties of line segments outlining the section plane's intersection surface. **Color**, **Linetype**, **Linetype Scale**, and **Lineweight** specify the color, linetype, linetype scale, and lineweight of the boundary lines.

Intersection Fill

These options specify properties of the fill inside the section plane's intersection:

Show — toggles the display of the fill.

Face Hatch — selects the type of hatching: none, predefined, custom, or user-defined (exactly the same as with the HATCH command).

Angle, **Hatch Scale**, and **Hatch Spacing** — specify the angle, scale, and spacing of the hatch pattern. The Hatch Spacing option is available only with user-defined patterns.

Color, **Linetype**, **Linetype Scale**, and **Lineweight** — specify the color, linetype, linetype scale, and lineweight of the hatch pattern.

Surface Transparency — varies the translucency of the fill surface; ranges from 0 to 100.

Cut-away Geometry

These options determine the properties of the cut-away geometry:

Show — toggles the display of the cut-away geometry.

Color, **Linetype**, **Linetype Scale**, and **Lineweight** — specify the color, linetype, linetype scale, and lineweight of the geometry.

Face Transparency — varies the translucency of the geometry's faces; ranges from 0 to 100.

Edge Transparency — varies the translucency of the geometry's edges; ranges from 0 to 100.

☑ **Apply Settings to All Section Objects** — applies properties to all section objects in the drawing. When off, applies to the selected section object only.

Reset — resets properties to default values.

CREATING SECTION PLANE OBJECTS

The section plane can be separated from the object. Here's how:

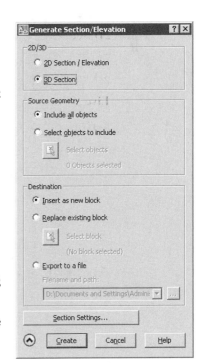

1. Select a section object.
2. Right-click, and then select **Generate 2D/3D Section** from the shortcut menu. Notice the dialog box of options:

2D/3D

These options are always displayed in the dialog box:

○ **2D Section/Elevation** — generates 2D sections.

⊙ **3D Section** — generates 3D sections.

Source Geometry

These options are displayed only when the ⊙ button is clicked:

⊙ **Include All Objects** — includes all 3D solids, surfaces, and regions, including those in xrefs and blocks.

○ **Select Objects to Include** — selects the objects from which to generate sections; prompts you to select objects.

Destination

⦿ **Insert as New Block** — inserts generated sections as blocks; displays the same prompts as the INSERT command.

○ **Replace Existing Block** — selects the section blocks to replace; works only if a block already exits.

○ **Export to a File** — saves sections as *.dwg* drawing files.

Filename and Path — specifies the path and file name for the saved section drawing.

Section Settings — displays Section Settings dialog box described above.

Create — creates the section as an independent object in block form.

3. Accept the default options (as listed above), and then click **Create**. Notice that the 3D section plane is now a separate object.

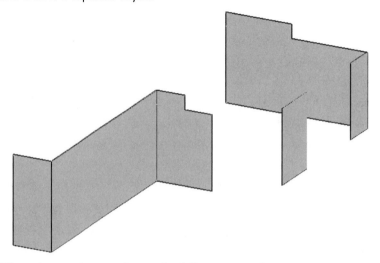

The **2D Section/Elevation** option produces the following result:

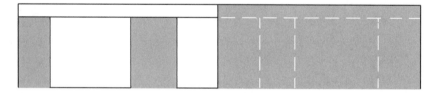

SOLID MODELING TUTORIAL

Let's draw a solid model using the commands from this chapter. The figure illustrates the finished model.

The dimensions of the 3D object are given below. You may refer to it as you construct the base model from solid primitives.

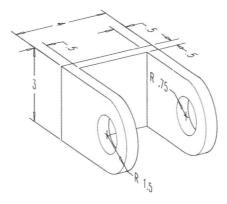

Before drawing any object, analyze it to determine the best primitives to use. In this tutorial, you use solid modeling commands and some 3D principles to construct the model. The figure below identifies the building components used to "assemble" the model.

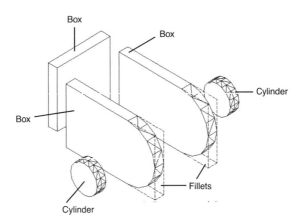

1. Start AutoCAD with a new drawing in the 3D Modeling workspace.

   Set snap to 0.5.

   Set visual style to Conceptual.

2. Select **Box** 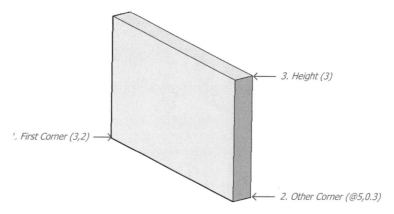 from the Dashboard palette, and enter the points indicated in the following command sequence.

   Command: **box**

   Select the first corner of the box.

   Specify first corner or [Center]: **3,2**

   Define the location of the opposite corner of the box.

   Specify other corner or [Cube/Length]: **@5,0.5**

   → 3. Height (3)

   '. First Corner (3,2) →

   → 2. Other Corner (@5,0.3)

   Now define the height of the box.

   Specify height or [2Point]: **3**

3. If the box looks small on the screen, zoom in with the Window option to make it look larger.

   Command: **zoom**

   Enter option [All/Extents/Window/Previous] <real time>: **w**

   Specify first corner: *(Pick a point.)*

   Specify opposite corner: *(Pick another point.)*

4. You will be working behind the box, so it can be helpful to see through it. In the Dashboard palette, click the **Xray Mode** button.

5. Repeat the **BOX** command. Refer to the figure for the points to enter.

   Command: **box**

   Specify first corner or [Center]: *(Select Point 1.)*

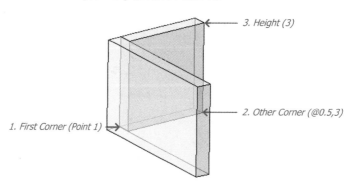

   3. Height (3)

   2. Other Corner (@0.5,3)

   1. First Corner (Point 1)

Specify other corner or [Cube/Length]: **@0.5,3**

Specify height or [2Point]: **3**

6. The next step is to drill a hole through the side. To ensure the hole is located accurately, first relocate the UCS icon to the face of the box.

Command: **ucs**

Current ucs name: *WORLD*

Specify origin of UCS or [Face/NAmed/OBject/Previous/View/World/X/Y/Z/ZAxis]

<World>: **n**

Specify origin of new UCS or [ZAxis/3point/OBject/Face/View/X/Y/Z] <0,0,0>: **3**

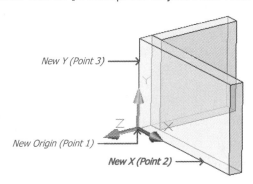

Specify new origin point <0,0,0>: *(Select point 1.)*

Specify point on positive portion of X-axis <1.0,0.0,0.0>: **nea**

of *(Select point 2.)*

Specify point on positive-Y portion of the UCS XY plane <3.0,3.0,0.0>: **nea**

of *(Select point 3.)*

7. Now add the drill hole: from the Dashboard, select **Cylinder**.

Command: **cylinder**

Specify the center point of the cylinder. Note that the absolute coordinates are relative to the origin of the new UCS.

Specify center point for base of cylinder or [Elliptical] <0,0,0>: **3.5,1.5**

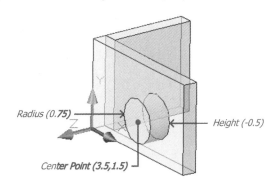

Specify the radius of the cylinder.

Specify radius for base of cylinder or [Diameter]: **0.75**

Designate the extrusion height of the cylinder. Since you want the extrusion to extend in the negative z-direction, the value is negative.

Specify height of cylinder or [Center of other end]: **–0.5**

8. Subtract the cylinder from the box. In the Dashboard, click 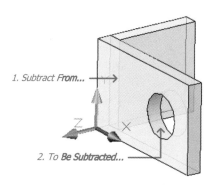 **Subtract**.

> Command: **subtract**
> Select solids and regions to subtract from...
> Select objects: *(Select the box with the cylinder.)*
> Select objects: *(Press* ENTER.*)*

1. Subtract From... →

Z   ×

2. To *Be Subtracted*... →

> 1 solid selected Select solids and regions to subtract...
> Select objects: *(Select the cylinder.)*
> Select objects: *(Press* ENTER.*)*

AutoCAD removes the cylinder from the box, creating the hole.

9. The box has rounded ends, which are created by the **FILLET** command. It works equally well with 2D and 3D objects.

You may find it easier to select the edges with snap turned off, and the view zoomed in closer.

> Command: *(Click* **SNAP** *on the status bar)* <Snap off>
> Command: **fillet**
> Current settings: Mode = TRIM, Radius = 0.0
> Select first object or [Undo/Polyline/Radius/Trim/Multiple]: *(Select point 1.)*
> Enter fillet radius <0.0>: **1.5**

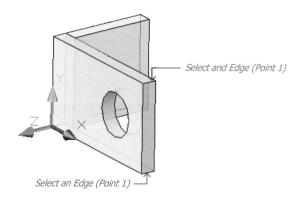

Select and Edge (Point 1)

Y

Z   ×

Select an Edge (Point 1)

> Select an edge or [Chain/Radius]: *(Select point 2.)*
> Select an edge or [Chain/Radius]: *(Press* ENTER.*)*

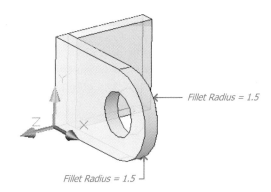

Fillet Radius = 1.5

Fillet Radius = 1.5

10. Rather than model the second leg, copy the first leg (the one with the hole):

   Command: **copy**

   Select objects: *(Select the leg.)*

   Select objects: *(Press* ENTER.*)*

   Specify base point or [Displacement] <Displacement>: **0,0,0**

   Specify second point or <use first point as displacement>: **0,0,-3.5**

   Specify second point or [Exit/Undo] <Exit>: *(Press* ENTER.*)*

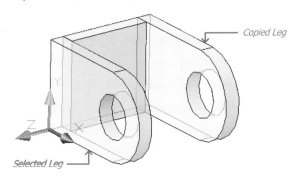

Copied Leg

Selected Leg

11. Combine the solids into a single composite body. In the Dashboard, click ⊙⊙ **Union**.

   Command: **union**

   Select objects: CTRL+A

   Select objects: *(Press* ENTER.*)*

   AutoCAD combines the three boxes into a single solid object. Note that the edge lines between the boxes are no longer a part of the object.

12. Save your work with the **SAVE** command, naming it "ch19-model.dwg."

SOLID MODELING TUTORIAL II

Solid modeling allows you to create the same object in many ways. One approach may be preferable to another, because it requires fewer steps, or uses commands with which you are more familiar. The previous tutorial described one approach; here is a completely different approach developed by technical editor Bill Fane.

1. Start a new drawing.

   Set the grid and snap to 0.5
2. Draw the profile with the **PLINE** command (lines and arcs) and a circle.

   As an alternative, you can use the **LINE** command, and then create the rounded end with the **FILLET** command.

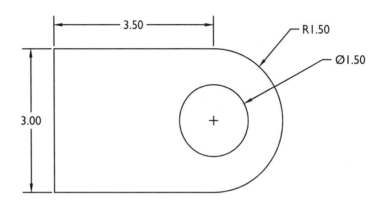

2. Convert the profile into a region with the **REGION** command.

   Command: **region**

   Select all objects.

   Select objects: **all**

   Select objects: *(Press ENTER to end object selection.)*

   2 loops extracted.

   2 Regions created.

   The objects don't look any different, but now consist of two region objects.
3. Subtract the circle from the outside region with the **SUBTRACT** command.

   Command: **subtract**

   First select the outside region:

   Select solids and regions to subtract from ...

   Select objects: *(Pick the outside region.)*

   Select objects: *(Press ENTER to end object selection.)*

   And then the circle region:

   Select solids and regions to subtract ...

   Select objects: *(Pick the circle.)*

   Select objects: *(Press ENTER to end object selection.)*

   Again, the objects look no different, but now consist of a region with a hole (island).
4. Use the **VPOINT** command to rotate the view so you can see your work in 3D.

   Command: **vpoint**

   Current view direction: VIEWDIR=0.0,0.0,1.0

   Specify a view point or [Rotate] <display compass & tripod>: **r**

Enter angle in XY plane from X axis <0>: **290**

Enter angle from XY plane <0>: **23**

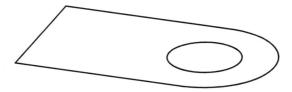

5. To "tilt" the region up, rotate it about the x axis:

    Command: *(Select the region.)*

    Command: **rotate3d**

    Specify first point on axis or define axis by [Object/Last/View/Xaxis/Yaxis/Zaxis/2points]: **x**

    Specify a point on the X axis <0,0,0>: *(Pick a corner of the region.)*

    Specify rotation angle or [Reference]: **90**

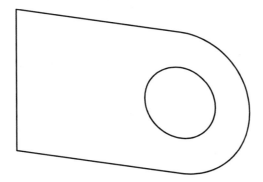

6. To convert the region into a solid model, use the **EXTRUDE** command. Give the model a thickness of 4.

    Command: **extrude**

    Current wire frame density: ISOLINES=4

    Select objects: *(Select region object.)*

    Select objects: *(Press **ENTER** to end object selection.)*

    Specify height of extrusion or [Path]: **4**

    Specify angle of taper for extrusion <0>: *(Press **ENTER** for 0 degrees of taper.)*

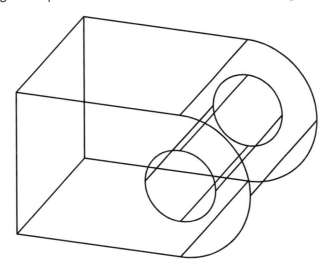

7. Before removing the "inside" of the part, set the UCS to the top face. That makes the model easier to work with.

Command: **ucs**

Enter an option [New/Move/orthoGraphic/Prev/Restore/Save/Del/Apply/?/World] <World>: **n**

Specify origin of new UCS or [ZAxis/3point/OBject/Face/View/X/Y/Z] <0,0,0>: **f**

Select face of solid object: *(Pick the top face, shown highlighted in the figure.)*

Enter an option [Next/Xflip/Yflip] <accept>: *(Press* ENTER.*)*

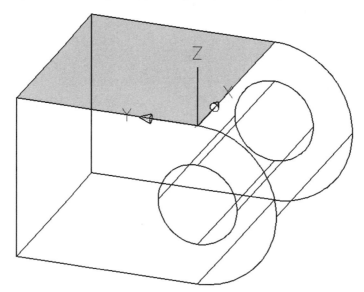

8. To create the "removal volume," draw a 3D box on the top face with a negative depth. The box is offset by 0.5 units from three edges. Use tracking mode to locate the corner of the box. It's easier with ortho turned on.

Command: *(Click* **ORTHO** *on the status bar.)* <Ortho on>

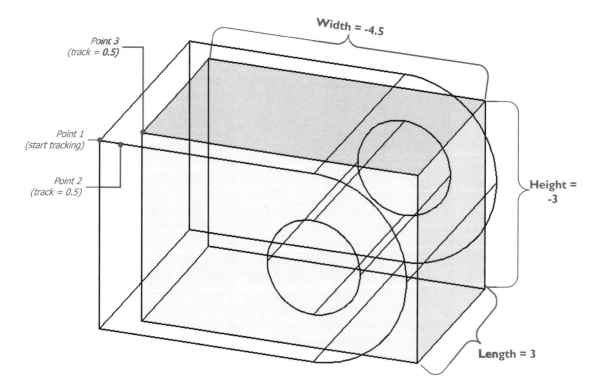

Command: **box**

Specify corner of box or [CEnter] <0,0,0>: **tk**

First tracking point: *(Pick point 1.)*

Next point (Press ENTER to end tracking): *(Move cursor to point 2, and then type **0.5**.)*

Next point (Press ENTER to end tracking): *(Move cursor to point 3, and then type **0.5**.)*

Next point (Press ENTER to end tracking): *(Press **ENTER**.)*

Tracking has located the starting point of the box, which you now draw:

Specify corner or [Cube/Length]:   **l**

Specify length: **3**

Specify width: **-4.5**

Specify height: **-3**

9. Use the **SUBTRACT** command to subtract the box from the model.

Command: **subtract**

First select the model:

Select solids and regions to subtract from ...

Select objects: *(Pick the model.)*

Select objects: *(Press **ENTER** to end object selection.)*

And then the box:

Select solids and regions to subtract ...

Select objects: *(Pick the box.)*

Select objects: *(Press **ENTER** to end object selection.)*

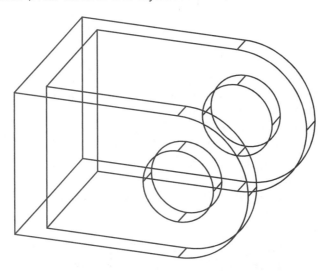

EXERCISES

1. From the Companion CD, open the *Ch19Fillet.dwg* file, a box with two holes. Apply a fillet radius of 0.1 to all edges. Save the result, which should be similar to the figure below.

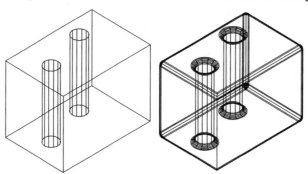

2. From the Companion CD, open the *Ch19Chamfer.dwg* file, a wedge. Apply a chamfer of 0.1 to the two edges shown in the figure below.

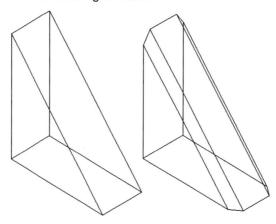

3. Using the commands you have learned in this book, draw the dishwasher rack wheel, illustrated below, as a solid model.

   The wheel is shown approximately full-size; some dimensions are:

          Outer diameter = 1-3/4"

          Wall thickness = 1/16"

          Axle hole = 5/32"

          Width of wheel = 3/4"

4. Draw the door latch cover, illustrated below, as a solid model.

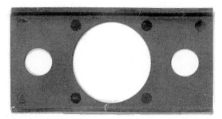

The cover is shown approximately full-size; some dimensions are:
   Overall length = 2-1/4"
   Overall width = 1-1/8"
As an aid, the profile of the cover, shown below, is available on the Companion CD as *Ch19Latch.dwg*.

5. From the Companion CD, open the *Ch19-SectionRotate.dwg* file, a drawing of a 3D part with a countersunk hole.

   Use the **3DROTATE** command to "flip" the part by 180 degrees. The before and after views are shown below.

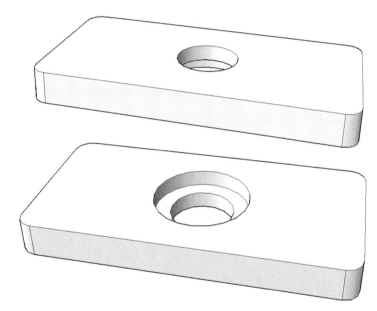

6. From the Companion CD, open the *Ch19HistoryEx.dwg* file, a drawing of a 3D part.
   a. What editing actions were performed on this part?
   b. Which 3D primitives were used to make the part?
   c. What command or system variable helps you determine the history of the part?

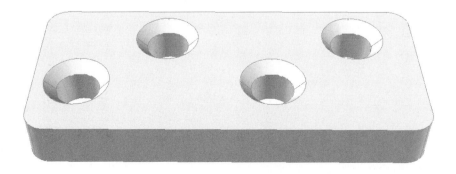

7. Continuing with the 3D part illustrated above, draw a section plane that cuts through all four holes.
    a. Which command did you use to place the section plane?
    b. Which option do you use to draw the section plane?
    c. Which object snaps (if any) did you use to place the section plane accurately?

8. Continuing with the sectioned 3D part from above, activate the section plane.

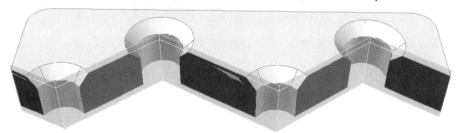

    a. What command did you use to active the section plane?
    b. Reverse the section view.
    c. Display the cut-away geometry, and then move it away from the 3D model.

CHAPTER REVIEW

1. Name two properties of solid objects that the **PROPERTIES** command can change.
2. Can the **FILLET** command create fillets between two solid objects?
3. Can the **CHAMFER** command create chamfers on more than one edge of a solid model at a time?
4. Briefly describe the purpose of the following commands when applied to 3D solids:
   a. **ALIGN**
   b. **ERASE**
   c. **MOVE**
   d. **ROTATE3D**
5. What must you do before filleting or chamfering between two solid objects?
6. Which key do you hold down to select the face of a body?
7. When chamfering solid objects, what is meant by the *base surface*?
8. What happens when the **EXPLODE** command is applied to:
   a. Solid models
   b. Regions
9. What is a *chain*?
   A *loop*?
10. How many edges does a rectangular face have?
11. What are *isolines*?
12. How would you create a lip around the top of a solid model?
13. Identify the commands that display the following grip tools, and their constituent parts:

Grip tool displayed by the _____ command:

Grip tool displayed by the _____ command:

14. When 3D solids are selected, they display two kinds of grips. Briefly describe the action of each grip shape:

■ _____

▲ _____

15. Label the action of the 3D model grips illustrated below:

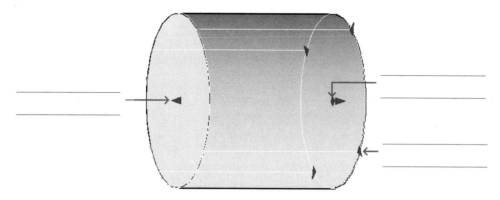

16. When do the grip tools align with selected faces?
17. Can grip tools be relocated in the drawing?
    If so, how?
18. Describe how the **GTDEFAULT** system variable affects the **MOVE** and **ROTATE** commands.
19. The grip tools are normally colored red, green, and blue. What is meant when one part of the tool turns yellow?
20. Can grips be used to change the base of 3D solid cylinders between round and elliptical?
    Slant their orientation in space?
21. What is a "hole"?
    Can holes be manipulated using grips?
22. What is the purpose of the **SOLIDHIST** system variable?
23. Describe a benefit of applying section planes to 3D models.
24. Briefly explain the function of the three section commands:
    SECTIONPLANE
    LIVESECTION
    JOGSECTION
25. What is the "cut-away" geometry?
    Can cut-away geometry be displayed by AutoCAD?
26. How can you remove section planes from drawings?
27. What is "live sectioning"?
28. Label the actions of the section plane grips illustrated below:

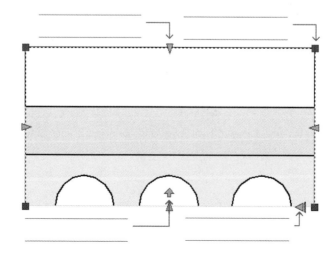

CHAPTER 20

Rendering 3D Models

Three-dimensional models are often viewed in wireframe mode, as if they were transparent. But many faces and isolines can make the models hard to visualize, especially for non-drafters. For this reason, AutoCAD removes hidden lines and renders 3D models realistically with lights, surface materials, and shadows.

In this chapter, you are introduced to visual styles and rendering through the following commands:

VSCURRENT applies the named visual style to 3D objects.

VISUALSTYLES creates new visual styles through a palette.

RENDERCROP generates photorealistic renderings of windowed 3D views.

RENDEREXPOSURE adjusts brightness and contrast of renderings (new to AutoCAD 2008).

VIEW defines backgrounds for visual styles and renderings.

SUNPROPERTIES adjusts the properties of the sun light.

RENDERENVIRONMENT creates fog effects.

LIGHT inserts lighting into 3D scenes.

ANIPATH creates animations along paths.

NEW TO AUTOCAD 2008 IN THIS CHAPTER

- The new **RENDEREXPOSURE** command adjust the brightness and contrast of renderings.
- The **NEWVIEW** command goes directly to the **VIEW** command's New View dialog box.

FINDING THE COMMANDS

On the Dashboard's **Visual Styles** panel:

Global Visual Styles — Control Commands
Named Visual Styles
Face Style — Facet/Smooth Shading
Isoline Color
Overhang
Jitter
Sihlouette
Obscured Edges Color — Intersection Color

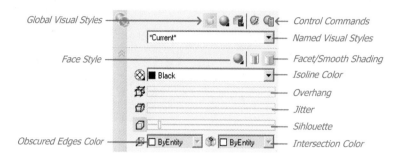

On the Dashboard's **Materials** panel:

Toggle Materials and Textures
Materials Palette
Mapping — Attach Materials by Layer

On the Dashboard's **Lights** panel:

Toggle User Lights and Sun Light — Light List palette
Date
Time
Create Point, Spot, and Distant Lights — Geography, Toggle Light Glyhphs, and Edit the Sun.
Brightness
Contrast
Midtones

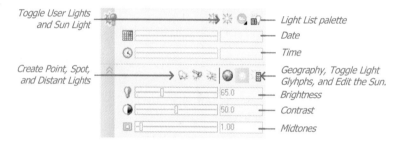

On the Dashboard's **Render** panel:

RenderCrop
RenderExposure — RenderEnvironment

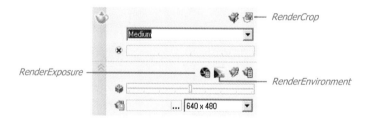

On the **Visual Styles** and **Render** toolbars:

VisualStyles

Render   Light   Materials   Render Environment

VsCurrent

Sun Properties

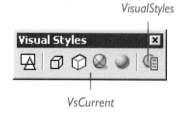

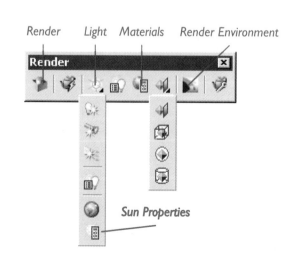

VSCURRENT

The **VSCURRENT** command sets the current visual style for 3D objects.

This command renders 3D models in real-time. (It replaces **SHADEMODE**, which is available as the renamed **-SHADEMODE** command.) AutoCAD includes five visual styles, and you can make your own:

 2D Wireframe — simulates AutoCAD's 2D drafting environment. Models are displayed in wireframe mode, the grid appears as dots, and the background color is white/black.

 3D Wireframe — displays models in wireframe mode also, but the grid appears as lines, the background color is gradated, and perspective viewing mode is allowed.

 3D Hidden — removes hidden lines from models; otherwise the same as 3D wireframe mode.

 Conceptual — renders models with a warm and cool set of colors, named "Gooch."

Realistic — rendes 3D models with realistic colors. If materials are attached to models, they are displayed.

(*History*: In the early days of AutoCAD, when desktop computers were slow, it could take all day to perform hidden-line removal from complex 3D drawings. This task was often left for overnight or for weekends. Rendering was unavailable. With today's fast computers, hidden-line removal occurs in real-time, while renderings take just a few seconds.)

BASIC TUTORIAL: APPLYING VISUAL STYLES

In this tutorial, you learn how to apply visual styles using two methods. First, open *vstyles.dwg* from the Companion CD, a 3D drawing of a donut and a gear sitting on a rectangular plate.

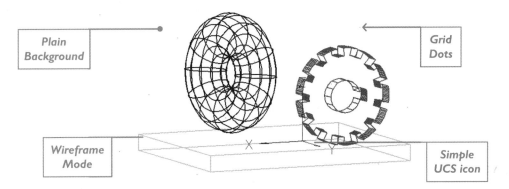

1. To change the visual style of 3D objects, start the **VSCURRENT** command:
   * From the **View** menu, choose **Visual Styles**, and then choose **3D Wireframe**.
   * In the Visual Styles toolbar, click the **3D Wireframe** button.
   * In the Dashboard's Visual Styles panel, select **3D Wireframe** from the droplist.
   * At the 'Command:' prompt, enter the **vscurrent** command.

     Command: **vscurrent** *(Press* ENTER.*)*
   * Alternatively, enter the alias **vs** at the keyboard.

2. If you start at the command prompt, AutoCAD prompts you as follows. Enter **3** for the 3D wireframe option:

   Enter an option [2dwireframe/3dwireframe/3dHidden/Realistic/Conceptual/Other]
   <2dwireframe>: **3**

Notice that the display changes: the grid changes to lines, and the UCS icon changes to the rendered style.

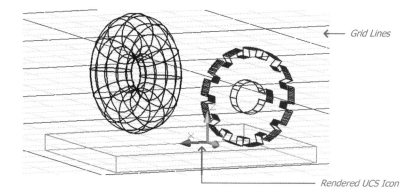

Grid Lines

Rendered UCS Icon

3. Turn on perspective viewing mode with the **PERSPECTIVE** system variable, as follows:

   Command: **perspective**

   Enter new value for PERSPECTIVE <0>: **1**

Gradiated Background Color

Perspective Model and Grid Lines

Notice that the model and grid take on the perspective effect, and that the background changes to gradated gray.

4. Repeat the **VSCURRENT** command, and then enter the **3dHidden** option to remove hidden lines from the model:

   Command: **vscurrent** *(Press* ENTER.*)*

   Enter an option [2dwireframe/3dwireframe/3dHidden/Realistic/Conceptual/Other]
   <2dwireframe>: **h**

Instantly, hidden lines are removed, although the grid lines continue to be displayed on top of the box.

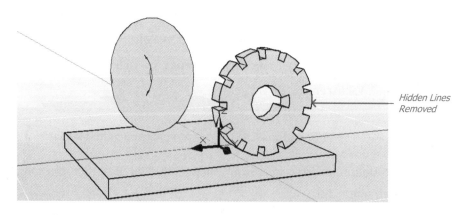

Hidden Lines Removed

5. Let's change the visual style to **Conceptual** using the Dashboard. (If the Dashboard palette is not displayed, use the **DASHBOARD** command to open it.)
    In the Visual Style control panel, click the **Visual Styles** droplist. (It currently displays "3D Hidden.") Notice that five or more icons represent visual styles.

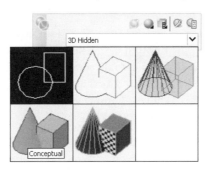

Select the **Conceptual** icon. Notice that the model changes colors, taking on a variety of hues. White models, for instance, take on the hues of blue, yellow, and brown.

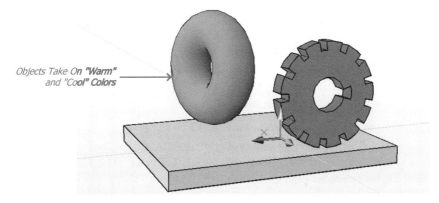

Objects Take On *"Warm"* and "Cool" *Colors*

6. Let's change the visual style to Realistic using the Dashboard again. From the Visual Styles droplist, select the **Realistic** icon. Notice that the model changes colors again, this time taking on solid colors. It doesn't look very realistic; we change that in the next section.

Rendered With Isolines

Rendered With Edges

ATTACHING MATERIALS

Materials define the look of rendered objects. They consist of colors or bitmaps (pictures). By varying the amount of shininess, roughness, and/or reflection, colors make CAD models look as if they were made of physical materials, such as plastic, concrete, and wood.

New drawings contain a single material definition, called *GLOBAL*, which holds default parameters for color, reflection, roughness, and ambience. It looks like gray plastic. Fortunately, AutoCAD includes an extensive library of predefined materials, and you can create your own with the **MATERIALS** command, if you have the fortitude. The easier alternative is to take digital photographs of everyday objects, like paving stones and plywood sheets, and then apply them to objects.

Materials appear when the visual style is set to Realistic, and when rendering commands, such as **RENDER** and **RENDERCROP**, are used. Materials are strictly for looks; they don't do anything useful, such as define the density of 3D solids.

Applying materials to drawings takes these steps:

Step 1. Open the Tool palette to the material tabs.

Step 2. Drag icons from the palette onto objects in drawings.

Step 3. Turn on the Realistic visual style to see the effect.

BASIC TUTORIAL: APPLYING MATERIALS

For this tutorial, continue with the *vstyles.dwg* drawing.

1. If the Tool palette is not visible, open it with the **CTRL+3** shortcut.
   If none of the material tabs is visible, right-click the lower end of the tabs. From the shortcut menu, select **Doors and Windows - Materials Sample**. The Tool palette displays several tabs with samples of materials.

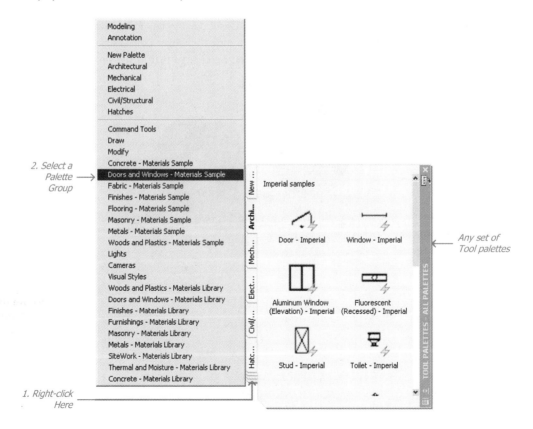

2. From the Doors and Windows - Materials Sample tab, select **Doors - Windows. Wood Doors.Ash**.

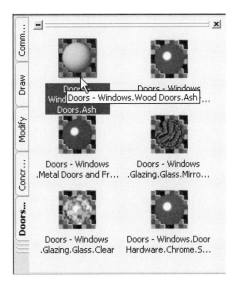

3. Notice that the cursor turns into a paintbrush with a selection box.

On the command line, AutoCAD prompts you:

Select objects: *(Select the box.)*

Notice that the box immediately changes appearance. It now looks as if it were crafted from ash wood.

4. You can use the brush cursor to "paint" other objects with this material, but not in this tutorial. Press **ENTER** to exit the prompt:

Select objects or [Undo]: *(Press **ENTER** to exit the command.)*

Note: You can use the **U** command to remove the material from the object immediately. To remove the material at a later time, use the **PROPERTIES** command to change the **Material** property to *GLOBAL*.

5. Apply materials to the donut and gear in a slightly different manner:
In the Tool palette, drag the "Doors - Windows.Metal Doors and Frames.Steel.Galvanized" icon onto the gear.

Instead of the brush cursor, you see the icon itself as the cursor.

Doors - Windows
.Metal Doors and Fr...

This drag'n drop method eliminates the need to answer and dismiss command-line prompts. The gear now looks as if it is made from galvanized steel.

6. Finally, drag the "Doors - Windows.Glazing.Glass.Mirrored" icon onto the donut. It takes on a gray shiny finish.

    Even though it has the mirror finish, the donut does not reflect anything, because the rendering technology used by visual styles does not handle reflections. This will have to wait for the **RENDERCROP** command, which is more sophisticated than **VSCURRENT**.

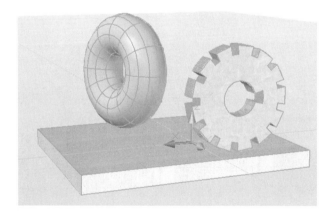

CLEANING UP THE RENDERED VISUAL STYLE

The objects in the *vstyle.dwg* drawing now have materials attached. They look more realistic, but there are still AutoCAD artifacts in the way. Let's remove the isolines, edge and grid lines, and the UCS icon.

1. Remove the UCS icon with the **UCSICON** command's **Off** option:

    Command: **ucsicon**

    Enter an option [ON/OFF/All/Noorigin/ORigin/Properties] <ON>: **off**

2. Turn off the grid by clicking **GRID** on the status bar.

    Command: *(Click* **GRID***.)* <Grid off>

3. Use the Dashboard palette to remove the isolines and edge lines. From the flyout menu, select **No Edges**.

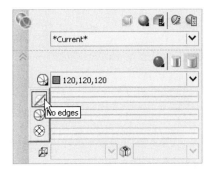

4. To remove the gradient gray background quickly, change the **PERSPECTIVE** system variable back to 0:

> Command: **perspective**
>
> Enter new value for PERSPECTIVE <1>: **0**

The viewport looks much cleaner now.

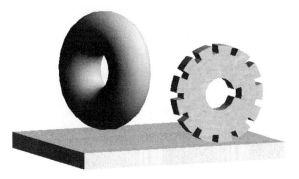

 **RENDERCROP**

RENDERCROP is one of several commands that generate photorealistic renderings of 3D views. Of them, I find that this is the quickest way to render drawings, because it renders areas of viewports, which is faster than rendering the entire viewport. (While AutoCAD can display visual styles in paper space, it cannot render in that mode.)

BASIC TUTORIAL: RENDERING DRAWINGS

Before using this command, open the *vstyle-complete.dwg* drawing from the Companion CD. The drawing contains 3D objects with materials already assigned .

1. First use the **RPREF** command to set the rendering preferences to **Medium**.
   This setting is a good compromise between quality (which renders slowly) and speed (which renders poorly).

   • From the **View** menu, choose **Render**, and then **Advanced Render Settings**.

   • In the Render toolbar, click the **Advanced Render Settings** button.

   • In the Dashboard's Render panel, select the **Advanced Render Settings** button.

   • At the 'Command:' prompt, enter the **rpref** command.

   > Command: **rpref** *(Press ENTER.)*

   • Alternatively, enter the **rpr** alias.

AutoCAD displays the Advanced Render Settings palette.

2. From Render Presets droplist, select **Medium**.

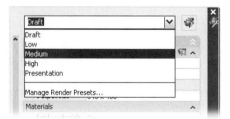

3. To render the 3D drawing, start the **RENDERCROP** command:

    Command: **rendercrop** *(Press ENTER.)*

4. AutoCAD prompts you to pick two points:

    Pick crop window to render: *(Pick a point.)*

    please enter the second point: *(Pick another point.)*

    The drawing is rendered.

    But the background is black, which makes it hard to see objects. You can the background color with the **VIEW** command, as described in a later next tutorial

5. To turn off the rendering, use the **REGEN** command:

    Command: **regen**

 **RENDEREXPOSURE**

The **RENDEREXPOSURE** commands adjusts the brightness and contrast of renderings. When you first try the command, AutoCAD complains, "Exposure control is not enabled," which is a mysterious way of telling you to change the value of the **LIGHTINGUNITS** system variable from 0 to 1 (metric units) or 2 (Imperial units).

1. To adjust the exposure of renderings, first ensure the **LIGHTINGUNITS** system variable is turned on:

    Command: **lightingunits** *(Press ENTER.)*

    Enter new value for LIGHTINGUNITS <0>: **2**

2. Start the **RENDEREXPOSURE** command:

    Command: **renderexposure** *(Press ENTER.)*

    AutoCAD displays the Adjust Rendered Exposure dialog box, and then re-renders the scene in a small preview window.

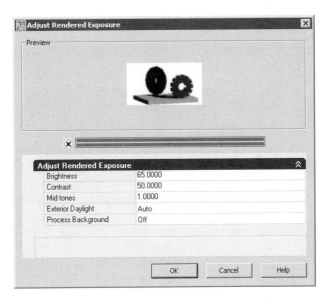

3. Try changing the Brightness and Contrast settings to see their effect.

4. When done, click **OK** to exit the dialog box.

5. Unfortunately, the effects are not interactive, and so you have to rerun the **RENDERCROP** command each time.

The following parameters can be adjusted in the dialog box.:

- **Brightness** — determines the global brightness of photometric lighting; ranges from 0 to 200 (Value is stored in the **LOGEXPBRIGHTNESS** system variable.)

- **Contrast** — determines the global contrast; ranges from 0 to 100 (**LOGEXPCONTRAST**).

- **Mid Tones** — determines the global midtones; ranges from 0 to 20 (**LOGEXPMIDTONES**).

- **Exterior Daylight** — toggles global daylight lighting (sun); switches between auto, off, and on (**LOGEXPDAYLIGHT**).

- **Process Background** — toggles processing of the background by exposure control; when off, the background is not rendered, and keeps the color or effect assigned by the **VIEW** command.

 **VIEW BACKGROUND**

The **VIEW** command defines the backgrounds of visual styles and renderings. (In older releases of AutoCAD, this was handled by the **BACKGROUND** command.) The background can consists of a solid color, two or three gradient colors, or an image. For the rendered image, create a white background.

1. Start the **VIEW** command using one of the following methods:
   - From the **View** menu, choose **Named Views**.

   - In the View toolbar, click the **Named Views** button.

   - In the Dashboard's 3D Navigate panel, click the View Names droplist, and then choose **Manage Views**.

   - At the 'Command:' prompt, enter the **view** command.

     Command: **view** *(Press* ENTER.*)*

   - Alternatively, enter the **v** alias.

AutoCAD displays the View Manager dialog box:

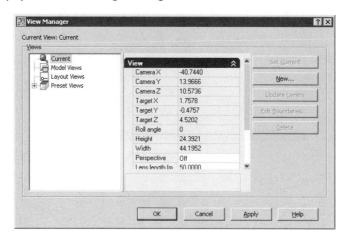

2. Backgrounds are attached to named views. To create one for this drawing, click **New**.
   AutoCAD displays the New View dialog box.

 Note: You can get directly to this dialog box with the **NEWVIEW** command (new to AutoCAD 2008).

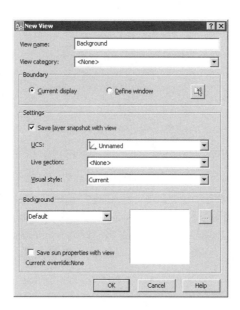

3. Give the view a name:

   View Name   **Background**.

4. To specify the background, select **Solid** from the Background droplist. AutoCAD displays
   the Background dialog box.

5. From the Type droplist, select **Solid**.
6. To specify the color, click the sample color rectangle under **Color**. AutoCAD displays the familiar Select Color dialog box.

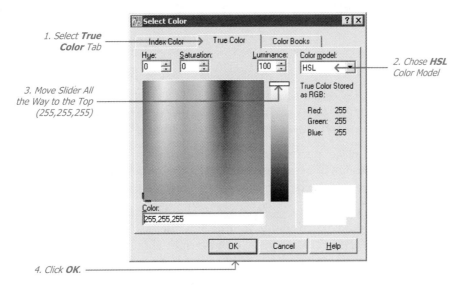

*1. Select **True Color** Tab*

3. Move Slider All the Way to the Top (255,255,255)

*2. Chose **HSL** Color Model*

*4. Click **OK**.*

To specify white as the color, follow these steps:

a. Choose the **True Color** tab.

b. Select the **HSL** color model.

c. Move the slider all the way up so that the color turns white. (It doesn't matter which color you start with.) The color number should read **255,255,255**.

d. Click **OK** to exit the dialog box.

7. Click **OK** to exit the Background dialog box.

Click **OK** to exit the New View dialog box.

8. Back in the View Manager dialog box, click **Set Current** to select the new named view.

Click **Apply** to apply the background to the viewport.

Click **OK** to exit the dialog box.

9. Repeat the **RENDERCROP** command. You can see the gear and the wooden box reflected in the donut.

CASTING SHADOWS WITH VISUAL STYLES

Objects can cast shadows in visual styles. AutoCAD can display three kinds of shadow, which are controlled by the **VSSHADOWS** system variable. Despite the capability, I recommend you not use shadows in visual styles, because their generation takes a large amount of computing power, slowing down AutoCAD.

The three settings are, as follows:

- **None** (**VSSHADOWS** = 0) — objects cast no shadows.

- **Ground shadows** (1) — objects cast shadows onto the "shadow plane," whose height is controlled by the **SHADOWPLANELOCATION** system variable.

- **Full shadows** (2) — objects cast shadows on the ground and on each other. Not all graphics boards can cast full shadows, so use the **3DCONFIG** command's **Manual Tune** option to determine whether your computer's graphics board has this capability.

In addition to the setting of these system variables and the capabilities of your graphics board, shadows are also affected by the types of lights in drawings. Before lights are added to drawings, there are two default lights positioned over your left and right shoulders; the shadows are cast as if the lights were directly overhead. Curious, but true.

(For the rendering commands, like RENDER and RENDERCROP, shadow casting is handled differently. Shadows and their display qualities are determined by the RPREF command's options. Once lights are placed in drawings, shadows are cast correctly: in the opposite direction from the light source.)

The Dashboard palette makes it easy to switch among the three shadow states: select the appropriate button from the droplist in the Visual Styles control panel.

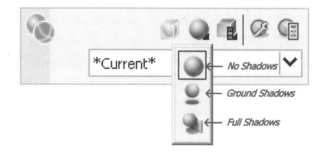

TUTORIAL: CASTING SHADOWS

Use the REGEN command to switch from rendered mode to visual style mode, which should be set to "Realistic" for this tutorial.

1. To display shadows in the current visual style, start the **VSSHADOWS** system variable:
   - At the 'Command:' prompt, enter the **vsshadows** system variable.

     Command: **vsshadows** (*Press* ENTER.)
   - Alternatively, in the Dashboard dialog box, choose the **Ground Shadows** button.

2. AutoCAD displays a prompt at the command line:
   Enter new value for VSSHADOWS <0>: **1**

   Enter **1** (one) to cast shadows on the ground, as illustrated below.

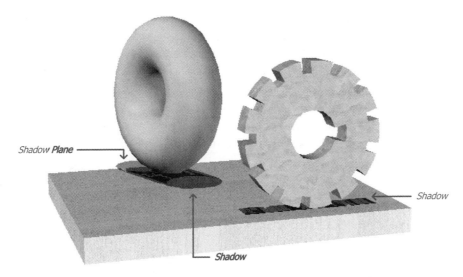

Shadow **Plane**

Shadow

Shadow

3. To see how the elevation of the shadow plane can be changed, enter the
SHADOWPLANELOCATION system variable, and then change it to -2 units:

Command: **shadowplanelocation**

Enter new value for SHADOWPLANELOCATION <1.0000>: **-2**

The shadows now appear under the wooden base. (See figure at left, below.)

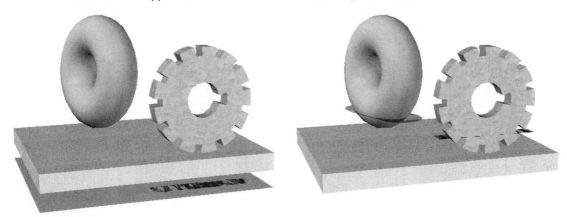

4. Change the value of **SHADOWPLANELOCATION** to 1. The shadows now appear to cut
through the donut and gear (see figure at right, above).

SUN LIGHT

Every 3D viewport has two default lights, placed over each of your shoulders. The lights are handy,
because they follow the viewpoint wherever you move it. The light over your right shoulder is brighter
than the one over your left. For more specific needs, AutoCAD can place five kinds of light, each
more specific in its nature:

- **Distance lights** shine light in parallel beams with constant intensity. Typically, you place a single
  distant light to simulate the sun.

- **Sun light** is a specific type of distant light. You can adjust the sun light by specifying a geographic
  location, day of the year, and time of day. To simulate a setting sun, change the color of the light
  to orange-red.

- **Point lights** shine light in all directions, with inverse linear, inverse square, or constant intensity.
  These lights work well as light bulbs in lamps.

- **Spot lights** shine light in a specific direction, in a cone shape. These lights work well as spotlights to beam from light locations to targets, such as high-intensity desk lamps and vehicle-mounted spotlights. When you place spotlights in drawings, you specify the hotspot of the light (where the light is brightest) and the falloff, where the light diminishes in intensity.

- **Web lights** shine light in a pattern defined by *.ies* files. They are also known as *photometric* lights, and have nothing to do with the World Wide Web. The *.ies* files are supplied by lighting manufacturers, and create renderings that are much more realistic. For Web lights to work, the **LIGHTINGUNITS** system variable must be turned on.

Of all these type of lights, the easiest to place is the sun light: just click the **Sun Status** button on the Dashboard palette's Lights control panel. This disables the two default lights, and turns on the sun light.

When the sun light is turned on, the shadows will probably change, because AutoCAD automatically locates the sun light according to the current date and time. See the figures below.

 Note: I find that the quality of shadows varies greatly, depending on the model of graphics board, whether hardware acceleration is turned on (through the **3DCONFIG** command's **Manual Tuning** option), and how well the board supports shadows. The figure at left shows dithered shadows when hardware acceleration is turned off. At right, hardware acceleration is turned on, but the shadows are incomplete.)

The sun is technically a "distant" light whose position is affected by date, time, and geography (location on earth.) Each of these factors can be adjusted in the Light panel of the Dashboard palette:

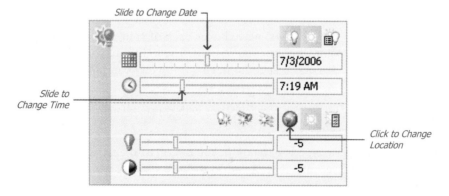

- **Change Date** — locates the sun by the day of the year. In the northern hemisphere, the longest shadows occur in winter, while the shortest are in summer; the reverse is true for the southern hemisphere. Shadow lengths do not change around the Equator.

- **Change Time** — locates the sun by time of day. The shortest shadows occur around noon, longer ones towards the morning and evening hours. No shadows are cast at night — AutoCAD has no moon shadows.

As you drag the sliders in the Dashboard panel, the shadows change interactively, as illustrated by the composite image below. Alternatively, you can enter exact dates and times in the text entry boxes.

- **Change Location** — locates the sun by geography; displays the Geographic Location dialog box.

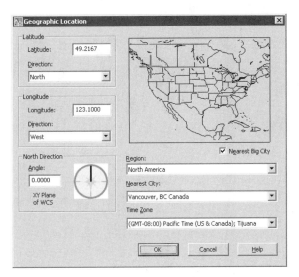

Latitude sets the latitude from 0 to 90 degrees between the Equator and the poles. (The value is stored in the **LATITUDE** system variable.) Its **Direction** option selects north or south of the equator.

Longitude sets the longitude from 0 to 180 between East and West. (The value is stored in the **LONGITUDE** system variable.) Its **Direction** option selects East or West of the Prime Meridian, which goes through Greenwich, England.

Angle specifies the direction of north; default is the positive y axis. (The value is stored in **NORTHDIRECTION** system variable.)

You can click points in the map to locate latitudes and longitudes; the red cross shows the current location. The **Nearest Big** City option displays the name of the nearest city to the latitude and longitude values entered. **Region** selects a region or the entire world, which then displays the related map.

Nearest City selects a city listed in AutoCAD's database, while **Time Zone** lets you select or change the time zone (stored in the **TIMEZONE** system variable).

This dialog box uses names and coordinates of cities found in the *sitename.txt* file found in the *C:\Documents and Settings\<username>\Application Data\Autodesk\AutoCAD 2008\R17.1\enu \Support* folder. You can add other locations by editing this text file with Notepad.

 **SUNPROPERTIES**

The **SUNPROPERTIES** command displays the Sun Properties palette, which allows fine-tuning of the sun light, such as the color and the harshness of shadows.

TUTORIAL: TUNING THE SUN LIGHT

1. To tune the sun light, start the **SUNPROPERTIES** command:
   - From the **View** menu, choose **Render**, choose **Light**, and then **Sun Properties**.
   - In the Render toolbar, select the **Sun Properties** button.
   - In the Dashboard palette's Light control panel, click the **Edit the Sun** button.
   - At the 'Command:' prompt, enter the **sunproperties** command.

     Command: **sunproperties** (Press ENTER.)

   In all cases, AutoCAD displays the Sun Properties palette.

To turn the sun on and off, select **On** or **Off** from the **Status** property. When off, all properties retain their values for the next time the sun is turned on.

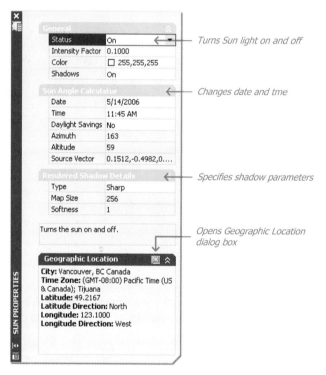

The **Intensity Factor** property allows you to make the sun brighter or dimmer. The range is from 0 (no light) to 9.999E+99, a very large number.

The **Color** property allows you to change the color of sunlight, such as light pink for early mornings and deep red for polluted sunsets. You can choose any AutoCAD color.

The **Shadows** parameters determine whether the sun light casts shadows.

The **Sun Angle Calculator** parameters are similar to those found on the Dashboard palette. They allow you to specify the date and time, and toggle daylight savings time. You can't change the Azimuth (the compass bearing), Altitude (height of the sun above the horizon), and Source Vector parameters; they are generated by AutoCAD based on the date and time.

When you click the **...** button in the Date field, a calendar pops up. Select a date by double-clicking it.

The **Rendered Shadow Details** section is available only when the Shadows parameter is turned on. These parameters apply to renderings only, and not to visual styles.

The **Type** parameter specifies whether shadows have hard (sharp) or soft edges. If soft, then two more parameters come into play. The **Map Size** specifies the size of shadow mapping, and ranges from 64 to 4096. Larger values mean more detailed shadows, which take longer to render. The **Softness** parameter ranges from 0 to 10; larger values mean softer shadow edges.

The graphics board in your computer determines whether AutoCAD can cast these shadows are cast, and the quality of the shadow simulation.

RENDERENVIRONMENT (FOG)

The **RENDERENVIRONMENT** command variable specifies an optional color and intensity for the atmosphere in renderings. (This command was known as **FOG** in AutoCAD 2007 and earlier.)

AutoCAD simulates fog by increasing the amount of gray in the rendering over distance. The farther away, the more dense the gray and the thicker the fog.

The color need not be gray. The subtle use of black, for example, enhances the illusion of depth, because objects farther away tend to be darker. Autodesk calls fog "depth cue." A limited application of yellow creates the illusion of glowing lamps; the reckless use of green fog simulates a Martian invasion.

Fog does not work in visual styles; it applies to renderings only.

TUTORIAL: ADDING FOG

1. To add the fog effect to renderings, start the **RENDERENVIRONMENT** command:
   - From the **View** menu, choose **Render**, and then **Render Environment**.

   - In the Dashboard palette's Render control panel, click the **Render Environment** button.

   - At the 'Command:' prompt, enter the **renderenvironment** command.

   - Alternatively, enter the **fog** alias. (No need to type R-E-N-D-E-R-E-N-V-I-R-O-N-M-E-N-T!).

      Command: **fog** *(Press* ENTER.*)*

   In all cases, AutoCAD displays the Fog/Depth Cue dialog box.
2. To turn on the fog effect, change **Enable Fog** to **On**. (To disable the fog effect temporarily, turn off this option; AutoCAD remembers all the fog settings the next time you turn it on.)

3. Turn on **Fog Background**. This toggle determines whether the fog affects the background. For example, if the background color in your rendering is normally white, but you choose black for the fog color, then the background becomes black.

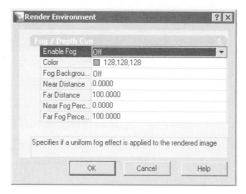

4. Select a color for the fog. Gray is the default. Choose a color from the droplist, or else select **Select Color** to display the Select Color dialog box.

5. Set the extent of the fog. The **Near Distance** and **Far Distance** properties determine where the fog begins and ends.

   • **Near Distance** positions the start of the fog effect. This property can be tricky to understand, because it represents a relative distance from the "camera" (your eye) to the back clipping plane. Try starting with a value of **0.45**.

   • **Far Distance** positions the end of the fog effect. The value also represents the percentage distance from the camera to the back clipping plane. Try starting with a value of **0.55**.

6. Specify the strength of the fog. The **Near Fog Percentage** and **Far Fog Percentage** properties determine the percentage of fog effect at near and far distances. For a stronger fog effect, increase the value of **Near Fog**; for a weaker effect, reduce the value of **Far Fog**.

7. Click **OK**, and then use the RENDERCROP command to see the fog effect.

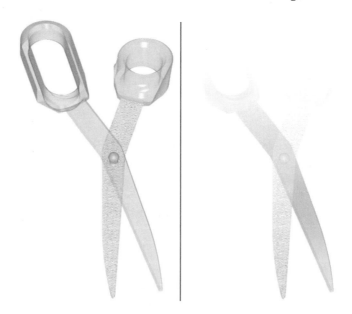

Left: Rendering without fog effect.
Right: Rendering with white fog effect.

I find I have to adjust the fog distance and strength parameters a number of times, each time executing a render, until the effect looks right.

 **ANIPATH**

The **ANIPATH** command makes movies from 3D drawings (short for "animation path"). This command moves a "camera" along a path, taking multiple snapshots, which become movie frames.

To create an animation, you specify (1) the camera path, (2) the target path, (3) the visual style, and (4) the output file format. The *path* is a line, arc, elliptical arc, circle, 2D or 3D polyline, or spline. To create a path that winds its way through the model, place a polyline with straight and curved segments. The *target* is the point that the camera looks at, which can be a path or a point. You can specify the duration of the movie, the number of frames, and/or its speed in fps (frames per second).

BASIC TUTORIAL: ANIMATING DRAWINGS

In this tutorial, you create an animation that circles about a 3D object. Before using this command, open *Ch19-SectionPlaneEx.dwg* from the Companion CD, a drawing containing the 3D model.

I. Draw a circle around the 3D object. The circle becomes the path for the camera.

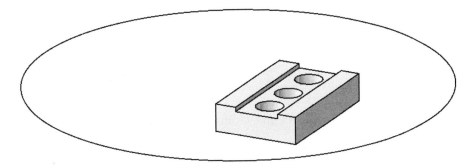

2. To create the animation, start the **ANIPATH** command:
   - From the **View** menu, choose **Motion Path Animations**.

   - At the 'Command:' prompt, enter the **anipath** command.

   Command: **anipath** *(Press* ENTER.*)*

AutoCAD opens the Motion Path Animation dialog box.

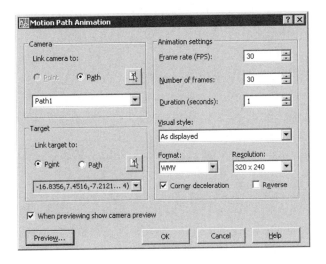

3. Select the following options:

Camera	Path
⬚	*(Pick the circle.)*
Target	Point
⬚	*(Pick one corner of the 3D model.)*
Frame rate	30
Duration	10
Visual Style	Conceptual
Format	AVI
Resolution	640x480
When Previewing Show Camera Preview	☑

Camera

When the target is a point, then the camera must be linked to a path — otherwise no animation would take place!

○ **Point** — links camera to a 3D point in a drawing; prompts you to name the point.

⊙ **Path** — links camera to path object; prompts you to name the path.

⬚ **Pick** — selects the point or path object.

Target

○ **Point** — links target to a 3D point in a drawing; prompts you to name the point.

⊙ **Path** — links target to path object; prompts you to name the path.

⬚ **Pick** — selects the point or path object.

Animation Settings

The number of frames and duration are linked through the frame rate: changing one changes the other.

Frame Rate **=** Frames **x** Seconds

Frame Rate (FPS) — specifies the animation speed in frames per second; range is 1 to 60. The speed most commonly used for movies is 30fps, although 15fps works if the file size needs to be smaller.

Number of Frames — specifies the number of frames to record during the animation. Changing this value changes the duration, based on fps.

Duration (seconds) — specifies the duration of the animation. Changing this value changes the number of frames, based on fps.

Visual Style — selects the visual style or preset rendering style in which to record the animation; the selected style does not show in the viewport. When materials and textures are attached, they will appear in the movie.

Format — specifies the animation file format: AVI (recommended), Apple MOV, MPG, or Microsoft WMV.

Resolution — specifies the width and height (in pixels) of the animation; range is from 160x120 to 1024x768. The higher the resolution, the better the viewing quality but the larger the file size; 640x480 is a good compromise.

☑ **Corner Deceleration** — slows the camera around sharp corners of the path.

Reverse — reverses the animation.

☐ **When Previewing Show Camera Preview** — displays the Animation Preview dialog box.

Preview — previews the animation in the Animation Preview dialog box.

4. Click **Preview**.

    Notice that AutoCAD displays the Animation Preview window. The camera circles the 3D model once, and the camera's view is displayed in the window. The target is at the center of the image in the preview window.

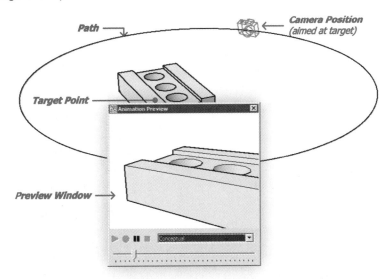

With its controls, the Animation Preview window lets you replay the animation. **Play** and **Pause** start and stop the movie. (The **Record** and **Save** buttons don't work here.) The slider lets you see specific movie frames; a tooltip reports the frame number.

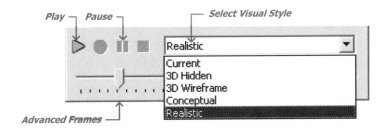

5. Close the preview window by clicking the **x** at the right end of the title bar.

    Notice that the Motion Path Animation dialog box returns.

6. Click **OK** to generate the movie file.

    AutoCAD displays the Save As dialog box. (The **Animation Settings** button lets you make last minute changes to the visual style, frame rate, resolution, and format.)

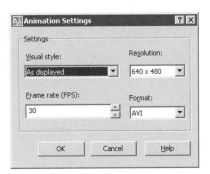

7. Select a folder, and then give the movie file a name, such as "AniPath."
   Click **Save**. Notice that AutoCAD runs the camera around the path again, this time
   generating the movie file. You can click **Cancel** to terminate the process.

8. When AutoCAD is finished, it quietly exits the command.
   To play back the movie, use Windows Explorer to look for the *.avi* file in the folder.
   Double-click the file to play back.

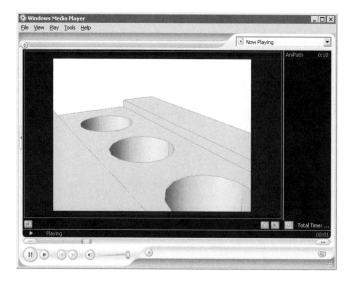

ALL ABOUT VISUAL STYLES

At the beginning of this chapter, you were introduced to visual styles. In this section, you learn all there is to know about modifying them.

Visual styles can be accessed in these ways: through the **VSCURRENT** command, the Dashboard palette's Visual Styles control, the **View | Visual Styles** menu, and the Visual Styles palette. There are many, many options and system variables that control visual styles. These are described in the following sections.

VISUAL STYLES CONTROL PANEL

The Visual Styles control on the Dashboard palette provides convenient access to many visual style settings. Behind the scenes, its buttons change values of commands and system variables. For example, clicking the **X-Ray Mode** button changes the value of the **VSFACEOPACITY** system variable between -60 (opaque) and 60 (translucent).

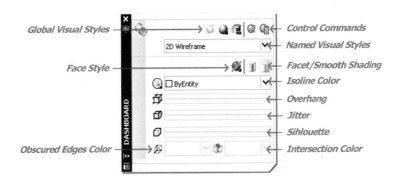

Global Visual Styles

The global visual style buttons affect every visual style.

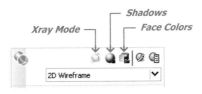

X-Ray Mode

The **X-Ray Mode** button toggles faces between opaque and translucent. You can use **VSFACEOPACITY** to change the percentage of opacity manually, which ranges from 1 (fully transparent) to 100 (fully opaque), as illustrated by the figure below.

Left: X-ray mode off.
Right: X-ray mode on.

Contrary to Autodesk's documentation, the value of 0 is not fully transparent, but is also fully opaque.

For maximum transparency, use a value of 1, instead. Use negative values, such as -60, to turn off transparency yet remember the percentage original setting.

Shadows

Shadows are cast by lights placed in the drawing, or by the ambient light. Shadows are not displayed in x-ray mode. The **Shadows** button is a flyout with these options:

- **No shadows.** Keep shadows turned off, because their generation greatly slows down AutoCAD. Turn on shadows only for final output of drawings.

- **Ground shadows.** Objects cast shadows on the shadow plane (ground), but not on each other. (The elevation of the shadow plane can be adjusted with the **SHADOWPLANELOCATION** system variable.) By adding the sun light to the drawing, you can adjust the angle of the shadows.

- **Full shadows.** Objects cast shadows on the shadow plane and on other objects. This option works only when (1) hardware acceleration is turned on, and (2) the graphics board supports shadows.

Face Colors

The color of faces can be changed to create special effects, such as less color (desaturate) or monochrome. The **Face Colors** button is a flyout with these options, from top to bottom:

- **Regular face colors.** No change is made to the faces.

- **Monochrome mode.** The face colors change to a single hue. You can manually change the monochrome hue through the **VSMONOCOLOR** system variable. It has a default value of "RGB:255,255,255" (white).

- **Tint mode.** The hue and saturation of the color specified by **VSMONOCOLOR** changes.

- **Desaturate mode.** The saturation of colors is reduced by 30%. (This setting is unaffected by **VSMONOCOLOR**.)

Control Buttons

The control buttons bring up dialog boxes.

Tuning Dialog

The **Tuning Dialog** button displays the Adaptive Degradation and Performance Tuning dialog box. (This is the same as entering the **3DCONFIG** command.) Click the **Manual Tune** button to change the settings from those automatically selected by AutoCAD. The technical editor notes that AutoCAD's default settings are very conservative.

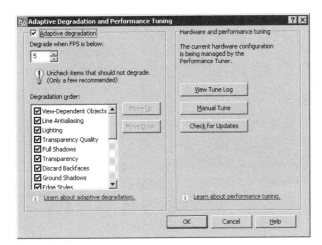
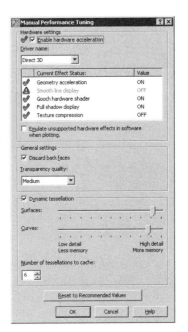

Visual Styles Manager

The **Visual Styles Manager** button displays the Visual Styles Manager palette. This is like entering the **VISUALSTYLES** command. The **VSSTATE** system variable reports whether or not the palette is open.

Predefined Visual Styles

The Visual Styles droplist displays preview images of predefined visual styles. Select one to apply to the entire viewport.

You can add custom styles to the droplist. To do so, create the new style in the Visual Styles Manager palette. The new style is added automatically to this droplist.

Face Styles

The second row of buttons toggles the style of faces.

Face Styles

The **Face Styles** button is a flyout with these options, from top to bottom:

- **No face style.** No styles are applied to the 3D model in the viewport.

- **Realistic face style.** This visual style uses dark and light colors.

- **Gooch face style.** This visual style uses cool and warm colors. Gooch uses cool and warm colors for rendering. That's because realistic renderings sometimes make certain faces too dark or too light. I recommend using this style.

From left to right: None, Realistic, and Gooch face styles.

Facets or Smooth

The **Facets** and **Smooth** buttons toggle the display of facets, which are the flat faces that AutoCAD uses to approximate curved surfaces.

Turning on **Smooth** removes the facets for better looking visual styles. (The button's background turns orange, when turned on.) These buttons have no effect on flat surfaces, such as the wedge illustrated below.

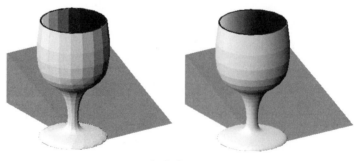

Left: Facets.
Right: Smooth.

The size of the facets is controlled by the FACETRES system variable, which ranges from 0.01 (coarse facets) to 10 (smooth).

Edge Styles

The **Edge styles** droplist determines the look of the edges, isolines, and facets. Facet edges apply to the edges and facets of curved and flat surfaces. Isolines apply only to curved surfaces. Only one can be active at a time, isolines or facet edges.

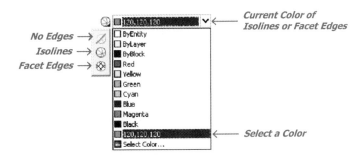

- **No edges.** Turns off the display of edges and isolines.

- **Isolines.** Displays isolines and specifies their color. (The **ISOLINES** system variable determines the number of isolines displayed.)

- **Facet edges.** Displays facets and specifies their color. This setting overrides the **Smooth** button; when you assign a color to facet edges, they are displayed, even though Smooth (no facets) is turned on.

From left to right: *None, Isolines, and Facet edge styles.*

The **Facet Edges/Isolines** droplist controls the color used to display facet edges and isolines.

Artistic Edge Effects

The artistic edge effects make 3D computer-generated drawings appear as if drawn by humans. None of these settings has an effect on donuts, spheres, and other curved surfaces, such as sweeps and lofts. The effects do not work at all when **No Edges** is selected for facet edges/isolines.

Edge Overhang

AutoCAD can extend the ends of edge lines to give drawings a casual look, called "pencil lines." The **Edge Overhang** button toggles the display of overhangs, while the accompanying slider adjusts the amount of overhang.

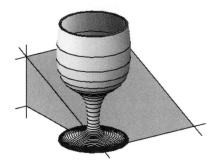

The amount of overhang ranges from 1 to 100 pixels, although the overhang effect seems to change, depending on the angle of the edge lines. When the value is zero or negative, the extensions are turned off. The setting is stored in the **VSEDGEOVERHANG** system variable.

Edge Jitter

Another pencil effect is called "jitter." The **Edge Jitter** button toggles the display of jitter, while the slider adjusts the amount of jitter.

The amount of jitter ranges between 1 = low, 2 = medium, and 3 = high. When the value is 0 or negative, the jitter effect is turned off. The setting is stored in the **VSEDGEJITTER** system variable.

Silhouette Edges

The silhouette effect fattens the outlines of objects. (This is different from lineweight, which affects the width of *all* lines and edges.) The effect is often used in technical documentation. The **Silhouette Edges** button toggles the display of silhouetting, while the slider adjusts the thickness of silhouette.

The range is between 1 and 25 pixels; the default value is 5. This control has no effect in Xray mode and does not affect isolines.

Hidden & Intersection Line Colors

At the bottom are controls for hidden and intersection lines. (Autodesk calls hidden lines "obscured.") The two droplists work only when the **3D Hidden** or **Conceptual** styles is selected. They specify the color to be used; AutoCAD can also specify linetypes, as described later.

Obscured Edges ⟶   ⟵ **Intersection Edges**

Color

Obscured Edges

The **Obscured Edges** control tells AutoCAD to display hidden (obscured) lines in a color different from visible lines and edges. The hidden lines are shown in the color you specify; AutoCAD displays them using a dashed line pattern. The pattern is zoom-dependent, so at some zoom levels the hidden lines look solid.

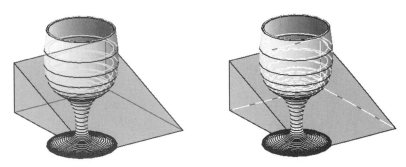

Left: *Hidden lines displayed.*
Right: *Hidden lines with dashed linetype.*

To change the linetype for hidden lines, use the **VSOBSCUREDLTYPE** system variable with the values listed below. You can use the same values to change the linetype for intersection lines with the **VSINTERSECTIONLTYPE** system variable; default is 1, solid.

0 - Off	6 - Long Dash
1 - Solid	7 - Double Short Dash
2 - Dashed (default)	8 - Double Medium Dash
3 - Dotted	9 - Double Long Dash
4 - Short Dash	10 - Medium Long Dash
5 - Medium Dash	11 - Sparse Dot

Intersection Edges

The **Intersection Edges** control displays lines and arcs where solids intersect. You can specify the color to be used for the intersection edges. The intersection edges are shown in the color you specify; AutoCAD displays them using a solid line pattern.

VISUAL STYLES PALETTE

The Visual Styles palette creates and modifies visual styles. Unlike the Dashboard, this palette provides access to *all* visual style parameters.

The **VISUALSTYLES** command opens the palette. The **VSSTATE** system variable reports whether or not the Visual Styles palette is open. Curiously, there doesn't seem to be a system variable that reports the name of the current visual style.

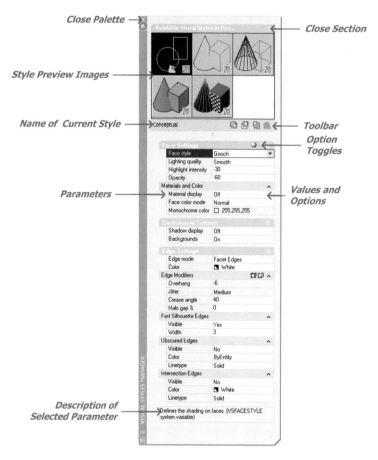

Close Palette → ... → *Close Section*

Style Preview Images →

Name of Current Style → *Conceptual* ← *Toolbar*

Parameters → ← *Values and Options*

← *Option Toggles*

Description of Selected Parameter →

Shortcut Menu Options

Right-click a visual style image in the preview area to display the shortcut menu with these options:

Create New Visual Style displays the New Visual Style dialog box. Enter a name for the visual style and (optionally) a description, and then click **OK**.

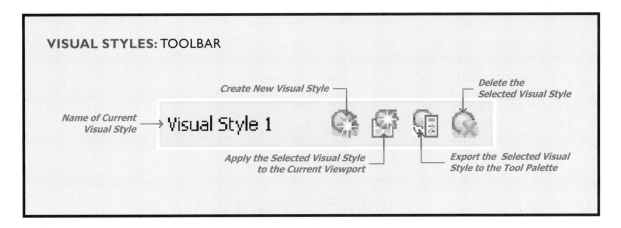

Apply to Current Viewport applies the selected style to the current model space viewport. **Apply to All Viewports** applies the selected style to all viewports of the current drawing. Visual styles cannot be applied to paper space.

Edit Name and Description allows you to change the name of the visual style and its description. Autodesk-defined styles cannot be renamed. **Delete** removes the visual style from the drawing. If the style is not being used, it is removed without warning; you can also use the PURGE command. In-use and Autodesk-defined styles cannot be erased.

Export to Active Tool Palette places the selected style in the current tool palette. You can also drag the style icon onto the tool palette.

Copy copies the visual style and its parameters to the Clipboard, where it can be pasted back into the Visual Styles palette to create a copy, or into the tool palette with **Paste**. The name of the copy is given the prefix of "Copy 1 of."

Size displays a submenu that lets you choose the size of preview icons:

I find that the **Small** option is big enough, and allows more preview images to be shown at a time. The **Full size** option adds buttons for scrolling through the preview images.

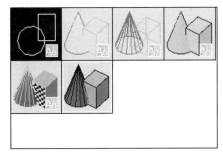
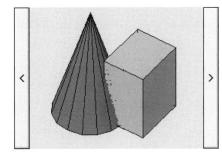

Left: *Small icons.*
Right: *Full icons with scroll buttons.*

(The Autodesk icon in the preview images indicates default visual styles that cannot be deleted.)

Reset to Default changes all parameters to their default values. This option applies to Autodesk-defined styles only. This is handy when you are experimenting with styles, and want to revert to their original settings. It's a good idea to make a copy, and then mess with it.

Face Settings

The **Face Settings** section changes the appearance of faces.

Face Settings Toolbar

Highlight Intensity reverses the value of the highlight intensity between positive and negative, or on and off. (The value is stored in the **VSFACEHIGHLIGHT** system variable.)

Opacity reverses the value of opacity between positive and negative, called "Xray mode" in the Visual Styles control panel (**VSFACEOPACITY**).

Face Style

Face Style specifies the style of face shading (**VSFACESTYLE**):

- **Real** — renders face realistically, like rendered mode; displays materials, if applied.

- **Gooch** — substitutes warm and cool colors for light and dark; allows faces to be seen more easily than with Real.

- **None** — applies no style to faces, like wireframe mode. This option cannot be selected when Edge Mode is set to None.

Lighting Quality toggles between **Faceted** and **Smooth**; this affects curved surfaces only (**VSLIGHTINGQUALITY**).

ADJUSTING SHADOWS

The **SHADOWPLANELOCATION** system variable changes the elevation of the "ground" (shadow plane) upon which shadows are cast. The default is 0 units, and is measured in the z direction. The plane is independent of everything else, and can result in some incorrect shadows, such as ones that intersect objects.

The **CSHADOW** system variable determines how each 3D object displays shadows. This allows you to determine the shadow-casting of individual objects:

> **0** — object casts onto and receives shadows from other objects (default setting).
> **1** — object casts shadows only.
> **2** — object receives shadows only.
> **3** — object ignores shadows.

Use the Dashboard palette's Light control panel to adjust the location and angle of shadow-casting. For instance, move the Date and Time sliders to rotate and lengthen the shadows (if the Sun is active). The Brightness and Contrast sliders make objects brighter and darker, but have no effect on the shadows, and no effect in Conceptual and wireframe visual styles.

Highlight Intensity specifies the size of highlights on faces. This affects faces without assigned materials, and applies equally to all objects in the viewport (VSFACEHIGHLIGHT). Highlighting is turned off when set to 0 or a negative value, such as -80. Larger numbers generate larger "intensity" areas.

 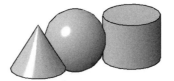 

Left to right: Highlight intensity = 0, 1, and 99.

Opacity specifies the amount of transparency of faces (VSFACEOPACITY); it applies equally to all objects in the viewport. The translucency is fully opaque when set to 0, 100, or a negative value (such as -27). Smaller numbers create more transparency; 1 = fully transparent (invisible). Entering a negative number turns off transparency, but retains the value. In macros, use the absolute value function to turn it back on.

From left to right: Opacity =1, 25, and 60.

Materials and Color

Materials toggles the display of materials and textures if they have been previously applied to objects with the MATERIALASSIGN command (name is stored in the CMATERIAL system variable).

When this parameter is set to a value other than **Off**, the earlier settings for highlights and opacity do not apply, because materials have their own values for these parameters (VSMATERIALMODE):

- **Off** — neither materials nor textures are displayed; visual style settings are applied to the model.

- **Materials** — materials are displayed, if assigned.

- **Materials** and Textures — materials and textures are displayed, if applied.

Face Color Mode specifies how colors are displayed (VSFACECOLORMODE).

- **Normal** — face colors are displayed normally.

- **Monochrome** — face colors are displayed in a monochrome color, as specified by the Monochrome Color option, described below.

- **Tint** — face colors have their hue and saturation changed.

- **Desaturate** — face colors have their saturation reduced by 30%.

Monochrome Color becomes available when **Monochrome** is selected from the Face Color Mode droplist. You select a color from the Select Color dialog box. **Tint Color** appears instead when "Tint" is selected. Both options "colorize" the model with a single color, as specified by the **VSMONOCOLOR** system variable (default value = "RGB:255,255,255" or white).

Environment Settings

The environment settings determine whether shadows and the background are displayed. You do not need to place any lights in the drawing for the visual style to generate shadows.

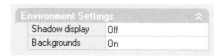

Shadow Display determines whether shadows are displayed. Shadow-casting slows down AutoCAD's display speed, so leave it turned off unless you are saving images to file or plotting them (**VSSHADOWS**).

- **Off** — objects cast no shadows, the preferred option.

- **Ground shadows** — objects cast shadows on the shadow plane, but not on each other.

- Full **shadows** — objects cast shadows on the ground and on each other. For full shadows, AutoCAD requires that hardware acceleration be turned on with the **3DCONFIG** command, and that the graphics board supports full shadow-casting.

Backgrounds toggles the display of a background in the viewport; the setting is saved in the **VSBACKGROUNDS** system variable. Backgrounds can consist of a solid color, a gradient of two or three colors, or a raster image — as specified with the **VIEW** command. The raster image can be in one of these formats: Targa (*.tga*), Bitmap (*.bmp*), PNG (*.png*), JPEG (*.jpg*), and TIFF (*.tif*).

Edge Settings

The **Edge Settings** parameters determine how edges are displayed.

Edge Modifiers Toolbar

Overhang Edges toggles the extension of edges beyond their boundaries. This button turns on the overhang effect by changing the **VSEDGEOVERHANG** system variable to a positive value.

Jitter Edges skews edge lines to mimic a hand drawn effect. This button turns on the jitter effect by changing the **VSEDGEJITTER** system variable to a positive value.

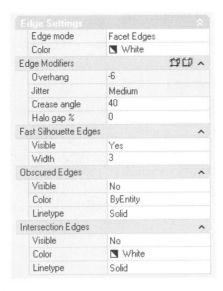

Edge Settings	⌃
Edge mode	Facet Edges
Color	◪ White
Edge Modifiers	⛶⛶ ⌃
Overhang	-6
Jitter	Medium
Crease angle	40
Halo gap %	0
Fast Silhouette Edges	⌃
Visible	Yes
Width	3
Obscured Edges	⌃
Visible	No
Color	ByEntity
Linetype	Solid
Intersection Edges	⌃
Visible	No
Color	◪ White
Linetype	Solid

Edge Settings

Edge Mode specifies which style of edge to display (**VSEDGES**).

- **None** — facets, isolines, and edges are not displayed. This setting cannot be used when the Face Style set to None.

- **Isolines** — isolines and edges are displayed.

- **Facet Edges** — facets and edges are displayed on objects with corners, when **VSEDGESMOOTH** is small enough.

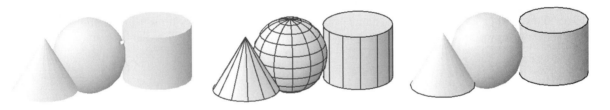

Left to right: Edge Mode = None, Facet Edges, and Isolines.

Color specifies the color for edges. The Select Color dialog box lets you select colors not listed in the droplist. One color applies to all edges (**VSEDGECOLOR**); default = 7 (white).

Edge Mode = Isolines

The following parameters appear only when Edge Mode = Isolines:

Number of Lines specifies the number of isolines drawn on curved surfaces. The value is stored in system variable **ISOLINES**; range is 0 to 2047. The default value of 4 is too low; a good number is either 0 (no isolines) or 12. Too many isolines completely blacken the curved objects.

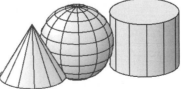

Left to right: Isolines = 2, 12, and 120.

This parameter does not go into effect until after you execute the **REGEN** command.

Always on Top toggles hidden-line removal of isolines (**VSISOONTOP**). This setting also affects flat surfaces.

- **Off** — all isolines are displayed.

- **On** — only foreground isolines are displayed; isolines located "around the back" are hidden.

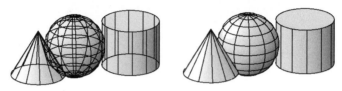

Left: *Isolines on top = Off.*
Right: *Isolines on top = On.*

Edge Modifiers

The **Edge Modifiers** settings apply to isolines and facet edges, but not when Edges are set to None.

Overhang extends edges beyond their boundaries (**VSEDGEOVERHANG**); range is 0 to 100 pixels.

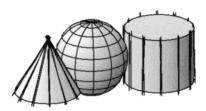

Jitter skews edge lines to mimic a hand drawn effect by skewing the lines that define edges. The **VSEDGEJITTER** system variable has the following settings:

- **0** — Off.

- **1** — Low jitter.

- **2** — Medium jitter.

- **3** — High jitter (or "caffeine mode," according to the copy editor).

Edge Mode = Facet Edges

The following parameters appear only when Edge Mode = Facet Edges:

Crease Angle specifies the angle beyond which facet edges are not shown on curved surfaces, removing the edge lines (**VSEDGESMOOTH**). The angle ranges from 0 to 180 degrees. Higher values turn off the display of facet edges. Because this affects only curved surfaces, it allows AutoCAD to show edges on planar surfaces but not curved ones, as illustrated below.

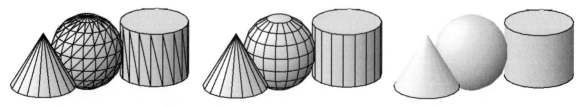

Left to right: *Crease angle = 0, 4, and 21 degrees.*

Halo Gap % specifies the gap generated between visually overlapping objects. The **Gap** option is not available when you are working with styles based on wireframe modes (Face Style = None). The value is stored in the **VSHALOGAP** system variable, which ranges from 0 to 100 pixels (not percent).

When greater than 0, silhouette edges are not displayed.

Fast Silhouette Edges

The **Fast Silhouette Edges** section defines the fattening of outlines around models. Both settings apply to all objects in the viewport equally:

Visible toggles the display of silhouette edges (**VSSILHEDGES**); 0 = off, 1 = on.

Width determines the width of the silhouette edges (**VSSILHWIDTH**); range is from 1 to 25 pixels.

AutoCAD does not display these edges when halo gap > 0, opacity > 0 (translucent objects), and during wireframe visual styles.

Edge Mode = Facet Edge

The following parameters appear only when Edge Mode = Facet Edges.

Obscured Edges

The **Obscured Edges** section determines what happens to obscured (hidden) edges and facets; it does not apply to isolines.

Visible toggles the visibility of obscured edges and facets. When on, hidden lines are shown "through" the model. Even when turned on, obscured edges are invisible on curved faces when the Crease Angle is large enough not to show facet lines (**VSOBSCUREDEDGES**); 0 = off, 1 = on.

Color specifies the color for visible obscured edges and facets (**VSOBSCUREDCCOLOR**); default = "byentity." Select a color from the droplist, or from the Select Colors dialog box.

Linetype specifies the line pattern for obscured (hidden) edges and facets (**VSOBSCUREDLTYPE**). You can select from a set of hardwired linetypes; you cannot use the linetypes accessed by the **LINETYPE** command or stored in the *acad.lin* file. The available linetypes are:

0 - Off	6 - Long Dash
1 - Solid	7 - Double Short Dash
2 - Dashed (default)	8 - Double Medium Dash
3 - Dotted	9 - Double Long Dash
4 - Short Dash	10 - Medium Long Dash
5 - Medium Dash	11 - Sparse Dot

No scale factor can be applied to these linetypes. Make sure the color differs from Edge Color; otherwise the linetype is invisible.

Edge Mode = Facet Edges

The following parameters appear only when Edge Mode = Facet Edges:

Intersection Edges

The **Intersection Edges** section determines what happens to edge lines along the intersection of two solid objects.

Visible toggles the display of intersection edges (**VSINTERSECTIONEDGES**). Autodesk recommends leaving this setting turned off to increase performance.

Color specifies the color of intersection lines. Select a color from the droplist or from the Select Color dialog box (**VSINTERSECTIONCOLOR**). The default is white (ACI #7), but can be any RGB color as well as ByBlock, ByLayer (color assigned by the layer), or ByEntity (color of the 3D solid).

Linetype specifies the linetype for intersection lines. The **VSINTERSCETIONLTYPE** system variable stores the linetype number. Select from one of the hardwired line patterns listed earlier. The scale is zoom dependent: when zoomed out too far, the linetypes look solid.

EXERCISES

1. From the Companion CD, open the *Teapot.dwg* file, a 3D drawing of a teapot.

    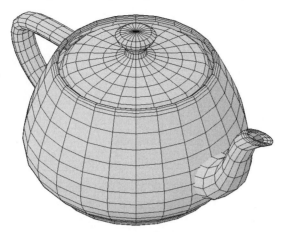

    a. Use the **VSCURRENT** command to remove hidden lines. Which option did you enter?

    b. Change the visual style to conceptual. Describe the changes to the teapot:

    _____

    _____

    (*History*: The teapot was used for testing purposes by 3D graphics programming pioneers as a complex 3D model. Hence, you see the teapot icon in AutoCAD toolbars for rendering commands.)

2. Continuing from exercise #1, use the **RENDERCROP** command. Does the image look different from that generated by the **VSCURRENT** command?

    Use the **REGEN** command. What happens to the lines?

3. From the Tool palette, select the "Doors - Windows.Glazing.Glass.Clear" material in the **Doors and Windows - Material Sample** tab. Apply it to the teapot, and then reapply the **RENDERCROP** command to make the teapot look like glass.

    Which settings make the rendering finish faster? Slower?

4. Use the **VIEW** command to place an image behind the teapot. (You can use the *IMG_8076.JPG* file from the Companion CD.) Change the visual style to conceptual. The result should look something like the figure illustrated below.

Use the **3DORBIT** command to rotate the teapot. What happens to the background?

CHAPTER REVIEW

1. Briefly explain the purpose of the **VSCURRENT** command.
2. How would you toggle perspective viewing mode?
3. What is the purpose of materials?
4. What are *obscured lines*?

   What is a halo gap?
5. When do materials appear on 3D objects?
6. Do materials define the density of 3D objects?
7. What effect does **FACETRES** have on renderings of 3D solid models?
8. What are the four types of lights available for rendering?

   a.

   b.

   c.

   d.
9. Can AutoCAD render just part of a drawing?
10. How would you place a picture in the background of a rendering?
11. What color creates the effect of "fog"?

    The illusion of increased depth?

    Which command creates fog effects?
12. What steps would you take to create the white color?
13. How would you make a 3D model look like glass?
14. Describe the two types of shadows:

    Ground shadows —

    Full shadows —
15. Is it advisable to have shadowcasting turned on?

    Why or why not?
16. Which system variable controls the height of the shadowplane?
17. What kind of light is the Sun light?
18. Describe the purpose of the **ANIPATH** command.

    How does the command work?
19. Explain the abbreviation "FPS."
20. What is a *target*?
21. How do you turn on XRay mode?

    What is the purpose of this mode?
22. What is edge jitter?

    Edge overhang?

UNIT VI

Plotting Drawings

CHAPTER 21

Working with Layouts

Up until now, you have been drawing and editing drawings in model space. *Models* are drawn in *model* space; drawings are *laid* out for plotting in *layouts*. Layouts determine the arrangement and views of 2D and 3D models on the plot sheet.

In this chapter, you find the following commands and options:

TILEMODE switches drawings between model and layout tabs.

ZOOM Xp scales models relative to layouts.

PSLTSCALE scales model-space linetypes relative to layouts.

SPACETRANS scales model-space distances relative to layouts.

Annotative scaling is tied to the scale factor of each viewport (new to AutoCAD 2008).

LAYER Current VP Freeze freezes layers in selected viewports.

Viewport overrides override layer properties on a per-viewport basis (new to AutoCAD 2008).

LAYOUTWIZARD steps through the layout creation process.

LAYOUT creates and modifies layouts.

VIEWPORTS creates rectangular viewports.

-VPORTS and **VPCLIP** create and change polygonal viewports.

VPMAX and **VPMIN** maximize and minimize the active viewport.

NEW TO AUTOCAD 2008 IN THIS CHAPTER

- Viewport scaling is set through the new drawing status bar, and controls the visibility of objects with the annotative property.
- **VPLAYEROVERRIDES** and **VPLAYEROVERRIDES** report on overridden layer properties in viewports.

FINDING THE COMMANDS

On the **Layouts** and **Viewports** toolbars:

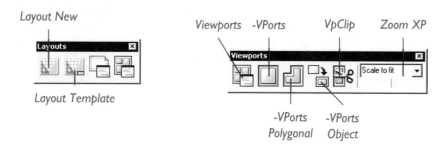

On the layout tabs:

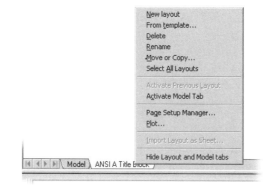

On the status bar:

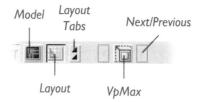

ABOUT LAYOUTS

Two of AutoCAD's drafting environments are known as model space and layouts.

Model space is where you draw the models full scale, 1:1. Recall from earlier chapters that you had to calculate a scale factor for text, linetypes, and hatch patterns, so that they would not appear too small when printed.

Layouts solve the scaling problem. Here is where you arrange the models as if they were on sheets of paper. In layouts, the paper is full size (1:1), so you don't need to figure out the scale factor for text, linetypes, and hatch patterns — they are all drawn full size.

Instead, the model is scaled down to fit the paper with these two steps: (1) creating *viewports* in layouts, which act like windows into model space, and then (2) using the ZOOM command to scale the model to the scale factor. The figure below illustrates a layout with a viewport containing a model.

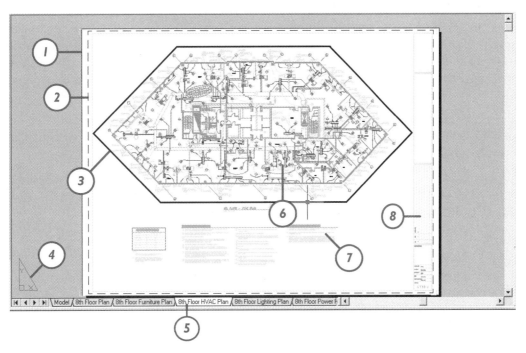

*1: Edge of "paper." **2:** Printer's margins. **3:** A polygonal viewport.*
*4: Paper space UCS icon. **5:** Model and layout tabs.*
*6: The model (scaled to fit the paper). **7:** Text placed in layout. **8:** Title block placed in layout.*

Depending on how your copy of AutoCAD is configured, the layout tabs may be stored on the status bar, along with the viewport and annotation scale controls, as illustrated below.

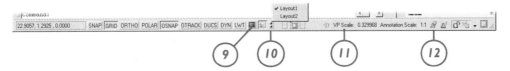

*9: Switch to model space. **10:** Access layouts. **11:** Viewport scale factor.*
12: Annotation controls.

Every drawing has just one model space, and one or more layouts. New drawings start off with two blank layouts named "Layout1" and "Layout2." Layouts can show the overall plan and zoomed-in details — all at the same time.

Each layout represents a drawing sheet to be plotted. Every layout can have the same or different paper sizes and orientation; each can be assigned the same or a different printer.

You can think of layouts as *interactive plot previews*. You can create, move, and delete viewports, the model, and other elements at your whim.

Each layout needs at least one viewport through which shows some or all of the model. But layouts often have two or more viewports, each showing details or overall views. (AutoCAD can display the contents of up to 64 viewports.) Each viewport can show the model at a different scale factor, and, in 3D drawings, show different views, such as the top, side, and front. (Layouts themselves are strictly 2D.)

Layouts typically have drawing borders and title blocks. (The viewports fit inside the drawing border.) You can add dimensions, text, and hatch patterns to layouts. A command is available that automatically translates linetype scales between the model scale factor and the layout.

New layouts are created from scratch or imported from other drawings, and can be moved, resized, and deleted. The order in which layouts appear can be changed. As well, each viewport can have a different set of layers visible, and have distinctive plot styles and visual styles. While layouts are always rectangular, viewports can be any shape.

Exploring Layouts

To experience layouts, open the *8th floor.dwg* file from the Companion CD, a floor plan drawing that used to be provided with AutoCAD. If necessary, select the **Model** tab. You see the entire model, but without any notes, drawing borders, and title blocks.

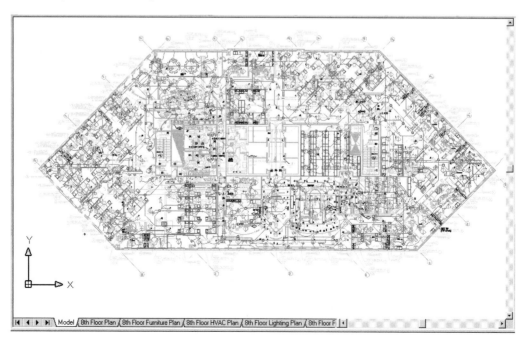

From among the layout tabs, select the **8th Floor Plan** tab by clicking on its name. The view changes dramatically. You see only the floor plan of the model, surrounded by the title block and drawing border. This illustrates how layers can be frozen in layouts to show details selectively.

The drawing is shown on a white rectangle, which represents the paper. In this case, the selected paper is "Architectural E1" size, which works out to 42" wide x 30" high. The dashed rectangle represents the printer's *margin*, the unprintable area along the four edges of the paper.

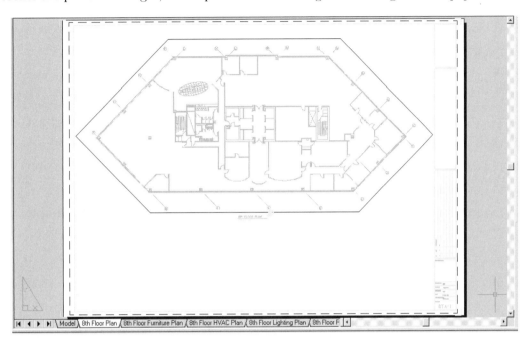

Select the **8th Furniture Plan** tab. The view changes again. In this layout, layers are turned on to show the floorplan with the furniture — desks, chairs, and so on.

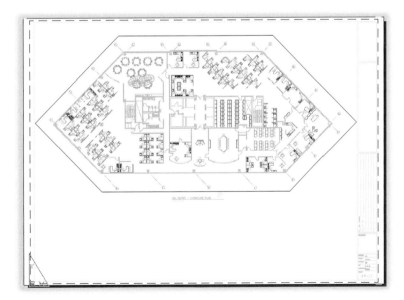

Select the other layout tabs, working your way through the HVAC plan (heating, ventilation, air conditioning), lighting plan, power plan, and plumbing plan. As an alternative to clicking tabs with the cursor, press **CTRL+PGUP** and **CTRL+PGDN** to switch between layouts.

When drawings have many layouts, an additional set of controls becomes useful. These "VCR" buttons take you to the next and previous layouts, as well as to the first and last layouts.

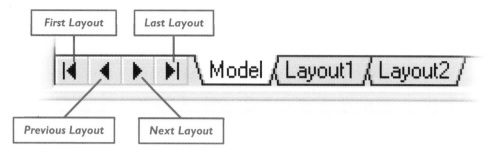

Hiding Layout Tabs

Layout tabs can be "removed" with the **OPTIONS** command, the default command for the 3D Modeling workspace. In the Layout Elements section of the Display tab, uncheck **Display Layout and Model tabs**. I don't see much point to this option, because the tabs share space with the horizontal scroll bar, which I find useful for panning, and so don't take up any extra room in AutoCAD's user interface.

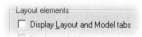

AutoCAD doesn't hide the tabs, but relocates them as buttons and shortcut menus on the status bar, replacing the MODEL/PAPER button. To access layout tabs, click the **Additional Layouts** button, and then select a layout name from the shortcut menu.

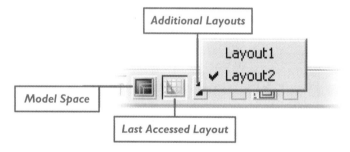

 — displays model space.

— displays the last-accessed layout.

— displays a list of layout names. (The check mark indicates the currently displayed layout.) Select a name from the shortcut menu to go to that layout.

To return the tabs to the scroll bar, right-click any of these three buttons, and then select **Display Layout and Model tabs** from the shortcut menu.

> **Note:** To switch between model and paper mode, double-click the layout, as follows:
> Double-click **inside** the viewport to enter model mode; or click the **VsMax** button.
> Double-click **outside** the viewport to return to paper mode.

TUTORIAL: WORKING WITH LAYOUTS

This tutorial takes you through some aspects of working with layouts.

1. From the Companion CD, open the *layout.dwg* file, the drawing of a streetscape.
2. Select the **Layout1** tab by either method described above.
   Notice that the cityscape drawing appears on a sheet of paper with two rectangles. The inner rectangle is the viewport, while the dashed rectangle is the printer margin.

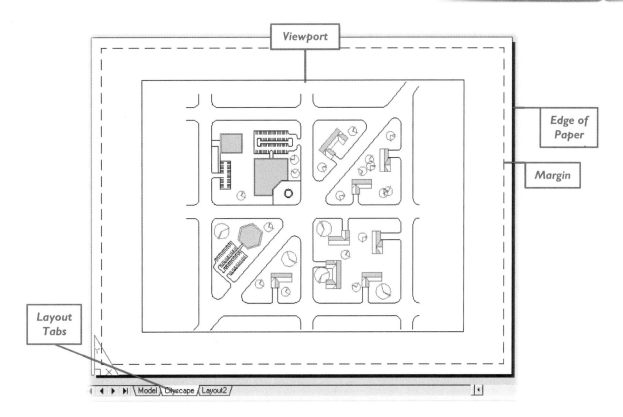

Viewport

Edge of Paper

Margin

Layout Tabs

Note: You can make many kinds of changes to viewports using the Properties palette. Some of the properties, however, don't work. For example, it appears that the Properties palette allows you to apply linetypes and lineweight to the viewport border, but these changes have no effect.

Properties that apply to the viewport border are:

Color — changes the color of the viewport border.

Layer — specifies the border's layer name.

Plot Style — specifies the plot style, if enabled.

Hyperlink — activates a link when the border is clicked.

Geometry — specifies the the center coordinates, and the size.

Properties that apply to the content of the viewport:

On — toggles the display of the view in the viewport.

Display Locked — prevents view changes, such as zooms and pans.

Standard Scale — scales the view using factors specified by **SCALELISTEDIT**.

Custom Scale — specifies other scale factors.

Shade Plot — specifies shading or rendering mode during plotting.

Properties that are controlled by other commands:

    Clipped — **MVIEW** and **-VPORTS** commands.

    UCS Per Viewport — **UCS** command.

    Visual Style — **VSCURRENT** system variable; can't set in paper space.

    Linked to Sheet View — **SHEETSET** command.

Linetype, Linetype scale, and Lineweight properties have no effect on viewports.

Working with Paper Mode

In paper mode, you work only with the "paper," not the model. "Working with the paper" includes drawing on the paper or page, manipulating viewports and their content, and plotting.

1. Try selecting the model. Notice that you can't, because the layout is in "paper" mode. (On the status bar, the indicator reads PAPER.)

2. Click the viewport border. Notice that you can select it, unlike the model.

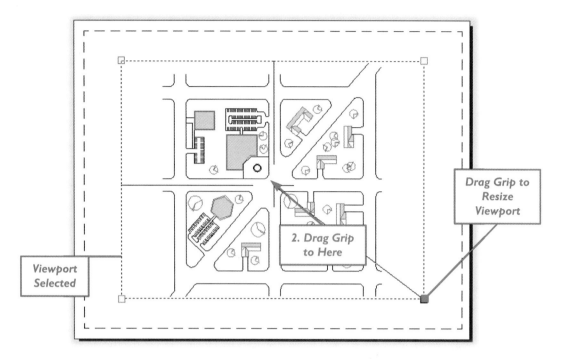

3. Grab a grip, and then make the viewport smaller by dragging. Notice what happens to the model: it has become "cropped" — part of the model is cut off visually. This illustrates how you can selectively show details of the model in layouts. Layouts need not show the entire model.

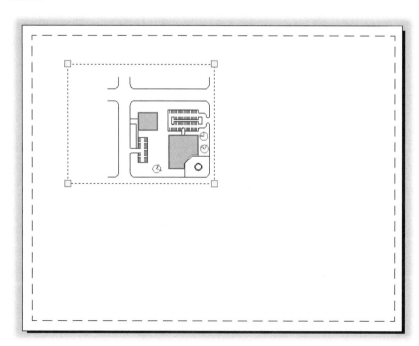

4. You can move viewports: grab an edge (not a handle), and then drag the viewport to a new position. Notice that the cropped model view moves with the viewport.

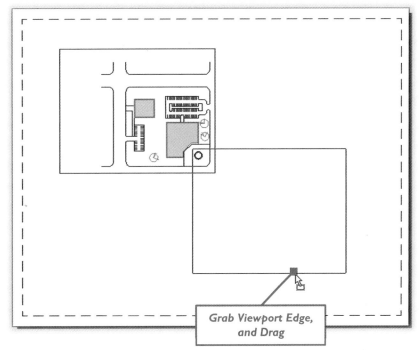

Grab Viewport Edge, and Drag

You can apply all of the usual grip editing commands to viewports: stretch, move, copy, scale (resize), rotate, and mirror.

5. Use the **COPY** command, for example, to copy viewports. Notice that the copy of the viewport contains exactly the same model view as the original. And, as the figure below illustrates, viewports can overlap.

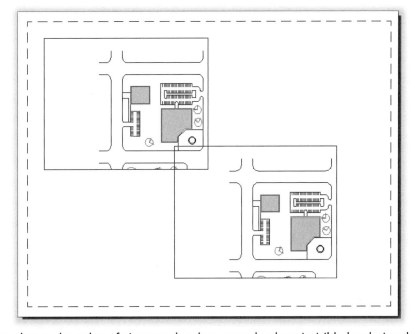

You can change the color of viewport borders, or make them invisible by placing them on a layer of their own, and then change the color or freeze the layer. The rest of the model is unaffected. (You cannot apply linetypes or lineweights to viewports borders.) To see viewport borders in the drawing, but not plot them, change the layer to be non-plotting.

6. The **ERASE** command deletes viewports.

    Other editing commands do not, however, work with viewports, including **OFFSET**, **TRIM**, **EXPLODE**, and **FILLET**.

Hiding Viewport Borders

Use this trick to hide viewport borders: assign viewports their own layer, and then freeze the layer. This makes the viewport borders invisible, but the content remains visible.

1. With the **LAYER** command, create a new layer named "VPorts."
2. Select the viewport borders, and from the Layer toolbar or Dashboard panel, click the **Layer Control** droplist to change the layer to "VPorts."

3. Freeze layer "VPorts." Notice that the viewport border disappears, but that the model view remains.

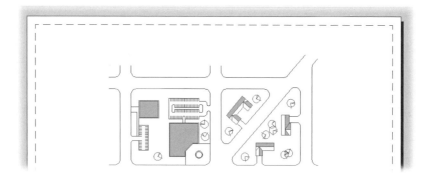

4. Thaw layer "VPorts" to bring back the viewport borders.

Switching from Paper to Model Mode

Until now, you have been manipulating viewports. Let's begin working with the model inside the viewports.

1. If the word PAPER or MODEL appears on the status bar, then, click PAPER. The button changes its name to MODEL. This means that you are now working with the model instead of the layout (paper) — yet you are still in layout mode.

    If the words PAPER or MODEL do not appear, then double-click inside the viewport to enter model space.

2. Either way, notice that one viewport has a heavy border. This is the *active* viewport, the one in which you can manipulate the model.

    To make another viewport active, simply select it. Alternatively, you can press **CTRL+R** to cycle through viewports.

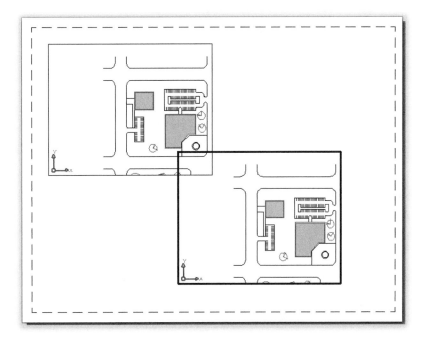

3. Select an object in the model. Notice that the grips and highlight appear in both viewports. This confirms that you are seeing two images of the same model.

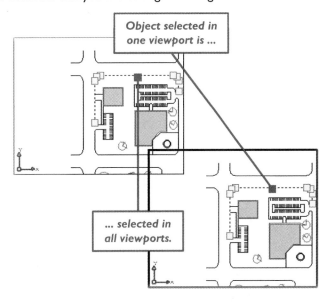

Object selected in one viewport is ...

... selected in all viewports.

SCALING MODELS IN VIEWPORTS

While objects are not independent of each other in different viewports, their visibility is. You can show models at a different scale in every viewport, hide and show layers, and override layer properties, such as color and linetype, and toggle the grid — all on a per-viewport basis. In this section, we look at the important topic of scaling models inside viewports.

VP Scale

The easiest way to scale models in viewports is to use the **VP Scale** droplist found on the status bar.

1. Select a viewport.
2. Click the **VP Scale** list, and then choose a scale factor.

   Notice that the model changes its size inside the viewport. You can choose a different scale factor for each viewport — or use the same one for all.

Not sure which scale factor to choose? Read ahead to the section on the ZOOM XP command.

The scale factors listed by the droplist can be edited with the SCALELISTEDIT command. See Chapter 22, "Plotting Drawings." The command allows you to add and remove scale factors; you can access it through the **Custom** item on the VP Scale droplist.

(This droplist does not work if the viewport is locked, either through the Properties palette's **Display Locked** or the MIVEW and -VPORTS commands' **Lock** options.)

Zoom XP

AutoCAD's two primary visibility commands are ZOOM and LAYER. In model mode, AutoCAD commands apply to the active viewport only; all other viewports are ignored.

1. The **ZOOM** command changes the size of the model in viewports.

   Enter the **ZOOM** command, and then use the **Extents** option. You should see the entire model in the active viewport.

Scale to fit
1:1
1:2
1:4
1:5
1:8
1:10
1:16
1:20
1:30
1:40
1:50
1:100
2:1
4:1
8:1
10:1
100:1
1/128" = 1'-0"
1/64" = 1'-0"
1/32" = 1'-0"
1/16" = 1'-0"
3/32" = 1'-0"
1/8" = 1'-0"
3/16" = 1'-0"
1/4" = 1'-0"
3/8" = 1'-0"
1/2" = 1'-0"
3/4" = 1'-0"
1" = 1'-0"
1-1/2" = 1'-0"
3" = 1'-0"
6" = 1'-0"
1'-0" = 1'-0"
Custom...

VP Scale: 0.664041 ▾   Annotation Scale: 1:1 ▾

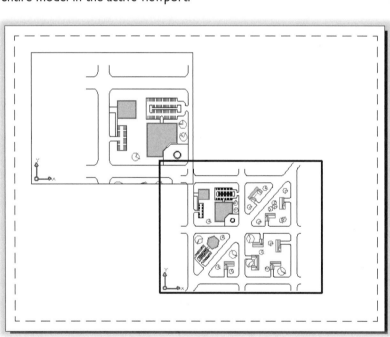

2. The **ZOOM** command's **XP** option is meant for use in layout viewports (short for "multiple in paper"). This option scales the model view relative to the layout. This is important: *You use ZOOM XP to scale models in layout viewports.*

Out of so many scale factors, which one is appropriate for this drawing? City plans have typically large scale factors, such as 1:1000, 1:2000, 1:5000, and so on. This model is approximately 470 feet across, while the viewport is about 4.7 inches across. Doing the math shows that the nearest standard scale factor is 1:2000:

(470' x 12 inches/foot) ÷ 4.7" = 1200

...so we select the next-highest standard scale factor.

3. Apply the scale factor of 1:2000 by using it with the **XP** suffix, as follows:

   Command: **zoom**

   Specify corner of window, enter a scale factor (nX or nXP), or

   [All/Center/Dynamic/Extents/Previous/Scale/Window/Object] <real time>: **1/2000xp**

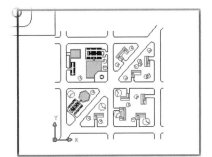

Notice that the layout now has two views of the model: a large one and a small one. This illustrates that viewports can have independent views of models.

Zoom Through Properties

As an alternative to **ZOOM XP**, use the **PROPERTIES** command to set the model scale. Here's how:

1. Switch back to PAPER mode.
2. Select the viewport border.
3. Right-click, and then select **Properties** from the shortcut menu.
4. In the **Misc** section of the Properties palette, click **Custom Scale**.

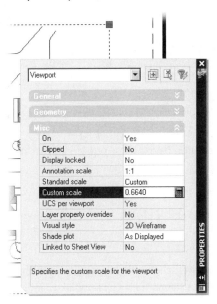

5. Enter a new scale factor, such as **0.015**, and then press **ENTER**. Notice that the size of the model changes in the viewport.

Here is an alternative method for having AutoCAD perform the calculation of the scale factor. Enter the following in the **Custom Scale** field:

   = 15/1000

...and then press the **END** key (not the **ENTER** key). AutoCAD returns the answer of 0.015.

Notes: You can use the **PAN** command to position the model inside the viewport.

You can lock the viewport to prevent anyone from changing the view of the model, such as by zooming or panning. In paper mode, select the viewport border. In the Properties palette, change the **Display Locked** property to **Yes**.

As an alternative, click the **Maximize Viewport** button to enter model space without messing up the viewport scale.

SELECTIVELY DISPLAY DETAILS IN VIEWPORTS

Earlier in this chapter, you saw how it was possible to make the viewport border invisible by freezing its layer. The LAYER command also allows you to change what is displayed *in* viewports. You can:

> **Freeze layers** to hide parts of the drawing; thaw the layers to make the parts visible again.

> **Override layer properties** to display parts of drawings differently.

The "ByLayer" properties that can be overridden in each viewport are color, linetype, lineweight, and plot style. Whether freezing or overriding, the process is the same:

> 1. Select a viewport.

> 2. Use the **LAYER** command to freeze layers or override properties.

> 3. Repeat for other viewports.

Let's look at this in some detail.

Freezing Layers

The LAYER command freezes layers in each viewport. This is done with the **Current VP Freeze** column.

1. Ensure the *layout.dwg* drawing is open. Select one of the viewports.
2. With the **LAYER** command, open the Layer Properties Manager dialog box.

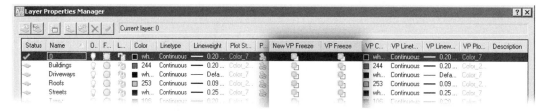

Towards the right are two columns you have not yet used:

- **Current VP Freeze** column freezes the selected layers in the current viewport (a.k.a. "VP") — as soon as you click **Apply** or exit this dialog box.

- **New VP Freeze** column freezes the selected layers — whenever a new viewport is created.

These two options have no effect in model tab, or in other layouts and other viewports.

3. Hold down the **CTRL** key, and then select names of the **Buildings**, **Roofs**, and **Trees** layers, as illustrated below.
4. Let go of the **CTRL** key, and then click a highlighted icon in the **Current VP Freeze** column. Notice that all selected suns turn to snowflakes, indicating the layers will be frozen (hidden from view).

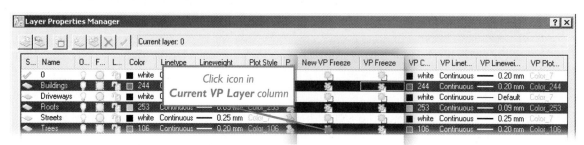

5. Choose the **OK** button. Notice that the buildings, roofs, and trees disappear from the viewport. The other viewport, however, is unaffected. This shows that viewports can have their layers frozen independently of each other.

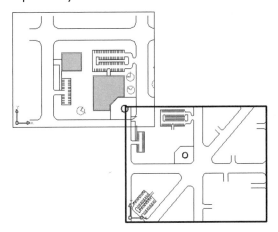

 NEW IN 2008 **Overriding Layer Properties**

You can override layer properties with commands like **COLOR** and **LINETYPE**, but these overrides affect all viewports equally. The **LAYER** command can override properties in each viewport.

For example, you might have some layers colored light gray in one viewport but black in others. Some viewports might show layers with the Hidden linetype, while others show the Continuous linetype, as illustrated below.

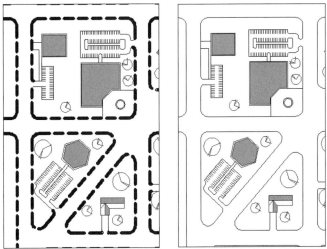

Left: *Streets in hidden linetype and heavy lineweight overrides.*
Right: *Streets displayed normally (ByLayer).*

The **LAYER** command overrides properties with the VP Color, VP Linetype, VP Lineweight, and VP Plot Style columns.

1. Continuing with the *layout.dwg* drawing, double-click inside a viewport to enter model space.
   Use the **U** command to undo the layer changes from the previous tutorial.

2. Open the Layer Properties Manager dialog box with the **LAYER** command.

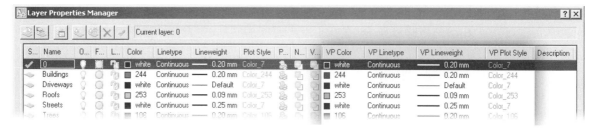

Towards the right are four columns that we are interested in:

- **VP Color** column overrides the ByLayer color in the current viewport.
- **VP Linetype** column overrides the ByLayer linetype.
- **VP Lineweight** column overrides the ByLayer lineweight.
- **VP Plot Style** column overrides the ByLayer plot style, if plot styles are enabled.

3. Select names of the **Streets** layer, as illustrated below.
4. In the **VP Lineweight** column, select 2.11mm.

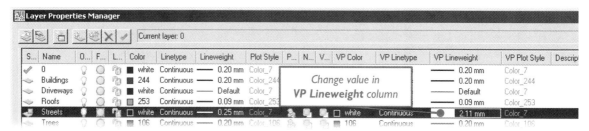

5. Choose the **OK** button.

   Notice that the streets are displayed with heavy lines in the current viewport; the other viewport is unaffected by the change.

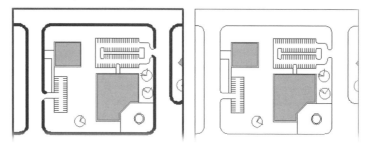

Left: Streets in heavy lineweight override.
Right: Streets displayed normally (ByLayer).

 Note: The **VPLAYEROVERRIDES** system variable reports whether any viewports contain layer property overrides. When 1, at least one does; when 0, none do. (This sysvar is read-only).

The same information is provided by the Properties palette. Select a viewport border, and then check whether the value of **Layer Property Overrides** is Yes or No. Like the sysvar, this setting is read-only.

VpOverrideMode — Toggling Properties Overrides

You can toggle property overrides with the **VPOVERRIDEMODE** system variable. It has two settings:

1 — overrides are displayed and plotted.

0 — overrides are not displayed or plotted.

This sysvar is handy for temporarily turning off overrides. To "permanently" turn them off, see the next section.

Reverting Properties to ByLayer

After you've overridden the layer properties of several viewports, you may want to change them back to "normal" (ByLayer). Autodesk provides three (!) ways to do this: through the **LAYER**, **-VPORTS**, and **MVIEW** commands.

The **LAYER** command's dialog box indicates overridden properties by highlighting them in cyan (light blue). In addition, a different icon appears in the Status column; pausing the cursor over the icon displays a yellow tooltip that reports the override value(s).

To remove overrides, follow these steps:

1. In the Layer dialog box, right-click any layer name.
2. From the shortcut menu, select **Remove Viewport Overrides For**.

    Notice the options:
- **Selected Layers** — removes overrides from selected layers only.
- **All Layers** — removes overlays from all layers.
- For both options, overrides can be removed from the current viewport or all viewports.

3. Select an option. Notice that the overrides and cyan highlights disappear.

The **MVIEW** and **-VPORTS** commands operate identically, curiously enough. Both have the LAyer option that removes the overrides. Here are the prompts for **MVIEW**:

> Command: **mview**
>
> Specify corner of viewport or [ON/OFF/Fit/.../LAyer/2/3/4] <Fit>: **la**
>
> Reset viewport layer property overrides back to global properties? [Yes/No]: **y**
>
> Select objects: *(Select one or more viewports.)*
>
> Select objects: *(Press ENTER to exit object selection.)*

Notice that the overrides are removed.

INSERTING TITLE BLOCKS

Layout mode is where you place scale-dependent objects, such as drawing notes and title blocks. In this tutorial, you add a title block and border to a drawing.

1. Ensure that Layout1 is in PAPER mode (no viewport with a heavy outline).
2. Create a new layer called "Titleblock," and then make it current.
3. Save the drawing with the **QSAVE** command.
4. Open Design Center (press **CTRL+2**).
5. In the Folder List, go to the folder holding the template drawings (not in the \template folder, which is empty):

    c:\documents and settings\\<*username*>\local settings\application
    data\autodesk\autocad 2008\r17.1\enu\template

    Replace <*username*> with your Windows login name, such as "administrator."

 Note: The folder is inconveniently hidden by Windows, and so it might not show up in DesignCenter. Here is one way to access it: enter the **ADCNAVIGATE** command and then provide the folder path listed above. DesignCenter immediately displays the content of the *template* folder.

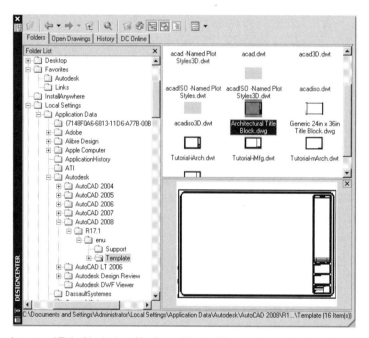

6. Drag the *Architectural Title Block.dwg* file from DesignCenter into the drawing.
7. When AutoCAD prompts you for the insertion point, pick the lower-left corner of the margin.

    Command: _-INSERT Enter block name or [?] <A$C27FF03C9>: "ANSI A title block.dwg"

    Specify insertion point or [Scale/X/Y/Z/Rotate/PScale/PX/PY/PZ/PRotate]: *(Pick the lower left corner of the margin.)*

8. Press **ENTER** at each of the remaining prompts.

    Enter X scale factor, specify opposite corner, or [Corner/XYZ] <1>: *(Press* **ENTER** *to keep scale factor at 1.0.)*

    Enter Y scale factor <use X scale factor>: *(Press* **ENTER** *to keep scale factor at 1.0.)*

    Specify rotation angle <0>: *(Press* **ENTER** *to keep angle at 1.0.)*

9. You may need to move the viewports so that they do not interfere with the title block or border.

USING TILEMODE

At the start of this chapter, I mentioned the **TILEMODE** system variable, which has been replaced by the layout tabs. For completeness, here is how to use it:

Command: **tilemode**
Enter new value for TILEMODE <0>: **1**

A value of **1** returns to the model tab, while a value of **0** returns to the last active layout tab.

 **VPMAX** AND  **VPMIN**

The **VPMAX** command maximizes the size of the current viewport (short for "viewport maximize"). This command is useful for increasing the size of small viewports. Conversely, the **VPMIN** command returns the viewport to its normal size. (As an alternative, you can double-click viewport borders to maximize and minimize them.)

Using this pair of commands is faster than switching back and forth between paper and model space, because AutoCAD automatically switches to model space at the same viewing scale as the layout. Another advantage is that these two commands do not affect model/paper space zoom scales.

The **VMPAX** and **VPMIN** commands work in layouts only, and not in Model tab. Drawings cannot be plotted or published while the viewport is maximized.

Left: *Maximize Viewport button.*
Right: *Minimize Viewport button surrounded by the Previous and Next buttons.*

When the layout has more than one viewport, two more buttons appear, as illustrated above at right. These button allow you to move to the next and previous viewport, in the order in which they were created. To return the viewport to its normal size, click the **Minimize Viewport** button.

TUTORIAL: MAXIMIZING AND MINIMIZING VIEWPORTS

1. For this tutorial, open the *vpmax.dwg* file from the CD.
   Click the **Layout** tab to ensure the drawing displays the viewports in layout mode.

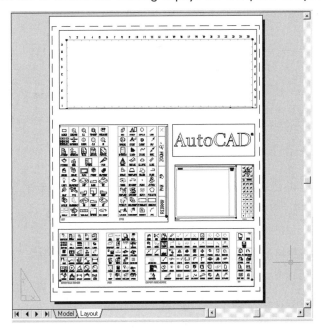

2. To maximize viewports, start the **VPMAX** command:
- In paper space, double-click a viewport border.

- In paper space, select a viewport. Right-click, and then select **Maximize Viewport** from the shortcut menu.

- On the status bar, click the **Maximize Viewport** button.

- At the 'Command:' prompt, enter the **vpmax** command.

    Command: **vpmax** *(Press* ENTER.*)*

3. If the layout contains more than one viewport, AutoCAD asks you to select one of them:

    Select a viewport to maximize: *(Select a viewport.)*

Notice that AutoCAD does the things illustrated in the figure below:

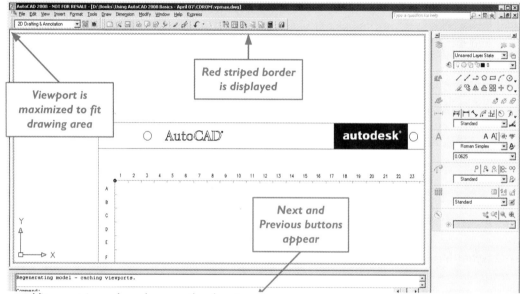

- Viewport is made as large as the drawing area.
- Viewport has a red striped border.
- **Previous** and **Next** buttons appear on the status bar, if the layout contains more than one viewport. And the **Maximize Viewport** button turns into the **Minimize Viewport** button.
- Space changes from paper to model.

4. To see other viewports in their maximized state, click the **Maximize Previous Viewport** and **Maximize Next Viewport** buttons.

AutoCAD displays the next viewport every time you click the button.

5. When done editing, minimize the viewport using one of these methods:
- Double-click the red striped border.

- Right-click anywhere inside the viewport, and then select the **Minimize Viewport** command from the shortcut menu.

- On the status bar, click the **Minimize Viewport** button.

- At the 'Command:' prompt, enter the **vpmin** command.

    Command: **vpmin** *(Press* ENTER.*)*

AutoCAD returns the viewport to its natural size.

Note: Here's how to maximize a viewport, but remain in paper space: select a viewport, and then use the **ZOOM** command with its **Object** option. AutoCAD maximizes the viewport.

 'SPACETRANS

The **SPACETRANS** command converts distances between model space units and paper space units (short for "space translation").

This command automatically determines the height of text being placed in model mode of a layout, but should be scaled appropriately for paper mode. The command is meant to be used transparently, during other commands when AutoCAD asks for heights, lengths, or distances. However, **SPACETRANS** only operates transparently after it has been run normally at least once.

The command may only be used in layouts — either in paper or model mode; attempting to use it in model space results in this complaint:

> ** Command not allowed in Model Tab **

(You may find it easier to use annotative scaling to scale text than this command.)

TUTORIAL: DETERMINING TEXT HEIGHT IN LAYOUTS

1. To specify that the height of text in model space matches a given height in a layout, ensure AutoCAD is displaying model space through the viewport in the layout, and then enter the following:

   Command: **text**

   Specify start point of text or [Justify/Style]: *(Pick a point.)*

   Specify height <0.2>: **'spacetrans**

2. You can enter a fraction, and AutoCAD calculates it. For example, 1/8" (0.125) is a typical height for text in drawings.

   >>Specify paper space distance <1.000>: **1/8**

   Press **ENTER**, and AutoCAD resumes the **TEXT** command, displaying the calculated height (0.125 x viewport scale factor):

   Resuming TEXT command.

   Specify height <0.2000>: 0.075783521943615

3. Continue with the command's other prompts:

   Specify rotation angle of text <0>: *(Press ENTER.)*

   Enter text: **SpaceTrans to the Rescue!** *(Press ENTER.)*

   Enter text: *(Press ENTER.)*

   The text is placed in the model at a size of 0.075... — instead of 0.125.

Notes: Use **SPACETRANS** to create notes in model space that appear at the correct height in the layout, and are plotted at the correct size.

Use it together with the **SCALETEXT** command to change existing text. Start the **SCALETEXT** command, and then enter **'SPACETRANS** to specify the new height. AutoCAD scales the text correctly.

SPACETRANS cannot be used with the **MTEXT** command. Instead use it as a calculator at the 'Command:' prompt to find the scaled text height, and then enter the height manually in the mtext editor:

> Command: **spacetrans**
>
> Specify model space distance <0.1250>: **1.0**
>
> 0.663080708963838

When the viewport scale changes, the text also changes size, depending on the mode in which it was placed:

When text is placed in PAPER mode
- It is not visible in model space.
- It does not change size when viewport scale changes.
- It issues the following prompts (note the word "model"):

> Select a viewport: *(Select a viewport border.)*
>
> Specify model space distance <1.0>: *(Enter a value.)*

When the text is placed in MODEL mode
- It is visible in model space.
- Its apparent size changes as seen from the layout.
- It issues the following single prompt (note the word "paper"):

> Specify paper space distance <1.000>: *(Enter a distance.)*

PSLTSCALE

The **PSLTSCALE** system variable matches linetype scaling to viewport scaling, so that linetypes drawn in the model appear the correct size (short for "paper space line type scale").

> Command: **psltscale**
>
> Enter new value for PSLTSCALE <1>: **0**

When set to **1** (the default), the viewport scale factor (set by **ZOOM Xp** or the viewport properties) controls linetype scaling. The lengths of dashes and gaps are based on paper space units — for objects drawn in model space and in layouts. The advantage is that viewports can have different zoom levels (scale factors), yet the linetypes look the same.

When set to **0**, the linetype scale in model space is independent of layout scale factors.

Note: After changing the value of **PSLTSCALE**, you need to use the **REGENALL** command to update linetype scaling in all viewports.

TUTORIAL: MAKING LINETYPE SCALES UNIFORM

1. From the CD, open the *psltscale.dwg* file, a drawing of objects with several linetypes.
   In this drawing, **PSLTSCALE** is turned off (set to **0**). Notice that the linetypes in the two viewports have different scales, because the viewports have different scales.

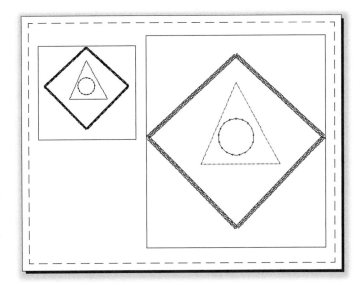

2. To make linetype scales the same in all layout viewports, start the **PSLTSCALE** system variable:

    Command: **psltscale** *(Press* **ENTER**.*)*

3. AutoCAD prompts you to change the value:

    Enter new value for PSLTSCALE <0>: **1**

4. The change to **PSLTSCALE** has no effect until you regenerate all the viewports:

    Command: **regenall** *(Press* **ENTER**.*)*

    Notice that the linetypes have the same scale factor in both viewports.

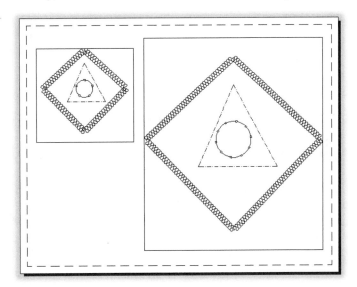

5. Change the linetype scale:

    Command: **ltscale**

    Enter new linetype scale factor <0.2500>: **.75**

    Regenerating layout. Regenerating model.

    The linetype scale changes in both viewports after AutoCAD automatically regenerates them. **VPMAX** shows the correct scale, whereas the Model tab does not.

LAYOUTWIZARD

The LAYOUTWIZARD command takes you through the steps of creating new layouts.

TUTORIAL: MANAGING LAYOUTS

1. To manage layouts, start the **LAYOUT** command:

    *   From the **Insert** menu, choose **Layouts**, and then **Create Layout Wizard**.

    *   From the **Layouts** toolbar, choose **New Layout**.

    *   At the 'Command:' prompt, enter the **layoutwizard** command.

        Command: **layoutwizard** *(Press* ENTER.*)*

2. In all cases, AutoCAD displays this first dialog box:

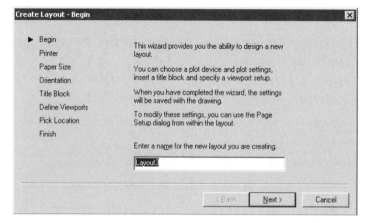

    Enter a name for the layout, which will appear on the tab.
    Click **Next**.

3. Select a printer from the list of Windows system printers and AutoCAD HDI printer drivers. (You can select a different printer later, if need be.)

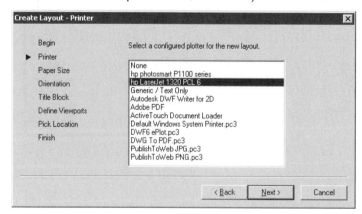

    Click **Next**.

4. Select a paper size from the droplist. Only those sizes supported by the printer are listed.

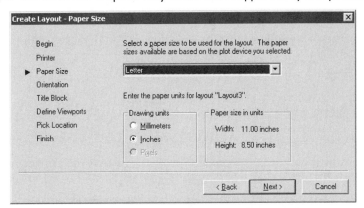

Drawing Units: Select the drawing units — inches or mm.
Click **Next**.

5. Select an orientation for the paper — landscape or portrait.

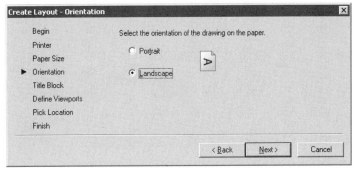

Click **Next**.

6. Select a drawing border/title block from the list.
(For A-size drawings, make sure the orientation of the border matches the orientation of the page you selected in the previous setup.)

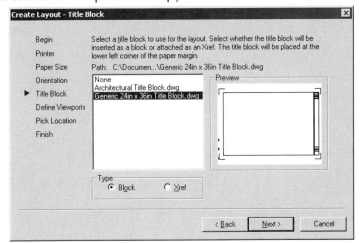

Type: Decide whether you want the drawing border to be inserted as a block, or attached as an externally-referenced drawing (xref). If you are not sure, select **Block**.
The following table provides you with the pros and cons of each:

Type	Pros	Cons
Block	Border and title block are part of the drawing, making it complete and easier to transport, such as by email.	Drawing size is larger. Blocks cannot be as readily updated as xrefs.
Xref	Drawing size is smaller. All xref'ed title blocks can be easily changed.	When sending the drawing to another office or client, you must ensure the xref is packaged with the drawing file.

7. **Viewport Setup:** Select the number and style of viewports. If you are not sure, select **Single**.

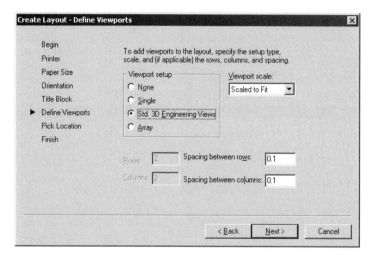

For engineering drawings, you would choose **Std. 3D Engineering Views,** which sets up the viewports and viewpoints illustrated below.

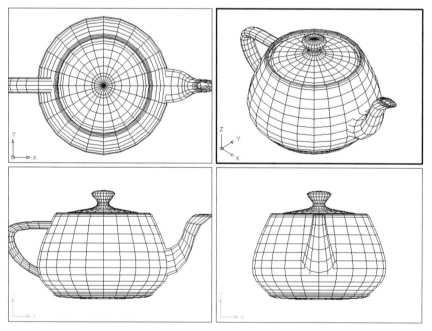

The four standard engineering viewpoints:
Top view — Isometric view
Side view — Front view.

The **Array** option sets up the viewports a rectangular array. You can specify the number of rows and columns, such as the 3x2 illustrated below.

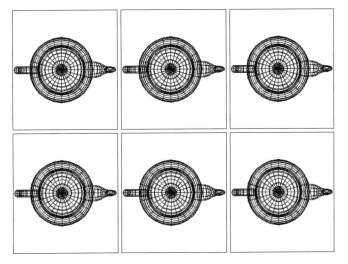

A 3-column x 2-row array of viewpoints, each with a gap of 0.1 units from its neighbor.

Viewport Setup	Comments
None	No viewport is created.
Single	One viewport is created.
Std. 3D Engineering Views	Four viewports are created adjusted to show the standard engineering viewpoints of 3D drawings: front, side, top, and isometric.
Array	*Rows* x *columns* array of viewports is created.

Viewport Scale: Select the viewport scale, which ranges from 100:1 to 1:100, and from 1/128"=1' to 1'=1'.

If you are not sure of the scale, select the **Scaled to Fit** option. (You can change it later by right-clicking the viewport border, selecting **Properties**, and then changing the value of **Custom Scale**.)

Optionally, you can also specify the gap between viewports; 0.1 units is the default.

8. Click the **Select Location** button to position the viewport(s).

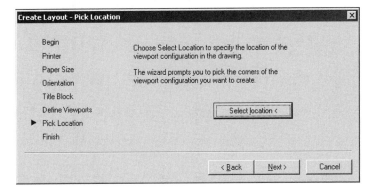

AutoCAD prompts you at the command line:

Specify first corner: *(Pick point 1.)*

Specify opposite corner: *(Pick point 2.)*

Pick two points to form a rectangle. AutoCAD later fits the viewport(s) to the rectangle.

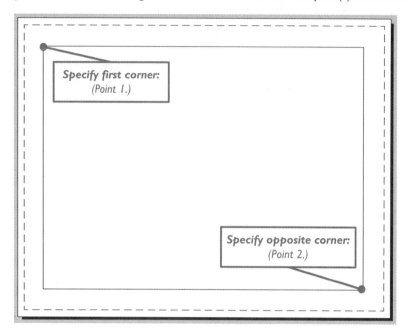

Click **Next**.

9. Click **Finish**.

AutoCAD creates the viewport, which allows the model to show through into the layout, the elements of which are illustrated below. (The title block and number of viewports varies, depending on the options you selected during the wizard.)

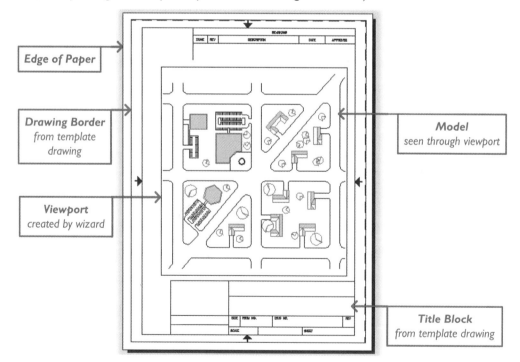

LAYOUT

The **LAYOUT** command manages layouts.

TUTORIAL: MANAGING LAYOUTS

1. To manage viewports, start the **LAYOUT** command:
   - From the **Insert** menu, choose **Layout**, and then **New Layout**.

   - In the **Layouts** toolbar, choose **New Layout**.

   - At the 'Command:' prompt, enter the **layout** command.

     Command: **layout** *(Press* ENTER.*)*

   - As an alternative, enter the **lo** alias.

2. In all cases, AutoCAD displays the prompts at the command line:

   Enter layout option [Copy/Delete/New/Template/Rename/SAveas/Set/?] <set>: *(Enter option.)*

Layout	Comment
Copy	Prompts you for the name of the layout to copy: Enter name of layout to copy:
Delete	Prompts you for the name of the layout to remove: Enter name of layout to delete: Note that the Model tab cannot be deleted. When you access this command through right-clicking the layout tab, the tab is immediately deleted.
New	Prompts you for a name, and then creates a new layout: Enter name of new layout: Names can be up to 255 characters long, but only a maximum of 31 are shown on the tab, fewer if there is less room. New tab are named "Layout*n*."
Rename	Prompts you for the name of the layout to rename: Enter name of layout to rename: Enter new layout name:
Saveas	Prompts you for the name of the layout to save as a *.dwt* template file: Enter layout to save to template:
Set	Prompts you for the name of the layout to make current. Enter layout to make current:
?	Lists the names of layouts.
Template	Creates new layouts based on *.dwt*, *.dwg*, and *.dxf* files. Displays the Insert Layout(s) dialog box (illustrated below), and then inserts objects and layouts into the drawing.

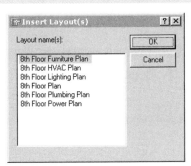

Note: Right-click any layout tab, if they are visible:

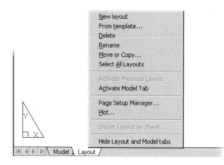

Several more options are available through the shortcut menu, which are not found in the **LAYOUT** command:

Activate Previous Layout — returns to the previously-accessed layout, useful when a drawing has many layouts.

Activate Model Tab — returns to the Model tab.

Page Setup Manager — executes the **PAGESETUP** command.

Plot — executes the **PLOT** command.

Import Layout as Sheet — adds the current layout as a sheet; available only when the Sheet Set Manager palette is open (new to AutoCAD 2008).

Hide Layout and Model Tabs — "hides" the tabs by moving them to the status bar.

 ## VIEWPORTS

The **VIEWPORTS** command creates and merges (removes) rectangular viewports.

When you switch to a layout for the first time, AutoCAD creates a single viewport automatically, which you can copy, move, and resize. To create additional viewports, use the **VIEWPORTS** command.

To create non-rectangular viewports, or convert objects into viewports, use **-VPORTS**, as described later. While both commands work in model space and in layouts, the non-rectangular viewports are available only in layouts.

These commands need to know two pieces of information: (1) the number of viewports, and (2) the location of the viewports.

TUTORIAL: CREATING RECTANGULAR VIEWPORTS

1. To create one or more rectangular viewports, start the **VIEWPORTS** command:
   * From the **View** menu, choose **Viewports**, and then **New Viewport**.
   * From the Viewports toolbar, choose the **Display Viewports Dialog** button.
   * At the 'Command:' prompt, enter the **viewports** command.

     Command: **viewports** *(Press* ENTER.*)*
   * Alternatively, enter the **vports** alias at the 'Command:' prompt.

   In all cases, AutoCAD displays the Viewports dialog box.

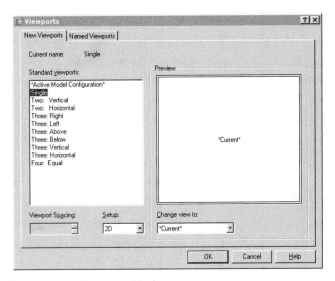

2. Select a style of viewport, as illustrated below.

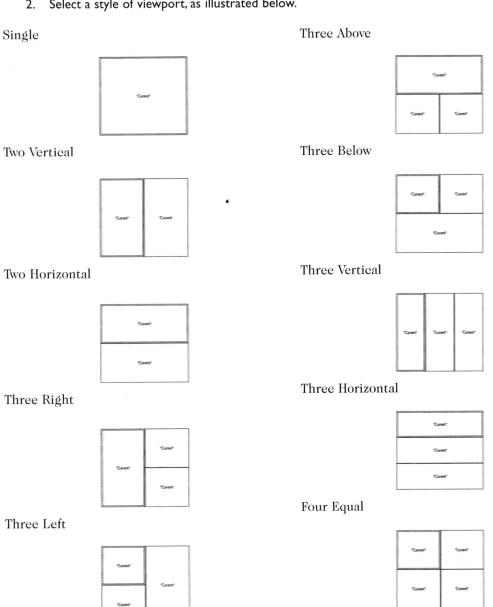

Single

Two Vertical

Two Horizontal

Three Right

Three Left

Three Above

Three Below

Three Vertical

Three Horizontal

Four Equal

3. Click **OK**. AutoCAD prompts you for the location of the viewports:

   Specify first corner or [Fit] <Fit>: *(Press* ENTER *to fit the viewports within the display area.)*

4. Press ENTER, and AutoCAD fits the viewports to the display area.

   Or, you can pick two points (forming a rectangle), and AutoCAD fits the viewports.

Notes: Viewports created in layouts are called "floating," because they can be moved. Each viewport is independent of the others. Polygonal (non-rectangular) viewports and viewports converted from objects can only be created in layouts.

Viewports created in model space are called "tiled," because they stick together as tightly as tiles. They cannot be moved or copied. You can merge and erase them with the **-VPORTS** command.

-VPORTS AND VPCLIP

The **-VPORTS** command creates non-rectangular viewports, while the **VPCLIP** command clips viewports.

You are not limited to viewports with rectangular or square shapes. AutoCAD can also create polygonal (non-rectangular) and circular viewports. These are sometimes called "clipped viewports," because they usually hide a portion of the model.

-VPORTS converts closed objects into viewports, including closed polylines, circles, ellipses, splines, and regions. Polylines can be made of line and arc segments, and may intersect themselves.

TUTORIAL: CREATING POLYGONAL VIEWPORTS

1. To create polygonal viewports, ensure AutoCAD is displaying a layout; this action cannot be carried out in model space.

2. Start the **-VPORTS** command:

   • From the **View** menu, choose **Viewports**, and then **Polygonal Viewport**.

   • At the 'Command:' prompt, enter the **-vports** command.

      Command: **-vports** *(Press* ENTER.*)*

3. In all cases, AutoCAD displays the prompts at the command line. Enter **P** to draw the outline of a polygonal viewport:

   Specify start point or

   [ON/OFF/Fit/Shadeplot/Lock/Object/Polygonal/Restore/LAyer/2/3/4] <Fit>: **p**

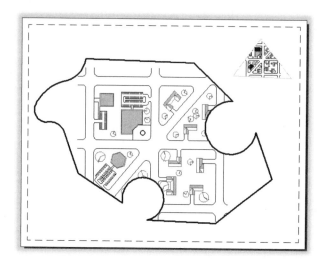

4. Pick a point to start.

    Specify start point: *(Pick a point.)*

5. The prompts that follow are exactly like those of the **PLINE** command. You can draw line segments and arcs. You must pick a minimum of three points, and then close the polyline.

An example of what is possible with polygonal viewport construction is illustrated earlier. There are two viewports: the larger one is made with line and arc segments; the smaller one in the upper right is made with the minimum of three line segments.

Note: In addition to viewports, AutoCAD can also clip externally-referenced drawings and other attached objects, but they each require their own command:

Xrefs — **XREFCLIP**
Image files — **IMAGECLIP**
DWF files — **DWFCLIP**
DGN files — **DGNCLIP**

Viewports created as polygons and from objects cannot have their scales locked.

TUTORIAL: CREATING VIEWPORTS FROM OBJECTS

Unusually-shaped viewports can be made from objects. For this tutorial, ensure AutoCAD is displaying a layout; this action cannot be carried out in model space.

1. Draw a polyline, circle, ellipse, spline, or region. If drawing a polyline or spline, use the **Close** option to ensure the object is closed.

    The figure below illustrates a self-intersecting polyline drawn with the **PLINE** command, and then smoothed with the **PEDIT** command.

 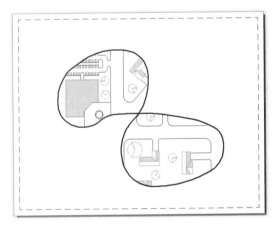

Left: *Object drawn in paper space.*
Right: *Object converted to viewport boundary.*

2. Start the **-VPORTS** command:

    *   From the **View** menu, choose **Viewports**, and then **Objects**.

    *   From the **Viewports** toolbar, choose the **Convert Object to Viewport** button.

    *   At the 'Command:' prompt, enter the **-vports** command.

        Command: **-vports** *(Press ENTER.)*

3. Enter **O** to convert a closed object into a viewport:

   Specify corner of viewport or

   [ON/OFF/Fit/Shadeplot/Lock/Object/Polygonal/Restore/2/3/4] <Fit>: **p**

4. Pick the object to convert.

   Select object to clip viewport: *(Select a closed object.)*

   Notice that AutoCAD converts the object into a viewport, and the model shows through, as illustrated above.

Once the viewport is in place, you can use grips editing to change the border, as illustrated by the figure below.

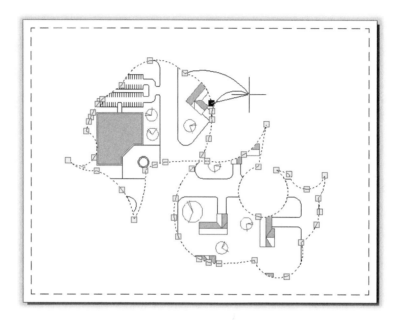

VPCLIP

The **VPCLIP** command operates in a manner similar to **-VPORTS**: it clips rectangular viewports with polylines or objects. In addition, it can further clip a clipped viewport. Better than **-VPORTS**, it converts clipped viewports back into rectangular ones.

> Command: **vpclip** *(Press ENTER.)*
>
> Select viewport to clip: *(Pick a viewport.)*
>
> Select clipping object or [Polygonal] <Polygonal>:

If you select a viewport previously clipped by this command, the **Delete** option also shows. The options are:

VpClip Option	Comment
Clipping object	Selects a closed object to be converted into a viewport boundary.
Polygonal	Picks points to designate line and arc segments for the viewport boundary.
Delete	Deletes the clipped viewport, and restores the rectangular viewport.

EXERCISES

1. From the Companion CD, open the *grader.dwg* file, the 2D drawing of a grader.

   Click the **Layout1** tab, and create two viewports on an A0- or E-size sheet of paper:

   a. The first viewport showing the entire grader, without the drawing border.

   b. The second viewport showing a detail of the air intake.

   The result should look similar to the figure below. (All figures in these exercises are courtesy of Autodesk, Inc.)

   Save the drawing as *layout1.dwg*.

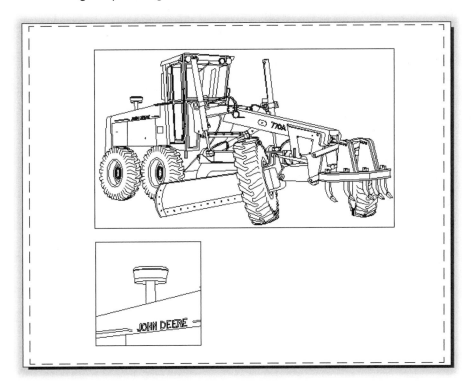

2. Start a new drawing, and then use the Layout Wizard to create a viewport with the following options:

Name	**Practice Layout**
Plotter	**DWF 6**
Paper Size	**ANSI B**
Drawing Units	**Inches**
Orientation	**Landscape**
Title Block	**ANSI B**
Viewport Setup	**Single**
Viewport Scale	**Scaled to Fit**

   Save the result as *layout2.dwg*.

3. From the CD, open the *langer.dwg* file, a drawing detailing sump pumps.

   Create four layout tabs of the following names, which match the names of the details.

   • Pump Connection Detail

   • Branch Pipe Support Detail

   • Duplex Ejector Pumps

   • Air Handler Unit Hanging Detail

In each layout, place the namesake detail.

What is the viewport scale factor?

Save the drawing as *layout3.dwg*. The result should look similar to the figure below.

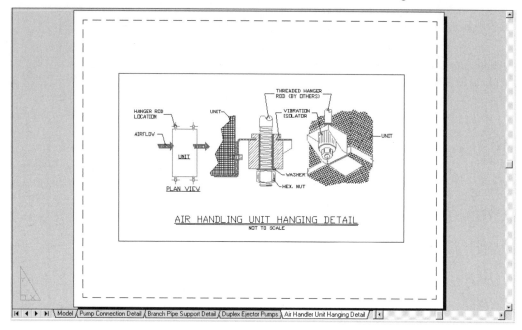

4. Use the **VIEWPORTS** command to create viewports with the following arrangements:

a.

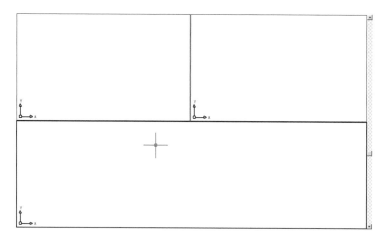

b.

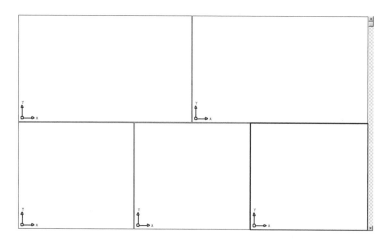

c.

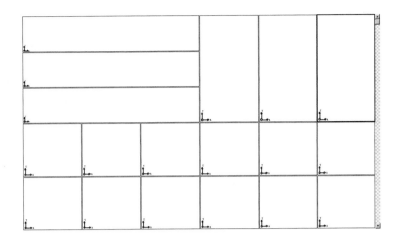

5. From the Companion CD, open the *wright.dwg* file, a floor plan and elevation of a house designed by architect Frank Lloyd Wright.

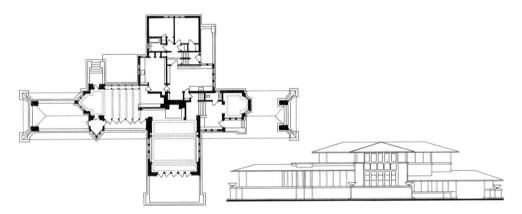

6. Create a pair of clipped viewports that show the two views independently, as illustrated below. Save the drawing as *layouts5.dwg*.

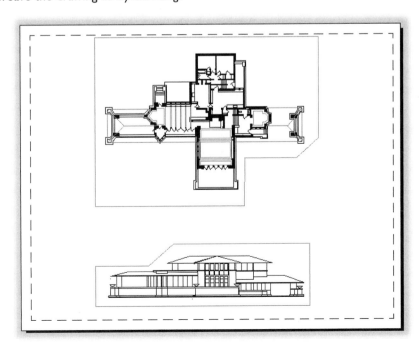

CHAPTER REVIEW

1. Explain the purpose of layouts.
2. Describe two ways to switch between model and layout mode:
   a.
   b.
3. In which tab are the following elements drawn full size?
   The model
   Title block
4. What does the layout represent?
5. Can a drawing have more than one layout?
6. How do you see the model in layouts?
7. Can viewports be modified?
   Can viewports be copied?
8. Name the parts of the layout:
   a.
   b.
   c.
   d.

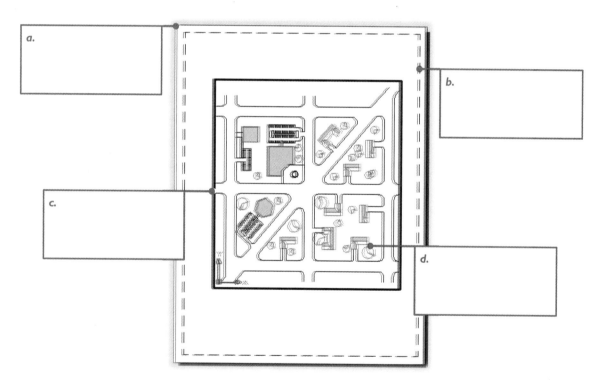

9. A layout has two viewports, of which one has a heavier border.
   What is the significance of the heavy border?
   What mode is AutoCAD in?
10. Explain the function of the following keystrokes when used with layout tabs:
    CTRL+R
    CTRL+PGDN

11. Describe the kind of drafting you can do when the layout is in:

    PAPER

    MODEL

12. Describe what happens when you resize a viewport in a layout?

    When you copy a viewport.

    When you select an object in one viewport.

13. Describe how to hide the viewport border.

14. What are the differences between the two viewports shown below? Explain a possible reason for the differences.

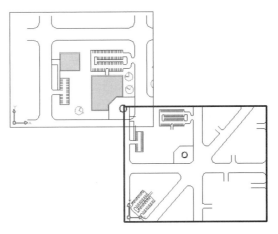

15. What are two ways to scale models in layouts?

    a.

    b.

16. What happens when you double-click a viewport?

17. Calculate the scale factor for the following viewport:

    Drawing of automobile = 12', bumper to bumper length

    Width of viewport = 10"

18. How would you move the model in the viewport?

19. Explain the difference between the **LAYER** command's **Current VP Freeze** and **New VP Freeze** options.

20. What does the snowflake mean in the Layer Properties Manager dialog box?

21. In which mode would you insert a title block?

    PAPER mode of layout tabs.

    MODEL mode of layout tabs.

    Model tab.

22. What advantage does the **VIEWPORTS** command have over the **-VPORTS** command?

    -VPORTS command over **VIEWPORTS** ?

23. Must viewports be rectangular?

    If no, in what way?

    If yes, why?

24. Describe the difference between:

    Tiled viewports

    Floating viewports

25. Briefly explain the function of the **PSLTSCALE** command.

26. When does **PSLTSCALE** take effect?

27. When is the **SPACETRANS** command commonly used?
28. What scale factor does **SPACETRANS** calculate?
29. Which command maximizes a viewport and also

    Switches to model space?

    Stays in paper space?
30. Describe one or more benefits to using the **VPMAX** command.
31. How can you add and remove scale factors from the VP Scale list?
32. When might you want to override layer properties in viewports?
33. List at least two ways of turning off layer property overrides.

CHAPTER 22

Plotting Drawings

The end product of CAD drafting is typically the drawing plotted on paper. In this chapter, you learn to plot drawings with these commands:

PREVIEW previews drawings before plotting.

PLOT plots drawings.

OPTIONS specifies default settings for plotting.

AUTOPUBLISH simultaneously saves drawings in DWG and DWF (new to AutoCAD 2008).

SCALELISTEDIT customizes lists of plot scales.

VIEWPLOTDETAILS reports on successful and failed plots.

PLOTSTAMP stamps plots with information about drawings and plotting.

PAGESETUP assigns plotters to layouts.

PUBLISH creates and plots drawing sets.

PLOTTERMANAGER creates and edits plotter configurations.

STYLESMANAGER creates and edits plot style tables.

CONVERTPSTYLES and **CONVERTCTB** convert color-based and named plot style tables.

NEW TO AUTOCAD 2008 IN THIS CHAPTER

- The **PLOT** command now emulates rendering and 3D effects not supported by the graphics board.
- The **AUTOPUBLISH** command determines whether drawings are simultaneously saved in DWF format during the **SAVE** and **CLOSE** commands.
- The **SCALELISTEDIT** command also determines the scale factors used with annotative scaling.

FINDING THE COMMANDS

On the **Standard Annotation** toolbar:

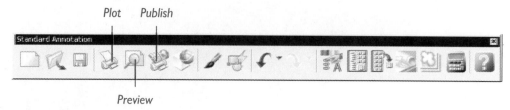

On the **Standard** toolbar:

ALL ABOUT PLOTTERS AND PRINTERS

AutoCAD works with any printer connected to your computer and your network; it checks automatically for all system printers registered with Windows.

SYSTEM PRINTERS

System printers are any local and network printers recognized by Windows. To see the list of system printers for your computer, choose the **Start** button on the Windows Taskbar, and then select **Settings | Printers** (varies with the version of Windows). Windows opens the Printers window, which lists your computer's system printers.

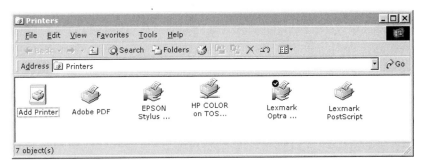

The figure illustrates the Printers window for the author's computer. From left to right, the icons indicate:

Add Printer — double-click this icon to add new printers to the computer. You only need to do this when Windows does not automatically recognize the new printer or plotter plugged into your computer. The automatic recognition is known as "Plug and Play."

Adobe PDF — the unadorned printer icon indicates a local printer. In this specific case, it creates PDF files from any document by "printing" to a file.

EPSON Stylus — the hand icon represents "sharing," and means the printer can be accessed by other computers on the network. To allow network access to printer(s), right-click the printer icon, and then select **Sharing**; in the **Properties | Sharing** dialog box, select the **Shared As** radio button, and then choose **OK**.

HP COLOR on TOS — the "wires" attached to the bottom of the printer icon indicate a network printer. "On" means this printer is attached to another computer on the network. "TOS" is the name of the other computer.

Lexmark Optra — the check mark on this icon indicates the default printer. AutoCAD and other Windows programs use this printer automatically, unless you specify a different one. To select a different printer as the default, right-click another of the other printer icons, and then select **Default Printer** from the shortcut menu.

To change the properties of a system printer, right-click its icon, and then select **Properties**. The content of the Properties dialog box varies, depending on the printer's capabilities. Commonly, though, you can set the default resolution, paper size and source, color management, and so on.

Nontraditional Printers

AutoCAD also works with nontraditional printers, the most common today being Adobe Acrobat PDF (short for "portable document format"). You can create PDF files of drawings by plotting with the "DWG to PDF.pc3" printer driver.

Autodesk promotes the DWF format for sharing drawings over the Internet. While the **DWFOUT** command is used to create *.dwf* files from drawings, Autodesk also provides the free DWF Writer software to create *.dwf* files from any other application.

Less popular today are faxes. Most computers have fax capability included with their modem. Windows lets you fax from any software program, including AutoCAD, provided the computer is connected to the telephone system. The process is as simple as selecting the fax as the printer. The drawback to faxing is that AutoCAD typically splits large drawings onto multiple A- or A4-size sheets of paper (roughly 8" x 11" each).

LOCAL AND NETWORK PRINTER CONNECTIONS

Printers are usually connected to computers through USB (universal serial bus) connections. Computers can theoretically have up to 128 USB ports, but two to eight is common. To support older printers, some computers have parallel ports; serial ports are used only for the oldest printers and pen plotters, which operate slowly and are difficult to configure.

Network printers are connected *directly* or *indirectly* to the network. If connected directly, the printer contains its own network card; if indirectly, the printer is connected to a computer, and then through the computer's network card to the network.

Computers connected to networks can print to any network printer — provided the computers have been given permission to access the network printers. This works in reverse, too: you can give other networked computers permission to use your computer's printer.

Differences Between Local and Network Printers

The primary difference between local and network printers is how Windows sees them. During printer setup, you need to tell Windows whether the printer is Local or Network, so that it knows whether to search for it on your computer's local ports, or along the network.

Another difference is that local printers are typically more available. A network printer might be inaccessible, because the network is down or because too many other computers are sending it files.

The advantage to networking printers is that everyone in the office can share all printers. If the printer attached to your computer breaks down (through mechanical failure, lack of paper or ink, and so on), you can easily access another one.

ABOUT DEVICE DRIVERS

A term you will occasionally read is "device driver." Without device drivers, software programs cannot communicate with printers, graphics boards, and other computer hardware — collectively known as "peripherals."

To understand the role of device drivers, it helps to know their history. In the early days of personal computers, there were very few choices for external peripherals, and the peripherals had very few abilities. For instance, early graphics boards and printers only output text — no graphics!

As personal computers became popular in the late 1980s, more and more companies began inventing add-ons for computers. Graphics boards and printers began outputting graphics, and then gained color. Varieties of input devices flourished, such as digitizing tablets and 100-button keypads.

Peripherals vendors typically took one of two approaches to make their products work with computers. One approach was to mimic a competitor's protocol that already worked with computers, such as HPGL (Hewlett-Packard Graphics Language) for plotters and TIGA (Texas Instruments Graphics Architecture) for graphics boards. If the standard lacked the desired capabilities, then the second approach was to create a new standard.

Even with standards in place, peripherals still needed a way to communicate with the computer. Initially, each peripheral vendor wrote their own software, called a "device driver" — software that drives devices, or "drivers" for short.

When Microsoft began marketing Windows, they decided to include as many drivers as possible to reduce the resistance to switching from DOS to Windows. Microsoft undertook the thankless job of writing thousands of drivers for nearly any peripheral on the market — albeit in many cases, one device driver served several related hardware products.

Initially, Autodesk undertook the same approach. The company wrote AutoCAD drivers for all peripherals, and even boasted of how long their list was. By the mid-1980s, however, the job became overwhelming, and so Autodesk began to require that peripheral vendors write the drivers themselves using the ADI kit supplied by Autodesk — the AutoCAD Device Interface. In the 1990s, the capabilities were expanded by the enhanced HDI kit — Heidi Device Interface.

(HEIDI is short for "HOOPS Extended Immediate-mode Drawing Interface." HOOPS, in turn, is short for "Hierarchical Object-Oriented Picture System." So really, HDI is short for Hierarchical Object-Oriented Picture System Extended Immediate-mode Drawing Interface Device Interface — whew!)

Drivers Today

Today, Autodesk provides drivers for a number of plotters and two graphics board standards, OpenGL and DirectX. It relies on Microsoft to provide drivers for common peripherals, such as mice, and on hardware vendors, for specialized peripherals.

From time to time, hardware companies update device drivers to fix bugs and enhance capabilities. They typically provide free downloads of the latest drivers. If plots don't work correctly, the first thing to suspect is a device driver that needs updating. On the other hand, if the driver is working well, *don't upgrade*, because sometimes upgrades can cause new problems.

To see the list of drivers installed on your computer, click **Start | Settings | Control Panel** on the Windows taskbar. Double-click the **System** icon, and then select the **Hardware** tab. Click **Device Manager** to see a list similar to that illustrated at right.

With this background information in place, let's see how AutoCAD handles the plotting of drawings.

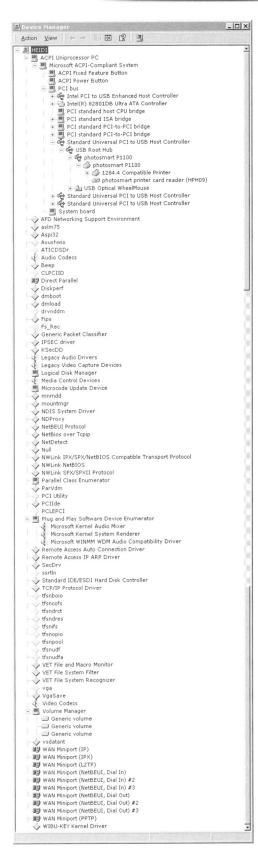

PREVIEW

The PREVIEW command lets you see how drawings will be plotted.

Previewing plots is important, because it ensures drawings will be plotted as you expect. The preview shows you whether all drawing elements will appear on the paper correctly, and provides visual checks of settings, for example to determine whether the drawing is centered or not. This saves you time and money, because previewing is much, much faster than plotting and costs nothing, except for a few extra seconds of your time.

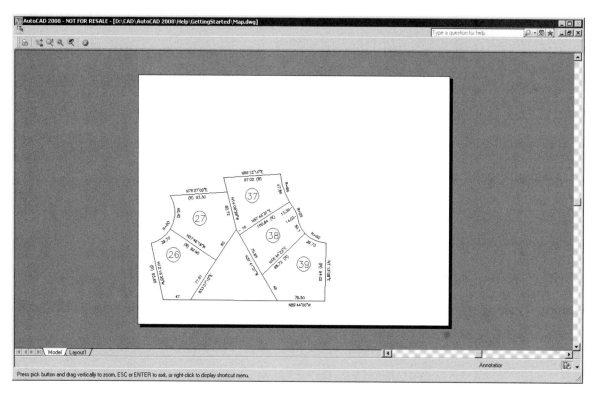

TUTORIAL: PREVIEWING PLOTS

1. To preview drawings before plotting, start the **PREVIEW** command:
   - From the **File** menu, choose **Plot Preview**.
   - From the Standard or Standard Annotative toolbars, choose the **Plot Preview** button.
   - In the Plot dialog box, click the **Preview** button.
   - Or, at the 'Command:' prompt, enter the **preview** command:

     Command: **preview** *(Press ENTER.)*
   - Alternatively, enter the **pre** alias at the 'Command:' prompt.

Note: Sometime, a plotter is not assigned to the drawing automatically, and AutoCAD complains:
No plotter is assigned. Use Page Setup to assign a plotter to the current Layout.

Other times, a plotter was assigned to the drawing that is not available from your computer, and AutoCAD complains:
The selected layout has an invalid hardcopy configuration.

One solution is to use the **PAGESETUP** command to assign a printer or plotter to the layout. Another solution is to modify the *.dwt* template files by assigning the system printer to Model tab and all layout tabs.

2. Once the preview is generated, AutoCAD displays it in the preview window. (If the drawing is complex, a dialog box may appear showing AutoCAD's progress in generating the preview image.

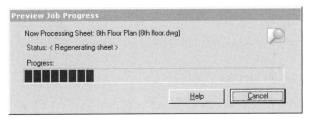

Cancel cancels the preview, and returns to the drawing editor.)

3. Press **ESC** to return to the drawing.

PREVIEWING PLOTS: ADDITIONAL METHODS

The preview shows the drawing on a white rectangle that represents the paper. (You can change the color of the area that represents the sheet of paper; use the **OPTIONS** command's **Display | Colors** button, and then select **Plot Preview** from the **Context** list.)

Controlling the Preview Image

Once in the preview window, you can zoom and pan the image. You can use the toolbar, or else right-click the drawing for a shortcut menu of viewing choices:

The cursor initially looks like a magnifying glass. As you drag the cursor up and down (by holding down the left mouse button), the preview zooms in and out.

When you select **Pan** from the menu or shortcut menu, the cursor changes to a hand. Hold down the left mouse button and the view moves (pans) as you move the mouse around.

The **Zoom Window** option operates identically to the regular **ZOOM** command's **Window** option.

The **Zoom Original** option returns the original view, much like the **ZOOM Extents** command.

Icon	Option	Cursor	Meaning
	Exit	...	Exits preview.
	Plot	...	Plots the drawing.
	Pan		Switches from zoom to pan mode.
	Zoom		Switches from pan to zoom mode.
	Zoom Window		Zooms into a windowed area.
	Zoom Original	...	Returns to normal view.

To exit the preview window, press **ENTER**, the spacebar, or **ESC**.

PLOT

The **PLOT** command displays the Plot dialog box, and then plots the drawing.

The dialog box lists all of the options AutoCAD needs to know before outputting drawings on paper or to files. Drawings can be plotted to printers, plotters, or to files.

TUTORIAL: PLOTTING DRAWINGS

1. To plot drawings, start the **PLOT** command:
    * From the **File** menu, choose **Plot**.

    * From the Standard or Standard Annotative toolbars, choose the **Plot** button.

    * On the keyboard, press the **CTRL+P** shortcut.

    * Right-click a layout tab, and select **Plot** from the shortcut menu.

    * At the 'Command:' prompt, enter the **plot** command.

      Command: **plot** (*Press* ENTER.)

    * Alternatively, enter the aliases **print** or **dwfout** (older commands that have been integrated into plotting) at the 'Command:' prompt.

    In all cases, AutoCAD displays the Plot dialog box.

    The Plot dialog box has many options, and it can be confusing to navigate. For your first plots, I recommend you specify the following information as a minimum:

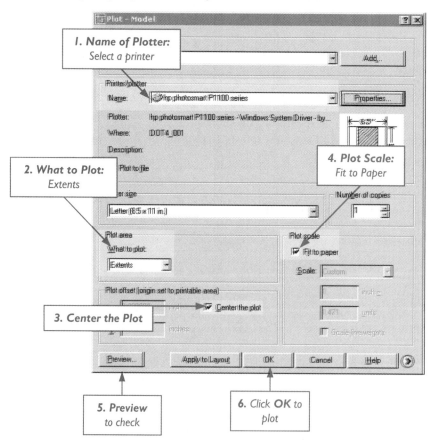

2. **Printer/plotter**: Select the name of a printer or plotter. In many cases, this is whatever printer is hooked up to your computer (the "system" or "default" printer).

3. **What to Plot**: select the **Extents** to ensure that every object is included in the plot.

4. **Plot Offset**: select **Center the plot.** (If the plot is not centered, it may be because **What to Plot** is set to **Display**.)

5. **Plot Scale**: select **Fit to Paper** so that nothing is printed off the edge of the page.

6. Click the **Preview** button to check that the plot will turn out.

7. Click the **OK** button to start the plot. AutoCAD plots the drawing.

   (If you wish only to set up the plotting parameters, but not actually plot, then click **Apply to Layout** to save the changes, and then **Cancel** to return to the drawing editor.)

Notes: AutoCAD does not plot layers that are frozen or have the No Plot property. As well, AutoCAD does not plot the Defpoints layer. If part of your drawing does not plot, it could be because you drew on the Defpoints layer.

To create 2D DWF files of drawings, use the **PLOT** command with its **DWF ePlot** "printer/plotter"; to create a multi-sheet 2D DWF file, use **PUBLISH** and its **DWF File** option. 3D versions of DWF files can only be created by the **3DDWF** command.

Here in greater detail are the steps to plot successfully.

Select a Plotter/Printer

Your computer may have two or more different printers available for plotting. The **Printer/Plotter** droplist contains the list of printers available to you — local and networked, as well as special drivers for plotting to files, such as PDF and DWF.

From the **Name** droplist, select a printer name. If the printer you want is not listed here, read about the **PLOTTERMANAGER** command later in this chapter.

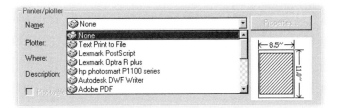

To change properties specific to the printer, such as its resolution or color management, choose the **Properties** button.

Plot to File

The **Plot to File** option saves drawings to files. The *plot file* can be read by plotters, or imported into other software. For example, many brands of plotters and graphics programs can import files created by HPGL plotter drivers, usually meant for Hewlett-Packard brand plotters.

Here's how to create plot files:

1. Click **Plot to File** so that a check mark appears.

2. Click **Plot**.

3. Instead of plotting the drawing, AutoCAD displays the Browse for Plot File dialog box. Select a folder, and if necessary change the file name.

4. Click **Save**.

AutoCAD generates the plot file, saving it under the name of the drawing and using the file extension of *.plt*.

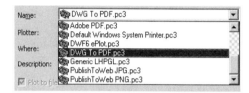

Some "plotters" are designed only to plot to file. These include DWF6, DWG to PDF, and Publish to Web. For them, the **Plot to File** option is turned on automatically and grayed-out, because it cannot be turned off.

DWG To PDF creates PDF files from drawings. These can be viewed by the free Acrobat Reader software from Adobe.

Select the Paper Size

Paper Size specifies the dimensions of the paper (media) used by printers. AutoCAD knows the appropriate media sizes for the selected printer. Thus the list of sizes you see here varies according to the printer. Make your selection from the droplist, and ensure the same size of paper is loaded into the printer.

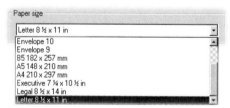

Most desktop printers use Letter paper in North America; drafters call it "A size." It measures 8.5x11". In Europe and other parts of the world that use metric sizes, A4 paper is the equivalent; it measures 8.26x11.7".

Sometimes printers will not print drawings, because the drawings have been set up with a printer or size of paper that is unavailable. If the paper size you need is not listed, you can add it with the **PLOTTERMANAGER**'s **Custom Paper Sizes** option.

Select the Plot Area

In many cases, you want to plot all of the drawing — known as the "extents." In some cases, you may want to plot specific parts of drawings. AutoCAD offers several options for determining how much of the drawing to plot:

Plot Area	Comments
Display	Plots the current view of drawings — the "what you see is what you get" plot.
Extents	Plots the extents, a rectangle that encompasses every part of the drawing containing objects; like performing a ZOOM **Extents**, and then plotting with the **Display** option.
Limits	Plots the limits of model space, as set by the LIMITS command.
Window	Plots a rectangular area identified by two windowed picks. After clicking **Window**, you are prompted to pick the two corners of a rectangle: Specify first corner: *(Pick a point or type X,Y coordinates.)* Specify opposite corner: *(Pick another point or type X,Y coordinates.)*
Layout	Plots the extents of the page in layouts (available only when plotting layouts).
View	Plots named views; available only when the drawing contains named views (created with the VIEW command). Select **View**, and then select a view name.

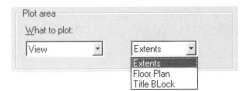

Plot Offset

The *plot origin* is at the lower left corner of the media for most plotters, just as it is for AutoCAD drawings. Some plotters, however, locate their origin at the center of the media; AutoCAD normally adjusts for them.

Sometimes you need to shift the plot on the media, for example to avoid a preprinted title block. Use the **Plot Offset** section of the Plot dialog box:

- Positive values shift the plot to the right and up.
- Negative values shift the plot to the left and down.

In most cases, however, you will probably choose the **Center the plot** option to center the plot on the media.

Note: Changing the plot's offset may result in a *clipped* plot, where part of the drawing is not plotted, because it extends beyond the edge of the media's margin.

Number of Copies

Most times, you want just one copy of the plot. If you need more, however, increase the **Number of Copies** counter. The largest number you can dial in is 99. Plot to file is limited to one copy.

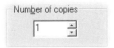

Select Plot Scale

To plot the drawing at a specific scale, select one of the predefined scale factors provided in the **Plot Scale** list box. These range from 1:1 to 1:100 and 1/128"=1'.

Plot Scale

For draft plots, select **Fit to Paper**. Together with a plot area of Extents, this ensures your entire drawing fits whatever size of media you select.

Another choice is **Custom**, where you specify the scale factor. Specify the number of *inches* (or millimeters) on the paper to match the number of drawing *units* to be plotted. For example, a scale of 1" = 8'-0" means that 1" of paper contains 8'-0" (or 96") of drawing.

To set the custom scale correctly, enter:

- Inches   **1"**
- Units   **9' 6"**

This is the same as a scale of 1/8" = 1'-0". The example assumes that the drawing units are set to architectural units. You can customize the list of plot scales with the **SCALELISTEDIT** command. (This command also affects annotative scale factors and the viewport scale list in the Properties palette.)

Preview the Plot

You should preview the plot to see the area of the paper on which the plot appears. AutoCAD's plot preview allows two types of preview: partial and full.

Partial Preview

The *partial* preview does not show the drawing elements, but rather the position of the plot on the paper. It's called "partial," because it shows only the basics. (*History*: this option was more desirable in decades past, because slow computers took a long time to generate full previews.)

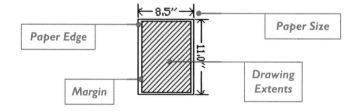

The paper's edge is shown by the outer rectangle. To see the paper size and printable area (a.k.a margin), pause the cursor over the partial preview image.

The margin is the non-printing area outside of the drawing; its size varies according to the capabilities of the printer and the settings in the Plot dialog box: Plot area, Plot offset, and Plot scale.

The area covered by the drawing is shown by the diagonal hatching.

Note: Use plot previews to discover plotting results before committing time, paper, and ink (toner) to a drawing that may be set up incorrectly. If the **Preview** button is grayed out, you have not yet selected a plotter for the layout.

Full Preview

The *full* preview shows the drawing as it would appears as a final plot on paper. See the PREVIEW command earlier in this chapter.

Save the Settings

If you plan to use the same plot parameters again, save them. In the Page Setup Name section (near the top of the Plot dialog box), choose the **Add** button. (As an alternative, you can click **Import**, and then select a drawing file containing plot setups.)

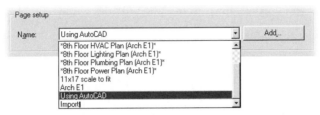

AutoCAD displays the User Defined Page Setups dialog box:

Enter a descriptive name, and choose **OK**. The next time you plot this drawing, you can select the page setup name from the **Page setup Name** droplist, and all of the dialog box's parameters changes to those of the saved page setup.

Plot the Drawing

When you have completed these steps, make sure the plotter is ready, and then choose the **OK** button at the bottom of the Plot dialog box.

(As an alternative, click **Apply to Layout**, and then **Cancel**. This saves the settings, but does not plot the drawing.)

AutoCAD displays a dialog box indicating its progress plotting each layout.

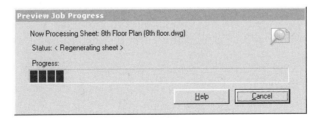

In the tray at the right end of the status bar, look for the tiny pen plotter icon moving a sheet of paper back and forth – even though pen plotters hardly exist any more, notes the technical editor, kind of like the diskette icon use for the Save button.

Right-click the plotter icon to see additional options:

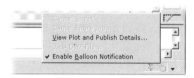

Cancel Sheet — cancels the current sheet being plotted.

Cancel Entire Job — cancels multi-sheet plot jobs.

View Plot and Publish Details — displays the dialog box discussed in the next section.

View DWF File — opens the *.dwf* file in the DWF Viewer, if drawings are published in DWF format.

Enable Balloon Notification — toggles the display of the balloon described below.

VIEWPLOTDETAILS

The **VIEWPLOTDETAILS** command reports on successful and failed plots.

When the plot is complete, AutoCAD displays a yellow balloon at the right end of the status bar. It reports whether (or not) the plot was successful.

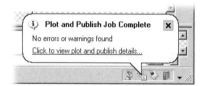

When the plotter is next to your desk, you don't need to see this balloon to know that the drawing has been plotted. But when the plotter is located elsewhere, it's good to know the status without having to consume precious calories by getting up from your desk and walking over to it. ("This is so tempting," muses the copy editor. "Although," adds the technical editor, "it's the only exercise some people get.")

Click the blue underlined text to view a full report on the plot. AutoCAD displays the Plot and Publish Details dialog box.

(As alternatives, you can enter the **VIEWPLOTDETAILS** command, or right-click the plotter icon on the status bar, and then select **View Plot and Publish Details**. To dismiss the balloon, click the x in the upper right corner.)

The **View** droplist displays all reports or just those containing errors. The **Copy to Clipboard** button copies the reports to the Clipboard; use the **Edit | Paste** command in a word processor to capture the report in a document.

Click **Close** to close the dialog box.

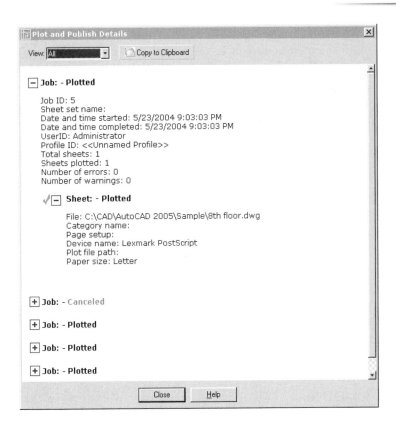

 PLOTTING DRAWINGS: ADDITIONAL METHODS

The **PLOT** command's dialog box sports a **More Options** button that expands to show more options. Click the button, or press **ALT+>**.

Let's look at each, except for **Plot Stamp On**, which is discussed later in this chapter.

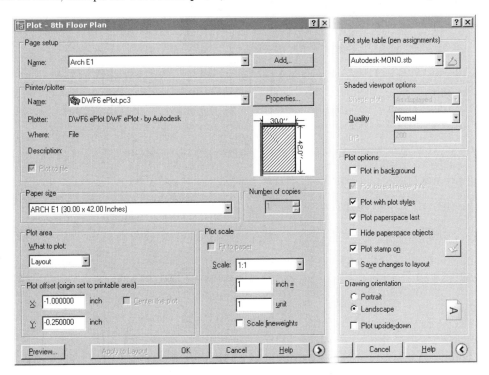

Select a Plot Style Table

The word "pen" in Pen Assignments derives from an earlier age when plotters plotted with actual pens — commonly ink pens, but also ballpoint pens, felt pens, and even pencils and knives (for cutting signs). The term has carried over to today's non-pen plotters that use laser and inkjet technology; now, it refers to varying widths, colors, and shades of gray.

Early releases of AutoCAD assigned colors of objects to pens. For example, all objects colored red (either by layer or by object) were plotted by a specific pen; those colored blue plotted with another pen, and so on. The pens didn't need to contain red or blue ink; typically they contained black ink and had tips of varying widths, such as 0.1" or 0.05".

Plot styles today assign plotter-specific properties to layers and objects, such as widths, colors, line-end capping, and patterns.

Select one of the preassigned plot style tables from the list. To create a new style, select **New** from the droplist.

To edit an existing style, click the **Edit** 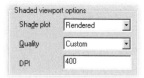 button. You learn more about plot styles under the **STYLESMANAGER** command later in this chapter.

Select Viewport Shade Options

Shaded Viewport Options controls plotting of viewports. Viewports typically display drawings with 2D and 3D wireframe views, but sometimes you want 3D models plotted as renderings or with hidden-lines removed. This option lets you specify the type and quality of shading.

Although the option has "viewport" in its name, it also applies to drawings plotted in Model tab. You can check the effect of changing **Shade plot** in the preview window; **Quality** and **DPI** don't show up until you plot on paper. Sometimes these options are unavailable:

Shade plot — unavailable when at least one viewport in layout mode has a Shade Plot other than "As Displayed" assigned to it by the Properties palette. This option is always available when plotting from the Model tab.

DPI — available only when **Quality** is set to "Custom."

Shade Plot Options

The **Shade plot** droplist replaces the "Shaded" option by visual styles. Even if your graphics board cannot display all visual styles effects, such as shadows and materials, AutoCAD can make the plotter simulate the effects in the plotted output.

 Note: If a viewport does not plot, it may be that its display has been turned off. You can turn it back on with the **PROPERTIES** command: in the Misc section, change the **On** option to "Yes."

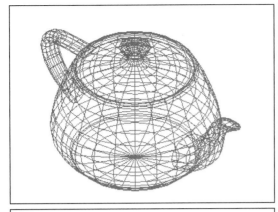
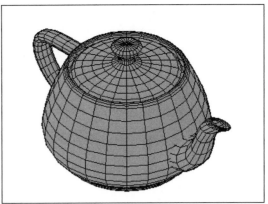
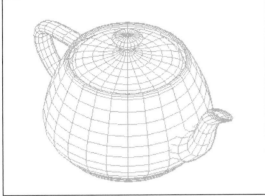

Top left: *3D drawing displayed in wirefame.*
Bottom left: *3D drawing with hidden lines removed.*

Top right: *Gooch shaded (conceptual visual style).*
Bottom right: *3D drawing rendered.*

Shade Plot	*Comment*
As Displayed	Plots objects as displayed in model or layout tabs. If shaded with the VISUALSTYLES command or rendered with the RENDER command, then the drawing is plotted that way.
Wireframe	Plots objects in wireframe, regardless of viewport display mode.
Hidden	Plots objects with hidden lines removed, regardless of display mode.
Visual Styles	
3D Hidden	Plots 3D objects with hidden lines removed.
3D Wireframe	Plots 3D objects in wireframe.
Conceptual	Plots 3D objects using Gooch rendering.
Realistic	Plots 3D objects in a realistic mode.
	If the drawing has other visual styles defined, they are included in this list.
Rendered	
Rendered	Plots 3D objects rendered, regardless of display mode.
Draft	Plots 3D objects at the lowest quality.
Low	Plots 3D objects...
Medium	....at varying levels...
High	...of quality.
Presentation	Plots objects at the highest quality.
	If the drawing has other rendering modes defined, they are included in this list.

Quality and DPI Options

The **Quality** droplist specifies the resolution for shaded and rendered plots, because they are raster images. (Wireframe and hidden-line-removed are plotted as vectors.) DPI is short for "dots per inch."

Quality	Meaning
Draft	Plotted at the output device's lowest resolution.
Preview	Plotted at a maximum resolution of 150 dpi.
Normal	Plotted at a maximum of resolution 300 dpi.
Presentation	Plotted at a maximum resolution of 600 dpi.
Maximum	Plotted at the output device's highest resolution.
Custom	Plotted at the resolution specified in the **DPI** text box

Note: Two-dimensional drawings cannot be rendered, or have hidden lines removed or visual styles applied.

You can see the effect of shading and hidden-line removal on 3D drawings in the plot preview window. Note that applying visual styles and renderings to drawings takes longer, and can slow the preview display and increase the plotting time.

Plot Options

The **Plot Options** area lists miscellaneous options.

Plot in Background

When the **Plot in Background** option is turned on, AutoCAD plots the drawing "in the background," allowing you to return to editing the drawing more quickly. (This is a replacement for autospooling, discussed later in this chapter.)

Always leave this option turned on, unless AutoCAD slows down unacceptably.

Plot Object Lineweights and Plot with Plot Styles

The **Plot with Lineweights** and **Plot with Plot Styles** options toggle each other; you can have one or the other, neither, but not both:

- **Plot with Lineweights** plots drawings with lineweights, *if* lineweights are turned on, and *if* lineweights have been assigned to layers and objects.

- **Plot with Plot Styles** plots drawings with plot styles, again only if defined and turned on.

Lineweights are described in Chapter 7, "Changing Object Properties," and plot styles are discussed later in this chapter.

Plot Paperspace Last

AutoCAD normally plots paper space objects before model space objects. Model space objects include viewport borders and title blocks. The **Plot Paperspace Last** option reverses the order.

Why does this matter? We're not sure, but perhaps it has to do with how color printers lay down the ink. This option is not available when plotting from the model tab, because it has no paperspace elements.

The technical editor notes that the following options also affect the plotted output: there are differences in how Hidden modes hide circles with thickness (some look like a can, others like a pipe), and the how values of the **DISPSILH** system variable create different looks.

Hide Paperspace Objects

The **Hide Paperspace Objects** option determines if hidden-line removal applies to 3D objects in paper space. This allows drawings not to show hidden-line removal, yet have hidden-line removed for the plot. The technical editor is unsure why anyone would want 3D objects in paper space, but here's how to do it:

1. In model space, create a 3D solid model with internal details, such as slots and holes.
2. From the **Edit** menu, select **Copy**, and then copy the entire model.
3. Switch to a layout tab, and the paste the model (**Edit | Paste**).
4. Use the preview command to notice the differences:
   - **Hide Paperspace Objects** = off — solids in paper space appear as wireframes, showing internal details.
   - **Hide Paperspace Objects** = on — solids in paper space have hidden lines removed.

This option is overridden by the **Shade plot** option of viewports, and is not available when plotting from the model tab.

Plot Stamp On

The **Plot Stamp On** option toggles plot stamping on plotted drawings. See the **PLOTSTAMP** command later in this chapter.

Save Changes to Layout

When turned on, the **Save Changes to Layout** option stores the changes made in this dialog box with the current layout.

The next time you plot from this layout tab, the same settings appear in the Plot dialog box. Select a different layout or drawing, and the settings are different.

Select the Orientation

Larger CAD drawings are usually plotted in *landscape* mode, with the long edge of the page oriented horizontally. To do this, choose the **Landscape** button in the **Drawing Orientation** area.

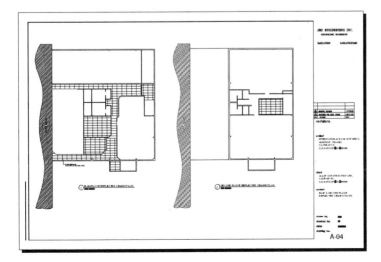

Left: D-size drawing plotted in landscape mode.
Right: A-size drawing plotted in portrait mode.

Smaller drawings are often printed in *portrait* mode, where the long edge is upright. Choose the **Portrait** button.

The **Plot upside-down** option is handy if you stuck the paper with the title block the wrong way around in the printer.

AUTOSPOOL

AutoCAD has an "autospool" option that lets it *spool* plots as an alternative to background plots. Spooling is set up through the Add a Plotter wizard's Port page.

Spooling is a technique (short for "simultaneous peripheral operations online") that speeds up printing jobs. When printing a document or plotting a drawing, the print data is sent to a file on disk; shortly thereafter, the spooling software starts up automatically, and sends the print data from the file to the printer. It is faster to save a file to disk than to print, so the application finishes the print job faster. (Spooling is also known as *buffering*, and in Windows it is handled by the Print Manager.)

AutoCAD allows you to use independent software for plot spooling. Before using the software, however, you must set up AutoCAD:

1. Install and configure the spooler software according to the vendor's instructions.
2. AutoCAD needs to know the name of the spooler, as well as the folder in which to place the spool files.

   From the **Tools** menu, select **Options**, and then the **Files** tab:
   - Open **Print File, Spooler, and Prolog Section Names**, and in **Print Spool Executable**, specify the spooler program name.

   - Open **Printer Support File Path**, and in **Print Spool File Location**, specify the folder name.

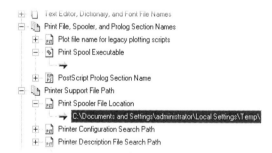

   Choose **OK**.
3. Set up the plotter by selecting **File | Plotter Manager**, and then working through the wizard.
4. Start the **PLOT** command, and then select the plotter from the **Plotter Configuration** droplist.

OPTIONS

The **OPTIONS** command allows you to set the default for plotting. From the **Tools** menu, select **Options**, and then choose the **Plot and Publish** tab.

PLOT AND PUBLISH TAB

This tab has settings specific to plotting and publishing drawings. "Plotting"means to plot one drawing at a time, while "publishing" means plotting sets of drawings (two or more).

Default Plot Settings for New Drawings

Here you specify the default plotter:

⊙ **Use As Default Output Device** selects the default printer from a list of system printers and devices defined by AutoCAD's *.pc3* (printer configuration) files. It will be used for new drawings, as well as those saved in versions prior to AutoCAD 2000.

○ **Use Last Successful Plot Setting**s selects the last-used plotter settings, instead of the default printer.

Click the **Add or Configure Plotters** button to display the Plotter Manager window; see the **PLOTTERMANAGER** command later in this chapter.

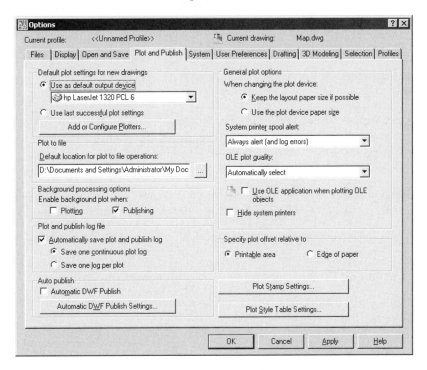

Plot To File

Default Location for Plot to File Operations specifies the drive and folder for storing drawings plotted to files.

Type the path, or click the **...** button to display the Select Default Location for All Plot-to-File Operations dialog box.

Background Processing Options

Background plotting lets you work on drawings while AutoCAD generates the plot. When **PLOT**, -**PLOT**, **PUBLISH**, and **-PUBLISH** are used in scripts, this setting is ignored and drawings are plotted and published in the foreground.

☐ **Plotting** performs background plot jobs; see the **PLOT** command in this chapter.

☑ **Publishing** performs background publish jobs; see the **PUBLISH** command.

Plot and Publish Log File

The log file saves plotting and publishing results in a CSV text file (comma-separated value) that can be imported into spreadsheet and database programs. Data include Job ID, Job name, Sheet set name, Category name, Date and time started and completed, Sheet name, Full file path, Selected layout name, Page setup name, Named page setup path, Device name, Paper size name, and Final status.

☑ **Automatically Save Plot and Publish Log** creates log files:

⊙ **Save One Continuous Plot Log** stores all plot data in a single log file.

○ **Save One Log File Per Plot** stores data about each plot in a separate log file.

 **AutoPublish**

AutoCAD can save the drawing in DWF format at the same time as it is saved in DWG format. Turn on this option when you work with DWF files, and need them to be up to date. Otherwise, leave this option turn off, because prolongs the save-file process.

☐ **Automatic DWF Publish** saves drawings in DWF format during the SAVE and CLOSE commands.

Click **Automatic DWF Publish** button to configure this feature. You can toggle auto-DWF publishing with the AUTOPUBLISH command.

General Plot Options

When changing the plot device:

⊙ **Keep the Layout Paper Size If Possible** uses the paper size specified by the plot setup. When the printer cannot handle the specified paper size, AutoCAD displays warning message, and then uses the size of paper specified by the PC3 plotter configuration file, or by the system printer's settings.

○ **Use the Plot Device Paper Size** ignores the size specified by the layout, and uses the paper size specified by the plotter's PC3 configuration file or the default system settings.

System Printer Spool Alert determines the type of alert displayed by port conflicts during plot spooling:

• **Always Alert (And Log Errors)** displays an alert and records the error to the log file.

• **Alert First Time Only (And Log Errors)** displays an alert the first time only, and then records the error to the log file.

• **Never Alert (And Log First Error)** never displays alerts, but records error to log files.

• **Never Alert (Do Not Log Errors)** never displays alerts and does not record to log files.

OLE Plot Quality specifies the quality of plotted OLE objects:

• **Monochrome** is suitable for black text.

• **Low Graphics** is suitable for colored test and simple graphics.

• **High Graphics** is meant for photographs.

• **Automatically Select** allows AutoCAD to choose the setting.

    ☐ **Use OLE Application When Plotting OLE Objects** launches the application that created the OLE object in order to improve the plot quality.

☐ **Hide System Printers** toggles the display of Windows system printers. System printers configured by the Add-a-Plotter wizard are always displayed in the Plot dialog box.

Specify Plot Offset Relative To

⊙ **Printable Area** offsets the plot relative to the printable area.

○ **Edge of Paper** offsets the plot relative to the edge of the paper.

Other Options

Plot Stamp Settings button opens the Plot Stamp dialog box; see the **PLOTSTAMP** command.

Plot Style Table Settings button opens the Plot Style Table Settings dialog box; see the **STYLESMANAGER** command.

FILE TAB

The Options dialog box's **File** tab specifies the default locations for plot-related files, as illustrated below:

SCALELISTEDIT

The **SCALELISTEDIT** command customizes lists of plot scales. The command also controls the list of scale factors displayed for viewports and page layouts and annotative scales. Use this command to shorten the list of unused scale factors and to add custom scales.

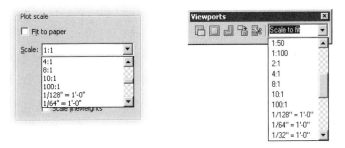

Left: *Scale factors listed in Plot dialog box.*
Right: *The same scale factors listed in Viewports toolbar.*

1. To edit scale factors, start the **SCALELISTEDIT** command:
   • From the **Format** menu, choose **Scale List**.

   • At the 'Command:' prompt, enter the **scalelistedit** command.

   Command: **scalelistedit** *(Press* ENTER.*)*

   AutoCAD displays the Edit Scale List dialog box, which lists the scale factors currently defined.

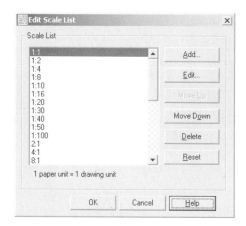

2. To add a new scale factor, click **Add**; to edit existing scale factors, select one, and then click **Edit**.

Name Appearing in Scale List — provides descriptive names that appear in place of scale factors; alternatively, enter an actual scale factor, such as 1:10.

Paper Units — specifies the size of objects on paper. This value is usually 1 (one) for drawings larger than the paper.

Drawing Units — specifies the size of objects in the drawing. This value is usually 1 for drawings smaller than the paper.

3. The **Move Up** and **Move Down** buttons move selected scale factors up and down the list. That allows you to place the frequently-used scales at the top of the list.

4. The **Delete** button removes the selected scale, while the **Reset** button deletes all custom scales, and then restores the default list.

5. Click **OK** to exit the dialog box.

The **-SCALELISTEDIT** command performs the same functions at the command line:

Command: **-scalelistedit**

Enter option [?/Add/Delete/Reset/Exit] <Add>: *(Enter an option.)*

 **PLOTSTAMP**

The **PLOTSTAMP** command stamps the plot with information about the drawing and plot.

When you plot sets of drawings, it sometimes becomes difficult to determine which set was plotted most recently, or which drawing files produced the plots. To identify the plots more readily, apply *plot stamps*. A plot stamp consists of one or two lines of text that list the drawing name, date and time plotted, the plotter device, plot scale, name of the computer that generated the plot, and so on. In addition, you may specify two custom pieces of data.

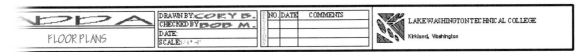

D:\CAD\AutoCAD 2007\Sample\Blocks and Tables - Imperial.dwg, D-size Plot, 5/13/2006 9:29:15 PM,
Administrator, DWF6 ePlot.pc3, 1:1

Plot stamp (below the title block) that provides information about the drawing and the plot.

Plot stamps are often placed along the edge of the drawing, as illustrated above. You can change the location, as well as the font and size of text. The plot stamp parameters can be saved to files, which is useful for billing clients, seeing who is hogging the plotter, and so on.

You access plot stamps with the **PLOTSTAMP** command, or from the **Plot Stamp Settings** button of the Plot and Publish dialog boxes.

TUTORIAL: STAMPING PLOTS

1. To apply a stamp to the plotted drawing, start the **PLOTSTAMP** command:
   - In the Plot or Publish dialog boxes, choose the **Plot Stamp Settings** button.

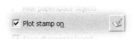

   - At the 'Command:' prompt, enter the **plotstamp** command.

     Command: **plotstamp** *(Press ENTER.)*
   - Alternatively, enter the **ddplotstamp** (the command's old name) alias at the 'Command:' prompt.

   In all cases, AutoCAD displays the Plot Stamp dialog box.

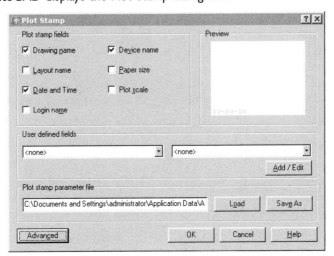

2. In the **Plot Stamp Fields** area, select the text you wish stamped on the plot:

Plot Stamp Fields	Example	Comments
Drawing Name	C:\17_22.DWG	Full path and file name of the drawing.
Layout Name	Model	Layout name; "Model" if plotted from model space.
Date and Time	6/26/2003 10:24:42 AM	Date and time of the plot; format is determined by the Regional Settings dialog box of the Control Panel.
Login Name	Administrator	Windows login name, as stored in the LOGINNAME sysvar.
Device Name	Lexmark Optra R+	Name of the plotting device.
Paper Size	Letter 8 ½ x 11 in	Size of the paper.
Plot Scale	1:0.8543125	Plot scale factor; when you see an unusual scale factor, such as that shown in the example, this means the drawing was probably scaled to fit the margins.

3. To define your own fields, click the **Add/Edit** button. AutoCAD displays the User Defined Fields dialog box.

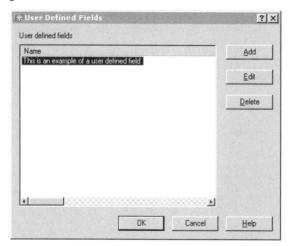

Click **Add**, and then enter text. You can add as many user-defined fields as you wish, but AutoCAD specifies a maximum of two per plot stamp. When done, click **OK**.

To include a user-defined field with the plot stamp, select them from the droplist under **User defined** fields.

4. Click **Save As** to save the settings to file. This allows you to reuse the settings with other drawings, or swap with friends.

AutoCAD displays the Plotstamp Parameter File Name dialog box. Enter a file name, and then click **Save**. AutoCAD saves the data in a *.pss* (plot stamp parameter) file.

5. Click **OK**.

Notes: When the options of the Plot Stamp dialog box are grayed out, this means that the *inches.pss* or *mm.pss* file in AutoCAD's *\support* folder is set to read-only. To change the setting with Windows Explorer: (1) right-click the *.pss* file, (2) select **Properties**, (3) uncheck **Read-only**, and (4) click **OK**.

The plot stamp data is not saved with the drawing, but is generated anew with each plot. Plot stamps are plotted with color 7 (on raster plotters) or pen 7 (on pen plotters).

STAMPING PLOTS: ADVANCED OPTIONS

The Advanced Options dialog box provides additional options for locating the stamp on the plot. In the Plot Stamp dialog box, click **Advanced**.

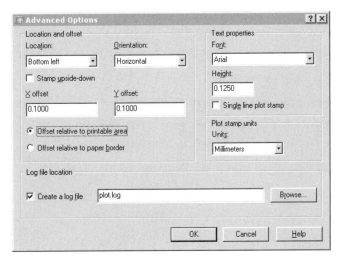

The **Location and Offset** area positions the stamp on the plot. The preview, unfortunately, is not located in this dialog box, so you need to go through a cycle: change settings, click **OK** to get out of this dialog box, check the plot preview, and return.

The **Stamp upside-down** option is useful, because otherwise the stamp can be confused for other text in the drawing. Also, some filing systems work better with upside-down plot stamps.

 Note: A large offset value may position the plot stamp text beyond the plotter's printable area, which may cause the text to be cut off. To prevent this, use the **Offset relative to printable area** option.

The **Text Properties** section selects the font and size of text. I recommend a narrow font, such as Arial Narrow or Future Condensed, to ensure all the plot stamp text fits the page. The default size, 0.2", tends to be too large; I recommend 0.1" or smaller.

Plot stamp units defines the offset and height numbers used by this dialog box. Select from inches, millimeters, and pixels.

The **Log File Location** area lets you specify the name and folder for the plot logging file. The *.log* file is in ASCII format, and contains exactly the same information as stamped on the plot, such as:

C:\17_22.DWG,Model,6/26/2003 12:34:48 PM,Administrator,Lexmark Optra R plus.

Click **OK** to exit the dialog box.

 ## PAGESETUP

The **PAGESETUP** command assigns plotters to layouts.

AutoCAD automatically assigns the default Windows system printer to new drawings. Right before plotting the drawing, you can select another printer in the **PLOT** command's dialog box. Or, you can assign a specific printer, such as your office's networked large-format plotter, as part of the initial setup for template drawings. This is done with the **PAGESETUP** command.

Page setups are lists of instructions that tell AutoCAD how to plot pages: model space, layouts, and sheets. (I think Autodesk uses the word "page," because AutoCAD prints these onto pages of paper.) The flexibility of page setups means that you can have a different page setup for every layout or sheet in a drawing. Each could be plotted by a different plotter on different size paper, and at different scales and orientations — in color or in monochrome. Page setup data are stored in the *.dwg* file.

The Page Setup Manager lists the layouts to which the page setup will be applied. This list can consist of a single layout name, or of all the layouts in the current drawing. (When opened from the Sheet Set Manager, it displays the name of the current sheet set.)

Page setups can be accessed in a number of places: in the Plot dialog box, from the Sheet Set Manager palette, and from the layout tabs.

TUTORIAL: PREPARING LAYOUTS FOR PLOTTING

1. To prepare drawings for plotting, start the **PAGESETUP** command:

    • From the **File** menu, choose **Page Setup Manager**.

    • Right-click a layout tab, and then select **Page Setup Manager** from the shortcut menu.

    • In the Sheet Set Manager, right-click a sheet, and then select **Publish | Manage Page Setups** from the shortcut menu.

    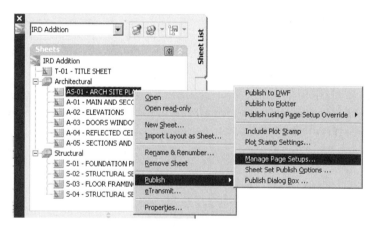

    • At the 'Command:' prompt, enter the **pagesetup** command.

    Command: **pagesetup** (*Press* ENTER.)

    AutoCAD displays the Page Setup Manager, which lists page setups, if any are already defined.

The **Display When Creating a New Layout** option displays this dialog box each time you create a new layout tab.

2. Look in the lower half of the dialog box. Notice that no plotter has been selected for this layout.

   To specify a plotter, click **Modify**.

   Notice the Page Setup dialog box, which looks almost identical to the Plot dialog box.

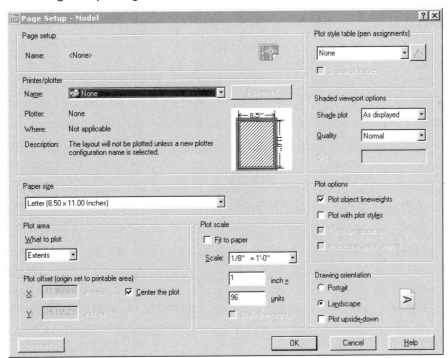

3. In this dialog box, you can preset many options for future plotting. The most important, however, is specifying the printer or plotter to use.

   In the **Printer/Plotter** area, select a printer from the **Name** droplist.

   The names of printers in the droplist vary with every computer, and include (1) Windows printer drivers, (2) AutoCAD plotter drivers, and (3) attached network printer drivers.

4. Set other parameters, if you wish. If you are unsure which settings to use, I recommend these:

Plot Area	**Extents**
Plot Offset	**Center the Plot**
Plot Scale	**Fit to Paper**

5. Click **OK**. Your drawing is now ready to plot.

PUBLISH

The **PUBLISH** command plots drawing sets. (In contrast, the **PLOT** command plots single drawings.)

A *drawing set* consists of one or more drawings and layouts that are plotted, in order, at one time. Engineering and architectural offices often use drawing sets to combine all drawings belonging to a project. This makes it easier to plot many drawings at once, such as at specific stages large projects. Alternatively, drawing sets can be generated as *.dwf* or *.plt* plot files for archiving or plotting later.

This command allows you to select the drawing files and layouts, and then reorder, rename, or copy them. This command also plots drawings. Lists of drawing sets can be saved as *.dsd* files (drawing set description) for later reuse. (The **BATCHPLT** utility has been removed from AutoCAD; its functions are now performed by the **PUBLISH** command.)

TUTORIAL: PUBLISHING DRAWING SETS

1. To create a drawing set, start the **PUBLISH** command:
   - From the **File** menu, choose **Publish**.

   - From the Standard or Standard Annotative toolbars, choose the **Publish** button.

   - Right-click two or more layout tabs, and then select **Publish Selected Layouts** from the shortcut menu.

   - At the 'Command:' prompt, enter the **publish** command.

     Command: **publish** *(Press ENTER.)*

   In all cases, AutoCAD displays the Publish Drawing Sheets dialog box. Initially, the dialog box lists the names of the current drawing(s) and layout(s) open in AutoCAD.

2. If you want to add more drawings and layouts to the list, click the **Add Sheets** to add more drawings. AutoCAD displays the Select Drawings dialog box. Select one or more drawings in the manner of the **OPEN** command, and then click **Select**.
   When AutoCAD adds drawings and their layouts to the list, each layout is listed separately, with a suffix:
   - **Model** indicates the model tab (model space view).
   - *Layout name* indicates the layout tabs (paper space view).

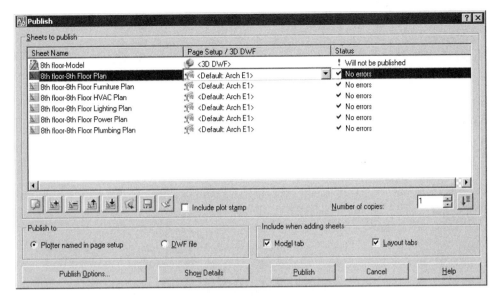

If layouts already appear in the list, AutoCAD prompts you to change their names:

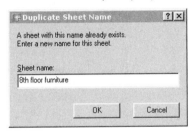

Change the name, and then click **OK**. (When you click **Cancel**, the layout is not added to the list.)

Note: Buttons with icons have these meanings:

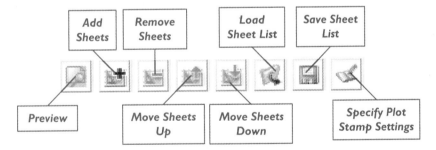

3. To remove layouts from the list, select one or more, and then click the **Remove** button.
   - Hold down the **SHIFT** key to select a range of layout names.
   - Hold down the **CTRL** key to select nonconsecutive layout names.

   AutoCAD removes the selected layouts without asking if you're sure.

4. Drawing sets are usually plotted in a specific order. Use the **Move Up** and **Move Down** buttons to change the order of the layouts.

5. The drawing sets can be "published" to the printer, or saved as a *.dwf* file for transmittal by email. Make the selection in the **Publish To** section:
   - **Plotter named in page setups** sends all of the layouts to the plotter. As the option name indicates, the layouts might not end up being plotted by the printer you expect. Prior to pressing the **Publish** button, ensure each layout is set up for the correct printer. This complexity is

necessary, because drawing sets sometimes need to be plotted to a variety of printers.

- **DWF file** saves the layouts in a *.dwf* file. You have the option of protecting the file with a password.

6. Click **Save List** to save the list of layouts for later reuse.

7. Click **Publish**.

   AutoCAD loads each drawing, and then generates the plot or the *.dwg* file. A dialog box notes its progress.

   If a drawing or layout cannot be found, AutoCAD skips it and carries on publishing the next layout. If errors are found, AutoCAD generates a *.csv* (comma separated value) log file, which you can read with a spreadsheet program.

When done, you are asked if you want to view the *.dwf* file in DWF Viewer, which is provided with AutoCAD.

BASIC OPTIONS

The Publish dialog box's user interface places many commands in toolbar icons, shortcut menus, and additional dialog boxes. Some commands are duplicated within these locations, as illustrated below.

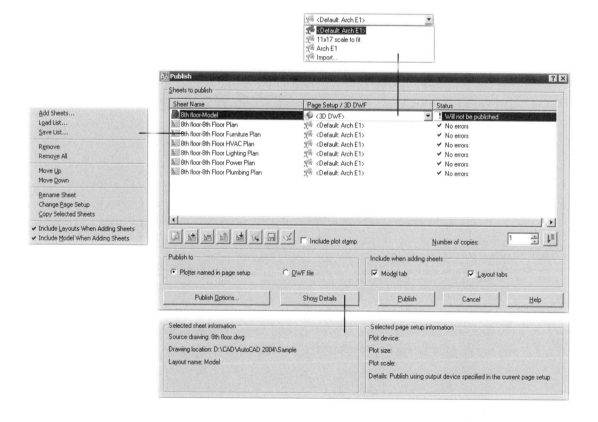

Shortcut Menu

Right-click the names of drawing layouts to access commands on the shortcut menu.

Shortcut Menu	Comment
Add Sheets	Adds "sheets" from other drawing files, which includes layouts and model-space drawings.
Load List	Loads a previously-saved *.dsd* drawing set description file.
Save List	Saves the current list of sheets as a *.dsd* file.
Remove	Removes the selected sheets from the list.
Remove All	Removes all sheets from the list.
Move Up	Moves the selected sheet up in the list.
Move Down	Moves the selected sheet down the list.
Rename Sheet	Changes the name of the selected sheet.
Change Page Setup	Selects another page setup name.
Copy Selected Sheets	Copies the selected sheets, adding the "-Copy(1)" suffix.
Include Layouts When Adding Sheets	Includes all layouts when adding drawings.
Include Model When Adding Sheets	Includes the model tab when adding drawings.

Page Setups Droplist

AutoCAD normally publishes each layout with its default page setup. You can change page setups: click a page setup name, and then select another one from the droplist.

Click **Import** to import page setups from other drawings. (Recall that page setups are stored in *.dwg* files.)

Plot Order

AutoCAD normally publishes the drawing set in the order listed by the dialog box. Some plotters output pages in reverse order. Click the **Currently publishing in default order** button to reverse the order in which layouts are plotted.

More Details

Click the **More Details** button to see more information about each sheet and its page setup.

After solemnly inspecting the details, click **Hide Details** to, you know, hide the details.

ADVANCED OPTIONS

Before publishing the drawing list in DWF format, click the **Publish Options** button. The Publish Options dialog box controls output for 2D and 3D DWF files.

Current User

The name that appears next to the icon in the upper-left corner varies. When this dialog box is accessed from the **PUBLISH** command, you see the Windows userid of the currently logged-in user, such as "Administrator"; when accessed from the Sheet Set Manager palette, you see the sheet set's name.

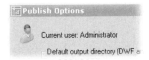 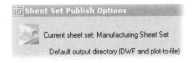

Left: *Current user name when dialog box accessed through* PUBLISH *command.*
Right: *Current sheet set name when dialog box accessed through* SHEETSET *command.*

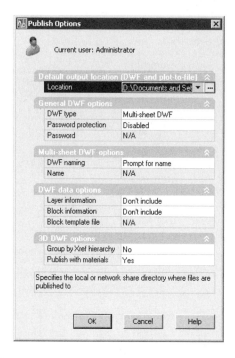

Default Output Directory

The **Default Output Directory** section determines where DWF and plot files are saved when you publish drawing sheets. This option is available only when saving in DWF format and when plotting to file.

Enter a different drive and folder name, or click the **...** button to display the Select Folder for Generated Files dialog box.

General DWF Options

The **General DWF Options** section selects whether files are published as a series of single-sheet DWFs, or as a single multisheet DWF.

DWF Type

Single-Sheet DWF — each sheet generates its own *.dwf* file. Use this option for compatibility with old DWF viewer software that does not accept DWF files in multisheet format.

Multi-Sheet DWF — all sheets are stored in a single DWF-format file. Use this option for the convenience of publishing to a single file.

Password Protection

DWF files are often distributed by email or on CD; adding a password prevents unauthorized people from viewing the files. You can use letters, numbers, punctuation, and non-ASCII characters for the password. The password is *case-sensitive*, which means that "autocad" is not the same password as "AutoCAD."

Disabled — password protection is turned off.

Specify Password — allows you to specify the password.

Prompt for Password — AutoCAD will prompt you for the password later, after you click the **Publish** button.

Password

The recipients must have the password to open the file; how they receive the password is up to you. If you forget the password, the DWF file cannot be recovered, although you can generate another DWF file from the original drawings.

Password — specifies the password, if the **Specify Password** option is enabled.

Multi-Sheet DWF Creation

When AutoCAD creates a single-sheet DWF, the *.dwf* file takes on the drawing's (sheet's) name. But what name should AutoCAD use when publishing more than one drawing? Here you specify the file name.

DWF Naming

Specify Name — allows you to enter the path and file name for the *.dwf* file.

Prompt for Name — AutoCAD will prompt you for the path and file name, after you click the **Publish** button.

Name — specifies the file name.

DWF Data Options

The **DWF Data** section includes options specific to DWF files.

Layer Information

Include — layer information is included; layers can be manipulated in DWF viewer software.

Don't Include — layer information is removed, and all objects appear to be on a single "layer."

Block Information

Include — blocks and attributes are included in the *.dwf* file.

Don't Include — blocks are included as exploded objects.

Block Template File

The *.blk* file defines which blocks and which attributes are included in the DWF file.

Create — creates a new *.blk* block template file; displays the Publish Block Template dialog box.

Edit — edits existing *.blk* files; displays the Publish Block Template dialog box.

3D DWF Options

The **3D DWF Options** section is available only when "<3D DWF>" is selected from the Publish dialog box's **Page Setup/3D DWF** column. And the <3D DWF> option is available only in model space. (AutoCAD does not generate 3D DWFs from layouts, because they are 2D only.)

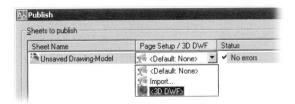

Layer Information

Group by Xref Hierarchy — objects are listed under each xref name, if externally-referenced drawings are linked to the sheet.

Publish with Materials — materials are included with objects. Materials are used by DWF Viewer.

Note: The technical editor suggests setting up the DWF plotter with a very large paper size. He finds that this slightly increases the file size with the benefit of dramatically higher resolution.

PLOTTERMANAGER

The **PLOTTERMANAGER** command creates and edits plotter configurations.

AutoCAD plots to any printer connected to your computer. This includes printers found in your office, large-format plotters used in engineering and architectural offices, and files on disk. (Large-format plotters create D- and E-size plots that are roughly three to four feet, or one meter, across.)

Most software relies on Windows to provide device drivers. The problem is that the device drivers provided by Microsoft are not accurate and flexible enough for AutoCAD's high-precision plotting. For this reason, Autodesk includes its own set of drivers, known as HDI.

To plot drawings with HDI drivers, first run the Plotter Manager. This configures in great detail how the plotters should print the drawings.

TUTORIAL: CREATING PLOTTER CONFIGURATIONS

1. To create configurations for plotters, start the **PLOTTERMANAGER** command:
   - From the **Tools** menu, choose **Wizards,** and then **Add Plotter.**
   - From the **File** menu, select **Plotter Manager,** and then double-click the **Add-A-Plotter Wizard** icon.

2. AutoCAD starts the *addplwiz.exe* program, which displays the Introduction Page.

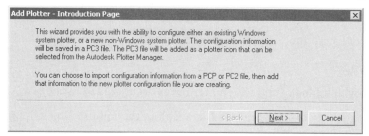

3. Choose **Next.**

4. In the Begin page, set up AutoCAD with one of these styles of printer:

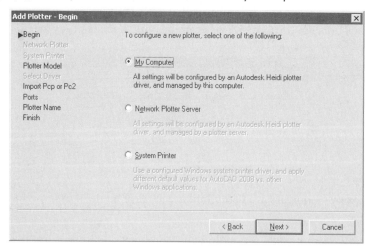

- **My Computer** are local printers and plotters controlled by AutoCAD's plotter drivers.
- **Network Plotter Server** are printers located on the network, also controlled by AutoCAD's plotter drivers.
- **System Printer** are local and network printers controlled by Windows printer drivers.

Unless you have a reason to do otherwise, choose **My Computer,** and then click **Next.**

5. AutoCAD displays a list of printer and plotter drivers provided by Autodesk.

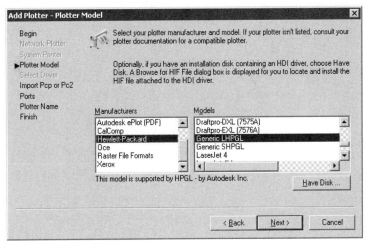

From the **Manufacturers** list, select the brand name of the plotter.

From the **Models** list, select the specific model number.

If your new plotter comes with AutoCAD-specific drivers on a diskette or CD, choose the **Have Disk** button.

(If you had selected Network Plotter Server, AutoCAD would have prompted you for the network location of the server. If System Printer, AutoCAD would have prompted you to select one.)

Choose **Next**.

Notes: If you do not see the name of your plotter's manufacturer, check the documentation. Often it lists brand names and model numbers of compatible plotters. If you cannot find this information, try selecting Adobe for PostScript printers, and Hewlett-Packard for large-format inkjet plotters.

To plot the drawing in a raster format, select **Raster File Formats**, and then a specific format:

Format	Color Depths	File Extension
Independent JPEG Group JFIF	Gray, RGB	.jpg
Portable Network Graphics PNG	Bitonal, gray, indexed, RGB, RGBA	.png
TIFF Uncompressed	Bitonal, indexed, gray, RGB, RGBA	.tif
TIFF Compressed	Bitonal, indexed, gray, RGB, RGBA	.tif
CALS MIL-R-28002A Type 1	Bitonal	.cal
Dimensional CALS Type 1	Bitonal	.cal
MS-Windows BMP Uncompressed	Bitonal, gray, indexed, RGB	.bmp
TrueVision TGA 2.0 Uncompressed	Indexed, gray, RGB, RGBA	.tga
Z-Soft PC Paintbrush PCX	Indexed, RGB	.pcx

6. The Import PCP or PC2 page is important only if you created plotter configuration files with AutoCAD Release 13 (*.pcp*) and 14 (*.pc2*). Here you import these files for use with newer releases of AutoCAD (as *.pc3* files).

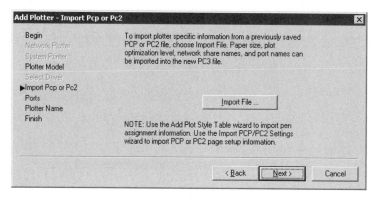

Choose **Next**.

7. The Ports page is the most difficult step for some users. Here you select the *port* to which the plotter is connected — or no port at all.

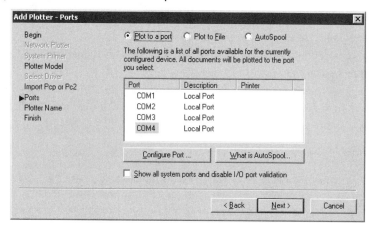

Your choices are:

- **Plot to a Port** — AutoCAD sends the drawing to the plotter through a local port or network port, such as a parallel port (designated LPT), serial port (designated COM), or USB port. Select this option for drawings plotted by printers and plotters.

- **Plot to File** — AutoCAD sends the drawing to a file on disk. Select this option to save the drawing to disk as a file in both plotter and raster formats.

- **AutoSpool** — AutoCAD sends the drawing to a file in a specified folder (defined by AutoCAD's Options dialog box), where another program processes the file. Select this option only if you know what you are doing. (If you need to ask, then you don't know.)

8. Some ports have further options. If necessary, choose the **Configure Port** button:

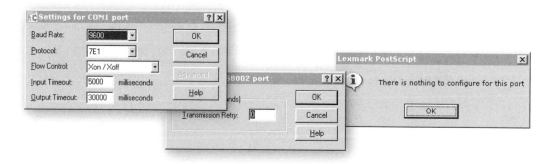

- **Serial** ports specify communications settings.

- **Parallel** and **USB** ports specify the transmission retry time.

- **Network** ports specify nothing, curiously enough.

Change settings, and then click **OK**.
Choose **Next**.

9. In the Plotter Name page, give the plotter configuration a name.

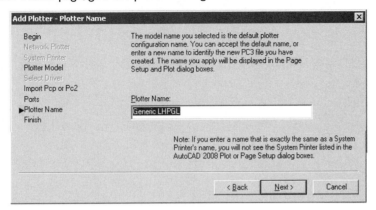

Enter a descriptive name, and then choose **Next**.

10. The Finish page has three buttons:
    - **Edit Plotter Configuration** — Displays AutoCAD's Plotter Configuration Editor dialog box, which allows you to specify options, such as media source, type of paper, type of graphics, and initialization strings.

    - **Calibrate Plotter** — calibrates the plotter. This allows you to confirm, for example, that a ten-inch line is indeed plotted 10.0000000 inches long.

    - **Finish** — completes the plotter configuration.

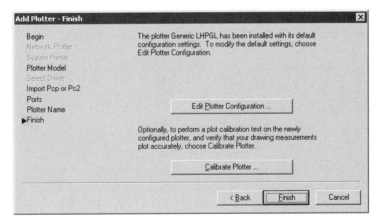

11. Choose **Finish**.

This configures the new plotter for AutoCAD. When you next use the **PLOT** command, this configuration appears in the list of available plotters.

IMPORTING OLDER CONFIGURATION FILES

As Autodesk added more capabilities to plotting, it also provided a way to import older plotter configuration files into new releases of AutoCAD. The table below, taken from an AutoCAD dialog box, summarizes the paths that configuration files can take:

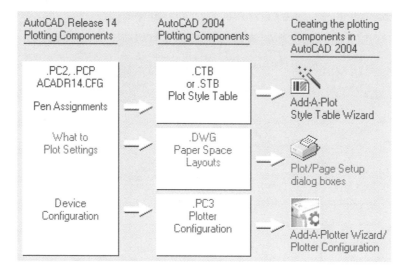

TUTORIAL: EDITING PLOTTER CONFIGURATIONS

1. To edit a plotter configuration, start the **PLOTTERMANAGER** command:
    - From the **File** menu, choose **Plotter Manager**.
    - At the 'Command:' prompt, enter the **plottermanager** command.

      Command: **plottermanager** (Press ENTER.)

    In all cases, AutoCAD requests that Windows display the Plotters window, which lists all HDI plotter configurations as *.pc3* files.

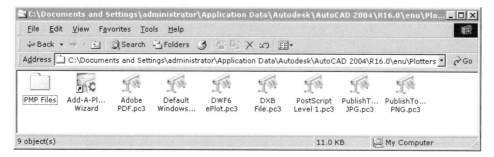

2. Double-click any *.pc3* icon, except *Default Windows System Printer.pc3*, because it must be edited through Windows.

3. Windows displays the Plotter Configuration Editor dialog box.
    Click the **Device and Settings** tab.

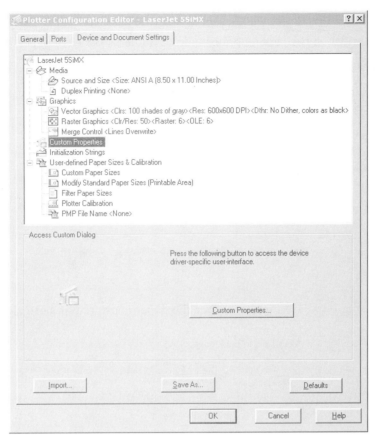

The available settings vary, depending on the capabilities of the plotter. (Shown below are settings for the HP LaserJet 5SiMX printer.)

Some settings are unavailable when the plotter does not support them. Other settings are handled through the **Custom Properties** button.

Selected options are noted in angle brackets, <like this>. When a change is made to a setting, AutoCAD places a red checkmark in front of it.

4. Change the settings, and then press **OK**.

THE PLOTTER CONFIGURATION EDITOR OPTIONS

The Plotter Configuration Editor lists features and capabilities that AutoCAD takes advantage of. Not all settings listed below are available for all printers and plotters. Some features must be set through Windows: select the **Custom Properties** node, and then click the **Custom Properties** button.

Media

The **Media** option specifies the paper source and destination, as well as the size and type of paper.

Source specifies the location of the paper. Examples include trays, sheet feed, and roll feed.

Width specifies the width of the paper; applies only to roll-fed sources of paper.

☐ **Automatic** means that the printer determines the paper source.

Size lists the standard and custom paper sizes the printer is capable of handling. **Printable Bounds** lists the margin measurement for the selected paper size.

Media Type specifies the type of paper.

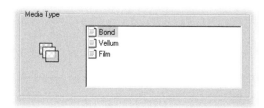

Duplex Printing

The **Duplex Printing** option determines whether the paper is printed on both sides.

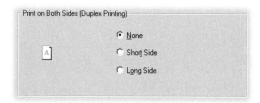

⊙ **None** turns off double-sided printing.

O **Short Side** sets the binding margin on the short edge.

O **Long Side** sets the binding margin on the long edge.

Media Destination

The **Media Destination** option handles paper activities after the plot, such as cutting, collating, and stapling.

Physical Pen Configuration

The **Physical Pen Configuration** option is for pen plotters only; it is largely obsolete, or nearly so. It controls the actions of each pen.

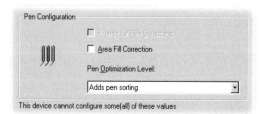

☐ **Prompt for Pen Swapping** forces AutoCAD to pause the plot so that you can change pens; available for single-pen plotters only.

☐ **Area Fill Correction** forces AutoCAD to plot filled areas by half-a-pen width narrower, so as not to draw filled areas too wide.

Pen Optimization Level optimizes pen motion to reduce total plotting time. This setting is not effective with very slow computers and slow plotters.

Physical Pen Characteristics

Physical Pen Characteristics allows you to specify the color, width, and manufacturer-recommended top speed of each pen (in inches or millimeters per second).

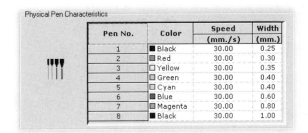

Graphics

Graphics options specify options for vector and raster graphics and TrueType fonts. The settings here vary depending on the capabilities of each printer and plotter.

Installed Memory reports to AutoCAD the added memory (RAM or hard disk) residing in the printer.

Total Installed Memory reports the total amount of memory installed in the printer.

Resolution and Color Depths / Vector Graphics

The **Resolution and Color Depths** section (a.k.a. Vector Graphics) specifies options for color depth (number of colors), resolution (dpi), and dithering (simulating a color by mixing two colors). Some printers trade off fewer colors for higher resolution.

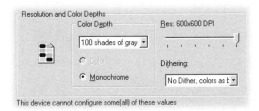

The **Raster Images** option trades off plotting speed for output quality (for plotters without pens).

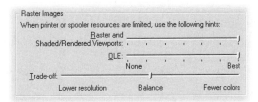

The **TrueType Text** option specifies whether to plot TrueType text as:

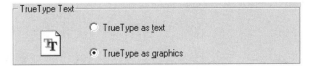

- **TrueType as text** — plotted as vector outlines; faster, but may not print with correct fonts. Only topmost lines are visible at intersections.

- **TrueType as graphics** — plotted as graphic images; slower, but guarantees the look of the fonts.

The **Merge Control** option specifies the look of crossing lines. This option is not valid when printing all colors as black, when using PostScript printers, and when the printer does not support merge control.

- **Lines Overwrite** — only topmost lines are visible at intersections.
- **Lines Merge** — colors of crossing lines are merged.

Custom Properties

The **Custom Properties** option displays the **Custom Properties** button. Click the button to view the Windows dialog boxes that control the printer. An example is illustrated below at right.

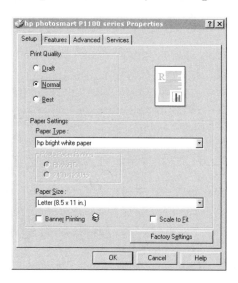

Initialization Strings

The **Initialization Strings** option sets control codes for non-system printers, including those for pre-initialization, post-initialization, and termination. These codes allow you to use plotters and printers not supported entirely by AutoCAD.

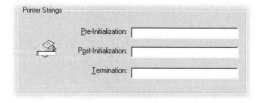

Note: Use the backslash (\ ) to emulate escape strings. For example, **\27** is sent as the escape character, and **\10** as the line-feed character.

User-Defined Paper Sizes and Calibration

The **User-Defined Paper Sizes and Calibration** option allows you to specify custom paper sizes (those not recognized by AutoCAD) and calibrates the printer. To help you define custom paper sizes, AutoCAD runs the Custom Paper Size wizard.

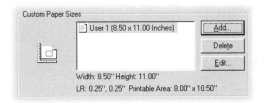

 Note: Calibration ensures drawings are plotted accurately. Autodesk recommends, "If your plotter provides a calibration utility, it is recommended that you use it instead of the AutoCAD utility."

Following printer calibration, AutoCAD stores the calibration data in a *.pmp* (plotter model parameter) file. The *.pmp* file must be attached to the printer's *.pc3* file, unless it was created during the Calibration stage of the Add-a-Plotter wizard.

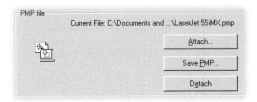

Save As

To save the settings, click the **Save As** button. Share the *.pc3* file with other AutoCAD users through the **Import** button.

STYLESMANAGER

The **STYLESMANAGER** command creates and edits plot style tables.

Plot styles control the printing properties of objects. Depending on the capabilities of the printer, plot styles control color or grayscale, dithering and screening, pen number and virtual pens, linetype and lineweight, line end and join styles, and fill style. Plot style *tables* are collections of plot styles assigned to layouts and the model tab.

AutoCAD works with two types of plot style table: color-dependent and named.

Color-dependent plot style tables (*.ctb* files) assign plot styles by the color of the layer or the objects, if ByLayer color has been overridden. (Prior to AutoCAD 2000, this was the only way to control printing. During the old **PLOT** command, the CAD operator assigned AutoCAD colors to pen numbers. For example, blue may have been assigned pen #3. During plotting, AutoCAD instructed the plotter to plot all blue-colored objects with whatever pen was in pen holder #3. Pens could be assigned linetypes, colors, and widths, depending on the capabilities of the plotter.)

Color-dependent plot style tables have exactly 256 plot styles, one for each ACI color (AutoCAD Index Color). AutoCAD cannot accurately plot true colors (16.7 million) and color books (DIC, Pantone, and RAL) with these style tables; if you must, however, use the **Use Entity Color** option to plot with the ACI color nearest to the true color

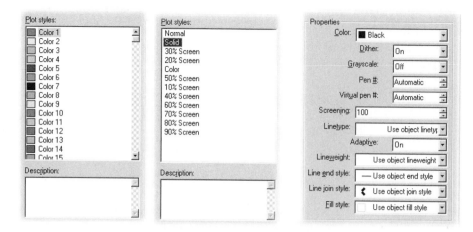

Left: *Color-based plot style names that are limited to the 256 ACI colors.*
Center: *Named plot styles that are unlimited.*
Right: *Properties that can be set for each plot style.*

Note: If you are not sure whether a drawing uses color-dependent or named plot styles, look at the **Plot Style Control** droplist on the Styles toolbar. If it is grayed out, color-dependent styles are in effect; if not, named plot styles are in effect.

Left: *Grayed-out droplist indicating color-based plot styles are in effect.*
Right: *Indicating named plot styles are available.*

Named plot style tables (*.stb* files) contain named plot styles assigned to layers and objects. This is much more flexible than color-dependent styles, because it is not limited to controlling objects by their color. Instead, each and every layer and object can have its own plot style. You can assign a different named STB to each layout in the drawing.

CONVERTCTB AND CONVERTPSTYLES

You can switch drawings between color-dependent and named plot style tables with the CONVERTPSTYLES command (short for "convert plot styles").

- **From color-dependent to named** — removes CTB color-dependent plot style tables from layouts, and replaces them with named ones. Color-dependent plot style tables should be first converted using these steps:

  Step 1: The **CONVERTCTB** command converts color-dependent plot style tables to named tables. Select a *.ctb* file from the dialog box, and then specify the name for the *.stb* file. The 256 converted plot styles are given generic names: Style1, Style2, and so on.

  Step 2: The **STYLESMANAGER** command renames the generic plot styles. Styles should be renamed *before* being attached to layouts.

  Step 3: The **CONVERTPSTYLES** command switches the drawing to named plot styles. The command has no options, but displays a warning:

Click **OK**, and a moment later AutoCAD confirms the conversion:

Drawing converted from Named plot style mode to Color Dependent mode.

- **From named to color-dependent** — any STB plot style names assigned to layers and objects in the drawing are erased. Using the **CONVERTPSTYLES** command results in a message similar to that shown above, along with the confirmation message:

Drawing converted from Color Dependent plot style mode to Named mode.

Although it appears that the UNDO command reverses the conversion, named plot styles are labeled "missing" and are not returned.

Setting Default Plot Styles

In almost all cases, your drawings should use named plot styles, because they have more benefits than color-based ones. This comparison lists the benefits:

Color-based Tables	Named Tables
256 plot styles.	Only as many plot styles as your drawings need.
	Benefit: Fewer plot styles to deal with.
Generic style names, such as "Color1" and "Color2."	Descriptive style names, such as "50% Screening."
	Benefit: Easier to understand style names.
Object colors control the plots.	Object color independent of plotting.
	Benefit: More versatile use of colors in drawings.
Single plot style for all layouts.	Every layout can have its own plot style.
	Benefit: Drawing can use layouts for different plotters.

Despite the drawbacks to color-based plot styles, they remain the default setting for new drawings. To change the default plot style table for new drawings, enter the **OPTIONS** command, and then select the **Plotting** tab. Select color or named plot style, and then a default table. Click **OK** to exit the dialog box, and then save the template file.

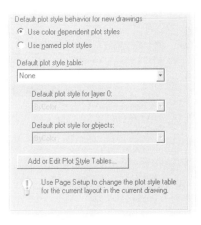

Assigning Plot Styles

Named plot styles can be assigned to objects, layers, and layouts. Like colors and linetypes, plot styles can be **ByLayer** and **ByBlock**. *ByLayer* means objects take on the plot style assigned to the layer; change the plot style, and all objects on the layer change. *ByBlock* means objects in a block take on the plot style assigned to the block. In addition, a plot style can be *Default*, which means objects take on the plot style assigned to the layout.

Objects

To assign a plot style to one or more objects, select them in the drawing, and then select the named plot style from the **Plot Style Control** droplist on the Properties toolbar. Click **Other** to see a list of all plot styles in the table.

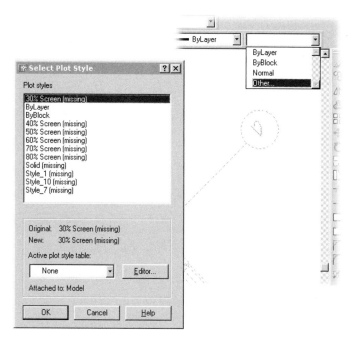

Alternatively, you can assign plot styles to objects with the Properties window through the **Plot Style** droplist.

Layers

To assign a plot style to layers, start the LAYER command.

In the dialog box, select one or more layers, and then select a named plot style from the **Plot Style** column. Repeat for other layers and plot styles.

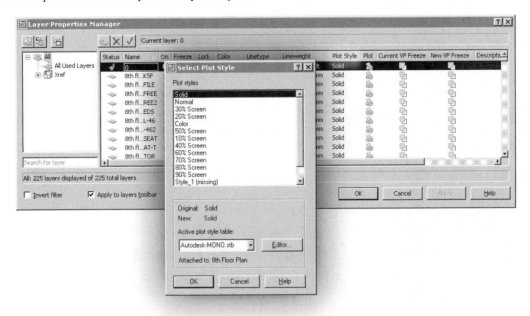

Click **OK** to exit the dialog box. All objects assigned to that layer now take on the same plot style.

Layouts

Assigning plot styles to layouts takes two steps. In the first step, right-click the Model tab, and then select **Page Setup Manager** from the shortcut menu (PAGESETUP command). In the Page Setup Manager, click **Modify**.

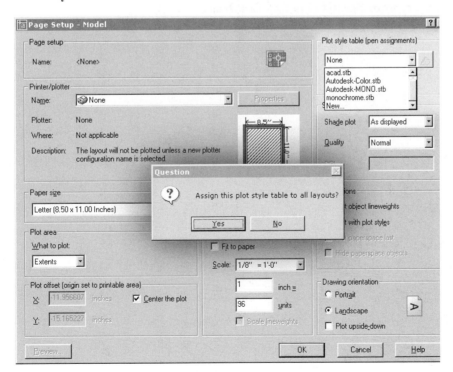

Select a **Plot Style Table** droplist. AutoCAD asks whether to apply the table to all layouts. You can answer yes or no:

- **Yes:** the same plot style table is assigned to the model tab and all layouts in the drawing.

- **No:** this plot style table is assigned only to the model tab.

Click **OK** to exit the dialog box.

If you want different plot style tables for each layout, then you have to take the second step. Select a layout tab. If AutoCAD does not automatically display the Page Setup Manager again, right-click the layout tab, and select **Page Setup Manager**. Notice that the **Plot style table** is "None." Select a table from the **Name** droplist, and then turn on the **Display plot styles** option. Click **OK** to exit the dialog box.

Repeat for each layout tab.

TUTORIAL: CREATING PLOT STYLES

1. To create new plot styles, start the **stylesmanager** command:
   - From the **File** menu, choose **Plot Style Manager**, and then select **Add-A-Plot Style Table Wizard**.

   - From the **Tools** menu, choose **Wizard**, and then **Add Plot Style Table**.

   - At the 'Command:' prompt, enter the **stylesmanager** command:

     Command: **stylesmanager***(Press* ENTER.*)*

     In the window, double-click **Add-A-Plot Style Table Wizard**.

   In all cases, AutoCAD runs the Add Plot Style Table "wizard," which leads you through the steps for creating a new plot style table.
   Click **Next**.

2. In the Begin dialog box, AutoCAD asks for the kind of plot style table to create. Select **Start from Scratch**, and then click **Next**.

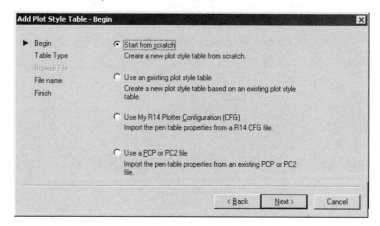

3. In the Table Type dialog box, select **Named Plot Style Table**.

Click **Next**.

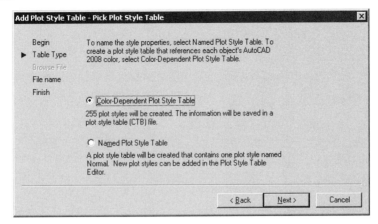

4. In the File Name dialog box, enter a descriptive name for the style table.
   Click **Next**.

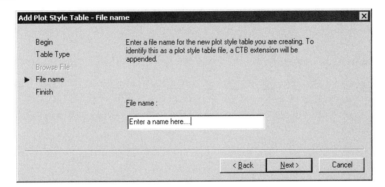

5. In the Finish dialog box, click **Finish**.

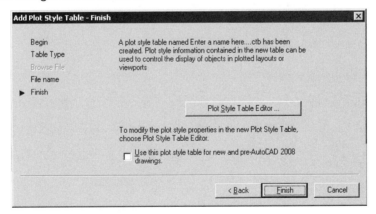

AutoCAD adds the *.stb* style table file to its collection stored in the *\plot styles* folder.

TUTORIAL: EDITING PLOT STYLES

1. To edit plot styles, start the **stylesmanager** command:
   * From the **File** menu, choose **Plot Style Manager**.

   * At the 'Command:' prompt, enter the **stylesmanager** command.

     Command: **stylesmanager** *(Press* ENTER.*)*

   In both cases, AutoCAD requests that Windows display the Plot Styles window.

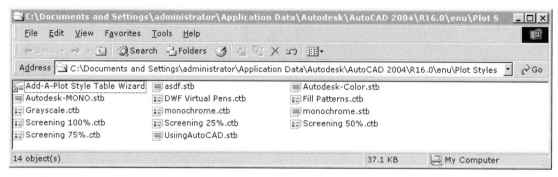

2. Double-click a plot style table file:
    *   **CTB**: color-based plot style table files.
    *   **STB**: named plot style table files.

    Notice that AutoCAD opens the Plot Style Table Editor.

3. Click the **Form View** tab, which lists plot style names on the left, and the associated plot style properties on the right.

4. Under **Plot styles**, select a plot style name — other than **Normal**. Notice the plot style properties at the right.
    Edit the properties, as required.

Note: The first plot style in every named plot style table is called "NORMAL." It lists the object's default properties, so that you can see its properties with no plot style applied. The NORMAL style cannot be edited or removed.

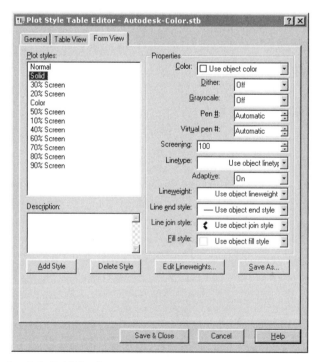

5. To add a plot style, click **Add Style**.
    Name the style, and then edit its properties. (You cannot add plot style to, or delete them from, color-based tables.)

6. To delete a style, select it, and then click **Delete Style**.

7. To save the plot style table, click **Save As**, and then name the table file.

8. When done, click **Save and Close**.

Plot Style Property	Comment
Color	Plotted color of objects; plot color overrides object color. *Default:* **Use Object Color**.
Dither	Toggles dithering to approximate colors with dot patterns. *Default:* **Off** for *.stb* files; **On** for *.ctb* files.
Grayscale	Converts object colors to grayscale during printing. *Default:* **Off**
Pen #	Specifies pen #; ranges from 0 to 32 (available for pen plotters only). *Default:* **Automatic** (pen #0).
Virtual Pen #	Simulates pen plotters for non-pen plotters; ranges from 0 to 255. *Default:* **Automatic** (pen #0).
Screening	Color intensity; ranges from 0 (white) to 100. Selecting a value other than 100 turns on Dithering. *Default:* **100**.
Linetype	Overrides object linetype with selected linetype. *Default:* **Use Object Linetype**.
Adaptive	Adjusts scale of the linetype to complete the pattern. Turn off if linetype scale is crucial. *Default:* **On**.
Lineweight	Overrides object lineweight with selected lineweight. *Default:* **Use Object Lineweight**.
Line End Style	Assigns a style to the end of object lines: Butt, Square, Round, and Diamond. *Default:* **Use Object End Style**.
Line Join Style	Assigns a style to the intersection of object lines: Miter, Bevel, Round, and Diamond. *Default:* **Use Object Join Style**.
Fill Style	Overrides object fill style with the following: Solid, Crosshatch, Diamonds, Horizontal Bars, Slant Left, Slant Right, Square Dots, Checkerboard, and Vertical Bar. *Default:* **Use Object Fill Style**.

EXERCISES

1. From the Companion CD, open the *edit4.dwg* file, a landscape drawing.

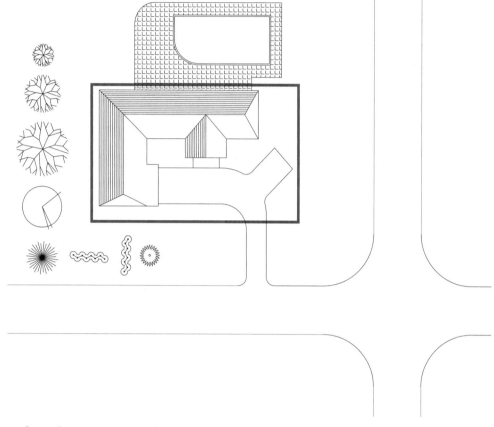

   a. Start the **PLOT** command, and assign a printer, if necessary.
   b. Plot the drawing at a scale factor of "Scaled to Fit."
   Does the drawing fit the page?

2. Use a ruler to measure the house's longest side. Assume the length represents 50'.
At what scale was the house plotted?

3. Plot the drawing again, using both landscape or portrait orientation.

4. Using the same drawing, set the plot scale factor to 1:1.
Preview the drawing.
Is the drawing larger or smaller?

5. Plot the drawing again, but this time use the **Window** option, and window the area
shown above by the blue rectangle.

6. Use the **PLOTSTAMP** command to attach a stamp to the plot.

7. Use the **PUBLISH** command to create a drawing set of the three drawings listed below:
   15_43.dwg
   15_44.dwg
   15_45.dwg

8. Create a plotter configuration for the printer attached to your computer. Name the configuration "UsingAutoCAD."

9. Create a plotter configuration for outputting compressed TIFF files with the **PLOT** command. Test the configuration by plotting a drawing, and then viewing the *.tif* file with AutoCAD's **REPLAY** command.

10. Calibrate the printer attached to your computer. How inaccurate was it before calibration?

11. Create a named plot style table with the following plot styles:
    50% screening.
    Checkerboard fill.
    Diamond join style.
    0.02 lineweight.
    Attach the plot style table to a drawing, and then plot it.

CHAPTER REVIEW

1. What is the purpose of plotting?

2. What part of the drawing is plotted when plotted to the extents of the drawing?

3. Describe how to plot only a portion of drawings.

4. How do you rotate the plot on the paper?

5. Why would you plot drawings to file?

6. Can you plot a drawing without first assigning a plotter or printer?

7. Explain the benefit to using the **PREVIEW** command.

8. Briefly describe the purpose of *device drivers*.

9. Does AutoCAD plot objects on frozen layers?

10. What kinds of plotters and printers does AutoCAD work with?

11. How large is a drawing plotted when the scale factor is
    a. Scaled to Fit?
    b. 1:1?
    c. 1/128" = 1?

12. Name an advantage to using full preview.
    To using partial preview.

13. What is the benefit of background plotting?

14. When might you want to offset the drawing on the paper?
    What is the drawback of offsetting?

15. Can AutoCAD create rendered plots?
    Can 2D drawings be rendered?

16. How do you instruct AutoCAD to plot with hidden lines removed in model space?
    In a layout?

17. Describe how plot stamps are useful.

18. When a plot stamp indicates the scale factor is 1:0.97531864, what might this indicate?

19. What are *drawing sets*?
    When might you want to use them?

20. Must all layouts in a drawing set come from the same drawing?

21. Explain the purpose of the **PUBLISH** command.
    Of the **PLOTTERMANAGER** command.

22. Can AutoCAD plot to printers located on networks?
    Connected via USB ports?

23. Explain the meaning of the parts of the printer icon:

    a.
    b.
    c.

24. What is the purpose of the *.pc3* file?
    The *.stb* file?
    The *.ctb* file?

25. Can AutoCAD print on both side of the paper?

26. Explain the purpose of the **STYLESMANAGER** command.

27. Of the two lists of plot style names shown below, which are color-based names and which are named styles?

a.   b.

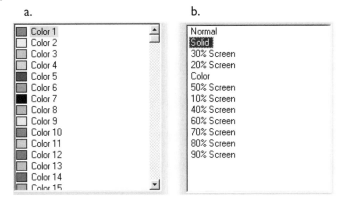

    a.

    b.

28. Describe two advantages of named plot styles over color-based ones:

    a.

    b.

29. Can named plot styles be applied to layers?

Layouts?

Views?

30. Explain the purpose of the **SCALELISTEDIT** command.

31. Describe how to create a PDF file of a drawing.

32. Why can 3D DWF files be created only from model tab?

33. If a plotted drawing does not fit the paper, what can you do to correct the situation?

    a.

    b.

    c.

34. What might be some reasons for plots failing to show lineweights?

    a.

    b.

    c.

UNIT VII

Appendices

APPENDIX

A

AutoCAD, Computers, and Windows

The Windows operating system is the foundation on which AutoCAD and other programs run. Windows loads automatically when you start the computer. (You might notice operating system messages and version numbers when your computer starts.) AutoCAD 2008 operates on the Windows 2000, XP, TabletPC, 2003 Server, and Vista operating systems manufactured by Microsoft. For XP and Vista, you can install separate versions of AutoCAD for 32- and 64-bit systems. (The 64-bit version can handle larger drawing files.)

HARDWARE OF AN AUTOCAD SYSTEM

CAD systems run on computers; *peripherals* connect to computers, as discussed later in this chapter. From smallest to largest, the five primary categories of computer are as follows:

Palm computers are the smallest of computers, fitting in the palm of your hand. Typically, data is input by writing with a stylus on the computer's screen; handwriting recognition software translates writing to text and graphics.

Palm computers can be used to view and edit CAD drawings, although AutoCAD itself is not available. The CAD software shown on the PalmOS computer at right is meant for use at job sites; drawings are exported in DXF format to AutoCAD.

Notebook computers are the smallest practical computers for running AutoCAD. They are the size of a notebook (hence the name), making them transportable. They are generally more expensive, but somewhat less powerful, than the equivalent desktop computer; most current notebook computers are powerful enough to run AutoCAD in 2D ,and in 3D for most features.

Notebook computers usually consist of a case that contains everything, except the printer.

Desktop computers are most commonly seen on (or under) desks. These versatile machines are sometimes referred to as "personal" computers, because they are primarily designed for use by one person. Desktop computers are often connected to other computers through networks and the Internet for sharing files and programs.

Personal computers usually consist of a case that contains the central processing unit, memory, one or more disk drives, and additional adapters, such as for sound and video. The monitor, keyboard, mouse, and output devices (such as printers) are external to the case.

Workstations are larger, faster, and more expensive than desktop computers. They typically contain two or more fast processors, much more memory, and much larger storage space. This class of computer is meant for running larger and more sophisticated programs than personal computers typically handle. For example, the special effects of Hollywood movies are created on workstations; some specific brands of CAD software run best on workstations.

Mainframe computers are the largest and most powerful of computers. They are meant for processing huge amounts of data, and are used by governments and companies to process data, such as weather forecasts and credit card transactions. These computers are not typically used for running CAD.

COMPONENTS OF COMPUTERS

Computers are made up of several essential parts. Let's look at some of them.

System Board

The system board (sometimes call the "motherboard") is a fiberglass board that holds most of the computer's chips and *expansion slots*. The CPU (central processing unit), RAM memory chips, ROM (read-only memory), drive controllers, and other parts are mounted on this board.

In addition, the board contains slots into which expansion boards can be mounted, such as video capture cards and drive adapters. These are the types of expansion slots:

>**ISA** (industry standard architecture) is the oldest type of expansion slot. The newest computers no longer have these slots.

>**PCI** (peripheral connect interface) is the most common slot in today's computers, found in both PC and Macintosh computers.

>**AGP** (advanced graphics port) is designed for high-speed data transfer between the CPU and graphics boards.

>**PC Card** or **PCMCIA** (personal computer memory card interface adapter; also known in jest as "people can't memorize confusing industry acronyms") are designed for notebook computers. The adapter card looks like a thick credit card, and might contain a modem, network interface, or even a disk drive. The newest notebook computers use **ExpressCard** slots with a smaller and faster interface.

>**SD** and **miniSD** (secure digital) are designed for Palm-size computers and notebook computers. These cards most often contain memory, but also include wireless networking, GPS (global positioning system), and Bluetooth communications capabilities.

>**Dedicated Memory** slots are designed for adding more memory to desktop and notebook computers. They are designed to access memory extremely quickly. External memory slots read memory cards from digital cameras and other portable devices.

Internal expansion slots are becoming less important as computer designers integrate most functions with the motherboard, including sound, networking, graphics, and drive interfaces. At the same time, high-speed USB2 and FireWire800 connectors are replacing expansion slots, making it possible to add peripherals without opening the computer case.

Central Processing Unit

The central processing unit (or CPU) is the "brain" of the computer. This is where the software is processed; and then the CPU sends instructions to the graphics board, printer, plotter, and other peripherals. With a few exceptions, all information passes through the central processing unit. To speed up calculations and access to memory, CPUs are usually packaged with *cache* memory, a very fast form of memory.

Physically, the central processor is a large computer chip mounted on the motherboard. Because of the heat generated by CPUs, it is common to have heat sinks (which look like a series of black fins) and small fans mounted on the chips. AMD and Intel manufacture the CPUs commonly used for AutoCAD.

CPUs are also located in other parts of the computer. The computer's graphics board has its own dedicated CPU for processing graphical images, often called the "GPU" (graphics processing unit). AutoCAD can make use of the GPU for processing visual styles and materials; this is known as "hardware acceleration," because the GPU is faster than the CPU at these tasks. Printers and plotters also contain CPUs for processing print and plot data.

Memory

Computer memory can be divided into two categories: ROM (read-only memory) and RAM (random access memory). ROM memory is contained in preprogramed chips mounted on the motherboard. ROM stores basic sets of commands for the computer, such as the instructions for starting the computer.

Random access memory (RAM) is what most people mean when they refer to "computer memory." It stores nearly all the information used by the computer while running; it is *temporary*, because all data in RAM is lost when the computer is turned off or when the operating system freezes up.

Software programs (such as AutoCAD) require a certain amount of RAM to run the program. This amount is given in megabytes. A common number might be "512MB," meaning 512 megabytes. A megabyte is 1,024KB (kilobytes); a kilobyte is 1,024 *bytes*.

Hard drive manufacturers often round down these values to make their products appear to have larger capacities, claiming that there are only 1,000 bytes in a kilobyte.

Disk Drives

Disk drives store large amounts of data, such as application programs, the Windows operating system, drawings, and other documents. Disk drives are either non-removable (*fixed*) or *removable*.

In almost all cases, fixed drives are hard drives. Today's desktop hard drives hold 240GB (gigabytes) and more of data. A gigabyte is 1,024 megabytes.

Removable drives include diskettes, USB data keys, CDs and DVDs, and even tape drives. The now-rarely-used 3-1/2" diskette holds 1.44MB, a relatively small amount of data; some computers support 2.88MB of data on these diskettes.

USB data keys hold 32MB to 16GB of flash memory, and connect to the USB ports of computers. They are small but expensive relative to other forms of data storage.

CD drives (short for "compact disc, read-only memory") are a common method for distributing software today. They are the same discs that hold music, but use a different format for coding data. CD discs can be read, but not written to. This makes the data secure from accidental erasure and attack from computer viruses.

Two types of CDs can be written to (called "burning CDs"): CD-R discs can be written to, but not erased, while CD-RW discs can be written to and erased. All CDs allow you to store up to 700MB of data.

DVDs (short for "digital versatile disc") hold 4.7GB of data. Your copy of AutoCAD probably arrived on a DVD. Dual-layer DVDs allow twice as much data to be recorded, while BlueRay and HD-DVDs have a capacity of 27GB - 54GB. DVD drives can read CDs, but CD drives cannot read DVDs.

As with CDs, there are three types of DVD: "DVD" discs can only be read; "DVD-R" can be written to (recorded); "DVD-RW" can be written to and erased.

Tape drives were the original backup medium, because they were relatively cheap. They suffer, however, from slow access speed. Today's tape drives can store 40GB or more data on tapes similar to those used by 8mm digital video cameras.

DISK CARE

All of your work is recorded to disk, so caring for disks is very important. If a disk becomes damaged, you will lose your work! Frequent backing-up (copying files to a second disk, tape, or recordable CD-R) and proper handling of disks minimize the possibility of file loss. AutoCAD includes the ability to back up drawings automatically every few minutes to your computer's hard disk, or safer yet, to the drive on another computer.

Hard disks are installed inside or outside the computer. Today's hard drives have self-parking heads that move to an area that does not have stored data; this helps protect disks from damage when moved. When the hard drive is on, do not move or tilt the computer. This is particularly important with notebook computers. A hard drive can be damaged by shock. If you move the computer or the drive, move it gently.

CD-ROMs and DVDs are relatively sturdy forms of data storage, but should also be handled with care. Avoid touching either surface of the discs; handle them by their centers and edges. When not in use, keep the discs in cases or protective sleeves; these protect the surfaces from which the laser reads data. Write on the label side with soft-tip pens only; damaging the label with a ballpoint pen damages the data on the disc. Keep discs out of the sun and out of high humidity areas.

Write-Protecting Data

Normally, disks can be both read and written to (new data saved on it). But you can *write-protect* some disks to preserve what is already stored on them. This allows the disk to be read by the computer, but not written to. Windows allows you to set *read-only* status for files, folders, and entire drives.

CD-ROMs and DVDs cannot be written to, so the data is always safe — except when damaged physically. Some hard drives include utility software that electronically *locks* the drive, preventing misuse.

 Some drives can erase data from CD-R and DVD-R discs, while all DVD-RW and CD-RW discs can have data erased and added. These discs have no write-protect switch, so it is possible to lose data. SD memory cards have a slider switch to prevent data from being overridden, as illustrated by "LOCK" at left.

PERIPHERAL HARDWARE

Peripheral hardware consists of add-on devices that perform specific functions. A properly-equipped CAD station consists of the following peripheral devices.

PLOTTERS

Plotters produce "hard" copies of your drawings on paper, vinyl sticker material, cloth, and other media. The two primary types of plotter today are inkjet plotters and laser printers. Some offices still have older pen plotters.

Inkjet Plotters

The most common plotter today is the *inkjet plotter*, which prints by sending an electrical current to the print head. In the print head, electricity heats up ink, causing it to squirt out the end of dozens of nozzles, hitting the paper and soaking in. Typical print heads have 48 nozzles lined up vertically. Print heads in large-format inkjet plotters can have as many as 512 nozzles.

For color plots, there are at least four print heads: one each for cyan (light blue), magenta (pink), yellow, and black. By combining these four colors using a process called *dithering*, inkjet plotters seem to produce 256 or more colors. For higher quality output, specialty inkjet plotters use seven or more colors of ink.

Desktop inkjet printers produce plots of A (8.5" x 11") or B-size (11" x 17"), while floor models plot up to E-size (36" x 48"). Inkjet printers have a typical resolution of 360 dots per inch (*dpi*), but some are capable of 1440 dpi and even 2880 dpi. Higher resolutions are generated by squirting smaller droplets of ink, and placing the droplets closer together. Inkjet plotters can use normal paper, but specially-coated paper results in a cleaner looking print with longer lasting and brighter colors.

When you need paper plots to last many years, make sure your inkjet printer uses pigment inks and archival-quality paper.

Laser Printers

Laser printers create plots using a process similar to photocopying: using a laser beam, the printer "paints" the image of text and drawings on its internal drum. This creates an electrostatic charge on the drum. As the drum rotates, it picks up black toner particles that stick to the electrically-charged portions of the drum. The toner particles are transferred to the paper, which then passes over a hot fuser to melt the toner onto the paper.

The laser-copy process produces sharp, clear prints. Although some laser printers have excellent graphics capabilities, they are usually restricted to a maximum of B-size. Resolution is typically 600 dpi, but some models boast 1200 dpi or higher.

GRAPHICS BOARDS

Graphics boards convert the images generated by AutoCAD into what you see on the monitor. Just about any graphics board made in the last five years works with AutoCAD, but may not support all of its 3D capabilities.

Some of the advanced renderings and visual styles in AutoCAD require advanced capabilities in graphics boards — such as shadow casting and high quality transparency. Autodesk maintains a Web site that lists graphics boards it has tested: <u>usa.autodesk.com/adsk/servlet/hc?siteID=123112 &id=6711853&linkID=9240618</u>. Autodesk ranks each on three levels: Not Supported; Supported, but not Recommended; and Supported and Recommended.

At the time of writing this book, only certain graphics boards from ATI and nVidia are fully compatible with the graphic demands of AutoCAD's imaging. Boards from Intel and XGI are not supported at all.

You can test the graphics board in your computer by running the **3DCONFIG** command, which determines its capabilities for:

- Geometry acceleration.
- Smooth line display.
- Gooch hardware shader.
- Full shadow display.
- Texture compression.

The technical editor observes that AutoCAD tends to be very conservative in rating graphics boards. He has overridden the default values without problems. The author, however, has experienced AutoCAD crashing when the recommended values are exceeded; for truly problematic cases, AutoCAD's **/nohardware** startup switch overrides hardware acceleration.

Monitors

Monitors display the work in progress. A monitor is sometimes referred to as a "display device," "LCD" (liquid crystal display), or "CRT" (cathode ray tube). CAD operators prefer monitors that measure 19 inches (diagonally) or larger, because they can see more of the drawing at a time.

The quality of the image on the monitor is determined by its resolution, which depends on the number of dots (pixels) displayed on the screen. The pixels make up the image. A typical 1280 x 1024-resolution display contains 1,280 pixels horizontally and 1,024 pixels vertically. Many professional CAD users prefer higher resolution displays, such as 1600 x 1200 or higher. The device driver, the operating system, the graphics board, and the monitor work together to support the high resolution.

A few graphics boards support two or even four monitors. The advantage to multi-monitor setups is that one monitor can display the drawing, while the second contains AutoCAD's toolbars, palettes, and other user interface elements. The third and fourth monitors could be used to display other software, or simply spread out the AutoCAD window. The drawback to multiple monitors is the added cost and loss of desk space.

INPUT DEVICES

Just as word processing programs require keyboards to input letters, numbers, and symbols, CAD programs require input devices to make and manipulate drawing elements. In addition to the keyboard, the most common input devices for CAD drawing are the mouse, digitizer, and scanner.

Wheel

Left (primary) button

Right button

Mouse

The mouse is an input device used for pointing and positioning the cursor. (The name "mouse" comes from its mouse-like appearance; cordless mice dispense with the tail, relying on a small transmitter to connect with the computer.) A mouse can be used for command input and screen interaction; it cannot digitize drawings. There are primarily two types of mouse: mechanical and optical.

A *mechanical* mouse has a ball under its housing. As the ball rolls, the mouse measures its motion, and then transmits the relative movement to the computer. An *optical* mouse uses optical technology to sense movement over almost any surface — except glass table tops and other shiny surfaces.

Today's mouse usually sports three or more buttons, with one button disguised as a roller wheel. More buttons are better, because they can be used for additional functions; AutoCAD supports up to 15 buttons. For example, the middle button can be made to issue a double-click, reducing by half the number of clicks. The roller wheel allows real-time zooms and pans in AutoCAD.

Digitizers

The digitizer is an electronic input device that transmits the *absolute* x and y location of the *puck* resting on its sensitized pad. Digitizers have a fine grid of wires sandwiched between glass layers. When the puck is moved across the pad, the absolute location is read and transmitted to the computer.

Digitizers can be used in two ways: as *pointing devices* to move the cursor around the screen (like a mouse), or as *tracing devices* to copy drawings into the computer at scale and in proper proportion. In "tablet" mode, the digitizing pad is calibrated to the actual absolute coordinates of the drawing. When used as a pointing device, the tablet is not calibrated.

Digitizers can be used with a stylus (similar in appearance to a pencil) or with a multi-button puck. Digitizers come in several sizes, ranging from a few square inches, to very large. Small pads can be used to digitize large drawings: moving the drawing on the pad and then recalibrating. This becomes annoying when you frequently work with large-scale drawings. In that case, the expense of a large-format digitizer (36" x 24" or larger) may be justified.

Digitizers are now very rare, used primarily for tracing existing maps.

Scanners

Scanners are used to "read" paper drawings and maps into computers. The scanner works by shining a bright light at the paper, while a head moves across the paper, measuring the amount of reflected light. (The head is made of many CCDs, the same charge couple devices used by digital cameras). Bright areas are the paper, while dark areas are lines.

The scanner sends data to the computer in raster format, where it can be imported by AutoCAD with the **IMAGEATTACH** command, or else converted to vector files by specialized software. Scanners are available in a variety of sizes, ranging from very small units that read business cards to E-size scanners that read large engineering drawings.

DRIVES AND FILES

Windows names the computer's disk drives with letters of the alphabet, followed by the colon symbol (**:**), such as A: and C:. **A:** and **B:** are reserved for two floppy disk drives — even when your computer has no floppy drives. **C:** almost always designates the first hard drive, while **D:** and subsequent letters are for all other drives, including additional hard drives, network drives, CD and DVD drives, and removable drives.

Computers access *Network drives* over networks. Computers must be connected with a network

The location of disk drives in typical desktop and notebook computers.

cable or a wireless network connection to share drawings and other files. The names of network drives are usually prefixed by a double backslash and their computer's name, such as \\Kat\C. The networked version of AutoCAD allows a single copy to be used by two or more operators, depending on the terms of the network license.

Removable drives are any disk drives from which the medium can be removed. When reading or writing using removable disks, you must be careful that the correct disk is inserted. (Removable drives are sometimes called "backup drives.")

The position of the drives in the computer can differ according to the cabinet design. The figure above illustrates typical locations.

FILE NAMES, EXTENSIONS, AND PATHS

Files have names of up to 255 letters and characters. At the end of the file name you often find a dot (.) and a three-letter code, called the *file extension*. This denotes the type of file. For example, AutoCAD drawing files have the *.dwg* extension, while AutoCAD template files use the *.dwt* extension.

Many programs, such as AutoCAD, add file extensions to file names automatically, while others require you to specify the extension. The display of file extensions is normally turned off by Microsoft. Because it is useful to identify files by their extension, I recommend that you turn on the display: In Windows Explorer, select **View | Folder Options**. In the dialog box, click the **View** tab, and then uncheck the **Hide extensions for known file types** option.

The full file name includes the *path*. The path specifies the names of the drive and folders in which the file is stored. For example, an AutoCAD drawing named *widget* stored on the C: drive in the *\autocad\drawings* folders is fully named by its path as:

$$c:\backslash autocad\backslash drawings\backslash widget.dwg$$

Sometimes you might see file names that are just eight characters long and perhaps end in ~1. Older operating systems, such as Windows 3.1 and DOS, limited file names to eight characters. When a file name of nine or more characters is copied to a computer running one of these older operating systems, the file name is *truncated* (chopped off) to six characters, and ~1 is added. For example, *houseplan.dwg* becomes *housep~1.dwg*. Similar names are truncated to ~2, and so on.

Wild-Card Characters

In all file dialog boxes, as well as some Windows functions and certain AutoCAD commands, you can use *wildcard characters* to specify groups of files. The two most common wildcard characters are the question mark (?) and the asterisk (*).

The **question mark** (?) fills in for *any single* character, even if the character is nothing. By using one or more question marks, you set an upper limit to the number of characters. When you enter:

car??

you refer to variations on "car...", such as:

car   cars   carts   card2

But not "cardindex," because it has more than five characters.

The **asterisk** (*)represents *any number of* characters. For example, when you enter:

*.dwg

you are referring to all files with the "*....dwg*" file extension, such as

floorplan.dwg   my drawing 1.dwg   8th floor.dwg

Alternately, when you type

floorplan.*

you refer to all files named "Floorplan..." regardless of their file extension, as well as to those with no extension:

> floorplan.dwg   floorplan.bak   floorplan.dwf   floorplan.

DISPLAYING FILES

Windows provides several ways to display and manipulate files on your computer.

Windows Explorer

Windows Explorer is the "official" method to view files in Windows. Not to be confused with Internet Explorer, Windows Explorer lets you view files in all folders on all drives, including on computers networked to yours.

File-Related Dialog Boxes

Almost all of AutoCAD's file-related dialog boxes — such as those opened by the **OPEN**, **SAVE AS**, and **IMPORT** commands — let you view and manipulate files. This is a handy shortcut that obviates the need to switch to Explorer.

DOS Session

The oldest method is to start a DOS session, and then type in commands — such as the **DIR**, **TREE**, **COPY**, **MOVE**, and **DELETE** commands— together with parameters, such as /**s**, *.*, and **a:** .

This approach is often employed by power users, because Windows Explorer cannot perform some file-related tasks, such as renaming all files in a folder, or copying files to devices, such as **lpt:**.

WINDOWS EXPLORER

Windows Explorer lets you view and manipulate files on your computer's drives and all networked drives that you have permission to access. You can change the way Explorer looks, but the most useful configuration has two panes, as illustrated by the figure below.

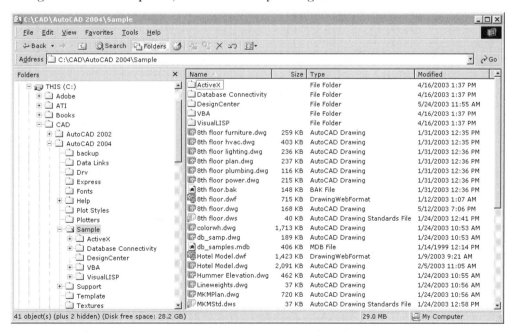

Windows Explorer showing two panes and file details.

Viewing Files

In Windows Explorer, the left *pane* displays the *tree* of drives (including networked drives) and folders (replacement name for subdirectories). This is called the "Folder view." To see more of the tree, use the scroll bar. The + (plus) sign beside a folder means it has subfolders. Click + to open the folder; the + changes to a − (minus) sign. Click − to close the folder.

In the right pane, Windows Explorer displays the files and subfolders contained in the folder currently selected in the left window pane. This is called the "Contents view." You change the sorting order of the **Name**, **Size**, **Type**, and **Modified** by clicking on the sort bar. (To see the sort bar, set the **View** to display file **Details**.) Click a second time and Explorer displays the list in reverse order, which is useful for displaying files sorted alphabetically or by size, beginning with either largest or smallest.

Windows Explorer also displays files and folders in different views, each giving you more detail. The **View** item on the menu bar lets you switch between large icons, small icons, list, detail, and thumbnail, as illustrated by the figure below. **Thumbnail** view is useful, because it previews AutoCAD drawing files.

Manipulating Files

By right-clicking on the name of any file or folder, you can change its name and perform other tasks. (Alternatively, use the **File** and **Edit** items on the menu bar.) When you right-click, Windows Explorer displays a shortcut menu with the following options, which differ depending on whether you right-click a folder or file:

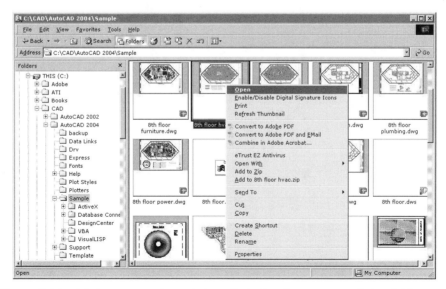

Open (folders) brings up another window displaying the contents of that folder.

Open (files) opens the file with the associated program. For example, a *.lsp* (AutoLISP source code) file is opened by Notepad. An *.exe* file, such as *acad.exe*, starts to run.

Explore (folders only) displays the contents of the folder in the pane at right.

Find (folders only) displays the Find:All Files dialog box, which lets you find files anywhere on the computer and (if connected) the network. You can search for text in *.dwg* files.

Sharing (folders only) allows you to give others on a network access to drives and folders. You can set three types of permission: (1) full read-write, (2) read only, and (3) access with password.

Send to sends the folder or file to one of several destinations, such as to the floppy drive or the Briefcase folder. Although the printer is listed, you cannot send folders directly to the printer; Windows Explorer first opens the associated application.

Create Shortcut places *shortcuts* as icons on the Windows desktop. Shortcuts are small files that point to the original file's location. They are excellent for accessing files without wandering through the folders to locate them each time.

Delete erases folders and files. Windows first asks, "Are you sure?" Deleted files and folders are stored in the Recycle Bin for a limited time, from which they can be recovered, if originally on a hard drive. The files are gone for good on floppy diskettes and Zip disks.

Rename changes the names of folders and files. As an alternative, click twice (slowly, do not double-click) on the name and Explorer lets you change the name directly.

Properties sets the attributes for the folder or file, including archive, read-only, hidden, and system.

Copying And Moving Files

Copying or moving a file using Windows Explorer takes three steps:

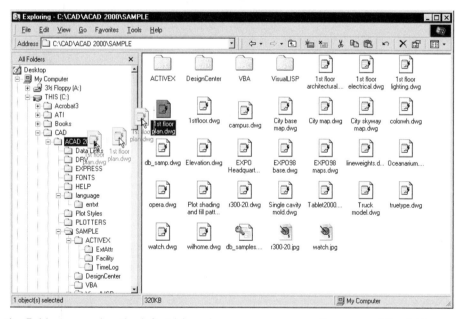

1. In the Folders pane (on the left side), make sure you can see the name of the folder you will be copying the file to. If you cannot see it, scroll the list to see more folders or open the appropriate closed folder by clicking the + sign.
2. In the Contents pane (on the right side), drag the file to the folder. (Drag means to click on the file and move the cursor without letting go of the mouse button until you reach its destination.)
3. To copy the file, hold down the **CTRL** key while dragging the file. As a reminder that you are copying the file, Windows Explorer displays a small + sign near the dragged icon.

If you drag the file on top of an application, the application will attempt to open the file. For example, dragging a *.dwg* file onto the *acad.exe* icon causes AutoCAD to launch with that drawing.

AUTOCAD FILE DIALOG BOXES

Windows Explorer is the official way to view and manipulate files in Windows. You can, however, perform some of these functions from within AutoCAD's file-related dialog boxes, which makes sense, because when saving and opening files, you probably want to work with them. Those dialog boxes are displayed when you use the OPEN, SAVEAS, EXPORT, and other file-related commands.

AutoCAD's File dialog box provides the following functions:

Switch Views. Choose the **Views** item to see the files listed in different views, such as a compact listing or with details (size, type, and modified date-time). Click the column headers to sort in alphabetical, size, date, etc. order; click a second time to sort in reverse order.

Select a Different Folder. Select a different folder by several methods: (1) the **Look in** list box lets you move directly to another folder, (2) single-click the **Up One Level** icon to move up the folder tree one folder at a time; double-click the folder to move down into it, or (3) click the **Back** button to return to the previous folder.

Create a New Folder. Click the **Create New Folder** icon to create a new folder in the current folder. As an alternative, press **ALT+5**.

Delete a File. Select the file name and press the **DEL** key. Windows displays a dialog box that asks, "Are you sure you want to send 'file name' to the Recycle Bin?" When you respond **Yes**, the file is not erased but moved to the Recycle Bin folder. If you delete a file by accident, you can go to the Recycle Bin and retrieve the file.

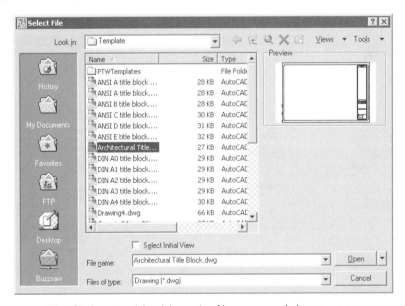

Rename a File. Click twice (slowly) on the file name, and then type a new name.

Move or Copy a File. Drag the file to another folder. Hold down the **CTRL** key to copy the file; hold down the **SHIFT** key to move a file.

Context-Sensitive Menu. Right-click a file or folder to bring up the same context-sensitive menu as seen in Windows Explorer.

Open or save the File. Double-click a file name to load it without choosing **OK**.

Places List. Located on the left hand side, the Places list provides convenient access to frequently used folders. To add a folder, drag it from the file list into the Places list.

AUTOCAD FILE EXTENSIONS

AutoCAD uses many types of files, the contents of which are often described by the file's extension. For example, linetypes are stored in files that end with *.lin*. Many of the file extensions used by AutoCAD are listed here.

Drawing Files

Extension	Description
.bak	Backup drawing files.
.dwf	Drawing Web format files.
.dwg	AutoCAD drawing files.
.dws	CAD standards files.
.dwt	Drawing template files.
.dxb	AutoCAD binary drawing interchange files.
.dxf	AutoCAD drawing interchange files.

Support Files

Extension	Description
.$ac	Temporary files created by AutoCAD.
.aws	AutoCAD workspace files.
.blk	Block template files.
.cfg	Configuration files.
.chm	HTML format help files.
.cui	Customize User Interface files.
.cus	Custom dictionary files.
.dct	Dictionary files.
.dst	Sheet set data files.
.dsd	Drawing set description files.
.err	Error log files.
.fdc	Field catalog files.
.fmp	Font mapping files.
.hlp	Windows-format help files.
.ies	Illumination distribution data files.
.lin	Linetype definition files.
.lli	Landscape libraries.
.log	Log files created by the LOGFILEON command.
.mli	Rendering material library files.
.mln	Multiline library files.
.mnc	Compiled menu files (obsolete as of AutoCAD 2006).
.mnd	Uncompiled menu files containing macros (obsolete).
.mnl	AutoLISP routines used by AutoCAD menus (obsolete).
.mns	AutoCAD-generated menu source files (obsolete).
.mnu	Menu source files (obsolete as of AutoCAD 2006).
.msg	Message files.
.pat	Hatch pattern definition files.
.pgp	Program parameters files (external commands and aliases).
.scr	Script files.
.shp	Shape and font definition files.
.shx	Compiled shape and Autodesk font files.
.ttf	Microsoft font files.
.xpg	Xml-format tool palette group files.

AutoCAD Program Files

Extension	Description
.arx	ObjectARx (AutoCAD Runtime eXtension) program files.
.dll	Dynamic link library files.
.exe	Executable files, such as AutoCAD itself.

Plotting Support Files

Extension	Description
.ctb	Color-table based plot parameter files.
.pcp	Plot configuration parameters files for AutoCAD Release 14.
.pc2	Plot configuration parameters files for AutoCAD 2000.
.pc3	Plot configuration parameters files since AutoCAD 2000i.
.plt	Plot files.
.stb	Style-table based plot parameter files.

Import-Export Files

Extension	Description
.3ds	3D Studio files.
.bmp	Windows raster files (device-independent bitmap).
.cdf	Comma delimited files (created by EATTEXT).
.dgn	MicroStation V8 design files (created by DGNEXPORT).
.dxe	Data extraction files (created by DATAEXTRACTION).
.dxx	DXF files (created by ATTEXT).
.pcx	Raster format file.
.png	Portable Network Graphics raster files.
.sat	ACIS solid object files (short for "Save As Text").
.slb	Slide library files.
.sld	Slide files.
.stl	Solid object stereo-lithography files (solids modeling).
.tif	Raster format files (Tagged image file format).
.tga	Raster format files (Targa).
.txt	Space delimited files (created by EATTEXT).
.wmf	Windows metafile files.
.xls	Excel spreadsheet files (created by EATTEXT).
.xml	DesignXML format files.

Miscellaneous Files

Extension	Description
.css	Cascading style sheet files.
.html	HTML files.
.htt	HTML template files.
.ini	Initialization files.
.js	JavaScript files.
.map	Map files displayed by GEOGRAPHICLOCATION.
.xls	XML style sheet.
.xml	Extended markup language files.
.xmx	External messages files.

LISP and ObjectARX Programming Files

Extension	Description
.cpp	ObjectARX source code files.
.dcl	Dialog control language descriptions of dialog box files.
.dvb	Visual Basic for Applications program files.
.def	ObjectARX definition files.
.fas	AutoLISP fast load programs.
.frm	VBA form definition files.
.h	ADS and ObjectARX function definitions files.
.lib	ObjectARX function library files.
.lsp	AutoLISP program files.
.mak	ObjectARX make files.
.rx	Lists of ObjectARX applications that load automatically.
.tlb	ActiveX Automation type library files.
.unt	Unit definition files.
.vlx	Compiled Visual LISP files.

REMOVING TEMPORARY FILES

While it is working, AutoCAD creates "temporary" files. When you exit AutoCAD, these temporary files are erased. If, however, your computer crashes, these files might be left on the hard disk. You can erase temporary files, but only when AutoCAD is not running. Never erase files with these extensions while AutoCAD is still running:

> . $$$.$a .ac$ .dwk .dwl .ef$ .sv$ .swr

These files can be erased to free up disk space when AutoCAD is not running. Alternatively, the **DRAWINGRECOVERY** and **RECOVERALL** commands help you select which of these files can be reused, following a crash.

B

AutoCAD Commands, Aliases, and Keyboard Shortcuts

AUTOCAD COMMANDS

The following commands are available in AutoCAD. Commands prefixed with - (hyphen) display their prompts at the command line only, and not in a dialog box. Shortcut keystrokes are indicated as CTRL+F4.

This icon indicates the command is new in AutoCAD 2008.

The Illustrated AutoCAD 2008 Quick Reference by author Ralph Grabowski provides a complete reference guide to all of AutoCAD's commands, and is available through Autodesk Press.

Command	Description
A	
About	Displays an AutoCAD information dialog box that includes version and serial numbers.
AcisIn	Imports ASCII-format ACIS files into the drawing, and then creates 3D solids, 2D regions, or body objects.
AcisOut	Exports AutoCAD 3D solids, 2D regions, or bodies as *.sat* ASCII-format ACIS files.
AdcClose	CTRL+2: Closes the DesignCenter palette.
AdCenter	CTRL+2: Opens DesignCenter; manages AutoCAD content.
AdcNavigate	Directs DesignCenter to the file name, folder, or network path you specify.
AecToAcad	Explodes AEC objects and adds them to a new drawing file.
Align	Uses three pairs of 3D points to move and rotate (align) 3D objects.
AmeConvert	Converts drawings made with AME v2.0 and v2.1 into ACIS solid models.
AniPath	Creates animations of drawings along paths.
AnnoUpdate	Updates annotative objects when styles have changed in the drawing.
AnnoReset	Resets the location of all scale representations of an annotative object to that of the current scale representation
Aperture	Adjusts the size of the target box used with object snap.

Command	Description
AppLoad	Displays a dialog box that lets you list AutoLISP, Visual Basic, and ObjectARX program names for easy loading into AutoCAD.
Arc	Draws arcs by a variety of methods.
Archive	Archives sheet sets in DWG or DWF format.
Area	Computes the area and perimeter of polygonal shapes.
Array	Makes multiple copies of objects.
Arx	Loads and unloads ObjectARX programs. Also displays the names of ObjectARX program command names.
AttachURL	Attaches hyperlinks to objects and areas.
AttDef	Creates attribute definitions.
AttDisp	Controls whether attributes are displayed.
AttEdit	Edits attributes.
AttExt	Extracts attribute data from drawings, and writes them to files for use with other programs.
2008 AttIpEdit	Lanches the in-place attribute editor.
AttReDef	Assigns existing attributes to new blocks, and new attributes to existing blocks.
AttSync	Synchronizes changed attributes with all blocks.
Audit	Diagnoses and corrects errors in drawing files.
2008 AutoPublish	Toggles the simultaneous saving of drawings in DWG and DWF formats.

B

Command	Description
Base	Specifies the origin for inserting one drawing into another.
BAttMan	Edits all aspects of attributes in a block; short for Block Attribute Manager.
BEdit	Switches to the block editor; makes the following commands available:
	BAction — adds actions to parameters in dynamic block definitions.
	BActionSet — associates actions with selected objects within dynamic blocks.
	BActionTool — adds actions to dynamic block definitions.
	BAassociate — associates actions with parameters in dynamic blocks.
	BAttOrder — controls the order in which attributes are displayed in dynamic blocks.
	BAuthorPalette — opens the Block Authoring palette in the Block Editor.
	BAuthorPaletteClose — closes the Block Authoring palette.
	BClose — closes the Block Editor.
	BCycleOrder — changes the cycling order of grips within dynamic block references.
	BGgripSet — creates, repositions, resets, and deletes grips associated with parameters.
	BLookupTable — creates and edits lookup tables associated with dynamic blocks.
	BParameter — adds parameters and grips to dynamic blocks.
	BSave — saves changes to dynamic blocks.
	BSaveAs — saves dynamic blocks by another name.
	BvHide — hides objects that have been assigned the state of invisibility.
	BvShow — shows objects assigned invisibility.
	BvState — assigns visibility states to objects in dynamic blocks.
BHatch	Fills an automatically-defined boundary with hatch patterns, solid colors, and gradient fills; previews and adjusts patterns.
Blipmode	Toggles display of marker blips.
Block	Creates symbols from groups of objects.
BlockIcon	Generates preview images for blocks created with AutoCAD Release 14 and earlier.
BmpOut	Exports selected objects from the current viewport to raster *.bmp* files.
Boundary	Draws closed boundary polylines.

Command	Description
Box	Creates 3D solid boxes and cubes.
Break	Erases parts of objects, breaks objects in two.
BRep	Removes construction history from 3D solid models.
Browser	Launches your computer's default Web browser with the URL you specify.

C

Command	Description
Cal	Runs a geometry calculator that evaluates integer, real, and vector expressions.
Camera	Sets the camera and target locations.
Chamfer	Trims intersecting lines, connecting them with a chamfer.
Change	Permits modification of an object's characteristics.
CheckStandards	Compares the settings of layers, linetypes, text styles, and dimension styles with those set in another drawing.
ChProp	Changes properties (linetype, color, and so on) of objects.
ChSpace	Moves and scales objects from model to paper space.
Circle	Draws circles by a variety of methods.
ClassicImage	Replaces the IMAGE command dialog box.
ClassicXref	Replaces the XREF command dialog box.
CleanScreenOn	CTRL+0: Maximizes the drawing area by turning off toolbars, title bar, and window borders.
CleanScreenOff	CTRL+0: Turns on toolbars, title bar, and window borders.
Close	CTRL+F4: Closes the current drawing.
CloseAll	Closes all open drawings; keeps AutoCAD open.
Color	Sets new colors for subsequently-drawn objects.
CommandLine	CTRL+9: Displays the command line palette.
CommandLineHide	CTRL+9: Hides the command line palette.
Compile	Compiles shapes and *.shp* and *.pfb* font files.
Cone	Creates 3D solid cones.
Convert	Converts 2D polylines and associative hatches in pre-AutoCAD Release 14 drawings to the "lightweight" format to save on memory and disk space.
ConvertCTB	Converts drawings from plot styles to color-based tables.
ConvertOldLights	Converts light blocks in drawings created by AutoCAD 2006 and older.
ConvertOldMaterials	Converts material definitions in drawings created by AutoCAD 2006 and older.
ConvertPStyles	Converts drawings from color-based tables to plot styles.
ConvToSolid	Converts closed polylines and circles to extruded 3D solids.
ConvToSurface	Converts 2D solids, regions, lines, polylines, arcs, and flat 3D faces to surfaces.
Copy	Makes one or more copies objects.
CopyBase	CTRL+SHIFT+C: Copies objects with a specified base point.
CopyClip	CTRL+C: Copies selected objects to the Clipboard in several formats.
CopyHist	Copies Text window text to the Clipboard.
CopyLink	Copies all objects in the current viewport to the Clipboard in several formats.
CopyToLayer	Copies selected objects to another layer through a dialog box.
Cui	Opens the Customize User Interface dialog box.
CuiExport	Exports customization settings to *.cui* files in disk.
CuiImport	Imports customization settings from *.cui* files.
CuiLoad	Loads *.cui* files.

Command	Description
CuiUnload	Unloads *.cui* files.
Customize	Customizes tool palette groups.
CutClip	CTRL+X: Cuts selected objects from the drawing to the Clipboard in several formats.
Cylinder	Creates 3D cylinders.
D	
Dashboard	Opens the Dashboard palette.
DashboardClose	Closes the Dashboard palette.
DataExtraction	Extracts attribute data and drawing information to spreadsheets or tables.
DataLink	Links tables with external spreadsheet files.
DataLinkUpdate	Updates linked tables.
DbcClose	CTRL+6: Closes the dbConnect Manager.
DbConnect	CTRL+6: Connects objects in drawings with tables in external database files.
DbList	Provides information about all objects in drawings.
DdEdit	Edits text, paragraph text, attribute text, and dimension text.
DdPType	Specifies the style and size of points.
DdVPoint	Sets 3D viewpoints.
Delay	Creates a delay between operations in a script file.
DetachURL	Removes hyperlinks from objects.
DgnAdjust	Adjusts the contrast of attached DGN files.
DgnAttach	Attaches MicroStation V8 2D DGN files as underlays.
DgnClip	Clips attached DGN files.
DgnExport	Exports the current drawing as a DGN file.
DgnImport	Imports MicroStation V8 2D DGN files as AutoCAD entities.
Dist	Computes the distance between two points.
DistantLight	Places distant lights in drawings.
Divide	Divides objects into an equal number of parts, and places specified blocks or point objects at the division points.
Donut	Constructs solid filled circles and doughnuts.
Dragmode	Toggles display of dragged objects.
DrawingRecovery	Opens the Drawing Recovery palette.
DrawingRecoveryHide	Closes the Drawing Recovery palette.
DrawOrder	Changes the order in which objects are displayed: selected objects and images are placed above or below other objects.
DSettings	Specifies drawing settings for snap, grid, polar, and object snap tracking.
DsViewer	Opens the Aerial View palette.
DView	Displays 3D views dynamically.
DwfAdjust	Changes the contrast, fade, and color of DWF files.
DwfAttach	Attaches *.dwf* files as uneditable underlays with a dialog box.
DwfClip	Clips DWF underlays.
DwfLayers	Toggles the display of layers of DWF underlays.

Command	Description
DwgProps	Sets and displays the properties of the current drawing.
DxbIn	Imports binary drawing interchange files.

DIMENSIONS

Command	Description
Dim	Specifies semi-automatic dimensioning capabilities.
Dim1	Executes a single AutoCAD Release 12-style dimension command.
DimAligned	Draws linear dimensions aligned to objects.
DimAngular	Draws angular dimensions.
DimArc	Draws arc dimensions along the circumference of arcs.
DimBaseline	Draws linear, angular, and ordinate dimensions that continue from baselines.
DimBreak	Breaks extension and dimension lines where they cross other objects.
DimCenter	Draws center marks on circles and arcs.
DimContinue	Draws linear, angular, and ordinate dimensions that continue from the last dimension.
DimDiameter	Draws diameter dimensions on circles and arcs.
DimDisassociate	Removes associativity from dimensions.
DimEdit	Edits the text and extension lines of associative dimensions.
DimInspect	Adds inspection text to dimensions.
DimJogged	Draws radial dimensions with jogged leaders.
DimJogLine	Jogs linear dimension lines.
DimLinear	Draws linear dimensions.
DimOrdinate	Draws ordinate dimensions in the x and y directions.
DimOverride	Overrides current dimension variables to change the look of selected dimensions.
DimRadius	Draws radial dimensions for circles and arcs.
DimReassociate	Associates dimensions with objects.
DimRegen	Updates associative dimensions.
DimSpace	Spaces two or more linear dimensions evenly.
DimStyle	Creates, names, modifies, and applies named dimension styles.
DimTEdit	Moves and rotates text in dimensions.

E

Command	Description
EAttEdit	Enhanced attribute editor.
Edge	Changes the visibility of 3D face edges.
EdgeSurf	Draws edge-defined 3D meshes.
Elev	Sets current elevation and thickness.
Ellipse	Constructs ellipses and elliptical arcs.
Erase	Removes objects from drawings.
eTransmit	Packages the drawing and related files for transmission by email or courier.
Explode	Breaks down blocks into individual objects; reduces polylines to lines and arcs.
Export	Exports drawings in several file formats.
Extend	Extends objects to meet boundary objects.
ExternalReferences	Displays the External References palette for attaching drawings, images, and *.dwf* files.
ExternalReferencesClose	Closes the External References palette; the longest command name in AutoCAD.

Command	Description
Extrude	Extrudes 2D closed objects into 3D solid objects.
F	
Field	Inserts updatable field text in drawings.
Fill	Toggles the display of solid fills.
Fillet	Connects two lines with an arc.
Filter	Creates selection sets of objects based on their properties.
Find	Finds and replaces text.
FlatShot	Creates 2D views of 3D models.
FreePoint	Places target-less point lights in drawings.
FreeSpot	Places target-less spot lights in drawings.
FreeWeb	Places target-less web lights in drawings.
G	
GeographicLocation	Specifies the sun position through date, time, city, longitude, and latitude.
GoToURL	Links to hyperlinks
Gradient	Opens the Hatch and Gradient dialog box to the Gradient tab.
GraphScr	F2: Switches to the drawing window from the Text window.
Grid	F7 and CTRL+G: Displays grid of specified spacing.
Group	Creates named selection sets of objects. (CTRL+SHIFT+A toggles group selection style.)
H	
Hatch	Opens the Hatch and Gradient dialog box.
HatchEdit	Edits associative hatch patterns.
Helix	Draws 3D helix objects (spirals).
Help	? and F1: Displays a list of AutoCAD commands with detailed information.
Hide	Removes hidden lines from the currently-displayed view.
HlSettings	Specifies properties for hidden line removal.
HyperLink	CTRL+K: Attaches hyperlinks to objects, or modifies existing hyperlinks.
HyperLinkOptions	Controls the visibility of the hyperlink cursor and the display of hyperlink tooltips.
I	
Id	Describes the position of a point in x,y,z coordinates.
-Image	Controls the insertion of raster images through the command line.
ImageAdjust	Controls the brightness, contrast, and fading of raster images.
ImageAttach	Attaches raster images to the current drawing.
ImageClip	Places rectangular or irregular clipping boundaries around images.
ImageFrame	Toggles the display and plotting of image frames.
ImageQuality	Controls the display quality of images.
Import	Imports a variety of file formats into drawings.
Imprint	Imprints 2D edges onto the faces of 3D models.
Insert	Inserts blocks and other drawings into the current drawing.
InsertObj	Inserts objects generated by another Windows application.
Interfere	Determines the interference of two or more 3D solids; displays dialog box with options.

Command	Description
Intersect	Creates a 3D solid or 2D region from the intersection of two or more 3D solids or 2D regions.
Isoplane	F5 and CTRL+E: Switches to the next isoplane.

J

Command	Description
JogSection	Adds 90-degree jogs to section planes.
Join	Unifies individual lines, arcs, splines, and elliptical arcs.
JpgOut	Exports views as JPEG files.
JustifyText	Changes the justification of text.

L

Command	Description
Layout	Creates a new layout and renames, copies, saves, or deletes existing layouts.
LayoutWizard	Designates page and plot settings for new layouts.
Leader	Draws leader dimensions.
Lengthen	Lengthens or shortens open objects.
Light	Creates, names, places, and deletes "lights" used by the RENDER command.
LightList	Edits light parameters in the Lights In Model palette.
LightListClose	Closes the Lights In Model palette.
Limits	Sets drawing boundaries.
Line	Draws straight line segments.
Linetype	Lists, creates, and modifies linetype definitions; loads them for use in drawings.
List	Displays database information for selected objects.
LiveSection	Activates section planes.
Load	Loads shape files into drawings.
Loft	Creates 3D surfaces and solids by lofting two or more curves.
LogFileOff	Closes the *.log* keyboard logging file.
LogFileOn	Writes the text of the 'Command:' prompt area to *.log* log file.
LWeight	Sets the current lineweight, lineweight display options, and lineweight units.

LAYER

Command	Description
LayCur	Changes the layer of selected objects to the current layer.
LayDel	Erases all objects from the specified layer; purges layer from drawing.
Layer	Creates and changes layers; toggles the state of layers; assigns linetypes, lineweights, plot styles, colors, and other properties to layers.
LayerP	Displays the previous layer state.
LayerPMode	Toggles the availability of the LAYERP command.
⊙ LayerState	Displays the Layer State dialog box for creating and editing layer states.
LayFrz	Freezes the layers of the selected objects.
LayIso	Turns off all layers except ones holding selected objects.
LayLck	Locks the layer of the selected object.
LayMch	Changes the layers of selected objects to that of a selected object.
LayMCur	Makes the selected object's layer current, like the AI_MOLC command.
LayMrg	Moves objects to another layer; purges the original layer from the drawing.
LayOff	Turns off the layer of the selected object.
LayOn	Turns on all layers.

Command	Description
LayThw	Thaws all layers.
LayTrans	Translates layer names from one space to another.
LayULk	Unlocks the layer of a selected object.
LayUnIso	Turns on layers that were turned off with the last LAYISO command.
LayVpi	Isolates the selected object's layer in the current viewport.
LayWalk	Displays objects on selected layers.

M

Command	Description
Markup	CTRL+7: Reviews and changes the status of marked-up *.dwf* files.
MarkupClose	CTRL+7: Closes the Markup Set Manager palette.
MassProp	Calculates and displays the mass properties of 3D solids and 2D regions.
MatchCell	Copies the properties from one table cell to other cells.
MatchProp	Copies properties from one object to other objects.
MaterialAttach	Attaches materials to objects by layer name.
MaterialMap	Maps materials to selected objects.
Materials	Manages, applies, and edits materials; opens the Materials palette.
MaterialsClose	Closes the Materials palette.
Measure	Places points or blocks at specified distances along objects.
Menu	Loads *.cui* and *.mnu* menu files of AutoCAD commands into the menu area.
MInsert	Inserts arrays of blocks.
Mirror	Creates mirror images of objects.
Mirror3D	Creates mirror images of objects rotated about a plane.
MLeader	Creates mult-line leaders.
MLeaderAlign	Aligns two or more mleaders.
MLeaderCollect	Collects two or more mleaders with block annotations into a single mleader with multiple annotations.
MLeaderEdit	Adds and removes leaders from mleaders.
MLeaderStyle	Specifies mleader styles.
MlEdit	Edits multilines.
MLine	Draws multiple parallel lines (up to 16).
MlStyle	Defines named mline styles, including color, linetype, and endcapping.
Model	Switches from layout tabs to Model tab.
Move	Moves objects.
MRedo	Redoes more than one undo operation.
MSlide	Creates *.sld* slide files of the current display.
MSpace	Switches to model space.
MtEdit	Edits mtext.
MText	Places formatted paragraph text inside a rectangular boundary.
Multiple	Repeats commands.
MView	Creates and manipulates viewports in paper space.
MvSetup	Sets up new drawings.

Command	Description
N	
Netload	Loads *.net* files.
New	CTRL+N: Creates new drawings.
NewView	Shortcut to the VIEW command's New View dialog box.
NewSheetSet	Creates a new sheet set.
O	
ObjectScale	Adds and removes scale factors from annotative objects.
Offset	Constructs parallel copies of objects.
OleConvert	Converts OLE (object linking and embedding) objects to other formats.
OleLinks	Controls objects linked to drawings.
OleOpen	Opens the source OLE file and application.
OleReset	Resets the OLE object to its original form.
OleScale	Displays the OLE Properties dialog box.
Oops	Restores objects accidentally erased by the previous command.
Open	CTRL+O: Opens existing drawings.
OpenDwfMarkup	Opens marked-up *.dwf* files.
OpenSheetset	Opens *.dst* sheet set files.
Options	Customizes AutoCAD's settings.
Ortho	F8 and CTRL+L: Forces lines to be drawn orthogonally or at an angle.
OSnap	F3 and CTRL+F: Locates geometric points of objects.
P	
PageSetup	Specifies the layout page, plotting device, paper size, and settings for new layouts.
Pan	Moves the view within the current viewport.
PartiaLoad	Loads additional geometry into partially-opened drawings.
-PartialOpen	Loads geometry from a selected view or layer into drawings.
PasteAsHyperlink	Pastes objects as hyperlinks.
PasteBlock	CTRL+SHIFT+V: Pastes copied block into drawings.
PasteClip	CTRL+V: Pastes objects from the Clipboard into the upper left corner of drawings.
PasteOrig	Pastes copied objects from the Clipboard into new drawings using the coordinates from the original.
PasteSpec	Controls the format of objects pasted from the Clipboard.
PcInWizard	Imports *.pcp* and *.pc2* configuration files of plot settings.
PEdit	Edits polylines and polyface objects.
PFace	Constructs polygon meshes defined by the location of vertices.
Plan	Returns to the plan view of the current UCS.
PlaneSurf	Creates flat surface areas.
PLine	Creates connected lines, arcs, and splines of specified width.
Plot	CTRL+P: Plots drawing to printers and plotters.
PlotStamp	Adds information about the drawing to the edge of the plot.
PlotStyle	Sets plot styles for new objects, or assigns plot styles to selected objects.
PlotterManager	Launches the Add-a-Plotter wizard and the Plotter Configuration Editor.
PngOut	Exports views as PNG files (portable network graphics format).

Command	Description
Point	Draws points.
PointLight	Places point lights in drawing.
Polygon	Draws regular polygons with a specified number of sides.
PolySolid	Converts lines, arcs, circles, and polylines into 3D solids with thickness.
PressPull	Presses and pulls bounded areas.
Preview	Provides a Windows-like plot preview.
Properties	CTRL+1: Displays and changes the properties of existing objects.
PropertiesClose	CTRL+1: Closes the Properties palette.
PSetUpIn	Imports user-defined page setups into new drawing layouts.
PSpace	Switches to paper space (layout mode).
Publish	Plots one or more drawings as a drawing set, or exports them in DWF format.
PublishToWeb	Creates a Web page from one or more drawings in DWF, JPEG, or PNG formats.
Purge	Deletes unused blocks, layers, linetypes, and so on.
Pyramid	Creates 3D solid pyramids and cones.

Q

Command	Description
QcClose	CTRL+8: Closes the Quick Calc palette.
QDim	Creates continuous dimensions quickly.
QLeader	Creates leaders and leader annotation quickly.
QNew	Starts new drawings based on template files.
QSave	CTRL+S: Saves drawings without requesting a file name.
QSelect	Creates selection sets based on filtering criteria.
QText	Redraws text as rectangles with the next regeneration.
QuickCalc	CTRL+8: Opens the Quick Calc palette.
QuickCui	Opens the Customize User Interface dialog box partially collapsed.
Quit	ALT+F4 and CTRL+Q: Exits AutoCAD.

R

Command	Description
Ray	Draws semi-infinite construction lines.
Recover	Attempts to recover corrupted or damaged files.
RecoverAll	Recovers and updates drawings and all attached reference files.
Rectang	Draws rectangles.
Redefine	Restores AutoCAD's definition of undefined commands.
Redo	CTRL+Y: Restores the operations changed by the previous UNDO command.
Redraw	Cleans up the display of the current viewport.
RedrawAll	Redraws all viewports.
RefClose	Saves or discards changes made during in-place editing of xrefs and blocks.
RefEdit	Selects references for editing.
RefSet	Adds and removes objects from a working set during in-place editing of references.
Regen	Regenerates the drawing in the current viewport.
RegenAll	Regenerates all viewports.
RegenAuto	Controls whether drawings are regenerated automatically.
Region	Creates 2D region objects from existing closed objects.

Command	Description
Reinit	Reinitializes the I/O ports, digitizer, display, plotter, and the *acad.pgp* file.
Rename	Renames blocks, linetypes, layers, text styles, views, and so on.
Render	Renders 3D objects.
RenderCrop	Renders a rectangular area of the drawing.
RenderEnvironment	Sets the fog options; displays the Render Environment dialog box.
2008 **RenderExposure**	Changes the brightness and contrast of renderings.
RenderPresets	Specifies the parameters for renderings; displays the Render Presets Manager dialog box.
RenderWin	Displays the Render window, which lists and displays the current and previous renderings and parameters.
ResetBlock	Resets dynamic blocks to their original definitions.
Resume	Continues playing a script file that had been interrupted by the ESC key.
RevCloud	Draws revision clouds.
Revolve	Creates 3D solids by revolving 2D closed objects around an axis.
RevSurf	Draws revolved 3D meshes.
Rotate	Rotates objects about specified center points.
Rotate3D	Rotates objects about a 3D axis.
RPref	Sets preferences for renderings; displays the Render Settings palette.
RPrefClose	Closes the Render Settings palette.
RScript	Restarts scripts.
RuleSurf	Draws ruled 3D meshes.

S

Command	Description
Save / SaveAs	CTRL+SHIFT+S: Saves the current drawing by a specified name.
SaveImg	Saves the current rendering in BMP, TGA, or TIF formats.
Scale	Changes the size of objects equally in the x, y, z directions.
ScaleListEdit	Adds and removes scale factors used by PLOT and other commands.
ScaleText	Resizes text.
Script	Runs script files in AutoCAD.
Section	Creates 2D regions from 3D solids by intersecting a plane through the solid.
SectionPlane	Creates section objects independently of the 3D model.
SecurityOptions	Sets up passwords and digital signatures for drawings.
Select	Preselects objects to be edited; CTRL+A selects all objects.
2008 **SetByLayer**	Resets properties to those specified by layers.
SetiDropHandler	Specifies how to treat i-drop objects when dragged into drawings from Web sites.
SetVar	Views and changes AutoCAD's system variables.
-ShadeMode	Shades 3D objects in the current viewport.
Shape	Places shapes from shape files into drawings.
Sheetset	CTRL+4: Opens the Sheet Set Manager palette.
SheetsetHide	CTRL+4: Closes the Sheet Set Manager palette.
Shell	Runs other programs outside of AutoCAD.
ShowMat	Reports the material definition assigned to selected objects.
SigValidate	Displays digital signature information in drawings.
Sketch	Allows freehand sketching.

Command	Description
Slice	Slices 3D solids with a plane.
Snap	F9 and CTRL+B: Toggles snap mode on or off, changes the snap resolution, sets spacing for the X- and Y-axis, rotates the grid, and sets isometric mode.
SolDraw	Creates 2D profiles and sections of 3D solid models in viewports created with the SOLVIEW command.
Solid	Draws filled triangles and rectangles.
SolidEdit	Edits faces and edges of 3D solid objects.
SolProf	Creates profile images of 3D solid models.
SolView	Creates viewports in paper space of orthogonal multi- and sectional view drawings of 3D solid model.
SpaceTrans	Converts length values between model and paper space.
Spell	Checks the spelling of text in the drawing.
Sphere	Draws 3D spheres.
Spline	Draws NURBS (spline) curve.
SplinEdit	Edits splines.
Spotlight	Places spotlights in drawings.
Standards	Compares CAD standards between two drawings.
Status	Displays information about the current drawing.
StlOut	Exports 3D solids to *.stl* files, in ASCII or binary format, for use with stereolithography.
Stretch	Moves selected objects while keeping connections to other objects unchanged.
Style	Creates and modifies text styles.
StylesManager	Displays the Plot Style Manager.
Subtract	Creates new 3D solids and 2D regions by subtracting one set of objects from a second.
SunProperties	Sets the properties of the Sun light; opens the Sun Properties palette.
SunPropertiesClose	Closes the Sun Properties palette.
Sweep	Creates 3D solids and surfaces by sweeping 2D curves along paths.
SysWindows	CTRL+TAB: Controls the size and position of windows.

T

Command	Description
Table	Inserts tables in drawings.
TablEdit	Edits the content of table cells.
TableExport	Exports tables as comma-separated text files.
TableStyle	Specifies table styles.
Tablet	F4: Aligns digitizers with existing drawing coordinates; operates only when a digitizer is connected.
TabSurf	Draws tabulated 3D meshes.
TargetPoint	Places target point lights in drawings.
Text	Places text in the drawing.
TextScr	F2: Displays the Text window.
TextToFront	Displays text and/or dimensions in front of all other objects.
Thicken	Creates 3D solids by thickening surfaces.
TifOut	Exports views as TIFF files (tagged image file format).
Time	Tracks time.
TInsert	Inserts blocks and drawings in table cells.
Tolerance	Selects tolerance symbols.

Command	Description
-Toolbar	Controls the display of toolboxes.
ToolPalettes	CTRL+3: Opens the Tool palette.
ToolPalettesClose	CTRL+3: Closes the Tool palette.
Torus	Draws doughnut-shaped 3D solids.
2008 **TpNavigate**	Sets the default tab for the Tools palette.
Trace	Draws lines with width.
Transparency	Toggles the background of bilevel images between transparent and opaque.
TraySettings	Specifies options for commands operating from the tray (right end of status bar).
TreeStat	Displays information on the spatial index.
Trim	Trims objects by defining other objects as cutting edges.

U

Command	Description
U	CTRL+Z: Undoes the effect of commands.
UCS	Creates and manipulates user-defined coordinate systems.
UCSicon	Controls the display of the UCS icon.
UcsMan	Manages user-defined coordinate systems.
Undefine	Disables commands.
Undo	Undoes several commands in a single operation.
Union	Creates new 3D solids and 2D regions from two solids or regions.
Units	Selects the display format and precision of units and angles.
UpdateField	Updates the contents of field text.
UpdateThumbsNow	Updates the preview images in sheet sets.

V

Command	Description
VbaIDE	ALT+F8: Launches the Visual Basic Editor.
VbaLoad	Loads VBA projects into AutoCAD.
VbaMan	Loads, unloads, saves, creates, embeds, and extracts VBA projects.
VbaRun	ALT+F11: Runs VBA macros.
VbaStmt	Executes VBA statements at the command prompt.
VbaUnload	Unloads global VBA projects.
View	Saves the display as a view; displays named views.
ViewPlotDetails	Reports on successful and unsuccessful plots.
ViewRes	Controls the fast zoom mode and resolution for circle and arc regenerations.
VisualStyles	Defines and edits visual styles used for renderings; displays the Visual Styles Manager palette.
VisualStylesClose	Closes the Visual Styles Manager palette.
VLisp	Launches the Visual LISP interactive development environment.
VpClip	Clips viewport objects.
VpLayer	Controls the independent visibility of layers in viewports.
VpMin	Restores the maximized viewport.
VpMax	Maximizes the current viewport (in layout mode).
VPoint	Sets the viewpoint from which to view 3D drawings.
VPorts	CTRL+R: Sets the number and configuration of viewports.
VsCurrent	Applies a named visual style to the current viewport.

Command	Description
VSlide	Displays *.sld* slide files created with MSLIDE.
VsSave	Saves a visual style by name in model space only.
VtOptions	Specifies options for visual transitions during the ZOOM command.
W	
WalkFlySettings	Specifies the settings for walk and fly animations; displays the Walk and Fly Settings dialog box.
WBlock	Writes objects to drawing files.
WebLight	Places web lights in drawings; these use *.ies* files to define light distribution.
Wedge	Creates 3D solid wedges.
WhoHas	Displays ownership information for opened drawing files.
WipeOut	Creates blank areas in drawings.
WmfIn	Imports *.wmf* files into drawings as blocks.
WmfOpts	Controls how *.wmf* files are imported.
WmfOut	Exports drawings as *.wmf* files.
WorkSpace	Creates, modifies, and saves workspaces; sets the current workspace.
WsSave	Saves the current user interface as a workspace.
WsSettings	Specifies options for workspaces through a dialog box.
X	
XAttach	Attaches externally-referenced drawing files to the drawing.
XBind	Binds externally-referenced drawings; converts them to blocks.
XClip	Defines clipping boundaries; sets the front and back clipping planes.
XEdges	Creates wireframe geometry from the edges of 3D solids and surfaces, as well as 2D regions.
XLine	Draws infinite construction lines.
XOpen	Opens externally-referenced drawings in independent windows.
Xplode	Breaks compound objects into component objects, with user control.
-Xref	Places externally-referenced drawings into drawings.
Z	
Zoom	Increases and decreases the viewing size of drawings.
3	
3D	Draws 3D objects of polygon meshes (boxes, cones, dishes, domes, meshes, pyramids, spheres, tori, and wedges).
3dAlign	Similar to the ALIGN command, but with different prompts.
3dArray	Creates 3D arrays.
3dClip	Switches to interactive 3D view, and opens the Adjust Clipping Planes palette.
3dConfig	Configures the 3D graphics system in a dialog box.
3dCOrbit	Switches to interactive 3D view, and sets objects into continuous motion.
3dDistance	Switches to interactive 3D view, and makes objects appear closer or farther away.
3dDWF	Exports drawings as 3D DWF files; displays the Export 3D DWF dialog box.
3dDwfPublish	Exports drawings as 3D *.dwf* files.
3dFace	Creates 3D faces.

Command	Description
3dFly	Navigates 3D drawings in fly-through mode; displays the Position Locator palette.
3dFOrbit	Displays an arcball for realtime view changes in 3D.
3dMesh	Draws 3D meshes.
3dMove	Displays the move handle on selected objects, so that they can be moved in 3D space.
3dOrbit	Controls the interactive 3D viewing.
3dOrbitCtr	Centers the view.
3dOrbitTransparent	Changes the 3D view in real time; an undocumented transparent command.
3dPan	Invokes interactive 3D view to drag the view horizontally and vertically.
3dPanTransparent	Pans 3D views in real time; an undocumented transparent command.
3dPoly	Draws 3D polylines.
3dRotate	Displays the rotate handle on selected objects so that they can be rotated in 3D space.
3dsIn	Imports 3D Studio geometry and rendering data.
3dSwivel	Switches to interactive 3D view, simulating the effect of turning the camera.
3dSwivelTransparent	Swivels the 3D viewpoint in real time; an undocumented transparent command.
3dWalk	Enters Walk mode; accesses the Walk Mode right-click menu.
3dZoom	Switches to interactive 3D view to zoom in and out.
3dZoomTransparent	Zooms the 3D view in real time; an undocumented transparent command.

COMMAND ALIASES

Aliases are shortened versions of command names. They are defined in the *acad.ppg* file. Those listed below in italics can be used only in the Block Editor.

Command	Aliases
A	
AdCenter	adc, dcenter, dc, content
Align	al
AppLoad	ap
Arc	a
Area	aa
Array	ar
-Array	-ar
AttDef	att, ddattdef
-AttDef	-att
AttEdit	ate, ddatte
-AttEdit	-ate, atte
AttExt	ddattext
AttIpEdit	ati
B	
BAction	*ac*
BClose	*cc*
BEdit	be
Block	b, bmake, bmod, acadblockdialog
-Block	-b
Boundary	bo, bpoly
-Boundary	-bo
BParameter	*param*
Break	br
BSave	*bs*
BvState	*bvs*
C	
Camera	cam
Chamfer	cha
Change	-ch
CheckStandards	chk
Circle	c
Color	col, colour, ddcolor
CommandLine	cli
Copy	cp, co

Command	Aliases
CTableStyle	ct
Cui	toolbar, to, tbconfig
CuiLoad	menuload
CuiUnload	menuunload
Cylinder	cyl
D	
DataExtraction	dx, eattext
DataLink	dl
DataLinkUpdate	dlu
DbConnect	dbc, ase, aad, aex, asq, ali, aro
DdEdit	ed
DdVpoint	vp
Dist	di
Divide	div
Donut	doughnut
DrawingRecovery	drm
Draworder	dr
DSettings	ds, se, ddrmodes, osnap, os, ddosnap
DsViewer	av
DView	dv
DIMENSIONS	
DimAligned	dal, dimali
DimAngular	dan, dimang
DimArc	dar
DimBaseline	dba, dimbase
DimCenter	dce
DimContinue	dco, dimcont
DimDiameter	ddi, dimdia
DimDisassociate	dda
DimEdit	dimed, ded
DimJogged	Jog
DimJogLine	djl
DimLinear	dli, dimlin

Command	Aliases
DimOrdinate	dor, dimord
DimOverride	dov, dimover
DimRadius	dra, dimrad
DimReassociate	dre
DimStyle	d, dimsty, dst, ddim
DimTEdit	dimted
E	
Ellipse	el
Erase	e
Explode	x
Export	exp
-ExportToAutoCAD	aectoacad
Extend	ex
ExternalReferences	er, image, im, xref, xr
Extrude	ext
F	
Fillet	f
Filter	fi
FlatShot	fshot
G	
GeographicLocation	geo, north, northdir
Gradient	dd
Group	g
-Group	-g
H	
Hatch	bhatch, h, bh
-Hatch	-h
Hatchedit	he
Hide	hi
I	
-Image	-im
ImageAdjust	iad
ImageAttach	iat
ImageClip	icl
Import	imp
Insert	i, inserturl, ddinsert
-Insert	-i
Insertobj	io
Interfere	inf
Intersect	in

Command	Aliases
J	
Join	j
L	
Layer	la, ddlmodes
-Layer	-la
LayerState	las, lman
-Layout	lo
Leader	lead
Lengthen	len
Line	l
Linetype	ltype, lt, ddltype
-Linetype	-ltype, -lt
List	ls, li
LtScale	lts
Lweight	lineweight, lw
M	
Markup	msm
MatchProp	ma, painter
MaterialMap	setuv
Materials	mat, finish, rmat
Measure	me
Mirror	mi
Mirror3D	3dmirror
MLeader	mld
MLeaderAlign	mla
MLeaderCollect	mlc
MLeaderEdit	mle
MLeaderStyle	mls
MLine	ml
Move	m
MSpace	ms
MText	t, mt
-MText	-t
MView	mv
O	
Offset	o
Open	openurl, dxfin
Options	op, preferences, ddselect, ddgrips, gr
-OSnap	-os

Command	Aliases
P	
Pan	p
-Pan	-p
-PartialOpen	partialopen
PasteSpec	pa
PEdit	pe
PLine	pl
Plot	print, dwfout
PlotStamp	ddplotstamp
Point	po
PointLight	freepoint
Polygon	pol
PolySolid	psolid
Preview	pre
Properties	props, pr, mo, ch, ddchprop, ddmodify,
PropertiesClose	prclose
PSpace	ps
PublishToWeb	ptw
Purge	pu
-Purge	-pu
Pyramid	pyr
Q	
QLeader	le
QuickCalc	qc
QuickCui	qcui
Quit	exit
R	
Rectang	rec, rectangle
Redraw	r
RedrawAll	ra
Regen	re
RegenAll	rea
Region	reg
Rename	ren
-Rename	-ren
Render	rr
RenderCrop	rc
RenderEnvironment	fog
RenderPresets	rp, rfileopt
RenderWin	rw, rendscr
Revolve	rev

Command	Aliases
Rotate	ro
RPref	rpr
S	
Save	saveurl
SaveAs	dxfout
Scale	sc
Script	scr
Section	sec
SectionPlane	splane
SetVar	set
Sheetset	ssm
Shell	sh
Slice	sl
Snap	sn
Solid	so
Spell	sp
Spline	spl
SplinEdit	spe
Standards	sta
Stretch	s
Style	st, ddstyle
Subtract	su
T	
Table	tb
TableStyle	ts
Tablet	ta
Text	dt, dtext
Thickness	th
Tilemode	ti, tm
Tolerance	tol
ToolPalettes	tp
Torus	tor
Tracking	tk, track
Trim	tr
U	
Ucs	dducs
UcsMan	uc, dducs, dducsp
Union	uni
Units	un, ddunits
-Units	-un

Command	Aliases
V	
View	v, ddview
-View	-v
VisualStyles	vsm
-VisualStyles	-vsm
VLisp	vlide
VPoint	-vp
VPorts	viewports
VsCurrent	vs, shademode, sha
W	
Wblock	w, acadwblockdialog
-Wblock	-w
Wedge	we
X	
XAttach	xa
XBind	xb
-XBind	-xb
XClip	xc
XLine	xl
Xplode	xp
-XRef	-xr

Command	Aliases
Z	
Zoom	Z
3	
3dAlign	3al
3dArray	3a
3dDwf	3dDwfPublish
3dFace	3f
3dMove	3m
3dOrbit	3do, orbit
3dPoly	3p
3dRotate	3r
3dWalk	3dw, 3dnavigate

KEYBOARD SHORTCUTS

CONTROL KEYS

Ctrl-key	Meaning
CTRL	Unlocks locked toolbars and windows temporarily; selects a face when picking 3D solid models.
CTRL+0	Toggles clean screen.
CTRL+1	Toggles the Properties palette.
CTRL+2	Toggles the AutoCAD DesignCenter palette.
CTRL+4	Toggles Sheetset Manager palette.
CTRL+6	Launches dbConnect.
CTRL+7	Toggles Markup Set Manager palette.
CTRL+8	Toggles QuickCalc palette.
CTRL+9	Toggles Command Line palette.
CTRL+A	Selects all objects in the current model or layout space.
CTRL+SHIFT+A	Toggles group selection mode.
CTRL+B	Turns snap mode on or off.
CTRL+C	Copies selected objects to the Clipboard.
CTRL+SHIFT+C	Copies selected objects with a base point to the Clipboard.
CTRL+D	Changes the coordinate display mode.
CTRL+E	Switches to the next isoplane.
CTRL+F	Toggles object snap on and off.
CTRL+G	Turns the grid on and off.
CTRL+H	Toggles pickstyle mode.
CTRL+K	Creates a hyperlink.
CTRL+L	Turns ortho mode on and off.
CTRL+N	Starts a new drawing.
CTRL+O	Opens a drawing.
CTRL+P	Prints the drawing.
CTRL+Q	Quits AutoCAD.
CTRL+R	Switches to the next viewport.
CTRL+S	Saves the drawing.
CTRL+SHIFT+S	Displays the Save Drawing As dialog box.
CTRL+T	Toggles tablet mode.
CTRL+V	Pastes from the Clipboard into the drawing or to the command prompt area.
CTRL+SHIFT+V	Pastes with an insertion point.
CTRL+X	Cuts selected objects to the Clipboard.
CTRL+Y	Performs the REDO command.
CTRL+Z	Performs the U command.
CTRL+TAB	Switches to the next drawing.

ALTERNATE KEYS

Alt-key	Meaning
ALT	Accesses keystroke shortcuts on menus.
ALT+TAB	Switches to the next application.

COMMAND LINE KEYSTROKES

These keystrokes are used in the command-prompt area and the Text window.

Keystroke	Meaning
left arrow	Moves the cursor one character to the left.
right arrow	Moves the cursor one character to the right.
HOME	Moves the cursor to the beginning of the line of command text.
END	Moves the cursor to the end of the line.
DEL	Deletes the character to the right of the cursor.
BACKSPACE	Deletes the character to the left of the cursor.
INS	Switches between insert and typeover modes.
up arrow	Displays the previous line in the command history.
down arrow	Displays the next line in the command history.
PGUP	Displays the previous screen of command text.
PGDN	Displays the next screen of command text.
CTRL+V	Pastes text from the Clipboard into the command line.
ESC	Cancels the current command.
TAB	Cycles through command names.

FUNCTION KEYS

Function Key	Meaning
F1	Calls up the help window.
F2	Toggles between the graphics and text windows.
F3	Toggles object snap on and off.
F4	Toggles tablet mode on and off; tablet must first be calibrated.
F5	Switches to the next isometric plane when in iso mode: left, top, and right.
F6	Toggles dynamic UCS on and off.
F7	Toggles grid display on and off.
F8	Toggles ortho mode on and off.
F9	Toggles snap mode on and off; to temporarily override, hold down F9 while selecting objects.
F10	Toggles polar tracking on and off.
F11	Toggles object snap tracking on and off.
F12	Toggles dynamic input.
ALT+F4	Exits AutoCAD.
ALT+F8	Displays the Macros dialog box.
ALT+F11	Starts Visual Basic for Applications editor.
CTRL+F4	Closes the current drawing.

TEMPORARY OVERRIDE (SHIFT) KEYS

Shift-key	Meaning
SHIFT	Toggles orthogonal mode;
	Selects 3D faces, edges, and vertices;
	Activates **3DORBIT** command;
	Switches between trim and extend modes during the EXTEND and TRIM commands.
SHIFT+A	Overrides object snap (osnap) mode.
SHIFT+C	Overrides CENter osnap.
SHIFT+D	Disables all snap modes and tracking.
SHIFT+E	Overrides ENDpoint osnap.
SHIFT+J	Overrides osnap tracking mode.
SHIFT+L	Disables all snap modes and tracking.
SHIFT+M	Overrides MIDpoint osnap.
SHIFT+P	Overrides ENDpoint osnap.
SHIFT+Q	Overrides osnap tracking mode.
SHIFT+S	Enables osnap enforcement.
SHIFT+V	Overrides MIDpoint osnap.
SHIFT+X	Overrides polar mode.
SHIFT+'	Overrides osnap mode.
SHIFT+,	Overrides CENter osnap.
SHIFT+.	Overrides polar mode.
SHIFT+;	Enables osnap enforcement.

MOUSE AND DIGITIZER BUTTONS

Button #	Mouse Button	Meaning
...	Wheel	Zooms or pans.
1	Left	Selects objects.
3	Center, wheel	Displays object snap menu.
2	Right	Displays shortcut menus.
4		Cancels command.
5		Toggles snap mode.
6		Toggles orthographic mode.
7		Toggles grid display
8		Toggles coordinate display.
9		Switches to isometric plane.
10		Toggles tablet mode.
SHIFT+1	SHIFT+Left	Toggles cycle mode.
SHIFT+2	SHIFT+Left	Displays object snap shortcut menu.
CTRL+2	CTRL+Right	Displays object snap shortcut men

MTEXT EDITOR SHORTCUT KEYS

Shortcut	Meaning
TAB	Moves cursor to the next tab stop, or creates the next indent level.
SHIFT+TAB	Outdents.
CTRL+A	Selects all text.
CTRL+B	**Boldfaces**.
CTRL+I	*Italicize*.
CTRL+U	<u>Underlines</u>.
CTRL+C	Copies selected text to the Clipboard.
CTRL+X	Cuts text from the editor and send it to the Clipboard.
CTRL+V	Pastes text from the Clipboard.
CTRL+Y	Redoes last undo.
CTRL+Z	Undoes last action.
CTRL+F	Inserts field text; displays Field dialog box.
CTRL+R	Finds and replaces text.
CTRL+SPACE	Removes formatting from selected text.
CTRL+SHIFT+SPACE	Inserts non-breaking space.
CTRL+SHIFT+U	Converts selected text to all UPPERCASE.
CTRL+SHIFT+L	Converts selected text to all lowercase.

C

AutoCAD System Variables

AutoCAD stores information about its current state, the drawing and the operating system in over 400 *system variables*. Those variables help programmers — who often work with menu macros and AutoLISP — to determine the state of the AutoCAD system.

CONVENTIONS

The following pages list all documented system variables, plus several more not documented by Autodesk. The listing uses the following conventions:

Bold	System variable is documented by Autodesk in AutoCAD 2008.
Bold Italic	System variable is listed neither by the **SETVAR** command nor in Autodesk's documentation.
🖭	System variable must be accessed through the **SETVAR** command; otherwise, the equivalent command is executed.
Default	Default value, as set in the *acad.dwg* prototype drawing; other template files may have different default values.
(R/O)	System variable is read-only (cannot be changed by the user).
Toggle	System variable has one of two values: 0 (off, closed, disabled) or 1 (on, open, enabled).
ACI color	System variable takes on one of AutoCAD's 255 colors; ACI is short for "AutoCAD color index."
🗗	System variable is new to AutoCAD 2008.

Variable Name	Default Value	Meaning
_Server (R/O)	0	Specifies the network authorization code.
_ToolPalettePath	*varies*	*Specifires the path to the Tool palette folder.*

A

Variable Name	Default Value	Meaning
AcadLspAsDoc	0	Controls whether *acad.lsp* is loaded into: **0** The first drawing only. **1** Every drawing.
AcadPrefix (R/O)	*varies*	Specifies paths used by AutoCAD search in Options | Files dialog box.
AcadVer (R/O)	"17.1"	Specifies the AutoCAD version number.
AcisOutVer	70	Controls the ACIS version number; values are 15, 16, 17, 18, 20, 21, 30, 40, or 70.
AdcState (R/O)	0	Toggle: reports if DesignCenter is active.
AFlags	0	Controls the default attribute display mode: **0** No mode specified. **1** Invisible. **2** Constant. **4** Verify. **8** Preset. **16** Lock position in block. **32** Multiple-line attributes.
AngBase	0	Controls the direction of zero degrees relative to the UCS.
AngDir	0	Controls the rotation of positive angles: **0** Clockwise. **1** Counterclockwise.
AnnoAllVisible	1	Toggles display of annotative objects at the current scale: **0** Displays only objects matching percent scale. **1** Displays all annotative objects (default).
AnnoAutoScale	4	Updates annotative objects when the annotation scale is changed: **1** Adds the new annotation scale to annotative objects (except on layers that are off, frozen, locked, or have viewports set to freeze). **2** As above, but includes objects on locked layers (excludes objects on layers that are off, frozen, or have viewports set to freeze). **3** The opposite of 2: applies to annotative objects on all layers, except objects on locked layers. **4** Applies to all objects regardless of layer status. **-1** Same as 1, but turned off. **-2** Same as 2, but turned off. **-3** Same as 3, but turned off. **-4** Same as 4, but turned off.
AnnotativeDwg	0	Toggles whether the drawing acts like an annotative block when inserted into other drawings: **0** Nonannotative block. **1** Annotative block. (This sysvar is read-only when drawing contains annotative text.)
ApBox	0	Toggles the display of the AutoSnap aperture box cursor.
Aperture	10	Controls the object snap aperture in pixels: **1** Minimum size. **50** Maximum size.
ApState	*0*	*Toggle: reports the state of the* **APBOX** *variable.*
Area (R/O)	0.0	Reports the area measured by the last **SREA** command.
AttDia	0	Controls the user interface for entering attributes: **0** Command-line prompts. **1** Dialog box.

Variable Name	Default Value	Meaning
AttIpe	0	Toggles display of in-place attribute text editor's toolbar: **0** Reduced-function toolbar. **1** Full-function toolbar.
AttMode	1	Controls the display of attributes: **0** Off. **1** Normal. **2** On; displays invisible attributes.
AttMulti	1	Toggles creation of multi-line attributes: **0** Attributes can be single-line only. **1** Attributes can be multi-line
AttReq	1	Toggles attribute values during insertion: **0** Uses default values. **1** Prompts user for values.
AuditCtl	0	Toggles creation of *.adt* audit log files: **0** File not created. **1** File created.
AUnits	0	Controls the type of angular units: **0** Decimal degrees. **1** Degrees-minutes-seconds. **2** Grads. **3** Radians. **4** Surveyor's units.
AUPrec	0	Controls the number of decimal places displayed by angles; range is 0 to 8.
AutoDwfPublish	1	Toggles automatic publishing of DWF Files when drawings are saved or closed: **0** DWF are not published. **1** DWF are published.
AutoSnap	55	Controls the AutoSnap display (sum): **0** Turns off all AutoSnap features. **1** Turns on marker. **2** Turns on SnapTip. **4** Turns on magnetic cursor. **8** Turns on polar tracking. **16** Turns on object snap tracking. **32** Turns on tooltips for polar tracking and object snap tracking.

B

Variable Name	Default Value	Meaning
BackgroundPlot	2	Controls background plotting and publishing (ignored during scripts): **0** Plot foreground; publish foreground. **1** Plot background; publish foreground. **2** Plot foreground; publish background. **3** Plot background; publish background.
BackZ	0.0	Controls the location of the back clipping plane offset from the target plane.
BAction Color	"7"	Specifies ACI text color for actions in Block Editor.
BDependencyHighlight	1	Toggles highlighting of dependent objects when parameters, actions, or grips selected in Block Editor: **0** Not highlighted. **1** Highlighted.
BGripObjColor	"141"	Specifies ACI color of grips in Block Editor.
BGripObjSize	8	Controls the size of grips in Block Editor; range is 1 to 255.
BgrdPlotTimeout	*20*	*Controls the timeout for failed background plots; ranges from 0 to 300 seconds.*
BindType	0	Controls how xref names are converted when being bound or edited: **0** From *xref|name* to *xref\$0\$name*. **1** From *xref|name* to *name*.

Variable Name	Default Value	Meaning
⌨ **BlipMode**	0	Toggles the display of blip marks.
BlockEditLock	0	Toggles the locking of dynamic blocks being edited: **0** Unlocked. **1** Locked.
BlockEditor (R/O)	0	Toggle: reports whether Block Editor is open.
BParameterColor	"7"	Specifies ACI color of parameters in the Block Editor.
BParameterFont	"Simplex.shx"	Controls the font used for parameter and action text in the Block Editor.
BParameterSize	12	Controls the size of parameter text and features in Block Editor: **1** Minimum. **255** Maximum.
BTMarkDisplay	1	Controls the display of value set markers: **0** Unlocked. **1** Locked.
BVMode	0	Controls the display of invisible objects in the Block Editor: **0** Invisible. **1** Visible and dimmed.

. . . .

C

Variable Name	Default Value	Meaning
CalcInput	1	Controls how formulas and global constants are evaluated in dialog boxes: **0** Not evaluated. **1** Evaluated after pressing ALT+ENTER.
CameraDisplay	0	Toggles the display of camera glyphs.
CameraHeight	0	Specifies the default height of cameras.
ⓐ **CAnnoScale**	"1:1"	Names the current annotative scale for the current viewport.
ⓐ **CAnnoScaleValue**(R/O)	1.0	Reports the current annotation scale.
CDate (R/O)	*varies*	Specifies the current date and time in the format YyyyMmDd.HhMmSsDd, such as 20080503.18082328
CeColor	"BYLAYER"	Controls the current color.
CeLtScale	1.0	Controls the current linetype scaling factor.
CeLType	"BYLAYER"	Controls the current linetype.
CeLWeight	-1	Controls the current lineweight in millimeters; valid values are 0, 5, 9, 13, 15, 18, 20, 25, 30, 35, 40, 50, 53, 60, 70, 80, 90, 100, 106, 120, 140, 158, 200, and 211, plus the following: **-1** BYLAYER. **-2** BYBLOCK. **-3** Default, as defined by LwDdefault.
CenterMT	0	Controls how corner grips stretch uncentered multiline text: **0** Center grip moves in same direction; opposite grip stays in place. **1** Center grip stays in place; both side grips move in direction of stretch.
ChamferA	0.0	Specifies the current value of the first chamfer distance.
ChamferB	0.0	Specifies the current value of the second chamfer distance.
ChamferC	0.0	Specifies the current value of the chamfer length.
ChamferD	0	Specifies the current value of the chamfer angle.
ChamMode	0	Toggles the chamfer input mode: **0** Chamfer by two lengths. **1** Chamfer by length and angle.
ⓐ *CipMode*	*0*	*Toggles Customer Involvement Program.*
CircleRad	0.0	Specifies the most-recent circle radius.
CLayer	"0"	Specifies name of current layer.
CleanScreenState (R/O)	0	Toggle: reports whether cleanscreen mode is active.
CliState (R/O)	1	Reports the command line palette.

Variable Name	Default Value	Meaning
CMaterial	"ByLayer"	Sets the name of the current material.
CmdActive (R/O)	1	Reports the type of command currently active (used by programs): **1** Regular command. **2** Transparent command. **4** Script file. **8** Dialog box. **16** Dynamic data exchange. **32** AutoLISP command. **64** ARX command.
CmdDia	1	Toggles **QLEADER**'s inplace text editor.
CmdEcho	1	Toggles AutoLISP command display: **0** No command echoing. **1** Command echoing.
CmdInputHistoryMax	20	Controls the maximum command input items stored; works with **INPUTHISTORYMODE**.
🅰 **CMleaderStyle**	"Standard"	Reports name of current mleader style.
CmdNames (R/O)	*varies*	Reports the name of the command currently active, such as "SETVAR."
CMLJust	0	Controls the multiline justification mode: **0** Top. **1** Middle. **2** Bottom.
CMLScale	1.0	Controls the scale of overall multiline width: *-n* Flips offsets of multiline. **0** Collapses to single line. *n* Scales by a factor of *n*.
CMLStyle	"STANDARD"	Specifies the current multiline style name.
Compass	0	Toggles the display of the 3D compass.
Coords	1	Controls the coordinate display style: **0** Updated by screen picks. **1** Continuous display. **2** Polar display upon request.
🅰 **CopyMode**	0	Toggles whether the **COPY** command repeats itself: **0** Command repeats copying. **1** Command makes one copy, then exits.
CPlotStyle	"ByColor"	Specifies the current plot style; options for named plot styles are: ByLayer, ByBlock, Normal, and User Defined.
CProfile (R/O)	"<<Unnamed Profile>>"	Specifies the name of the current profile.
CrossingColor	3	ACI color of crossing rectangle.
CShadow	0	Shadows cast by 3D objects, if graphics board is capable: **0** Casts and receives shadows. **1** Casts shadows. **2** Receives shadows. **3** Ignores shadows.
CTab	"Model"	Specifies the name of the current tab.
CTableStyle	"Standard"	Specifies the name of the current table style name.
CursorSize	5	Controls the cursor size as a percentage of the viewport size. **1** Minimum size. **100** Full viewport.
CVPort	2	Specifies the current viewport number.

. . . .

D

DashboardState	0	Toggle: reports whether Dashboard is open.

Variable Name	Default Value	Meaning
⊡ **DataLinkNotify**	2	Controls reporting on data links: **0** All notifications disabled. **1** Displays data link icon in tray. **2** Displays the icon and a warning balloon in the tray.
Date (R/O)	*varies*	Reports the current date in Julian format, such as 2448860.54043252.
DbcState (R/O)	0	Toggles: specifies whether dbConnect Manager is active.
DblClkEdit	1	Toggles editing by double-clicking objects.
DBMod (R/O)	4	Reports how the drawing has been modified: **0** No modification since last save. **1** Object database modified. **2** Symbol table modified. **4** Database variable modified. **8** Window modified. **16** View modified. **32** Field modified.
DctCust	"sample.cus"	Specifies the name of custom spelling dictionary.
DctMain	"enu"	Controls the code for spelling dictionary: **enu** American English **eng** British English (ise) **enc** Canadian English **cat** Catalan **csy** Czech **dan** Danish **nld** Dutch (primary) **fin** Finnish **fra** French (accented capitals) **frc** French (unaccented capitals) **deu** German (post-reform) **deo** German (pre-reform) **ita** Italian **nor** Norwegian (Bokmal) **ptb** Portuguese (Brazilian) **ptg** Portuguese (Iberian) **rus** Russian **esp** Spanish **sve** Swedish
DefaultLighting	1	Toggles distant lighting.
DefaultLightingType	0	Toggles between new (1) and old (0) type of lights.
DefaultViewCategory	*""*	*Specifies the default name for View Category in the* VIEW *command's New View dialog box*
DefLPlStyle	"ByColor"	Reports the default plot style for layer 0.
DefPlStyle	"ByColor"	Reports the default plot style for new objects.
DelObj	1	Toggles the deletion of source objects: **-2** Users are prompted whether to erase all defining objects. **-1** Users are prompted whether to erase profiles and cross sections. **0** Objects retained. **1** Profiles and cross sections erased. **2** All defining objects erased.
DemandLoad	3	Controls application loading when drawing contains proxy objects: **0** Apps not demand-loaded. **1** Apps loaded when drawing opened. **2** Apps loaded at first command. **3** Apps loaded when drawing is opened or at first command.
⊡ **DgnFrame**	0	Controls display of DGN underlay frame: **0** Does not display frame. **1** Displays and plots frame. **2** Displays but does not plot frame.

Variable Name	Default Value	Meaning
ⓐ **DgnOsnap**	1	Toggles object snapping to DGN elements:
		0 Does not osnap.
		1 Osnaps.
DiaStat (R/O)	0	Reports whether user exited dialog box by clicking:
		0 **Cancel** button.
		1 **OK** button.

Dimension Variables

Variable Name	Default Value	Meaning
DimADec	0	Controls angular dimension precision:
		-1 Use DimDec setting (default).
		0 Zero decimal places (minimum).
		8 Eight decimal places (maximum).
DimAlt	Off	Toggles alternate units:
		On Enabled.
		Off Disabled.
DimAltD	2	Controls alternate unit decimal places.
DimAltF	25.4	Controls alternate unit scale factor.
DimAltRnd	0.0	Controls rounding factor of alternate units.
DimAltTD	2	Controls decimal places of tolerance alternate units; range is 0 to 8.
DimAltTZ	0	Controls display of zeros in alternate tolerance units:
		0 Zeros not suppressed.
		1 All zeros suppressed.
		2 Include 0 feet, but suppress 0 inches.
		3 Includes 0 inches, but suppress 0 feet.
		4 Suppresses leading zeros.
		8 Suppresses trailing zeros.
DimAltU	2	Controls display of alternate units:
		1 Scientific.
		2 Decimal.
		3 Engineering.
		4 Architectural; stacked.
		5 Fractional; stacked.
		6 Architectural.
		7 Fractional.
		8 Windows desktop units setting.
DimAltZ	0	Controls the display of zeros in alternate units:
		0 Suppresses 0 ft and 0 in.
		1 Includes 0 ft and 0 in.
		2 Includes 0 ft; suppress 0 in.
		3 Suppresses 0 ft; include 0 in.
		4 Suppresses leading 0 in dec dims.
		8 Suppresses trailing 0 in dec dims.
		12 Suppresses leading and trailing zeroes.
ⓐ **DimAnno** (R/O)	0	Reports whether current dimstyle is:
		0 Not annotative.
		1 Annotative.
DimAPost	""	Specifies the prefix and suffix for alternate text.
DimArcSym	0	Specifies the location of the arc symbol:
		0 Before dimension text.
		1 Above the dimension text.
		2 Not displayed.
DimAso	On	Toggles associative dimensions:
		On Dimensions are created associative.
		Off Dimensions are not associative.

Variable Name	Default Value	Meaning
DimAssoc	2	Controls how dimensions are created: **0** Dimension elements are exploded. **1** Single dimension object, attached to defpoints. **2** Single dimension object, attached to geometric objects.
DimASz	0.18	Controls the default arrowhead length.
DimAtFit	3	Controls how text and arrows are fitted when there is insufficient space between extension lines (leader is added when DimTMove = 1): **0** Text and arrows outside extension lines. **1** Arrows first outside, then text. **2** Text first outside, then arrows. **3** Either text or arrows, whichever fits better.
DimAUnit	0	Controls the format of angular dimensions: **0** Decimal degrees. **1** Degrees.Minutes.Seconds. **2** Grads. **3** Radians. **4** Surveyor units.
DimAZin	0	Controls the display of zeros in angular dimensions: **0** Displays all leading and trailing zeros. **1** Suppresses 0 in front of decimal. **2** Suppresses trailing zeros behind decimal. **3** Suppresses zeros in front and behind the decimal.
DimBlk	""	Specifies the name of the arrowhead block: Architectural tick: "Archtick" Box filled: "Boxfilled" Box: "Boxblank" Closed blank: "Closedblank" Closed filled: "" (default) Closed: "Closed" Datum triangle filled: "Datumfilled" Datum triangle: "Datumblank" Dot blanked: "Dotblank" Dot small: "Dotsmall" Dot: "Dot" Integral: "Integral" None: "None" Oblique: "Oblique" Open 30: "Open30" Open: "Open" Origin indication: "Origin" Right-angle: "Open90"
DimBlk1	""	Specifies the name of first arrowhead's block; uses same list of names as under DimBlk. **.** No arrowhead.
DimBlk2	""	Specifies name of second arrowhead block.
DimCen	0.09	Controls how center marks are drawn: *-n* Draws center lines. **0** No center mark or lines drawn. *+n* Draws center marks of length *n*.
DimClrD	0	ACI color of dimension lines: **0** BYBLOCK (default). **256** BYLAYER.
DimClrE	0	Specifies ACI color of extension lines and leaders.
DimClrT	0	Specifies ACI color of dimension text.
DimDec	4	Controls the number of decimal places for the primary tolerance; range is 0 to 8.
DimDLE	0.0	Controls the length of the dimension line extension.

Arrowhead legend (alongside DimBlk):
- Closed filled
- Closed blank
- Closed
- Dot
- Architectural tick
- Oblique
- Open
- Origin indicator
- Origin indicator 2
- Right angle
- Open 30
- Dot small
- Dot blank
- Dot small blank
- Box
- Box filled
- Datum triangle
- Datum triangle filled
- Integral
- None

Variable Name	Default Value	Meaning
DimDLI	0.38	Controls the increment of the continued dimension lines.
DimDSep	"."	Specifies the decimal separator (must be a single character).
DimExe	0.18	Controls the extension above the dimension line.
DimExO	0.0625	Specifies the extension line origin offset.

DimFrac	0	Controls the fraction format when DimLUnit set to 4 or 5:
		  0 Horizontal.
		  1 Diagonal.
		  2 Not stacked.

DimFXL	1	Default length of fixed extension lines.
DimFxlOn	0	Toggles fixed extension lines.
DimGap	0.09	Controls the gap between text and the dimension line.
DimJogAngle	45	Specifies default angle for jogged dimension lines.

DimJust	0	Controls the positioning of horizontal text:
		  0 Center justify.
		  1 Next to first extension line.
		  2 Next to second extension line.
		  3 Above first extension line.
		  4 Above second extension line.

DimLdrBlk	""	Specifies the name of the block used for leader arrowheads; same as DimBlock. **.** Suppresses display of arrowhead.
DimLFac	1.0	Controls the linear unit scale factor.
DimLim	Off	Toggles the display of dimension limits.
DimLtEx1	""	Specifies linetype for the first extension line.
DimLtEx2	""	Specifies linetype for second extension line.
DimLtype	""	Specifies linetype name for dimension line.

DimLUnit	2	Controls dimension units (except angular); replaces DimUnit:
		  1 Scientific.
		  2 Decimal.
		  3 Engineering.
		  4 Architectural.
		  5 Fractional.
		  6 Windows desktop.

DimLwD	-2	Controls the dimension line lineweight; valid values are BYLAYER, BYBLOCK, or integer multiples of 0.01mm.
DimLwE	-2	Controls the extension lineweight.
DimPost	""	Specifies the default prefix or suffix for dimension text (maximum 13 characters): **""** No suffix.
DimRnd	0.0	Controls the rounding value for dimension distances.

DimSAh	Off	Toggles separate arrowhead blocks:
		  Off Use arrowheads defined by DimBlk.
		  On Use arrowheads defined by DimBlk1 and DimBlk2.

DimScale	1.0	Controls the overall dimension scale factor:
		  0 Value is computed from the scale between current model space viewport and paper space.
		  >0 Scales text and arrowheads.

DimSD1	Off	Toggles display of the first dimension line:
		  On First dimension line is suppressed.
		  Off Not suppressed.

DimSD2	Off	Toggles display of the second dimension line:
		  On Second dimension line is suppressed.
		  Off Not suppressed.

Variable Name	Default Value	Meaning
DimSE1	Off	Toggles display of the first extension line: **On** First extension line is suppressed. **Off** Not suppressed.
DimSE2	Off	Toggles display of the second extension line: **On** Second extension line is suppressed. **Off** Not suppressed.
DimSho	On	Toggles dimension updates while dragging: **On** Dimensions are updated during drag. **Off** Dimensions are updated after drag.
DimSOXD	Off	Toggles display of dimension lines outside of extension lines: **On** Dimension lines not drawn outside extension lines. **Off** Are drawn outside extension lines.
⌨ **DimStyle** (R/O)	"STANDARD"	Reports the current dimension style.
DimTAD	0	Controls the vertical position of text: **0** Centered between extension lines. **1** Above dimension line, except when dimension line not horizontal and DimTIH = 1. **2** On side of dimension line farthest from the defining points. **3** Conforms to JIS.
DimTDec	4	Controls the number of decimal places for primary tolerances; range is 0 to 8.
DimTFac	1.0	Controls the scale factor for tolerance text height.
DimTFill	0	Toggles background fill color for dimension text.
DimTFillClr	0	Specifies background color for dimension text.
DimTIH	On	Toggles alignment of text placed inside extension lines: **Off** Text aligned with dimension line. **On** Text is horizontal.
DimTIX	Off	Toggles placement of text inside extension lines: **Off** Text placed inside extension lines, if room. **On** Text forced between the extension lines.
DimTM	0.0	Controls the value of the minus tolerance.
DimTMove	0	Controls how dimension text is moved: **0** Dimension line moves with text. **1** Adds a leader when text is moved. **2** Text moves anywhere; no leader.
DimTOFL	Off	Toggles placement of dimension lines: **Off** Dimension lines not drawn when arrowheads are outside. **On** Dimension lines drawn, even when arrowheads are outside.
DimTOH	On	Toggles text alignment when outside of extension lines: **Off** Text aligned with dimension line. **On** Text is horizontal.
DimTol	Off	Toggles generation of dimension tolerances: **Off** Tolerances not drawn. **On** Tolerances are drawn.
DimTolJ	1	Controls vertical justification of tolerances: **0** Bottom. **1** Middle. **2** Top.
DimTP	0.0	Specifies the value of the plus tolerance.
DimTSz	0.0	Controls the size of oblique tick strokes: **0** Arrowheads. **>0** Oblique strokes.
DimTVP	0.0	Controls the vertical position of text when DimTAD = 0: **1** Turns on DimTAD (=1). **>-0.7** *or* **<0.7** Dimension line is split for text.

Variable Name	Default Value	Meaning
DimTxSty	"STANDARD"	Specifies the dimension text style.
DimTxt	0.18	Controls the text height.
DimTZin	0	Controls the display of zeros in tolerances:
		0 Suppresses 0 ft and 0 in.
		1 Includes 0 ft and 0 in.
		2 Includes 0 ft; suppress 0 in.
		3 Suppresses 0 ft; include 0 in.
		4 Suppresses leading 0 in decimal dim.
		8 Suppresses trailing 0 in decimal dim.
		12 Suppresses leading and trailing zeroes.
DimUPT	Off	Controls user-positioned text:
		Off Cursor positions dimension line.
		On Cursor also positions text.
DimZIN	0	Controls the display of zero in feet-inches units:
		0 Suppresses 0 ft and 0 in.
		1 Includes 0 ft and 0 in.
		2 Includes 0 ft; suppress 0 in.
		3 Suppresses 0 ft; include 0 in.
		4 Suppresses leading 0 in decimal dim.
		8 Suppresses trailing 0 in decimal dim.
		12 Suppresses leading and trailing zeroes.
DispSilh	0	Toggles the silhouette display of 3D solids.
Distance (R/O)	0.0	Reports the distance last measured by the Dist command.
DonutId	0.5	Controls the inside diameter of donuts.
DonutOd	1.0	Controls the outside diameter of donuts.
⌨ **DragMode**	2	Controls the drag mode:
		0 No drag.
		1 On if requested.
		2 Automatic.
DragP1	10	Controls the regen drag display; range is 0 to 32767.
DragP2	25	Controls the fast drag display; range is 0 to 32767.
DragVs	""	Specifies the default visual style when 3D solids are created by dragging the cursor; disabled when visual style is 2D wireframe.
DrawOrderCtrl	3	Controls the behavior of draw order:
		0 Draw order not restored until next regen or drawing reopened.
		1 Normal draw order behavior.
		2 Draw order inheritance.
		3 Combines options 1 and 2.
DrState	0	Toggles the Drawing Recovery palette.
DTextEd	2	Controls the user interface of TEXT command:
		0 In-place text editor.
		1 Edit text dialog box.
		2 Can click elsewhere in drawing to start new text string.
DwfFrame	2	Controls display of the frame around DWF overlays:
		0 Frame is turned off.
		1 Frame is displayed and plotted.
		2 Frame is displayed but not plotted.
DwfOsnap	1	Toggles osnapping of the DWF frame.
DwgCheck	0	Toggles checking of whether drawing was edited by software other than AutoCAD:
		0 Suppresses dialog box.
		1 Displays warning dialog box.
		2 Warning appears on command line.
DwgCodePage (R/O)	*varies*	Same value as SysCodePage.

Variable Name	Default Value	Meaning
DwgName (R/O)	*varies*	Reports the current drawing file name, such as "drawing1.dwg".
DwgPrefix (R/O)	*varies*	Reports the drawing's drive and folder, such as "d:\acad 2008\".
DwgTitled (R/O)	0	Reports whether the drawing file name is: **0** "drawing1.dwg". **1** User-assigned name.
DxEval	12	Controls which commands cause AutoCAD to check for changed external data files (sum): **0** Never. **1** Open. **2** Save. **4** Plot. **8** Publish. **16** eTransmit and Archive. **32** Save with automatic update. **64** Plot with automatic update. **128** Publish with automatic update. **256** eTransmit/Archive with automatic update.
DynDiGrip	31	Controls the dynamic dimensions displayed during grip stretch editing (DynDiVis =2): **0** None. **1** Resulting dimension. **2** Length change dimension. **4** Absolute angle dimension. **8** Angle change dimension. **16** Arc radius dimension.
DynDiVis	1	Controls the dynamic dimensions displayed during grip stretch editing: **0** First (in the cycle order). **1** First two (in the cycle order). **2** All (as specified by DynDiGrip).
DynMode	0	Controls dynamic input features. (Click DYN on status bar to turn on hidden modes.) **-3** Both on hidden. **-2** Dimensional input on hidden. **-1** Pointer input on hidden. **0** Off. **1** Pointer input on. **2** Dimensional input on. **3** Both on.
DynPiCoords	0	Toggles pointer input coordinates: **0** Relative. **1** Absolute.
DynPiFormat	0	Toggles pointer input coordinates: **0** Polar. **1** Cartesian.
DynPiVis	1	Controls when pointer input is displayed: **0** When user types at prompts for points. **1** When prompted for points. **2** Always.
DynPrompt	1	Toggles display of prompts in Dynamic Input tooltips.
DynToolTips	1	Toggles which tooltips are affected by tooltip appearance settings: **0** Only Dynamic Input value fields. **1** All drafting tooltips.

E

Variable Name	Default Value	Meaning
EdgeMode	0	Toggles edge mode for the TRIM and EXTEND commands: **0** Does not extend. **1** Extends cutting edge.

Variable Name	Default Value	Meaning
Elevation	0.0	Specifies the current elevation, relative to current UCS.
EnterpriseMenu (R/O)	"."	Reports the path and *.cui* file name.
ErrNo (R/O)	0	Reports error numbers from AutoLISP, ADS, & Arx.
ErState (R/O)	0	Toggle: reports display of the External References palette.
Expert	0	Controls the display of prompts: **0** Normal prompts. **1** 'About to regen, proceed?' and 'Really want to turn the current layer off?' **2** 'Block already defined. Redefine it?' and 'A drawing with this name already exists. Overwrite it?' **3** Linetype command messages. **4** UCS Save and VPorts Save. **5** DimStyle Save and DimOverride.
ExplMode	1	Toggles whether the **EXPLODE** and **XPLODE** commands explode non-uniformly scaled blocks: **0** Does not explode. **1** Explodes.
ExtMax (R/O)		Specifies upper-right coordinate of drawing extents.
	-1.0E+20, -1.0E+20, -1.0E+20	
ExtMin (R/O)		Specifies lower-left coordinate of drawing extents.
	1.0E+20, 1.0E+20, 1.0E+20	
ExtNames	1	Controls the format of named objects: **0** Names are limited to 31 characters, and can include A - Z, 0 - 9, dollar ($), underscore (_), and hyphen (-). **1** Names are limited to 255 characters, and can include A - Z, 0 - 9, spaces, and any characters not used by Windows or AutoCAD for special purposes.

. . . .

F

Variable Name	Default Value	Meaning
FaceTRatio	0	Controls the aspect ratio of facets on rounded 3D bodies: **0** Creates an *n* by 1 mesh. **1** Creates an *n* by *m* mesh.
FaceTRres	0.5000	Controls the smoothness of shaded and hidden-line objects; range is 0.01 to 10.
FieldDisplay	1	Toggles background to field text: **0** No background. **1** Gray background.
FieldEval	31	Controls how fields are updated: **0** Not updated **1** Updated with Open. **2** Updated with Save. **4** Updated with Plot. **8** Updated with eTransmit. **16** Updated with regeneration.
FileDia	1	Toggles the user interface for file-access commands, such as **OPEN** and **SAVE**: **0** Displays command-line prompts. **1** Displays file dialog boxes.
FilletRad	0.0	Specifies the current fillet radius.
FillMode	1	Toggles the fill of solid objects, wide polylines, fills, and hatches.
FontAlt	"simplex.shx"	Specifies the font used for missing fonts.
FontMap	"acad.fmp"	Specifies the name of the font mapping file.
FrontZ (R/O)	0.0	Reports the front clipping plane offset.
FullOpen (R/O)	1	Reports whether the drawing is: **0** Partially loaded. **1** Fully open.

Variable Name	Default Value	Meaning
FullPlotPath	1	Specifies the format of file name sent to plot spooler: **0** Drawing file name only. **1** Full path and drawing file.

. . . .

G

Variable Name	Default Value	Meaning
GfAng	*0*	*Controls the angle of gradient fill; 0 to 360 degrees.*
GfClr1	*"RGB 000,000,255"*	*Specifies the first gradient color in RGB format.*
GfClr2	*"RGB 255,255,153"*	*Specifies the second gradient color in RGB format.*
GfClrLum	*1.0*	*Controls the level of gray in one-color gradients:* *0 Black.* *1 White.*
GfClrState	*1*	*Specifies the type of gradient fill:* *0 Two-color.* *1 One-color.*
GfName	*1*	*Specifies the style of gradient fill:* *1 Linear.* *2 Cylindrical.* *3 Inverted cylindrical.* *4 Spherical.* *5 Inverted spherical.* *6 Hemispherical.* *7 Inverted hemispherical.* *8 Curved.* *9 Inverted curved.*
GfAShift	*0*	*Controls the origin of the gradient fill:* *0 Centered.* *1 Shifted up and left.*
GlobCheck	*0*	*Controls reporting on dialog boxes:* *-1 Turn off local language.* *0 Turn off.* *1 Warns if larger than 640x400.* *2 Also reports size in pixels.* *3 Additional information.*
GridDisplay	2	Determines grid display (sum of bitcodes): **0** Grid restricted to area specified by the LIMITS command. **1** Grid is infinite. **2** Adaptive grid, with fewer grid lines when zoomed out. **4** Generates more grid lines when zoomed in. **8** Grid follows the x,y plane of the dynamic UCS.
GridMajor	5	Specifies number of minor grid lines per major line.
GridMode	0	Toggles the display of the grid.
GridUnit	0.5,0.5	Controls the x, y spacing of the grid.
GripBlock	0	Toggles the display of grips in blocks: **0** At block insertion point. **1** Of all objects within block.
GripColor	160	Specifies ACI color of unselected grips.
GripDynColor	140	Specifies ACI color of grips in dynamic blocks.
GripHot	1	Specifies ACI color of selected grips.
GripHover	3	Specifies ACI grip color when cursor hovers.
GripObjLimit	100	Controls the maximum number of grips displayed; range is 1 to 32767; 0 = grips never suppressed.
Grips	1	Toggles the display of grips.
GripSize	5	Controls the size of grip; range 1 - 255 pixels.
GripTips	1	Toggles the display of grip tips when cursor hovers over custom objects.

Variable Name	Default Value	Meaning
GtAuto	1	Toggles display of grip tools.
GtDefault	0	Toggles which commands are the default commands in 3D views: **0** MOVE and ROTATE. **1** 3DMOVE and 3DROTATE
GtLocation	0	Controls location of grip tools: **0** Aligns grip tool with UCS icon. **1** Aligns with the last selected object.

. . . .

Variable Name	Default Value	Meaning
H		
HaloGap	0	Controls the distance by which haloed lines are shortened in 2D wireframe visual style; a percentage of 1 unit.
HidePrecision	0	Controls the precision of hide calculations in 2D wireframe visual style: **0** Single precision, less accurate, faster. **1** Double precision, more accurate, but slower (recommended).
HideText	On	Controls the display of text during HIDE: **On** Text is neither hidden nor hides other objects, unless text object has thickness. **Off** Text is hidden and hides other objects.
Highlight	1	Toggles object selection highlighting.
HPAng	0	Specifies current hatch pattern angle.
HpAssoc	1	Toggles associativity of hatches: **0** Not associative. **1** Associative.
HpBound	1	Controls the object created by the HATCH and BOUNDARY commands: **0** Region. **1** Polyline.
HpDouble	0	Toggles double hatching.
HpDrawOrder	3	Controls draw order of hatches and fills: **0** None. **1** Behind all other objects. **2** In front of all other objects. **3** Behind the hatch boundary. **4** In front of the hatch boundary.
HpGapTol	0	Controls largest gap allowed in hatch boundaries; ranges from 0 to 5000 units.
HpInherit	0	Toggles how MATCHPROP copies the hatch origin from source object to destination objects: **0** As specified by HpOrigin. **1** As specified by the source hatch object.
ⓐ **HpMaxlines**	1000000	Controls the maximum number of hatch lines; range is 100 - 10,000,000.
HpName	"ANSI31"	Specifies default hatch name.
HpObjWarning	10000	Specifies the maximum number of hatch boundaries that can be selected before AutoCAD flashes warning message; range is 1 to 1073741823.
HpOrigin	0,0	Specifies the default origin for hatch objects.
HpOrigMode	0	Controls the default hatch origin point:. **0** Specified by HpOrigin. **1** Bottom-left corner of hatch's rectangular extents. **2** Bottom-right corner of hatch's rectangular extents. **3** Top-right corner of hatch's rectangular extents. **4** Top-left corner of hatch's rectangular extents. **5** Center of hatch's rectangular extents.
HpScale	1.0	Specifies the current hatch scale factor; cannot be zero.

Variable Name	Default Value	Meaning
HpSeparate	0	Controls the number of hatch objects made from multiple boundaries: **0** Creates single hatch objects. **1** Creates separately hatch objects.
HpSpace	1.0	Controls the default spacing of user-defined hatches; cannot be zero.
HyperlinkBase	""	Specifies the path for relative hyperlinks.

. . . .

I

Variable Name	Default Value	Meaning
ImageHlt	0	Toggles image frame highlighting when raster images are selected.
ImpliedFace	1	Toggles detection of implied faces.
IndexCtl	0	Controls creation of layer and spatial indices: **0** No indices created. **1** Layer index created. **2** Spatial index created. **3** Both indices created.
InetLocation	"http://www.autodesk.com"	Specifies the default URL for Browser.
InputHistoryMode	15	Controls the content and location of user input history (bitcode sum of): **0** No history displayed. **1** Displayed at the command line, and in dynamic prompt tooltips accessed with Up and Down arrow keys. **2** Current command displayed in the shortcut menu. **4** All commands in the shortcut menu. **8** Blipmark for recent input displayed in the drawing.
InsBase	0.0,0.0,0.0	Controls the default insertion base point relative to the current UCS for **INSERT** and **XREF** commands.
InsName	""	Specifies the default block name: . Set to no default.
InsUnits	1	Specifies drawing units of blocks dragged into drawings: **0** Unitless. **1** Inches. **2** Feet. **3** Miles. **4** Millimeters. **5** Centimeters. **6** Meters. **7** Kilometers. **8** Microinches. **9** Mils. **10** Yards. **11** Angstroms. **12** Nanometers. **13** Microns. **14** Decimeters. **15** Decameters. **16** Hectometers. **17** Gigameters. **18** Astronomical Units. **19** Light Years. **20** Parsecs.
InsUnitsDefSource	4	Controls source drawing units value; ranges from 0 to 20; see above.
InsUnitsDefTarget	4	Controls target drawing units; see list above.
IntelligentUpdate	20	Controls graphics refresh rate in frames per second; range is 0 (off) to 100 fps.
InterfereColor	"1"	Specifies color of interference objects.
InterfereObjVs	"Realistic"	Specifies visual style of interference objects.

Variable Name	Default Value	Meaning
InterfereVpVs	"3d wireframe"	Specifies visual style during interference checking.
IntersectionColor	257	Specifies ACI color of intersection polylines in 2D wireframe visual style: **0** ByBlock. **256** ByLayer. **257** ByEntity.
IntersectionDisplay	Off	Toggles display of 3D surface intersections during HIDE command in 2D wireframe visual mode: **Off** Does not draw intersections. **On** Draws polylines at intersections.
ISaveBak	1	Toggles creation of *.bak* backup files.
ISavePercent	50	Controls the percentage of waste in saved *.dwg* file before cleanup occurs: **0** Slower full saves. **>0** Faster partial saves. **100** Maximum.
IsoLines	4	Controls the number of contour lines on 3D solids; range is 0 to 2047.

L

Variable Name	Default Value	Meaning
LastAngle (R/O)	0	Reports the end angle of last-drawn arc.
LastPoint	0,0,0	Reports the x,y,z coordinates of the last-entered point.
LastPrompt (R/O)	"*varies*"	Reports the last string on the command line.
Latitude	"37.7950"	Specifies last-used angle of latitude.
LayerEval	1	Controls when newly-added (unreconciled) layers are detected: **0** Never. **1** When xref layers are added to the drawing **2** When new layers are added to drawings and xrefs.
LayerFilterAlert	2	Controls the deletion of layer filters in excess of 99 filters *and* the number of layers: **0** Off. **1** Deletes all filters without warning, when layer dialog box opened. **2** Recommends deleting all filters when layer dialog box opened. **3** Displays dialog box for selecting filters to erase, upon opening the drawing.
LayerNotify	15	Controls when alerts are displayed about unreconciled layers (bitcode): **0** Off. **1** Plotting. **2** Opening drawings. **4** Loading, reloading, and attaching xrefs. **8** Restoring layer states. **16** Saving drawings. **32** Inserting blocks, etc.
LayLockFadeCtl	90	Controls the fading of locked layers during the LAYISO command: **0** Not faded. **>0** Faded up to 90 percent. **<0** Not faded, but the value is saved.
LayoutRegenCtl	2	Controls display list for layouts: **0** Display-list regenerated with each tab change. **1** Display-list is saved for model tab and last layout tab. **2** Display list is saved for all tabs.
LegacyCtrlPick	0	Toggles function of Ctrl+pick: **0** Selects faces, edges, and vertices of 3D solids. **1** Cycles through overlapping objects.
LensLength (R/O)	50.0	Reports perspective view lens length, in mm.
LightGlyphDisplay	1	Toggles display of light glyph.

Variable Name	Default Value	Meaning
LightingUnits	2	Controls the type of lighting used: **0** Generic lighting. **1** International units of photometric lighting. **2** American units of photometric lighting.
LightListState	0	Toggles display of Light List palette.
LightsInBlocks	1	Toggles use of lights in blocks: **0** Off. **1** On.
LimCheck	0	Toggles drawing limits checking.
LimMax	12.0,9.0	Controls the upper right drawing limits.
LimMin	0.0,0.0	Controls the lower left drawing limits.
LinearBrightness	0	Controls the overall brightness of generic lighting in renderings; range is -10 to 10.
LinearContrast	0	Controls the overall contrast of generic lighting in renderings; range is -10 to 10.
LispInit	1	Toggles AutoLISP functions and variables: **0** Preserved from drawing to drawing. **1** Valid in current drawing only.
Locale (R/O)	"enc"	Reports ISO language code; see DctMain.
LocalRootPrefix (R/O)	"d:\docume..."	Reports the path to folder holding local customizable files.
LockUi	0	Controls the position and size of toolbars and palettes; hold down Ctrl key to unlock temporarily (bitcode sum): **0** Toolbars and palettes unlocked. **1** Docked toolbars locked. **2** Docked palettes locked. **4** Floating toolbars locked. **8** Floating palettes locked.
LoftAng1	90	Angle of loft to first cross-section; range is 0 to 359.9 degrees.
LoftAng2	90	Specifies angle of loft to second cross-section.
LoftMag1	0.0	Specifies magnitude of loft at first cross-section; range is 1 to 10.
LoftMag2	0.0	Specifies magnitude of loft at last cross-section.
LoftNormals	1	Specifies location of loft normals: **0** Ruled. **1** Smooth. **2** First normal. **3** Last normal. **4** Ends normal. **5** All normal. **6** Use draft angle and magnitude.
LoftParam	7	Specifies the loft shape: **1** Minimizes twists between cross-sections. **2** Aligns start-to-end direction of each cross-section. **4** Generates simple solids and surfaces, instead of spline solids and surfaces. **8** Closes the surface or solid between the first and last cross-sections.
LogExpBrightness	65.0	Controls the overall brightness of photometric lighting in renderings; range is 0 to 200.
LogExpContrast	50.0	Controls the overall contrast of photometric lighting in renderings; range is 0 to 100.
LogExpDaylight	2	Controls how daylight is displayed photometric renderings: **0** Off. **1** On. **2** Same as sun status.

Variable Name	Default Value	Meaning
⊡ LogExpMidtones	1.00	Controls the overall midtones of photometric lighting in renderings; range is 0.01 to 20.
⊡ *LogExpPhysicalScale*	*1500.000*	*Physically scales photometric lights.*
LogFileMode	0	Toggles writing command prompts to *.log* file.
LogFileName (R/O)	"...\Drawing1.log"	Reports file name and path for *.log* file.
LogFilePath	"d:\acad 2008\"	Specifies path to the *.log* file.
LogInName (R/O)	"*username*"	Reports user's login name; truncated after 30 characters.
Longitude	-122.3940	Specifies current angle of longitude.
⌨ LTScale	1.0	Controls linetype scale factor; cannot be 0.

LUnits	2	Controls linear units display:
		  1 Scientific.
		  2 Decimal.
		  3 Engineering.
		  4 Architectural.
		  5 Fractional.

LUPrec	4	Controls decimal places (or inverse of smallest fraction) of linear units; range is 0 to 8.
LwDefault	25	Controls the default lineweight, in millimeters; must be one of the following values: 0, 5, 9, 13, 15, 18, 20, 25, 30, 35, 40, 50, 53, 60, 70, 80, 90, 100, 106, 120, 140, 158, 200, or 211.
LwDisplay	0	Toggles whether lineweights are displayed; setting saved separately for Model space and each layout tab.

LwUnits	1	Toggles units used for lineweights:
		  0 Inches.
		  1 Millimeters.

M

Variable Name	Default Value	Meaning
MacroTrace	*0*	*Toggles diesel debug mode.*
MatState	0	Toggles display of Materials palette.
MaxActVP	64	Controls the maximum number of viewports to display; range is 2 to 64.
MaxSort	1000	Controls the maximum names sorted alphabetically; range is 0 to 32767.

MButtonPan	1	Toggles the behavior of the wheel mouse:
		  0 As defined by AutoCAD *.cui* file.
		  1 Pans when dragging with wheel.

MeasureInit	0	Toggles drawing units for default drawings:
		  0 English.
		  1 Metric.

Measurement	0	Toggles current drawing units:
		  0 English.
		  1 Metric.

MenuCtl	1	Toggles the display of submenus in side menu:
		  0 Only with menu picks.
		  1 Also with keyboard entry.

MenuEcho	0 ...	Controls menu and prompt echoing (sum):
		  0 Displays all prompts.
		  1 Suppresses menu echoing.
		  2 Suppresses system prompts.
		  4 Disables ^P toggle.
		  8 Displays all input-output strings.

MenuName (R/O)	"acad"	Reports path and file name of *.cui* file.

MirrText	0	Toggles text handling by MIRROR command:
		  0 Retains text orientation.
		  1 Mirrors text.

ModeMacro	""	Invokes Diesel macros.

Variable Name	Default Value	Meaning
MsLtScale	1	Controls how linetypes are scaled: **0** Not scaled by the annotation scale. **1** Scaled by the annotation scale.
MsmState (R/O)	0	Specifies if Markup Set Manager is active: **0** No. **1** Yes.
MsOleScale	1.0	Controls the size of text-containing OLE objects when pasted in model space: **-1** Scales by value of PLOTSCALE. **0** Scales by value of DIMSCALE. **>0** Scale factor.
MTextEd	"Internal"	Controls the name of the MText editor: **.** Uses default editor. **0** Cancels the editing operation. **-1** Uses the secondary editor. **"blank"** MTEXT internal editor. **"Internal"** MTEXT internal editor. **"oldeditor"** Previous internal editor. **"Notepad"** Windows Notepad editor. **":lisped"** Built-in AutoLISP function. *string* Name of editor fewer than 256 characters long using this syntax: *:AutoLISPtextEditorFunction#TextEditor.*
MTextFixed	2	Controls the mtext editor appearance: **0** Mtext editor is used. **1** Mtext editor remembers its location. **2** Difficult-to-read text is displayed horizontally at a larger size.
MTJigString	"abc"	Specifies the sample text displayed by mtext editor; maximum 10 letters; enter . for no text.
MyDocumentsPrefix (R/O)		Reports path to the *my documents* folder of the currently logged-in user.
	"C:\Documents and Settings*username*\My Documents"	

. . . .

N

NodeName (R/O)	*"AC$"*	*Reports the name of the network node; range is one to three characters.*
NoMutt	0	Toggles display of messages (a.k.a. muttering) during scripts, LISP, macros: **0** Displays prompt, as normal. **1** Suppresses muttering.
NorthDirection	0	Specifies angle of sun relative to positive y axis.
NwfState	*1*	*Reports whether New Features Workshop displays when AutoCAD starts.*

. . . .

O

ObscuredColor	257	Specifies ACI color of objects obscured by HIDE in 2D wireframe visual style.
ObscuredLtype	0	Specifies linetype of objects obscured by HIDE in 2D wireframe visual mode: **0** Invisible. **1** Solid. **2** Dashed. **3** Dotted. **4** Short dash. **5** Medium dash. **6** Long dash. **7** Double short dash. **8** Double medium dash. **9** Double long dash.

Variable Name	Default Value	Meaning
		10 Medium long dash.
		11 Sparse dot.
OffsetDist	-1.0	Controls current offset distance:
		<0 Offsets through a specified point.
		>0 Default offset distance.
OffsetGapType	0	Controls how polylines reconnect when segments are offset:
		0 Extends segments to fill gap.
		1 Fills gap with fillet (arc segment).
		2 Fills gap with chamfer (line segment).
OleFrame	2	Controls the visibility of the frame around OLE objects:
		0 Frame is not displayed and not plotted.
		1 Frame is displayed and is plotted.
		2 Frame is displayed but is not plotted.
OleHide	0	Controls display and plotting of OLE objects:
		0 All OLE objects visible.
		1 Visible in paper space only.
		2 Visible in model space only.
		3 Not visible.
OleQuality	3	Controls the quality of display and plotting of embedded OLE objects:
		0 Monochrome.
		1 Low quality graphics.
		2 High quality graphics.
		3 Automatically selected mode.
OleStartup	0	Toggles loading of OLE source applications to improve plot quality:
		0 Does not load OLE source application.
		1 Loads OLE source application when plotting.
OpmState	0	Toggles whether Properties palette is active.
OrthoMode	0	Toggles orthographic mode.
OsMode	20517	Controls current object snap mode (sum):
		0 NONe.
		1 ENDpoint.
		2 MIDpoint.
		4 CENter.
		8 NODe.
		16 QUAdrant.
		32 INTersection.
		64 INSertion.
		128 PERpendicular.
		256 TANgent.
		512 NEARest.
		1024 QUIck.
		2048 APPint.
		4096 EXTension.
		8192 PARallel.
		16383 All modes on.
		16384 Object snap turned off via OSNAP on the status bar.
OSnapCoord	2	Controls keyboard overrides object snap:
		0 Object snap overrides keyboard.
		1 Keyboard overrides object snap.
		2 Keyboard overrides object snap, except in scripts.
OSnapHatch	0	Toggles whether hatches are snapped:
		0 Osnaps ignore hatches.
		1 Hatches are snapped.
OSnapNodeLegacy	1	Toggles whether osnap snaps to mtext insertion points.
OsnapZ	0	Toggles osnap behavior in z direction:
		0 Uses the z-coordinate.
		1 Uses the current elevation setting.

Variable Name	Default Value	Meaning
OsOptions	3	Determines when objects with negative z values are osnaped:
		0 Uses the actual z coordinate.
		1 Substitutes z coordinate with the elevation of the current UCS.

. . . .

P

Variable Name	Default Value	Meaning
PaletteOpaque	2	Controls transparency of palettes:
		0 Turned off by user.
		1 Turned on by user.
		2 Unavailable, but turned on by user.
		3 Unavailable, and turned off by user.
PaperUpdate	0	Toggles how AutoCAD plots layouts with paper size different from plotter's default:
		0 Displays a warning dialog box.
		1 Changes paper size to that of the plotter configuration file.
PDMode	0	Controls point display style (sum):
		0 Dot.
		1 No display.
		2 +-symbol.
		3 x-symbol.
		4 Short line.
		32 Circle.
		64 Square.

Variable Name	Default Value	Meaning
PDSize	0.0	Controls point display size:
		>0 Absolute size, in pixels.
		0 5% of drawing area height.
		<0 Percentage of viewport size.
PEditAccept	0	Toggles display of the **PEDIT** command's 'Object selected is not a polyline. Do you want to turn it into one? <Y>:' prompt.
PEllipse	0	Toggles object used to create ellipses:
		0 True ellipse.
		1 Polyline arcs.
Perimeter (R/O)	0.0	Reports perimeter calculated by the last **AREA**, **DBLIST**, and **LIST** commands.
Perspective	0	Toggles perspective mode; not available in 2D wireframe visual mode.
PerpectiveClip	5.0	Controls position of eyepoint clipping, as a percentage; range is 0.01% - 10.0%.
PFaceVMax (R/O)	4	Reports the maximum vertices per 3D face.
PickAdd	1	Toggles meaning of Shift key on selection sets:
		0 Adds to selection set.
		1 Removes from selection set.
PickAuto	1	Toggles selection set mode:
		0 Single pick mode.
		1 Automatic windowing and crossing.
PickBox	3	Controls selection pickbox size; range is 0 to 50 pixels..
PickDrag	0	Toggles selection window mode:
		0 Pick two corners.
		1 Pick a corner; drag to second corner.
PickFirst	1	Toggles command-selection mode:
		0 Enter command first.
		1 Select objects first.

Variable Name	Default Value	Meaning
PickStyle	1	Controls how groups and associative hatches are selected: **0** Includes neither. **1** Includes groups. **2** Includes associative hatches. **3** Includes both.
Platform (R/O)	*"varies"*	Reports the name of the operating system.
PLineGen	0	Toggles polyline linetype generation: **0** From vertex to vertex. **1** From end to end.
PLineType	2	Controls the automatic conversion and creation of 2D polylines by **PLINE**: **0** Not converted; creates old-format polylines. **1** Not converted; creates optimized lwpolylines. **2** Polylines in older drawings are converted on open; **PLINE** creates optimized lwpolyline objects.
PLineWid	0.0	Controls current polyline width.
PlotOffset	0	Toggles the plot offset measurement: **0** Relative to edge of margins. **1** Relative to edge of paper.
PlotRotMode	2	Controls the orientation of plots: **0** Lower left = 0,0. **1** Lower left plotter area = lower left of media. **2** X, y-origin offsets calculated relative to the rotated origin position.
PlQuiet	0	Toggles display during batch plotting and scripts (replaces CmdDia): **0** Plot dialog boxes and nonfatal errors are displayed. **1** Logs nonfatal errors; plot dialog boxes are not displayed.
PolarAddAng	""	Holds a list of up to 10 user-defined polar angles; each angle can be up to 25 characters long, each separated with a semicolon (;). For example: 0;15;22.5;45.
PolarAng	90	Controls the increment of polar angle; contrary to Autodesk documentation, you may specify any angle.
PolarDist	0.0	Controls the polar snap increment when SnapStyl is set to 1 (isometric).
PolarMode	0	Controls polar and object snap tracking: **0** Measure polar angles based on current UCS (absolute), track orthogonally; don't use additional polar tracking angles; and acquire object tracking points automatically. **1** Measure polar angles from selected objects (relative). **2** Use polar tracking settings in object snap tracking. **4** Use additional polar tracking angles (via PolarAng). **8** Press Shift to acquire object snap tracking points.
PolySides	4	Controls the default number of polygon sides; range is 3 to 1024.
Popups (R/O)	1	Reports display driver support of AUI: **0** Not available. **1** Available.
PreviewEffect	2	Controls the visual effect for previewing selected objects: **0** Dashed lines. **1** Thick lines. **2** Thick dashed lines.
PreviewFilter	7	Controls the exclusion of objects from selection previewing (sum): **0** No objects excluded. **1** Objects on locked layers. **2** Objects in xrefs. **4** Tables. **8** Multiline text. **16** Hatch patterns. **32** Groups.
Product (R/O)	"AutoCAD"	Reports the name of the software.

Variable Name	Default Value	Meaning
Program (R/O)	"acad"	Reports the name of the software's executable file.
ProjectName	""	Controls the project name of the current drawing; searches for xref and image files.
ProjMode	1	Controls the projection mode for TRIM and EXTEND commands: **0** Does not project. **1** Projects to x,y-plane of current UCS. **2** Projects to view plane.
ProxyGraphics	1	Toggles saving of proxy images in drawings: **0** Not saved; displays bounding box. **1** Image saved with drawing.
ProxyNotice	0	Toggles warning message displayed when drawing contains proxy objects.
ProxyShow	1	Controls the display of proxy objects: **0** Not displayed. **1** All displayed. **2** Bounding box displayed.
ProxyWebSearch	0	Toggles checking for object enablers: **0** Does not check for object enablers. **1** Checks for object enablers when an Internet connection is present.
PsLtScale	1	Toggles paper space linetype scaling: **0** Uses model space scale factor. **1** Uses viewport scale factor.
PSolHeight	4.0	Specifies default height of polysolid objects.
PSolWidth	0.25	Specifies default width of polysolid objects.
PsProlog	""	Specifies the PostScript prologue file name.
PsQuality	75	Controls resolution of PostScript display, in pixels: **<0** Display as outlines; no fill. **0** Displays no fills. **>0** Displays filled.
PStyleMode	1	Toggles the plot color matching mode of the drawing: **0** Uses named plot style tables. **1** Uses color-dependent plot style tables.
PStylePolicy (R/O)	1	Reports whether the object color is associated with its plot style: **0** Not associated. **1** Associated.
PsVpScale	0.0	Controls the view scale factor (ratio of units in paper space to units in newly-created model space viewports) 0 = scaled to fit.
PublishAllSheets	1	Determines which sheets (model space and layouts) are loaded automatically into the PUBLISH command's list: **0** Current drawing only. **1** All open drawings.
PublishCollate	1	Controls how sheets are published: **0** Sheet sets are processed one sheet at a time; a separate plot is created for each sheet. **1** Sheets sets are processed as a single job and single plot file.
PUcsBase (R/O)	""	Reports name of UCS defining the origin and orientation of orthographic UCS settings; in paper space only.

. . . .

Q

QcState	0	Toggles whether QuickCalc palette is open.
QTextMode	0	Toggles quick text mode.

. . . .

R

RasterDpi	300	Controls the conversion of millimeters or inches to pixels, and vice versa; range is 100 to 32767.

Variable Name	Default Value	Meaning
`a` *RasterPercent*	*20*	*Percentage of system memory to allocate to plotting raster images; range is 0 - 100%.*
RasterPreview (R/O)	1	Toggles creation of BMP preview image.
`a` *RasterThreshold*	*20*	*Amount of RAM to allocate to plotting raster images; range is 0 - 2000MB.*
RecoveryMode	2	Controls recording of drawing recovery information after software failure:
		0 Note recorded.
		1 Recorded; Drawing Recovery palette not displayed automatically.
		2 Recorded, and Drawing Recovery palette displays automatically.
RefEditName	""	Specifies the reference file name when in reference-editing mode.
RegenMode	1	Toggles regeneration mode:
		0 Regens with each view change.
		1 Regens only when required.
Re-Init	0	Controls the reinitialization of I/O devices:
		1 Digitizer port.
		4 Digitizer.
		16 Reloads PGP file.
RememberFolders	1	Toggles the path search method:
		0 Path specified in desktop AutoCAD icon is default for file dialog boxes.
		1 Last path specified by each file dialog box is remembered.
RenderPrefsState	0	Toggles display of the Render Preferences palette.
`a` **RenderUserLights**	1	Determines which lights are rendered:
		0 Default *or* user-defined lights are rendered.
		1 Default *and* user-defined lights are rendered.
ReportError	1	Determines if AutoCAD sends an error report to Autodesk:
		0 No error report created.
		1 Error report is generated and sent to Autodesk.
RoamableRootPrefix (R/O)		Reports the path to the root folder where roamable customized files are located.
	"d:\documents and settings*username*\application data\aut..."	
RTDisplay	1	Toggles raster display during real-time zoom and pan:
		0 Displays the entire raster image.
		1 Displays raster outline only.

· ·

S

Variable Name	Default Value	Meaning
`a` **SaveFidelity**	0	Toggles how annotative objects are translated to earlier releases (bitcode):
		0 No changes made.
		1 Each scale representation is saved to a separate layer.
SaveFile (R/O)	""	Reports the automatic save file name.
SaveFilePath	"...\temp\"	Specifies the path for automatic save files.
SaveName (R/O)	""	Reports the drawing's save-as file name.
SaveTime	10	Controls the automatic save interval, in minutes; 0 = disable auto save.
ScreenBoxes (R/O)	0	Reports the maximum number of menu items supported by display; 0 = screen menu turned off.
ScreenMode (R/O)	3	Reports the state of AutoCAD display:
		0 Text screen.
		1 Graphics screen.
		2 Dual-screen display.
ScreenSize (R/O)	*varies*	Reports the current viewport size, in pixels, such as 719.0,381.0.
SDI	*0*	*Controls the multiple-document interface (SDI is "single document interface"):*
		0 Turns on MDI.
		1 Turns off MDI (only one drawing may be loaded into AutoCAD).
		2 Disables MDI for apps that cannot support MDI; read-only.
		3 (R/O) Disables MDI for apps that cannot support MDI, even when SDI= 1.

Variable Name	Default Value	Meaning
ⓐ SelectionAnnoDisplay	1	Toggles how alternate scale representations are displayed when annotative objects are selected: **0** Not displayed. **1** Displayed according to the value of XFadeCtl.
SelectionArea	1	Toggles use of colored selection areas.
SelectionAreaOpacity	25	Controls the opacity of color selection areas; range is 0 (transparent) to 100 (opaque).
SelectionPreview	3	Controls selection preview: **0** Off. **1** On when commands are inactive. **2** On when commands prompt for object selection.
ⓐ SetBylayerMode	127	Controls which properties are affected by the **SETBYLAYER** command: **0** None. **1** Colors. **2** Linetypes. **4** Lineweights. **8** Materials. **16** Plot styles. **32** ByBlock is changed to Bylayer. **64** Blocks are changed from ByBlock to ByLayer.
ShadEdge	3	Controls shading by **SHADE** command: **0** Faces only shaded. **1** Faces shaded, edges in background color. **2** Only edges, in object color. **3** Faces in object color, edges in background color.
ShadeDif	70	Controls percentage of diffuse to ambient light; range is 0 to 100%.
ShadowPlaneLocation	0.0	Default height of the shadow plane.
ShortcutMenu	11	Controls display of shortcut menus (sum): **0** Disables all. **1** Default shortcut menus. **2** Edit shortcut menus. **4** Command shortcut menus when commands are active. **8** Command shortcut menus only when options available at command line. **16** Shortcut menus when the right button held down longer.
ShowHist	1	Toggles display of history in solids: **0** Original solids are not displayed. **1** Display of original solids depends on **SHOWHISTORY** property settings. **2** Displays all original solids.
ShowLayerUsage	1	Toggles layer-usage icons in Layers dialog box.
ShpName	""	Specifies the default shape name: **.** Set to no default.
SigWarn	1	Toggles display of dialog box when drawings with digital signatures are opened: **0** Only when signature is invalid. **1** Always.
SketchInc	0.1	Controls the **SKETCH** command's recording increment.
SkPoly	0	Toggles sketch line mode: **0** Record as lines. **1** Record as a polyline.
SnapAng	0	Controls rotation angle for snap and grid; when not 0, grid lines are not displayed.
SnapBase	0.0,0.0	Controls current origin for snap and grid.

Variable Name	Default Value	Meaning
SnapIsoPair	0	Controls current isometric drawing plane: **0** Left isoplane. **1** Top isoplane. **2** Right isoplane.
SnapMode	0	Toggles snap mode.
SnapStyl	0	Toggles snap style: **0** Normal. **1** Isometric.
SnapType	0	Toggles snap for the current viewport: **0** Standard snap. **1** Polar snap.
SnapUnit	0.5,0.5	Controls x,y spacing for snap distances.
SolidCheck	1	Toggles solid validation.
SolidHist	1	Toggles retention of history in solids.
SortEnts	127	Controls object display sort order: **0** Off. **1** Object selection. **2** Object snap. **4** Redraw. **8** Slide generation. **16** Regeneration. **32** Plot. **64** PostScript output.
SplFrame	0	Toggles polyline and mesh display: **0** Polyline control frame not displayed; display polygon fit mesh; 3D faces invisible edges not displayed. **1** Polyline control frame displayed; display polygon defining mesh; 3D faces invisible edges displayed.
SplineSegs	8	Controls number of line segments that define splined polylines; range is -32768 to 32767. **<0** Drawn with fit-curve arcs. **>0** Drawn with line segments.
SplineType	6	Controls type of spline curve: **5** Quadratic Bezier spline. **6** Cubic Bezier spline.
SsFound	""	Specifies path and file name of sheet sets.
SsLocate	1	Toggles whether sheet set files are opened with drawing: **0** Not opened. **1** Opened automatically.
SsmAutoOpen	1	Toggles whether the Sheet Set Manager is opened with drawing (SsLocate must be 1): **0** Not opened. **1** Opened automatically.
SsmPollTime	60	Controls time interval between automatic refreshes of status data in sheet sets; range is 20 to 600 seconds (SsmSheetStatus = 2).
SsmSheetStatus	2	Controls refresh of status data in sheet sets: **0** Not automatically refreshed. **1** Refresh when sheet set is loaded or updated. **2** Also refresh as specified by SsmPollTime.
SsmState (R/O)	0	Toggles whether Sheet Set Manager is open.
StandardsViolation	2	Controls whether alerts are displayed when CAD standards are violated: **0** No alerts. **1** Alert displayed when CAD standard violated. **2** Displays icon on status bar when file is opened with CAD standards, and when non-standard objects are created.

Variable Name	Default Value	Meaning
Startup	0	Controls which dialog box is displayed by the **NEW** and **QNEW** commands:
		0 Displays Select Template dialog box.
		1 Displays Startup and Create New Drawing dialog box.
StepSize	6.0	Specifies length of steps in walk mode; range is 1E-6 to 1E+6.
StepsPerSec	2	Specifies speed of steps in walk mode; range is 1-30.
SunPropertiesState	0	Toggles display of Sun Properties palette.
SunStatus	0	Toggles display of light by the sun.
SurfTab1	6	Controls density of m-direction surfaces and meshes; range is 5 to 32766.
SurfTab2	6	Density of n-direction surfaces and meshes; range is 2 to 32766.
SurfType	6	Controls smoothing of surface by **PEDIT**:
		5 Quadratic Bezier spline.
		6 Cubic Bezier spline.
		8 Bezier surface.
SurfU	6	Controls surface density in m-direction; range is 2 to 200.
SurfV	6	Specifies surface density in n-direction; range is 2 to 200.
SysCodePage (R/O)	"ANSI_1252"	Reports the system code page; set by operating system.

T

Variable Name	Default Value	Meaning
TableIndicator	1	Toggles display of column letters and row numbers during table editing.
TableToolbar	1	Toggles display of cell editing toolbar:
		0 Not displayed.
		1 Displayed.
TabMode	0	Toggles tablet mode.
Target (R/O)	0.0,0.0,0.0	Reports target coordinates in the current viewport.
Tbaskbar	*1*	*Toggles whether each drawing appears as a button on the Windows taskbar.*
TbCustomize	1	Toggles whether toolbars can be customized.
TDCreate (R/O)	*varies*	Reports the date and time that the drawing was created, such as 2448860.54014699.
TDInDwg (R/O)	*varies*	Reports the duration since the drawing was loaded, such as 0.00040625.
TDuCreate (R/O)	*varies*	Reports the universal date and time when the drawing was created, such as 2451318. 67772165.
TDUpdate (R/O)	*varies*	Reports the date and time of last update, such as 2448860.54014699.
TDUsrTimer (R/O)	*varies*	Reports the decimal time elapsed by user-timer, such as 0.00040694.
TDuUpdate (R/O)	*varies*	Reports the universal date and time of the last save, such as 2451318.67772165.
TempOverrides	1	Toggles temporary overrides.
TempPrefix (R/O)	"d:\temp"	Reports the path for temporary files set by Temp variable.
TextEval	0	Toggles the interpretation of text input during the **-TEXT** command:
		0 Literal text.
		1 Read **(** and **!** as AutoLISP code.
TextFill	1	Toggles the fill of TrueType fonts when plotted:
		0 Outline text.
		1 Filled text.
TextOutputFileFormat	0	Controls Unicodes for plot and text window log files:
		0 ANSI format.
		1 UTF-8 (Unicode).
		2 UTF-16LE (Unicode).
		3 UTF-16BE (Unicode).

Variable Name	Default Value	Meaning
TextQlty	50	Controls the resolution of TrueType fonts when plotted; range is 0 to 100.
TextSize	0.2	Controls the default height of text (2.5 in metric units).
TextStyle	"Standard"	Specifies the default name of text style.
Thickness	0.0	Controls the default object thickness.
TileMode	1	Toggles the view mode: **0** Displays layout tab. **1** Displays model tab.
TimeZone	-80000	Specifies current time zone.
ToolTipMerge	0	Toggles the merging of tooltips during dynamic display.
ToolTips	1	Toggles the display of tooltips.
TpState (R/O)	0	Reports if Tool Palettes palette is open.
TraceWid	0.0500	Specifies current width of traces.
TrackPath	0	Controls the display of polar and object snap tracking alignment paths: **0** Displays object snap tracking path across the entire viewport. **1** Displays object snap tracking path between the alignment point and "From point" to cursor location. **2** Turns off polar tracking path. **3** Turns off polar and object snap tracking paths.
TrayIcons	1	Toggles the display of the tray on status bar.
TrayNotify	1	Toggles service notifications displayed by the tray.
TrayTimeout	0	Controls length of time that tray notifications are displayed; range is 0 to 10 seconds.
TreeDepth	3020	Controls the maximum branch depth (in *xxyy* format): *xx* Model-space nodes. *yy* Paper-space nodes. *>0* 3D drawing. *<0* 2D drawing.
TreeMax	10000000	Controls the memory consumption during drawing regeneration.
TrimMode	1	Toggles trims during CHAMFER and FILLET: **0** Leaves selected edges in place. **1** Trims selected edges.
TSpaceFac	1.0	Controls the mtext line spacing distance measured as a factor of "normal" text spacing; ranges from 0.25 to 4.0.
TSpaceType	1	Controls the type of mtext line spacing: **1** At Least: adjusts line spacing based on the height of the tallest character in a line of mtext. **2** Exactly: uses the specified line spacing; ignores character height.
TStackAlign	1	Controls vertical alignment of stacked text: **0** Bottom aligned. **1** Center aligned. **2** Top aligned.
TStackSize	70	Controls size of stacked text as a percentage of the current text height; range is 1 to 127%.

U

Variable Name	Default Value	Meaning
UcsAxisAng	90	Controls the default angle for rotating the UCS around an axis (via the UCS command using the X, Y, or Z options); valid values limited to: 5, 10, 15, 18, 22.5, 30, 45, 90, or 180.
UcsBase	""	Specifies name of UCS that defines the origin and orientation of orthographic UCS settings.
UcsDetect	0	Toggles dynamic UCS mode.
UcsFollow	0	Toggles view displayed with new UCSs: **0** No change. **1** Automatically aligns UCS with new view.

1112

Variable Name	Default Value	Meaning
▦ **UcsIcon**	3	Controls display of the UCS icon: **0** Off. **1** On. **2** Displays at UCS origin, if possible. **3** On, and displayed at origin.
UcsName (R/O)	""	Reports the name of current UCS view: **""** Current UCS is unnamed.
UcsOrg (R/O)	0.0,0.0,0.0	Reports the origin of current UCS relative to WCS.
UcsOrtho	1	Controls whether the related orthographic UCS settings are restored automatically: **0** UCS setting remains unchanged when orthographic view is restored. **1** Related ortho UCS is restored automatically when an ortho view is restored.
UcsView	1	Toggles whether the current UCS is saved with a named view.
UcsVp	1	Toggles whether the UCS in active viewports remains fixed (locked) or changes (unlocked) to match the UCS of the current viewport.
UcsXDir (R/O)	1.0,0.0,0.0	Reports the x-direction of current UCS relative to WCS.
UcsYDir (R/O)	0.0,1.0,0.0	Reports the y-direction of current UCS relative to WCS.
UndoCtl (R/O)	21	Reports the status of undo: **0** Undo disabled. **1** Undo enabled. **2** Undo limited to one command. **4** Auto-group mode. **8** Group currently active. **16** Combines zooms and pans.
UndoMarks (R/O)	0	Reports the number of undo marks.
UnitMode	0	Toggles the type of units display: **0** As set by UNITS command. **1** As entered by user.
UpdateThumbnail	15	Controls how thumbnails are updated (sum): **0** Thumbnail previews not updated. **1** Sheet views updated. **2** Model views updated. **4** Sheets updated. **8** Updated when sheets or views are created, modified, or restored. **16** Updated when the drawing is saved.
UserI1 *thru* **UserI5**	0	Five user-definable integer variables.
UserR1 *thru* **UserR5**	0.0	Five user-definable real variables.
UserS1 *thru* **UserS5**	""	Five user-definable string variables; values are not saved.

V

Variable Name	Default Value	Meaning
ViewCtr (R/O)	*varies*	Reports x,y,z-coordinate of center of current view, such as 15,9,56.
ViewDir (R/O)	*varies*	Reports current view direction relative to UCS (0.0,0.0,1.0 = plan view).
ViewMode (R/O)	0	Reports the current view mode: **0** Normal view. **1** Perspective mode on. **2** Front clipping on. **4** Back clipping on. **8** UCS-follow on. **16** Front clip not at eye.
ViewSize (R/O)	*varies*	Reports the height of current view in drawing units.
ViewTwist (R/O)	0	Reports the twist angle of current view.

Variable Name	Default Value	Meaning
VisRetain	1	Controls xref drawing's layer settings: **0** Xref-dependent layer settings are not saved in the current drawing. **1** Xref-dependent layer settings are saved in the current drawing, and take precedence over settings in the xref'ed drawing the next time the current drawing is loaded.
VpLayerOverrides (R/O)	0	Reports whether VP layer properties are overridden in current viewport: **0** None. **1** At least one layer.
VpLayerOverridesMode	1	Controls display and plot of VP layer property overrides: **0** Not displayed or plotted. **1** Displayed and plotted.
VpMaximizedState (R/O)	0	Toggle: reports whether viewport has been maximized by **VPMAX**.
VsBackgrounds	1	Toggles display of backgrounds in visual styles.
VsEdgeColor	"ByEntity"	Specifies the edge color.
VsEdgeJitter	-2	Specifies the level of pencil effect: **0** *or -n* None. **1** Low. **2** Medium. **3** High.
VsEdgeOverhang	-6	Specifies extension of pencil lines beyond edges; range is 1 to 100 pixels; *-n* = none.
VsEdges	1	Specifies types of edges to display: **0** No edges displayed. **1** Isolines displayed. **2** Facets and edges displayed.
VsEdgeSmooth	1	Specifies crease angle; range is 0 to 180 degrees.
VsFaceColorMode	0	Determines color of faces.
VsFaceHighlight	-30	Specifies the size of highlights; range is -100 to 100. Ignored when VsMaterialMode is 1 or 2 and objects have materials attached.
VsFaceOpacity	-60	Controls the transparency/opacity of faces; range is -100 to 100 (fully opaque).
VsFaceStyle	0	Determines how faces are displayed: **0** None. **1** Real. **2** Gooch.
VsHaloGap	0	Specifies halo gap; range is 0 to 100 pixels.
VsHidePrecision	0	Toggles accuracy of hides and shades.
VsIntersectionColor	"7 (white)"	Specifies the color of intersecting polylines.
VsIntersectionEdges	0	Toggles the display of intersecting edges.
VsIntersectionLtype	1	Specifies the linetype for intersecting polylines: **0** Off. **1** Solid. **2** Dashed. **3** Dotted. **4** Short dash. **5** Medium dash. **6** Long dash. **7** Double-short dash. **8** Double-medium dash. **9** Double-long dash. **10** Medium-long dash. **11** Sparse dot.
VsIsoOntop	0	Toggles whether isolines are displayed.

Variable Name	Default Value	Meaning
VsLightingQuality	1	Toggles the quality of lighting: **0** Facets displayed. **1** Facets smoothed.
VsMaterialMode	0	Controls the display of material finishes: **0** No materials displayed. **1** Materials displayed. **2** Materials and textures displayed.
VSMax (R/O)	*varies*	Reports the upper-right corner of virtual screen, such as 37.4,27.0,0.0.
VSMin (R/O)	*varies*	Reports the lower-left corner of virtual screen, such as -24.9,-18.0,0.0.
VsMonoColor	"RGB:255,255,255"	Specifies the monochrome tint.
VsObscuredColor	"ByEntity"	Specifies the color of obscured lines.
VsObscuredEdges	1	Toggles the display of obscured edges.
VsObscuredLtype	1	Specifies the linetype of obscured lines.
VsShadows	0	Determines the quality of shadows: **0** No shadows displayed. **1** Ground shadows displayed. **2** Full shadows displayed.
VsSilhEdges	0	Toggles the display of silhouette edges.
VsSilhWidth	5	Specifies the width of silhouette edge lines; range is 1 to 25 pixels.
VsState	0	Toggles the Visual Styles palette.
VtDuration	750	Controls the duration of smooth view transition; range is 0 to 5000 secs.
VtEnable	3	Controls smooth view transitions for: **0** Turned off. **1** Pan and zoom. **2** View rotation. **3** Pan, zoom, and view rotation. **4** During scripts only. **5** Pan and zoom during scripts. **6** View rotation during scripts. **7** Pan, zoom, and view rotation during scripts.
VtFps	7	Controls minimum speed for view transitions; range is 1 to 30 fps.

W

Variable Name	Default Value	Meaning
WhipArc	0	Toggles display of circular objects: **0** Displays as connected vectors. **1** Displays as true circles and arcs.
WhipThread	1	Controls multithreaded processing on two CPUs (if present) during redraws and regens: **0** Single-threaded calculations. **1** Regenerations multi-threaded. **2** Redraws multi-threaded. **3** Regens and redraws multi-threaded.
WindowAreaColor	5	Specifies ACI color of windowed selection area.
WmfBkgnd	Off	Toggles background of *.wmf* files: **Off** Background is transparent. **On** Background is same as AutoCAD's background color.
WmfForegnd	Off	Toggles foreground colors of exported WMF images: **Off** Foreground is darker than background. **On** Foreground is lighter than background.
WorldUcs (R/O)	1	Toggles matching of WCS with UCS: **0** Current UCS does not match WCS. **1** UCS matches WCS.
WorldView	1	Toggles view during 3DORBIT, DVIEW, and VPOINT commands: **0** Current UCS. **1** WCS.

Variable Name	Default Value	Meaning
WriteStat (R/O)	1	Toggle: reports whether *.dwg* file is read-only: **0** Drawing file cannot be written to. **1** Drawing file can be written to.
WsCurrent	"*varies*"	Controls name of current workspace.

X

Variable Name	Default Value	Meaning
XClipFrame	0	Toggles visibility of xref clipping boundary.
XEdit	1	Toggles editing of xrefs: **0** Cannot in-place refedit. **1** Can in-place refedit.
XFadeCtl	50	Controls faded display of objects not being edited in-place: **0** No fading; minimum value. **90** 90% fading; maximum value.
XLoadCtl	2	Controls demand loading: **0** Demand loading turned off; entire drawing is loaded. **1** Demand loading turned on; xref file opened. **2** Demand loading turned on; a *copy* of the xref file is opened.
XLoadPath	"...\temp"	Specifies path for storing temporary copies of demand-loaded xref files.
XRefCtl	0	Toggles creation of *.xlg* xref log files.
XrefNotify	2	Controls notification of updated and missing xrefs: **0** No alert displayed. **1** Icon indicates xrefs are attached; a yellow alert indicates missing xrefs. **2** Also displays balloon messages when an xref is modified.
XrefType	0	Toggles xrefs: **0** Attached. **1** Overlaid.

Z

Variable Name	Default Value	Meaning
ZoomFactor	60	Controls the zoom level via mouse wheel; range from 3 to 100.
ZoomWheel	0	Switches the zoom direction when mouse wheel is rotated forward: **0** Zooms in. **1** Zooms out.

3

Variable Name	Default Value	Meaning
3dConversionMode	1	Controls how material and light definitions are converted when pre-AutoCAD 2008 drawings are opened: **0** None converted. **1** Converted automatically. **2** Converted after prompted.
3dDwfPrc	2	Level of precision in drawings exported as *.dwf* files: **1** 1 **2** 0.5 **3** 0.2 **4** 0.1 **5** 0.01 **6** 0.001
3dSelectionMode	1	Toggles how visually overlapping objects are selected (other than 2D and 3D wireframe): **0** Uses traditional 3D selection. **1** Uses line-of-sight selection.

APPENDIX

D

AutoCAD Toolbars and Menus

AUTOCAD TOOLBARS

Autodesk provides AutoCAD 2008 with the toolbars illustrated below.

CAD STANDARDS

CAMERA ADJUSTMENT

DIMENSION

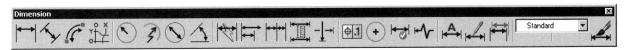

DRAW

DRAWORDER

The icon indicates the toolbar is new to AutoCAD 2008, or is an existing toolbar with new buttons.

INQUIRY

 INSERT

LAYERS

LAYERS II

LAYOUTS

LIGHTS

MAPPING

MODELING

MODIFY

MODIFY II

MULTILEADER

OBJECT SNAP

ORBIT

PROPERTIES

REFEDIT

REFERENCE

RENDER

SOLID EDITING

STANDARD

STANDARD ANNOTATION

STYLES

TEXT

UCS

UCS II

VIEW

VIEWPORTS

VISUAL STYLES

WALK AND FLY

WEB

WORKSPACES

ZOOM

3D NAVIGATION

EXPRESS TOOLS: BLOCKS

EXPRESS TOOLS: STANDARD

EXPRESS TOOLS: TEXT

IMPRESSION: AUTODESK IMPRESSION

AUTOCAD MENUS

FILE MENU

EDIT MENU

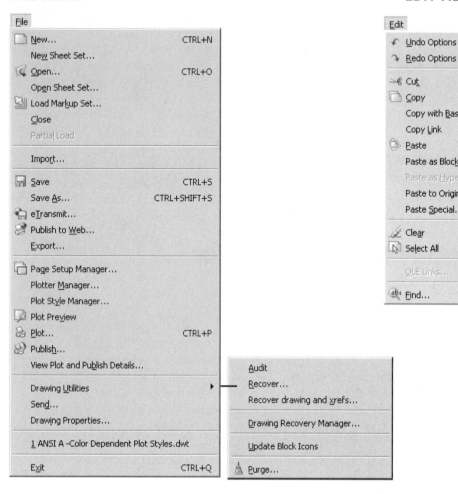

VIEW MENU

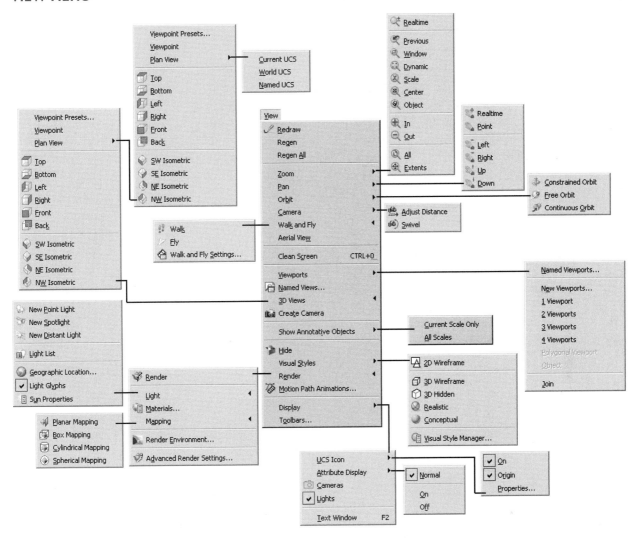

INSERT MENU

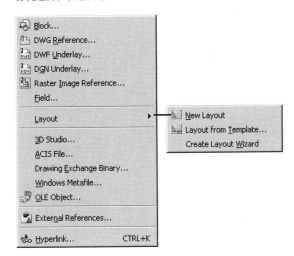

FORMAT MENU

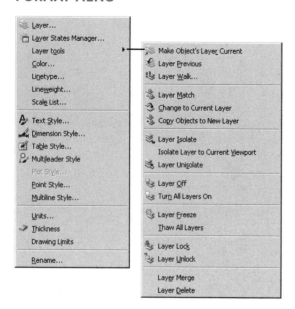

TOOLS MENU

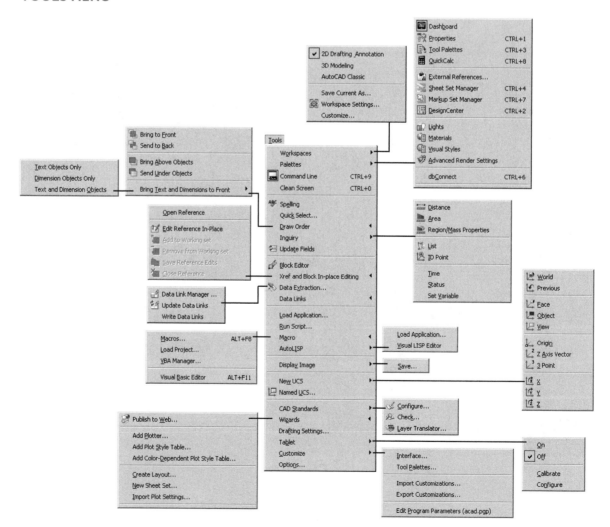

DRAW MENU

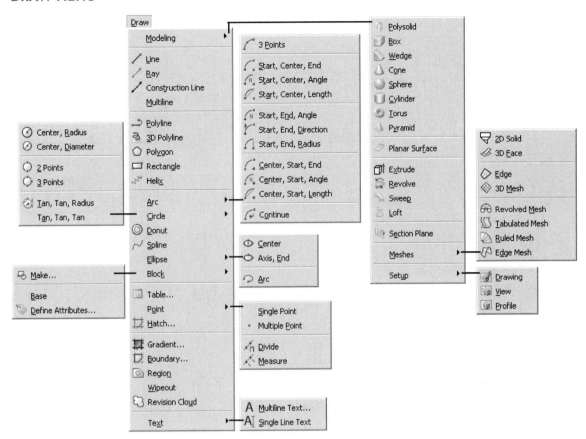

DIMENSION MENU

MODIFY MENU

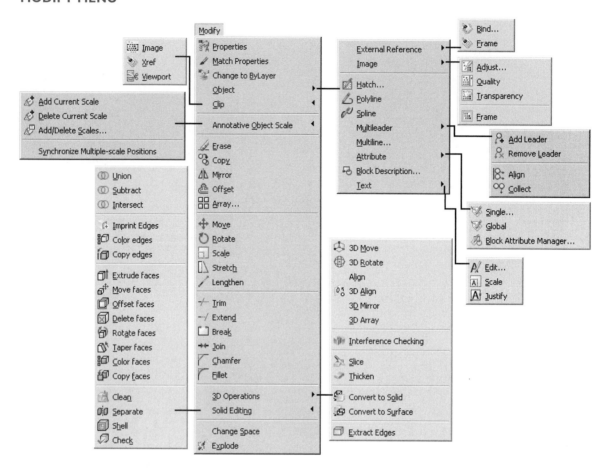

WINDOW MENU

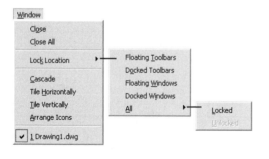

HELP MENU

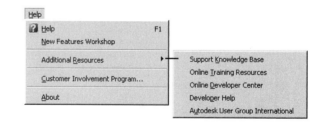

WINDOW CONTROL MENU

EXPRESS MENU

DATA VIEW MENU

This menu is available only after the **DBCONNECT** command is used for the first time.

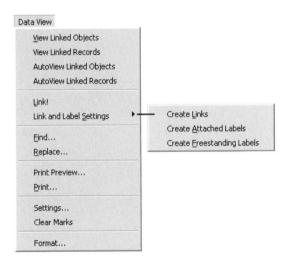

AutoCAD Fonts, Linetypes, Hatch Patterns, Gradients & Lineweights

FONTS

The fonts shown below are just some of the many provided with AutoCAD in SHX and TrueType format. SHX fonts are based on Autodesk's design, while TrueType are based on Microsoft technology. In addition to using the fonts provided by Autodesk, AutoCAD can use any TrueType font found on your computer. SHX fonts are stored in the \autocad 2008\fonts folder, while TTF fonts are stored in the \windows\fonts folder.

Complex
ABCDEFGHIJKLMNOPQRSTUVWZXYZ
abcdefghijklmnopqrstuvwzxyz
0123456789 ° ± ø
!@#$%^&*()_+−=,./<>?;':"[]{}\|'~

Country Blueprint
ABCDEFGHIJKLMNOPQRSTUVWZXYZ
abcdefghijklmnopqrstuvwzxyz
0123456789 ° ±Ø
!@#$%^&*()_+−=,./<>?;':"[]{}\|'~

Euroroman
ABCDEFGHIJKLMNOPQRSTUVWZXYZ
abcdefghijklmnopqrstuvwzxyz
0123456789 ⫿ ± ÿ
!@#$%^&*()_+−=,./<>?;':"⫿⫿\|'~

GDT
ABCDEFGHIJKLMNOPQRSTUVWZXYZ
∠⊥◻⌢⌒//⫽⌀⟋⌇=⌖⌒ⓁⓂØ⌘Ⓟ¢ⓄⓈ⟋⟋—⎵∪∨⊳⊽⊳⊳
0123456789 ° ± ø
!@#$%⌒&*()_+−=,./<>?;':"[]{}\|±˙

Monotxt
ABCDEFGHIJKLMNOPQRSTUVWZXYZ
abcdefghijklmnopqrstuvwzxyz
0123456789 ° ± ø
! @#$%^&*()_+-=,. /<>?;' ¦ "[] ( )\|` ~

RomanC (Complex)
ABCDEFGHIJKLMNOPQRSTUVWZXYZ
abcdefghijklmnopqrstuvwzxyz
0123456789 ° ± ø
!@#$%^&*()_+-=,./<>?;':"[]{}\|'~

RomanD (Duplex)
ABCDEFGHIJKLMNOPQRSTUVWZXYZ
abcdefghijklmnopqrstuvwzxyz
0123456789 ° ± ø
!@#$%^&*()_+-=,./<>?;':"[]{}\|'~

RomanS (Simplex)
ABCDEFGHIJKLMNOPQRSTUVWZXYZ
abcdefghijklmnopqrstuvwzxyz
0123456789 ° ± ø
!@#$%^&*()_+-=,./<>?;':"[]{}\|'~

RomandT (Triplex)
ABCDEFGHIJKLMNOPQRSTUVWZXYZ
abcdefghijklmnopqrstuvwzxyz
0123456789 ° ± ø
!@#$%^&*()_+-=,./<>?;':"[]{}\|'~

Romantic
ABCDEFGHIJKLMNOPQRSTUVWZXYZ
abcdefghijklmnopqrstuvwzxyz
0123456789 ° ± Ø
!@#$%^&*()_+-=,./<>?;':"[]{ }\|`~

ScriptC (Complex)
ABCDEFGHIJKLMNOPQRSTUVWXYZ
abcdefghijklmnopqrstuvwzxyz
0123456789 ° ± ø
!@#$%^&*()_+-=,./<>?;':" []{}\|'~

ScriptS (Simplex)
ABCDEFGHIJKLMNOPQRSTUVWZXYZ
abcdefghijklmnopqrstuvwzxyz
0123456789 ° ± ø
!@#$%^&*()_+-=,./<>?;':" []{}\|'~

Simplex
ABCDEFGHIJKLMNOPQRSTUVWZXYZ
abcdefghijklmnopqrstuvwzxyz
0123456789 ° ± ø
!@#$%^&*()_+-=,./<>?;':"[]{}\|'~

SyAstro (Astronomical Symbols)
⊙☿♀⊕♂♃♄♅Ψ♇☌☍✶☊♈♉☟♋♌♍♎♏♐♑♒♓☄
✶"∪∪∩∈→↑←↓∇^´`˘✗§†‡∃©ℒ®©
0123456789 ° ± ø
!@#$%^&*()_+-=,./<>?;':"[]{}\|'~

SyMap (Mapping Symbols)
⬠△✕✿⋈✠⊢⊣⟶⟵✗♁♆✦✚⬡◯✿⬟
†‡✢☙"⋯◯◯◯◯◯◯◯◯◯◯◯◯◯⊥∴⋮♤♡◇♣✿✿✿
0123456789 ° ± ø
!@#$%^&*()_+-=,./<>?;':"[]{}\|'~

SyMath (Mathematical Symbols)
ℵ'‖±∓×·÷=≠≡<>≦≧∝~√∪∪∩↑∈→↑
←↓∂∇√∫∮∞§†‡∃∏∑()[]{}{}√≅∫≈≅
0123456789 ° ± ø
!@#$%^&*()_+-=,./<>?;':"[]{}\|'~

SyMeteo (Weather Symbols)
〰〰〰〰〰〰〰✳〰───⋀⋁─╱
╲─⋀〰〰〰〰〰〰〰〰◯
0123456789 ° ± ø
!@#$%^&*()_+-=,./<>?;':"[]{}\|'~

SyMusic (Music Symbols)
·♭♪◦◦●#♮♭━━✗♩𝄞𝄢:‖•─⌐⌐⋀─═⟶
·♭♪◦◦●#♮♭━━♩♪𝄞𝄢:‖⊙♀♂⊕♂♃♄♇♃Ψ♇
0123456789 ° ± ø
!@#$%^&*()_+-=,./<>?;':"[]{}\|'~

TXT
ABCDEFGHIJKLMNOPQRSTUVWZXYZ
abcdefghijklmnopqrstuvwzxyz
0123456789 ° ± ø
!@#$%^&*()_+-=,./<>?;'¦"[]()\|`~

LINETYPES

The linetypes shown below are provided with AutoCAD in the *acad.lin* and *acadiso.lin* files. Linetype files are stored in a hidden folder named *C:\Documents and Settings\<user login name>\Application Data\Autodesk\AutoCAD 2008\R17.1\enu\support*, or in a folder defined by the Support File Location item in the Options | Files dialog box.

Linetype	Description
ACAD_ISO02W100	ISO dash __ __ __ __ __ __ __ __ __ __ __
ACAD_ISO03W100	ISO dash space __ __ __ __ __ __
ACAD_ISO04W100	ISO long-dash dot ____ . ____ . ____ . ____ . _
ACAD_ISO05W100	ISO long-dash double-dot ____ .. ____ .. ____ .
ACAD_ISO06W100	ISO long-dash triple-dot ____ ... ____ ... ____
ACAD_ISO07W100	ISO dot . . . .
ACAD_ISO08W100	ISO long-dash short-dash ____ __ ____ __ ____ __
ACAD_ISO09W100	ISO long-dash double-short-dash ____ __ __ ____
ACAD_ISO10W100	ISO dash dot __ . __ . __ . __ . __ . __ . __ .
ACAD_ISO11W100	ISO double-dash dot __ __ . __ __ . __ __ . __
ACAD_ISO12W100	ISO dash double-dot __ .. __ .. __ .. __ ..
ACAD_ISO13W100	ISO double-dash double-dot __ __ .. __ __ ..
ACAD_ISO14W100	ISO dash triple-dot __ ... __ ... __ ...
ACAD_ISO15W100	ISO double-dash triple-dot __ __ ... __ __ .
BATTING	Batting SSSSSSSSSSSSSSSSSSSSSSSSSSSSSSSSSSSSSS
BORDER	Border __ __ . __ __ . __ __ . __ __ .
BORDER2	Border (.5x) __ __ . __ __ . __ __ . __ __ .
BORDERX2	Border (2x) ____ ____ . ____ ____ . ____
CENTER	Center ____ _ ____ _ ____ _ ____ _ ____
CENTER2	Center (.5x) ____ _ ____ _ ____ _ ____ _ ____
CENTERX2	Center (2x) _____ __ _____ __ _____
DASHDOT	Dash dot __ . __ . __ . __ . __ . __ . __
DASHDOT2	Dash dot (.5x) __.__.__.__.__.__.__.__.__.
DASHDOTX2	Dash dot (2x) ____ . ____ . ____ . ____
DASHED	Dashed __ __ __ __ __ __ __ __ __ __ _
DASHED2	Dashed (.5x) _ _ _ _ _ _ _ _ _ _ _ _ _ _ _ _
DASHEDX2	Dashed (2x) ____ ____ ____ ____ ____ ____
DIVIDE	Divide ____ .. ____ .. ____ .. ____ .. ____
DIVIDE2	Divide (.5x) __.._.__.._.__.._.__.._.__
DIVIDEX2	Divide (2x) _____ .. _____ .. _
DOT	Dot . . . .
DOT2	Dot (.5x) . . . .
DOTX2	Dot (2x) . . . .
FENCELINE1	Fenceline circle ----O-----O----O-----O----O---
FENCELINE2	Fenceline square ----[]-----[]----[]-----[]----
GAS_LINE	Gas line ----GAS----GAS----GAS----GAS----GAS---
HIDDEN	Hidden __ __ __ __ __ __ __ __ __ __ _
HIDDEN2	Hidden (.5x) _ _ _ _ _ _ _ _ _ _ _ _ _ _ _ _
HIDDENX2	Hidden (2x) ____ ____ ____ ____ ____ ____
HOT_WATER_SUPPLY	Hot water supply ---- HW ---- HW ---- HW ----
PHANTOM	Phantom _____ __ __ _____ __ __ _____
PHANTOM2	Phantom (.5x) ____ _ _ ____ _ _ ____ _ _ ____
PHANTOMX2	Phantom (2x) _____ ____ ____ _
TRACKS	Tracks -I-I-I-I-I-I-I-I-I-I-I-I-I-I-I-I-I-I
ZIGZAG	Zig zag /\/\/\/\/\/\/\/\/\/\/\/\/\/\/\/\/

HATCH PATTERNS

The hatch patterns shown below are provided by Autodesk with AutoCAD in the *acad.pat* and *acadiso.pat* files, as well as generated as solid-color and gradient fills. Pattern files are stored in a hidden folder named *C:\Documents and Settings\<user login name>\Application Data \Autodesk\AutoCAD 2008\R17.1\enu\support*, or in a folder defined by the Support File Location item in the Options | Files dialog box.

OTHER PATTERNS

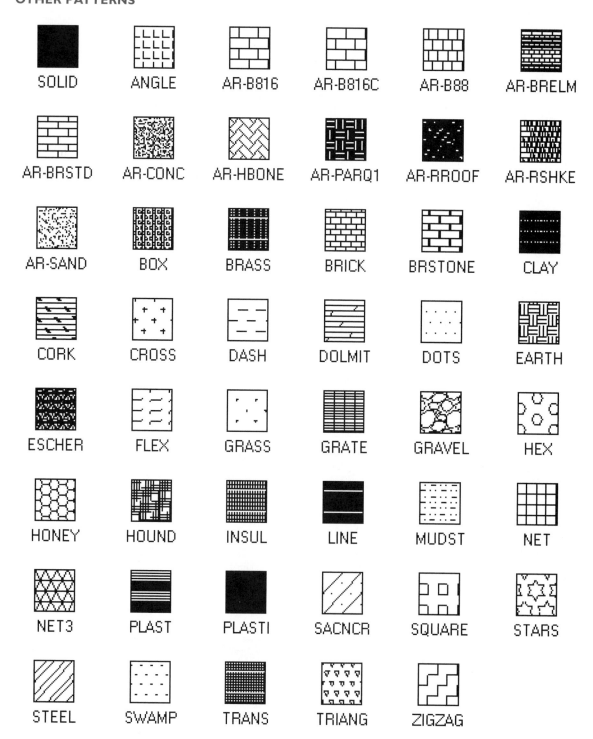

ANSI PATTERNS

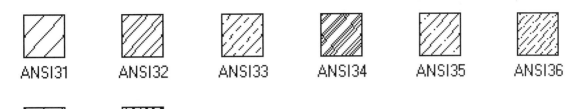

ANSI31   ANSI32   ANSI33   ANSI34   ANSI35   ANSI36

ANSI37   ANSI38

ISO PATTERNS

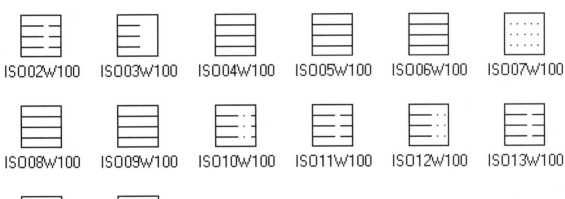

ISO02W100   ISO03W100   ISO04W100   ISO05W100   ISO06W100   ISO07W100

ISO08W100   ISO09W100   ISO10W100   ISO11W100   ISO12W100   ISO13W100

ISO14W100   ISO15W100

GRADIENT PATTERNS

Centered at 0 Degrees

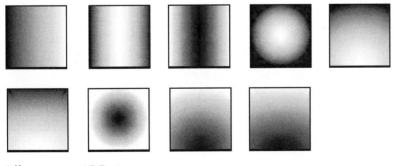

Off-center at 45 Degrees

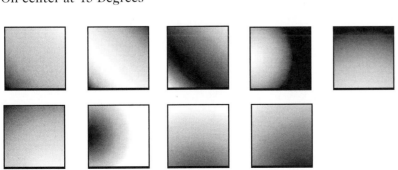

LINEWEIGHTS

The lineweights shown below are "hardwired" into AutoCAD; you cannot add, change, or remove them. Lineweights are measured in inches or millimeters.

Millimeters	Sample
0.00 mm*	————————————
0.05 mm	————————————
0.09 mm	————————————
0.13 mm	————————————
0.15 mm	————————————
0.18 mm	————————————
0.20 mm	————————————
0.25 mm	————————————
0.30 mm	————————————
0.35 mm	————————————
0.40 mm	————————————
0.50 mm	————————————
0.53 mm	————————————
0.60 mm	————————————
0.70 mm	————————————
0.80 mm	————————————
0.90 mm	————————————
1.00 mm	————————————
1.06 mm	————————————
1.20 mm	————————————
1.40 mm	————————————
1.58 mm	————————————
2.00 mm	————————————
2.11 mm	————————————

* When lineweight = 0.0, AutoCAD displays the objects at the screen and plotter resolution. For the screen, this is 1 pixel; for the printer, this is $1/\text{dpi}$ resolution (dpi = dot per inch).

APPENDIX

F

DesignCenter Symbols & Dynamic Blocks

DESIGNCENTER BLOCKS

The symbols (blocks) shown on the following pages are provided by Autodesk with AutoCAD, and can be accessed with DesignCenter. These DWG-format files are stored in the \autocad 2008\ sample\designcenter folder, and are often available in metric or Imperial versions.

AEC: HOME - SPACE PLANNER

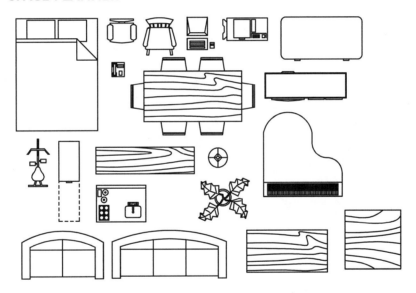

AEC: HOUSE DESIGNER

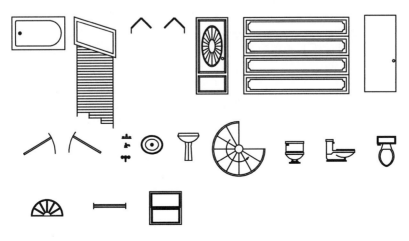

AEC: KITCHENS

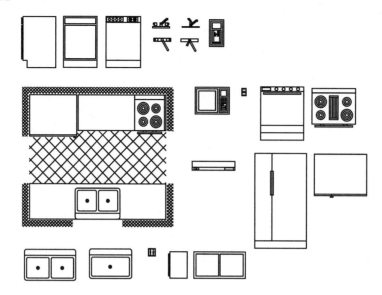

AEC: LANDSCAPING

MECHANICAL FASTENERS - METRIC

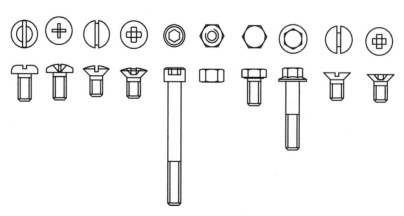

MECHANICAL FASTENERS - US

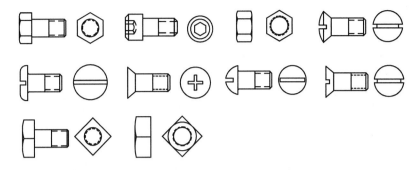

MECHANICAL VAC - HEATING VENTILATION AIR CONDITIONING

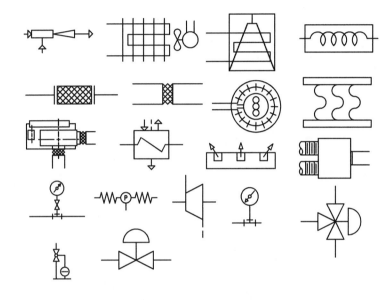

MECHANICAL HYDRAULIC - PNEUMATIC

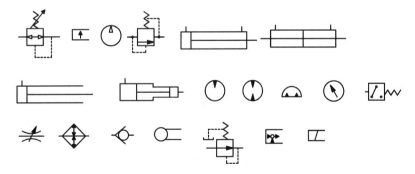

MECHANICAL PIPE FITTINGS

MECHANICAL PLANT PROCESS

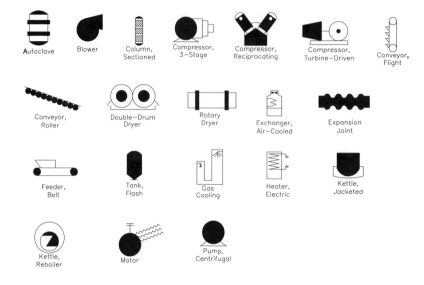

Autoclave   Blower   Column, Sectioned   Compressor, 3-Stage   Compressor, Reciprocating   Compressor, Turbine-Driven   Conveyor, Flight

Conveyor, Roller   Double-Drum Dryer   Rotary Dryer   Exchanger, Air-Cooled   Expansion Joint

Feeder, Belt   Tank, Flash   Gas Cooling   Heater, Electric   Kettle, Jacketed

Kettle, Reboiler   Motor   Pump, Centrifugal

MECHANICAL WELDING

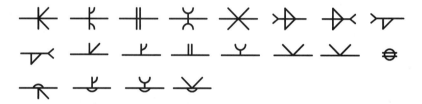

ELECTRONICS: ANALOG INTEGRATED CIRCUITS

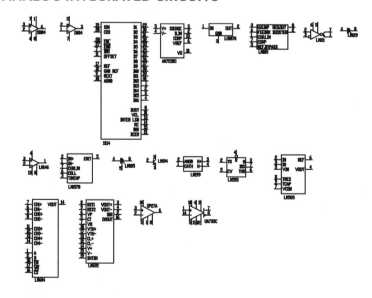

ELECTRONICS: BASIC ELECTRONICS

ELECTRONICS: CMOS INTEGRATED CIRCUITS

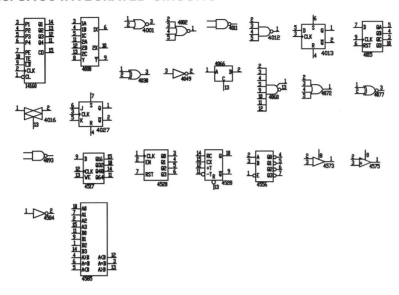

ELECTRONICS: ELECTRICAL POWER

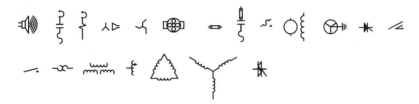

DYNAMIC BLOCK LIBRARIES

The dynamic blocks shown on the following pages are found in the *\autocad 2008\sample\dynamic blocks* folder. A single dynamic block can represent many similar symbols. After being placed in drawings with the **INSERT** command, some blocks can be modified to change their shape; others change their size.

ANNOTATION

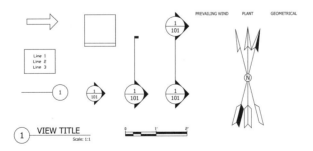

ARCHITECTURAL

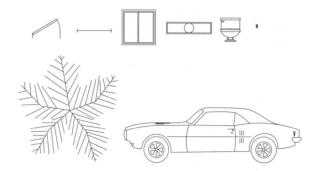

MECHANICAL

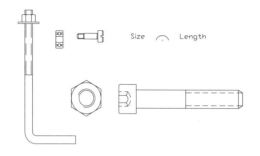

Size ⌒ Length

ELECTRICAL

1 ○ ○ 2   1 ○ ○ 2   1 ○—○ 2   1 ○—○ 2

1 WHITE 2   1 | | 2

CIVIL/STRUCTURAL

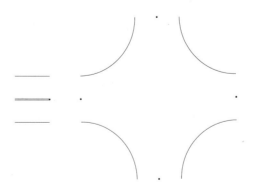

 80      I I

Command and system variable names are shown in **SMALLCAPS BOLDFACE**.